AF357831

Emerging Translational Research in Neurological and Psychiatric Diseases: From In Vitro to In Vivo Models, from Animals to Humans, from Qualitative to Quantitative Methods 2.0

Emerging Translational Research in Neurological and Psychiatric Diseases: From In Vitro to In Vivo Models, from Animals to Humans, from Qualitative to Quantitative Methods 2.0

Editors

Masaru Tanaka
Lydia Giménez-Llort
Simone Battaglia
Chong Chen
Piril Hepsomali

Basel • Beijing • Wuhan • Barcelona • Belgrade • Novi Sad • Cluj • Manchester

Editors
Masaru Tanaka
HUN-REN-SZTE
Neuroscience Research
Group, Hungarian Research
Network, University of
Szeged (HUN-REN-SZTE)
Szeged
Hungary

Lydia Giménez-Llort
Institute of Neuroscience
(INc)
Barcelona
Spain

Simone Battaglia
Centre for Studies and
Research in Cognitive
Neuroscience, Department of
Psychology, University of
Bologna, Department of
Psychology, University of
Turin
Bologna/Turin
Italy

Chong Chen
Yamaguchi University
Graduate School of Medicine
Yamaguchi
Japan

Piril Hepsomali
University of Reading
Reading
UK

Editorial Office
MDPI AG
Grosspeteranlage 5
4052 Basel, Switzerland

This is a reprint of articles from the Topic published online in the open access journals *Biomedicines* (ISSN 2227-9059), *Brain Sciences* (ISSN 2076-3425), *Cells* (ISSN 2073-4409), *International Journal of Molecular Sciences* (ISSN 1422-0067), *Neurology International* (ISSN 2035-8377), and *Pharmaceuticals* (ISSN 1424-8247) (available at: https://www.mdpi.com/topics/Neurological_Psychiatric_Diseases_2).

For citation purposes, cite each article independently as indicated on the article page online and as indicated below:

Lastname, A.A.; Lastname, B.B. Article Title. *Journal Name* **Year**, *Volume Number*, Page Range.

ISBN 978-3-7258-1741-2 (Hbk)
ISBN 978-3-7258-1742-9 (PDF)
doi.org/10.3390/books978-3-7258-1742-9

Contents

About the Editors

Masaru Tanaka

Masaru Tanaka, M.D., Ph.D., is a Senior Research Fellow in the Danube Neuroscience Research Laboratory, HUN-REN-SZTE Neuroscience Research Group, Hungarian Research Network, University of Szeged (HUN-REN-SZTE). His scientific interests include depression, anxiety, pain in dementia, their comorbid nature, and translational research in neurological diseases and psychiatric disorders. He is the author of over 80 papers in peer-reviewed journals and is ranked in D1 in the global publication rankings. He is an Editorial Board Member of *Frontiers in Neuroscience, Anesthesia Research*, the *Journal of Integrative Neuroscience, Advances in Clinical Experimental Medicine, Biology and Life Sciences*, and *Biomedicines*, among others. He obtained a PhD in Medicine and an MD in general Medicine from the University of Szeged, and a Bachelors' degree in Biophysics from the University of Illinois, Urbana-Champaign.

Lydia Giménez-Llort

Lydia Giménez-Llort, B.Sc., Ph.D., is a Full Professor in Psychiatry at the Autonomous University of Barcelona (UAB) and the Senior Research Fellow of Translational Behavioral Neuroscience group, Institut of Neuroscience, UAB. Her scientific and clinical interests include aging, neurodegenerative disorders (Alzheimer's disease, Parkinson's disease, and multiple sclerosis), misophonia, grief, pain in dementia, and gender medicine.

She holds a degree in Biology (University of Barcelona), a Master's in Biochemistry and Molecular Biology (UB), a PhD in Anatomical Pathology (CSIC, Spanish Research Council and UB), a Postdoc in Neuroscience (Karolinska Institutet, Sweden), a Postdoc in Psychiatry (UAB), the Ramón y Cajal Tenure Track (UAB), and a Master's in Psychology—Problem Solving Strategies and Advanced Problem Solving Strategies—SBT (CTs di Arezzo, Italy and Barcelona, Spain).

She is an Editorial Board Member of *Geriatrics* and *Behavioral Sciences*, a Guest Associate Editor for *Frontiers in Neuropharmacology, Public Mental Health, Cognitive Neuroscience, Brain Disease Mechanisms*, and *Alzheimer's Disease and Related Dementias* and a Review Editor for *Cognitive Neuroscience*.

Simone Battaglia

Simone Battaglia, Ph.D., is an Assistant Professor in Cognitive Neuroscience at the Centre for Studies and Research in Cognitive Neuroscience, in the Department of Psychology at the University of Bologna (Italy) and holds a Research Fellowship at the Department of Psychology of the University of Turin (Italy).

His research experience focuses on investigating the intricate functional interplay of different brain areas involved in emotional learning, action control, brain plasticity, decision-making, and a variety of cognitive tasks. To this end, his research activities primarily revolve around the utilization of non-invasive brain stimulation techniques such as transcranial magnetic stimulation (TMS) and transcranial direct current stimulation (tDCS), in addition to employing various neuroscientific techniques to record physiological measures, including EEG, SCR, HRV, and EMG. He has conducted extensive research involving both healthy individuals, in whom he examines intra/inter-individual differences, and patients with acquired brain injuries. In his investigations, he employs a multimodal approach that integrates behavioral assessments, electrophysiological measurements, and neurostimulation techniques. The overarching aim of his research is to develop innovative therapeutic protocols, with a particular focus on utilizing the cortico-cortical paired associative stimulation (ccPAS) method to facilitate neuroplasticity and enhance functional recovery. Recently, in 2023, he was recognized as one of the World's 2% Top Scientists by Stanford University and Elsevier, and in 2024, he won the 'Young Investigator Award' granted by the European Society for Cognitive and Affective Neuroscience (ESCAN).

Chong Chen

Chong Chen, Ph.D., is an Assistant Professor in the Division of Neuropsychiatry, Department of Neuroscience, Yamaguchi University Graduate School of Medicine. His primary research interest is understanding the cognitive and neural computational mechanisms of human decision-making and learning, and how these processes are dysregulated in neuropsychiatric diseases such as depressive and anxiety disorders. He is also focused on developing behavioral and psychological interventions, such as physical exercise and contact with natural environments, to enhance cognitive functions, build resilience, promote mental health, and provide therapeutic effects for neuropsychiatric conditions. He serves as an Associator Editor for *Psychological Research, Brain and Cognition*, and *Frontiers in Psychology* and is a member of the Editorial Boards of several journals, including *Scientific Reports* and *Discover Psychology*.

Piril Hepsomali

Piril Hepsomali, Ph.D., is a lecturer (i.e., Assistant Professor) in Psychology at the School of Psychology and Clinical Language Sciences, University of Reading, United Kingdom. Her research focuses on nutritional psychiatry/neuroscience, where she has contributed to expanding the scientific understanding of the complex interactions between diet/nutrients and mental, sleep, and cognitive wellbeing, with a specific focus on their biological and neural correlates in different populations. She has also been involved in undertaking clinical trials with nutritional interventions for mental, cognitive, and sleep health outcomes and related biomarkers. She obtained a PhD in Psychology from the University of Southampton, United Kingdom, a Master's degree in Cognitive Sciences from the Middle East Technical University, Turkiye, and a Bachelors' degree in Psychology from the Izmir University of Economics, Turkiye.

Editorial

Innovation at the Intersection: Emerging Translational Research in Neurology and Psychiatry

Masaru Tanaka [1,*,†], Simone Battaglia [2,3,*,†], Lydia Giménez-Llort [4,5], Chong Chen [6], Piril Hepsomali [7], Alessio Avenanti [2,8,‡] and László Vécsei [1,9,‡]

[1] HUN-REN-SZTE Neuroscience Research Group, Hungarian Research Network, University of Szeged (HUN-REN-SZTE), Danube Neuroscience Research Laboratory, Tisza Lajos krt. 113, H-6725 Szeged, Hungary; vecsei.laszlo@med-u-szeged.hu

[2] Center for Studies and Research in Cognitive Neuroscience, Department of Psychology "Renzo Canestrari", Cesena Campus, Alma Mater Studiorum Università di Bologna, 47521 Cesena, Italy; alessio.avenanti@unibo.it

[3] Department of Psychology, University of Turin, 10124 Turin, Italy

[4] Institut de Neurociències, Universitat Autònoma de Barcelona, Cerdanyola del Vallès, 08193 Barcelona, Spain; lidia.gimenez@uab.cat

[5] Department of Psychiatry & Forensic Medicine, Faculty of Medicine, Campus Bellaterra, Universitat Autònoma de Barcelona, Cerdanyola del Vallès, 08193 Barcelona, Spain

[6] Division of Neuropsychiatry, Department of Neuroscience, Yamaguchi University Graduate School of Medicine, Yamaguchi 755-8505, Japan; cchen@yamaguchi-u.ac.jp

[7] School of Psychology and Clinical Language Sciences, University of Reading, Reading RG6 6ET, UK; p.hepsomali@reading.ac.uk

[8] Neuropsychology and Cognitive Neuroscience Research Center (CINPSI Neurocog), Universidad Católica del Maule, Talca 3460000, Chile

[9] Department of Neurology, Albert Szent-Györgyi Medical School, University of Szeged, Semmelweis u. 6, H-6725 Szeged, Hungary

* Correspondence: tanaka.masaru.1@med.u-szeged.hu (M.T.); simone.battaglia@unibo.it (S.B.); Tel.: +36-62-342-847 (M.T.)

† These authors contributed equally to this work.

‡ These authors contributed equally to this work.

Citation: Tanaka, M.; Battaglia, S.; Giménez-Llort, L.; Chen, C.; Hepsomali, P.; Avenanti, A.; Vécsei, L. Innovation at the Intersection: Emerging Translational Research in Neurology and Psychiatry. *Cells* 2024, *13*, 790. https://doi.org/10.3390/cells13100790

Received: 16 April 2024
Accepted: 28 April 2024
Published: 7 May 2024

1. Introduction

Translational research in neurological and psychiatric diseases is a rapidly advancing field that promises to redefine our approach to these complex conditions [1–9]. The Topic "Emerging Translational Research in Neurological and Psychiatric Diseases" is a testament to this evolution, showcasing 22 pioneering papers that span a diverse range of topics and methodologies. This Editorial aims to provide a cohesive overview of the collection, emphasizing the synergy between various research domains and the translational potential of the findings presented. These papers fall into the following subtopics: Advancements in Neuroimaging and Neurostimulation, Insights into Neurodegenerative and Neuroinflammatory Disorders, and Exploring Neuropsychiatric Disorders and Treatments.

The first subtopic opens with a series of papers that explore the frontiers of neuroimaging and neurostimulation. These studies not only enhance our understanding of brain morphology and function, but also introduce groundbreaking techniques for modulating brain activity [10–16]. Non-invasive brain stimulation techniques, including transcranial magnetic stimulation (TMS), transcranial direct current stimulation, neurofeedback, and deep brain stimulation (DBS), are frequently used to influence neural activity in specific brain regions, providing therapeutic benefits for conditions such as depression and anxiety [17–23]. The implications for diagnosis and therapy are profound, offering new pathways for addressing neurological and psychiatric disorders that were once considered intractable [24–32].

A central theme of the second subtopic is the in-depth examination of neurodegenerative and neuroinflammatory disorders such as Alzheimer's disease (AD); this includes exploring strategies for neural regeneration and repair. Neuroplasticity, the brain's ability to reorganize itself by forming new neural connections, is particularly relevant in neurodegenerative diseases [33–38]. Interventions targeting neural plasticity aim to slow cognitive decline by promoting adaptive changes in the brain's structure through cognitive training, physical exercise, or environmental enrichment [39–45]. The papers delve into the pathophysiology of diseases such as AD, highlighting the latest strategies for neural regeneration and repair. The inclusion of research on rare and genetic neurological disorders broadens the scope, providing insights into the genetic factors that influence neurology and the potential for personalized medicine [46–51].

The third subtopic also casts a spotlight on neuropsychiatric disorders, offering a nuanced look at conditions such as depression and their treatments [52–57]. Topics include the role of antidepressants, the importance of neural oscillations in cognitive functions, and the interplay between neurotrophic factors and genetic variants. These papers underscore the complexity of mental health and the multifaceted nature of effective treatment strategies. Each paper in this collection contributes to a mosaic of knowledge that enhances our understanding of the brain and its disorders. Employing both qualitative and quantitative methods, the researchers provide a rich tapestry of data that not only reflects the current state of the field, but also lays the groundwork for future innovations [58–61]. As we look ahead, the insights garnered from this collection will undoubtedly inform the next generation of translational research, ultimately leading to improved outcomes for patients worldwide.

2. Topic Articles

2.1. Neuroimaging and Neurostimulation Techniques

Neuroimaging and neurostimulation techniques encompass a diverse range of studies that explore the intricate workings of the brain and the potential therapeutic interventions for neurological conditions [10,11,18,27,62–64]. From a systematic review of brain morphometric changes throughout life to the nuanced effects of ultrasound on cortical activity, these articles shed light on the dynamic field of neuroimaging [12,20,22,28,29,65,66]. Furthermore, the authors explore the field of neurostimulation, specifically examining the potential therapeutic applications of arginine–vasopressin receptor antagonists in the treatment of stroke, the transformative capabilities of DBS in the context of neurodegenerative disorders, and the safety implications associated with transcranial electrical stimulation. Additionally, cutting-edge research on electroencephalography (EEG) biomarkers using machine learning underscores innovative approaches being undertaken to understand and predict outcomes in disorders of consciousness. Collectively, these studies represent the forefront of neuroscientific research, offering insights and raising questions about the future of brain health and therapy.

2.1.1. Neuroimaging Techniques for Brain Morphometry and Function

One of the main challenges in neuroscience is understanding how the brain works in health and disease [67–69]. To achieve this goal, researchers use various methods to measure the structure and activity of the brain, such as magnetic resonance imaging (MRI), EEG, and positron emission tomography (PET). These methods provide information about the anatomy, function, metabolism, and connectivity of the brain regions involved in different cognitive processes, emotions, and behaviors. They are also used to diagnose and monitor various neurological and psychiatric disorders, such as stroke, AD, schizophrenia (SCZ), and depression [70–74]. In this subsection, we review two papers that explore the use of different neuroimaging techniques for measuring the structure and activity of the brain.

Statsenko et al. propose a protocol for the systematic review and meta-analysis of longitudinal changes in brain morphology across the lifespan, from infancy to old age, using MRI data from different cohorts and studies [75]. The authors aim to search for characteristic features of non-pathological development and degeneration in distinct brain structures and devise a precise descriptive model of brain morphometry in age groups. They expect to find age-related patterns of brain morphometry changes that can serve as normative references for clinical studies and diagnoses of neurodegenerative diseases. The authors also plan to explore the potential factors that may influence brain morphometry, such as sex, genetics, lifestyle, and environmental exposures. They hope that their study will provide a comprehensive overview of the current knowledge and gaps in the field of brain morphometry and aging, and stimulate further research on this topic (Table 1).

Table 1. Major subtopics covering the Topic "Emerging Translational Research in Neurological and Psychiatric Diseases".

Subtopics	Ref.
1. Neuroimaging and Neurostimulation Techniques	
a. Neuroimaging Techniques for Brain Morphometry and Function	[75,76]
b. Neurostimulation Techniques for Brain Modulation and Therapy	[77–80]
2. Neurodegenerative and Neuroinflammatory Disorders	
a. Alzheimer's Disease and Related Disorders	[81–84]
b. Neural Regeneration and Repair	[85–87]
c. Rare and Genetic Neurological Disorders	[88]
3. Neuropsychiatric Disorders and Treatments	
a. Depression and Antidepressants	[89,90]
b. Neural Oscillations and Cognitive Functions	[91,92]
c. Neurotrophic Factors and Genetic Variants	[93]
d. Neurotransmission and Neuroprotection	[94]
e. Neuromodulation and Neuroregeneration	[95,96]

Di Gregorio et al. evaluated the accuracy of EEG biomarkers in predicting the clinical outcome of patients with disorders of consciousness following severe acquired brain injury, using a machine learning approach to analyze EEG signals recorded during resting state and auditory stimulation [76]. The authors enrolled 20 patients with disorders of consciousness and 20 healthy controls, and performed EEG recordings at baseline and after one month. They extracted several features from the EEG signals, such as spectral power, coherence, entropy, and complexity, and used a support vector machine classifier to discriminate between patients and controls, and between patients with different levels of consciousness. The authors found that the EEG features had high accuracy in classifying the patients and the controls, and were correlated with the clinical outcome of the patients. They concluded that EEG biomarkers serve as a reliable and objective tool for assessing the level of consciousness and prognosis of patients with disorders of consciousness.

2.1.2. Neurostimulation Techniques for Brain Modulation and Therapy

Neurostimulation techniques are promising approaches for modulating the brain activity and function in various neurological and psychiatric conditions, such as stroke, Parkinson's disease (PD), depression, and chronic pain [10,11,18,20,97–100]. They can also be used to enhance cognitive performance, learning, and memory in healthy individuals [101–103]. These techniques involve applying different types of stimuli, such as ultrasounds, electric currents, magnetic fields, or drugs, to specific brain regions or networks, in order to alter their excitability, connectivity, or plasticity [104–111]. In this subtopic, we review four papers that explore the use of different neurostimulation techniques for brain modulation and therapy.

The first paper investigates the effects of weak ultrasound on the rat motor cortex, and suggests that it can induce neuromodulation without causing tissue damage [77]. The authors used functional MRI (fMRI) and electrophysiological recordings to measure the changes in blood-oxygen-level-dependent (BOLD) signals and local field potentials (LFPs) in response to ultrasound stimulation. They found that ultrasound increased the BOLD signals and LFPs in the stimulated area, as well as in the contralateral motor cortex and the thalamus. They also observed that ultrasound enhanced the motor-evoked potentials elicited by TMS, indicating an increase in corticospinal excitability. The authors concluded that weak ultrasound can be used as a non-invasive and focal neurostimulation technique for modulating the motor cortex.

The second paper reviews the role of arginine–vasopressin (AVP) in stroke, and proposes that its type 1 receptor antagonists (V1RAs) can be used as a novel neuroprotective strategy [78]. The authors summarized the evidence that AVP is involved in the pathophysiology of stroke, such as increasing blood pressure, promoting inflammation, inducing cerebral edema, and impairing cerebral blood flow. They also discussed the potential benefits of V1RAs in reducing these deleterious effects and improving neurological outcomes after stroke. They highlighted the results of animal studies and clinical trials that showed that V1RAs can decrease the infarct size, attenuate brain damage, and enhance the recovery of motor and cognitive functions. The authors suggested that V1RAs can be used as an adjunctive therapy for stroke, especially in combination with thrombolysis or mechanical thrombectomy.

The third paper discusses the challenges and opportunities of DBS for the treatment of PD and AD, and highlights the need for personalized and adaptive stimulation paradigms [79]. The authors reviewed the current state of the art and the future directions of DBS for these neurodegenerative disorders, focusing on the selection of optimal stimulation targets, parameters, and patterns. They emphasized the importance of tailoring the stimulation to the individual patient's symptoms, disease stage, and neural activity, as well as adjusting the stimulation in real time based on the feedback from biomarkers, such as EEG, LFPs, or neurochemicals. They also explored the potential of DBS to modulate the neural circuits and networks involved in motor, cognitive, and emotional functions, and to restore the balance between excitation and inhibition in the brain.

The fourth paper evaluates the safety of a special waveform of transcranial electrical stimulation (TES) in vivo, and demonstrates that it can modulate cortical excitability without inducing seizures or neuronal damage [80]. The authors used a novel TES waveform, called the alternating current square wave (ACSW), which consists of alternating positive and negative pulses with a fixed duration and amplitude. They applied ACSW to the rat somatosensory cortex and measured the changes in cortical excitability, seizure susceptibility, and histological alterations. They found that ACSW increased the cortical excitability, as measured by the amplitude of the somatosensory evoked potentials (SEPs), but did not induce seizures, even at high intensities. Furthermore, ACSW did not cause any neuronal damage, inflammation, or apoptosis in the stimulated cortex. The authors concluded that ACSW is a safe and effective TES waveform for modulating cortical excitability (Table 1).

2.2. Neurodegenerative and Neuroinflammatory Disorders

2.2.1. Alzheimer's Disease and Related Disorders

AD is the most common cause of dementia, a progressive and irreversible decline in cognitive functions such as memory, language, reasoning, and judgment [112–115]. AD is characterized by the accumulation of amyloid plaques and neurofibrillary tangles in the brain, leading to neuronal loss and synaptic dysfunction [116–118]. Other forms of dementia include vascular dementia, caused by impaired blood flow to the brain; frontotemporal dementia, caused by degeneration of the frontal and temporal lobes; and Lewy body dementia, resulting from abnormal deposits of alpha-synuclein protein in the brain [119–122]. In this subsection, we review four papers that investigate the etiology, pathogenesis, biomarkers, and therapeutic strategies for AD and related disorders.

The first paper explores the potential role of the liver–brain axis in AD and aging, and how it is influenced by sex, isolation, and obesity [81]. The authors used a mouse model of AD and wildtype counterparts with normal aging where males exhibit obesity. They found that hepatic oxi-inflammation was associated with worse cognitive and behavioral impairments in the mice, and that this effect was more pronounced in male mice and in isolated and obese mice. They also observed that hepatic oxi-inflammation increased the brain AD–neuropathological levels and reduced the expression of neurotrophic factors and synaptic proteins. The authors suggested that hepatic oxi-inflammation and neophobia, a fear of novelty, could be potential targets for preventing or delaying AD and aging.

The second paper examines the risk factors and mechanisms for psychotic symptoms in AD, such as hallucinations, delusions, and agitation, using electronic medical records and deep learning models [82]. The authors analyzed the data of over 300,000 patients with AD, and identified several clinical and demographic variables that were associated with psychotic symptoms, such as age, sex, race, comorbidities, medications, and cognitive and functional status. They also used a convolutional neural network to extract features from the text of the medical records, which improved the prediction of psychotic symptoms. The authors concluded that electronic medical records and deep learning models could provide valuable insights into the etiology and management of psychotic symptoms in AD.

The third paper investigates the role of stromal interaction molecule (STIM) 1 and STIM2, two calcium sensors that regulate intracellular calcium homeostasis, in the pathophysiology of AD, using a mouse model of the disease [83]. The authors measured the expression and localization of STIM1 and STIM2 in the hippocampus of the mice, and found that both were reduced and in different places in the neurons of the Alzheimer's mice compared to the control mice. Electrophysiological recordings and calcium imaging revealed that the Alzheimer's mice exhibited impaired synaptic transmission and calcium signaling in the hippocampal neurons. Overexpressing STIM1 or STIM2 in the hippocampus of the Alzheimer's mice improved their synaptic function and memory performance. The authors suggested that STIM1 and STIM2 could be novel targets for restoring calcium homeostasis and synaptic plasticity in AD.

The fourth paper evaluates the effect of intraoperative hypothermia, a common complication during surgery, on the vascular function and integrity of the rat hippocampus, a brain region that is vulnerable to ischemia and AD [84]. The authors induced hypothermia in the rats by lowering their body temperature to 28 °C for 2 h during surgery, and measured changes in the blood–brain barrier permeability, cerebral blood flow, and vascular reactivity in the hippocampus. They found that hypothermia increased the blood–brain barrier permeability and reduced the cerebral blood flow and vascular reactivity in the CA1 region of the hippocampus; these effects persisted for 24 h post surgery. Additionally, they observed that hypothermia impaired the spatial learning and memory in the rats, and that these deficits correlated with vascular dysfunction. The authors concluded that intraoperative hypothermia could induce vascular dysfunction and cognitive impairment in the hippocampus, potentially influencing the development and progression of AD (Table 1).

2.2.2. Neural Regeneration and Repair

The nervous system is composed of billions of neurons and glial cells that communicate and cooperate to perform various functions, such as sensation, movement, cognition, and emotion [123–127]. However, the nervous system is also vulnerable to various forms of injury and disease, such as trauma, stroke, infection, degeneration, and malformation, which can impair or destroy neural functions and structures [128–130]. Consequently, there is a significant need to develop effective methods to enhance the recovery and restoration of neural function after injury or disease, such as stem cell therapy, gene therapy, tissue engineering, and neurotrophic factors [131–135]. In this subsection, we review three papers that explore the use of different methods for neural regeneration and repair.

The first paper reviews the techniques, mechanisms, potential applications, and challenges of somatic cell reprogramming for nervous system diseases [85]. Somatic cell reprogramming is a process that converts a differentiated cell into another cell type, such as a pluripotent stem cell or a neural cell, by introducing specific factors or stimuli. The authors summarize the current state of the art and the future directions of somatic cell reprogramming for generating neural cells, such as neurons, astrocytes, oligodendrocytes, and microglia, and for modeling and treating various nervous system diseases, such as AD, PD, spinal cord injury, and brain tumors. They also discuss the advantages and limitations of somatic cell reprogramming, including the efficiency, safety, scalability, and ethical issues, and propose possible solutions and strategies.

The second paper evaluates the effect of the autologous genetically enriched leucoconcentrate (AGEL) on the lumbar spinal cord morpho-functional recovery in a mini pig with thoracic spine contusion injury [86]. AGEL is a cell-based therapy consisting of autologous leukocytes that are genetically modified to overexpress the neurotrophin-3 gene, a potent factor promoting neural survival, differentiation, and regeneration. The authors induced a thoracic spine contusion injury in a mini pig, and injected AGEL into the lumbar spinal cord at 24 h and 7 days after injury. They measured changes in spinal cord morphology, electrophysiology, and locomotor function at different time points after injury and treatment. They found that AGEL significantly improved the spinal cord morphology, such as reducing the cavity size, increasing the tissue sparing, and enhancing the axonal growth and myelination. They also found that AGEL significantly improved the spinal cord electrophysiology, such as by increasing the amplitude and decreasing the latency of SEPs and motor-evoked potentials. Furthermore, AGEL significantly enhanced locomotor function, increasing Basso–Beattie–Bresnahan and grid walk scores. The authors concluded that AGEL is a promising therapy for spinal cord injury, and that the mini pig is a suitable animal model for preclinical studies.

The third paper investigates the role of the extracellular-signal-regulated kinase (ERK)1/2 signaling pathway in regulating the tubulin-binding cofactor B (TBCB) expression and affecting the astrocyte process formation after acute fetal alcohol exposure [87]. ERK1/2 is a key kinase that mediates various cellular processes, such as proliferation, differentiation, and survival. TBCB is a protein that regulates microtubule dynamics and stability, which are essential for the cytoskeleton and morphology of cells. The authors exposed rat fetuses to ethanol on the 12th day of gestation, and isolated the astrocytes from the cortex and hippocampus of the pups on postnatal day 1. They measured changes in the ERK1/2 activation, TBCB expression, and astrocyte process formation in vitro. They found that ethanol exposure decreased the ERK1/2 activation and TBCB expression, and impaired the astrocyte process formation in both regions. They also found that pharmacological inhibition or the genetic knockdown of ERK1/2 mimicked the effects of ethanol exposure, while pharmacological activation or the genetic overexpression of ERK1/2 reversed the effects of ethanol exposure. The authors suggested that the ERK1/2 signaling pathway regulates TBCB expression and affects astrocyte process formation after acute fetal alcohol exposure, and that this pathway could be a potential target for preventing or treating fetal alcohol spectrum disorders (Table 1).

2.2.3. Rare and Genetic Neurological Disorders

The paper by Sivananthan et al. introduces the Buffy Coat Score (BCS) as a novel biomarker for assessing treatment response in neuronal ceroid lipofuscinosis type 2 (NCL2) [88]. The BCS is a quantitative measure derived from blood samples, specifically the buffy coat layer, which can reflect the cellular changes in response to therapy. This advancement is significant as it provides a less invasive, cost-effective, and timely method for monitoring disease progression and therapeutic efficacy in NCL2 patients, potentially improving patient management and treatment outcomes (Table 1).

2.3. Neuropsychiatric Disorders and Treatments

These papers explore the diagnosis, pathophysiology, and pharmacological or non-pharmacological interventions for various mental health conditions, such as depression, bipolar disorder (BD), anxiety, SCZ, and substance abuse.

2.3.1. Depression and Antidepressants

Depression is a serious mood disorder that affects millions of people worldwide. It is characterized by persistent feelings of sadness, hopelessness, and loss of interest in daily activities, as well as physical symptoms such as fatigue, insomnia, and appetite changes [136–138]. Depression can impair the quality of life and increase the risk of suicide and other health problems [139–142]. The causes of depression are complex and multifactorial, involving genetic, biological, psychological, and environmental factors [143–147]. Treatment usually involves a combination of psychotherapy and pharmacotherapy, with antidepressants being the most commonly prescribed drugs [148–151]. In this subsection, we review two papers that explore the diagnosis, pathophysiology, and pharmacological interventions for major depressive disorder.

The first paper by Vasiliu provides a narrative review of the efficacy, tolerability, and safety of toludesvenlafaxine, a novel antidepressant that belongs to the class of serotonin and norepinephrine reuptake inhibitors (SNRIs) [89]. The author summarizes the results of preclinical and clinical studies that evaluated the pharmacokinetics, pharmacodynamics, and therapeutic effects of toludesvenlafaxine in comparison with other SNRIs, such as venlafaxine, duloxetine, and desvenlafaxine. The author also discusses the potential advantages and disadvantages of toludesvenlafaxine, such as its lower risk of drug interactions, its longer half-life, and its higher incidence of adverse events. The author concludes that toludesvenlafaxine is a promising antidepressant that may offer some benefits over existing SNRIs; however, more studies are needed to confirm its efficacy and safety in different populations and settings.

The second paper by Kalkman proposes a novel in vitro screen to detect new antidepressant principles based on the inhibition of microglial glycogen synthase kinase-3 beta (GSK3β) activity [90]. The author explains that microglia, the immune cells of the brain, play a key role in neuroinflammation, which is implicated in the pathogenesis of depression. GSK3β is a kinase that regulates various cellular processes, such as metabolism, proliferation, and apoptosis; its inhibition has antidepressant-like effects in animal models. The author suggests that different kinds of antidepressants, such as selective serotonin reuptake inhibitors, SNRIs, monoamine oxidase inhibitors, and ketamine, share a common mechanism of action: the inhibition of microglial GSK3β activity. The author proposes a simple and rapid assay that measures the GSK3β activity in microglial cells exposed to different compounds, which could be used to screen for novel antidepressant principles. This assay is hoped to facilitate the discovery and development of new and more effective antidepressants.

2.3.2. Neural Oscillations and Cognitive Functions

Neural oscillations are rhythmic patterns of electrical activity that occur in the brain, and reflect the synchronization of neural populations [152–155]. Neural oscillations can be measured by various techniques, such as EEG or magnetoencephalography (MEG), and can be classified into different frequency bands, such as delta, theta, alpha, beta, and gamma [156–159]. Neural oscillations are involved in various cognitive functions, such as memory, attention, and decision making, and are modulated by various factors, such as sensory input, task demands, and emotional states [160–163]. Neural oscillations are also affected by neuropsychiatric disorders, including SCZ, BD, and autism, and may serve as biomarkers or targets for diagnosis and intervention [164–167]. In this subsection, we review two papers that explore the role of alpha oscillations, a type of brain wave that occurs in the range of 8–12 Hz, in various cognitive functions, and how they are affected by neuropsychiatric disorders.

The first paper provides a comprehensive review of the role of alpha oscillations among key neuropsychiatric disorders in the adult and developing human brain, based on evidence from the last 10 years of research [91]. The authors summarize the findings from EEG, MEG, and fMRI studies that investigated the alpha oscillations in SCZ, BD, major depressive disorder, anxiety disorders, obsessive compulsive disorder, post-traumatic stress disorder, attention-deficit/hyperactivity disorder, autism spectrum disorder, and Tourette syndrome. The authors discuss the similarities and differences in alpha oscillations across these disorders, and how they relate to the clinical symptoms, cognitive impairments, and neurodevelopmental trajectories. The authors also highlight the potential applications of alpha oscillations for the diagnosis, prognosis, and treatment of these disorders, and suggest some future directions and challenges for the field.

The second paper examines the cost of imagined actions in a reward-valuation task, and how it is modulated by alpha oscillations. The authors conducted an EEG experiment in which healthy participants had to choose between two options that differed in the amount of reward and the number of actions required to obtain it [92]. Participants had to either perform or imagine the actions, and the authors measured the alpha oscillations during the choice and the action phases. The authors found that the participants preferred the option with fewer actions, regardless of the reward amount, and that this preference was stronger when they had to imagine the actions rather perform them. The authors also found that alpha oscillations increased during the choice phase, and that this increase correlated with the preference for fewer actions. The authors suggested that alpha oscillations reflect the inhibition of irrelevant or costly actions, and that imagined actions have a higher cost than performed actions (Table 1).

2.3.3. Neurotrophic Factors and Genetic Variants

Neurotrophic factors are molecules that support the survival, growth, and differentiation of neurons and glial cells in the nervous system. They also regulate various aspects of neural function, such as synaptic transmission, plasticity, and neurogenesis [168–171]. One of the most widely studied neurotrophic factors is nerve growth factor (NGF), which binds to its receptor, nerve growth factor receptor (NGFR), also known as p75NTR [172–174]. NGF and NGFR play crucial roles in the development and maintenance of the cholinergic system, which is important for learning, memory, and cognition [163,175–177]. They are also implicated in the pathophysiology of various neuropsychiatric disorders, such as SCZ, BD, and AD, which are characterized by cognitive impairment, neuroinflammation, and neurodegeneration [161,178–180]. In this subsection, we review one paper that investigates the role of the NGFR gene and its single-nucleotide polymorphisms (SNPs), rs2072446 and rs11466162, in psychiatric disorders.

The paper investigates the role of the NGFR gene and its SNPs, rs2072446 and rs11466162, in psychiatric disorders, such as SCZ, BD, and AD [93]. The authors performed a meta-analysis of 18 studies that examined the association between these SNPs and the risk or severity of these disorders. The authors found that the NGFR gene and its SNPs were significantly associated with psychiatric disorders, especially SCZ and BD. The authors also found that the NGFR gene and its SNPs interacted with other genes and environmental factors to influence the pathogenesis and progression of these disorders. The authors suggested that the NGFR gene and its SNPs could be potential biomarkers or therapeutic targets for psychiatric disorders.

2.3.4. Neurotransmission and Neuroprotection

Neurotransmission is the process of communication between neurons and other cells in the nervous system, mediated by chemical messengers called neurotransmitters. Neurotransmitters can modulate the neural activity and synaptic plasticity, which are essential for learning, memory, and cognition [160,162,181,182]. Neurotransmitters can also confer neuroprotection, which is the ability of the nervous system to resist or recover from damage caused by various insults, such as oxidative stress, excitotoxicity, inflammation, and degeneration [183–187]. One of the most important neurotransmitters in the nervous system is glutamate, which is the main excitatory neurotransmitter that mediates rapid synaptic transmission and long-term potentiation [188]. However, excessive glutamate can also cause excitotoxicity, the overstimulation and subsequent death of neurons [189]. Therefore, maintaining a balance between glutamate and its antagonists, such as kynurenic acid, is crucial for neural function and integrity [190]. In this subsection, we review one paper that explores the role of kynurenic acid (KYNA) in neurotransmission and neuroprotection.

The paper investigates the effect of KYNA on memory enhancement and its mechanisms in neurotransmission [94]. KYNA is an endogenous metabolite of tryptophan that acts as an antagonist of glutamate receptors, such as N-methyl-D-aspartate, α-amino-3-hydroxy-5-methyl-4-isoxazolepropionic acid, and kainate receptors. The authors administered KYNA to rats and mice and measured changes in their memory performance, synaptic transmission, and neurotransmitter levels. The authors found that KYNA improved the memory performance of the animals in various tasks, such as passive avoidance, object recognition, and the Morris water maze. KYNA also increased synaptic transmission and the levels of acetylcholine, dopamine, and serotonin in the hippocampus and the cortex, which are brain regions involved in memory formation and consolidation. The authors suggest that KYNA enhances memory by modulating the glutamatergic and cholinergic systems, and increasing the levels of other neurotransmitters involved in memory and cognition. They also propose that KYNA may have neuroprotective effects against oxidative stress and neuroinflammation, which are associated with neurodegenerative diseases such as AD (Table 1).

2.3.5. Neuromodulation and Neuroregeneration

Neuromodulation is the process of altering neural activity and function by applying external stimuli such as light, electric currents, magnetic fields, or drugs to specific brain regions or networks [14,191]. Neuroregeneration involves restoring neural structures and functions by promoting the survival, growth, and differentiation of neurons and glial cells [192,193]. Both processes are important for enhancing the recovery and restoration of neural function after injury or disease, such as trauma, stroke, infection, degeneration, and malformation [194]. In this subsection, we review two papers that explore the use of different methods for neuromodulation and neuroregeneration.

The first paper investigates the effect of heterologous fibrin biopolymer and photobiomodulation on morphofunctional improvements of the facial nerve and muscles after injury [95]. Heterologous fibrin biopolymer, a biomaterial derived from horse blood, acts as a scaffold and hemostatic agent for tissue repair. Photobiomodulation, a technique using low-level laser therapy, modulates cellular metabolism and function. The authors induced a facial nerve injury in rats and treated them with heterologous fibrin biopolymer, photobiomodulation, or both. They measured the changes in facial nerve morphology, electrophysiology, and functionality, as well as facial muscle histology at different time points after injury and treatment. They found that the combination of heterologous fibrin biopolymer and photobiomodulation significantly improved recovery of the facial nerve and muscles compared to the control or the single treatments. This combination treatment also increased the expression of neurotrophic factors, such as NGF and brain-derived neurotrophic factor, which are crucial for neural survival, differentiation, and regeneration. The authors concluded that heterologous fibrin biopolymer and photobiomodulation are effective methods for enhancing neuromodulation and neuroregeneration of the facial nerve and muscles after injury.

The second paper examines the establishment of a mouse model of recurrent primary dysmenorrhea, a common gynecological disorder characterized by painful menstrual cramps [96]. The authors used a chemical agent, zymosan, to induce an inflammatory response in the uterus of female mice. They measured changes in uterine contraction, blood flow, and pain behavior, as well as the expression of inflammatory mediators such as prostaglandins and cytokines. They found that zymosan injection caused significant increases in uterine contraction, blood flow, and pain behavior, along with the expression of inflammatory mediators in the mice. These effects were repeated in the subsequent estrous cycles, mimicking the clinical features of recurrent primary dysmenorrhea. The authors suggested that this mouse model could be used to study the pathophysiology and pharmacology of recurrent primary dysmenorrhea and to test the efficacy of potential treatments, such as neuromodulators and neuroprotective agents (Table 1).

3. Conclusions

The topic "Emerging Translational Research in Neurological and Psychiatric Diseases: From In Vitro to In Vivo Models, from Animals to Humans, from Qualitative to Quantitative Methods 2.0" represents a significant advancement in our quest to understand and treat neurological and psychiatric disorders. The 22 papers in this collection span a broad spectrum of research, from the molecular bases of neurodegenerative diseases to innovative therapeutic approaches for neuropsychiatric conditions. They reflect the collaborative efforts of scientists and clinicians who are committed to unraveling the complexities of the brain and mind. The insights gleaned from these studies underscore the importance of multidisciplinary but integrative approaches, integrating neuroimaging, neurostimulation, genetic analysis, and computational modeling [195–200]. As we synthesize knowledge from in vitro experiments, animal models, and human clinical trials, we pave the way for transformative breakthroughs that can be translated into effective interventions. This collection not only highlights the current achievements in the field, but also illuminates the path forward. It encourages ongoing dialogue and research collaboration, fostering an environment where scientific curiosity and clinical need converge to inspire innovation. As we continue to push the boundaries of our understanding, we remain hopeful that the work encapsulated in these pages will lead to improved outcomes for individuals affected by neurological and psychiatric diseases worldwide.

Author Contributions: Conceptualization, M.T. and S.B.; writing—original draft preparation, M.T.; writing—review and editing, M.T., S.B., L.G.-L., C.C., P.H., A.A. and L.V.; supervision, M.T. and S.B.; project administration, M.T. and S.B.; funding acquisition, M.T. and S.B. All authors have read and agreed to the published version of the manuscript.

Funding: This work was supported by the National Research, Development, and Innovation Office—NKFIH K138125, SZTE SZAOK-KKA No: 2022/5S729, and the HUN-REN Hungarian Research Network to L.V. and M.T. This work was also supported by #NEXTGENERATIONEU (NGEU) and funded by the Ministry of University and Research (MUR), National Recovery and Resilience Plan (NRRP), project MNESYS (PE0000006)—A Multiscale integrated approach to the study of the nervous system in health and disease (DN. 1553 11.10.2022) to S.B. and A.A. This work was also funded by the UAB-GE-260408 and ArrestAD H2020 Fet-OPEN-1-2016-2017-737390, European Union's Horizon 2020 research and innovation program under grant agreement No. 737390 to L.G.-L.

Institutional Review Board Statement: Not applicable.

Informed Consent Statement: Not applicable.

Data Availability Statement: Data sharing is not applicable to this article.

Conflicts of Interest: The authors declare no conflicts of interest.

Abbreviations

AD	Alzheimer's disease
ACSW	Alternating current square wave
AGEL	Autologous genetically enriched leucoconcentrate
AVP	Arginine–vasopressin
BCS	Buffy Coat Score
BOLD	Blood-oxygen-level-dependent
BD	Bipolar disorder
DBS	Deep brain stimulation
EEG	Electroencephalography
ERK	Extracellular-signal-regulated kinase
fMRI	Functional magnetic resonance imaging
GSK3β	Glycogen synthase kinase-3 beta

KYNA	Kynurenic acid
LFPs	Local field potentials
NGF	Nerve growth factor
NGFR	Nerve growth factor receptor
MEG	Magnetoencephalography
MRI	Magnetic resonance imaging
NCL2	Neuronal ceroid lipofuscinosis type 2
PD	Parkinson's disease
SCZ	Schizophrenia
SEPs	Somatosensory evoked potentials
SNPs	Single-nucleotide polymorphisms
SNRIs	Serotonin and norepinephrine reuptake inhibitors
STIM	Stromal interaction molecule
TBCB	Tubulin-binding cofactor B
TMS	Transcranial magnetic stimulation
TES	Transcranial electrical stimulation
V1RAs	Vasopressin type 1 receptor antagonists

References

1. Machado-Vieira, R. Tracking the impact of translational research in psychiatry: State of the art and perspectives. *J. Transl. Med.* **2012**, *10*, 175. [CrossRef] [PubMed]
2. Aragona, M. The Impact of Translational Neuroscience on Revisiting Psychiatric Diagnosis: State of the Art and Conceptual Analysis. *Balk. Med. J.* **2017**, *34*, 487–492. [CrossRef] [PubMed]
3. Tanaka, M.; Szabó, Á.; Vécsei, L.; Giménez-Llort, L. Emerging Translational Research in Neurological and Psychiatric Diseases: From In Vitro to In Vivo Models. *Int. J. Mol. Sci.* **2023**, *2023*, 15739. [CrossRef] [PubMed]
4. Sullivan, J.A.; Dumont, J.R.; Memar, S.; Skirzewski, M.; Wan, J.; Mofrad, M.H.; Ansari, H.Z.; Li, Y.; Muller, L.; Prado, V.F. New frontiers in translational research: Touchscreens, open science, and the mouse translational research accelerator platform. *Genes Brain Behav.* **2021**, *20*, e12705. [CrossRef] [PubMed]
5. Tanaka, M.; Vécsei, L. From Lab to Life: Exploring Cutting-Edge Models for Neurological and Psychiatric Disorders. *Biomedicines* **2024**, *2024*, 613. [CrossRef] [PubMed]
6. Stieglitz, T. Of man and mice: Translational research in neurotechnology. *Neuron* **2020**, *105*, 12–15. [CrossRef] [PubMed]
7. Tanaka, M.; Szabó, Á.; Vécsei, L. Preclinical modeling in depression and anxiety: Current challenges and future research directions. *Adv. Clin. Exp. Med.* **2023**, *32*, 505–509. [CrossRef] [PubMed]
8. Baker, M.; Hong, S.-I.; Kang, S.; Choi, D.-S. Rodent models for psychiatric disorders: Problems and promises. *Lab. Anim. Res.* **2020**, *36*, 9. [CrossRef] [PubMed]
9. Martos, D.; Lőrinczi, B.; Szatmári, I.; Vécsei, L.; Tanaka, M. Impact of C-3 Side Chain Modifications on Kynurenic Acid: A Behavioral Analysis of Its Analogs in the Motor Domain. *Int. J. Mol. Sci.* **2024**, *2024*, 3394. [CrossRef]
10. Tanaka, M.; Diano, M.; Battaglia, S. Editorial: Insights into structural and functional organization of the brain: Evidence from neuroimaging and non-invasive brain stimulation techniques. *Front. Psychiatry* **2023**, *14*, 1225755. [CrossRef]
11. Battaglia, S.; Schmidt, A.; Hassel, S.; Tanaka, M. Editorial: Case reports in neuroimaging and stimulation. *Front. Psychiatry* **2023**, *14*, 1264669. [CrossRef] [PubMed]
12. Tervo-Clemmens, B.; Marek, S.; Barch, D.M. Tailoring Psychiatric Neuroimaging to Translational Goals. *JAMA Psychiatry* **2023**, *80*, 765–766. [CrossRef] [PubMed]
13. Margolis, K.G.; Cryan, J.F.; Mayer, E.A. The microbiota-gut-brain axis: From motility to mood. *Gastroenterology* **2021**, *160*, 1486–1501. [CrossRef] [PubMed]
14. McCormick, D.A.; Nestvogel, D.B.; He, B.J. Neuromodulation of brain state and behavior. *Annu. Rev. Neurosci.* **2020**, *43*, 391–415. [CrossRef] [PubMed]
15. Mujica-Parodi, L.R.; Amgalan, A.; Sultan, S.F.; Antal, B.; Sun, X.; Skiena, S.; Lithen, A.; Adra, N.; Ratai, E.-M.; Weistuch, C. Diet modulates brain network stability, a biomarker for brain aging, in young adults. *Proc. Natl. Acad. Sci. USA* **2020**, *117*, 6170–6177. [CrossRef]
16. Gonzalez-Escamilla, G.; Muthuraman, M.; Ciolac, D.; Coenen, V.A.; Schnitzler, A.; Groppa, S. Neuroimaging and electrophysiology meet invasive neurostimulation for causal interrogations and modulations of brain states. *Neuroimage* **2020**, *220*, 117144. [CrossRef]
17. Battaglia, S.; Avenanti, A.; Vécsei, L.; Tanaka, M. Neurodegeneration in Cognitive Impairment and Mood Disorders for Experimental, Clinical and Translational Neuropsychiatry. *Biomedicines* **2024**, *2024*, 574. [CrossRef]
18. Antal, A.; Luber, B.; Brem, A.-K.; Bikson, M.; Brunoni, A.R.; Kadosh, R.C.; Dubljević, V.; Fecteau, S.; Ferreri, F.; Flöel, A. Non-invasive brain stimulation and neuroenhancement. *Clin. Neurophysiol. Pract.* **2022**, *7*, 146–165. [CrossRef]
19. Tanaka, M.; Chen, C. Towards a mechanistic understanding of depression, anxiety, and their comorbidity: Perspectives from cognitive neuroscience. *Front. Behav. Neurosci.* **2023**, *17*, 1268156. [CrossRef]

20. Bhattacharya, A.; Mrudula, K.; Sreepada, S.S.; Sathyaprabha, T.N.; Pal, P.K.; Chen, R.; Udupa, K. An overview of noninvasive brain stimulation: Basic principles and clinical applications. *Can. J. Neurol. Sci.* **2022**, *49*, 479–492. [CrossRef]

21. Battaglia, S.; Avenanti, A.; Vécsei, L.; Tanaka, M. Neural Correlates and Molecular Mechanisms of Memory and Learning. *Int. J. Mol. Sci.* **2024**, *25*, 2724. [CrossRef] [PubMed]

22. Begemann, M.J.; Brand, B.A.; Ćurčić-Blake, B.; Aleman, A.; Sommer, I.E. Efficacy of non-invasive brain stimulation on cognitive functioning in brain disorders: A meta-analysis. *Psychol. Med.* **2020**, *50*, 2465–2486. [CrossRef] [PubMed]

23. Orendáčová, M.; Kvašňák, E. Effects of transcranial alternating current stimulation and neurofeedback on alpha (EEG) dynamics: A review. *Front. Hum. Neurosci.* **2021**, *15*, 628229. [CrossRef] [PubMed]

24. Tajti, J.; Szok, D.; Csáti, A.; Szabó, Á.; Tanaka, M.; Vécsei, L. Exploring Novel Therapeutic Targets in the Common Pathogenic Factors in Migraine and Neuropathic Pain. *Int. J. Mol. Sci.* **2023**, *2023*, 4114. [CrossRef]

25. Tanaka, M.; Szabó, Á.; Körtési, T.; Szok, D.; Tajti, J.; Vécsei, L. From CGRP to PACAP, VIP, and Beyond: Unraveling the Next Chapters in Migraine Treatment. *Cells* **2023**, *2023*, 2649. [CrossRef] [PubMed]

26. Jászberényi, M.; Thurzó, B.; Bagosi, Z.; Vécsei, L.; Tanaka, M. The Orexin/Hypocretin System, the Peptidergic Regulator of Vigilance, Orchestrates Adaptation to Stress. *Biomedicines* **2024**, *12*, 448. [CrossRef] [PubMed]

27. Padberg, F.; Bulubas, L.; Mizutani-Tiebel, Y.; Burkhardt, G.; Kranz, G.S.; Koutsouleris, N.; Kambeitz, J.; Hasan, A.; Takahashi, S.; Keeser, D. The intervention, the patient and the illness–personalizing non-invasive brain stimulation in psychiatry. *Exp. Neurol.* **2021**, *341*, 113713. [CrossRef] [PubMed]

28. Dell'Osso, B.; Di Lorenzo, G. *Non Invasive Brain Stimulation in Psychiatry and Clinical Neurosciences*; Springer: Berlin/Heidelberg, Germany, 2020.

29. Kan, R.L.; Zhang, B.B.; Zhang, J.J.; Kranz, G.S. Non-invasive brain stimulation for posttraumatic stress disorder: A systematic review and meta-analysis. *Transl. Psychiatry* **2020**, *10*, 168. [CrossRef] [PubMed]

30. Lefaucheur, J.P.; Aleman, A.; Baeken, C.; Benninger, D.H.; Brunelin, J.; Di Lazzaro, V.; Filipović, S.R.; Grefkes, C.; Hasan, A.; Hummel, F.C.; et al. Evidence-based guidelines on the therapeutic use of repetitive transcranial magnetic stimulation (rTMS): An update (2014–2018). *Clin. Neurophysiol.* **2020**, *131*, 474–528. [CrossRef]

31. Edwards, J.D.; Dominguez-Vargas, A.U.; Rosso, C.; Branscheidt, M.; Sheehy, L.; Quandt, F.; Zamora, S.A.; Fleming, M.K.; Azzollini, V.; Mooney, R.A.; et al. A translational roadmap for transcranial magnetic and direct current stimulation in stroke rehabilitation: Consensus-based core recommendations from the third stroke recovery and rehabilitation roundtable. *Int. J. Stroke* **2024**, *19*, 145–157. [CrossRef]

32. Turrini, S.; Bevacqua, N.; Cataneo, A.; Chiappini, E.; Fiori, F.; Candidi, M.; Avenanti, A. Transcranial cortico-cortical paired associative stimulation (ccPAS) over ventral premotor-motor pathways enhances action performance and corticomotor excitability in young adults more than in elderly adults. *Front. Aging Neurosci.* **2023**, *15*, 1119508. [CrossRef] [PubMed]

33. Bássoli, R.; Audi, D.; Ramalho, B.; Audi, M.; Quesada, K.; Barbalho, S. The Effects of Curcumin on Neurodegenerative Diseases: A Systematic Review. *J. Herb. Med.* **2023**, *42*, 100771. [CrossRef]

34. Schaefers, A.T.; Teuchert-Noodt, G. Developmental neuroplasticity and the origin of neurodegenerative diseases. *World J. Biol. Psychiatry* **2016**, *17*, 587–599. [CrossRef] [PubMed]

35. Sasmita, A.O.; Kuruvilla, J.; Ling, A.P.K. Harnessing neuroplasticity: Modern approaches and clinical future. *Int. J. Neurosci.* **2018**, *128*, 1061–1077. [CrossRef] [PubMed]

36. Toricelli, M.; Pereira, A.A.R.; Abrao, G.S.; Malerba, H.N.; Maia, J.; Buck, H.S.; Viel, T.A. Mechanisms of neuroplasticity and brain degeneration: Strategies for protection during the aging process. *Neural Regen. Res.* **2021**, *16*, 58–67. [PubMed]

37. Weerasinghe-Mudiyanselage, P.D.; Ang, M.J.; Kang, S.; Kim, J.-S.; Moon, C. Structural plasticity of the hippocampus in neurodegenerative diseases. *Int. J. Mol. Sci.* **2022**, *23*, 3349. [CrossRef] [PubMed]

38. Gatto, R.G. Molecular and microstructural biomarkers of neuroplasticity in neurodegenerative disorders through preclinical and diffusion magnetic resonance imaging studies. *J. Integr. Neurosci.* **2020**, *19*, 571–592. [CrossRef] [PubMed]

39. Battaglia, S.; Nazzi, C.; Thayer, J. Heart's tale of trauma: Fear-conditioned heart rate changes in post-traumatic stress disorder. *Acta Psychiatr. Scand.* **2023**, 1–4. [CrossRef] [PubMed]

40. Buglio, D.S.; Marton, L.T.; Laurindo, L.F.; Guiguer, E.L.; Araújo, A.C.; Buchaim, R.L.; Goulart, R.d.A.; Rubira, C.J.; Barbalho, S.M. The role of resveratrol in mild cognitive impairment and Alzheimer's disease: A systematic review. *J. Med. Food* **2022**, *25*, 797–806. [CrossRef]

41. Yuan, T.-F.; Li, W.-G.; Zhang, C.; Wei, H.; Sun, S.; Xu, N.-J.; Liu, J.; Xu, T.-L. Targeting neuroplasticity in patients with neurodegenerative diseases using brain stimulation techniques. *Transl. Neurodegener.* **2020**, *9*, 44. [CrossRef]

42. Battaglia, S.; Nazzi, C.; Thayer, J.F. Genetic differences associated with dopamine and serotonin release mediate fear-induced bradycardia in the human brain. *Transl. Psychiatry* **2024**, *14*, 24. [CrossRef] [PubMed]

43. Camandola, S.; Plick, N.; Mattson, M.P. Impact of coffee and cacao purine metabolites on neuroplasticity and neurodegenerative disease. *Neurochem. Res.* **2019**, *44*, 214–227. [CrossRef] [PubMed]

44. Svensson, M.; Lexell, J.; Deierborg, T. Effects of physical exercise on neuroinflammation, neuroplasticity, neurodegeneration, and behavior: What we can learn from animal models in clinical settings. *Neurorehabilit. Neural Repair* **2015**, *29*, 577–589. [CrossRef] [PubMed]

45. Mercerón-Martínez, D.; Ibaceta-González, C.; Salazar, C.; Almaguer-Melian, W.; Bergado-Rosado, J.A.; Palacios, A.G. Alzheimer's disease, neural plasticity, and functional recovery. *J. Alzheimer's Dis.* **2021**, *82*, S37–S50. [CrossRef]

46. Kölker, S.; Gleich, F.; Mütze, U. Rare disease registries are key to evidence-based personalized medicine: Highlighting the European experience. *Front. Endocrinol.* **2022**, *13*, 832063. [CrossRef]
47. Özdinler, P.H. Expanded access: Opening doors to personalized medicine for rare disease patients and patients with neurodegenerative diseases. *FEBS J.* **2021**, *288*, 1457–1461. [CrossRef]
48. Smoller, J.W. Psychiatric genetics and the future of personalized treatment. *Depress. Anxiety* **2014**, *31*, 893. [CrossRef]
49. Levchenko, A.; Nurgaliev, T.; Kanapin, A.; Samsonova, A.; Gainetdinov, R.R. Current challenges and possible future developments in personalized psychiatry with an emphasis on psychotic disorders. *Heliyon* **2020**, *6*, e03990. [CrossRef]
50. Gregorio, F.; Battaglia, S. The intricate brain-body interaction in psychiatric and neurological diseases. *Adv. Clin. Exp. Med. Off. Organ Wroc. Med. Univ.* **2024**. [CrossRef]
51. Papadopoulou, E.; Pepe, G.; Konitsiotis, S.; Chondrogiorgi, M.; Grigoriadis, N.; Kimiskidis, V.K.; Tsivgoulis, G.; Mitsikostas, D.D.; Chroni, E.; Domouzoglou, E. The evolution of comprehensive genetic analysis in neurology: Implications for precision medicine. *J. Neurol. Sci.* **2023**, *447*, 120609. [CrossRef]
52. Cuijpers, P.; Stringaris, A.; Wolpert, M. Treatment outcomes for depression: Challenges and opportunities. *Lancet Psychiatry* **2020**, *7*, 925–927. [CrossRef] [PubMed]
53. Marwaha, S.; Palmer, E.; Suppes, T.; Cons, E.; Young, A.H.; Upthegrove, R. Novel and emerging treatments for major depression. *Lancet* **2023**, *401*, 141–153. [CrossRef]
54. Ormel, J.; Hollon, S.D.; Kessler, R.C.; Cuijpers, P.; Monroe, S.M. More treatment but no less depression: The treatment-prevalence paradox. *Clin. Psychol. Rev.* **2022**, *91*, 102111. [CrossRef] [PubMed]
55. Gold, S.M.; Köhler-Forsberg, O.; Moss-Morris, R.; Mehnert, A.; Miranda, J.J.; Bullinger, M.; Steptoe, A.; Whooley, M.A.; Otte, C. Comorbid depression in medical diseases. *Nat. Rev. Dis. Primers* **2020**, *6*, 69. [CrossRef]
56. Nemeroff, C.B. The State of Our Understanding of the Pathophysiology and Optimal Treatment of Depression: Glass Half Full or Half Empty? *Am. J. Psychiatry* **2020**, *177*, 671–685. [CrossRef]
57. Li, Q.; Fu, Y.; Liu, C.; Meng, Z. Transcranial direct current stimulation of the dorsolateral prefrontal cortex for treatment of neuropsychiatric disorders. *Front. Behav. Neurosci.* **2022**, *16*, 893955. [CrossRef] [PubMed]
58. Lucas, P.J.; Baird, J.; Arai, L.; Law, C.; Roberts, H.M. Worked examples of alternative methods for the synthesis of qualitative and quantitative research in systematic reviews. *BMC Med. Res. Methodol.* **2007**, *7*, 4. [CrossRef]
59. Östlund, U.; Kidd, L.; Wengström, Y.; Rowa-Dewar, N. Combining qualitative and quantitative research within mixed method research designs: A methodological review. *Int. J. Nurs. Stud.* **2011**, *48*, 369–383. [CrossRef]
60. Palinkas, L.A.; Mendon, S.J.; Hamilton, A.B. Innovations in Mixed Methods Evaluations. *Annu. Rev. Public Health* **2019**, *40*, 423–442. [CrossRef]
61. Desale, P.; Dhande, R.; Parihar, P.; Nimodia, D.; Bhangale, P.N.; Shinde, D.; Dhande, R.; Bhangale, P.N., Jr. Navigating Neural Landscapes: A Comprehensive Review of Magnetic Resonance Imaging (MRI) and Magnetic Resonance Spectroscopy (MRS) Applications in Epilepsy. *Cureus* **2024**, *16*, e56927. [CrossRef]
62. Bestmann, S.; Feredoes, E. Combined neurostimulation and neuroimaging in cognitive neuroscience: Past, present, and future. *Ann. N. Y Acad. Sci.* **2013**, *1296*, 11–30. [CrossRef]
63. Peng, S.; Dhawan, V.; Eidelberg, D.; Ma, Y. Neuroimaging evaluation of deep brain stimulation in the treatment of representative neurodegenerative and neuropsychiatric disorders. *Bioelectron. Med.* **2021**, *7*, 4. [CrossRef] [PubMed]
64. Yen, C.; Lin, C.L.; Chiang, M.C. Exploring the Frontiers of Neuroimaging: A Review of Recent Advances in Understanding Brain Functioning and Disorders. *Life* **2023**, *13*, 1472. [CrossRef] [PubMed]
65. Scheepens, D.S.; van Waarde, J.A.; Lok, A.; de Vries, G.; Denys, D.A.J.P.; van Wingen, G.A. The Link Between Structural and Functional Brain Abnormalities in Depression: A Systematic Review of Multimodal Neuroimaging Studies. *Front. Psychiatry* **2020**, *11*, 485. [CrossRef] [PubMed]
66. Fomenko, A.; Chen, K.S.; Nankoo, J.F.; Saravanamuttu, J.; Wang, Y.; El-Baba, M.; Xia, X.; Seerala, S.S.; Hynynen, K.; Lozano, A.M.; et al. Systematic examination of low-intensity ultrasound parameters on human motor cortex excitability and behavior. *Elife* **2020**, *9*, e54497. [CrossRef] [PubMed]
67. Thompson, P.M.; Jahanshad, N.; Ching, C.R.K.; Salminen, L.E.; Thomopoulos, S.I.; Bright, J.; Baune, B.T.; Bertolín, S.; Bralten, J.; Bruin, W.B.; et al. ENIGMA and global neuroscience: A decade of large-scale studies of the brain in health and disease across more than 40 countries. *Transl. Psychiatry* **2020**, *10*, 100. [CrossRef] [PubMed]
68. Kringelbach, M.L.; Deco, G. Brain States and Transitions: Insights from Computational Neuroscience. *Cell Rep.* **2020**, *32*, 108128. [CrossRef] [PubMed]
69. Barabási, D.L.; Bianconi, G.; Bullmore, E.; Burgess, M.; Chung, S.; Eliassi-Rad, T.; George, D.; Kovács, I.A.; Makse, H.; Nichols, T.E. Neuroscience needs network science. *J. Neurosci.* **2023**, *43*, 5989–5995. [CrossRef] [PubMed]
70. Rokicki, J.; Wolfers, T.; Nordhøy, W.; Tesli, N.; Quintana, D.S.; Alnaes, D.; Richard, G.; de Lange, A.G.; Lund, M.J.; Norbom, L.; et al. Multimodal imaging improves brain age prediction and reveals distinct abnormalities in patients with psychiatric and neurological disorders. *Hum. Brain Mapp.* **2021**, *42*, 1714–1726. [CrossRef]
71. Livint Popa, L.; Dragos, H.; Pantelemon, C.; Verisezan Rosu, O.; Strilciuc, S. The Role of Quantitative EEG in the Diagnosis of Neuropsychiatric Disorders. *J. Med. Life* **2020**, *13*, 8–15. [CrossRef]
72. Kreisl, W.C.; Kim, M.J.; Coughlin, J.M.; Henter, I.D.; Owen, D.R.; Innis, R.B. PET imaging of neuroinflammation in neurological disorders. *Lancet Neurol.* **2020**, *19*, 940–950. [CrossRef]

73. Meyer, J.H.; Cervenka, S.; Kim, M.J.; Kreisl, W.C.; Henter, I.D.; Innis, R.B. Neuroinflammation in psychiatric disorders: PET imaging and promising new targets. *Lancet Psychiatry* **2020**, *7*, 1064–1074. [CrossRef]

74. Shusharina, N.; Yukhnenko, D.; Botman, S.; Sapunov, V.; Savinov, V.; Kamyshov, G.; Sayapin, D.; Voznyuk, I. Modern methods of diagnostics and treatment of neurodegenerative diseases and depression. *Diagnostics* **2023**, *13*, 573. [CrossRef] [PubMed]

75. Statsenko, Y.; Habuza, T.; Smetanina, D.; Simiyu, G.L.; Meribout, S.; King, F.C.; Gelovani, J.G.; Das, K.M.; Gorkom, K.N.-V.; Zaręba, K. Unraveling lifelong brain morphometric dynamics: A protocol for systematic review and meta-analysis in healthy neurodevelopment and ageing. *Biomedicines* **2023**, *11*, 1999. [CrossRef]

76. Di Gregorio, F.; La Porta, F.; Petrone, V.; Battaglia, S.; Orlandi, S.; Ippolito, G.; Romei, V.; Piperno, R.; Lullini, G. Accuracy of EEG biomarkers in the detection of clinical outcome in disorders of consciousness after severe acquired brain injury: Preliminary results of a pilot study using a machine learning approach. *Biomedicines* **2022**, *10*, 1897. [CrossRef] [PubMed]

77. Chu, P.-C.; Huang, C.-S.; Chang, P.-K.; Chen, R.-S.; Chen, K.-T.; Hsieh, T.-H.; Liu, H.-L. Weak ultrasound contributes to neuromodulatory effects in the rat motor cortex. *Int. J. Mol. Sci.* **2023**, *24*, 2578. [CrossRef]

78. Chojnowski, K.; Opiełka, M.; Gozdalski, J.; Radziwon, J.; Dańczyszyn, A.; Aitken, A.V.; Biancardi, V.C.; Winklewski, P.J. The Role of Arginine-Vasopressin in Stroke and the Potential Use of Arginine-Vasopressin Type 1 Receptor Antagonists in Stroke Therapy: A Narrative Review. *Int. J. Mol. Sci.* **2023**, *24*, 2119. [CrossRef]

79. Senevirathne, D.K.L.; Mahboob, A.; Zhai, K.; Paul, P.; Kammen, A.; Lee, D.J.; Yousef, M.S.; Chaari, A. Deep Brain Stimulation beyond the Clinic: Navigating the Future of Parkinson's and Alzheimer's Disease Therapy. *Cells* **2023**, *12*, 1478. [CrossRef] [PubMed]

80. Adeel, M.; Chen, C.-C.; Lin, B.-S.; Chen, H.-C.; Liou, J.-C.; Li, Y.-T.; Peng, C.-W. Safety of Special Waveform of Transcranial Electrical Stimulation (TES): In Vivo Assessment. *Int. J. Mol. Sci.* **2022**, *23*, 6850. [CrossRef]

81. Fraile-Ramos, J.; Garrit, A.; Reig-Vilallonga, J.; Giménez-Llort, L. Hepatic Oxi-Inflammation and Neophobia as Potential Liver–Brain Axis Targets for Alzheimer's Disease and Aging, with Strong Sensitivity to Sex, Isolation, and Obesity. *Cells* **2023**, *12*, 1517. [CrossRef]

82. Fan, P.; Miranda, O.; Qi, X.; Kofler, J.; Sweet, R.A.; Wang, L. Unveiling the Enigma: Exploring Risk Factors and Mechanisms for Psychotic Symptoms in Alzheimer's Disease through Electronic Medical Records with Deep Learning Models. *Pharmaceuticals* **2023**, *16*, 911. [CrossRef] [PubMed]

83. Skobeleva, K.; Shalygin, A.; Mikhaylova, E.; Guzhova, I.; Ryazantseva, M.; Kaznacheyeva, E. The STIM1/2-regulated calcium homeostasis is impaired in hippocampal neurons of the 5xFAD mouse model of Alzheimer's disease. *Int. J. Mol. Sci.* **2022**, *23*, 14810. [CrossRef]

84. Li, T.; Xu, G.; Yi, J.; Huang, Y. Intraoperative Hypothermia Induces Vascular Dysfunction in the CA1 Region of Rat Hippocampus. *Brain Sci.* **2022**, *12*, 692. [CrossRef]

85. Chen, J.; Huang, L.; Yang, Y.; Xu, W.; Qin, Q.; Qin, R.; Liang, X.; Lai, X.; Huang, X.; Xie, M. Somatic Cell Reprogramming for Nervous System Diseases: Techniques, Mechanisms, Potential Applications, and Challenges. *Brain Sci.* **2023**, *13*, 524. [CrossRef]

86. Garifulin, R.; Davleeva, M.; Izmailov, A.; Fadeev, F.; Markosyan, V.; Shevchenko, R.; Minyazeva, I.; Minekayev, T.; Lavrov, I.; Islamov, R. Evaluation of the Autologous Genetically Enriched Leucoconcentrate on the Lumbar Spinal Cord Morpho-Functional Recovery in a Mini Pig with Thoracic Spine Contusion Injury. *Biomedicines* **2023**, *11*, 1331. [CrossRef] [PubMed]

87. Zheng, Y.; Huo, J.; Yang, M.; Zhang, G.; Wan, S.; Chen, X.; Zhang, B.; Liu, H. ERK1/2 Signalling Pathway Regulates Tubulin-Binding Cofactor B Expression and Affects Astrocyte Process Formation after Acute Foetal Alcohol Exposure. *Brain Sci.* **2022**, *12*, 813. [CrossRef] [PubMed]

88. Sivananthan, S.; Lee, L.; Anderson, G.; Csanyi, B.; Williams, R.; Gissen, P. Buffy coat score as a biomarker of treatment response in neuronal ceroid lipofuscinosis type 2. *Brain Sci.* **2023**, *13*, 209. [CrossRef] [PubMed]

89. Vasiliu, O. Efficacy, Tolerability, and Safety of Toludesvenlafaxine for the Treatment of Major Depressive Disorder—A Narrative Review. *Pharmaceuticals* **2023**, *16*, 411. [CrossRef]

90. Kalkman, H.O. Inhibition of Microglial GSK3β Activity Is Common to Different Kinds of Antidepressants: A Proposal for an In Vitro Screen to Detect Novel Antidepressant Principles. *Biomedicines* **2023**, *11*, 806. [CrossRef]

91. Ippolito, G.; Bertaccini, R.; Tarasi, L.; Di Gregorio, F.; Trajkovic, J.; Battaglia, S.; Romei, V. The role of alpha oscillations among the main neuropsychiatric disorders in the adult and developing human brain: Evidence from the last 10 years of research. *Biomedicines* **2022**, *10*, 3189. [CrossRef]

92. Sellitto, M.; Terenzi, D.; Starita, F.; di Pellegrino, G.; Battaglia, S. The Cost of Imagined Actions in a Reward-Valuation Task. *Brain Sci.* **2022**, *12*, 582. [CrossRef] [PubMed]

93. Zhao, L.; Hou, B.; Ji, L.; Ren, D.; Yuan, F.; Liu, L.; Bi, Y.; Yang, F.; Yu, S.; Yi, Z. NGFR gene and single nucleotide polymorphisms, rs2072446 and rs11466162, playing roles in psychiatric disorders. *Brain Sci.* **2022**, *12*, 1372. [CrossRef] [PubMed]

94. Martos, D.; Tuka, B.; Tanaka, M.; Vécsei, L.; Telegdy, G. Memory enhancement with kynurenic acid and its mechanisms in neurotransmission. *Biomedicines* **2022**, *10*, 849. [CrossRef]

95. Bueno, C.R.d.S.; Tonin, M.C.C.; Buchaim, D.V.; Barraviera, B.; Junior, R.S.F.; Santos, P.S.d.S.; Reis, C.H.B.; Pastori, C.M.; Pereira, E.d.S.B.M.; Nogueira, D.M.B. Morphofunctional Improvement of the Facial Nerve and Muscles with Repair Using Heterologous Fibrin Biopolymer and Photobiomodulation. *Pharmaceuticals* **2023**, *16*, 653. [CrossRef] [PubMed]

96. Hong, F.; He, G.; Zhang, M.; Yu, B.; Chai, C. The establishment of a mouse model of recurrent primary dysmenorrhea. *Int. J. Mol. Sci.* **2022**, *23*, 6128. [CrossRef] [PubMed]

97. Ting, W.K.; Fadul, F.A.; Fecteau, S.; Ethier, C. Neurostimulation for Stroke Rehabilitation. *Front. Neurosci.* **2021**, *15*, 649459. [CrossRef] [PubMed]

98. Moisset, X.; Lanteri-Minet, M.; Fontaine, D. Neurostimulation methods in the treatment of chronic pain. *J. Neural Transm.* **2020**, *127*, 673–686. [CrossRef] [PubMed]

99. Avenanti, A.; Coccia, M.; Ladavas, E.; Provinciali, L.; Ceravolo, M.G. Low-frequency rTMS promotes use-dependent motor plasticity in chronic stroke: A randomized trial. *Neurology* **2012**, *78*, 256–264. [CrossRef]

100. Turrini, S.; Bevacqua, N.; Cataneo, A.; Chiappini, E.; Fiori, F.; Battaglia, S.; Romei, V.; Avenanti, A. Neurophysiological Markers of Premotor-Motor Network Plasticity Predict Motor Performance in Young and Older Adults. *Biomedicines* **2023**, *11*, 1464. [CrossRef]

101. Grimaldi, D.; Papalambros, N.A.; Zee, P.C.; Malkani, R.G. Neurostimulation techniques to enhance sleep and improve cognition in aging. *Neurobiol. Dis.* **2020**, *141*, 104865. [CrossRef]

102. Haneef, Z.; Gavvala, J.R.; Combs, H.L.; Han, A.; Ali, I.; Sheth, S.A.; Stinson, J.M. Brain Stimulation Using Responsive Neurostimulation Improves Verbal Memory: A Crossover Case-Control Study. *Neurosurgery* **2022**, *90*, 306–312. [CrossRef]

103. Ridgewell, C.; Heaton, K.J.; Hildebrandt, A.; Couse, J.; Leeder, T.; Neumeier, W.H. The effects of transcutaneous auricular vagal nerve stimulation on cognition in healthy individuals: A meta-analysis. *Neuropsychology* **2021**, *35*, 352. [CrossRef] [PubMed]

104. Yüksel, M.M.; Sun, S.; Latchoumane, C.; Boch, J.; Courtine, G.; Raffin, E.E.; Hummel, F.C. Low-intensity focused ultrasound neuromodulation for stroke recovery: A novel deep brain stimulation approach for neurorehabilitation? *IEEE Open J. Eng. Med. Biol.* **2023**, *4*, 300–318. [CrossRef] [PubMed]

105. Kim, E.; Kim, S.; Kwon, Y.W.; Seo, H.; Kim, M.; Chung, W.G.; Park, W.; Song, H.; Lee, D.H.; Lee, J. Electrical stimulation for therapeutic approach. *Interdiscip. Med.* **2023**, *1*, e20230003. [CrossRef]

106. Vecchio, P.F.F.M.; Iodice, R.; Ferreri, F.F.; Miraglia, M.B.C.; Orlando, E.J. 8.1 Transcranial magnetic stimulation. In *Magnetic Materials and Technologies for Medical Applications*; Woodhead Publishing: Sawston, UK, 2021; p. 227.

107. Carzoli, K.L.; Kogias, G.; Fawcett-Patel, J.; Liu, S.J. Cerebellar interneurons control fear memory consolidation via learning-induced HCN plasticity. *Cell Rep.* **2023**, *42*, 113057. [CrossRef] [PubMed]

108. Krishna, V.; Sammartino, F.; Rezai, A. A Review of the Current Therapies, Challenges, and Future Directions of Transcranial Focused Ultrasound Technology: Advances in Diagnosis and Treatment. *JAMA Neurol.* **2018**, *75*, 246–254. [CrossRef] [PubMed]

109. Borgomaneri, S.; Zanon, M.; Di Luzio, P.; Cataneo, A.; Arcara, G.; Romei, V.; Tamietto, M.; Avenanti, A. Increasing associative plasticity in temporo-occipital back-projections improves visual perception of emotions. *Nat. Commun.* **2023**, *14*, 5720. [CrossRef] [PubMed]

110. Valchev, N.; Tidoni, E.; Hamilton, A.F.C.; Gazzola, V.; Avenanti, A. Primary somatosensory cortex necessary for the perception of weight from other people's action: A continuous theta-burst TMS experiment. *Neuroimage* **2017**, *152*, 195–206. [CrossRef] [PubMed]

111. Doss, M.K.; Madden, M.B.; Gaddis, A.; Nebel, M.B.; Griffiths, R.R.; Mathur, B.N.; Barrett, F.S. Models of psychedelic drug action: Modulation of cortical-subcortical circuits. *Brain* **2022**, *145*, 441–456. [CrossRef]

112. Alzheimer, A. Die arteriosklerotische atrophie des gehirns. *Allg. Z. Für Psychiatr.* **1895**, *51*, 809–811.

113. Alzheimer, A. Über eine eigenartige Erkrankung der Hirnrinde. *Allg. Z. Fur Psychiatr. Und Psych.-Gerichtl. Med.* **1907**, *64*, 146–148.

114. Gainotti, G.; Quaranta, D.; Vita, M.G.; Marra, C. Neuropsychological predictors of conversion from mild cognitive impairment to Alzheimer's disease. *J. Alzheimers Dis.* **2014**, *38*, 481–495. [CrossRef]

115. Papanastasiou, C.A.; Theochari, C.A.; Zareifopoulos, N.; Arfaras-Melainis, A.; Giannakoulas, G.; Karamitsos, T.D.; Palaiodimos, L.; Ntaios, G.; Avgerinos, K.I.; Kapogiannis, D. Atrial fibrillation is associated with cognitive impairment, all-cause dementia, vascular dementia, and Alzheimer's disease: A systematic review and meta-analysis. *J. Gen. Intern. Med.* **2021**, *36*, 3122–3135. [CrossRef] [PubMed]

116. Blocq, P.M.G. Sur la lésion et la pathogénie de l'épilepsie dite essentielle. *Sem. Médicale* **1892**, *12*.

117. Glenner, G.G.; Wong, C.W. Alzheimer's disease: Initial report of the purification and characterization of a novel cerebrovascular amyloid protein. *Biochem. Biophys. Res. Commun.* **1984**, *120*, 885–890. [CrossRef] [PubMed]

118. Tzioras, M.; McGeachan, R.I.; Durrant, C.S.; Spires-Jones, T.L. Synaptic degeneration in Alzheimer disease. *Nat. Rev. Neurol.* **2023**, *19*, 19–38. [CrossRef] [PubMed]

119. Bir, S.C.; Khan, M.W.; Javalkar, V.; Toledo, E.G.; Kelley, R.E. Emerging Concepts in Vascular Dementia: A Review. *J. Stroke Cerebrovasc. Dis.* **2021**, *30*, 105864. [CrossRef] [PubMed]

120. Puppala, G.K.; Gorthi, S.P.; Chandran, V.; Gundabolu, G. Frontotemporal Dementia—Current Concepts. *Neurol. India* **2021**, *69*, 1144–1152. [CrossRef] [PubMed]

121. Milán-Tomás, Á.; Fernández-Matarrubia, M.; Rodríguez-Oroz, M.C. Lewy Body Dementias: A Coin with Two Sides? *Behav. Sci.* **2021**, *11*, 94. [CrossRef]

122. Kara, B.; Gordon, M.N.; Gifani, M.; Dorrance, A.M.; Counts, S.E. Vascular and nonvascular mechanisms of cognitive impairment and dementia. *Clin. Geriatr. Med.* **2023**, *39*, 109–122. [CrossRef]

123. Thau, L.; Reddy, V.; Singh, P. Anatomy, Central Nervous System. In *StatPearls*; StatPearls Publishing LLC.: Treasure Island, FL, USA, 2024.

124. Pistono, C.; Bister, N.; Stanová, I.; Malm, T. Glia-derived extracellular vesicles: Role in central nervous system communication in health and disease. *Front. Cell Dev. Biol.* **2021**, *8*, 623771. [CrossRef] [PubMed]

125. Bigbee, J.W. Cells of the Central Nervous System: An Overview of Their Structure and Function. In *Glycobiology of the Nervous System*; Springer: Cham, Switzerland, 2022; pp. 41–64.

126. Von Bartheld, C.S.; Bahney, J.; Herculano-Houzel, S. The search for true numbers of neurons and glial cells in the human brain: A review of 150 years of cell counting. *J. Comp. Neurol.* **2016**, *524*, 3865–3895. [CrossRef] [PubMed]

127. Xin, Y.; Tian, M.; Deng, S.; Li, J.; Yang, M.; Gao, J.; Pei, X.; Wang, Y.; Tan, J.; Zhao, F. The key drivers of brain injury by systemic inflammatory responses after sepsis: Microglia and neuroinflammation. *Mol. Neurobiol.* **2023**, *60*, 1369–1390. [CrossRef] [PubMed]

128. Rice, D.; Barone, S., Jr. Critical periods of vulnerability for the developing nervous system: Evidence from humans and animal models. *Environ. Health Perspect.* **2000**, *108* (Suppl. S3), 511–533. [CrossRef] [PubMed]

129. Snyder, J.M.; Gibson-Corley, K.N.; Radaelli, E. Nervous system. In *Pathology of Genetically Engineered and Other Mutant Mice*; Wiley Online Library: Hoboken, NJ, USA, 2021; pp. 462–492.

130. Howard, R.; Al-Mayhani, T.; Carr, A.; Leff, A.; Morrow, J.; Rossor, A. Toxic, metabolic and physical insults to the nervous system. In *Neurology: A Queen Square Textbook*; Wiley Online Library: Hoboken, NJ, USA, 2024; pp. 903–943.

131. Jiao, Y.; Liu, Y.W.; Chen, W.G.; Liu, J. Neuroregeneration and functional recovery after stroke: Advancing neural stem cell therapy toward clinical application. *Neural Regen. Res.* **2021**, *16*, 80–92. [CrossRef] [PubMed]

132. Chen, W.; Hu, Y.; Ju, D. Gene therapy for neurodegenerative disorders: Advances, insights and prospects. *Acta Pharm. Sin. B* **2020**, *10*, 1347–1359. [CrossRef] [PubMed]

133. Papadimitriou, L.; Manganas, P.; Ranella, A.; Stratakis, E. Biofabrication for neural tissue engineering applications. *Mater. Today Bio* **2020**, *6*, 100043. [CrossRef] [PubMed]

134. Liu, W.; Wang, X.; O'Connor, M.; Wang, G.; Han, F. Brain-Derived Neurotrophic Factor and Its Potential Therapeutic Role in Stroke Comorbidities. *Neural Plast.* **2020**, *2020*, 1969482. [CrossRef] [PubMed]

135. Li, R.; Li, D.-h.; Zhang, H.-y.; Wang, J.; Li, X.-k.; Xiao, J. Growth factors-based therapeutic strategies and their underlying signaling mechanisms for peripheral nerve regeneration. *Acta Pharmacol. Sin.* **2020**, *41*, 1289–1300. [CrossRef]

136. Maurer, D.M.; Raymond, T.J.; Davis, B.N. Depression: Screening and Diagnosis. *Am. Fam. Physician* **2018**, *98*, 508–515.

137. Khune, A.A.; Rathod, H.K.; Deshmukh, S.P.; Chede, S.B. Mental health, depressive disorder and its management: A review. *GSC Biol. Pharm. Sci.* **2023**, *25*, 001–013. [CrossRef]

138. Schulz, D. Depression development: From lifestyle changes to motivational deficits. *Behav. Brain Res.* **2020**, *395*, 112845. [CrossRef] [PubMed]

139. Lépine, J.P.; Briley, M. The increasing burden of depression. *Neuropsychiatr. Dis. Treat.* **2011**, *7*, 3–7. [CrossRef]

140. Fernandez-Rodrigues, V.; Sanchez-Carro, Y.; Lagunas, L.N.; Rico-Uribe, L.A.; Pemau, A.; Diaz-Carracedo, P.; Diaz-Marsa, M.; Hervas, G.; de la Torre-Luque, A. Risk factors for suicidal behaviour in late-life depression: A systematic review. *World J. Psychiatry* **2022**, *12*, 187. [CrossRef]

141. Obuobi-Donkor, G.; Nkire, N.; Agyapong, V.I. Prevalence of major depressive disorder and correlates of thoughts of death, suicidal behaviour, and death by suicide in the geriatric population—A general review of literature. *Behav. Sci.* **2021**, *11*, 142. [CrossRef]

142. Mann, F.; Wang, J.; Pearce, E.; Ma, R.; Schlief, M.; Lloyd-Evans, B.; Ikhtabi, S.; Johnson, S. Loneliness and the onset of new mental health problems in the general population. *Soc. Psychiatry Psychiatr. Epidemiol.* **2022**, *57*, 2161–2178. [CrossRef] [PubMed]

143. Filatova, E.V.; Shadrina, M.I.; Slominsky, P.A. Major Depression: One Brain, One Disease, One Set of Intertwined Processes. *Cells* **2021**, *10*, 1283. [CrossRef] [PubMed]

144. Kendall, K.; Van Assche, E.; Andlauer, T.; Choi, K.; Luykx, J.; Schulte, E.; Lu, Y. The genetic basis of major depression. *Psychol. Med.* **2021**, *51*, 2217–2230. [CrossRef] [PubMed]

145. Remes, O.; Mendes, J.F.; Templeton, P. Biological, psychological, and social determinants of depression: A review of recent literature. *Brain Sci.* **2021**, *11*, 1633. [CrossRef]

146. Botha, F.B.; Dozois, D.J. The influence of emphasizing psychological causes of depression on public stigma. *Can. J. Behav. Sci. /Rev. Can. Des Sci. Du Comport.* **2015**, *47*, 313. [CrossRef]

147. Bhatt, S.; Nagappa, A.N.; Patil, C.R. Role of oxidative stress in depression. *Drug Discov. Today* **2020**, *25*, 1270–1276. [CrossRef] [PubMed]

148. Cuijpers, P.; Noma, H.; Karyotaki, E.; Vinkers, C.H.; Cipriani, A.; Furukawa, T.A. A network meta-analysis of the effects of psychotherapies, pharmacotherapies and their combination in the treatment of adult depression. *World Psychiatry* **2020**, *19*, 92–107. [CrossRef] [PubMed]

149. Cuijpers, P.; Oud, M.; Karyotaki, E.; Noma, H.; Quero, S.; Cipriani, A.; Arroll, B.; Furukawa, T.A. Psychologic treatment of depression compared with pharmacotherapy and combined treatment in primary care: A network meta-analysis. *Ann. Fam. Med.* **2021**, *19*, 262–270. [CrossRef]

150. Guidi, J.; Fava, G.A. Sequential combination of pharmacotherapy and psychotherapy in major depressive disorder: A systematic review and meta-analysis. *JAMA Psychiatry* **2021**, *78*, 261–269. [CrossRef] [PubMed]

151. Henssler, J.; Alexander, D.; Schwarzer, G.; Bschor, T.; Baethge, C. Combining antidepressants vs antidepressant monotherapy for treatment of patients with acute depression: A systematic review and meta-analysis. *JAMA Psychiatry* **2022**, *79*, 300–312. [CrossRef] [PubMed]

152. Grover, S.; Nguyen, J.A.; Reinhart, R.M.G. Synchronizing Brain Rhythms to Improve Cognition. *Annu. Rev. Med.* **2021**, *72*, 29–43. [CrossRef] [PubMed]

153. Wilson, D.; Moehlis, J. Recent advances in the analysis and control of large populations of neural oscillators. *Annu. Rev. Control* **2022**, *54*, 327–351. [CrossRef]

154. Daffertshofer, A.; Pietras, B. Phase synchronization in neural systems. In *Synergetics*; Springer: New York, NY, USA, 2020; pp. 221–233.

155. Wang, Y.; Shi, X.; Cheng, B.; Chen, J. Synchronization and rhythm transition in a complex neuronal network. *IEEE Access* **2020**, *8*, 102436–102448. [CrossRef]

156. Lopes da Silva, F. EEG and MEG: Relevance to neuroscience. *Neuron* **2013**, *80*, 1112–1128. [CrossRef]

157. Donoghue, T.; Schaworonkow, N.; Voytek, B. Methodological considerations for studying neural oscillations. *Eur. J. Neurosci.* **2022**, *55*, 3502–3527. [CrossRef]

158. Balart-Sánchez, S.A.; Bittencourt-Villalpando, M.; van der Naalt, J.; Maurits, N.M. Electroencephalography, magnetoencephalography, and cognitive reserve: A systematic review. *Arch. Clin. Neuropsychol.* **2021**, *36*, 1374–1391. [CrossRef] [PubMed]

159. Kim, J.A.; Davis, K.D. Neural oscillations: Understanding a neural code of pain. *Neurosci.* **2021**, *27*, 544–570. [CrossRef] [PubMed]

160. Battaglia, M.R.; Di Fazio, C.; Battaglia, S. Activated Tryptophan-Kynurenine metabolic system in the human brain is associated with learned fear. *Front. Mol. Neurosci.* **2023**, *16*, 1217090. [CrossRef]

161. Mehterov, N.; Minchev, D.; Gevezova, M.; Sarafian, V.; Maes, M. Interactions Among Brain-Derived Neurotrophic Factor and Neuroimmune Pathways Are Key Components of the Major Psychiatric Disorders. *Mol. Neurobiol.* **2022**, *59*, 4926–4952. [CrossRef] [PubMed]

162. Battaglia, S.; Di Fazio, C.; Vicario, C.M.; Avenanti, A. Neuropharmacological modulation of N-methyl-D-aspartate, noradrenaline and endocannabinoid receptors in fear extinction learning: Synaptic transmission and plasticity. *Int. J. Mol. Sci.* **2023**, *24*, 5926. [CrossRef]

163. Bruno, F.; Abondio, P.; Montesanto, A.; Luiselli, D.; Bruni, A.C.; Maletta, R. The Nerve Growth Factor Receptor (NGFR/p75(NTR)): A Major Player in Alzheimer's Disease. *Int. J. Mol. Sci.* **2023**, *24*, 3200. [CrossRef]

164. Farzan, F. Transcranial Magnetic Stimulation-Electroencephalography for Biomarker Discovery in Psychiatry. *Biol. Psychiatry* **2024**, *95*, 564–580. [CrossRef] [PubMed]

165. Günther, A.; Hanganu-Opatz, I.L. Neuronal oscillations: Early biomarkers of psychiatric disease? *Front. Behav. Neurosci.* **2022**, *16*, 1038981. [CrossRef] [PubMed]

166. Lu, Z.; Wang, H.; Gu, J.; Gao, F. Association between abnormal brain oscillations and cognitive performance in patients with bipolar disorder: Molecular mechanisms and clinical evidence. *Synapse* **2022**, *76*, e22247. [CrossRef]

167. Ronconi, L.; Vitale, A.; Federici, A.; Pini, E.; Molteni, M.; Casartelli, L. Altered neural oscillations and connectivity in the beta band underlie detail-oriented visual processing in autism. *NeuroImage Clin.* **2020**, *28*, 102484. [CrossRef]

168. Ribeiro, F.F.; Xapelli, S. Intervention of Brain-Derived Neurotrophic Factor and Other Neurotrophins in Adult Neurogenesis. *Adv. Exp. Med. Biol.* **2021**, *1331*, 95–115. [CrossRef] [PubMed]

169. Van Hook, M.J. Brain-derived neurotrophic factor is a regulator of synaptic transmission in the adult visual thalamus. *J. Neurophysiol.* **2022**, *128*, 1267–1277. [CrossRef] [PubMed]

170. Yang, T.; Nie, Z.; Shu, H.; Kuang, Y.; Chen, X.; Cheng, J.; Yu, S.; Liu, H. The role of BDNF on neural plasticity in depression. *Front. Cell. Neurosci.* **2020**, *14*, 82. [CrossRef]

171. Leschik, J.; Gentile, A.; Cicek, C.; Peron, S.; Tevosian, M.; Beer, A.; Radyushkin, K.; Bludau, A.; Ebner, K.; Neumann, I. Brain-derived neurotrophic factor expression in serotonergic neurons improves stress resilience and promotes adult hippocampal neurogenesis. *Prog. Neurobiol.* **2022**, *217*, 102333. [CrossRef] [PubMed]

172. Mitra, S.; Gera, R.; Linderoth, B.; Lind, G.; Wahlberg, L.; Almqvist, P.; Behbahani, H.; Eriksdotter, M. A Review of Techniques for Biodelivery of Nerve Growth Factor (NGF) to the Brain in Relation to Alzheimer's Disease. *Adv. Exp. Med. Biol.* **2021**, *1331*, 167–191. [CrossRef] [PubMed]

173. Lorenzini, L.; Baldassarro, V.A.; Stanzani, A.; Giardino, L. Nerve growth factor: The first molecule of the neurotrophin family. In *Recent Advances in NGF and Related Molecules: The Continuum of the NGF "Saga"*; Springer: Berlin/Heidelberg, Germany, 2021; pp. 3–10.

174. Zha, K.; Yang, Y.; Tian, G.; Sun, Z.; Yang, Z.; Li, X.; Sui, X.; Liu, S.; Zhao, J.; Guo, Q. Nerve growth factor (NGF) and NGF receptors in mesenchymal stem/stromal cells: Impact on potential therapies. *Stem Cells Transl. Med.* **2021**, *10*, 1008–1020. [CrossRef] [PubMed]

175. Eva, C. Nerve growth factor: Influence on cholinergic neurons in the CNS. In *CNS Neurotransmitters and Neuromodulators*; CRC Press: Boca Raton, FL, USA, 2020; pp. 233–256.

176. Liu, Z.; Wu, H.; Huang, S. Role of NGF and its receptors in wound healing. *Exp. Ther. Med.* **2021**, *21*, 599. [CrossRef] [PubMed]

177. Geula, C.; Dunlop, S.R.; Ayala, I.; Kawles, A.S.; Flanagan, M.E.; Gefen, T.; Mesulam, M.M. Basal forebrain cholinergic system in the dementias: Vulnerability, resilience, and resistance. *J. Neurochem.* **2021**, *158*, 1394–1411. [CrossRef]

178. Cuello, A.C.; Pentz, R.; Hall, H. The brain NGF metabolic pathway in health and in Alzheimer's pathology. *Front. Neurosci.* **2019**, *13*, 441218. [CrossRef]

179. Lai, N.-S.; Yu, H.-C.; Huang Tseng, H.-Y.; Hsu, C.-W.; Huang, H.-B.; Lu, M.-C. Increased serum levels of brain-derived neurotrophic factor contribute to inflammatory responses in patients with rheumatoid arthritis. *Int. J. Mol. Sci.* **2021**, *22*, 1841. [CrossRef]

180. Beattie, E.; Zhou, J.; Grimes, M.; Bunnett, N.; Howe, C.; Mobley, W. A signaling endosome hypothesis to explain NGF actions: Potential implications for neurodegeneration. In Proceedings of the Cold Spring Harbor Symposia on Quantitative Biology; Cold Spring Harbor Laboratory Press: Cold Spring Harbor, NY, USA, 1996; pp. 389–406.
181. Battaglia, S.; Di Fazio, C.; Mazzà, M.; Tamietto, M.; Avenanti, A. Targeting Human Glucocorticoid Receptors in Fear Learning: A Multiscale Integrated Approach to Study Functional Connectivity. *Int. J. Mol. Sci.* **2024**, *25*, 864. [CrossRef] [PubMed]
182. Tortora, F.; Hadipour, A.L.; Battaglia, S.; Falzone, A.; Avenanti, A.; Vicario, C.M. The role of serotonin in fear learning and memory: A systematic review of human studies. *Brain Sci.* **2023**, *13*, 1197. [CrossRef] [PubMed]
183. Valotto Neto, L.J.; Reverete de Araujo, M.; Moretti Junior, R.C.; Mendes Machado, N.; Joshi, R.K.; dos Santos Buglio, D.; Barbalho Lamas, C.; Direito, R.; Fornari Laurindo, L.; Tanaka, M. Investigating the Neuroprotective and Cognitive-Enhancing Effects of Bacopa monnieri: A Systematic Review Focused on Inflammation, Oxidative Stress, Mitochondrial Dysfunction, and Apoptosis. *Antioxidants* **2024**, *13*, 393. [CrossRef] [PubMed]
184. Direito, R.; Barbalho, S.M.; Sepodes, B.; Figueira, M.E. Plant-Derived Bioactive Compounds: Exploring Neuroprotective, Metabolic, and Hepatoprotective Effects for Health Promotion and Disease Prevention. *Pharmaceutics* **2024**, *16*, 577. [CrossRef]
185. Fornari Laurindo, L.; Aparecido Dias, J.; Cressoni Araújo, A.; Torres Pomini, K.; Machado Galhardi, C.; Rucco Penteado Detregiachi, C.; Santos de Argollo Haber, L.; Donizeti Roque, D.; Dib Bechara, M.; Vialogo Marques de Castro, M. Immunological dimensions of neuroinflammation and microglial activation: Exploring innovative immunomodulatory approaches to mitigate neuroinflammatory progression. *Front. Immunol.* **2024**, *14*, 1305933. [CrossRef] [PubMed]
186. Matias, J.N.; Achete, G.; Campanari, G.S.d.S.; Guiguer, É.L.; Araújo, A.C.; Buglio, D.S.; Barbalho, S.M. A systematic review of the antidepressant effects of curcumin: Beyond monoamines theory. *Aust. N. Z. J. Psychiatry* **2021**, *55*, 451–462. [CrossRef] [PubMed]
187. de Souza, G.A.; de Marqui, S.V.; Matias, J.N.; Guiguer, E.L.; Barbalho, S.M. Effects of Ginkgo biloba on diseases related to oxidative stress. *Planta Medica* **2020**, *86*, 376–386.
188. Lutzu, S.; Castillo, P.E. Modulation of NMDA Receptors by G-protein-coupled receptors: Role in Synaptic Transmission, Plasticity and Beyond. *Neuroscience* **2021**, *456*, 27–42. [CrossRef] [PubMed]
189. Armada-Moreira, A.; Gomes, J.I.; Pina, C.C.; Savchak, O.K.; Gonçalves-Ribeiro, J.; Rei, N.; Pinto, S.; Morais, T.P.; Martins, R.S.; Ribeiro, F.F.; et al. Going the Extra (Synaptic) Mile: Excitotoxicity as the Road Toward Neurodegenerative Diseases. *Front. Cell Neurosci.* **2020**, *14*, 90. [CrossRef]
190. McGrath, T.; Baskerville, R.; Rogero, M.; Castell, L. Emerging Evidence for the Widespread Role of Glutamatergic Dysfunction in Neuropsychiatric Diseases. *Nutrients* **2022**, *14*, 917. [CrossRef]
191. Soleimani, G.; Nitsche, M.A.; Bergmann, T.O.; Towhidkhah, F.; Violante, I.R.; Lorenz, R.; Kuplicki, R.; Tsuchiyagaito, A.; Mulyana, B.; Mayeli, A.; et al. Closing the loop between brain and electrical stimulation: Towards precision neuromodulation treatments. *Transl. Psychiatry* **2023**, *13*, 279. [CrossRef] [PubMed]
192. Nagappan, P.G.; Chen, H.; Wang, D.Y. Neuroregeneration and plasticity: A review of the physiological mechanisms for achieving functional recovery postinjury. *Mil. Med. Res.* **2020**, *7*, 30. [CrossRef] [PubMed]
193. Burns, T.C.; Quinones-Hinojosa, A. Regenerative medicine for neurological diseases—Will regenerative neurosurgery deliver? *Bmj* **2021**, *373*, n955. [CrossRef] [PubMed]
194. Stevenson, R.; Samokhina, E.; Rossetti, I.; Morley, J.W.; Buskila, Y. Neuromodulation of Glial Function During Neurodegeneration. *Front. Cell Neurosci.* **2020**, *14*, 278. [CrossRef] [PubMed]
195. Török, N.; Török, R.; Molnár, K.; Szolnoki, Z.; Somogyvári, F.; Boda, K.; Tanaka, M.; Klivényi, P.; Vécsei, L. Single nucleotide polymorphisms of indoleamine 2, 3-dioxygenase 1 influenced the age onset of Parkinson's disease. *Front. Biosci.-Landmark* **2022**, *27*, 265. [CrossRef]
196. Ji, J.L.; Demšar, J.; Fonteneau, C.; Tamayo, Z.; Pan, L.; Kraljič, A.; Matkovič, A.; Purg, N.; Helmer, M.; Warrington, S. QuNex—An integrative platform for reproducible neuroimaging analytics. *Front. Neuroinform.* **2023**, *17*, 1104508. [CrossRef] [PubMed]
197. Duggineny, S. Neurological Disorders: A Comprehensive Review of Insights and Innovations in Treatment Development. *Unique Endeavor Bus. Soc. Sci.* **2023**, *2*, 28–36.
198. Everett, J.N.; Dettwyler, S.A.; Jing, X.; Stender, C.; Schmitter, M.; Baptiste, A.; Chun, J.; Kawaler, E.A.; Khanna, L.G.; Gross, S.A. Impact of comprehensive family history and genetic analysis in the multidisciplinary pancreatic tumor clinic setting. *Cancer Med.* **2023**, *12*, 2345–2355. [CrossRef] [PubMed]
199. Constant, A.; Badcock, P.; Friston, K.; Kirmayer, L.J. Integrating evolutionary, cultural, and computational psychiatry: A multilevel systemic approach. *Front. Psychiatry* **2022**, *13*, 763380. [CrossRef]
200. Büttenbender, P.C.; de Azevedo Neto, E.G.; Heckler, W.F.; Barbosa, J.L.V. A computational model for identifying behavioral patterns in people with neuropsychiatric disorders. *IEEE Lat. Am. Trans.* **2022**, *20*, 582–589. [CrossRef]

Study Protocol

Unraveling Lifelong Brain Morphometric Dynamics: A Protocol for Systematic Review and Meta-Analysis in Healthy Neurodevelopment and Ageing

Yauhen Statsenko [1,2,3,*], Tetiana Habuza [3,4], Darya Smetanina [1], Gillian Lylian Simiyu [1], Sarah Meribout [1,2,5], Fransina Christina King [6,7], Juri G. Gelovani [1,8,9,10], Karuna M. Das [1], Klaus N. -V. Gorkom [1], Kornelia Zaręba [11], Taleb M. Almansoori [1,*], Miklós Szólics [12,13], Fatima Ismail [14] and Milos Ljubisavljevic [6,7]

1   Radiology Department, College of Medicine and Health Sciences, United Arab Emirates University, Al Ain P.O. Box 15551, United Arab Emirates
2   Medical Imaging Platform, ASPIRE Precision Medicine Research Institute Abu Dhabi, Al Ain P.O. Box 15551, United Arab Emirates
3   Big Data Analytics Center, United Arab Emirates University, Al Ain P.O. Box 15551, United Arab Emirates
4   Department of Computer Science and Software Engineering, College of Information Technology, United Arab Emirates University, Al Ain P.O. Box 15551, United Arab Emirates
5   Internal Medicine Department, Maimonides Medical Center, New York, NY 11219, USA
6   Physiology Department, College of Medicine and Health Sciences, United Arab Emirates University, Al Ain P.O. Box 15551, United Arab Emirates
7   Neuroscience Platform, ASPIRE Precision Medicine Research Institute Abu Dhabi, Al Ain P.O. Box 15551, United Arab Emirates
8   Biomedical Engineering Department, College of Engineering, Wayne State University, Detroit, MI 48202, USA
9   Siriraj Hospital, Mahidol University, Nakhon Pathom 73170, Thailand
10 Provost Office, United Arab Emirates University, Al Ain P.O. Box 15551, United Arab Emirates
11 Obstetrics & Gynecology Department, College of Medicine and Health Sciences, United Arab Emirates University, Al Ain P.O. Box 15551, United Arab Emirates
12 Neurology Division, Medicine Department, Tawam Hospital, Al Ain P.O. Box 15258, United Arab Emirates
13 Internal Medicine Department, College of Medicine and Health Sciences, United Arab Emirates University, Al Ain P.O. Box 15551, United Arab Emirates
14 Pediatric Department, College of Medicine and Health Sciences, United Arab Emirates University, Al Ain P.O. Box 15551, United Arab Emirates
*   Correspondence: e.a.statsenko@gmail.com or e.a.statsenko@uaeu.ac.ae (Y.S.); taleb.almansoor@uaeu.ac.ae (T.M.A.)

Citation: Statsenko, Y.; Habuza, T.; Smetanina, D.; Simiyu, G.L.; Meribout, S.; King, F.C.; Gelovani, J.G.; Das, K.M.; Gorkom, K.N.-V.; Zaręba, K.; et al. Unraveling Lifelong Brain Morphometric Dynamics: A Protocol for Systematic Review and Meta-Analysis in Healthy Neurodevelopment and Ageing. *Biomedicines* **2023**, *11*, 1999. https://doi.org/10.3390/biomedicines11071999

Academic Editors: Lydia Giménez-Llort, Masaru Tanaka, Chong Chen, Simone Battaglia and Piril Hepsomali

Received: 16 February 2023
Revised: 27 June 2023
Accepted: 30 June 2023
Published: 14 July 2023

Abstract: A high incidence and prevalence of neurodegenerative diseases and neurodevelopmental disorders justify the necessity of well-defined criteria for diagnosing these pathologies from brain imaging findings. No easy-to-apply quantitative markers of abnormal brain development and ageing are available. We aim to find the characteristic features of non-pathological development and degeneration in distinct brain structures and to work out a precise descriptive model of brain morphometry in age groups. We will use four biomedical databases to acquire original peer-reviewed publications on brain structural changes occurring throughout the human life-span. Selected publications will be uploaded to Covidence systematic review software for automatic deduplication and blinded screening. Afterwards, we will manually review the titles, abstracts, and full texts to identify the papers matching eligibility criteria. The relevant data will be extracted to a 'Summary of findings' table. This will allow us to calculate the annual rate of change in the volume or thickness of brain structures and to model the lifelong dynamics in the morphometry data. Finally, we will adjust the loss of weight/thickness in specific brain areas to the total intracranial volume. The systematic review will synthesise knowledge on structural brain change across the life-span.

Keywords: brain; nerve degeneration; brain diseases; neurodevelopmental disorders; ageing; regression analysis; atrophy; neuroimaging; meta-analysis; cognition

1. Introduction

Brain structure continuously changes throughout life. In healthy individuals, age-related brain atrophy and neurodevelopment account for these changes. In patients with mental and psychological disorders, disease-related brain atrophy takes place. The atrophy can start in late adolescence and young adulthood and lead to abnormal brain development [1]. Differentiation between normal and abnormal structural changes remains a challenge [2–4]. The current study focuses on the age-specific anatomy of the brain and describes the structural evolution of the brain across the life-span. A descriptive model that recapitulates key features of brain development and ageing is a powerful medium for obtaining comprehensive knowledge on the abnormalities signalling neurodegeneration. The model is a potential tool assisting clinicians in the early diagnostics of dementia. Studies on the structural signs of both normal and abnormal brain development and ageing remain relevant today.

The necessity of continuous research on the aforementioned issue is paramount. Autism spectrum disorder puts a substantial socio-economic burden, and the average diagnostic delay after initial concerns does not differ among high-, medium- and low-income countries [5]. On average, autism incidence is around 80 cases per 100,000 children in developed economies. The UAE is in the top-10 list of countries with the highest autism rates, which reached 112.40 in 2021. The statistics on middle- and low-income countries is incomplete because of misreported cases [6].

Neurocognitive slowing is a typical functional outcome of normal brain ageing, while cognitive decline is a sign of cognitive deterioration. In high-income countries, the numbers of people with cognitive decline has been rising due to population ageing. The latest favourable trends in dementia incidence are typical for the Western countries. Meanwhile, wealthy countries of other regions show the opposite tendency [7]. It is expected that around 35.25 million people will be diagnosed with dementia in Asia by 2025, while 13.97 million people will be diagnosed in European countries [8]. Neurodegenerative diseases (ND) are among leading causes of life loss and disability among the elderly, and the number of deaths due to Alzheimer's disease has risen disproportionally in comparison to the top attributed cases of mortality (e.g., heart disease, cancer, and strokes) [9]. Individual causes of dementia are hard to detect and predict. Therefore, the disease is commonly diagnosed at late stages after the prominent manifestation of intellectual decline.

The identification of infants and elderly at risk of neurological disorders is important for optimal disease management [10]. For this reason, scientists look for highly sensitive screening and diagnosing tools, which enable early therapeutic strategies and targeted interventions. According to the recent updates from molecular biology studies, genetic tests can detect over 500 neurodevelopmental diseases [11]. The test specificity is sufficient for distinguishing dementia variants [12,13]. Still, the available diagnostic methods possess the following limitations. First, genetic tests are invasive, labour intensive, and time consuming. In addition, neurologists face difficulties in choosing an appropriate genetic test and interpreting the laboratory results [14]. Second, neurobehavioral and cognitive questionnaires are relatively short. Usually, they are well-tolerated by examinees, but the sensitivity and specificity are low in the paediatric population and in adults. For example, the Hammersmith Infant Neurological Examination predicts adverse neurophysiological outcomes at 1 year with a sensitivity of 50–64% and specificity of 73–77% [15]. The sensitivity and specificity of differentiation among dementia phenotypes with cognitive tests is around 71.92% and 70.06%, correspondently [16,17]. Third, molecular imaging with positron emission tomography can reveal early metabolic changes that are not visible on CT and MRI scans. The latter mainly document the brain macrostructure [18]. Although nuclear imaging is highly sensitive, the examination is invasive. The exposure to radiation and possible allergic reactions are the drawbacks of radiotracer injection. Additionally, the number of PET scanners is insufficient to arrange the routine screening of patients at risk of dementia [19,20]. MRI is the method of choice for detecting structural abnormalities in the brain, identifying disease-specific diagnostic signs at late stages of neurodegeneration, and reflecting their functional outcomes [4,21,22]. The early diagnostics of dementia necessitate the quantitative analysis of MRI findings along with bioengenering technologies. The creation of population norms will advance these technologies, thus meeting the demands of neuroscience for early and reliable diagnostics.

Scientists fail to describe the exact pathophysiological mechanisms in which age-related brain atrophy contributes to malfunctioning of the nervous system. Many publications cover either the structural or functional impairment. High variability in individual anatomy and physiology complicates studies on structure–function association [4,23–25]. Morphological studies with structural MRI revealed a marked variance across individuals in the extent of age-related brain change [26]. Research on brain functioning produced inconclusive findings on the onset and rate of episodic memory loss in the elderly. Different inherited and lifestyle factors account for these results [27,28]. A link between structural and functional impairment has not yet been explicitly explained.

Hypothetically, different brain parts have a common pace of age-related structural changes in the normal ageing, while a disproportion in the regional atrophy indicates accelerated brain ageing. Another hypothesis of the current study is the specificity of brain morphometry findings regarding normal growing and maturation in childhood and adolescence. In analogy to late life, morphometric changes in the pediatric population also remain an object of ongoing studies [29]. For this reason, we want to create a descriptive model of volumetric changes in the brain across the life-span.

The aim of our review is to find the characteristic features of non-pathological development and decline in distinct brain structures and to work out a precise descriptive model of brain morphometry among age groups. Specifically, we want to determine whether diverse brain compartments develop and decline proportionally throughout life and to calculate the annual rate of change.

2. Materials and Methods

We will perform a meta-analysis of available findings and modelled age-related changes in the brain. This approach is advantageous over a retrospective analysis or prospective observation, especially when researchers want to establish populational norms. A meta-analysis serves as a reliable source of data if performed in accordance with the standardized methodology [30,31]. Another reason for this study design is the intent to overcome limitations of original studies such as a small sample size, a narrow age range of cohorts, etc.

This protocol will be prepared in accordance with the Preferred Reporting Items for Systematic Review and Meta-Analyses Protocol (PRISMA-P) and registered in the PROSPERO International database of prospectively registered systematic reviews (protocol No: CRD42022354112). The PRISMA-P checklist is included in the supporting material (Supplementary File S1).

2.1. Eligibility Criteria

To perform the search, we will pre-define and list the inclusion and exclusion criteria for the literature (see Table 1). Given that we focus on the structural change of the healthy population, the articles reporting findings on the following pathologies will be excluded: impaired development, mental disorders, organic brain pathologies, injuries to the head, and neurodegenerative diseases. We will not limit the age and sex of the study participants. The literature sources will be required to report findings of in vivo MRI examinations of the total brain or specific brain structures with outcome measurements such as absolute or proportional change in volume and/or size (width, length, thickness). We will exclude reviews, case reports, theses, dissertations, conference abstracts, editorial letters, and protocol papers. Animal studies and interventional research will also be eliminated from the search query.

Table 1. Inclusion and exclusion criteria.

Inclusion Criteria	Exclusion Criteria	
	for Literature	for Subjects
1. Original peer-reviewed studies 2. Studies of the longitudinal and cross-sectional design 3. Studies on absolute or proportional change in volume, thickness, and other dimensions of the brain structures 4. Female and male participants of any age starting from birth 5. Individuals free from mental disorders, brain pathologies, and injuries	1. Grey literature 2. Editorial letters and protocol papers 3. Case studies and reviews 4. Studies performed on animals 5. Interventional studies (both therapeutic and surgical interventions) 6. Exposure of the participants to any factor that can potentially affect results.	Patients suffering from: 1. Mental and psychological disorders (F00–F99 in ICD-10) 2. Cerebrovascular diseases (I60– I69) 3. Organic pathology of the central nervous system (e.g., brain and meninges tumors: C71, D32–33) 4. Injury to the head (S00–S09)

2.2. Information Sources and Strategy

A comprehensive systematic search for literature will be conducted using four biomedical databases: PubMed, Embase, Scopus, and Web of Science. The pre-search was performed in September–October 2022 in PubMed and its Medical Subject Headings (MeSH). The review is set to start in November 2023. The strategy will be updated ahead of the manuscript submission. The names of 112 brain structures will be used as the keywords for the search strategy. The structures will be segmented from the brain MRI with the FreeSurfer software [32]. We will look for combinations of all the key terms in the "title", "abstract", and "MeSH"/"thesaurus" fields. MeSH terms will include "brain", "Magnetic Resonance Imaging", "organ size," "atrophy", "aging", and "age factors". Our search will be limited to the papers published in English from 1990 to 2023. We will conduct hand screening of reference lists in the retrieved papers. All records identified in the literature search will be uploaded to the systematic review software Covidence for automatic deduplication and blinded screening. Reproducible search strings for all databases will be appended to the review (Supplementary File S2).

2.3. Study Selection

Once the papers are uploaded to the Covidence software, two independent reviewers will subsequently screen the title/abstract and full text against the predefined criteria. A third reviewer will resolve the discrepancies identified with the software. The reasons for exclusion of full text records will be stored. A PRISMA flow diagram will be used to report the screening process.

2.4. Data Extraction

Two independent reviewers will extract the data from the final list of papers into a summary of findings table. These records will include the basic characteristics (the authors, country, publication year, journal, study design, and mean age of study participants) in addition to the targeted data (change in size and volume of a brain structure within a period of time). The third reviewer will resolve eventual discrepancies in the data collection.

2.5. Quality Assessment of Individual Studies

Two independent reviewers will critically appraise each individual study included into the analysis. They will assess the quality of individual studies with the Joanna Brigs Institute checklist for analytical cross-sectional studies [33]. The tool has 8 questions with multiple choice answers "yes", "no", "unclear", and "not applicable". If a study scores less than 4 "yes" answers, it will be excluded from the statistical analysis.

We will construct funnel plots for each brain region and visually assess them. In the diagrams, the effect size is plotted against the standard error of the effect size. Asymmetry of the graph indicates publication bias. In our review, the effect size corresponds to annual atrophy of a studied brain structure.

2.6. Data Analysis and Synthesis

Prior to statistical analysis, we will explore the heterogeneity level of the studies with the Higgins–Thompson I^2 test [34]. Potential sources of heterogeneity include the strength of the magnetic field of MRI scanners, the type of segmentation, sample size, and study design. If I^2 exceeds 75%, we will perform a narrative systematic review instead of the meta-analysis. The final manuscript will present the number of qualifying articles and give a description of the overall trend of structural changes in the brain. The systematic review will analyse the sample size, age, and sex of the participants, and it will derive average statistical data on annual changes in brain regions. The team statistician will calculate the following parameters: (1) the average annual pace of enlargement or atrophy of each brain structure and (2) the side-specific change of brain structures in size or volume. For the analysis, we will use descriptive statistics and machine learning.

We will calculate the mean value of the left and right side volumes to receive the average size. To compensate for the variability in head size, the data will be estimated on a normalised volume in percentage to the total intracranial volume. Afterwards, we will compute the percentage of relative change per year, which is the first derivative of the model divided by the initial value and provided in % per year. To model the lifelong evolution of volumetric data for specific brain areas, we will consider linear, quadratic, cubic, or higher degree equations (see Equations (1)–(4)). Scatterplots in Figure 1 depict the models that we built based on the results of a preliminary search for references covering age-related change in the hippocampal [2,35–41] and lateral ventricle volumes [2,3,42–44].

$$Vol = \beta_0 + \beta_1 Age + \epsilon \tag{1}$$

$$Vol = \beta_0 + \beta_1 Age + \beta_2 Age^2 + \epsilon \tag{2}$$

$$Vol = \beta_0 + \beta_1 Age + \beta_2 Age^2 + \beta_3 Age^3 + \epsilon \tag{3}$$

$$Vol = \beta_0 + \beta_1 Age + \beta_2 Age^2 + ... + \beta_k Age^k, k = 1, 10. \tag{4}$$

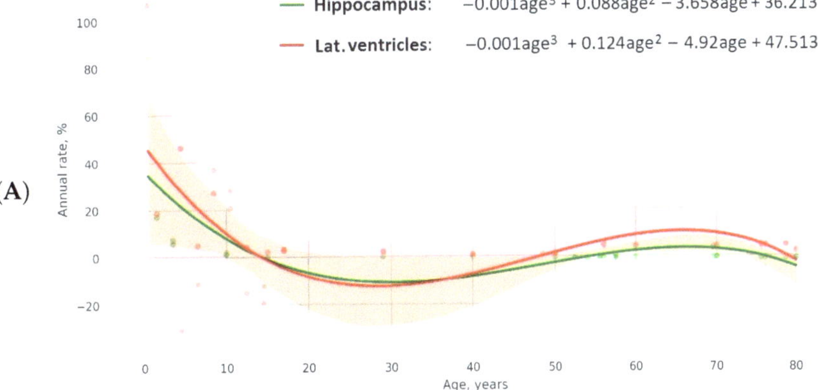

Figure 1. *Cont.*

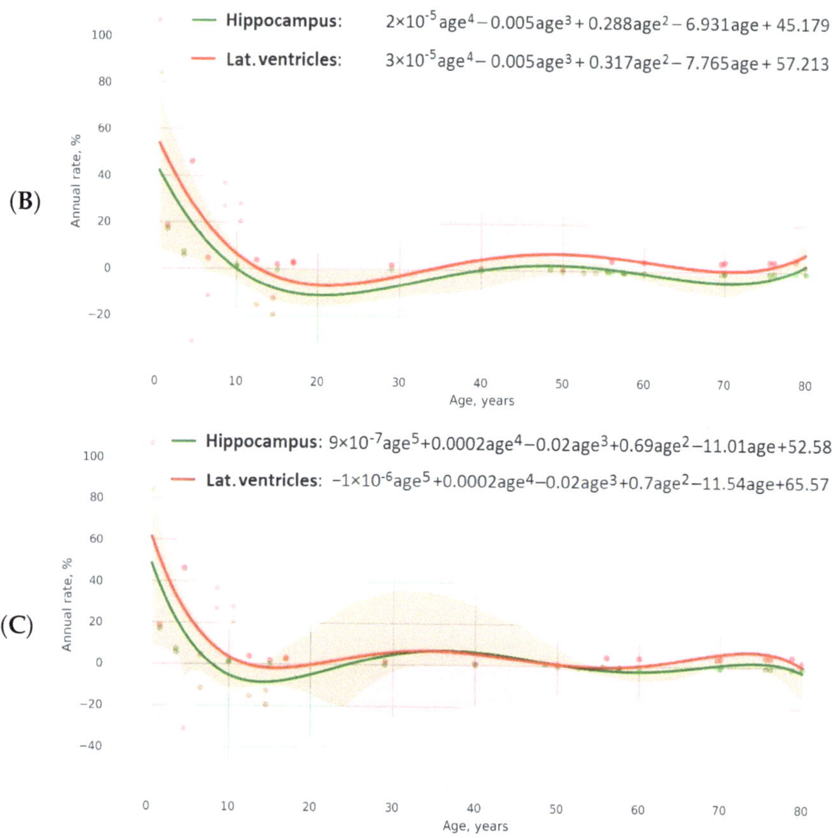

Figure 1. Cubic (**A**), fourth (**B**), and fifth order (**C**) models of lifelong change in hippocampal and lateral ventricle relative volumes.

We will also use hybrid models with exponential cumulative distributions for growth with the linear, quadratic, cubic, or higher degree equations (see Equations (5)–(7)).

$$Vol = \beta_4(1 - e^{-Age/\beta_5}) + \beta_0 + \beta_1 Age + \epsilon \tag{5}$$

$$Vol = \beta_4(1 - e^{-Age/\beta_5}) + \beta_0 + \beta_1 Age + \beta_2 Age^2 + \epsilon \tag{6}$$

$$Vol = \beta_4(1 - e^{-Age/\beta_5}) + \beta_0 + \beta_1 Age + \beta_2 Age^2 + \beta_3 Age^3 + \epsilon. \tag{7}$$

Then, we will select the model explaining most of the data with a minimum number of parameters. To identify the best one among the candidate models, we will use a Bayesian information criterion (see a scatterplot in Figure 2). Finally, we will assess the portion of brain atrophy in a specific brain area by calculating the percentage of atrophy, i.e., its absolute relative difference with the control model. As the control model, we will use the relative rate of change in the CSF volume, which is a marker of the total brain shrinkage. The study pipeline is shown in Figure 3.

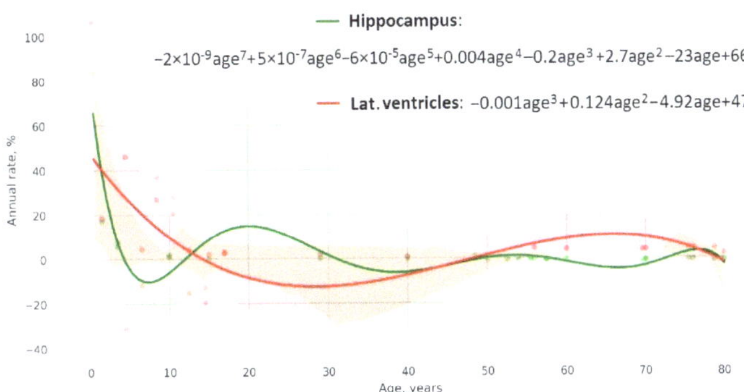

Figure 2. Best model types for hippocampus and lateral ventricles according to Bayesian information criterion.

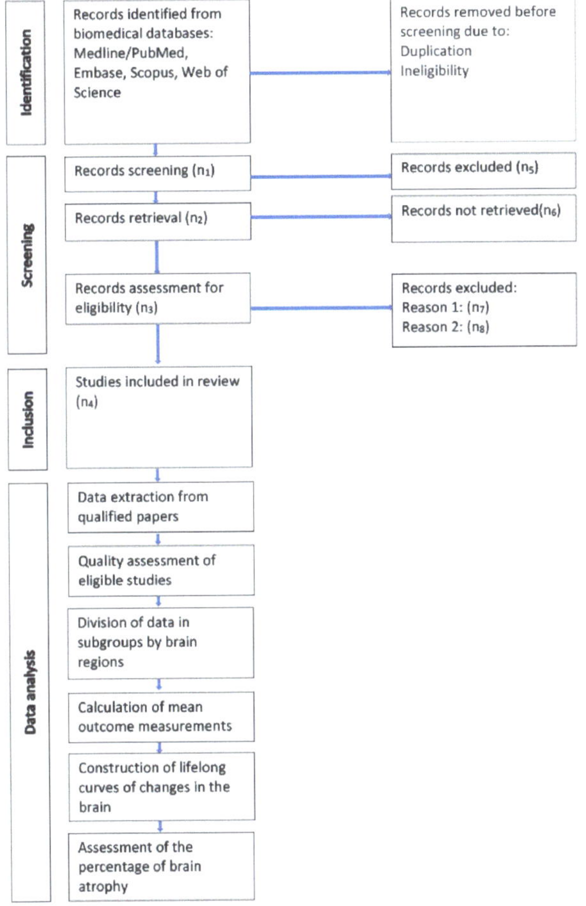

Figure 3. Study pipeline.

3. Discussion

3.1. Establishing Descriptive Model for Brain Ageing

A vast amount of literature is available on the volumetric decline of specific brain regions and neurocognitive slowing. Still, a thorough descriptive model of normal brain ageing is missing. This can be attributed to the following limitations that are typical for recent studies. First, researchers investigated specific age groups, and they did not study changes throughout life. Second, different methodological approaches used by the authors may account for incompatible findings [45]. For example, some studies used the cross-sectional design, and others—used the longitudinal one. Third, individual variations in brain structure limit our ability to establish population norms. Fourth, genetics, environmental, lifestyle, and cultural distinctions contribute to pronounced difference in brain morphology among nations. The current systemic review and meta-analysis investigates brain development and ageing in the global population.

MRI-based neuroimaging studies with voxel- and surface-based brain morphometry can detect a tiny change of brain structure. With these techniques, bioengineers help clinicians to quantify cortical and subcortical grey matter atrophy in terms of volume loss, macro-morphological changes, and cortical thinning [46,47]. One can evaluate structural damage to the white matter with voxel-based morphometry [46]. Recent studies reported the following outcomes of brain atrophy in normal ageing: volumetric reduction in the cortex and the sudden shrinkage of neuronal networks [48]. The latter describes the way in which brain atrophy impairs structural and functional connectivity. However, the aforementioned studies failed to provide a precise descriptive model of structural changes in cognitively preserved individuals, and future research should address this shortcoming. The evolution of brain structure in different life periods is briefly discussed in the next paragraphs.

3.1.1. Period of Development

The total brain volume trajectory is strongly associated with age and cognitive status both in children and the elderly [49,50]. Radiological findings may promote the early diagnostics of impaired neurodevelopment. Still, diagnostic criteria for autism and other neurodevelopmental disorders are not uniform, and the resources for proper examination are limited. Various intellectual, behavioral, and psychiatric disorders in children may result from abnormal cortical development. To identify atypical change during the maturation period, physicians need the normative values for the cortical brain structures in 1–6 year old healthy children. This is the peak brain development period. The early signs of behavioral and developmental disorders become apparent at this age [42]. Researchers try to find out which particular mechanisms of brain development result in dysfunction in socio-emotional and communication networks in infants and toddlers with autism [51,52].

3.1.2. Period of Maturation

The dynamics of the structural brain change in middle-aged adults, and older adults resemble the trajectory of functional performance throughout life [45,53]. From a neuro-physiological point of view, an increase in performance lasts up to the age of 40, and it is followed by a plateau in neurocognitive functioning in the midlife and a decline in older adults [54–57]. The speed and extent of change seems to accelerate with age. It remains unknown if the pace of age-related transformation is common for distinct brain regions or if atrophy is slightly more prominent in certain locations.

3.1.3. Period of Decline

Brain atrophy accounts for morphometric and functional changes in physiological and accelerated ageing. At the microscopic level, the atrophy presents with glial, myelin, axonal and/or neuronal loss. Macroscopically, brain atrophy results in brain shrinkage and compensatory enlargement of cerebrospinal fluid spaces: the ventricles and the sub-arachnoid space. A ventricular volume trajectory shows a strong association with age, because it is a summary marker of atrophy of the grey and white matter [58,59]. Specific structural markers serve as radiologic signs of disease-related atrophy, e.g., the hippocampal volume trajectory is clearly associated with amyloid angiopathy [58,60]. However, the sensitivity of the quantitative radiomical markers is too low to use them in screening for neurodegeneration, and their specificity is insufficient for differentiation among dementia subtypes. Large-scale standardised studies do not provide a comprehensive outline of the normal volumetric decline. In the absence of such studies, physicians make subjective judgements on the pace of brain ageing. This increases the chance of late or false diagnosis. Meanwhile, the therapeutic efficiency of anti-amyloid drugs is significantly higher at early stages of Alzheimer's disease, which also justifies the importance of research on normal and pathological brain ageing [61–66].

3.2. Developing Reference Norms with Meta-Analysis

The current article is a protocol of the future study. Once we complete the analysis, we will compare the findings with the results of studies on relevant issues. For the discussion, we will consider available systematic reviews covering the following research questions: (1) an atrophy rate of brain parts vulnerable to changes in neurodevelopmental delay and neurodegenerative disorders and (2) brain structural correlates of the aforementioned pathologies.

Relationships between environmental factors and brain structure are not the major research topic for the study. Still, the data on environmental risks may show the multi-dimensionality of the research question. The latter is not limited to brain changes across the life, but it also includes the impact of adverse life events, dietary patterns, physical exercise, and vitamin or mineral supplementation on cognitive function in children and the elderly [67–71]. The future manuscript will contain the results of the prospective meta-analysis discussed in the context of healthy and pathological transformations in the brain.

The necessity of the current systematic review arises from the absence of reliable markers of delayed neurodevelopment and accelerated ageing: no threshold criteria for abnormal annual change in the brain are established. Visual assessment of the MRI by the radiologist can confirm dementia diagnosis when cognitive decline becomes prominent. However, subtle changes in the brain remain misreported [21]. This evidences the necessity of supporting clinical decision making with the analysis of radiomical data or developing computer-aided decision tools for automatic image analysis.

Statistics for meta-analysis help to diminish the risk of unreliable research outcomes resulting from inter-study heterogeneity. The meta-analytic approach allows for combining evidence from numerous studies, thus enlarging the study sample, as well as consolidating complex and sometimes conflicting findings [72]. After receiving data from different populations, we will apply a random-effects model to minimise errors in data presentation [72]. The model assists in controlling for unobserved heterogeneity, which is constant over time and not correlated with independent variables.

Our meta-analysis will illustrate life-long trends in brain volumetric changes. We will use regression models to provide trendlines for the variables of interest [73,74]. The trendlines will reflect normative values for volumetric changes in brain parts. This approach will allow us to identify people at risk of brain disorders at a certain age [74]. A large individual deviation from the normative curves may indicate pathologic ageing.

We will focus on the studies of cognitively normal people and model the annual rate of change in their brains. A reason for narrowing the research question to the cognitively intact population is limitations of the concept of accelerated ageing, which should be critically reappraised. For example, it remains unknown whether neurodegeneration is a kind of accelerated ageing [75,76] or an outcome of pathological changes in the body [75,77–79].

When studying accelerated ageing, one should consider individual multi-component reserves in the brain. The structural reserve refers to the number of neurons and synapses, whereas the cognitive reserve determines the ability of the brain to cope with structural brain damage [80]. Supposedly, individual cognitive and structural reserves affect the quality of life in the elderly population and modulate the risk of developing Alzheimer's disease and other types of dementia. Variance in cognitive abilities and amount of neurons and synapses hardens the determination of the precise age of the organism. Calculation of the biological brain age is tricky, because indicators of normal ageing are still missing and the methodology of assessing reserves is not standardised [81]. Hence, we should rather focus on non-pathologic brain ageing.

The objective of the current study totally fits the idea of 'Precision Medicine', which is a concept of personalizing disease prevention and treatment. The concept integrates advanced statistical analysis into routine assessment of clinical findings along with the environmental, social, and behavioral factors impacting the individual [82]. Currently, radiological reports have limited value due to the semi-quantitative evaluation of structural changes by radiologists. Diagnostic errors arise from technical limitations of scanners and misinterpretation by physicians [83,84]. Therefore, bioengineering should create reliable tools for computerised diagnostics and automatic analysis of imaging findings.

We aim to improve the situation by applying quantitative assessment of radiological findings (radiomics) into practice instead of keeping it exceptionally for research. Radiomical findings will support early diagnostics and clinical decision making, thus meeting the demand of time. Radiomics mine high-dimensional data on organ structure and correlate this information with age and clinical endpoints [85]. Modelling structural changes in the healthy cohort will promote future studies on risk stratification of neurodevelopmental/neurodegenerative diseases with radiomics. These studies are currently performed and published, but their clinical applicability remains low [23–25,53,55,86–88].

4. Conclusions

This meta-analysis will help modelling the populational curves of brain development and ageing. For performing a comprehensive assessment of life-long structural changes, we will take into analysis multiple morphometry parameters, e.g., volume, thickness, length and width of brain regions. A systematic review and meta-analysis is the preferable study design for the formulated research question. Combining data from cross-sectional studies and longitudinal observations will allow us to acquire more statistical data on structural changes in the brain. The theoretical value of the future study is the implementation of highly sensitive screening and quantitative assessment of individual risks, which fully fits the idea of 'Precision Medicine'. The practical application of the study is establishing the reference norms that could be used in screening individuals for developmental delay and cognitive decline.

Supplementary Materials: The following supporting information can be downloaded at: https://www.mdpi.com/article/10.3390/biomedicines11071999/s1, Supplementary File S1: PRISMA checklist; Supplementary File S2: Search strategy for PubMed .

Author Contributions: Conceptualization, Y.S.; methodology, K.M.D. and K.Z.; software, T.H.; validation, F.I.; formal analysis, T.H.; investigation, D.S. and G.L.S.; data curation, M.S.; writing—original draft preparation, D.S., Y.S. and S.M.; writing—review and editing, Y.S. and F.C.K.; visualization, T.H.; supervision, K.N.-V.G., T.M.A. and F.I.; project administration, Y.S. and J.G.G.; funding acquisition, M.L. All authors have read and agreed to the published version of the manuscript.

Funding: This research is supported by ASPIRE, the technology program management pillar of Abu Dhabi's Advanced Technology Research Council (ATRC) via the ASPIRE Precision Medicine Research Institute of Abu Dhabi (ASPIREPMRIAD) award grant number VRI-20-10 (21R098).

Institutional Review Board Statement: The systematic review does not require an ethical approval. The findings of the study will be published in a peer-reviewed journal and presented at scientific conferences as a poster or presentation.

Informed Consent Statement: Not applicable.

Data Availability Statement: No datasets were generated or analysed during the current study. All relevant data from this study will be made available upon study completion.

Conflicts of Interest: The authors declare that the research was conducted in the absence of any commercial or financial relationships that could be construed as a potential conflict of interest.

Abbreviations

The following abbreviations are used in this manuscript:

MeSH   Medical Subject Headings
ND   Neurodegenerative disease
PRISMA-P   Preferred Reporting Items for Systematic Reviews and Meta-Analyses Protocol

References

1. Lebel, C.; Beaulieu, C. Longitudinal development of human brain wiring continues from childhood into adulthood. *J. Neurosci.* **2011**, *31*, 10937–10947. [CrossRef] [PubMed]
2. Fjell, A.M.; Walhovd, K.B.; Fennema-Notestine, C.; McEvoy, L.K.; Hagler, D.J.; Holland, D.; Brewer, J.B.; Dale, A.M. One-year brain atrophy evident in healthy aging. *J. Neurosci.* **2009**, *29*, 15223–15231. [CrossRef] [PubMed]
3. Tamnes, C.K.; Walhovd, K.B.; Dale, A.M.; Østby, Y.; Grydeland, H.; Richardson, G.; Westlye, L.T.; Roddey, J.C.; Hagler, D.J., Jr.; Due-Tønnessen, P.; et al. Brain development and aging: Overlapping and unique patterns of change. *Neuroimage* **2013**, *68*, 63–74. [CrossRef]
4. Statsenko, Y.; Meribout, S.; Habuza, T.; Almansoori, T.M.; Gorkom, K.N.V.; Gelovani, J.G.; Ljubisavljevic, M. Patterns of structure-function association in normal aging and in Alzheimer's disease: Screening for mild cognitive impairment and dementia with ML regression and classification models. *Front. Aging Neurosci.* **2023**, *14*, 943566. [CrossRef]
5. Matos, M.B.; Bara, T.S.; Cordeiro, M.L. Autism Spectrum Disorder Diagnoses: A Comparison of Countries with Different Income Levels. *Clin. Epidemiol.* **2022**, *2022*, 959–969. [CrossRef]
6. Autism Rates by Country. 2021. Available online: http://worldpopulationreview.com/country-rankings/autism-rates-by-country (accessed on 4 October 2021).
7. Roehr, S.; Pabst, A.; Luck, T.; Riedel-Heller, S.G. Is dementia incidence declining in high-income countries? A systematic review and meta-analysis. *Clin. Epidemiol.* **2018**, *2018*, 1233–1247. [CrossRef] [PubMed]
8. Numbers of People with Dementia Worldwide. 2015. Available online: https://www.alzint.org (accessed on 4 October 2021).
9. Mattson, M.P.; Arumugam, T.V. Hallmarks of brain aging: Adaptive and pathological modification by metabolic states. *Cell Metab.* **2018**, *27*, 1176–1199. [CrossRef]
10. Cioni, G.; Inguaggiato, E.; Sgandurra, G. Early intervention in neurodevelopmental disorders: Underlying neural mechanisms. *Dev. Med. Child. Neurol.* **2016**, *58*, 61–66. [CrossRef]
11. Chung, W.K.; Berg, J.S.; Botkin, J.R.; Brenner, S.E.; Brosco, J.P.; Brothers, K.B.; Currier, R.J.; Gaviglio, A.; Kowtoniuk, W.E.; Olson, C.; et al. Newborn screening for neurodevelopmental diseases: Are we there yet? *Am. J. Med. Genet.* **2022**, *190*, 222–230. [CrossRef]
12. Leidinger, P.; Backes, C.; Deutscher, S.; Schmitt, K.; Mueller, S.C.; Frese, K.; Haas, J.; Ruprecht, K.; Paul, F.; Stähler, C.; et al. A blood based 12-miRNA signature of Alzheimer disease patients. *Genome Biol.* **2013**, *14*, R78. [CrossRef]
13. Lovrecic, L.; Kastrin, A.; Kobal, J.; Pirtosek, Z.; Krainc, D.; Peterlin, B. Gene expression changes in blood as a putative biomarker for Huntington's disease. *J. Mov. Disord.* **2009**, *24*, 2277–2281. [CrossRef] [PubMed]
14. Huang, Y.; Yu, S.; Wu, Z.; Tang, B. Genetics of hereditary neurological disorders in children. *Transl. Pediatr.* **2014**, *3*, 108. [CrossRef]
15. Venkata, S.K.R.G.; Pournami, F.; Prabhakar, J.; Nandakumar, A.; Jain, N. Disability prediction by early Hammersmith neonatal neurological examination: A diagnostic study. *J. Child Neurol.* **2020**, *35*, 731–736. [CrossRef] [PubMed]
16. Gordon, P.H.; Wang, Y.; Doorish, C.; Lewis, M.; Battista, V.; Mitsumoto, H.; Marder, K. A screening assessment of cognitive impairment in patients with ALS. *Amyotroph. Lateral Scler.* **2007**, *8*, 362–365. [CrossRef] [PubMed]
17. Mackin, R.S.; Ayalon, L.; Feliciano, L.; Areán, P.A. The sensitivity and specificity of cognitive screening instruments to detect cognitive impairment in older adults with severe psychiatric illness. *J. Geriatr. Psychiatry Neurol.* **2010**, *23*, 94–99. [CrossRef] [PubMed]
18. Minoshima, S.; Mosci, K.; Cross, D.; Thientunyakit, T. Brain [F-18] FDG PET for clinical dementia workup: Differential diagnosis of Alzheimer's disease and other types of dementing disorders. *Semin. Nucl. Med.* **2021**, *51*, 230–240. [CrossRef]
19. Ossenkoppele, R.; Rabinovici, G.D.; Smith, R.; Cho, H.; Schöll, M.; Strandberg, O.; Palmqvist, S.; Mattsson, N.; Janelidze, S.; Santillo, A.; et al. Discriminative accuracy of [18F] flortaucipir positron emission tomography for Alzheimer disease vs other neurodegenerative disorders. *Jama* **2018**, *320*, 1151–1162. [CrossRef]
20. Tripathi, M.; Tripathi, M.; Damle, N.; Kushwaha, S.; Jaimini, A.; D'Souza, M.M.; Sharma, R.; Saw, S.; Mondal, A. Differential diagnosis of neurodegenerative dementias using metabolic phenotypes on F-18 FDG PET/CT. *Neuroradiol. J.* **2014**, *27*, 13–21. [CrossRef]
21. Koikkalainen, J.; Rhodius-Meester, H.; Tolonen, A.; Barkhof, F.; Tijms, B.; Lemstra, A.W.; Tong, T.; Guerrero, R.; Schuh, A.; Ledig, C.; et al. Differential diagnosis of neurodegenerative diseases using structural MRI data. *NeuroImage Clin.* **2016**, *11*, 435–449. [CrossRef]

22. Statsenko, Y.; Gorkom, K.; Ljubisabljevic, M.; Szolics, M.; Baylis, C.; Al Mansoori, T.M.; Das, K.M.; Soteriades, E.; Charykova, I.A.; Dotsenko, T.; et al. Psychological outcomes of age-related brain atrophy. *Neuroradiology* **2019**, *61*, 73.

23. Habuza, T.; Zaki, N.; Mohamed, E.A.; Statsenko, Y. Deviation from model of normal aging in alzheimer's disease: Application of deep learning to structural MRI data and cognitive tests. *IEEE Access* **2022**, *10*, 53234–53249. [CrossRef]

24. Habuza, T.; Zaki, N.; Statsenko, Y.; Elyassami, S. MRI and cognitive tests-based screening tool for dementia. *J. Neurol. Sci.* **2021**, *429*, 118964. [CrossRef]

25. Habuza, T.; Statsenko, Y.; Uzianbaeva, L.; Gorkom, K.; Zaki, N.; Belghali, M. Models of brain cognitive and morphological changes across the life: Machine learning-based approach. ESNR 2021. *Neuroradiology* **2021**, *63*, 42.

26. Gorbach, T.; Pudas, S.; Lundquist, A.; Orädd, G.; Josefsson, M.; Salami, A.; de Luna, X.; Nyberg, L. Longitudinal association between hippocampus atrophy and episodic-memory decline. *Neurobiol. Aging* **2017**, *51*, 167–176. [CrossRef] [PubMed]

27. Josefsson, M.; de Luna, X.; Pudas, S.; Nilsson, L.G.; Nyberg, L. Genetic and lifestyle predictors of 15-year longitudinal change in episodic memory. *J. Am. Geriatr. Soc.* **2012**, *60*, 2308–2312. [CrossRef]

28. Statsenko, Y.; Habuza, T.; Gorkom, K.N.V.; Zaki, N.; Almansoor, T.M. Applying the inverse efficiency score to visual-motor task for studying speed/accuracy performance while aging. *Front. Aging Neurosci.* **2020**, *12*, 452. [CrossRef] [PubMed]

29. Gilmore, J.H.; Knickmeyer, R.C.; Gao, W. Imaging structural and functional brain development in early childhood. *Nat. Rev. Neurosci.* **2018**, *19*, 123–137. [CrossRef] [PubMed]

30. Butler, A.; Hall, H.; Copnell, B. A guide to writing a qualitative systematic review protocol to enhance evidence-based practice in nursing and health care. *Worldviews Evid.-Based Nurs.* **2016**, *13*, 241–249. [CrossRef]

31. Pigott, T.D.; Polanin, J.R. Methodological guidance paper: High-quality meta-analysis in a systematic review. *Rev. Educ. Res.* **2020**, *90*, 24–46. [CrossRef]

32. Freesurfer Labels. 2013. Available online: https://www.slicer.org/wiki (accessed on 21 March 2022).

33. Aromataris, E.; Stern, C.; Lockwood, C.; Barker, T.H.; Klugar, M.; Jadotte, Y.; Evans, C.; Ross-White, A.; Lizarondo, L.; Stephenson, M.; et al. JBI series paper 2: Tailored evidence synthesis approaches are required to answer diverse questions: A pragmatic evidence synthesis toolkit from JBI. *J. Clin. Epidemiol.* **2022**, *150*,196–202. [CrossRef]

34. Higgins, J.P.; Thompson, S.G.; Deeks, J.J.; Altman, D.G. Measuring inconsistency in meta-analyses. *BMJ* **2003**, *327*, 557–560. [CrossRef]

35. Walhovd, K.B.; Fjell, A.M.; Reinvang, I.; Lundervold, A.; Dale, A.M.; Eilertsen, D.E.; Quinn, B.T.; Salat, D.; Makris, N.; Fischl, B. Effects of age on volumes of cortex, white matter and subcortical structures. *Neurobiol. Aging* **2005**, *26*, 1261–1270. [CrossRef] [PubMed]

36. Gilmore, J.H.; Shi, F.; Woolson, S.L.; Knickmeyer, R.C.; Short, S.J.; Lin, W.; Zhu, H.; Hamer, R.M.; Styner, M.; Shen, D. Longitudinal development of cortical and subcortical gray matter from birth to 2 years. *Cereb. Cortex* **2012**, *22*, 2478–2485. [CrossRef]

37. Raz, N.; Lindenberger, U.; Rodrigue, K.M.; Kennedy, K.M.; Head, D.; Williamson, A.; Dahle, C.; Gerstorf, D.; Acker, J.D. Regional brain changes in aging healthy adults: General trends, individual differences and modifiers. *Cereb. Cortex* **2005**, *15*, 1676–1689. [CrossRef]

38. Fjell, A.M.; Westlye, L.T.; Grydeland, H.; Amlien, I.; Espeseth, T.; Reinvang, I.; Raz, N.; Holland, D.; Dale, A.M.; Walhovd, K.B.; et al. Critical ages in the life course of the adult brain: Nonlinear subcortical aging. *Neurobiol. Aging* **2013**, *34*, 2239–2247. [CrossRef]

39. Walhovd, K.B.; Westlye, L.T.; Amlien, I.; Espeseth, T.; Reinvang, I.; Raz, N.; Agartz, I.; Salat, D.H.; Greve, D.N.; Fischl, B.; et al. Consistent neuroanatomical age-related volume differences across multiple samples. *Neurobiol. Aging* **2011**, *32*, 916–932. [CrossRef] [PubMed]

40. Uematsu, A.; Matsui, M.; Tanaka, C.; Takahashi, T.; Noguchi, K.; Suzuki, M.; Nishijo, H. Developmental trajectories of amygdala and hippocampus from infancy to early adulthood in healthy individuals. *PLoS ONE* **2012**, *7*, e46970. [CrossRef]

41. Scahill, R.I.; Frost, C.; Jenkins, R.; Whitwell, J.L.; Rossor, M.N.; Fox, N.C. A longitudinal study of brain volume changes in normal aging using serial registered magnetic resonance imaging. *Arch. Neurol.* **2003**, *60*, 989–994. [CrossRef]

42. Remer, J.; Croteau-Chonka, E.; Dean D.C., III; D'arpino, S.; Dirks, H.; Whiley, D.; Deoni, S.C. Quantifying cortical development in typically developing toddlers and young children, 1–6 years of age. *Neuroimage* **2017**, *153*, 246–261. [CrossRef] [PubMed]

43. Storsve, A.B.; Fjell, A.M.; Tamnes, C.K.; Westlye, L.T.; Overbye, K.; Aasland, H.W.; Walhovd, K.B. Differential longitudinal changes in cortical thickness, surface area and volume across the adult life span: Regions of accelerating and decelerating change. *J. Neurosci.* **2014**, *34*, 8488–8498. [CrossRef]

44. Takahashi, T.; Zhou, S.Y.; Nakamura, K.; Tanino, R.; Furuichi, A.; Kido, M.; Kawasaki, Y.; Noguchi, K.; Seto, H.; Kurachi, M.; et al. A follow-up MRI study of the fusiform gyrus and middle and inferior temporal gyri in schizophrenia spectrum. *Prog. Neuropsychopharmacol. Biol. Psychiatry* **2011**, *35*, 1957–1964. [CrossRef] [PubMed]

45. Statsenko, Y.; Habuza, T.; Smetanina, D.; Simiyu, G.L.; Uzianbaeva, L.; Gorkom, K.N.V.; Zaki, N.; Charykova, I.; Al Koteesh, J.; Ljubisavljevic, M.; et al. Brain morphometry and cognitive performance in normal brain aging: Age- and sex-related structural and functional changes. *Front. Aging Neurosci.* **2022**, *13*, 713680. [CrossRef]

46. Pini, L.; Pievani, M.; Bocchetta, M.; Altomare, D.; Bosco, P.; Cavedo, E.; Galluzzi, S.; Marizzoni, M.; Frisoni, G.B. Brain atrophy in Alzheimer's disease and aging. *Ageing Res. Rev.* **2016**, *30*, 25–48. [CrossRef]

47. Dieckmann, N.; Roediger, A.; Prell, T.; Schuster, S.; Herdick, M.; Mayer, T.E.; Witte, O.W.; Steinbach, R.; Grosskreutz, J. Cortical and subcortical grey matter atrophy in Amyotrophic Lateral Sclerosis correlates with measures of disease accumulation independent of disease aggressiveness. *Neuroimage Clin.* **2022**, *36*, 103162. [CrossRef] [PubMed]

48. Li, X.; Pu, F.; Fan, Y.; Niu, H.; Li, S.; Li, D. Age-related changes in brain structural covariance networks. *Front. Hum. Neurosci.* **2013**, *7*, 98. [CrossRef] [PubMed]

49. Fletcher, E.; Gavett, B.; Harvey, D.; Farias, S.T.; Olichney, J.; Beckett, L.; DeCarli, C.; Mungas, D. Brain volume change and cognitive trajectories in aging. *Neuropsychology* **2018**, *32*, 436. [CrossRef]

50. Pangelinan, M.M.; Zhang, G.; VanMeter, J.W.; Clark, J.E.; Hatfield, B.D.; Haufler, A.J. Beyond age and gender: Relationships between cortical and subcortical brain volume and cognitive-motor abilities in school-age children. *Neuroimage* **2011**, *54*, 3093–3100. [CrossRef]

51. Courchesne, E.; Pierce, K.; Schumann, C.M.; Redcay, E.; Buckwalter, J.A.; Kennedy, D.P.; Morgan, J. Mapping early brain development in autism. *Neuron* **2007**, *56*, 399–413. [CrossRef]

52. Shen, M.D.; Piven, J. Brain and behavior development in autism from birth through infancy. *Dialogues Clin. Neurosci.* **2017**, *19*, 325–333 . [CrossRef]

53. Statsenko, Y.; Habuza, T.; Neidl-Van Gorkom, K.; Zaki, N.; Almansoori, T.M.; Al Zahmi, F.; Ljubisavljevic, M.R.; Belghali, M. Proportional Changes in Cognitive Subdomains During Normal Brain Aging. *Front. Aging Neurosci.* **2021**, *13*, 673469. [CrossRef]

54. Statsenko, Y.; Habuza, T.; Charykova, I.; Gorkom, K.; Zaki, N.; Almansoori, T.; Ljubisavljevic, M.; Szolics, M.; Al Koteesh, J.; Ponomareva, A.; et al. AI models of age-associated changes in CNS composition identified by MRI. *J. Neurol. Sci.* **2021**, *429*, 118303. [CrossRef]

55. Statsenko, Y.; Habuza, T.; Uzianbaeva, L.; Gorkom, K.; Belghali, M.; Charykova, I. Correlation between lifelong dynamics of psychophysiological performance and brain morphology. ESNR 2021. *Neuroradiology* **2021**, *63*, 41–42.

56. Gorkom, K.; Statsenko, Y.; Habuza, T.; Uzianbaeva, L.; Belghali, M.; Charykova, I. Comparison of brain volumetric changes with functional outcomes in physiologic brain aging. ESNR 2021. *Neuroradiology* **2021**, *63*, 43–44.

57. Uzianbaeva, L.; Statsenko, Y.; Habuza, T.; Gorkom, K.; Belghali, M.; Charykova, I. Effects of sex age-related changes in brain morphology. ESNR 2021. *Neuroradiology* **2021**, *63*, 42–43.

58. Erten-Lyons, D.; Dodge, H.H.; Woltjer, R.; Silbert, L.C.; Howieson, D.B.; Kramer, P.; Kaye, J.A. Neuropathologic basis of age-associated brain atrophy. *JAMA Neurol.* **2013**, *70*, 616–622. [CrossRef] [PubMed]

59. Kuo, P.L.; Schrack, J.A.; Shardell, M.D.; Levine, M.; Moore, A.Z.; An, Y.; Elango, P.; Karikkineth, A.; Tanaka, T.; de Cabo, R.; et al. A roadmap to build a phenotypic metric of ageing: Insights from the Baltimore Longitudinal Study of Aging. *J. Intern. Med.* **2020**, *287*, 373–394. [CrossRef]

60. Nagaraja, N.; Wang, W.E.; Duara, R.; DeKosky, S.T.; Vaillancourt, D. Mediation of Reduced Hippocampal Volume by Cerebral Amyloid Angiopathy in Pathologically Confirmed Patients with Alzheimer's Disease. *J. Alzheimer's Dis.* **2023**, *93*, 495–507. [CrossRef]

61. Jack, C.R., Jr.; Knopman, D.S.; Jagust, W.J.; Shaw, L.M.; Aisen, P.S.; Weiner, M.W.; Petersen, R.C.; Trojanowski, J.Q. Hypothetical model of dynamic biomarkers of the Alzheimer's pathological cascade. *Lancet Neurol.* **2010**, *9*, 119–128. [CrossRef] [PubMed]

62. Jack, C.R., Jr.; Albert, M.S.; Knopman, D.S.; McKhann, G.M.; Sperling, R.A.; Carrillo, M.C.; Thies, B.; Phelps, C.H. Introduction to the recommendations from the National Institute on Aging-Alzheimer's Association workgroups on diagnostic guidelines for Alzheimer's disease. *Alzheimer's Dement.* **2011**, *7*, 257–262. [CrossRef] [PubMed]

63. Sperling, R.A.; Aisen, P.S.; Beckett, L.A.; Bennett, D.A.; Craft, S.; Fagan, A.M.; Iwatsubo, T.; Jack, C.R., Jr.; Kaye, J.; Montine, T.J.; et al. Toward defining the preclinical stages of Alzheimer's disease: Recommendations from the National Institute on Aging-Alzheimer's Association workgroups on diagnostic guidelines for Alzheimer's disease. *Alzheimer's Dement.* **2011**, *7*, 280–292. [CrossRef] [PubMed]

64. Jack, C.R., Jr.; Knopman, D.S.; Jagust, W.J.; Petersen, R.C.; Weiner, M.W.; Aisen, P.S.; Shaw, L.M.; Vemuri, P.; Wiste, H.J.; Weigand, S.D.; et al. Tracking pathophysiological processes in Alzheimer's disease: An updated hypothetical model of dynamic biomarkers. *Lancet Neurol.* **2013**, *12*, 207–216. [CrossRef] [PubMed]

65. Lozupone, M.; Solfrizzi, V.; D'Urso, F.; Di Gioia, I.; Sardone, R.; Dibello, V.; Stallone, R.; Liguori, A.; Ciritella, C.; Daniele, A.; et al. Anti-amyloid-β protein agents for the treatment of Alzheimer's disease: An update on emerging drugs. *Expert Opin. Emerg. Drugs* **2020**, *25*, 319–335. [CrossRef] [PubMed]

66. Perneczky, R.; Jessen, F.; Grimmer, T.; Levin, J.; Flöel, A.; Peters, O.; Froelich, L. Anti-amyloid antibody therapies in Alzheimer's disease. *Brain* **2023**, *146*, 842–849. [CrossRef] [PubMed]

67. Wrigglesworth, J.; Ward, P.; Harding, I.H.; Nilaweera, D.; Wu, Z.; Woods, R.L.; Ryan, J. Factors associated with brain ageing-a systematic review. *BMC Neurol.* **2021**, *21*, 312. [CrossRef]

68. de Melo Coelho, F.G.; Gobbi, S.; Andreatto, C.A.A.; Corazza, D.I.; Pedroso, R.V.; Santos-Galduróz, R.F. Physical exercise modulates peripheral levels of brain-derived neurotrophic factor (BDNF): A systematic review of experimental studies in the elderly. *Arch. Gerontol. Geriatr.* **2013**, *56*, 10–15. [CrossRef]

69. Colich, N.L.; Rosen, M.L.; Williams, E.S.; McLaughlin, K.A. Biological aging in childhood and adolescence following experiences of threat and deprivation: A systematic review and meta-analysis. *Psychol. Bull.* **2020**, *146*, 721. [CrossRef]

70. Milte, C.M.; McNaughton, S.A. Dietary patterns and successful ageing: a systematic review. *Eur. J. Nutr.* **2016**, *55*, 423–450. [CrossRef]

71. Sachdev, H.; Gera, T.; Nestel, P. Effect of iron supplementation on mental and motor development in children: Systematic review of randomised controlled trials. *Public Health Nutr.* **2005**, *8*, 117–132. [CrossRef]

72. Cronin, P.; Kelly, A.M.; Altaee, D.; Foerster, B.; Petrou, M.; Dwamena, B.A. How to perform a systematic review and meta-analysis of diagnostic imaging studies. *Acad. Radiol.* **2018**, *25*, 573–593. [CrossRef]

73. Grucza, R.A.; Sher, K.J.; Kerr, W.C.; Krauss, M.J.; Lui, C.K.; McDowell, Y.E.; Hartz, S.; Virdi, G.; Bierut, L.J. Trends in adult alcohol use and binge drinking in the early 21st-century United States: A meta-analysis of 6 National Survey Series. *Alcohol. Clin. Exp.* **2018**, *42*, 1939–1950. [CrossRef]

74. Koh, D.; Nam, J.; Graubard, B.I.; Chen, Y.; Locke, S.J.; Friesen, M.C. Evaluating temporal trends from occupational lead exposure data reported in the published literature using meta-regression. *Ann. Occup. Hyg.* **2014**, *58*, 1111–1125. [CrossRef] [PubMed]

75. Miller, M.W.; Sadeh, N. Traumatic stress, oxidative stress and post-traumatic stress disorder: Neurodegeneration and the accelerated-aging hypothesis. *Mol. Psychiatry* **2014**, *19*, 1156–1162. [CrossRef] [PubMed]

76. Ghosh, C.; De, A. Basics of aging theories and disease related aging-an overview. *PharmaTutor* **2017**, *5*, 16–23.

77. Wadhwa, R.; Gupta, R.; Maurya, P.K. Oxidative stress and accelerated aging in neurodegenerative and neuropsychiatric disorder. *Curr. Pharm. Des.* **2018**, *24*, 4711–4725. [CrossRef] [PubMed]

78. Bersani, F.S.; Mellon, S.H.; Reus, V.I.; Wolkowitz, O.M. Accelerated aging in serious mental disorders. *Curr. Opin. Psychiatry* **2019**, *32*, 381. [CrossRef]

79. Hou, Y.; Dan, X.; Babbar, M.; Wei, Y.; Hasselbalch, S.G.; Croteau, D.L.; Bohr, V.A. Ageing as a risk factor for neurodegenerative disease. *Nat. Rev. Neurol.* **2019**, *15*, 565–581. [CrossRef]

80. Stern, Y. Cognitive reserve in ageing and Alzheimer's disease. *Lancet Neurol.* **2012**, *11*, 1006–1012. [CrossRef]

81. Bethlehem, R.A.; Seidlitz, J.; White, S.R.; Vogel, J.W.; Anderson, K.M.; Adamson, C.; Adler, S.; Alexopoulos, G.S.; Anagnostou, E.; Areces-Gonzalez, A.; et al. Brain charts for the human lifespan. *Nature* **2022**, *604*, 525–533. [CrossRef]

82. Habuza, T.; Navaz, A.N.; Hashim, F.; Alnajjar, F.; Zaki, N.; Serhani, M.A.; Statsenko, Y. AI applications in robotics, precision medicine, and medical image analysis: An overview and future trends. *Inform. Med. Unlocked* **2021**, *24*, 100596. [CrossRef]

83. Bruno, M.A.; Walker, E.A.; Abujudeh, H.H. Understanding and confronting our mistakes: The epidemiology of error in radiology and strategies for error reduction. *Radiographics* **2015**, *35*, 1668–1676. [CrossRef]

84. Wang, S.; Summers, R.M. Machine learning and radiology. *Med. Image Anal.* **2012**, *16*, 933–951. [CrossRef] [PubMed]

85. Salvatore, C.; Castiglioni, I.; Cerasa, A. Radiomics approach in the neurodegenerative brain. *Aging Clin. Exp. Res.* **2021**, *33*, 1709–1711. [CrossRef] [PubMed]

86. Statsenko, Y.; Habuza, T.; Talako, T.; Kurbatova, T.; Simiyu, G.L.; Smetanina, D.; Sido, J.; Qandil, D.S.; Meribout, S.; Gelovani, J.G.; et al. Reliability of Machine Learning in Eliminating Data Redundancy of Radiomics and Reflecting Pathophysiology in COVID-19 Pneumonia: Impact of CT Reconstruction Kernels on Accuracy. *IEEE Access* **2022**, *10*, 120901–120921. [CrossRef]

87. Statsenko, Y.; Habuza, T.; Charykova, I.; Gorkom, K.N.V.; Zaki, N.; Almansoori, T.M.; Baylis, G.; Ljubisavljevic, M.; Belghali, M. Predicting age from behavioral test performance for screening early onset of cognitive decline. *Front. Aging Neurosci.* **2021**, *13*, 661514. [CrossRef]

88. Statsenko, Y.; Habuza, T.; Charykova, I.; Gorkom, K.; Zaki, N.; Almansoori, T.; Ljubisavljevic, M.; Szolics, M.; Al Koteesh, J.; Ponomareva, A.; et al. Predicting cognitive age for screening for neurodegeneration. *J. Neurol. Sci.* **2021**, *429*, 118994. [CrossRef]

Article

Accuracy of EEG Biomarkers in the Detection of Clinical Outcome in Disorders of Consciousness after Severe Acquired Brain Injury: Preliminary Results of a Pilot Study Using a Machine Learning Approach

Francesco Di Gregorio [1], Fabio La Porta [2,*], Valeria Petrone [2], Simone Battaglia [3,4], Silvia Orlandi [5], Giuseppe Ippolito [3], Vincenzo Romei [3], Roberto Piperno [2] and Giada Lullini [2]

[1] UO Medicina Riabilitativa e Neuroriabilitazione, Azienda Unità Sanitaria Locale, 40133 Bologna, Italy
[2] IRCCS Istituto delle Scienze Neurologiche di Bologna
[3] Centro Studi e Ricerche in Neuroscienze Cognitive, Dipartimento di Psicologia, Alma Mater Studiorum—Università di Bologna, Campus di Cesena, 47521 Cesena, Italy
[4] Dipartimento di Psicologia, Università di Torino, 10124 Torino, Italy
[5] Department of Electrical, Electronic and Information Engineering "Guglielmo Marconi", University of Bologna, Viale Risorgimento, 2, 40136 Bologna, Italy
* Correspondence: fabio.laporta@isnb.it

Citation: Di Gregorio, F.; La Porta, F.; Petrone, V.; Battaglia, S.; Orlandi, S.; Ippolito, G.; Romei, V.; Piperno, R.; Lullini, G. Accuracy of EEG Biomarkers in the Detection of Clinical Outcome in Disorders of Consciousness after Severe Acquired Brain Injury: Preliminary Results of a Pilot Study Using a Machine Learning Approach. *Biomedicines* **2022**, *10*, 1897. https://doi.org/10.3390/biomedicines10081897

Academic Editors: Carlo Santambrogio and Juan Sahuquillo

Received: 1 June 2022
Accepted: 29 July 2022
Published: 5 August 2022

Publisher's Note: MDPI stays neutral with regard to jurisdictional claims in published maps and institutional affiliations.

Abstract: Accurate outcome detection in neuro-rehabilitative settings is crucial for appropriate long-term rehabilitative decisions in patients with disorders of consciousness (DoC). EEG measures derived from high-density EEG can provide helpful information regarding diagnosis and recovery in DoC patients. However, the accuracy rate of EEG biomarkers to predict the clinical outcome in DoC patients is largely unknown. This study investigated the accuracy of psychophysiological biomarkers based on clinical EEG in predicting clinical outcomes in DoC patients. To this aim, we extracted a set of EEG biomarkers in 33 DoC patients with traumatic and nontraumatic etiologies and estimated their accuracy to discriminate patients' etiologies and predict clinical outcomes 6 months after the injury. Machine learning reached an accuracy of 83.3% (sensitivity = 92.3%, specificity = 60%) with EEG-based functional connectivity predicting clinical outcome in nontraumatic patients. Furthermore, the combination of functional connectivity and dominant frequency in EEG activity best predicted clinical outcomes in traumatic patients with an accuracy of 80% (sensitivity = 85.7%, specificity = 71.4%). These results highlight the importance of functional connectivity in predicting recovery in DoC patients. Moreover, this study shows the high translational value of EEG biomarkers both in terms of feasibility and accuracy for the assessment of DoC.

Keywords: disorders of consciousness; traumatic brain injury; electroencephalography; brain plasticity and connectivity; post-anoxic coma; severe acquired brain injury; acquired brain damage; linear discriminant analyses; brain functional impairment; neurocognitive disorders

1. Introduction

After acute brain injury and coma, a large number of surviving patients develop severe disorders of consciousness (DoC), such as unresponsive wakefulness syndrome (UWS) or minimally conscious state (MCS). UWS patients preserve basic functions, such as eye-opening and reflexive movements [1–3], but remain unresponsive to the external environment. MCS patients instead show minimal but reliable behavioral evidence (i.e., visual fixation or visual pursuit, verbalizations, etc.) of self and environmental consciousness [4,5]. UWS and MCS can be persistent states or can evolve toward varying degrees of recovery of consciousness. The assessment of the rehabilitative potential and the prediction of the possible clinical outcome of DoC patients are crucial for clinical and ethical reasons, and they allow for the identification of rehabilitative needs and the design of customized rehabilitative programs. Although standard clinical scales and innovative neurophysiological methods can help diagnose and predict the clinical outcome in DoC patients, these still represent complex challenges for clinicians. Indeed, even the distinction among UWS, MCS, and the emergence from MCS is based on clinical and behavioral evidence which might be hard to identify.

Behavioral and clinical scales such as the Coma Recovery Scale (CRS-R) and the Glasgow Outcome Scale (GOS) provide criteria for the diagnosis of DoC [6,7] and allow for longitudinal monitoring of the behavioral responsiveness of these patients. However, several fMRI and EEG studies [8–12] showed that ~15–20% of DoC patients with no evidence of overt behavioral responsiveness may nevertheless show signs in the brain activity of covert consciousness. Moreover, the prediction of patients' clinical outcomes, solely based on clinical scales, is not reliable (misdiagnosis rate up to 40% [13–15]), especially in DoC after traumatic etiologies, where the prediction of the clinical outcome can be even less accurate compared to post anoxic/ischemic etiologies [13]. Therefore, accurate diagnostic tools that rely on brain activity [12] and on patients' characteristics have become a critical need. In this sense, the EEG can be a useful tool. Indeed, the EEG already has numerous applications in clinical settings for the prediction of recovery in neurological patients [16] and after the application of specific protocols [17,18]. Moreover, new robust statistical methodologies, such as machine learning, have been already implemented in EEG studies to help with clinical and rehabilitative decision making [16,17,19–21]. Accordingly, recent studies investigated the sensitivity and accuracy of quantitative EEG (qEEG) and EEG-based functional connectivity measures to predict the clinical outcome in DoC patients [20–22]. Specifically, patients with reactive EEG signal to external stimuli, larger EEG amplitudes, and stronger activity in the higher-frequency bands (i.e., alpha 7–13 Hz and beta 14–25 Hz) are more likely to have a positive outcome after 3–6 months [23–25].

Furthermore, the complexity of the EEG signal in terms of diversity and integration is considered an important proxy of the consciousness level [26,27]. Indeed, measures that quantify the complexity of information content in the brain activity, such as the evoked EEG activity after transcranial magnetic stimulation [28–30] and the permutation entropy (PeEN), is reduced in patients with DoC [31–34] and in patients with worse clinical outcome [21,35]. Another important finding suggests that the level of functional connectivity in the brain, reflecting information sharing within different cortical areas [36], is an accurate index for discriminating different degrees of consciousness and predicting recovery in DoC patients [20–22]. In particular, EEG-based measures of connectivity are larger in healthy participants compared to DoC patients and in MCS compared to UWS patients [20,21,37,38]. Lastly, DoC patients with stronger global connectivity [21], anterior forebrain connectivity [39,40], and thalamic–cortical connectivity [41] can have a better long-term outcome in terms of disability and level of consciousness [42].

However, most of the studies reported above recorded brain activity using different paradigms and EEG configurations or high spatial sampling of scalp electrodes (i.e., high-density EEG). Albeit promising, these methodologies are often time-demanding, requiring specific protocols and apparatuses, thus preventing a large-scale implementation in clinical settings where, more likely, only standard EEGs with the 10–20 EEG montage density are available. Thus, implementing new methodologies with high feasibility in clinical settings becomes fundamental in terms of clinical usefulness and to exploit the potential of EEG-based measures in predicting the clinical outcome of DoC patients. From this perspective, the recent literature has started to highlight the relevance of quantitative EEG measures and specifically how the dominant frequency [23] and alpha power [22] extracted from standard clinical EEG can predict clinical outcome in DoC patients. On this basis, in the present study, we hypothesized that quantitative measures and measures of functional connectivity derived from standard clinical EEG can reliably predict the clinical outcome of DoC patients after traumatic and nontraumatic brain injury.

To test this hypothesis, we set two steps of analyses with the main goal of identifying EEG predictors of functional outcome in DoC patients. The first step of the study was to identify and characterize the role of EEG biomarkers in discriminating the etiology and the clinical outcome in DoC patients. Specifically, we identified those EEG measures which can discriminate DoC patients as a function of the etiology of their brain injury (i.e., traumatic or nontraumatic) and clinical outcome 6 months after the injury (i.e., improved patients or nonimproved patients) using a standard clinical resting-state EEG. The study's second step was to investigate the accuracy of the EEG biomarkers to predict clinical outcomes separately in traumatic and nontraumatic brain injury patients. To this end, we integrated etiological and outcome information using a machine learning approach. Specifically, the identification of EEG biomarkers of etiology serves to quantify the differential contribution of EEG parameters in the prediction of the clinical outcome separately in the two etiological groups. Therefore, in the two etiology groups, we investigated the accuracy of the EEG biomarkers, as identified in the first step of analyses, to discriminate DoC patients who improved vs. those who did not.

To summarize, the present study can offer several novel insights:

1. Translational value: We highlight the translational value and feasibility of EEG biomarkers based on standard clinical EEG in the assessment of functional outcome in DoC patients;
2. Methodological value: We directly compare DoC patients with different etiologies to identify those EEG biomarkers able to predict the clinical outcome in traumatic and nontraumatic DoC patients;
3. Computational value: we propose a machine learning model, based on those discriminative EEG biomarkers, for the classification of the functional outcome in DoC patients.

2. Materials and Methods

Step 1: To identify EEG biomarkers of different etiologies and clinical outcomes in DoC patients.

In this retrospective study, electrophysiological measures extracted from a standard clinical EEG recorded at 1 month after acute brain injury (T0) were used to discriminate the etiology of the brain injury and to predict clinical outcome 6 months after the injury (T1) in DoC patients. Specifically, we first investigated which EEG measures (i.e., qEEG and functional connectivity measures) can discriminate patients on the basis of their brain injury etiology (i.e., traumatic or nontraumatic) and the corresponding clinical outcome at 6 months (i.e., improved or nonimproved patients) to identify those EEG measures able to predict clinical changes (i.e., outcome) of the level of consciousness as measured by a specific clinical scale (i.e., the GOS) [6,43].

2.1. Participants

Inclusion criteria for all patients were as follows: (1) severe acquired brain injury after traumatic or nontraumatic etiologies; (2) diagnosis of disorder of consciousness (i.e., MCS or UWS) with the use of specific clinical scales (GOS < 3 or GOSE < 3 or CRS-R < 23); (3) EEG and clinical data availability at 1 and 6 months after brain injury; (4) age between 18 and 80 years old. Exclusion criteria were as follows: (1) diagnosis of locked-in syndrome (patients with LiS present total paralysis, but intact consciousness); (2) diagnosis of brain death, which implies a persistent state; (3) infectious lesions of the brain (i.e., abscess and encephalitis).

All procedures were conducted in accordance with the Declaration of Helsinki and approved by the Ethical Committee Area Vasta Emilia Centro (CE num. 841-2021-OSS-AUSLBO) Bologna, Italy. Data were acquired during routine clinical care by trained clinicians between 2005 and 2020.

2.2. Intervention: EEG Data Acquisition

Resting-state EEG data were recorded according to the Italian guidelines [44] for the clinical use of the EEG. A total of 19 Ag/AgCl-cup electrodes were positioned according to the 10/20 system and referenced to the linked ear lobes. Impedance for EEG and electrooculogram (EOG) electrodes were kept below 10 kΩ. EEG data were continuously recorded at a sampling rate of 1024 Hz. All electrodes were offline resampled to 500 Hz and filtered with a 1–30 Hz bandpass filter. A single EEG session lasted 20 min. Offline, EEG artefacts were eliminated using the pop_autorej function on EEGLAB [45], which automatically detects and eliminates artefact data. This function first identifies extremely large potential fluctuations in order to detect artefacts from scalp electrodes data or other unreasonably large amplitude events. Then, it rejects data epochs containing data values outside a given standard deviation (3 SD). Lastly, the EEG data were re-epoched in segments of 1 s with the function pop_repoch on EEGLAB [46], and linear trends were corrected with the function 'detrend' on EEGLAB.

2.3. Control

No control for the intervention was planned as all participants underwent EEG assessment. However, comparisons between patients with different etiologies and clinical outcomes were planned.

2.4. Outcome Measures

2.4.1. Clinical Measures

Patients' demographic information and the date of the brain injury were collected and reported in specific case report forms. Moreover, brain injury etiologies were used to distinguish between traumatic and nontraumatic brain injury patients.

The Glasgow Outcome Scale (GOS) [14,21,47–49], a specific scale used to discriminate between different levels of functional outcome in DoC patients, was administered 1 month and 6 months after the brain injury. All patients underwent usual care during the 6 months after brain injury. The difference between the GOS score at 1 month after brain injury (T0) and the score at 6 months after brain injury (T1) was used to estimate changes in the functional outcome (i.e., clinical outcome). Score differences were used to distinguish patients with an improvement in the clinical outcome (GOS score difference of at least +1 point between T0 and T1) from those with a lack of improvement in the clinical outcome (i.e., those who did not show any change in the GOS score between T0 and T1 or an impairment) [21,50]. As the GOS directly assesses the functional outcome in terms of consciousness and disability, it is a suitable clinical test to reach our goal to predict clinical outcome in DoC patients [14,48,51–55].

2.4.2. Quantitative EEG Measures

Four quantitative EEG (qEEG) measures were extracted: z-scored power spectral density, dominant frequency peak, permutation entropy, and mean amplitude. First, power spectral data from the averaged electrodes and epochs were calculated using the "pop_spectopo" function on EEGLAB. The power spectral data expressed in $\mu v^2/Hz$ were transformed in z points to identify the frequency peak (i.e., dominant frequency) in Hertz (Hz) and the power spectral density (z-scored PSD) divided into specific frequencies: delta (1–3 Hz), theta (4–7 Hz), alpha (8–13 Hz), and beta (14–30 Hz) [38,56]. The power was calculated as the mean value in each frequency band. Furthermore, in order to evaluate permutation entropy (i.e., PeEn) on single electrodes and on the average of all electrodes, the "pec" function of EEGLAB was applied to the EEG data [33]. Similarly, the "mean" function was applied to estimate the mean amplitude (Amp) in μv [23].

2.4.3. Functional Connectivity Measures

EEG-based functional connectivity estimates the association between electrode signals. There are several methods for quantifying functional connectivity on the basis of EEG data [20,21]. In particular, methods can be based on associations between phases (e.g., weighted phase lag index), between power in specific frequency bands (e.g., partial coherence), and in the complexity of the EEG signal (e.g., mutual information). For the connectivity analyses, the EEG signal was first filtered with a spatial filter (i.e., Laplacian filter) [57]. The Laplacian filter allows subtracting the activity of contiguous electrodes from each electrode and reduces the risk of false positive connectivity due to the effects of common neural sources on contiguous electrodes. The connectivity indices were calculated for each pair of electrodes, which resulted in 19 × 19 connectivity matrices. Connectivity measures were then extracted in specific regions of interest (ROI): right frontoparietal, left frontoparietal, frontal interhemispheric, central interhemispheric, and posterior interhemispheric ROIs.

In the present study, three functional connectivity measures were considered: (a) weighted phase lag index (wPLI), (b) partial coherence (PCoh), and (c) mutual information (MI).

(a)   For the weighted phase lag index (wPLI), the time–frequency data were first calculated via convolution with complex Morlet wavelets. Convolution was performed via frequency-domain multiplication [58,59]. In order to prevent the artefact of the "edges", the signal was re-epoched in epochs of 2 s. The wPLI evaluates the consistency of the phase differences between two timeseries [60] (e.g., EEG signal over specific electrodes). The wPLI was calculated on the individual EEG dominant frequency [46]. wPLI values can range from 0 to 1, where a higher value indicates a relationship between the phases of two signals. A correct threshold for each participant was set to detect residual false positive connectivity [61].

(b)   The partial coherence (PCoh) values were calculated on the entire EEG signal with the "pop_newcrossf" function on EEGLAB and on the individual EEG dominant frequency (to improve frequency resolution, the pad ratio parameter was set to 8) [62]. Absolute correlations were extracted for each pair of electrodes and corrected for multiple comparisons. The PCoh values can range from 0 to 1. Larger values indicate a stronger relationship between the two signals at a specific frequency.

(c)   The mutual information (MI) is a functional connectivity index that estimates the level of information shared between two variables or time series. The MI is calculated by adding the individual entropies (H) of the two timeseries and subtracting the joint entropy. For MI analyses, data were first divided into 10 bins on the basis of the Freedman–Diaconis rule [58]. The MI values were extracted with the "mutualinformationx" function on MATLAB for each pair of electrodes. Higher values indicate higher levels of information shared between two signals in terms of oscillations and similarity in the waveforms.

2.5. Statistical Analyses

For the first step, between-group statistical analyses were performed to identify EEG biomarkers of different etiologies and clinical outcomes. Specifically, analyses were separately performed on two between-subject factors: etiology (traumatic (TBI) vs. nontraumatic (non-TBI) etiologies) [21] and clinical outcome (improved vs. nonimproved patients) [21–23,63]. For both etiology and clinical outcome, we employed the same analytical strategy. In particular, mixed-model ANOVAs with repeated measures on the between-subject variable group (TBI vs. non-TBI for factor etiology and improved vs. nonimproved patients for the factor clinical outcome) were performed. For the functional connectivity indices, the additional within-subject variable ROI (right frontoparietal, left frontoparietal, frontal interhemispheric, central interhemispheric, and posterior interhemispheric) was considered. Between-group planned comparisons were performed using two-tailed *t*-tests with 1000 bootstrap corrections. To compensate for violations of sphericity in the ANOVA, Greenhouse–Geisser corrections were applied [64], and corrected *p*-values were reported. Effect sizes were estimated with partial eta squared (ηp^2) and Cohen's d (d) for between-group comparisons. A preliminary Shapiro–Wilk test for normality distribution was performed for all measures [65]. For similar statistical procedures, see also [66–70].

Step 2: Accuracy of EEG biomarkers to predict clinical outcome in traumatic and nontraumatic brain injury.

For the second step, we integrated results from step 1 using a machine learning procedure (see also [46]) to maximize the informative value provided by the combinations of EEG biomarkers in predicting the clinical outcome at the level of the individual patient for the TBI and non-TBI groups.

2.5.1. Features Extraction and Data Aggregation

The initial total number of features examined in aim 1 was 43. The most discriminative features were then selected on the basis of the statistical analyses and results of step 1. In particular, for the EEG feature extraction procedure, we extracted the following information from each participant's recording: qEEG features (dominant frequency, permutation entropy, mean amplitudem and delta, theta, alpha, and beta zPSD) and functional connectivity features (wPLI, PCoh, and MI). The selected features were aggregated in matrices where the rows represented participants (i.e., instances) and the columns represented the values of the features. The instances were used to feed a machine learning algorithm, as explained in the next section.

2.5.2. Classification Method

For this step of the analysis, a stepwise linear discriminant analysis (LDA) was applied with a leave-one-subject-out cross-validation. The goal of LDA is to discriminate two classes of data in low-dimensional space by retaining the features with the higher discriminative power. LDA was already used in previous research studies, and it was recommended by the International Federation of Clinical Neurophysiology for EEG research [71] and for the assessment of DoC [72,73]. However, we additionally used a traditional multivariate logistic regression as a control for the LDA as suggested in previous studies [74–76]. Within the leave-one-subject-out cross-validation, each feature array was used once as validation data, with the remaining data as the training data [3,77]. Then, the percentage of correctly classified instances was calculated. While this percentage reflects the classification accuracy, we also calculated sensitivity and specificity. For the classification, we used the clinical outcome (i.e., improved vs. nonimproved patients) as a categorical grouping variable and the etiology (i.e., TBI vs. non-TBI) as a factor. In this way, the accuracy of EEG biomarkers of the clinical outcome was estimated separately for TBI and non-TBI patients. Moreover, given the retrospective nature of the study, we could deal with imbalanced classifiers [78]; thus, we calculated additional metrics, such as the balanced accuracy (the average of sensitivity and specificity) and the precision (the number of positive class predictions divided by the sum of true-positive and false-positive instances). Only EEG variables that showed sensitivity to distinguish between groups in the analyses performed to address the first step entered the LDA.

It is important to note that, for the study purpose, we needed an independent clinical indicator (i.e., the GOS) which could discriminate between two groups of patients, with a relative lower and higher outcome. In other words, we needed a binary indicator to show the potential of EEG biomarkers to discriminate individuals on the basis of their outcome, without any pretense to make an accurate diagnosis of the level of consciousness for each individual. Thus, for the purpose of the study, we regard the GOS a sufficient and adequate clinical indicator [79].

2.5.3. Sample Size Estimation and Statistical Software Employed

The preliminary sample size for step 1 was calculated with the G*Power software, version 3.1 (Heinrich Heine University Düsseldorf, Düsseldorf, Germany) [80]. Parameters used in the analyses were derived from previous studies on the primary endpoint z-score PSD on the alpha band [22]. The following parameters were used: α (two-tailed) = 0.05 (threshold probability for rejecting the null hypothesis; type I error rate), β = 0.2 (probability of failing to reject the null hypothesis under the alternative hypothesis), and ηp^2 = 0.56 (effect size calculated on preliminary studies). The minimum estimated sample size, using these parameters, was 28 subjects.

All statistical analyses for steps 1 and 2 were performed with the SPSS software (IBM Corp., Armonk, NY, USA) (Version 13). The feature extraction was performed using custom-made routines in MATLAB 2015 b (The MathWorks, Inc., Natick, MA, USA) and EEGLAB (v. 13.0.1).

3. Results

Step 1: To identify EEG biomarkers of different etiologies and clinical outcomes in DoC patients.

3.1. Clinical Results

Thirty-three DoC patients (21 males aged 19–71) were included in this study. Descriptive statistics (Table 1) showed that 15 patients suffered from traumatic brain injury, while 18 patients had a nontraumatic etiology (i.e., vascular and anoxic brain injury). In particular, within the non-TBI group, nine patients suffered from an ischemic or hemorrhagic stroke while nine patients had a post-anoxic etiology. Lesion locations were extracted for each participant according to the most recent CT or MRI scan. The visual inspection of non-TBI patients' lesion profiles showed a maximal lesion overlap over the basal ganglia and the thalamus. Fifteen patients presented diffuse axonal damages after TBI; specifically, according to the Marshall classification of traumatic brain injury [81], eight patients were classified as "diffuse injury" (IV), and seven patients were classified as "evacuated mass lesion" (V). All patients underwent usual care and received symptomatic treatments. Whenever necessary, patients were treated with decompressive craniectomy to reduce high intracranial pressure. Demographic data show that non-TBI patients were about 15 years older than TBI patients ($t(31) = 2.62$, $p = 013$, d = 0.46). At baseline, no differences emerged between groups in the severity of brain injury as revealed by the analyses of the Glasgow Coma Scale (GCS) at admission (i.e., baseline) ($t(31) = 0.57$, $p = 575$, d = 0.11). Lastly, clinical changes were comparable between the two etiological groups as the proportion of patients improved at T1, albeit numerically larger in the TBI group, was not statistically different between TBI (50%) and non-TBI patients (29.5%; independent samples Mann–Whitney U test; $p = 215$).

Table 1. Demographic and clinical data.

	TBI	Non-TBI
N	15	18
Gender		
Males	11	10
Females	4	8
Mean age in years (SD)	34.3 (4.4)	49.1 (3.59)
Glasgow Coma Scale at baseline	4.73 (0.5)	4.22 (0.7)
Clinical outcome at T1		
Improved	8	5
Nonimproved	7	13

Patients' etiologies were divided into traumatic (TBI) and nontraumatic brain injury. Diagnostic classifications at 1 month after brain injury (T0) are reported as unresponsive wakefulness syndrome (UWS) and minimal conscious states (MCS). Clinical outcomes at 6 months after the injury (T1) are reported as improved (i.e., patients showing positive changes in the GOS > +1) or nonimproved patients (patients showing no changes or negative changes in the GOS < 0). Standard errors of the mean (SD) are in brackets.

3.2. Etiology Biomarkers

EEG data were analyzed for the between-subject factor etiology. As demographic results showed that TBI patients were younger than non-TBI patients (see above), the variable age was used as a covariate in the subsequent analyses of covariance (ANCOVAs). The ANCOVA (similar to the ANOVA) examined the effects of independent variables on dependent variables while factoring out the effect of the covariate "age". A visual assessment of Figure 1A,B reveals the generally slower dominant frequency in TBI patients compared to non-TBI, which additionally showed stronger residual connectivity expressed by the wPLI. Statistical analyses confirmed these impressions. Specifically, qEEG results showed a faster dominant frequency (one-way ANCOVA $F_{(2,30)} = 3.763$, one-tailed $p = 031$, $\eta p^2 = 111$) in the non-TBI group (M = 5.966 Hz, standard error of the mean SE = 0.606 Hz) compared to the TBI group (M = 3.997 Hz, SE = 0.621 Hz) ($t(31) = 2.26$, $p = 031$, d = 0.39), as shown in Figure 1A. Similar analyses on the other qEEG measures did not show further significant results (all $p \geq 228$).

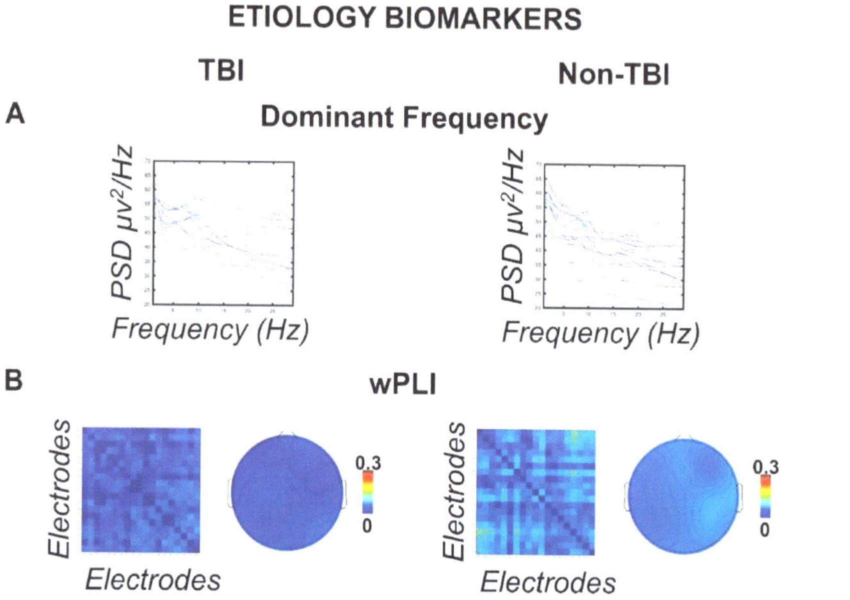

Figure 1. Etiology biomarkers. (**A**) Single-subject power spectral density (PSD) in the frequency range (1–30 Hz) divided into traumatic (TBI) and nontraumatic (non-TBI) etiologies. The circles identify the dominant frequency peaks showing a distribution of the peaks toward slower frequencies in the TBI group. (**B**) Connectivity matrices between scalp electrodes of the weighted phase lag index (wPLI) for TBI and non-TBI groups showing stronger global connectivity in the non-TBI group. Topographies show the grand mean wPLI connectomes for each electrode.

Analyses on the functional connectivity indices showed larger connectivity in the wPLI for the non-TBI group (M = 053 PLI, SE = 012 PLI) in all the considered ROIs (main effect of etiology in the repeated measures ANCOVA; $F_{(1,30)} = 6.862$, $p = 007$, $\eta p^2 = 186$) compared to the TBI group (M = 047 PLI, SE = 014 PLI) (Figure 1B). No further significant between groups or interaction effects were found in the other functional connectivity indices (all F < 0.344, all $p > 562$, all $\eta p^2 < 011$).

3.3. Outcome Biomarkers

EEG data were analyzed for the between-subject factor 'clinical outcome' (improved vs. nonimproved patients). A visual assessment of the Figure 2A,B shows that EEG-based connectivity (expressed by the PCoh and the MI) calculated at T0 was stronger for those patients who showed at T1 functional outcome improvements. Statistical analyses confirmed these impressions. Analyses on the functional connectivity indeed showed a general stronger connectivity both for the MI (main effect of group, $F_{(1,31)} = 5.36$, $p = 027$, $\eta p^2 = 147$) and for the PCoh (main effect of group, $F_{(1,31)} = 5.81$, $p = 022$, $\eta p^2 = 158$) indices in the improved outcome group (MI mean = 326, SE = 046; PCoh mean = 464, SE = 042) compared to the nonimproved outcome group (MI mean = 188, SE = 037; PCoh mean = 333, SE = 034) Figure 2A,B. To further confirm the relative improvement in DoC patients at 6 months (T1), we ran two control analyses on the factor clinical outcome, controlling for the initial severity of the brain injury. To this aim, we first used the GCS score at baseline and then the GOS score at 1 month (T0) as covariate variables in two separate analyses of covariance (ANCOVAs). The results of the ANCOVAs with the within-subject factor ROI, the between-subject factor clinical outcome, and the covariate GCS or GOS at T0 confirmed our main findings. In particular, both analyses showed a significant effect of the between-subject factor clinical outcome for the MI (all $F_{(1,30)} > 4.62$, $p < 04$, $\eta p^2 > 133$) and PCoh (all $F_{(1,30)} > 4.95$, $p < 034$, $\eta p^2 > 142$), thus confirming again that patients with higher general connectivity at T0 have a higher probability of a better clinical outcome at T1 regardless of the initial GCS or GOS levels. No further significant between groups or interaction effects were found for the functional connectivity indices (all $F < 1.138$, all $p > 339$, all $\eta p^2 < 035$). However, main effects of the factor ROI (all $F > 4.842$, all $p < 008$, all $\eta p^2 < 135$) were found for all the functional connectivity indices (wPLI, PCoh and MI), indicating a general stronger interhemispheric connectivity (i.e., frontal, central, and posterior interhemispheric ROIs) in all groups compared to the intrahemispheric connectivity (i.e., right and left frontoparietal ROIs).

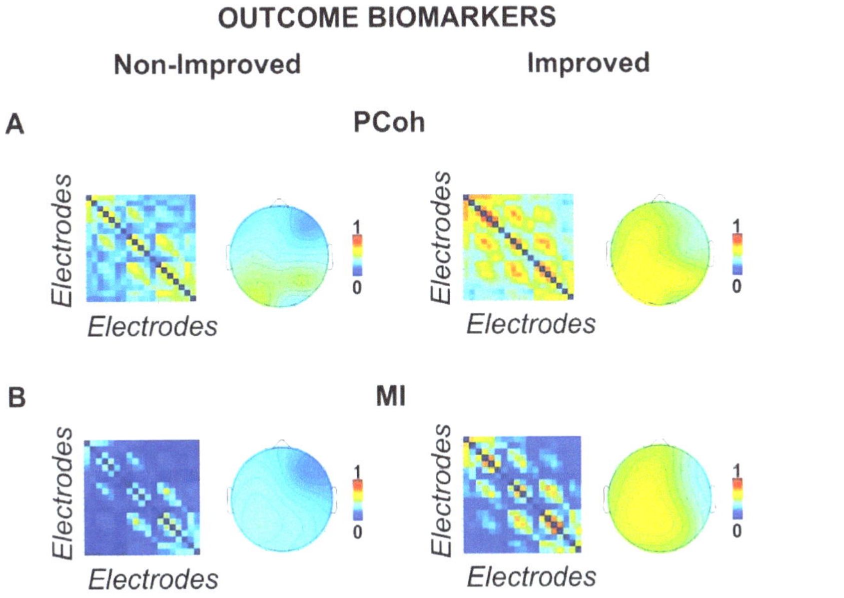

Figure 2. Outcome biomarkers. Connectivity matrices between scalp electrodes of the (**A**) partial coherence (PCoh) and the (**B**) mutual information (MI) for improved and nonimproved patients. Both figures show stronger connectivity for improved patients. Topographies show grand mean PCoh and MI connectomes for each electrode.

Lastly, between-group comparisons for the qEEG measures did not show any significant difference (all $p \geq 308$).

Step 2: Accuracy of EEG biomarkers to predict clinical outcome in traumatic and nontraumatic brain injury.

3.4. Linear Discriminant Analysis

Instances used for LDA analyses were 15 for TBI patients and 18 for non-TBI patients. Four features were selected for the LDA analyses on the basis of the results of step 1: one qEEG feature (i.e., dominant frequency) and three functional connectivity features (i.e., wPLI, MI, and PCoh) calculated as the mean values across ROIs. The data partition and LDA procedure are described in Figure 3. Results and ROC curves for sensitivity and specificity of the LDA are reported in Figure 4 and Table 2.

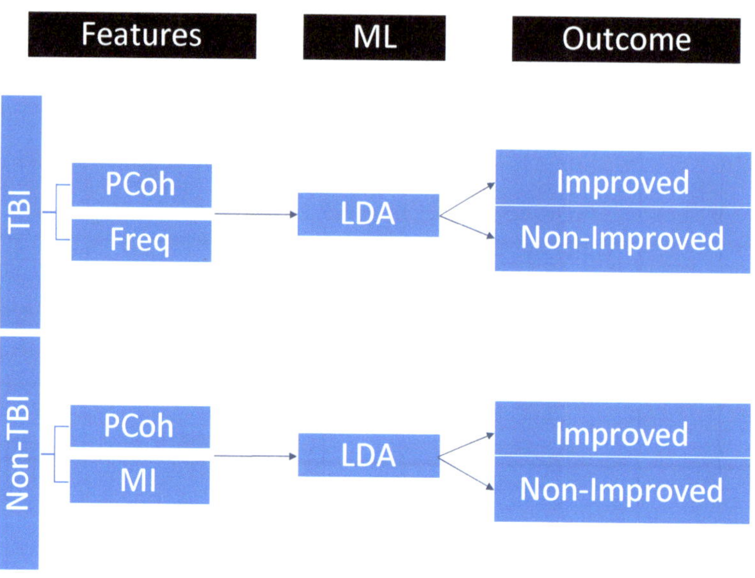

Figure 3. EEG features selected for traumatic (TBI) and nontraumatic (non-TBI) patients and included in the linear discriminant analysis (LDA) for clinical outcome prediction. LDA parameters were as follows: discriminant type = diagLinear (all classes had the same diagonal covariance matrix), gamma = 1, and delta = 0. PCoh = partial coherence, Freq = dominant frequency, MI = mutual information, ML = machine learning.

Table 2. LDA results.

Etiology	Features		LDA		Clinical Outcome [95% CI]	
		Acc	Sens	Spec	Non-Improved	Improved
TBI	Pcoh and Freq	80.0%	85.7%	71.4%	[−1.262, 0.306]	[−0.523, 1.420]
Non-TBI	PCoh and MI	83.3%	92.3%	60.0%	[−1.179, 0.304]	[−0.099, 2.199]

The best accuracy results in the discrimination of the clinical outcome between improved and nonimproved patients are reported separately for traumatic (TBI) and nontraumatic (non-TBI) etiologies. The 95% confidence intervals (CI) are reported for the feature combinations. LDA = linear discriminative analyses, Acc = accuracy, Sens = sensitivity, Spec = specificity, PCoh = partial coherence, Freq = dominant frequency, MI = mutual information.

The accuracy of the EEG features to discriminate between the improved and non-improved patients in the two etiological groups was analyzed. In the TBI group, the combination between global PCoh (i.e., mean of PCoh across ROIs) and the dominant frequency measures resulted in the best discrimination accuracy for the 6 month outcome (accuracy = 80%, balanced accuracy = 78.55%, and precision = 77.7%) with a sensitivity (i.e., patients correctly classified as nonimproved patients) of 85.7% and a specificity (i.e., patients correctly classified as improved patients) of 71.4%. The other considered measures, taken alone or in conjunction, resulted in an overall discrimination accuracy < 73.3%. In the non-TBI group, the best discrimination accuracy was evinced for the combination of two functional connectivity indices (MI and PCoh accuracy = 83.3%, balanced accuracy = 76.65%, and precision = 85.7%), calculated as the mean connectivity across ROIs, with sensitivity = 92.3% and specificity = 60.0%. The other considered measures, taken alone or in conjunction, resulted in an overall discrimination accuracy < 83.3%. Multivariate logistic regression on the same EEG features further confirmed the accuracy results of LDA in discriminating improved and nonimproved patients in both TBI (accuracy = 80%) and non-TBI patients (accuracy = 83.3%).

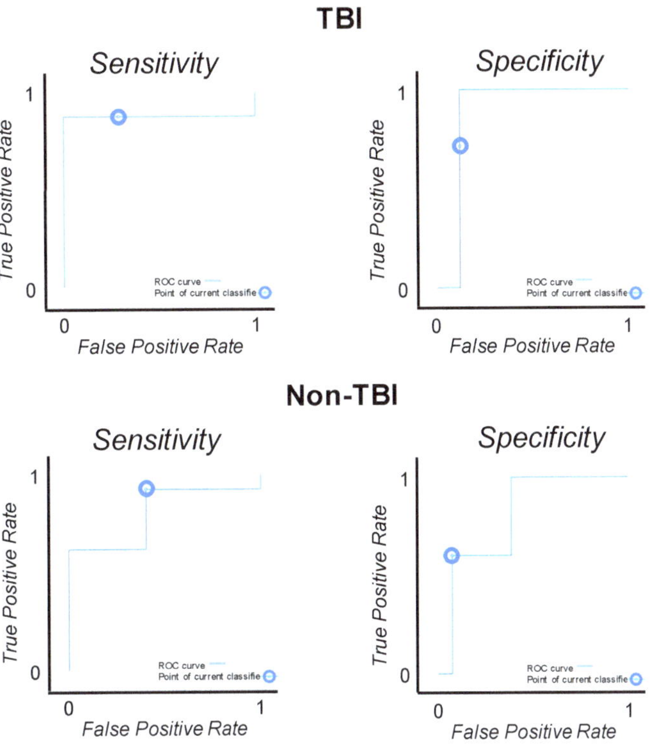

Figure 4. ROC curves. The ROC curves for sensitivity and specificity of the linear discriminant analyses on the patients' clinical outcome (TBI vs. non-TBI patients).

4. Discussion

The present study investigated the accuracy of psychophysiological measures extracted from standard clinical EEG for the prediction of the clinical outcome in traumatic and nontraumatic patients with DoC. Our findings described how EEG-derived measures of connectivity and qEEG measures are associated with clinical characteristics and 6 month outcomes. In particular, our results highlighted that DoC patients with a traumatic etiology show reduced dominant frequencies and lower connectivity based on wPLI compared to patients with vascular and anoxic etiologies. Furthermore, we showed that higher connectivity in the EEG brain network can predict behavioral changes in the functional outcome as measured by the GOS. Lastly, we demonstrated that the combination of different EEG features (quantitative and connectivity measures) can reliably predict 6 month clinical outcome of DoC patients after both TBI and non-TBI etiologies.

Previous studies have highlighted that quantitative EEG measures and functional connectivity within critical brain networks can provide diagnostic and predictive information for the assessment of DoC [20–22]. For instance, the most common EEG change associated with severe brain injury is frequency slowing with predominant delta and theta activity [45,82,83]. However, the neuropathological patterns of DoC after TBI and non-TBI are different. Several studies have reported that the pathologic substrate in DoC patients after TBI is a diffuse axonal damage, while, in non-TBI patients, lesions can be more focal especially for vascular etiologies [62,63]. In TBI, the diffuse axonal damage can cause a disconnection between cortical and subcortical structures, such as the brainstem, the thalamus, and the cerebral cortex, which are involved both in the emergence of consciousness and in the genesis of EEG rhythms [64,65], as in both cognitive (i.e., memory and learning) and social functioning [84,85]. Indeed, partial deafferentations of cortical and subcortical areas can also produce EEG rhythms within slower bands [66]. In contrast, non-TBI DoC patients with a vascular etiology show more focal cortical and subcortical lesions, with structural connectivity relatively preserved [62,63], but altered motor and behavioral abilities [86]. Thus, different neuropathological and connectivity profiles may be responsible for diverse EEG patterns. Accordingly, our results show differences in the brain patterns for TBI and non-TBI patients; thus, we can hypothesize that lower wPLI connectivity and dominant frequency in TBI presumably reflect the typical diffuse axonal damage.

Furthermore, our results confirmed that resting-state EEG-based connectivity is an accurate proxy of patients' long-term clinical outcomes [20,21]. The notion that connectivity is important for the recovery in DoC patients after brain injury is consistent with evidence from PET [50,87,88], functional MRI [89–91], and high-density EEG [20,21,56]. In particular, DoC patients often show severe functional and structural thalamocortical lesions and disconnections with a drop in the EEG coherence in the damaged hemispheres [92,93]. The decrease in EEG coherence seems to be partially determined by damages of the neuronal network implicated in the emergence of consciousness [27,31,92,93] and could reflect an impairment in information sharing within the brain networks [36,94–96]. Indeed, levels of EEG coherence, as measured by partial coherence (PCoh) and mutual information (MI), are lower in UWS and MCS patients and in patients with a poor clinical outcome [20,36,92,97]. Accordingly, our results show that a larger levels of PCoh and of distributed information sharing (as calculated by the MI) within brain networks can support the processes needed to achieve functional improvements in DoC patients. However, while previous studies identified thalamocortical [98,99] and frontoparietal [36,100] connectivity as crucial hubs for the emergence of consciousness, we did not replicate these results here. This is probably related to the low spatial resolution of the standard clinical EEG, which provides a reduced number of scalp electrodes. Most importantly however, standard EEG protocols are easier to implement, and the accuracy of global connectivity indices to predict patients' clinical outcome is still high (~80–83.3%) and comparable to the accuracy of other EEG settings and high-density EEG (~75–87%) [20,21,25,72]. However, the mentioned studies used different machine learning algorithms for classification and outcome prediction of DoC patients. Thus, our results cannot be directly compared with studies that used different EEG settings and statistical approaches. Importantly, here, we further validated those EEG biomarkers which can better predict functional outcome in DoC patients and excluded other biomarkers that show lower accuracy when a standard clinical EEG is used.

Lastly, it is important to notice that the etiology is an essential factor for clinicians evaluating EEG predictors of the clinical outcome in DoC patients. Specifically, while coherence in both etiological groups has a high predictive value, the global frequency of the EEG activity is a relevant feature for the prediction of the potential clinical outcome only in TBI patients. EEG features such as the dominant frequency, which presumably reflects structural damages as we and previous studies report [49], could influence the functional outcome of TBI patients [22]. Indeed, a recent study, using a standard EEG setting, showed that relative larger activity at faster frequencies, specifically in the alpha band, is a sign of potential functional improvement in TBI patients. Thus, we could hypothesize that diffuse axonal injury and the related measures are additional crucial factors to predict the outcome of DoC patients after TBI. This highlights the need for the integration of different indices and methodologies to optimize the predictive accuracy in different etiological groups.

Study Limitations

The results of this study should be considered in light of some limitations. Different studies applied LDA for the classification of clinical outcomes in neurological, cognitive, and psychiatric conditions [101–106]. However, replication of the analysis using a larger sample size collected within prospective multicentric studies and using multiple assessment timepoints (e.g., 3 months, 6 months, and 1 year follow-up) is desirable to support the stability and generalizability of the present results. In this sense, the applicability of the standard resting-state EEG in different clinical settings represents a great advantage in terms of feasibility and replicability of the data. Furthermore, although the GOS is a frequently used coma-to-community assessment tool and is largely used for outcome prediction [20,24,43,50], it is more prone to random errors of measurement being a single-item scale. Indeed, despite the GOS being able to provide a general evaluation of the level of consciousness, it is known to be less responsive to clinical changes than summative rating scale such as the Disability Rating Scale [107]. In particular, measurement errors can be larger in TBI patients compared to non-TBI [108–110]. Thus, the limitation of the GOS measure could explain the lower overall accuracy of the EEG features in the prediction of the clinical outcome in TBI patients as evinced by our results. Future studies could eventually highlight the diagnostic accuracy of standard EEG using different clinical scales, such as the Coma Recovery Scale—Revised [7], as the reference for the clinical outcome.

5. Conclusions

This study presented three main outcomes. The most important outcome of this study allowed us to describe the advantages and limitations of a standard EEG as a tool for clinical assessment and classification of patients with DoC after severe acquired brain injury. Secondly, measures based on standard EEG, such as qEEG and functional connectivity, are promising tools for predicting and classifying DoC patients in terms of rehabilitative potentials and functional recovery. Although spontaneous functional improvements can be observed in TBI and non-TBI patients, the identification of those EEG biomarkers able to predict possible improvements is a valuable effort. Lastly, our results demonstrated that different measures extracted from the standard EEG could be beneficially combined to discriminate the patients' potential clinical outcome using a machine learning approach. Hence, EEG biomarkers may prove highly relevant in supporting clinical and rehabilitative decision making.

Author Contributions: Conceptualization, F.D.G., V.P., F.L.P. and R.P.; methodology, F.D.G., V.P., F.L.P. and R.P.; software, F.D.G. and V.P.; validation, F.D.G., G.L., F.L.P. and R.P.; formal analysis, F.D.G.; investigation, F.D.G., V.P. and G.I.; resources, V.R., G.L., F.L.P. and R.P.; data curation, F.D.G., V.P. and S.O.; writing—original draft preparation, F.D.G.; writing—review and editing, F.D.G., V.R., G.L., G.I., F.L.P., R.P. and S.B.; visualization, F.D.G., V.P., F.L.P., S.B. and R.P.; supervision, F.D.G., G.L., F.L.P. and R.P.; project administration, F.D.G., V.R., G.L., F.L.P. and R.P. All authors have read and agreed to the published version of the manuscript.

Funding: This work was supported by the IRCCS Istituto delle Scienze Neurologiche di Bologna and by S.B. with private funding.

Institutional Review Board Statement: This study was approved by the Ethical Committee Area Vasta Emilia Centro (CE num. 841-2021-OSS-AUSLBO).

Informed Consent Statement: Informed consent was obtained from all subjects involved in the study.

Data Availability Statement: The datasets used and analyzed during the current study are available from the corresponding author on reasonable request. The anonymised EEG raw data are publicly available for download at https://doi.org/10.5281/zenodo.6951440.

Acknowledgments: The authors thanks all the staff of the Neurorehabilitation Unit of the ISNB of Bologna.

Conflicts of Interest: The authors declare no conflict of interest.

Abbreviations

DoC   Disorders of consciousness
EEG   Electroencephalography
TBI   Traumatic brain injury
wPLI   Weighted phase lag index
PCoh   Partial coherence
MI   Mutual information
LDA   Linear discriminant analyses

References

1. Laureys, S.; Celesia, G.G.; Cohadon, F.; Lavrijsen, J.; León-Carrión, J.; Sannita, W.G.; Sazbon, L.; Schmutzhard, E.; von Wild, K.R.; Zeman, A.; et al. Unresponsive Wakefulness Syndrome: A New Name for the Vegetative State or Apallic Syndrome. *BMC Med.* **2010**, *8*, 68. [CrossRef] [PubMed]
2. Giacino, J.T.; Fins, J.J.; Laureys, S.; Schiff, N.D. Disorders of Consciousness after Acquired Brain Injury: The State of the Science. *Nat. Rev. Neurol.* **2014**, *10*, 99–114. [CrossRef] [PubMed]
3. Höller, Y.; Thomschewski, A.; Bergmann, J.; Kronbichler, M.; Crone, J.S.; Schmid, E.V.; Butz, K.; Höller, P.; Nardone, R.; Trinka, E. Connectivity Biomarkers Can Differentiate Patients with Different Levels of Consciousness. *Clin. Neurophysiol. Off. J. Int. Fed. Clin. Neurophysiol.* **2014**, *125*, 1545–1555. [CrossRef] [PubMed]
4. Schnakers, C.; Vanhaudenhuyse, A.; Giacino, J.; Ventura, M.; Boly, M.; Majerus, S.; Moonen, G.; Laureys, S. Diagnostic Accuracy of the Vegetative and Minimally Conscious State: Clinical Consensus versus Standardized Neurobehavioral Assessment. *BMC Neurol.* **2009**, *9*, 35. [CrossRef]
5. Thibaut, A.; Bodien, Y.G.; Laureys, S.; Giacino, J.T. Minimally Conscious State "plus": Diagnostic Criteria and Relation to Functional Recovery. *J. Neurol.* **2020**, *267*, 1245–1254. [CrossRef]
6. Seel, R.T.; Sherer, M.; Whyte, J.; Katz, D.I.; Giacino, J.T.; Rosenbaum, A.M.; Hammond, F.M.; Kalmar, K.; Pape, T.L.-B.; Zafonte, R.; et al. Assessment Scales for Disorders of Consciousness: Evidence-Based Recommendations for Clinical Practice and Research. *Arch. Phys. Med. Rehabil.* **2010**, *91*, 1795–1813. [CrossRef]
7. La Porta, F.; Caselli, S.; Ianes, A.B.; Cameli, O.; Lino, M.; Piperno, R.; Sighinolfi, A.; Lombardi, F.; Tennant, A. Can We Scientifically and Reliably Measure the Level of Consciousness in Vegetative and Minimally Conscious States? Rasch Analysis of the Coma Recovery Scale-Revised. *Arch. Phys. Med. Rehabil.* **2013**, *94*, 527–535. [CrossRef]
8. Owen, A.M.; Coleman, M.R. Detecting Awareness in the Vegetative State. *Ann. N. Y. Acad. Sci.* **2008**, *1129*, 130–138. [CrossRef]
9. Monti, M.M.; Vanhaudenhuyse, A.; Coleman, M.R.; Boly, M.; Pickard, J.D.; Tshibanda, L.; Owen, A.M.; Laureys, S. Willful Modulation of Brain Activity in Disorders of Consciousness. *N. Engl. J. Med.* **2010**, *362*, 579–589. [CrossRef]
10. Naci, L.; Sinai, L.; Owen, A.M. Detecting and Interpreting Conscious Experiences in Behaviorally Non-Responsive Patients. *NeuroImage* **2017**, *145 Pt B*, 304–313. [CrossRef]
11. Claassen, J.; Doyle, K.; Matory, A.; Couch, C.; Burger, K.M.; Velazquez, A.; Okonkwo, J.U.; King, J.-R.; Park, S.; Agarwal, S.; et al. Detection of Brain Activation in Unresponsive Patients with Acute Brain Injury. *N. Engl. J. Med.* **2019**, *380*, 2497–2505. [CrossRef] [PubMed]
12. Luppi, A.I.; Cain, J.; Spindler, L.R.B.; Górska, U.J.; Toker, D.; Hudson, A.E.; Brown, E.N.; Diringer, M.N.; Stevens, R.D.; Massimini, M.; et al. Mechanisms Underlying Disorders of Consciousness: Bridging Gaps to Move Toward an Integrated Translational Science. *Neurocrit. Care* **2021**, *35* (Suppl. 1), 37–54. [CrossRef] [PubMed]
13. Carter, B.G.; Butt, W. Review of the Use of Somatosensory Evoked Potentials in the Prediction of Outcome after Severe Brain Injury. *Crit. Care Med.* **2001**, *29*, 178–186. [CrossRef] [PubMed]
14. Amantini, A.; Carrai, R.; Fossi, S.; Pinto, F.; Grippo, A. The Role of Early Electroclinical Assessment in Improving the Evaluation of Patients with Disorders of Consciousness. *Funct. Neurol.* **2011**, *26*, 7–14. [PubMed]
15. van Erp, W.S.; Lavrijsen, J.C.M.; van de Laar, F.A.; Vos, P.E.; Laureys, S.; Koopmans, R.T.C.M. The Vegetative State/Unresponsive Wakefulness Syndrome: A Systematic Review of Prevalence Studies. *Eur. J. Neurol.* **2014**, *21*, 1361–1368. [CrossRef] [PubMed]
16. Hussain, I.; Park, S.J. HealthSOS: Real-Time Health Monitoring System for Stroke Prognostics. *IEEE Access* **2020**, *8*, 213574–213586. [CrossRef]
17. Hussain, I.; Park, S.-J. Quantitative Evaluation of Task-Induced Neurological Outcome after Stroke. *Brain Sci.* **2021**, *11*, 900. [CrossRef]
18. Hussain, I.; Young, S.; Kim, C.H.; Benjamin, H.C.M.; Park, S.J. Quantifying Physiological Biomarkers of a Microwave Brain Stimulation Device. *Sensors* **2021**, *21*, 1896. [CrossRef]
19. Hussain, I.; Hossain, M.A.; Jany, R.; Bari, M.A.; Uddin, M.; Kamal, A.R.M.; Ku, Y.; Kim, J.-S. Quantitative Evaluation of EEG-Biomarkers for Prediction of Sleep Stages. *Sensors* **2022**, *22*, 3079. [CrossRef]
20. Engemann, D.A.; Raimondo, F.; King, J.-R.; Rohaut, B.; Louppe, G.; Faugeras, F.; Annen, J.; Cassol, H.; Gosseries, O.; Fernandez-Slezak, D.; et al. Robust EEG-Based Cross-Site and Cross-Protocol Classification of States of Consciousness. *Brain* **2018**, *141*, 3179–3192. [CrossRef]

21. Chennu, S.; Annen, J.; Wannez, S.; Thibaut, A.; Chatelle, C.; Cassol, H.; Martens, G.; Schnakers, C.; Gosseries, O.; Menon, D.; et al. Brain Networks Predict Metabolism, Diagnosis and Prognosis at the Bedside in Disorders of Consciousness. *Brain* **2017**, *140*, 2120–2132. [CrossRef] [PubMed]
22. O'Donnell, A.; Pauli, R.; Banellis, L.; Sokoliuk, R.; Hayton, T.; Sturman, S.; Veenith, T.; Yakoub, K.M.; Belli, A.; Chennu, S.; et al. The Prognostic Value of Resting-State EEG in Acute Post-Traumatic Unresponsive States. *Brain Commun.* **2021**, *3*, fcab017. [CrossRef] [PubMed]
23. Bagnato, S.; Boccagni, C.; Sant'Angelo, A.; Prestandrea, C.; Mazzilli, R.; Galardi, G. EEG Predictors of Outcome in Patients with Disorders of Consciousness Admitted for Intensive Rehabilitation. *Clin. Neurophysiol. Off. J. Int. Fed. Clin. Neurophysiol.* **2015**, *126*, 959–966. [CrossRef] [PubMed]
24. Estraneo, A.; Fiorenza, S.; Magliacano, A.; Formisano, R.; Mattia, D.; Grippo, A.; Romoli, A.M.; Angelakis, E.; Cassol, H.; Thibaut, A.; et al. Multicenter Prospective Study on Predictors of Short-Term Outcome in Disorders of Consciousness. *Neurology* **2020**, *95*, e1488–e1499. [CrossRef] [PubMed]
25. Scarpino, M.; Lolli, F.; Hakiki, B.; Lanzo, G.; Sterpu, R.; Atzori, T.; Portaccio, E.; Draghi, F.; Amantini, A.; Grippo, A. EEG and Coma Recovery Scale-Revised Prediction of Neurological Outcome in Disorder of Consciousness Patients. *Acta Neurol. Scand.* **2020**, *142*, 221–228. [CrossRef] [PubMed]
26. Tononi, G.; Boly, M.; Massimini, M.; Koch, C. Integrated Information Theory: From Consciousness to Its Physical Substrate. *Nat. Rev. Neurosci.* **2016**, *17*, 450–461. [CrossRef]
27. Koch, C.; Massimini, M.; Boly, M.; Tononi, G. Neural Correlates of Consciousness: Progress and Problems. *Nat. Rev. Neurosci.* **2016**, *17*, 307–321. [CrossRef]
28. Rosanova, M.; Gosseries, O.; Casarotto, S.; Boly, M.; Casali, A.G.; Bruno, M.-A.; Mariotti, M.; Boveroux, P.; Tononi, G.; Laureys, S.; et al. Recovery of Cortical Effective Connectivity and Recovery of Consciousness in Vegetative Patients. *Brain* **2012**, *135 Pt 4*, 1308–1320. [CrossRef]
29. Sarasso, S.; Rosanova, M.; Casali, A.G.; Casarotto, S.; Fecchio, M.; Boly, M.; Gosseries, O.; Tononi, G.; Laureys, S.; Massimini, M. Quantifying Cortical EEG Responses to TMS in (Un)Consciousness. *Clin. EEG Neurosci.* **2014**, *45*, 40–49. [CrossRef]
30. Casali, A.G.; Gosseries, O.; Rosanova, M.; Boly, M.; Sarasso, S.; Casali, K.R.; Casarotto, S.; Bruno, M.-A.; Laureys, S.; Tononi, G.; et al. A Theoretically Based Index of Consciousness Independent of Sensory Processing and Behavior. *Sci. Transl. Med.* **2013**, *5*, 198ra105. [CrossRef]
31. Demertzi, A.; Tagliazucchi, E.; Dehaene, S.; Deco, G.; Barttfeld, P.; Raimondo, F.; Martial, C.; Fernández-Espejo, D.; Rohaut, B.; Voss, H.U.; et al. Human Consciousness Is Supported by Dynamic Complex Patterns of Brain Signal Coordination. *Sci. Adv.* **2019**, *5*, eaat7603. [CrossRef] [PubMed]
32. Gosseries, O.; Schnakers, C.; Ledoux, D.; Vanhaudenhuyse, A.; Bruno, M.-A.; Demertzi, A.; Noirhomme, Q.; Lehembre, R.; Damas, P.; Goldman, S.; et al. Automated EEG Entropy Measurements in Coma, Vegetative State/Unresponsive Wakefulness Syndrome and Minimally Conscious State. *Funct. Neurol.* **2011**, *26*, 25–30.
33. Thul, A.; Lechinger, J.; Donis, J.; Michitsch, G.; Pichler, G.; Kochs, E.F.; Jordan, D.; Ilg, R.; Schabus, M. EEG Entropy Measures Indicate Decrease of Cortical Information Processing in Disorders of Consciousness. *Clin. Neurophysiol. Off. J. Int. Fed. Clin. Neurophysiol.* **2016**, *127*, 1419–1427. [CrossRef] [PubMed]
34. Claassen, J.; Velazquez, A.; Meyers, E.; Witsch, J.; Falo, M.C.; Park, S.; Agarwal, S.; Michael Schmidt, J.; Schiff, N.D.; Sitt, J.D.; et al. Bedside Quantitative Electroencephalography Improves Assessment of Consciousness in Comatose Subarachnoid Hemorrhage Patients. *Ann. Neurol.* **2016**, *80*, 541–553. [CrossRef] [PubMed]
35. Stefan, S.; Schorr, B.; Lopez-Rolon, A.; Kolassa, I.-T.; Shock, J.P.; Rosenfelder, M.; Heck, S.; Bender, A. Consciousness Indexing and Outcome Prediction with Resting-State EEG in Severe Disorders of Consciousness. *Brain Topogr.* **2018**, *31*, 848–862. [CrossRef] [PubMed]
36. King, J.-R.; Sitt, J.D.; Faugeras, F.; Rohaut, B.; El Karoui, I.; Cohen, L.; Naccache, L.; Dehaene, S. Information Sharing in the Brain Indexes Consciousness in Noncommunicative Patients. *Curr. Biol.* **2013**, *23*, 1914–1919. [CrossRef] [PubMed]
37. Naro, A.; Bramanti, A.; Leo, A.; Cacciola, A.; Manuli, A.; Bramanti, P.; Calabrò, R.S. Shedding New Light on Disorders of Consciousness Diagnosis: The Dynamic Functional Connectivity. *Cortex* **2018**, *103*, 316–328. [CrossRef]
38. Sitt, J.D.; King, J.-R.; El Karoui, I.; Rohaut, B.; Faugeras, F.; Gramfort, A.; Cohen, L.; Sigman, M.; Dehaene, S.; Naccache, L. Large Scale Screening of Neural Signatures of Consciousness in Patients in a Vegetative or Minimally Conscious State. *Brain* **2014**, *137 Pt 8*, 2258–2270. [CrossRef]
39. Fridman, E.A.; Beattie, B.J.; Broft, A.; Laureys, S.; Schiff, N.D. Regional Cerebral Metabolic Patterns Demonstrate the Role of Anterior Forebrain Mesocircuit Dysfunction in the Severely Injured Brain. *Proc. Natl. Acad. Sci. USA* **2014**, *111*, 6473–6478. [CrossRef]
40. Lant, N.D.; Gonzalez-Lara, L.E.; Owen, A.M.; Fernández-Espejo, D. Relationship between the Anterior Forebrain Mesocircuit and the Default Mode Network in the Structural Bases of Disorders of Consciousness. *NeuroImage Clin.* **2016**, *10*, 27–35. [CrossRef]
41. Redinbaugh, M.J.; Phillips, J.M.; Kambi, N.A.; Mohanta, S.; Andryk, S.; Dooley, G.L.; Afrasiabi, M.; Raz, A.; Saalmann, Y.B. Thalamus Modulates Consciousness via Layer-Specific Control of Cortex. *Neuron* **2020**, *106*, 66–75. [CrossRef] [PubMed]
42. Threlkeld, Z.D.; Bodien, Y.G.; Rosenthal, E.S.; Giacino, J.T.; Nieto-Castanon, A.; Wu, O.; Whitfield-Gabrieli, S.; Edlow, B.L. Functional Networks Reemerge during Recovery of Consciousness after Acute Severe Traumatic Brain Injury. *Cortex* **2018**, *106*, 299–308. [CrossRef] [PubMed]

43. Formisano, R.; Contrada, M.; Ferri, G.; Schiattone, S.; Iosa, M.; Aloisi, M. The Glasgow Outcome Scale Extended-Revised (GOSE-R) to Include Minimally Conscious State in the Vegetative State/Unresponsive Wakefulness Syndrome Category: A Correlation with Coma Recovery Scale-Revised (CRS-R). *Eur. J. Phys. Rehabil. Med.* **2019**, *55*, 139–140. [CrossRef]

44. Mecarelli, O.; Brienza, M.; Grippo, A.; Amantini, A. Disorders of Consciousness. In *Clinical Electroencephalography*; Mecarelli, O., Ed.; Springer: Cham, Switzerland, 2019; pp. 731–765.

45. Delorme, A.; Makeig, S. EEGLAB: An Open Source Toolbox for Analysis of Single-Trial EEG Dynamics Including Independent Component Analysis. *J. Neurosci. Methods* **2004**, *134*, 9–21. [CrossRef] [PubMed]

46. Trajkovic, J.; Di Gregorio, F.; Ferri, F.; Marzi, C.; Diciotti, S.; Romei, V. Resting State Alpha Oscillatory Activity Is a Valid and Reliable Marker of Schizotypy. *Sci. Rep.* **2021**, *11*, 10379. [CrossRef]

47. Lapitska, N.; Gosseries, O.; Delvaux, V.; Overgaard, M.; Nielsen, F.; Maertens de Noordhout, A.; Moonen, G.; Laureys, S. Transcranial Magnetic Stimulation in Disorders of Consciousness. *Rev. Neurosci.* **2009**, *20*, 235–250. [CrossRef] [PubMed]

48. Song, M.; Yang, Y.; He, J.; Yang, Z.; Yu, S.; Xie, Q.; Xia, X.; Dang, Y.; Zhang, Q.; Wu, X.; et al. Prognostication of Chronic Disorders of Consciousness Using Brain Functional Networks and Clinical Characteristics. *eLife* **2018**, *7*, e36173. [CrossRef]

49. Hebb, M.O.; McArthur, D.L.; Alger, J.; Etchepare, M.; Glenn, T.C.; Bergsneider, M.; Martin, N.; Vespa, P.M. Impaired Percent Alpha Variability on Continuous Electroencephalography Is Associated with Thalamic Injury and Predicts Poor Long-Term Outcome after Human Traumatic Brain Injury. *J. Neurotrauma* **2007**, *24*, 579–590. [CrossRef]

50. Stender, J.; Gosseries, O.; Bruno, M.-A.; Charland-Verville, V.; Vanhaudenhuyse, A.; Demertzi, A.; Chatelle, C.; Thonnard, M.; Thibaut, A.; Heine, L.; et al. Diagnostic Precision of PET Imaging and Functional MRI in Disorders of Consciousness: A Clinical Validation Study. *Lancet* **2014**, *384*, 514–522. [CrossRef]

51. Duclos, C.; Norton, L.; Laforge, G.; Frantz, A.; Maschke, C.; Badawy, M.; Letourneau, J.; Slessarev, M.; Gofton, T.; Debicki, D.; et al. Protocol for the Prognostication of Consciousness Recovery Following a Brain Injury. *Front. Hum. Neurosci.* **2020**, *14*, 582125. [CrossRef]

52. Stocchetti, N.; Zanier, E.R. Chronic Impact of Traumatic Brain Injury on Outcome and Quality of Life: A Narrative Review. *Crit. Care* **2016**, *20*, 148. [CrossRef] [PubMed]

53. Edlow, B.L.; Claassen, J.; Schiff, N.D.; Greer, D.M. Recovery from Disorders of Consciousness: Mechanisms, Prognosis and Emerging Therapies. *Nat. Rev. Neurol.* **2021**, *17*, 135–156. [CrossRef] [PubMed]

54. Pauli, C.S.; Conroy, M.; Vanden Heuvel, B.D.; Park, S.-H. Cannabidiol Drugs Clinical Trial Outcomes and Adverse Effects. *Front. Pharmacol.* **2020**, *11*, 63. [CrossRef] [PubMed]

55. Pugin, D.; Vargas, M.-I.; Thieffry, C.; Schibler, M.; Grosgurin, O.; Pugin, J.; Lalive, P.H. COVID-19-Related Encephalopathy Responsive to High-Dose Glucocorticoids. *Neurology* **2020**, *95*, 543–546. [CrossRef] [PubMed]

56. Chennu, S.; Finoia, P.; Kamau, E.; Allanson, J.; Williams, G.B.; Monti, M.M.; Noreika, V.; Arnatkeviciute, A.; Canales-Johnson, A.; Olivares, F.; et al. Spectral Signatures of Reorganised Brain Networks in Disorders of Consciousness. *PLoS Comput. Biol.* **2014**, *10*, e1003887. [CrossRef]

57. Cohen, M.X. Comparison of Different Spatial Transformations Applied to EEG Data: A Case Study of Error Processing. *Int. J. Psychophysiol. Off. J. Int. Organ. Psychophysiol.* **2015**, *97*, 245–257. [CrossRef]

58. Cohen, M.X. *Analyzing Neural Time Series Data: Theory and Practice*; MIT Press: Cambridge, MA, USA, 2014.

59. Cohen, M.X. Effects of Time Lag and Frequency Matching on Phase-Based Connectivity. *J. Neurosci. Methods* **2015**, *250*, 137–146. [CrossRef]

60. Stam, C.J.; Nolte, G.; Daffertshofer, A. Phase Lag Index: Assessment of Functional Connectivity from Multi Channel EEG and MEG with Diminished Bias from Common Sources. *Hum. Brain Mapp.* **2007**, *28*, 1178–1193. [CrossRef]

61. Hardmeier, M.; Hatz, F.; Bousleiman, H.; Schindler, C.; Stam, C.J.; Fuhr, P. Reproducibility of Functional Connectivity and Graph Measures Based on the Phase Lag Index (PLI) and Weighted Phase Lag Index (WPLI) Derived from High Resolution EEG. *PLoS ONE* **2014**, *9*, e108648. [CrossRef]

62. Di Gregorio, F.; Maier, M.E.; Steinhauser, M. Errors Can Elicit an Error Positivity in the Absence of an Error Negativity: Evidence for Independent Systems of Human Error Monitoring. *NeuroImage* **2018**, *172*, 427–436. [CrossRef]

63. Bareham, C.A.; Roberts, N.; Allanson, J.; Hutchinson, P.J.A.; Pickard, J.D.; Menon, D.K.; Chennu, S. Bedside EEG Predicts Longitudinal Behavioural Changes in Disorders of Consciousness. *NeuroImage Clin.* **2020**, *28*, 102372. [CrossRef] [PubMed]

64. Greenhouse, S.W.; Geisser, S. On Methods in the Analysis of Profile Data. *Psychometrika* **1959**, *24*, 95–112. [CrossRef]

65. Shapiro, S.S.; Wilk, M.B. An Analysis of Variance Test for Normality (Complete Samples). *Biometrika* **1965**, *52*, 591–611. [CrossRef]

66. Di Gregorio, F.; Maier, M.E.; Steinhauser, M. Are Errors Detected before They Occur? Early Error Sensations Revealed by Metacognitive Judgments on the Timing of Error Awareness. *Conscious. Cogn.* **2020**, *77*, 102857. [CrossRef]

67. Di Gregorio, F.; La Porta, F.; Casanova, E.; Magni, E.; Bonora, R.; Ercolino, M.G.; Petrone, V.; Leo, M.R.; Piperno, R. Efficacy of Repetitive Transcranial Magnetic Stimulation Combined with Visual Scanning Treatment on Cognitive and Behavioral Symptoms of Left Hemispatial Neglect in Right Hemispheric Stroke Patients: Study Protocol for a Randomized Controlled Trial. *Trials* **2021**, *22*, 24. [CrossRef] [PubMed]

68. Trajkovic, J.; Di Gregorio, F.; Marcantoni, E.; Thut, G.; Romei, V. A TMS/EEG Protocol for the Causal Assessment of the Functions of the Oscillatory Brain Rhythms in Perceptual and Cognitive Processes. *STAR Protoc.* **2022**, *3*, 101435. [CrossRef]

69. Di Gregorio, F.; La Porta, F.; Lullini, G.; Casanova, E.; Petrone, V.; Simoncini, L.; Ferrucci, E.; Piperno, R. Efficacy of Repetitive Transcranial Magnetic Stimulation Combined With Visual Scanning Treatment on Cognitive-Behavioral Symptoms of Unilateral Spatial Neglect in Patients With Traumatic Brain Injury: Study Protocol for a Randomized Controlled Trial. *Front. Neurol.* **2021**, *12*, 702649. [CrossRef]
70. Di Gregorio, F.; Maier, M.E.; Steinhauser, M. Early Correlates of Error-Related Brain Activity Predict Subjective Timing of Error Awareness. *Psychophysiology* **2022**, *59*, e14020. [CrossRef]
71. Babiloni, C.; Barry, R.J.; Başar, E.; Blinowska, K.J.; Cichocki, A.; Drinkenburg, W.H.I.M.; Klimesch, W.; Knight, R.T.; Lopes da Silva, F.; Nunez, P.; et al. International Federation of Clinical Neurophysiology (IFCN)-EEG Research Workgroup: Recommendations on Frequency and Topographic Analysis of Resting State EEG Rhythms. Part 1: Applications in Clinical Research Studies. *Clin. Neurophysiol. Off. J. Int. Fed. Clin. Neurophysiol.* **2020**, *131*, 285–307. [CrossRef]
72. Noirhomme, Q.; Brecheisen, R.; Lesenfants, D.; Antonopoulos, G.; Laureys, S. "Look at My Classifier's Result": Disentangling Unresponsive from (Minimally) Conscious Patients. *NeuroImage* **2017**, *145 Pt B*, 288–303. [CrossRef]
73. Corchs, S.; Chioma, G.; Dondi, R.; Gasparini, F.; Manzoni, S.; Markowska-Kacznar, U.; Mauri, G.; Zoppis, I.; Morreale, A. Computational Methods for Resting-State EEG of Patients with Disorders of Consciousness. *Front. Neurosci.* **2019**, *13*, 1–7. [CrossRef]
74. Liong, C.Y.; Foo, S.F. Comparison of Linear Discriminant Analysis and Logistic Regression for Data Classification. *AIP Conf. Proc.* **2013**, *1522*, 1159–1165. [CrossRef]
75. Lei, P.W.; Koehly, L.M. Linear Discriminant Analysis versus Logistic Regression: A Comparison of Classification Errors in the Two-Group Case. *J. Exp. Educ.* **2003**, *72*, 25–49. [CrossRef]
76. Fan, X.; Wang, L. Comparing Linear Discriminant Function with Logistic Regression for the Two-Group Classification Problem. *J. Exp. Educ.* **1999**, *67*, 265–286. [CrossRef]
77. Höller, Y.; Bergmann, J.; Thomschewski, A.; Kronbichler, M.; Höller, P.; Crone, J.S.; Schmid, E.V.; Butz, K.; Nardone, R.; Trinka, E. Comparison of EEG-Features and Classification Methods for Motor Imagery in Patients with Disorders of Consciousness. *PLoS ONE* **2013**, *8*, e80479. [CrossRef]
78. Krawczyk, B. Learning from Imbalanced Data: Open Challenges and Future Directions. *Prog. Artif. Intell.* **2016**, *5*, 221–232. [CrossRef]
79. Wright, J. The Glasgow Outcome Scale. Available online: http://www.tbims.org/combi/gos (accessed on 1 March 2022).
80. Faul, F.; Erdfelder, E.; Lang, A.-G.; Buchner, A. G*Power 3: A Flexible Statistical Power Analysis Program for the Social, Behavioral, and Biomedical Sciences. *Behav. Res. Methods* **2007**, *39*, 175–191. [CrossRef]
81. Marshall, L.F.; Marshall, S.B.; Klauber, M.R.; Van Berkum Clark, M.; Eisenberg, H.; Jane, J.A.; Luerssen, T.G.; Marmarou, A.; Foulkes, M.A. The Diagnosis of Head Injury Requires a Classification Based on Computed Axial Tomography. *J. Neurotrauma* **1992**, *9* (Suppl. 1), S287–S292.
82. Pietrelli, M.; Zanon, M.; Làdavas, E.; Grasso, P.A.; Romei, V.; Bertini, C. Posterior Brain Lesions Selectively Alter Alpha Oscillatory Activity and Predict Visual Performance in Hemianopic Patients. *Cortex* **2019**, *121*, 347–361. [CrossRef]
83. Di Gregorio, F.; Trajkovic, J.; Roperti, C.; Marcantoni, E.; Di Luzio, P.; Avenanti, A.; Thut, G.; Romei, V. Tuning Alpha Rhythms to Shape Conscious Visual Perception. *Curr. Biol.* **2022**, *32*, 988–998. [CrossRef]
84. Battaglia, S.; Fabius, J.H.; Moravkova, K.; Fracasso, A.; Borgomaneri, S. The Neurobiological Correlates of Gaze Perception in Healthy Individuals and Neurologic Patients. *Biomedicines* **2022**, *10*, 627. [CrossRef]
85. Battaglia, S. Neurobiological Advances of Learned Fear in Humans. *Adv. Clin. Exp. Med.* **2022**, *31*, 217–221. [CrossRef] [PubMed]
86. Sellitto, M.; Terenzi, D.; Starita, F.; Pellegrino, G.; Battaglia, S. The Cost of Imagined Actions in a Reward-Valuation Task. *Brain Sci.* **2022**, *12*, 582. [CrossRef] [PubMed]
87. Stender, J.; Kupers, R.; Rodell, A.; Thibaut, A.; Chatelle, C.; Bruno, M.-A.; Gejl, M.; Bernard, C.; Hustinx, R.; Laureys, S.; et al. Quantitative Rates of Brain Glucose Metabolism Distinguish Minimally Conscious from Vegetative State Patients. *J. Cereb. Blood Flow Metab. Off. J. Int. Soc. Cereb. Blood Flow Metab.* **2015**, *35*, 58–65. [CrossRef] [PubMed]
88. Stender, J.; Mortensen, K.N.; Thibaut, A.; Darkner, S.; Laureys, S.; Gjedde, A.; Kupers, R. The Minimal Energetic Requirement of Sustained Awareness after Brain Injury. *Curr. Biol.* **2016**, *26*, 1494–1499. [CrossRef] [PubMed]
89. Achard, S.; Delon-Martin, C.; Vértes, P.E.; Renard, F.; Schenck, M.; Schneider, F.; Heinrich, C.; Kremer, S.; Bullmore, E.T. Hubs of Brain Functional Networks Are Radically Reorganized in Comatose Patients. *Proc. Natl. Acad. Sci. USA* **2012**, *109*, 20608–20613. [CrossRef] [PubMed]
90. Vanhaudenhuyse, A.; Noirhomme, Q.; Tshibanda, L.J.-F.; Bruno, M.-A.; Boveroux, P.; Schnakers, C.; Soddu, A.; Perlbarg, V.; Ledoux, D.; Brichant, J.-F.; et al. Default Network Connectivity Reflects the Level of Consciousness in Non-Communicative Brain-Damaged Patients. *Brain* **2010**, *133 Pt 1*, 161–171. [CrossRef]
91. Balogh, L.; Tanaka, M.; Török, N.; Taguchi, S. Crosstalk between Existential Phenomenological Psychotherapy and Neurological Sciences in Mood and Anxiety Disorders. *Biomedicines* **2021**, *9*, 340. [CrossRef]
92. Cavinato, M.; Genna, C.; Manganotti, P.; Formaggio, E.; Storti, S.F.; Campostrini, S.; Arcaro, C.; Casanova, E.; Petrone, V.; Piperno, R.; et al. Coherence and Consciousness: Study of Fronto-Parietal Gamma Synchrony in Patients with Disorders of Consciousness. *Brain Topogr.* **2015**, *28*, 570–579. [CrossRef]
93. Davey, M.P.; Victor, J.D.; Schiff, N.D. Power Spectra and Coherence in the EEG of a Vegetative Patient with Severe Asymmetric Brain Damage. *Clin. Neurophysiol. Off. J. Int. Fed. Clin. Neurophysiol.* **2000**, *111*, 1949–1954. [CrossRef]

94. Imperatori, L.S.; Betta, M.; Cecchetti, L.; Canales-Johnson, A.; Ricciardi, E.; Siclari, F.; Pietrini, P.; Chennu, S.; Bernardi, G. EEG Functional Connectivity Metrics WPLI and WSMI Account for Distinct Types of Brain Functional Interactions. *Sci. Rep.* **2019**, *9*, 8894. [CrossRef] [PubMed]

95. Cooke, J.; Poch, C.; Gillmeister, H.; Costantini, M.; Romei, V. Oscillatory Properties of Functional Connections Between Sensory Areas Mediate Cross-Modal Illusory Perception. *J. Neurosci. Off. J. Soc. Neurosci.* **2019**, *39*, 5711–5718. [CrossRef] [PubMed]

96. Fotia, F.; Cooke, J.; Van Dam, L.; Ferri, F.; Romei, V. The Temporal Sensitivity to the Tactile-Induced Double Flash Illusion Mediates the Impact of Beta Oscillations on Schizotypal Personality Traits. *Conscious. Cogn.* **2021**, *91*, 103121. [CrossRef] [PubMed]

97. Schorr, B.; Schlee, W.; Arndt, M.; Bender, A. Coherence in Resting-State EEG as a Predictor for the Recovery from Unresponsive Wakefulness Syndrome. *J. Neurol.* **2016**, *263*, 937–953. [CrossRef] [PubMed]

98. Crone, J.S.; Bio, B.J.; Vespa, P.M.; Lutkenhoff, E.S.; Monti, M.M. Restoration of Thalamo-Cortical Connectivity after Brain Injury: Recovery of Consciousness, Complex Behavior, or Passage of Time? *J. Neurosci. Res.* **2018**, *96*, 671–687. [CrossRef]

99. Alkire, M.T.; Hudetz, A.G.; Tononi, G. Consciousness and Anesthesia. *Science* **2008**, *322*, 876–880. [CrossRef]

100. Dehaene, S.; Naccache, L. Towards a Cognitive Neuroscience of Consciousness: Basic Evidence and a Workspace Framework. *Cognition* **2001**, *79*, 1–37. [CrossRef]

101. Wang, T.; Bezerianos, A.; Cichocki, A.; Li, J. Multikernel Capsule Network for Schizophrenia Identification. *IEEE Trans. Cybern.* **2020**, *52*, 4741–4750. [CrossRef]

102. Rapcan, V.; D'Arcy, S.; Yeap, S.; Afzal, N.; Thakore, J.; Reilly, R.B. Acoustic and Temporal Analysis of Speech: A Potential Biomarker for Schizophrenia. *Med. Eng. Phys.* **2010**, *32*, 1074–1079. [CrossRef]

103. Lin, W.; Gao, Q.; Du, M.; Chen, W.; Tong, T. Multiclass Diagnosis of Stages of Alzheimer's Disease Using Linear Discriminant Analysis Scoring for Multimodal Data. *Comput. Biol. Med.* **2021**, *134*, 104478. [CrossRef]

104. Neto, E.; Biessmann, F.; Aurlien, H.; Nordby, H.; Eichele, T. Regularized Linear Discriminant Analysis of EEG Features in Dementia Patients. *Front. Aging Neurosci.* **2016**, *8*, 273. [CrossRef] [PubMed]

105. Gerez, M.; Tello, A. Selected Quantitative EEG (QEEG) and Event-Related Potential (ERP) Variables as Discriminators for Positive and Negative Schizophrenia. *Biol. Psychiatry* **1995**, *38*, 34–49. [CrossRef]

106. V, K.R.; Rajagopalan, S.S.; Bhardwaj, S.; Panda, R.; Reddam, V.R.; Ganne, C.; Kenchaiah, R.; Mundlamuri, R.C.; Kandavel, T.; Majumdar, K.K.; et al. Machine Learning Detects EEG Microstate Alterations in Patients Living with Temporal Lobe Epilepsy. *Seizure* **2018**, *61*, 8–13. [CrossRef] [PubMed]

107. Nunnally, J. Psychometric Theory—25 Years Ago and Now. *Educ. Res.* **1975**, *4*, 7–21.

108. Kasner, S.E. Clinical Interpretation and Use of Stroke Scales. *Lancet Neurol.* **2006**, *5*, 603–612. [CrossRef]

109. McMillan, T.; Wilson, L.; Ponsford, J.; Levin, H.; Teasdale, G.; Bond, M. The Glasgow Outcome Scale—40 Years of Application and Refinement. *Nat. Rev. Neurol.* **2016**, *12*, 477–485. [CrossRef]

110. Lieh-Lai, M.W.; Theodorou, A.A.; Sarnaik, A.P.; Meert, K.L.; Moylan, P.M.; Canady, A.I. Limitations of the Glasgow Coma Scale in Predicting Outcome in Children with Traumatic Brain Injury. *J. Pediatr.* **1992**, *120 Pt 1*, 195–199. [CrossRef]

International Journal of
Molecular Sciences

MDPI

Article

Weak Ultrasound Contributes to Neuromodulatory Effects in the Rat Motor Cortex

Po-Chun Chu [1], Chen-Syuan Huang [1], Pi-Kai Chang [2], Rou-Shayn Chen [3], Ko-Ting Chen [4,5], Tsung-Hsun Hsieh [5,6,7,*] and Hao-Li Liu [1,*]

1   Department of Electrical Engineering, National Taiwan University, Taipei 10617, Taiwan
2   Department of Pediatrics, School of Medicine, University of Utah, Salt Lake City, UT 84112, USA
3   Division of Movement Disorder, Department of Neurology, Chang Gung Memorial Hospital and Chang Gung University College of Medicine, Taipei 10507, Taiwan
4   Department of Neurosurgery, Chang Gung Memorial Hospital at Linkou, Taoyuan 33305, Taiwan
5   Neuroscience Research Center, Chang Gung Memorial Hospital, Linkou, Taoyuan 33305, Taiwan
6   School of Physical Therapy, Graduate Institute of Rehabilitation Science, Chang Gung University, Taoyuan 33302, Taiwan
7   Healthy Aging Research Center, Chang Gung University, Taoyuan 33302, Taiwan
*   Correspondence: hsiehth@mail.cgu.edu.tw (T.-H.H.); hlliu@ntu.edu.tw (H.-L.L.)

Abstract: Transcranial focused ultrasound (tFUS) is a novel neuromodulating technique. It has been demonstrated that the neuromodulatory effects can be induced by weak ultrasound exposure levels (spatial-peak temporal average intensity, $I_{SPTA} < 10 \ mW/cm^2$) in vitro. However, fewer studies have examined the use of weak tFUS to potentially induce long-lasting neuromodulatory responses in vivo. The purpose of this study was to determine the lower-bound threshold of tFUS stimulation for inducing neuromodulation in the motor cortex of rats. A total of 94 Sprague–Dawley rats were used. The sonication region aimed at the motor cortex under weak tFUS exposure (I_{SPTA} of 0.338–$12.15 \ mW/cm^2$). The neuromodulatory effects of tFUS on the motor cortex were evaluated by the changes in motor-evoked potentials (MEPs) elicited by transcranial magnetic stimulation (TMS). In addition to histology analysis, the in vitro cell culture was used to confirm the neuromodulatory mechanisms following tFUS stimulation. In the results, the dose-dependent inhibitory effects of tFUS were found, showing increased intensities of tFUS suppressed MEPs and lasted for 30 min. Weak tFUS significantly decreased the expression of excitatory neurons and increased the expression of inhibitory GABAergic neurons. The PIEZO-1 proteins of GABAergic neurons were found to involve in the inhibitory neuromodulation. In conclusion, we show the use of weak ultrasound to induce long-lasting neuromodulatory effects and explore the potential use of weak ultrasound for future clinical neuromodulatory applications.

Keywords: transcranial focused ultrasound; neuromodulation; weak ultrasound; rats; motor-evoked potentials; transcranial magnetic stimulation; in vivo; c-fos; GABAergic neurons; PIEZO-1 protein

Citation: Chu, P.-C.; Huang, C.-S.; Chang, P.-K.; Chen, R.-S.; Chen, K.-T.; Hsieh, T.-H.; Liu, H.-L. Weak Ultrasound Contributes to Neuromodulatory Effects in the Rat Motor Cortex. *Int. J. Mol. Sci.* **2023**, *24*, 2578. https://doi.org/10.3390/ijms24032578

Academic Editors: Masaru Tanaka, Lydia Giménez-Llort, Simone Battaglia, Chong Chen and Piril Hepsomali

Received: 7 December 2022
Revised: 23 January 2023
Accepted: 24 January 2023
Published: 30 January 2023

1. Introduction

Neuromodulation is currently a rapidly growing field where technology refers to interacting and interfering with the nervous system through chemical, mechanical, electrical, or electromagnetic methods to activate, inhibit, regulate or modify the neural activity for the treatment of neurological and neuropsychiatric disorders [1–3]. Neuromodulatory techniques have been developed through invasively or non-invasively stimulating cortical or subcortical regions for modulating brain activity [4–6]. For example, as a clinical tool, the deep brain stimulation technique (DBS) is a typical invasively intracranial neuromodulation approach that requires the surgical implantation of stimulating electrodes to deep brain area and has been proven to be effective for Parkinson's disease, essential tremor, dystonia, and epilepsy [7–11]. Non-invasive neuromodulation approaches such as repetitive transcranial magnetic stimulation (rTMS) or transcranial direct current stimulation (tDCS), are capable of inducing after-effects outlasting the stimulation period through plasticity-like mechanisms and have been approved for therapeutic purposes in neurological and psychiatric disorders [1,12,13]. However, the rTMS or tDCS have major limits. For example, the stimulation areas of rTMS or tDCS are relatively superficial and cannot penetrate deeper than the cortical layer of the brain, and their effects are variable both within and between individuals [13–15]. Hence, there is an unmet need for a novel transcranial brain stimulation technique for modulating a deeper area and for more consistent effects.

Transcranial focused ultrasound (tFUS), which converges the acoustic wave of ultrasound to a target point distant from the transducer, is a novel neuromodulating technology for noninvasively targeting deep brain tissue [4,16,17]. In particular, the main advantage is that the tFUS can stimulate much deeper brain structures and has a higher spatial resolution than rTMS or tDCS [18]. In contrast to high-intensity focused ultrasound which has shown its success in tissue ablation through a thermal effect, tFUS at a much lower intensity has shown the ability to modulate neuronal activity without causing tissue damage [19–23]. For instance, following ultrasound stimulation, suppressive neuromodulatory effects on human motor cortical excitability as measured by motor-evoked potentials (MEPs) elicited by transcranial magnetic stimulation (TMS) were found [24]. Another study demonstrated that tFUS delivered transiently increased excitability in the motor cortex [25]. The alternation of main effects from the facilitation or inhibition induced by tFUS could result from the application of differing tFUS parameters, such as duty cycle, amplitude, or duration of stimulation [26,27].

Concerning stimulation intensity of tFUS for neuromodulation, several in vivo and human studies utilized different exposure levels of tFUS from tens to thousands of milliwatts per square centimeter for identifying the neuromodulatory effects. Wang et al. [28] and Yuan et al. [29] found an increase in neural oscillations and cortical hemodynamic responses in the motor cortex by using tFUS with 80 to 400 mW/cm^2 spatial-peak temporal-average intensity (I_{SPTA}). In the animal study, Kim and associates demonstrated the suppressive responses to the sonication of either the primary motor cortex or the thalamus in conscious sheep at 3600 mW/cm^2 I_{SPTA} tFUS [30]. Dallapiazza et al. showed that 25,000 mW/cm^2 I_{SPTA} tFUS could functionally modulate to somatosensory evoke potential while delivering energy to the swine thalamus [21]. The findings of animal and human experiments have promoted the understanding of different biological effects based on different exposure levels of tFUS and show that the intensity of tFUS could be a key parameter for the induction of neuromodulation.

Recently, in contrast to the high intensity of tFUS for the neuromodulation, the application of weak-intensity ($I_{SPTA} < 10$ mW/cm^2) ultrasound to induce a neural modulatory effect has been reported in vitro and brain-slice experiments [31,32]. However, a physiological response from neuromodulation after weak-intensity sonication has not been demonstrated. To our knowledge, no study has directly investigated the cortex of the in vivo brain. We hypothesize that an in vivo and long-lasting neuromodulating effect can be generated by using weak ultrasound in a transcranial noninvasive manner. For this reason, a feasibility study of using weak ultrasound ($I_{SPTA} < 10$ mW/cm^2) to explore neuromodulatory effects is needed. The purpose of this study is trying to explore the feasibility to evoke the neuromodulation effects via designed increased intensity level of weak ultrasound ($I_{SPTA} < 10$ mW/cm^2) and with the assessment of motor-evoked potentials (MEPs) in the primary motor cortex (M1) of rat brains [24]. We also additionally assessed the selected ultrasound parameters with real-time calcium signals from primary neurons and histologic examinations to identify potential physiological mechanisms.

2. Results

The neuromodulatory effects of tFUS were evaluated by the changes of MEPs elicited by TMS on the motor cortex (M1) of rat brains [24,33,34]. Furthermore, to evaluate the safety of tFUS and verify the neuronal activity changes following weak tFUS, the glial fibrillary acidic protein (GFAP) staining and expression levels of c-Fos and GAD-65 were conducted. Furthermore, to identify potential physiological mechanisms, we additionally assessed the selected ultrasound parameters with real-time calcium signals in vitro from primary neurons through real-time calcium signal and immunostaining with the piezo inhibitor to confirm the activities of modulated GABAergic neurons.

2.1. The Changes in Motor-Evoked Potentials (MEPs) before and after Weak Sonication

To evaluate the neuromodulatory effects after ultrasound stimulation, animals were anesthetized and then mounted in a prone position in a stereotaxic apparatus. A homemade 0.7-MHz focused ultrasound transducer was used to deliver spatial-peak temporal average intensity (I_{SPTA}) from 0.014 mW/cm^2 to 50 mW/cm^2 for 5 min. MEPs were recorded for 10 min before tFUS to serve as the baseline, then MEP recording continued for 30 min to follow the post-tFUS effect. All animals exhibited 70.25 ± 1.16 (mean ± SEM) maximum machine output in resting motor threshold (RMT), and no significant differences between groups were observed (Supplementary Figure S1). Figure 1 illustrates a typical case of MEP change after various intensities of weak tFUS (from I_{SPTA} 0 mW/cm^2 (sham) to I_{SPTA} 12.15 mW/cm^2). The MEPs of the right limb for each time point after sham treatment (group 1) and I_{SPTA} of 0.014 mW/cm^2 exhibited no significant changes (Figure 1A,B). The MEPs decreased 10 min after I_{SPTA} of 0.338 mW/cm^2 and I_{SPTA} of 3.038 mW/cm^2 were applied (Figure 1C,D). Notably, following ultrasound exposure at I_{SPTA} of 12.15 mW/cm^2, the amplitude of MEPs decreased to 0.5 mV and persisted under that low level for 30 min (Figure 1E).

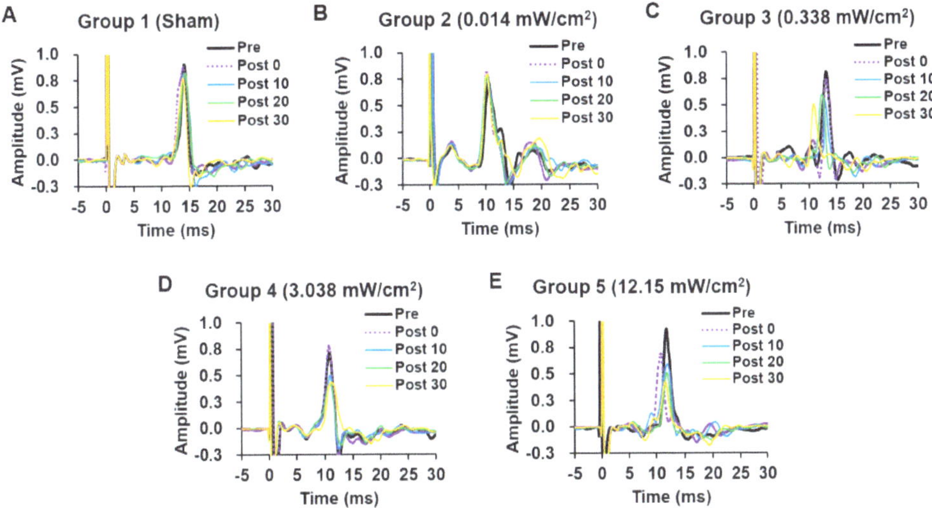

Figure 1. Representative MEPs of a rat before and after sham stimulation with various intensities of tFUS are presented at each measured period (preintervention and 0, 10, 20, 30 min after tFUS) with a colored line. (**A**) The MEPs of the right limb for each time point after sham treatment (**B**) The MEPs of the right limb for each time point after I_{SPTA} of 0.014 mW/cm^2 (**C**) The MEPs of the right limb for each time point after I_{SPTA} of 0.338 mW/cm^2 (**D**) The MEPs of the right limb for each time point after I_{SPTA} of 3.038 mW/cm^2 (**E**) The MEPs of the right limb for each time point after I_{SPTA} of 12.15 mW/cm^2.

We quantitatively evaluated the neuromodulation effect in each group by observing MEP responses in both limbs. Increased exposure levels monotonically decreased MEP responses (Figure 2). The normalized averaged amplitude of MEP data for left/right limbs from group 1 (sham) and group 2 (I_{SPTA} 0.014 mW/cm^2) exhibited no significant difference (Figure 2A,B). Following exposure to an I_{SPTA} of 0.338 mW/cm^2 (group 3, Figure 2C), the amplitude of MEPs for the right limb decreased from 0.94 ± 0.22 to 0.79 ± 0.28, which was significantly lower than the MEP for the left limb (from 1.00 ± 0.44 to 0.83 ± 0.49, $p < 0.05$ at all time points). Over the post-tFUS period (0–30 min), group 4 (I_{SPTA} 3.038 mW/cm^2) and group 5 (I_{SPTA} 12.15 mW/cm^2) exhibited declined MEPs of the right limb (30.5% and 31.4% decreases compared with pre-FUS values, respectively) (Figure 2D,E). In group 4 and group 5, the MEPs of the left limb also decreased by 20% and 24.3%, respectively, but no significant differences were observed between limbs ($p = 0.353$ and 0.356, respectively). This indicates that the higher intensity of tFUS ($I_{SPTA} > 3.038$ mW/cm^2) can lead to a suppression effect in not only the ipsilateral sonicated motor cortex but also in the contralateral region.

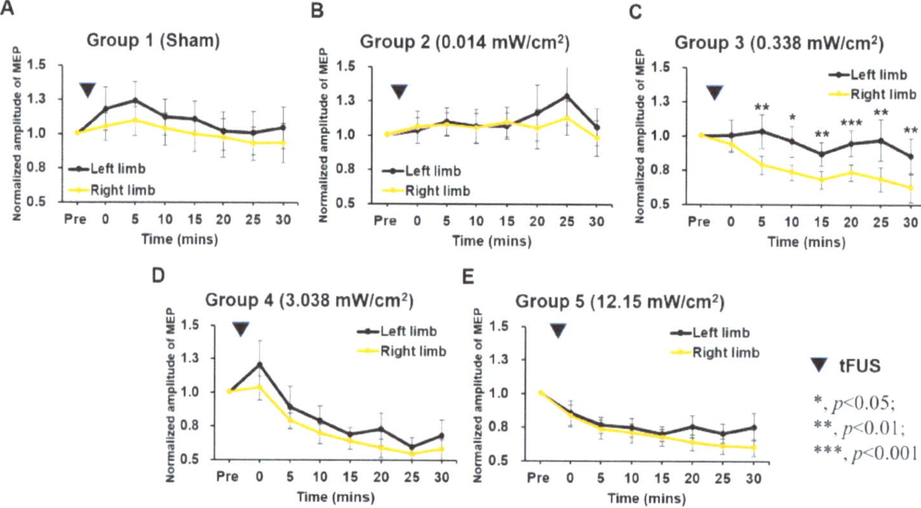

Figure 2. Comparison of the averaged MEP from both limbs in various groups 30 min after tFUS (*, **, and *** represent p values of <0.05, <0.01, and <0.001, respectively).

To compare the neuromodulatory effect among various sonication groups, 30-min longitudinal MEP responses were rearranged and colocalized (Figure 3). For the MEPs in the targeted limb, repeated-measures ANOVA indicated the significant main of time ($F_{7,525} = 14.104$, $p < 0.001$), ultrasound exposure dose ($F_{4,75} = 5.129$, $p = 0.001$), and time × dose interaction ($F_{28,525} = 2.609$, $p < 0.001$). The MEPs of both limb at each time point after sham tFUS (group 1) and tFUS I_{SPTA} 0.014 mW/cm^2 exhibited no significant changes (Figure 3A,B). At 30 min post-tFUS, tFUS reduced the MEP amplitude of the targeted limb as the intensity exceeded an I_{SPTA} of 0.338 mW/cm^2 (Figure 3A), but intensities of only 3.038 and 12.15 mW/cm^2 could be observed on the left side, suggesting that the ultrasound-exposure-induced neural suppression effect is target-side-dependent (Figure 3B). Conversely, differences in MEP amplitude changes in the untargeted limb were noticeable only in groups 4 (I_{SPTA} 3.038 mW/cm^2) and 5 (I_{SPTA} 12.15 mW/cm^2), which were observed 5–15 min after tFUS ($p < 0.05$, compared to group 1 (sham)). Similarly, changes in MEP amplitude did not recover to baseline levels in these two groups, indicating that the duration of neuromodulation depended on tFUS intensity. In terms of overall time (0–30 min), MEP amplitudes of 0.338 W/cm^2 (group 3) began to induce an MEP-suppressive effect in the targeted motor cortex (no effect was observed on the untargeted contralateral region), which appears to be the minimal exposure level required to induce a neuromodulatory suppressive effect (Figure 3C).

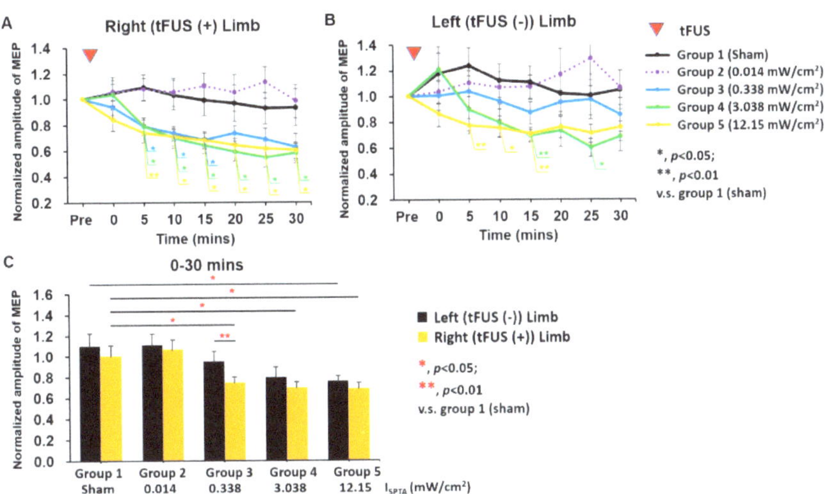

Figure 3. Overall changes in MEP in all groups. (**A**) Changes in right limbs (tFUS targeted) within 30 min after tFUS (**B**) Change in left limbs (tFUS untargeted) within 30 min after tFUS. (**C**) Change in both limbs at overall follow-up. (* and ** represent *p* values of <0.05 and <0.01, respectively).

2.2. Histological Examinations

To evaluate whether tFUS induces neuron damage, rats were sacrificed 90 min after tFUS at an I_{SPTA} of 50 mW/cm^2. The GFAP staining was conducted, and group 1 (sham) and animals that received tFUS at an I_{SPTA} of 50 mW/cm^2 were used as a comparison group. Upon GFAP immunohistochemical staining, no obvious astrogliosis at or near the sonicated sites in brains treated with tFUS was observed (Figure 4A). Quantification analysis demonstrated no significant change in GFAP immunoreactivity after I_{SPTA} = 50 mW/cm^2 sonication compared with group 1 treatment (sham; t = 1.066, *p* = 0.328) (Figure 4B). This suggests that the tFUS intensities employed in this study do not affect the normal phenotype of astrocytes in rat brains.

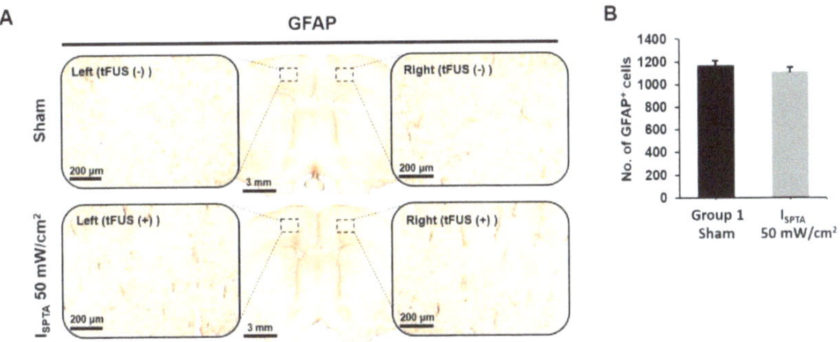

Figure 4. (**A**) Representative images of glial fibrillary acidic protein (GFAP) immunostaining and examples of GFAP images after sham tFUS and tFUS at an I_{SPTA} of 50 mW/cm^2. (**B**) No significant difference was observed between the two groups in the GFAP of the right hemisphere.

We further verified the expression levels of c-Fos and GAD-65 as markers of neuronal and synaptic activity to verify neuronal activity changes induced using weak ultrasound in two representative groups (group 3 [I_{SPTA} = 0.338 mW/cm^2] and group 5 [I_{SPTA} = 12.15 mW/cm^2]) (Figure 5). Compared with c-Fos markers of the contralateral brain in group 3 and group5, animals receiving tFUS at an I_{SPTA} of 0.338 mW/cm^2, exhibited a greater reduction in c-Fos-positive neurons (261 ± 13.7 to 235 ± 9.4 cell counts in the ipsilateral motor cortex versus sham FUS. tFUS at an I_{SPTA} of 12.15 mW/cm^2 further decreased c-Fos-positive neurons (215 ± 8.8 cell counts in the ipsilateral motor cortex versus sham FUS t = 4.87, $p = 0.008$). No reduction in cell counts in the contralateral cortex was observed). The animals with sham tFUS stimulation had a symmetric c-Fos-positive signal between hemispheres, indicating that weak tFUS reduced neuronal activity in the targeted motor cortex.

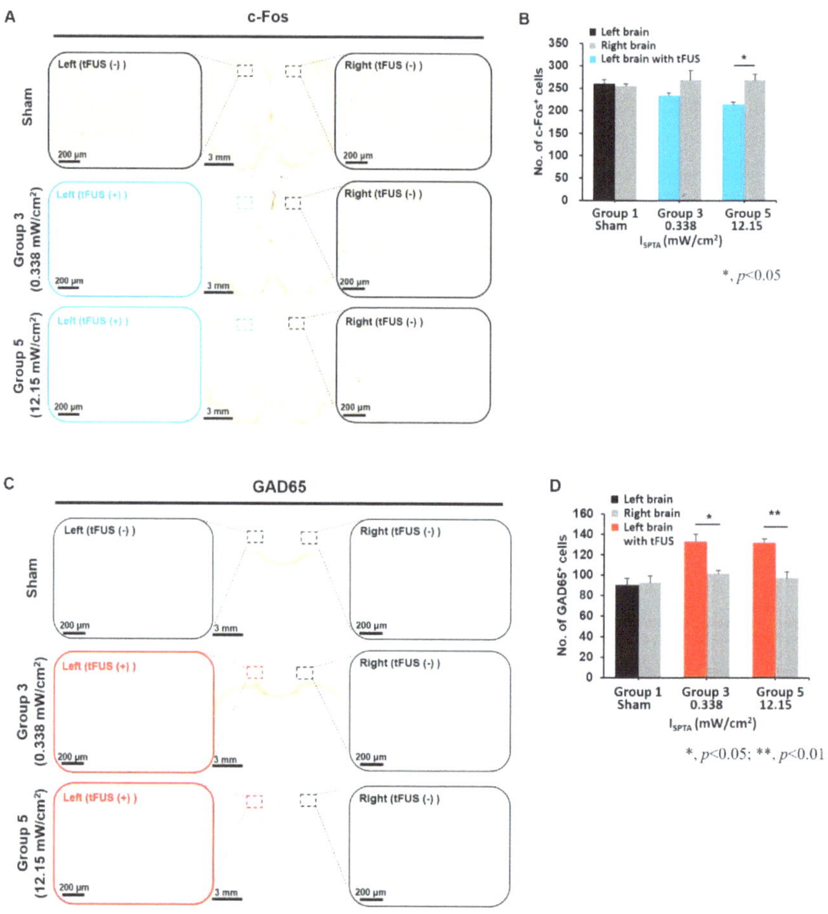

Figure 5. Comparison of c-Fos and GAD-65 signals in motor cortexes among the tested groups, including groups 1 (sham), 3, and 5. (**A**) A typical example of c-Fos staining (**B**) c-Fos-positive cell change in group 1 (sham), group 3 ($I_{SPTA} = 0.338$ mW/cm^2), and group 5 ($I_{SPTA} = 12.15$ mW/cm^2). Blue windows and bars represent the effects of tFUS. (**C**) A typical example of GAD-65 staining (**D**) GAD-65-positive cell changes in group 1 (sham), group 3 ($I_{SPTA} = 0.338$ mW/cm^2), and group 5 ($I_{SPTA} = 12.15$ mW/cm^2). Red windows and blue bars represent the effects of tFUS. (* and ** represent p values of <0.05 and <0.01, respectively).

According to a GAD-65 analysis, animals that received sham tFUS had symmetric positive cells in the ipsilateral and contralateral motor cortexes (93 ± 11.3 versus 92 ± 7.1, t = 0.4, $p = 0.71$). Exposure to tFUS of 0.338 mW/cm^2 strongly increased GAD-65-positive cell counts compared with those in the contralateral motor cortex (133 ± 11.85 versus 102 ± 5.4, t = 4.2, $p = 0.014$). Similarly, further increased GAD-65 cell counts (t = 5.9,

$p = 0.004$) were observed in the ipsilateral motor cortex after tFUS at 12.15 mW/cm^2 (131 ± 3.5 versus 95 ± 9.5 cell counts, respectively), indicating weak tFUS-induced GABA inhibitor generation in the sonicated area, and an ultrasound dose-dependent pattern was observed.

2.3. Primary Neuron Cells with Weak Sonication

To confirm the ultrasound-induced neuromodulatory pathway, calcium exchange caused by neural activity change, interfered with by ultrasound, was then observed in vitro from the primary neurons obtained from embryonic rodent cerebral cortices. The signal of calcium influx induced by weak ultrasound exposure (I_{SPTA} = 12.15 mW/cm^2) was recorded (Figure 6A), which was observed using the detected green fluorescence with the Fluo-4 calcium indicator. An increasing amplitude of calcium influx was detected from 0.09 ± 0.02 to 1 ± 0.13 15 min after 5-min sonication, suggesting weak-intensity ultrasound activated the calcium channels of primary neurons (Figure 6B). Images from Fluo-4 staining and anti-GABA staining were used to determine the subtype of ultrasound-activated neurons. We observed that GABAergic neurons had anti-GABA positive signals and also expressed Fluo-4-positive signals, suggesting that inhibitory type neurons were stimulated by weak sonication (Supplementary Figure S2), providing evidence to confirm the in vivo observation that weak-intensity ultrasound induced inhibitory neuromodulatory effect.

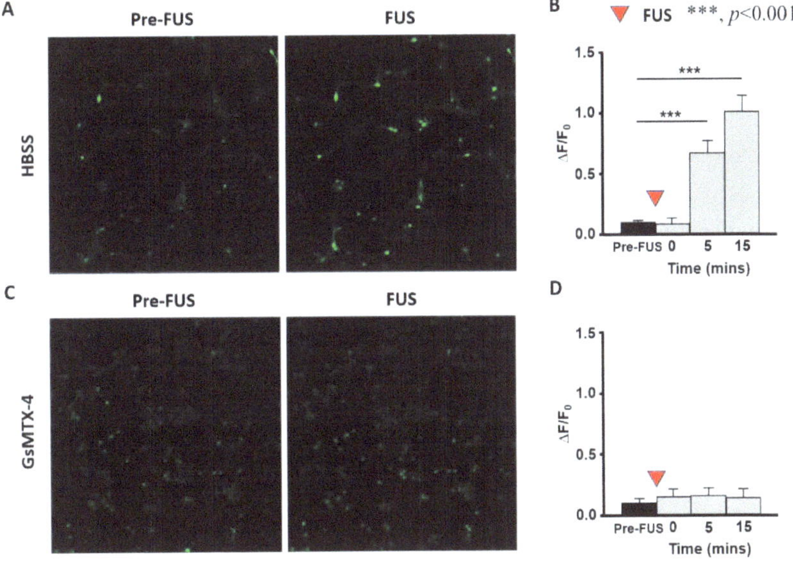

Figure 6. Calcium images of primary neurons following weak ultrasound sonication (I_{SPTA} = 12.15 mW/cm^2). (**A**) Representative images of Fluo-4-AM fluorescence presenting calcium signals in primary neurons before and after weak sonication. (**B**) Quantitative analysis indicates that weak sonication increases the levels of calcium signals in primary neurons within 20 min. Bars represent mean ± SD from two experiments. (n = 66, *** represents $p < 0.001$, Friedman test post hoc Tukey test). (**C**) Representative calcium images of primary neurons pretreated with 5 µM GsMTx-4 before and after weak sonication. (**D**) Quantitative analysis indicated no increased calcium signals in primary neurons pretreated with GsMTx-4 after weak sonication within 20 min. Bars represent the mean ± SD from three experiments. (n = 219. *** represents $p < 0.001$, one-way repeated-measures analysis post hoc Tukey test). ΔF/F0: fluorescence changes/before weak sonication.

Figure 6C,D demonstrates that the inhibitor GsMTX-4 of piezo channels blocked calcium signal activation from ultrasound weak exposure caused by the addition of HBSS-medium.

This suggests that the PIEZO-1 proteins of GABAergic neurons are highly involved in the cortical inhibitory neuromodulation process evoked by weak-intensity ultrasound.

3. Discussion

In the current study, we confirmed that weak ultrasound can induce neuromodulation effects in rat motor cortical regions and identified its major modulation types. In addition to the electrophysiological examination, the immunohistochemistry examinations in c-Fos, GAD-65, and primary neuron calcium quantification were conducted to confirm the ultrasound-induced neuromodulatory effects under weak exposure.

We have established a tFUS-TMS rat model to identify the neuromodulatory effects induced by weak tFUS exposure by measuring TMS-evoked MEPs in rats. We found that dose-dependent neuromodulation effects on the motor cortex were induced by different exposure levels of tFUS. We found that 5 min tFUS (I_{SPTA} of 0.338–12.15 mW/cm^2) decreased the amplitude of MEPs for 30 min, whereas the low intensity of tFUS (I_{SPTA} of 0.014 mW/cm^2) did not inhibit MEPs. Based on the results, we determined that ultrasound exposure levels as low as 0.338 mW/cm2 can induce neuromodulatory effects and may be the lower-bound exposure threshold. Although earlier studies have reported similar ideas in vitro [31,35], to our best knowledge, this work is the first to test the feasibility of using weak ultrasound exposure to induce the neuromodulatory effects in vivo. The findings of this study may also provide useful information for exploring the potential of using weak ultrasound stimulation for future clinical applications such as neuromodulation therapy for epilepsy, Parkinson's disease, essential tremors, depression, or neuropathic pain [18,36–39].

The current results confirm that tFUS pulsations induce neural excitatory (c-Fos, activation of cells) and inhibitory (GAD-65) synaptic transmission, indicating that the neuromodulatory effect induced by weak ultrasound stimulation originated in the inhibitory-type neurons. On the basis of increased GAD-65 expression and reduced c-Fos expression after weak ultrasound stimulation, the results reveal that weak tFUS exposure can generally suppress the neuron activity in the motor cortex. The decrease in c-Fos-positive cells and a strong increase in the GAD-65 signal indicated the deactivation of cortical activity and an increase in the activity of inhibitory interneurons. Similar results were also reported in the previous tFUS-epilepsy study, showing the decreased expression of c-Fos and increased GAD-65 levels in the cortex and hippocampus [37]. Furthermore, the finding of a strengthening of the GABAergic synapses by tFUS is supported by another noninvasive repetitive TMS (rTMS) study, showing that GAD-65 was increased following rTMS and demonstrated that rTMS affected the expression of activity-dependent proteins in cortical inhibitory interneurons [40]. Although the precise mechanism underlying the the tFUS induced neuromodulation effects remains unclear, our current results suggest that the activation in inhibitory neurons could play a critical role in the tFUS induced neuromodulation.

We further observed that weak (I_{SPTA} < 10 mW/cm^2) ultrasound significantly increased the calcium influxes in primary neurons, which responded to the increased Fluo-4 staining on GABAergic neurons. Although the detailed mechanisms of weak ultrasound–induced neuromodulatory effects have not yet been fully explored, earlier in vitro studies utilized low-intensity (I_{SPTA} < 100 mW/cm^2) showed that the calcium influx in primary cortical neurons is mediated from the specific mechanosensitive ion channels after ultrasound stimulation [41,42]. The mechanosensitive channels such as transient receptor potential channels, acid-sensing ion channels (ASICs), and Piezo channels were reported to play a role in mechanosensitive interaction with the aforementioned acoustic regimes [43,44]. A previous cell culture study reported that ASICs and cytoskeletal proteins induce a rapid calcium flow change (a few seconds after sonication) under weak ultrasound exposure (I_{SPTA} < 10 mW/cm^2) [31]. Unlike the rapid response of ASIC proteins, we found that the piezo-type mechanosensitive ion channels (e.g., PIEZO-1) exhibited relatively slow dynamics (with minutes after sonication). Because the observed neural modulatory dynamics in this study did not exhibit a rapid response pattern, it is suggested that the triggered proteins in this study were mechanosensitive ion channel proteins. Further investigations are required to provide definitive evidence supporting this hypothesis.

Considerable attention has been paid to the duration of neuromodulatory effects induced by tFUS. In the current study, the results show that 5 min tFUS can inhibit motor cortical excitability for 30 min. When compared with other studies, Kim et al. reported a reversely suppressive effect for 2 min in the motor cortex with 1-min tFUS [30], and another study reported that 40-s tFUS had a lasting effect on the supplementary motor area or frontal polar cortex in primates, producing even longer-lasting modulation of brain cortical activation and connectivity (>1 h) [45]. Yoon et al. reported a 5-min neural suppression effect in the somatosensory cortex after a 2-min sonication [46]. Dallapiazza and colleagues used 40-s tFUS to suppress somatosensory evoked potential in a swine brain for 5 min [21]. In a rabbit model, Yoo et al. reported that 18-s tFUS to the visual cortex suppressed visual evoked potential for 15 min [47]. tFUS suppressed pharmacal-induced epileptic signals for 30 min more than 1 week after 10-min tFUS [37] and 30-min FUS [48]. These results suggest that the suppression effect of tFUS may be longer than the exposure duration, and this long-term effect may depend on the exposure site and total exposure duration. Based on our results, we observed that 5-min weak tFUS reduced cortical excitability for up to 30 min. A procedure involving repeated 5-min sonication or prolonged sonication may help treat chronic conditions such as Parkinson's disease or epilepsy [49,50]

Regarding the interhemispheric neuromodulatory inhibition between ipsilateral and contralateral M1, the suppressive neuromodulation effects on the ipsilateral and contralateral M1 were found while increased tFUS intensity over 3 mW/cm^2 (group 4, 5). tFUS-altered functional connectivities have been reported through functional magnetic resonance imaging in a primate study [45]. The possible explanation for this bilateral effect is that ultrasound mechanical whole-brain vibration may have confounding auditory effects [51]. However, the weak tFUS protocols utilized in the current study were using pulsed mode and low pulse repetition frequency [52]. The confounding auditory effect can be eliminated in the current study. Further study could be needed to find the possible mechanisms in the interhemispheric neuromodulatory inhibition induced by tFUS.

Compared with other non-invasive brain neuromodulatory tools such as rTMS and tDCS, the weak-intensity tFUS ($I_{SPTA} < 10$ mW/cm^2) applied in the current study may have more clinical opportunities. We demonstrate the promising neuromodulatory effects by using a very low dose of tFUS in the motor cortex. For current FDA-approved guidelines, the clinical diagnosis ultrasound such as Transcranial Doppler (TCD) ($I_{SPTA} \leq 720$ mW/cm^2) may play a role in the application of neuromodulation. A review article shows that TCD presents several advantages including its high spatial resolution, easy administration and better clinical accessibility [53]. With delivering weak-intensity ultrasound from a diagnosis transducer, the tFUS can apply not only as diagnostic imaging but also can be a therapeutic tool for neurological disorders.

For future clinical applications of tFUS for neuromodulation, it could have a few limitations. For example, we did not characterize the long-term effects of tFUS in motor cortical excitability and neuroactivity over days. Another limitation is that the protocols of weak ultrasound delivered by the current animal setting would be needed to be modified for the future use of tFUS in humans. The adjustments of dose and related instruments of tFUS would be necessary. Despite these limitations, our findings indicate that the neuromodulatory effects induced by weak-intensity tFUS may serve as a reference for future clinical neuromodulatory studies.

In conclusion, we confirmed that weak ultrasound had long-lasting neuromodulatory effects in the motor cortical region of rats. A lower-bound threshold (I_{SPTA} 0.338 mW/cm^2) of tFUS to induce the suppressive neuromodulatory effect, which lasted for 30 min, was identified. The findings from histology and primary neuron cells confirmed that GABA synthesis in inhibitory neurons was stimulated artificially. This study provides the safety and feasibility of the use of weak ultrasound exposure to induce neuromodulatory effects, which can explore the potential use of weak ultrasound for future clinical neuromodulatory applications in neurological and neuropsychiatric disorders.

4. Materials and Methods

4.1. Animals and Preparations

All animal experiments were approved by the Institutional Animal Care and Use Committee of Chang Gung University (IACUC No. CGU107-231). Ninety-four male Sprague-Dawley rats (350–400 g; BioLASCO, Taipei, Taiwan) were used in the experiments. The rats were housed with a 12-h light/dark cycle at a temperature of 25 ± 1 °C and had ad libitum access to food and water. All animals were initially anesthetized with Zoletil (65 mg/kg; Vibac, Carros, France) and Rompun (10 mg/kg; Bayer, Leverkusen, Germany) and then mounted on a stereotaxic apparatus (Stoelting, Wood Dale, IL, USA). The body temperature of the experimental rats was kept constant using a circulating water heater. All animals were randomly assigned into groups for testing the effects of tFUS.

4.2. tFUS Setup and Study Design

The anesthetized test rats were mounted on a stereotaxic apparatus 60 min before focused ultrasound (FUS) to avoid the dampening of neuronal activity from anesthesia (Figure 7A). Anesthesia depth was adjusted for the absence of abdominal contractions to the tail pinch [54]. A self-manufactured FUS transducer (fundamental frequency = 0.7 MHz) was used. For ultrasound sonication, burst signals of 0.7 MHz were generated by a function generator (A33420; Agilent, Santa Clara, CA, USA) and amplified by a radiofrequency power amplifier (240L; E&I, NY, USA). Acoustic pressure was measured in a free field filled and rat skull bone with deionized degassed water using a needle-type hydrophone (HNA-0400; ONDA, Sunnyvale, CA, USA) to estimate the transcranial pressure and acoustic deformation [55]. The diameter and length of the half-maximum pressure amplitude of the ultrasound field and transcranial ultrasound field were within 2 and 2 mm, respectively (Figure 7B,C). During sonication, the ultrasonic gel was used to coat the tFUS inducer of the animal's head, and the sonication region was aimed at the left primary motor cortex (Figure 7D).

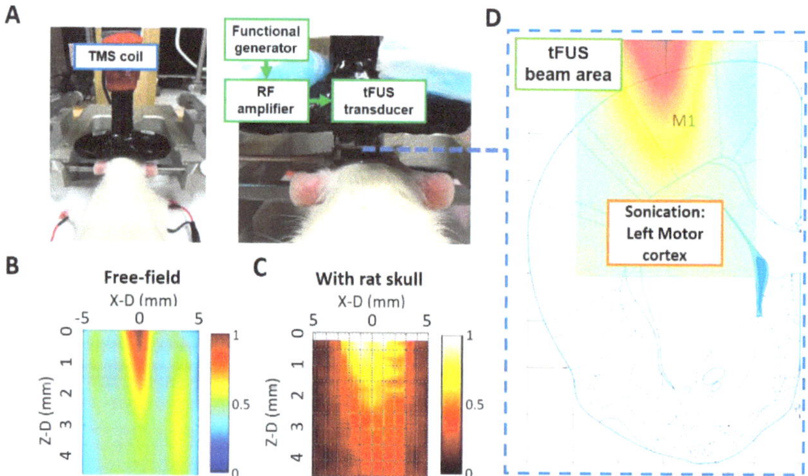

Figure 7. (**A**) The Center of the transcranial magnetic stimulation (TMS) coil and transcranial focused ultrasound (tFUS) transducer are positioned and focused on the motor cortex of the left hemisphere. (**B**) Characterization of the FUS pressure fields in sagittal (Y–Z plane) scans of ultrasound pressure distribution using a hydrophone-based ultrasound field mapping system. (**C**) Characterization of the tFUS pressure fields in sagittal (Y–Z plane) scans. (**D**) Conceptual schematic of the topographical sonication depth targeting the motor cortex area [56].

The ultrasound exposure parameters of the groups and experimental design flow are summarized in Figure 8A. tFUS exposure level was set from 0 mechanical index (MI) to 0.08 MI in groups 2 to 5. The duty cycle was set at 8% (pulse repetition frequency = 100 Hz; burst length = 0.8 ms), which produced weak exposure intensities (I_{SPTA} from 0.014 mW/cm^2 to 12.15 mW/cm^2). The total sonication duration was 5 min. Figure 8B illustrates the study protocol and time course. Initially, MEP was recorded for 10 min to serve as the baseline, and tFUS was then performed while the animals were under anesthesia. MEP recording continued for 30 min to measure the post-tFUS effects.

A

Group	Mechanical index (MI)	Duty cycle	I_{SPTA} (mW/cm^2)	Duration
1	0 (sham)	N.A	N.A	N.A
2	0.003		0.014	
3	0.013	8 %	0.338	5 minutes
4	0.04		3.038	
5	0.081		12.15	

B

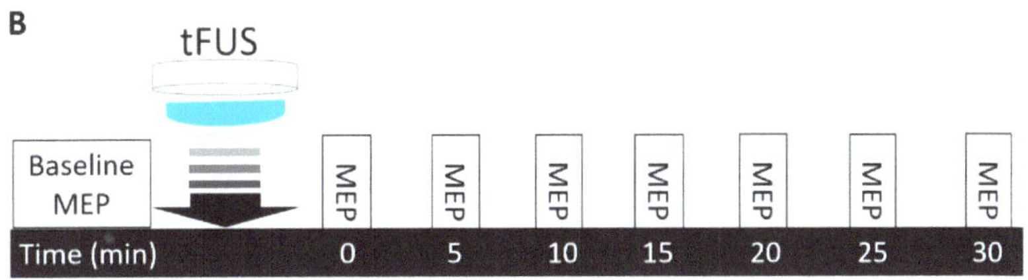

Figure 8. (**A**) Summary of tFUS parameters used in individual experimental groups. MI: Mechanical index; I_{SPTA} = spatial-peak temporal averaged intensity. Duty cycle = percentage of time divided by the total time of a stimulus train. (**B**) Motor-evoked potential (MEP) recording and tFUS sonication protocol.

4.3. Recordings and TMS Assessments

The assessment of changes in MEP using TMS is a common measurement tool in clinical practice involving cortical neuromodulation of the motor cortex [54,57]. Electromyographic activity (EMG) was recorded with monopolar uninsulated 27G stainless steel needle electrodes inserted into the belly of the bilateral brachioradialis muscle [33,54]. A reference electrode was positioned distally in the paw [33,54]. The EMG signal was amplified (1000×) and filtered using a 60-Hz notch and 10 Hz–1 kHz bandpass filters before digitization at 4 kHz (MP36, BIOPAC System, CA, USA) [54].

All TMS evaluation sessions were performed using a Rapid^2 magnetic stimulator (Magstim, Whitland, UK) and a figure-of-eight coil (external diameter = 55 mm, internal diameter = 10 mm; Magstim). To evaluate the effect of different FUS schemes on cortical excitability, MEPs were recorded using single-pulse TMS. tFUS sonication was performed on the left M1 region. On the basis of these results, we investigated the effect of weak tFUS using ipsilateral sonication input of the brain. The coil was held in the stereotaxic frame with a junction of rings placed over the rat's dorsal scalp; this is a position that can reliably elicit equally bilateral forelimb MEPs [54,58]. MEPs evoked using single-pulse TMS were recorded to evaluate the effect of tFUS on cortical excitability. The resting motor threshold (RMT) is defined as the minimal intensity of stimulation required for eliciting MEPs from a contralateral brachioradialis muscle of > 20 μV in 5 out of 10 consecutive trials [54,59]. For the assessment of MEP amplitudes, single TMS pulses were applied at 10-s intervals at 120% RMT, and the EMG signal was recorded for 200 ms after a single TMS pulse. The stimulus intensity remained unchanged throughout the experiment. Peak-to-peak amplitudes of

MEP were analyzed offline. To compare the effect of interventions on cortical excitability, all averaged MEP amplitudes were normalized to baseline before FUS intervention.

4.4. Histological and Immunohistochemistry Examinations

To confirm the safety of protocols, rats received I_{SPTA} 50 mW/cm^2 tFUS under high tFUS stimulation intensity and were sacrificed 90 min after receiving tFUS for histological investigation of the brain tissue. Immunohistochemical analysis was used to detect the expression of the GFAP, which indicates the activation of astrocytes following injury or stress [48,60–63]. Brain sections (30 µm) were cut free-floating in phosphate-buffered saline using Leica CM3050 S Research Cryostat (Leica Biosystems, Nussloch, Germany) for each tissue block, and five consecutive sections per well were collected in multiwell culture plates. Immunohistochemical staining was performed on formalin-fixed, frozen sections using rabbit polyclonal anti-GFAP antibody (ab7260; dilution 1:1000; Abcam, Cambridge, UK) as the primary antibody. To investigate the mechanism of neuromodulation from various weak tFUS intensities on the motor cortex in brain tissue, rats from group 1 (sham) to group 5 (I_{SPTA} = 12.15 mW/cm^2) were sacrificed 20 min after receiving tFUS. C-Fos (using anti-c-Fos [1:1000, AB190289; Abcam]) [64] and GAD-65 (using anti-GAD-65 [1:100, GAD-65-101AP; FabGennix, Frisco, TX, USA]) [37,65] were used to examine expression level changes in the motor cortex. Changes in GFAP-positive signals with FUS sonication were quantified, and the number of c-Fos-positive and GAD-65-positive cells were analyzed using Image J software (Media Cybernetics, Rockville, MD, USA).

4.5. Primary Neuron Cell Culture and Real-Time Calcium Signal following Weak-Intensity Sonication

In addition to assessing the neuromodulatory effects of tFUS in vivo, we examined the changes in neural activity under weak sonication in vitro in the primary neurons derived from rodent embryonic E14-15. The forebrain was isolated from embryonic E14-15 under a dissecting microscope. The forebrain tissue was digested with papain (1 mg/mL in Ca^{2+}, Mg^{2+}-free Hanks' balanced salt solution [HBSS]; HBSS w/o Ca^{2+}-Mg^{2+}: 5.33 mM KCl, 0.44 mM KH$_2$PO$_4$, 137.93 mM NaCl, 0.34 mM Na$_2$HPO$_4$, 8.34 mM NaHCO$_3$, 5.56 mM glucose, pH 7.3, and 300 mOSM) at 37 °C, gently shaken for 20 min, mechanically minced with a glass pipette, and then centrifuged at 100× g for 5 min. The clear supernatant was discarded, and cell pellets were resuspended in Ca^{2+}, Mg^{2+}-free HBSS. Cells were cultured at 10^6 cells/mL in 35-mm glass bottom dishes (ibidi, Martinsried, Germany), which were coated with poly-L-lysine and supplemented with Neurobasal medium containing B27 (2%; Thermo Fisher, Waltham, MA, USA), GlutaMax (0.5 mM; Thermo Fisher), glutamatic acid (25 µM; Thermo Fisher), and penicillin–streptomycin (1%; GeneDirexX) at 37 °C /5% CO$_2$ in a humidified incubator. Primary neurons were used for experiments after 11–14 days of culturing culture. Before the observation of calcium images, primary neurons were loaded with fluorescent calcium indicator Fluo-4-AM (2%; Thermo Fisher) for 20 min at room temperature [66].

A ring-shaped FUS transducer was used in the cultured dish during calcium imaging. The transducer was placed confocally with the objective of the confocal fluorescence microscope (LS780, Carl Zeiss, Oberkochen, Germany). The culture dish was filled with HBSS and covered with the ultrasound transducer. The FUS intensity applied for observation was the same as that used for group 5 (MI = 0.08, duty cycle = 8%, I_{SPTA} = 12.15 mW/cm^2, duration = 5 min). During the experiment, an ultrasound was performed on primary neurons for 5 min, and then the real-time neural calcium signals were recorded for 20 min.

To confirm the modulated neural activities using ultrasound, a piezo inhibitor, GsMTX-4 [67] was freshly diluted with distilled water, and 5 µmol/L (piezo inhibitor) was added to cultures 30 min before ultrasound treatment. After sonication, Fluo-4 in the primary neurons was first fixed with 1-ethyl-3-(3-dimethylaminopropyl) carbodiimide (40 mg/mL, EDAC; Sigma-Aldrich, St. Louis, MO, USA) for 30 min. Subsequently, the primary neurons in culture dishes were fixed using 4% formaldehyde for 20 min at room temperature. The primary and second antibodies used for immunostaining to examine activated GABAergic neurons from weak ultrasound stimulation were rabbit anti-GABA (1:2000; Sigma-Aldrich) and Alex Fluor 488-conjugated goat-anti-rabbit igG (1:1000, Abcam), respectively.

4.6. Statistical Analysis

All data are expressed as the average $\pm$ standard error of the mean (SEM). Data were analyzed using SPSS for Windows version 17.0 (IBM, Armonk, NY, USA). Comparisons of the tFUS-induced changes in MEPs, which were normalized to the last 5-min pre-FUS baseline, were performed using two-way repeated-measures analysis of variance (ANOVA) with sonication (from I_{SPTA} 0 [sham] to 12.15 mW/cm^2) as a between-subject factor and time (5, 10, 15, 20, 25, and 30 min after FUS) as a within-subject factor. The data of GFAP immunohistochemistry, c-Fos, and GAD-65 were analyzed using a one-way ANOVA followed by Dunnett's post hoc comparisons.

Supplementary Materials: The following supporting information can be downloaded at: https://www.mdpi.com/article/10.3390/ijms24032578/s1, Figure S1: Resting motor threshold (RMT) in each group. Data are presented as the mean $\pm$ standard error of the mean (n = 15 per group). Figure S2: Representative images of Fluo-4 and anti-GABA immunostaining of rodent primary cortical neurons at DIV 12 after weak ultrasound stimulation. Weak ultrasound stimulation–activated neurons were stained with Fluo-4 (green). The activated neurons were partially co-immunostained with anti-GABA.

Author Contributions: T.-H.H., H.-L.L. and P.-K.C. conceived and designed the experiments. P.-C.C., C.-S.H., P.-K.C. and T.-H.H. performed the experiments. T.-H.H. and H.-L.L. provided the equipment. P.-C.C., C.-S.H., T.-H.H. and H.-L.L. developed the methodology. P.-C.C., C.-S.H., P.-K.C. and T.-H.H. analyzed the data. P.-C.C., P.-K.C., R.-S.C., K.-T.C., T.-H.H. and H.-L.L. contributed to the writing and editing the manuscript. All authors have read and agreed to the published version of the manuscript.

Funding: This work was supported by the National Science and Technology Council (grant MOST 110-2321-B-002-010, MOST 111-2321-B-002-014, MOST 111-2622-B-182-002, MOST 109-2314-B-182-029-MY3, MOST 108-2314-B-182-015-MY3) and Chang Gung Medical Foundation, Taiwan (CM-RPD1M0701, CMRPD1M0251 and CMRPD1K0671).

Institutional Review Board Statement: The animal study was reviewed and approved by the Institutional Animal Care and Use Committee, Chang Gung University IACUC Approval No: CGU107-231, Period of Protocol: valid from 1 August 2019 to 31 July 2022.

Informed Consent Statement: Not applicable.

Data Availability Statement: The raw data supporting the conclusion of this study are available from the corresponding authors on reasonable request.

Acknowledgments: The authors would like to thank Ying-Zu Huang for his helpful advice and are grateful to the Neuroscience Research Center of Chang Gung Memorial Hospital at Linkou, Taiwan and the Technology Commons, College of Life Science, National Taiwan University, Taiwan.

Conflicts of Interest: H.-L.L. served as a technical consultant at NaviFUS Corp., Taiwan and currently holds several therapeutic ultrasound-related patents; P.-C.C. concurrently served as a part-time research and development scientist at NaviFUS Corp., Taiwan.

References

1. Johnson, M.D.; Lim, H.H.; Netoff, T.I.; Connolly, A.T.; Johnson, N.; Roy, A.; Holt, A.; Lim, K.O.; Carey, J.R.; Vitek, J.L.; et al. Neuromodulation for brain disorders: Challenges and opportunities. *IEEE. Trans. Biomed. Eng.* **2013**, *60*, 610–624. [CrossRef]
2. Parpura, V.; Silva, G.A.; Tass, P.A.; Bennet, K.E.; Meyyappan, M.; Koehne, J.; Lee, K.H.; Andrews, R.J. Neuromodulation: Selected approaches and challenges. *J. Neurochem.* **2013**, *124*, 436–453. [CrossRef]
3. Muresanu, D.F. Neuromodulation with pleiotropic and multimodal drugs–future approaches to treatment of neurological disorders. *Acta Neurochir Suppl.* **2010**, *106*, 291–294.
4. Darmani, G.; Bergmann, T.O.; Butts Pauly, K.; Caskey, C.F.; de Lecea, L.; Fomenko, A.; Fouragnan, E.; Legon, W.; Murphy, K.R.; Nandi, T.; et al. Non-invasive transcranial ultrasound stimulation for neuromodulation. *Clin. Neurophysiol.* **2022**, *135*, 51–73. [CrossRef]

5. Knotkova, H.; Hamani, C.; Sivanesan, E.; Le Beuffe, M.F.E.; Moon, J.Y.; Cohen, S.P.; Huntoon, M.A. Neuromodulation for chronic pain. *Lancet* **2021**, *397*, 2111–2124. [CrossRef]

6. Bormann, N.L.; Trapp, N.T.; Narayanan, N.S.; Boes, A.D. Developing Precision Invasive Neuromodulation for Psychiatry. *J. Neuropsychiatry Clin. Neurosci.* **2021**, *33*, 201–209. [CrossRef]

7. Rossi, S.; Santarnecchi, E.; Valenza, G.; Ulivelli, M. The heart side of brain neuromodulation. *Philos. Trans. A Math. Phys. Eng. Sci.* **2016**, *374*. [CrossRef]

8. Cleary, R.T.; Bucholz, R. Neuromodulation Approaches in Parkinson's Disease Using Deep Brain Stimulation and Transcranial Magnetic Stimulation. *J. Geriatr. Psychiatry Neurol.* **2021**, *34*, 301–309. [CrossRef]

9. Krauss, J.K.; Lipsman, N.; Aziz, T.; Boutet, A.; Brown, P.; Chang, J.W.; Davidson, B.; Grill, W.M.; Hariz, M.I.; Horn, A.; et al. Technology of deep brain stimulation: Current status and future directions. *Nat. Rev. Neurol.* **2021**, *17*, 75–87. [CrossRef]

10. Dietz, N.; Neimat, J. Neuromodulation: Deep Brain Stimulation for Treatment of Dystonia. *Neurosurg. Clin. N. Am.* **2019**, *30*, 161–168. [CrossRef]

11. Yan, H.; Ren, L.; Yu, T. Deep brain stimulation of the subthalamic nucleus for epilepsy. *Acta. Neurol. Scand.* **2022**, *146*, 798–804. [CrossRef]

12. Baur, D.; Galevska, D.; Hussain, S.; Cohen, L.G.; Ziemann, U.; Zrenner, C. Induction of LTD-like corticospinal plasticity by low-frequency rTMS depends on pre-stimulus phase of sensorimotor mu-rhythm. *Brain Stimul.* **2020**, *13*, 1580–1587. [CrossRef]

13. Huang, Y.Z.; Lu, M.K.; Antal, A.; Classen, J.; Nitsche, M.; Ziemann, U.; Ridding, M.; Hamada, M.; Ugawa, Y.; Jaberzadeh, S.; et al. Plasticity induced by non-invasive transcranial brain stimulation: A position paper. *Clin. Neurophysiol.* **2017**, *128*, 2318–2329. [CrossRef]

14. Lopez-Alonso, V.; Cheeran, B.; Rio-Rodriguez, D.; Fernandez-Del-Olmo, M. Inter-individual variability in response to non-invasive brain stimulation paradigms. *Brain Stimul.* **2014**, *7*, 372–380. [CrossRef]

15. Lopez-Alonso, V.; Fernandez-Del-Olmo, M.; Costantini, A.; Gonzalez-Henriquez, J.J.; Cheeran, B. Intra-individual variability in the response to anodal transcranial direct current stimulation. *Clin. Neurophysiol.* **2015**, *126*, 2342–2347. [CrossRef]

16. Darrow, D.P. Focused Ultrasound for Neuromodulation. *Neurotherapeutics* **2019**, *16*, 88–99. [CrossRef] [PubMed]

17. Mueller, J.K.; Ai, L.; Bansal, P.; Legon, W. Numerical evaluation of the skull for human neuromodulation with transcranial focused ultrasound. *J. Neural Eng.* **2017**, *14*, 066012. [CrossRef]

18. di Biase, L.; Falato, E.; Di Lazzaro, V. Transcranial Focused Ultrasound (tFUS) and Transcranial Unfocused Ultrasound (tUS) Neuromodulation: From Theoretical Principles to Stimulation Practices. *Front. Neurol.* **2019**, *10*, 549. [CrossRef]

19. Tyler, W.J.; Lani, S.W.; Hwang, G.M. Ultrasonic modulation of neural circuit activity. *Curr. Opin. Neurobiol.* **2018**, *50*, 222–231. [CrossRef]

20. Tyler, W.J.; Tufail, Y.; Finsterwald, M.; Tauchmann, M.L.; Olson, E.J.; Majestic, C. Remote excitation of neuronal circuits using low-intensity, low-frequency ultrasound. *PLoS ONE* **2008**, *3*, e3511. [CrossRef] [PubMed]

21. Dallapiazza, R.F.; Timbie, K.F.; Holmberg, S.; Gatesman, J.; Lopes, M.B.; Price, R.J.; Miller, G.W.; Elias, W.J. Noninvasive neuromodulation and thalamic mapping with low-intensity focused ultrasound. *J. Neurosurg.* **2018**, *128*, 875–884. [CrossRef] [PubMed]

22. Lee, W.; Lee, S.D.; Park, M.Y.; Foley, L.; Purcell-Estabrook, E.; Kim, H.; Fischer, K.; Maeng, L.S.; Yoo, S.S. Image-Guided Focused Ultrasound-Mediated Regional Brain Stimulation in Sheep. *Ultrasound Med. Biol.* **2016**, *42*, 459–470. [CrossRef]

23. Fini, M.; Tyler, W.J. Transcranial focused ultrasound: A new tool for non-invasive neuromodulation. *Int. Rev. Psychiatry* **2017**, *29*, 168–177. [CrossRef] [PubMed]

24. Legon, W.; Bansal, P.; Tyshynsky, R.; Ai, L.; Mueller, J.K. Transcranial focused ultrasound neuromodulation of the human primary motor cortex. *Sci. Rep.* **2018**, *8*, 10007. [CrossRef] [PubMed]

25. Gibson, B.C.; Sanguinetti, J.L.; Badran, B.W.; Yu, A.B.; Klein, E.P.; Abbott, C.C.; Hansberger, J.T.; Clark, V.P. Increased Excitability Induced in the Primary Motor Cortex by Transcranial Ultrasound Stimulation. *Front. Neurol.* **2018**, *9*, 1007. [CrossRef]

26. Baek, H.; Pahk, K.J.; Kim, H. A review of low-intensity focused ultrasound for neuromodulation. *Biomed Eng. Lett.* **2017**, *7*, 135–142. [CrossRef]

27. Blackmore, J.; Shrivastava, S.; Sallet, J.; Butler, C.R.; Cleveland, R.O. Ultrasound Neuromodulation: A Review of Results, Mechanisms and Safety. *Ultrasound Med. Biol.* **2019**, *45*, 1509–1536. [CrossRef]

28. Wang, X.; Yan, J.; Wang, Z.; Li, X.; Yuan, Y. Neuromodulation Effects of Ultrasound Stimulation under Different Parameters on Mouse Motor Cortex. *IEEE Trans. Biomed. Eng.* **2020**, *67*, 291–297. [CrossRef]

29. Yuan, Y.; Wang, Z.; Liu, M.; Shoham, S. Cortical hemodynamic responses induced by low-intensity transcranial ultrasound stimulation of mouse cortex. *NeuroImage* **2020**, *211*, 116597. [CrossRef]

30. Kim, H.C.; Lee, W.; Kunes, J.; Yoon, K.; Lee, J.E.; Foley, L.; Kowsari, K.; Yoo, S.S. Transcranial focused ultrasound modulates cortical and thalamic motor activity in awake sheep. *Sci. Rep.* **2021**, *11*, 19274. [CrossRef]

31. Lim, J.; Tai, H.H.; Liao, W.H.; Chu, Y.C.; Hao, C.M.; Huang, Y.C.; Lee, C.H.; Lin, S.S.; Hsu, S.; Chien, Y.C.; et al. ASIC1a is required for neuronal activation via low-intensity ultrasound stimulation in mouse brain. *eLife* **2021**, *10*, e61660. [CrossRef] [PubMed]

32. Lim, J.; Chu, Y.C.; Tai, H.H.; Chien, A.; Huang, S.S.; Chen, C.C.; Wang, J.L. Auditory independent low-intensity ultrasound stimulation of mouse brain is associated with neuronal ERK phosphorylation and an increase of Tbr2 marked neuroprogenitors. *Biochem. Biophys. Res. Commun.* **2022**, *613*, 113–119. [CrossRef] [PubMed]

33. Rotenberg, A.; Muller, P.A.; Vahabzadeh-Hagh, A.M.; Navarro, X.; Lopez-Vales, R.; Pascual-Leone, A.; Jensen, F. Lateralization of forelimb motor evoked potentials by transcranial magnetic stimulation in rats. *Clin. Neurophysiol.* **2010**, *121*, 104–108. [CrossRef] [PubMed]

34. Marufa, S.A.; Hsieh, T.H.; Liou, J.C.; Chen, H.Y.; Peng, C.W. Neuromodulatory effects of repetitive transcranial magnetic stimulation on neural plasticity and motor functions in rats with an incomplete spinal cord injury: A preliminary study. *PLoS ONE* **2021**, *16*, e0252965. [CrossRef]

35. Plaksin, M.; Kimmel, E.; Shoham, S. Cell-Type-Selective Effects of Intramembrane Cavitation as a Unifying Theoretical Framework for Ultrasonic Neuromodulation. *eNeuro* **2016**, *3*, ENEURO.0136-15.2016. [CrossRef]

36. Zhang, T.; Pan, N.; Wang, Y.; Liu, C.; Hu, S. Transcranial Focused Ultrasound Neuromodulation: A Review of the Excitatory and Inhibitory Effects on Brain Activity in Human and Animals. *Front. Hum. Neurosci.* **2021**, *15*, 749162. [CrossRef]

37. Chen, S.G.; Tsai, C.H.; Lin, C.J.; Lee, C.C.; Yu, H.Y.; Hsieh, T.H.; Liu, H.L. Transcranial focused ultrasound pulsation suppresses pentylenetetrazol induced epilepsy in vivo. *Brain Stimul.* **2020**, *13*, 35–46. [CrossRef]

38. Lescrauwaet, E.; Vonck, K.; Sprengers, M.; Raedt, R.; Klooster, D.; Carrette, E.; Boon, P. Recent Advances in the Use of Focused Ultrasound as a Treatment for Epilepsy. *Front. Neurosci.* **2022**, *16*, 886584. [CrossRef]

39. Yu, K.; Niu, X.; He, B. Neuromodulation Management of Chronic Neuropathic Pain in The Central Nervous system. *Adv. Funct. Mater.* **2020**, *30*. [CrossRef]

40. Trippe, J.; Mix, A.; Aydin-Abidin, S.; Funke, K.; Benali, A. theta burst and conventional low-frequency rTMS differentially affect GABAergic neurotransmission in the rat cortex. *Exp. Brain Res.* **2009**, *199*, 411–421. [CrossRef]

41. Yoo, S.; Mittelstein, D.R.; Hurt, R.C.; Lacroix, J.; Shapiro, M.G. Focused ultrasound excites cortical neurons via mechanosensitive calcium accumulation and ion channel amplification. *Nat. Commun.* **2022**, *13*, 493. [CrossRef] [PubMed]

42. Qiu, Z.; Kala, S.; Guo, J.; Xian, Q.; Zhu, J.; Zhu, T.; Hou, X.; Wong, K.F.; Yang, M.; Wang, H.; et al. Targeted Neurostimulation in Mouse Brains with Non-invasive Ultrasound. *Cell Rep.* **2020**, *32*, 108033. [CrossRef]

43. Chu, Y.C.; Lim, J.; Chien, A.; Chen, C.C.; Wang, J.L. Activation of Mechanosensitive Ion Channels by Ultrasound. *Ultrasound Med. Biol.* **2022**, *48*, 1981–1994. [CrossRef] [PubMed]

44. Shen, X.; Song, Z.; Xu, E.; Zhou, J.; Yan, F. Sensitization of nerve cells to ultrasound stimulation through Piezo1-targeted microbubbles. *Ultrason. Sonochem.* **2021**, *73*, 105494. [CrossRef]

45. Verhagen, L.; Gallea, C.; Folloni, D.; Constans, C.; Jensen, D.E.; Ahnine, H.; Roumazeilles, L.; Santin, M.; Ahmed, B.; Lehericy, S.; et al. Offline impact of transcranial focused ultrasound on cortical activation in primates. *eLife* **2019**, *8*, e40541. [CrossRef]

46. Yoon, K.; Lee, W.; Lee, J.E.; Xu, L.; Croce, P.; Foley, L.; Yoo, S.S. Effects of sonication parameters on transcranial focused ultrasound brain stimulation in an ovine model. *PLoS ONE* **2019**, *14*, e0224311. [CrossRef]

47. Yoo, S.S.; Bystritsky, A.; Lee, J.H.; Zhang, Y.; Fischer, K.; Min, B.K.; McDannold, N.J.; Pascual-Leone, A.; Jolesz, F.A. Focused ultrasound modulates region-specific brain activity. *NeuroImage* **2011**, *56*, 1267–1275. [CrossRef]

48. Chu, P.C.; Yu, H.Y.; Lee, C.C.; Fisher, R.; Liu, H.L. Pulsed-Focused Ultrasound Provides Long-Term Suppression of Epileptiform Bursts in the Kainic Acid-Induced Epilepsy Rat Model. *Neurotherapeutics* **2022**. [CrossRef] [PubMed]

49. Lee, K.S.; Clennell, B.; Steward, T.G.J.; Gialeli, A.; Cordero-Llana, O.; Whitcomb, D.J. Focused Ultrasound Stimulation as a Neuromodulatory Tool for Parkinson's Disease: A Scoping Review. *Brain Sci.* **2022**, *12*, 289. [CrossRef] [PubMed]

50. Lee, C.C.; Chou, C.C.; Hsiao, F.J.; Chen, Y.H.; Lin, C.F.; Chen, C.J.; Peng, S.J.; Liu, H.L.; Yu, H.Y. Pilot study of focused ultrasound for drug-resistant epilepsy. *Epilepsia* **2021**, *63*, 162–175. [CrossRef]

51. Guo, H.; Hamilton, M., 2nd; Offutt, S.J.; Gloeckner, C.D.; Li, T.; Kim, Y.; Legon, W.; Alford, J.K.; Lim, H.H. Ultrasound Produces Extensive Brain Activation via a Cochlear Pathway. *Neuron* **2018**, *98*, 1020–1030. [CrossRef]

52. Mohammadjavadi, M.; Ye, P.P.; Xia, A.; Brown, J.; Popelka, G.; Pauly, K.B. Elimination of peripheral auditory pathway activation does not affect motor responses from ultrasound neuromodulation. *Brain Stimul.* **2019**, *12*, 901–910. [CrossRef] [PubMed]

53. Iyer, P.C.; Madhavan, S. Non-invasive brain stimulation in the modulation of cerebral blood flow after stroke: A systematic review of Transcranial Doppler studies. *Clin. Neurophysiol.* **2018**, *129*, 2544–2551. [CrossRef] [PubMed]

54. Hsieh, T.H.; Huang, Y.Z.; Rotenberg, A.; Pascual-Leone, A.; Chiang, Y.H.; Wang, J.Y.; Chen, J.J. Functional Dopaminergic Neurons in Substantia Nigra are Required for Transcranial Magnetic Stimulation-Induced Motor Plasticity. *Cereb. Cortex* **2015**, *25*, 1806–1814. [CrossRef] [PubMed]

55. O'Reilly, M.A.; Muller, A.; Hynynen, K. Ultrasound insertion loss of rat parietal bone appears to be proportional to animal mass at submegahertz frequencies. *Ultrasound Med. Biol.* **2011**, *37*, 1930–1937. [CrossRef] [PubMed]

56. Paxinos, G.; Watson, C. *The Rat Brain in Stereotaxic Coordinates: Hard Cover Edition*; Elsevier: Amsterdam, The Netherlands, 2006.

57. Fomenko, A.; Chen, K.S.; Nankoo, J.F.; Saravanamuttu, J.; Wang, Y.; El-Baba, M.; Xia, X.; Seerala, S.S.; Hynynen, K.; Lozano, A.M.; et al. Systematic examination of low-intensity ultrasound parameters on human motor cortex excitability and behavior. *eLife* **2020**, *9*, e54497. [CrossRef]

58. Gersner, R.; Kravetz, E.; Feil, J.; Pell, G.; Zangen, A. Long-term effects of repetitive transcranial magnetic stimulation on markers for neuroplasticity: Differential outcomes in anesthetized and awake animals. *J. Neurosci.* **2011**, *31*, 7521–7526. [CrossRef]

59. Vahabzadeh-Hagh, A.M.; Muller, P.A.; Pascual-Leone, A.; Jensen, F.E.; Rotenberg, A. Measures of cortical inhibition by paired-pulse transcranial magnetic stimulation in anesthetized rats. *J. Neurophysiol.* **2011**, *105*, 615–624. [CrossRef]

60. Eng, L.F.; Ghirnikar, R.S.; Lee, Y.L. Glial fibrillary acidic protein: GFAP-thirty-one years (1969–2000). *Neurochem. Res.* **2000**, *25*, 1439–1451. [CrossRef]
61. Hsieh, T.H.; Kang, J.W.; Lai, J.H.; Huang, Y.Z.; Rotenberg, A.; Chen, K.Y.; Wang, J.Y.; Chan, S.Y.; Chen, S.C.; Chiang, Y.H.; et al. Relationship of mechanical impact magnitude to neurologic dysfunction severity in a rat traumatic brain injury model. *PLoS ONE* **2017**, *12*, e0178186. [CrossRef]
62. Kuo, C.W.; Chang, M.Y.; Liu, H.H.; He, X.K.; Chan, S.Y.; Huang, Y.Z.; Peng, C.W.; Chang, P.K.; Pan, C.Y.; Hsieh, T.H. Cortical Electrical Stimulation Ameliorates Traumatic Brain Injury-Induced Sensorimotor and Cognitive Deficits in Rats. *Front. Neural. Circuits.* **2021**, *15*, 693073. [CrossRef] [PubMed]
63. Yu, Y.W.; Hsieh, T.H.; Chen, K.Y.; Wu, J.C.; Hoffer, B.J.; Greig, N.H.; Li, Y.; Lai, J.H.; Chang, C.F.; Lin, J.W.; et al. Glucose-Dependent Insulinotropic Polypeptide Ameliorates Mild Traumatic Brain Injury-Induced Cognitive and Sensorimotor Deficits and Neuroinflammation in Rats. *J. Neurotrauma* **2016**, *33*, 2044–2054. [CrossRef] [PubMed]
64. Bullitt, E. Expression of *C-fos*-like protein as a marker for neuronal activity following noxious stimulation in the rat. *J. Comp. Neurol.* **1990**, *296*, 517–530. [CrossRef] [PubMed]
65. Patel, A.B.; de Graaf, R.A.; Martin, D.L.; Battaglioli, G.; Behar, K.L. Evidence that GAD65 mediates increased GABA synthesis during intense neuronal activity in vivo. *J. Neurochem.* **2006**, *97*, 385–396. [CrossRef] [PubMed]
66. Wu, M.P.; Kao, L.S.; Liao, H.T.; Pan, C.Y. Reverse mode Na^+/Ca^{2+} exchangers trigger the release of Ca^{2+} from intracellular Ca^{2+} stores in cultured rat embryonic cortical neurons. *Brain Res.* **2008**, *1201*, 41–51. [CrossRef]
67. Bae, C.; Sachs, F.; Gottlieb, P.A. The mechanosensitive ion channel Piezo1 is inhibited by the peptide GsMTx4. *Biochemistry* **2011**, *50*, 6295–6300. [CrossRef]

 International Journal of *Molecular Sciences*

Review

The Role of Arginine-Vasopressin in Stroke and the Potential Use of Arginine-Vasopressin Type 1 Receptor Antagonists in Stroke Therapy: A Narrative Review

Karol Chojnowski [1], Mikołaj Opiełka [1], Jacek Gozdalski [2,*], Jakub Radziwon [1], Aleksandra Dańczyszyn [1], Andrew Vieira Aitken [3,4], Vinicia Campana Biancardi [3,4] and Paweł Jan Winklewski [5,6,*]

1   Student Scientific Circle of the Department of Adult Neurology, Medical University of Gdansk, 17 Smoluchowskiego Street, 80-214 Gdansk, Poland
2   Department of Adult Neurology, Medical University of Gdansk, 17 Smoluchowskiego Street, 80-214 Gdansk, Poland
3   Department of Anatomy, Physiology, and Pharmacology, College of Veterinary Medicine, Auburn University, Auburn, AL 36849, USA
4   Center for Neurosciences Initiative, Auburn University, Auburn, AL 36849, USA
5   Department of Human Physiology, Medical University of Gdansk, 15 Tuwima Street, 80-210 Gdansk, Poland
6   2nd Department of Radiology, Medical University of Gdansk, 17 Smoluchowskiego Street, 80-214 Gdansk, Poland
*   Correspondence: gozdalski@gumed.edu.pl (J.G.); pawel.winklewski@gumed.edu.pl (P.J.W.)

Citation: Chojnowski, K.; Opiełka, M.; Gozdalski, J.; Radziwon, J.; Dańczyszyn, A.; Aitken, A.V.; Biancardi, V.C.; Winklewski, P.J. The Role of Arginine-Vasopressin in Stroke and the Potential Use of Arginine-Vasopressin Type 1 Receptor Antagonists in Stroke Therapy: A Narrative Review. *Int. J. Mol. Sci.* **2023**, *24*, 2119. https://doi.org/10.3390/ijms24032119

Academic Editors: Lydia Giménez-Llort, Masaru Tanaka, Chong Chen, Simone Battaglia and Piril Hepsomali

Received: 16 October 2022
Revised: 15 January 2023
Accepted: 16 January 2023
Published: 20 January 2023

Abstract: Stroke is a life-threatening condition in which accurate diagnoses and timely treatment are critical for successful neurological recovery. The current acute treatment strategies, particularly non-invasive interventions, are limited, thus urging the need for novel therapeutical targets. Arginine vasopressin (AVP) receptor antagonists are emerging as potential targets to treat edema formation and subsequent elevation in intracranial pressure, both significant causes of mortality in acute stroke. Here, we summarize the current knowledge on the mechanisms leading to AVP hyperexcretion in acute stroke and the subsequent secondary neuropathological responses. Furthermore, we discuss the work supporting the predictive value of measuring copeptin, a surrogate marker of AVP in stroke patients, followed by a review of the experimental evidence suggesting AVP receptor antagonists in stroke therapy. As we highlight throughout the narrative, critical gaps in the literature exist and indicate the need for further research to understand better AVP mechanisms in stroke. Likewise, there are advantages and limitations in using copeptin as a prognostic tool, and the translation of findings from experimental animal models to clinical settings has its challenges. Still, monitoring AVP levels and using AVP receptor antagonists as an add-on therapeutic intervention are potential promises in clinical applications to alleviate stroke neurological consequences.

Keywords: arginine-vasopressin; vasopressin receptors; copeptin; ischemic stroke; stroke pathophysiology; acute stress response; neuroendocrine dysfunction; cerebral edema; neuroinflammation; blood-brain barrier

1. Introduction

Stroke results from an acute central nervous system injury caused by the disruption of cerebral blood flow or bleeding within or around the brain, with consequent neurological damage and loss of function. Ischemic stroke due to cerebral blood vessel occlusion comprises most of the cases (~62% of all new strokes in 2019), followed by intracerebral hemorrhage stroke (28%), and subarachnoid hemorrhage stroke (10%) [1]. Independent of the cause, in 2019 stroke remained the second-leading cause of death globally and the third-leading cause of disability and all-death mortality combined [1].

The extent of the neurological injury in response to stroke, particularly ischemic stroke as the most common type, depends on various factors, including the severity, duration, and area affected. The pathological mechanisms are complex and the time window to intervene at each stage is limited. With the decrease of cerebral blood flow, oxygen and glucose deprivation that follow change the ionic environment, thus promoting excitotoxicity and neuronal loss [2]. The resulting blood-brain barrier (BBB) disruption, increased levels of reactive species, and neuroinflammation all contribute to poor outcomes in ischemic stroke. In the post-stroke phase, a chronic neuroinflammatory process might occur at the infarct core and distal regions, promoting long-term neuronal tissue damage. A subsequent stroke-induced secondary neurodegeneration may occur in the distal regions. Neurodegeneration may lead to loss or impairment of motor function, autonomic and/or cognitive systems depending on the functional location of neuronal loss and the structure affected [3].

Despite scientific advances in understanding the mechanisms of stroke, etiology, and prognosis, effective treatments are still deficient. For instance, though mitigating post-stroke neuroinflammation is promising in improving stroke outcomes, multiple neuroprotectants have failed clinical efficacy to date [2]. Hence, a constant need to search for novel or complementary targets to decrease the burden of stroke. Likewise, long-term prognosis-effective biomarkers remain a cause for concern.

In this narrative review, we present several pieces of evidence suggesting arginine vasopressin (AVP) neuropeptide receptors as a potential therapeutic target, and copeptin, a surrogate marker of AVP, as an effective biomarker for long-term prognosis and risk stratification in stroke.

AVP is a neuropeptide critically involved in the maintenance of body homeostasis. The primary physiological mechanism controlling AVP release to the circulation is osmotic stimuli, followed by nonosmotic stimuli, including blood volume and pressure changes [4]. However, AVP secretion also occurs during stress-related stimuli associated with acute onset diseases, including stroke, in a hypersecretion manner [5,6]. As a neurotransmitter and neuromodulator within the central nervous system (CNS), AVP plays significant roles in maintaining physiological cerebral fluid balance, electrolyte homeostasis, and vascular resistance [5,7–10]. All those functions are highly affected by a cerebrovascular accident, suggesting that dysregulated AVP release may play a significant role in stroke pathophysiology. Indeed, AVP hypersecretion is implicated in major stroke-related complications, including brain edema, vasoconstriction, oxidative stress, BBB disruption, and neuroinflammation [11–14].

The AVP precursor peptide (pre-proAVP) is synthesized mainly within neurosecretory neurons of the hypothalamus and subsequently transported along the axons to the neurohypophysis. Proteolytic cleavage during axonal transport forms the mature AVP alongside copeptin and dissociates their intracellular carrier protein neurophysin II in equimolar amounts, independent of the stimuli applied [15–17]. In contrast to AVP's short half-life and unstable nature, copeptin is stable in the serum and plasma of patients with elevated AVP, reliably reflecting AVP circulating concentrations [18]. Hence, copeptin has been used as a surrogate marker for AVP in multiple disease states, including myocardial infarction and stroke [18]. Recently, studies exploring the detrimental role of AVP in animal models of ischemic and hemorrhagic stroke have supported a positive correlation between copeptin level and stroke severity. Moreover, an association between increased levels of AVP in stroke patients and poor prognosis has been established [19–25]. Hence, from a clinical perspective, the importance of understanding further the AVP involvement in stroke-elicited damage progression as copeptin, as it is secreted in equimolar concentrations as AVP and has been considered a surrogate marker of AVP, is an effective biomarker in terms of risk stratification after ischemic stroke and in predicting vascular events after a transient ischemic attack (TIA) [21,26].

Another clinical implication in understanding AVP stroke pathology in stroke settings is using AVP receptor antagonists as an add-on or complementary target to mitigate AVP-induced edema and BBB disruption [26–33]. AVP receptor antagonists have been FDA-approved for hyponatremia, and hence are readily available for repurposing [34,35].

In this narrative review, we first briefly summarize the physiological role of AVP. We then discuss the underlying mechanisms contributing to AVP hypersecretion during the onset of the cerebrovascular accident and how this phenomenon causes detrimental effects on both systemic and local scales. Furthermore, recent data suggesting a promising role for copeptin use as a prognostic tool in stroke is introduced. Finally, we summarize current evidence for using selective and mixed AVP receptor antagonists in the acute phase of ischemic stroke and as an add-on therapy for mitigating post-stroke complications. The discussion is primarily focused on AVP in ischemic stroke, as most data has studied this type of stroke. Our goal is to provide the reader with a comprehensive, state-of-the-art narrative review highlighting AVP's critical role in stroke pathology and its potential for therapeutic targets.

2. The Physiological Role of Arginine Vasopressin

AVP is synthesized in magnocellular neurons located in the supraoptic and paraventricular nucleus of the hypothalamus (PVN). The AVP gene encodes the precursor protein pre-proAVP, which is proteolytically cleaved during axonal transport towards the pituitary gland to form AVP and copeptin in equal concentrations [15,16]. Physiological secretion of AVP to the circulation occurs primarily in response to osmotic stimuli, sensed mainly by osmoreceptors and through nonosmotic cues detected by baroreceptors and atrial stretch receptors [4]. Once released, circulating AVP acts as a hormone in widespread targets regulating many physiological functions to promote fluid homeostasis by controlling water balance and plasma ion concentration, and maintaining systemic blood pressure, by controlling the vascular tonus [4].

AVP acts primarily through AVP activation of G protein-coupled receptors, namely: V1a (V1aR) and V2 receptors (V2R), to promote its physiological functions [8]. AVP acting on V2R found in the distal tubules and collecting ducts of the kidney regulates proper fluid exchange and plasma osmolality within the renal system [36]. The AVP/V2R activation causes upregulation of aquaporin 2 (AQP2), resulting in ion gradient-driven water influx and subsequent urine concentration, plasma dilution, and plasma osmolality reduction [36].

Blood pressure regulation by AVP is complex and still not fully understood, given that AVP has interactions with other systems, such as the sympathetic and the renin-angiotensin system, both critical regulators of blood pressure. Still, AVP predominantly regulates blood pressure through V1aR and V2R, located on endothelial and vascular smooth muscles. Their activation by AVP causes either vasoconstriction or vasodilation, depending on the vascular bed [37], and aids blood pressure maintenance in normotensive and hypertensive subjects [38,39]. Moreover, AVP vasoconstriction response depends upon arteriolar diameters. As such, the vasoconstrictive response to AVP in large arterioles is more significant than that observed by norepinephrine treatment [40].

On the other hand, smaller arterioles do not show differences between treatments in rodents' microcirculation [40]. Hence, AVP has clinical importance and use in advanced hypovolemic or vasodilatory shock states [41].

In the CNS, AVP acts as a neurotransmitter and neuromodulator and participates in various physiological and behavioral functions, as summarized in the remainder of this section. In concert with the corticotropin-releasing hormone, AVP regulates corticosteroids' secretion within the CNS and periphery, contributing to the endocrine stress response [42,43]. In the CNS, AVP regulates the hypothalamic-pituitary-adrenal (HPA) axis through a third characterized receptor named V1b located in the corticotropic cells in the anterior pituitary gland (V1bR, also called V3) [44]. Additionally, AVP directly activates V1aR in the adrenocortical cells to release adrenocorticotropic hormone (ACTH), hence potentiating the release of cortisol, an essential hormone involved in stress response [45,46].

Moreover, AVP is present within the so-called intrinsic vasopressin fiber system composed of astrocytes and astrocyte-like cells. Although the precise function thereof remains unknown, it is postulated that the vasopressin fiber system facilitates neocortical water flux via AQP4 in response to V1aR activation. Thus, playing a role in brain water regulation and ion homeostasis [26,47].

In the medial preoptic area (MPO), a hypothalamic nucleus containing AVP-activated (AVP+) neurons, AVP stimulates γ-aminobutyric acid (GABA) release. Local, chemogenic stimulation of all MPO AVP+ neurons induces local hyperthermia while decreasing endotoxin or stress-activated fever with a generally antipyretic effect [48–50].

Within the suprachiasmatic nuclei (SCN), daily rhythm regulation is tightly connected to AVP signaling and HPA activity [51]. HPA-regulated cortisol increase might be inhibited when the AVP is derived from parvocellular SCN neurons or stimulated if derived from parvocellular PVN neurons [52,53]. Although AVP's role in daily rhythm regulation has not been clearly defined, based on recent bioluminescence findings, AVP signaling influences SCN period, precision, and organization [54].

AVP affects neuronal oscillations generated at various scales, from individual neurons through local networks to multiple neural systems across the brain [55]. Among other functions, AVP activation translates into fine-tuning neural responses behind emotional reactivity, risk acceptance, cognitive functioning, social interactions, and maternal behavior in mammals besides oxytocin [56,57].

Additionally, AVP is one of the modulators of neurogenesis. In rat models, brain AVP content gradually increases from 16 days of pregnancy, while pituitary AVP rapidly increases from 19 days until birth [58]. Trials on rodents showed that lower AVP concentration neonatally is associated with reduced brain weight, mainly affecting the cerebellum. In 10–15% of cases, these abnormalities persist throughout life [58].

3. Arginine Vasopressin Hypersecretion and Release Mechanisms during Stroke

Several studies indicate excess release of AVP in association with somatic and psychological stressors present in acute onset diseases, including myocardial infarction, sepsis, and stroke [59]. Interestingly, the stress-related release of AVP in response to pathological stimuli shows a much higher increase in magnitude than classical osmotic stimuli. For instance, in the baboon model of hemorrhagic shock, baseline copeptin level (reflecting AVP) increased approximately 36-fold (median level from 7.5 to 269 pmol/L) [60].

Indications of increased release of AVP in ischemic stroke were reported in animal models of stroke and stroke patients. AVP gene and protein expression were reported as elevated within the SON and PVN in an experimental model of cerebral ischemia and reperfusion, suggesting transcription and translation of AVP are increased in acute brain injury [61]. Patients with ischemic stroke show a significant increase of AVP in plasma and cerebrospinal fluid [62–65]. Elevated AVP levels in blood plasma were observed in patients with ischemic stroke during the 24 h period [63]. The observed AVP hypersecretion was independent of plasma osmolality and mean arterial pressure and correlated with the mean size of the lesion and level of neurological deficit. However, no correlation was observed between AVP release and vascular territory of ischemic injury [63].

During the acute phase of stroke, secretion of AVP occurs mainly from neurosecretory MNCs neurons located in the PVN and SON, as the AVP neurons are resilient to ischemia and are capable of secreting AVP after ischemic insult [66,67]. However, other local sources of AVP during brain injury were also identified, including activated microglia, choroid plexus, and to a lesser extent, brain endothelial cells [62,68]. Notably, AVP release in experimental stroke was found in the infarct and peri-infarct spatial location areas [62,66].

In the acute phase of ischemic stroke, overactivation of the angiotensin-converting-enzyme (ACE)/angiotensin (Ang II)/angiotensin receptor 1 (AT1R) axis, sympathetic and baroreceptor dysregulation are frequently observed. Such dysregulation causes blood pressure fluctuations, impairs cerebrovascular autoregulation, increases pro-inflammatory cytokine production in the parenchyma, and induces hyperglycemia [67,69–72]. The state of sympathetic predominance after stroke can cause catecholamines release, which activates the magnocellular neurons in PVN and SON, resulting in AVP release [73,74]. Moreover, the state of hyperglycemia following stroke (as a result of direct post-stroke hyperglycemia or exacerbation of pre-existing diabetes) may increase the number of AVP-expressing neurons in PVN and SON, resulting in increased release of AVP during stroke [75,76]. Another nonosmotic systemic stimulus contributing to AVP release during stroke is the increased intracranial pressure (ICP) [77]. ICP in stroke was classically attributed to cerebral edema formation after stroke [78]. However, recent evidence demonstrates that increased ICP is also associated with the acute phase of stroke before the edema formation [79]. In stroke, mechanical pressure created before/during edema formation exerts direct pressure on the hypothalamus, which results in AVP release from PVN and SON [77].

The local mechanisms contributing to AVP release during stroke include glutamate release, local hyperosmotic environment formation, structural and functional changes in astrocytes located in PVN and SON, and pro-inflammatory mediators' release. Excessive and prolonged glutamate release occurs during stroke in infarct and peri-infarct areas, leading to exaggerated depolarization of postsynaptic neurons and neuronal death through excitotoxicity [80]. After the stroke, elevated pools of circulating glutamate are detected in patients' plasma and cerebrospinal fluid, which can directly interact with AVP neurons in PVN and SON [81,82]. AVP neurons in the hypothalamus receive dense glutaminergic innervation, accounting for approximately 25% of the total number of synapses formed in PVN and SON. Previous work shows glutamate induces AVP release in conscious rats by stimulating non-n-methyl-d-aspartate (NMDA) receptors [83]. Furthermore, after middle cerebral artery occlusion (MCAO) in rats, glutamate levels rapidly increase in PVN and SON in the first 15 min after the procedure.

Moreover, ischemic stroke induces increased BBB permeability, enabling the passage of ions, molecules, and fluids into the brain parenchyma, disrupting ion homeostasis [14]. Particularly after ischemic injury, an increase in BBB Na+/H+ exchangers, Na+-K+-Cl− cotransporters, or the calcium-activated potassium channel KCa3.1, induces the uptake of the Na+ ions in the brain parenchyma [14,84]. Consequently, the formation of a local hyperosmotic environment in PVN and SON can enhance the release of AVP [85].

The release of AVP during stroke is also regulated by interactions of astrocytes and microglia with magnocellular neurons located in PVN and SON. In physiological conditions, the activity of AVP neurons is under negative feedback modulation from astrocytes [86,87]. Increased AVP secretion in PVN and SON increases the aquaporin 4 (AQP4) expression, which is associated with the extension of glial fibrillary acidic protein (GFAP) filaments. Therefore, astrocytic processes can reversely expand around AVP neurons and inhibit AVP secretion [87]. However, during stroke, the maladaptation of astrocyte plasticity causes the regulatory volume decrease of astrocytes, followed by the release of glutamate into the extracellular space [88]. It has been demonstrated that MCAO and basilar artery occlusion (BAO) reduce the GFAP and AQP4 in astrocytes located around AVP neurons, followed by the separation of GFAP from AQP4, thereby increasing the activity of AVP neurons [89,90].

Additionally, activated microglia and astrocytes can release many pro-inflammatory mediators during stroke [91]. Previous works indicate that AVP can be released during a cerebral injury in response to TNFa, IL-1, and IL-6. Furthermore, AVP acts synergistically with those cytokines and increases the local release of CXC and CC chemokines [92].

Following the hypersecretion of AVP during stroke onset, an early (1–2 h) increase of V1aR is observed in astrocytes, neurons during axonal beading, and brain endothelium [6,31,93]. An increase in V1aR expression is correlated with AQP4 upregulation. The V1aR is also redistributed from the astrocyte body to the astrocytic processes, where the AQP4s are localized [90]. Consequently, the injured brain parenchyma is more susceptible to AVP-induced effects, particularly brain edema formation.

4. Arginine Vasopressin Contribution to Ischemic Stroke Pathomechanism

4.1. Systemic Effects of AVP in Stroke

4.1.1. Arginine Vasopressin and Hyponatremia

The detrimental effects of AVP overproduction can be observed in patients with the syndrome of inappropriate antidiuretic hormone secretion (SIADH), which is commonly associated with stroke (3.9–45.3% and 40–45% on admission and during hospitalization, respectively) [94,95]. AVP's action to effectively lower the plasma osmolality is mainly focused on reducing the concentration of sodium, a predominant osmolyte. Thus, a pathological increase in AVP secretion results in hyponatremia. Of note, although AVP induces water conservation in the kidneys, the regulatory mechanisms, i.e., the renin-angiotensin system, prevent extracellular volume expansion. Therefore, the elevated levels of AVP observed in SIADH result in the pathological state of euvolemic hyponatremia [96].

The severity of CNS damage caused by hyponatremia is causally associated with the magnitude of sodium deficiency and the rate of change in plasma sodium concentration [96]. In severe cases of hyponatremia, the hypoosmotic state produces a gradient-driven water influx into the brain causing cerebral edema [97]. The edema and increased intracranial pressure (ICP) promote further AVP secretion followed by edema accretion [26,77]. For instance, posterior reversible encephalopathy syndrome (PRES), a clinicoradiological condition characterized by brain edema as a primary symptom was associated with an increased AVP secretion [98].

As mentioned above, nearly half of stroke patients develop hyponatremia either on admission or during hospitalization [94]. Moreover, sodium concentration below 135 mmol/L is associated with higher mortality and larger baseline intracerebral hemorrhage volume [99–102]. Although there are many overlapping causes of low sodium levels in stroke patients, AVP's role seems significant, especially in hemorrhagic stroke. Indeed, in ischemic stroke and mild/moderate subarachnoid hemorrhage cases, 7% and 71% of hyponatremia cases were attributed to SIADH, respectively [103,104].

4.1.2. Arginine Vasopressin, Autonomic Nervous System, Inflammation, and Immunosuppression Response

Strokes may provoke damage to anatomical structures of the autonomic nervous system, causing autonomic dysregulation. Likewise, AVP acts centrally, coordinating autonomic responses, i.e., AVP administered intraventricularly in rodent models causes sympathetic activation [105–107]. AVP-mediated sympathetic activation and stroke-elicited disruption of central control of the autonomic nervous system could account for catecholamine release, and thus immunosuppression. The diminished immune response can be detrimental in stroke patients due to the increased risk of infection complications and potential bystander autoimmune response [108].

AVP is secreted in response to IL-6, an inflammatory cytokine released during infection. This temporal relationship may explain the hyponatremia associated with high CRP values [109]. Moreover, AVP modulates immune response by increasing IFN-y and primary antibody production in inflammatory cells. Furthermore, AVP stimulates the release of prolactin, a hormone with pro-inflammatory properties [110]. This data suggests the involvement of AVP in inflammatory response exacerbation (caused by, e.g., hospital-acquired infection), which may be related to mid-term stroke outcome worsening.

4.1.3. Arginine Vasopressin and Stress Response

The release of AVP, and other stroke-related events, such as cytokine release and ischemic disruption of HPA inhibitory areas, can entail the HPA axis activation [111,112]. Indeed, one of the earliest events following a brain injury caused by ischemia or hemorrhage (depending on the type of stroke) is the activation of the HPA axis and subsequent hypercortisolism [57,112]. As previously mentioned, AVP, in concert with corticotrophin-releasing hormone (CRH), can stimulate the pituitary release of ACTH, subsequently activating adrenal glucocorticoid secretion [45,46]. AVP, in physiological conditions, is a weaker ACTH secretagogue compared to CRH. However, AVP acts synergistically with CRH and causes extensive ACTH release after stimulation by various acute stressors, including stroke [113].

The resulting increase in cortisol starts promptly after the stroke onset and persists from seven days to months after an insult [114]. The effect of cortisol on brain tissue is detrimental and is associated with structural changes in all brain regions, but predominantly gray matter [115]. Mounting evidence suggests that chronic cortisol elevation causes atrophic changes in the hippocampus, leading to hippocampus-dependent learning and memory impairment [116]. The loss of hippocampal volume may be attributable to brain-derived trophic factor expression changes in the hippocampus [117]. Moreover, the prolonged high cortisol levels are associated with an exacerbation of Alzheimer's disease, which is putatively a result of an increase in oxidative stress and amyloid beta peptide toxicity [118].

Regarding stroke, in the most recent systematic review involving a total of 1340 patients, high cortisol levels at admission were associated with greater dependency, morbidity, and mortality in patients with an ischemic stroke. Nevertheless, there is no evidence of cortisol's causal role in worsening stroke outcomes to date [114]. Moreover, the degree of involvement of AVP in ischemia-elicited cortisol rise has not yet been clarified.

Hypersecretion of AVP and consequent increase in plasma AVP concentration may lead to depression, including endogenous or major depressive disorder and anxiety due to HPA disorder and following hypercortisolemia [57,112]. The exact biochemical mechanism is not well known; however, there seems to be functional interaction between AVP (via V1aR) and serotonin (via 5-HT receptor) at the level of the hypothalamus. This association might be significant in aggressive behavior as well [119,120]. Such behavior often exists in stroke patients during the acute phase, who may experience depressive mood, aggression, or hostility even if they have never been diagnosed with depression or other mood disorders before [121].

4.1.4. Arginine Vasopressin and Platelet Aggregation

AVP-induced platelet aggregation is facilitated by V1R activation and subsequent thromboxane release [122]. However, the effect of AVP on platelets could only be observed in physiologically unattainable concentrations of AVP. Moreover, there is no link between AVP and platelet activation in human pathology [123]. Thus, the significant influence of AVP on platelet aggregation in stroke is theoretically unlikely and unsubstantiated.

4.2. Central Nervous System Local Effects of Arginine Vasopressin System in Stroke

Hydromineral disturbance and neurovascular unit damage are strongly associated with stroke. Moreover, phenomena such as BBB disruption, neuroinflammation, astrocyte, and neuronal swelling, take place during the acute phase of the ischemic stroke and account for brain edema formation (Table 1) [11]. Under the term "brain edema" associated with ischemic/hemorrhagic stroke, cytotoxic and vasogenic edema can be distinguished [11,124]. Although both forms of edema coexist simultaneously, the former dominates in ischemic stroke, whereas the latter is associated with hemorrhagic stroke [125]. The water exchange between vascular and perivascular space and between perivascular and cellular space is regulated by AQP4, a water channel expressed mainly on astrocytes and endothelial cells [126]. In cytotoxic edema, AQP4 upregulation causes water movement from extracellular space into the astrocytes aggravating astrocyte compensatory swelling, thus increasing brain edema [127–129]. However, in vasogenic edema, AQP4 upregulation aids water movement from brain interstitial fluid to the perivascular drainage system, i.e., the glymphatic system, thus attenuating the damaging effects of local edema [125,130–132]. The data on AVP's influence on AQP4 expression is unclear as studies show its up-regulatory and down-regulatory effects [31,33,133].

Regarding ischemic stroke, in one study, treatment with a V1R antagonist increased AQP4 expression [134]. Contrarily, in a study on intracranial hemorrhage animal models, V1aR antagonists led to the downregulation of the AQP4 expression [135]. Interestingly, in both of these studies, AVP receptor antagonism proved beneficial in brain edema attenuation [134,135]. It is clear that further studies are needed to address differences between ischemic and hemorrhagic stroke regarding AVP-mediated AQP4 expression modulation and edema formation.

AVP influences ion balance in astrocytes and endothelial cells. The ion imbalance caused by Na/K pump dysfunction is aggravated by V1aR-mediated luminal ion transporters' activity enhancement [26,136]. AVP activates endothelial V1R, which results in AMPK and ERK1/2 pathways activation leading to NKCC1, NHE 1, and NHE2 upregulation [26,136,137]. Moreover, AVP enhances sympathetic response, which also stimulates NKCC1 [138]. The resulting influx of sodium into the endothelial cells and astrocytes promotes osmosis-driven water movement into the extracellular space and astrocytes [139].

Stimulation of V1aRs found on cerebral vasculature leads to endothelin-1 overexpression and protein kinase C (PKC) pathway activation, which translates into oxidative stress exacerbation and BBB disruption. A study conducted by Faraco et al. has shown that water deprivation-induced AVP hypersecretion elicits AVP-mediated oxidative stress leading to cerebrovascular dysregulation in murine brains [140].

Moreover, AVP may damage BBB by stimulating matrix metallopeptidase 9 (MMP9) expression in the brain endothelium [92]. The activation of MMP9 causes proteolytic degradation of the basal lamina, increasing the vessel permeability [141,142]. Furthermore, MMP9 alters the expression of tight junction components, resulting in increased neutrophil and macrophage infiltration through BBB into the brain [143]. The influx of inflammatory cells is augmented by AVP-mediated neutrophil (CXCL1 and CXCL2) and monocyte (CCL2) chemoattractant production in the brain endothelium and astrocytes [92]. AVP also enhances the production of VEGF, a potent vascular permeability-increasing agent, in mesangial cells acting through V1aR, further enhancing BBB leakage [144,145].

AVP elicits the vasoconstriction of cerebral blood vessels via V1aR, causing an increase in cerebral perfusion pressure (CPP) [146]. In the setting of SAH-induced AVP release, vasoconstriction causes a decrease in cerebral perfusion [65]. Contrarily, the increase in CPP caused by AVP administration was associated with the improvement of cerebral blood flow in TBI patients [146]. Most probably, the elevated CPP counteracted the dysfunction of cerebral blood flow autoregulation elicited by sympathetic activation and arterioles distention [147,148]. These results support the rationale for further research regarding the influence of AVP on cerebral blood flow in other diseases associated with cerebral autoregulation impairment, including ischemic stroke [149].

In the rodent TBI model, AVP aggravated injury-elicited inflammatory response. Mechanistically, AVP amplifies CXC and CC chemokines synthesis in the endothelium and astrocytes. In line with this, AVP stimulates the expression of high-mobility group box1, a well-known inflammatory cytokine, in astrocytes during hypoxia/reoxygenation [150,151]. A recent study has shown AVP involvement in IL-6 production in murine hearts, which may play a role in myocardial inflammation in heart failure [152]. Taken together, local immunogenic effects of AVP in combination with enhanced immune cell infiltration caused by AVP-mediated BBB disruption and pro-inflammatory mediators' production may play a significant role in stroke-associated neuroinflammation.

As highlighted in Table 1, several neuropathological mechanisms associated with stroke pathology may benefit from the AVP receptor antagonism as a target to improve secondary mechanisms of stroke.

Table 1. Receptor-mediated effects of arginine vasopressin related to stroke pathophysiology.

Mechanism	Involved Receptors	Receptor Location in the Brain	Effects	Authors
Brain Edema Formation	V1aR	Neurons, Glia	■ Disruption of the hydromineral balance ■ Redistribution of V1 receptors in astrocytes ■ AQP-4 channel regulation—increase in brain water retention ■ Modulation of Na^+/K^+ ion channels ■ Development of Intracranial Hypertension	[6,33,133,153–156]
Time-dependent Arteries and arterioles vasodilation and vasoconstriction.	V1aR	Smooth muscle layer of brain vasculature; Endothelium	■ Reduced cerebral blood flow. Increase of CPP, but not CBF ■ Aggravation of cerebral ischemia.	[145,146,157–160]
Increased BBB permeability	V1aR	Astrocytes, podocytes, brain endothelium	■ AVP-mediated stimulation of cerebral microvascular endothelial cell NKCC and NHE ■ Astrocytic endfeet swelling	[144,145,156,161,162]
AVP potentiating action on the CRH-evoked ACTH secretion.	V1bR (V3)	Adenohypophysis	■ Cortisol release, hypercortisolism	[163,164]
Post-stroke, edema-associated inflammation	V1aR	Astrocytes	■ Stimulation of excessive AVP release by inflammatory cytokines including IL-6 ■ Hmgb1, Prx2, and Tlr2 upregulation by AVP ■ Synergistic AVP action with TNF-alpha on chemokines release	[92,150,165]

AVP, arginine vasopressin; V1aR, Vasopressin receptor 1A; V1bR (V3), Vasopressin receptor 1b (V3 pituitary receptor); AQP-4, Aquaporin-4; CPP, Cerebral perfusion pressure; CBF, Cerebral blood flow; BBB, Blood-brain barrier; CRH, Corticotropin-releasing hormone; ACTH, Adrenocorticotropic hormone; NKCC, Na-K-Cl cotransporter; NHE, Na/H exchanger; IL-6, Interleukin- 6; Hmgb1, High mobility group box 1 protein; Prx2, peroxiredoxin 2; Tlr2, Toll-like receptor-2; TNF-alpha, Tumor necrosis factor alpha.

5. The Prognostic Value of Copeptin in Acute Stroke Patients

As summarized above, increases in AVP are detrimental to stroke and have a considerable potential of being used as a therapeutic target as the mechanisms are directly associated with AVP receptors. Furthermore, although AVP is an unstable peptide, it is released in equimolar concentrations with copeptin and validated as a surrogate marker of AVP.

Clinically, copeptin measurements in acute ischemic stroke serve as a tool to predict functional outcomes [20,166]. A meta-analysis of the correlation between copeptin levels and functional outcomes in acute ischemic stroke has shown that a 10-fold increase in copeptin level was associated with a higher risk of poor 3-month and 1-year outcomes (OR = 2.56, 95% CI: 1.97–3.32) and all-cause mortality (OR = 4.16, 95% CI: 2.77–6.25) [24]. Moreover, the authors observed that using copeptin levels in addition to the National Institutes of Health Stroke Scale (NIHSS) score provided a better prediction for unfavorable outcomes than using the NIHSS score alone, showing the importance of such predictive biomarkers in everyday clinical practice [24]. Accordingly, De Marchis and colleagues recently proposed a scale for risk stratification based on clinical features and copeptin levels [167]. The CoRisk score, available for calculation online, considers the patient's age, NIHSS score, copeptin plasma concentration levels, and information on whether the patient received intravenous therapy [168]. A validation study of the CoRisk score scale showed good sensitivity in outcome prediction, with three in four patients classified correctly, with an area under the curve (AUC) of 0.819 [167].

Another promising application of copeptin measurement is to predict stroke after an episode of TIA. Previous studies have shown that patients with a higher copeptin level measured immediately after the TIA event are at a greater risk of developing a stroke or any cerebrovascular re-event [21,169]. In addition, in a long-term follow-up of patients after TIA or stroke, copeptin was a predictive factor of the recurrent vascular event [25]. Together, these data show that it may be beneficial to improve the ABCD2 score for TIA (age, blood pressure, clinical features, duration of TIA, and presence of diabetes) by adding copeptin values or monitoring the patients with large artery stenosis and high copeptin concentration [21,23].

In general, copeptin seems to be an effective biomarker in risk stratification after ischemic stroke and in predicting vascular events after TIA. However, copeptin is also associated with low specificity or low discrimination as a biomarker in other conditions. For instance, a few studies assessed the risk of stroke-associated infections using copeptin [20,170–172]. While higher levels of the biomarker were associated with a higher risk of complications, especially pneumonia, the results were unsatisfactory regarding diagnostic ability [20]. Copeptin alone had a similar predictive value observed with white blood count or C-reactive protein (CRP), and its addition to established scales resulted in a minor improvement in discrimination ability [170–172].

Likewise, applying copeptin as a diagnostic tool to distinguish stroke patients from stroke-free subjects has been tested with inconsistent results [19,22,173]. However, it may help to exclude vascular causes in patients with dizziness in the emergency department [173].

In summary, copeptin seems to be an effective biomarker opening an opportunity to improve stroke prognosis. It should be considered that the field of novel biomarkers regarding cerebrovascular disease management is growing rapidly. The combination of both circulating blood protein biomarkers and genetic testing can be used for cerebrovascular risk assessment and outcome stratification after an ischemic event [174–176]. Still, the interpretation of blood-based biomarkers should always be correlated with individual clinical judgment and the physician's best knowledge to avoid the danger of undertreatment of patients who are believed to be at a high risk of a poor outcome [177].

6. Targeting Arginine Vasopressin Receptors in the Acute Phase of Ischemic Stroke

The mechanisms mentioned above, especially brain edema exacerbation, prompted the rationale for AVP receptor antagonism in stroke. The use of a selective V1aR antagonist (SR49059) in MCAO rodent stroke models improved neurological outcomes and reduced infarct area [30,135,178]. The histopathological findings, such as decreased brain water content in the infarct area, sodium shift into the brain, reduced AQP4 expression, and reduction of BBB disruption, could be a plausible explanation for the beneficial outcome of SR49059-treated animals. Notably, both histological (reduced brain water content and Na+ accumulation) and neurobehavioral (higher modified Garcia score and better performance in beam balance test and wire hanging test) outcomes were achieved when the V1aR antagonist was administered no more than one hour after the procedure [29,81,135,178]. The drug administration after three or six hours since the MCAO proved ineffective [30].

OPC-31260, a V2R antagonist, prevented water and sodium accumulation in the brain in both general cerebral hypoxia and SAH rat models. Moreover, in both studies, V2R antagonism enhanced plasma AVP increase in subjected rodents [27,28].

Tolvaptan is a clinically approved V2R antagonist, successfully used in hyponatremia treatment [34]. However, in the ischemic stroke mouse model, Tolvaptan and other selective V2R antagonists, apart from the established aquaretic effect, did not produce the neuroprotective effects displayed by V1aR antagonists [134,179,180].

Conivaptan is an FDA-approved V1aR and V2R antagonist [35]. The main indication for Conivaptan use is euvolemic hyponatremia in specific disorders, including SIADH [181]. In animal stroke models, Conivaptan improved neurological outcomes, reduced brain water content, and attenuated BBB disruption [179]. In a different study, AVP receptor antagonism reduced brain edema and BBB disruption and improved neurological deficits in mice subjected to MCAO. Importantly, in this study, Conivaptan proved effective when administered three hours after an infarct induction [182], suggesting it may also be effective when administered within a similar time window of standard stroke treatment (i.e., thrombolysis). Recently, Can et al. have shown evidence for Conivaptan superiority over mannitol in diuretic activity in ischemic stroke achieved by a 30-min common carotid occlusion. Moreover, Conivaptan decreased serum 2-Phospho-D-Glycerate-Hydrolase (NSE) and increased progranulin. The former is a clinical biomarker associated with post-injury brain dysfunction, whereas the latter is known for its neuroprotective properties [183]. Regarding the safety of Conivaptan in stroke, the standard dose of the drug (20 mg) administered every 12 h for 2 days in combination with standardized intracranial hemorrhage (ICH) management proved safe and well tolerated by the patients [184]. In a case report published by Hedna et al. presenting a patient with post-operational ICH, the addition of Conivaptan to ineffective conventional anti-edema therapy resulted in both conscious level and motor function improvement, as well as radiologically-assessed brain edema resolution with no reported side effects [185].

Overall, antagonism of just V1R or both V1R and V2R provides a better outcome in rodent stroke models than using V2R antagonists alone. Recent literature regards Conivaptan as a safe, mixed AVP receptor antagonist that has the potential to supplement stroke treatment. Nevertheless, phase II studies are needed to test the effectiveness of Conivaptan in stroke.

7. Discussion

During the acute phase of stroke, AVP release is uncontrolled and larger than observed in physiological conditions. The mechanisms for AVP release in stroke are independent of plasma osmolality and involve a complex interplay and synergism between regulatory systems, systemic factors, and local components. Among them, AVP secretion underlying mechanisms include an interplay between the sympathetic and renin-angiotensin system; systemic changes in glycemia; and regional increase in glutamate excitability, as well as maladaptation of the neurovascular unit (reviewed in [11]). The resultant AVP hypersecretion is associated with many pathological detrimental phenomena, including

(1) brain edema caused by hyponatremia aggravation, AQP4 expression modulation, and ion transporters' function disruption, (2) BBB integrity disruption through MMP9 and VEGF upregulation (3) hypercortisolemia caused by HPA axis activation, (4) neuroinflammation through inflammatory cytokines upregulation and enhancement of inflammatory cells infiltration, and (5) vasoconstriction through direct V1a-mediated effect on cerebral blood vessels and the promotion of oxidative stress, causing cerebrovascular dysregulation (Figure 1, and Tables 1 and 2). Using the stable peptide copeptin as a surrogate marker of AVP, studies have found an association between a higher concentration of copeptin in stroke patients (hence, AVP) with a higher risk of poor outcomes and all-cause mortality. Furthermore, patients with higher copeptin levels at the time of TIA were at a greater risk of developing a stroke or any cerebrovascular re-event. Altogether, the revised individual literature establishes the potential of using copeptin as a biomarker for AVP plasma levels.

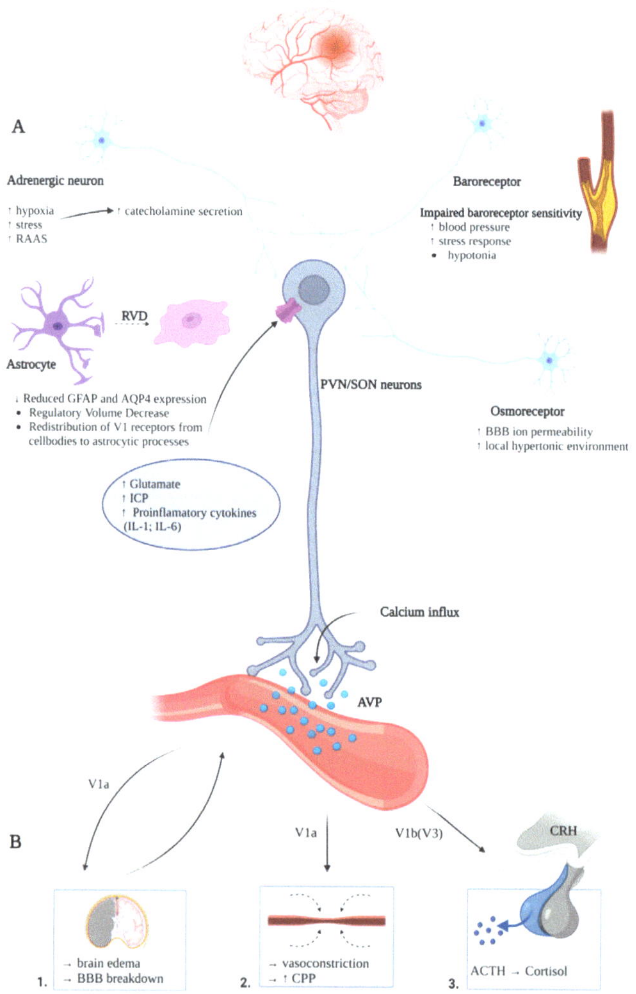

Figure 1. Main mechanisms causing arginine-vasopressin (AVP) release during stroke (**A**) and main local deteriorating effects caused by excessive AVP release (**B**). AVP neurons receive dense afferent input

from baroreceptors, osmoreceptors, adrenergic, and glutaminergic neurons. Stroke induces the dysregulation of baroreceptors and the renin-angiotensin-aldosterone system, and increases the sympathetic stimulation of the paraventricular nucleus of hypothalamus (PVN) neurons (**A**). After the stroke, the initial disruption of the blood-brain barrier causes local hyperosmotic environment formation and stimulation of osmoreceptors (**A**). The release of AVP is also regulated by the interaction of astrocytes with AVP-containing neurons. During a stroke, the regulatory volume decrease of astrocytes occurs, followed by the downregulation of GFAP and AQP4 (**A**). Additionally, stroke causes the release of glutamate and pro-inflammatory cytokines, which stimulate AVP release (**A**). After the release, AVP activates the V1aR and V1b(V3)R located in the brain, brain vasculature, and pituitary, respectively (**B**). Upon release, AVP exacerbates brain edema formation (1). Consequently, the expanding brain edema can increase the intracranial pressure and exert direct pressure on adenohypophysis, further increasing the AVP release (1). Vasopressin causes time-dependent arteries and arterioles vasoconstriction and increases cerebral perfusion pressure (CPP) but not cerebral blood flow (CBF) (2). Vasopressin also acts synergistically with corticotropin (CRH), causing excessive cortisol release (3). Abbreviations: ACTH, adrenocorticotropic hormone; AQP4, aquaporin-4; AVP, arginine vasopressin; BBB, blood-brain barrier; CBF, cerebral blood flow; CPP, cerebral perfusion pressure; CRH, corticotropin-releasing hormone; GFAP, glial fibrillary acidic protein; IL-1, interleukin 1; IL -6, interleukin 6; PVN, paraventricular nucleus; RAAS, renin–angiotensin–aldosterone system; and RVD, regulatory volume decrease.

Targeting the AVP receptor with pharmacological receptor antagonists, either individually or in combination, aims to prevent pathological events spurred by V1R and V2R activation. A mixed AVP antagonist, Conivaptan, achieved the best safety and neurological outcomes, providing the most significant degree of potential in stroke treatment.

Of note, most of the beneficial effects of AVP receptor antagonists were achieved in ischemia-reperfusion models. Thus, it is reasonable to infer that AVP receptor antagonism-oriented therapy is more efficient when applied together with the recanalization of the occluded vessel. In line with this, we believe that future studies involving AVP receptor antagonists should focus on using AVP antagonists as an add-on therapy for standard recanalization (thrombolysis/thrombectomy) treatment. It is crucial to recognize the potential of the anti-edematous effects of AVP receptor antagonists in post-stroke secondary brain damage alleviation. We strongly feel that treatment of post-stroke brain edema using AVP receptor antagonists in concert with standard brain edema treatment could be a viable means of stroke outcome improvement and mortality rate reduction, especially in patients with refractory brain edema [185]. As demonstrated previously, a phase I clinical trial on Conivaptan use, along with standard treatment for reduction of perihematomal edema, has proven its safety and tolerability in patients treated with ICH. In our view, relatively large amounts of data pointing to the efficacy of AVP receptor antagonists in ischemic stroke animal models supports the rationale for future clinical trials in ischemic stroke patients.

Table 2. The main results of selective and non-selective vaptans administration in animal and human stroke studies.

Experimental Model/Type of Study	Antagonist (Dose)	Time of Administration	Results	Author
Mouse; 60 min tMCAO	Conivaptan (i.v 0.2 mg, 0.2 mg) and Tolvaptan (p.o 0.2 mg)	At reperfusion	■ Both antagonists produced aquaresis; ■ Conivaptan: ↑neurological deficits, ↓ BWC, ↓ BBB disruption	[179]
Mouse; 60 min tMCAO	V1R antagonist (i. cv. 500 ng, 1000 ng) V2R antagonist (i. cv. 1000 ng)	10 min after surgery	■ V1R antagonist- ↓ infarct volume; ↓ BWC (1000 ng); ↑ AQP4 expression (1000 ng) ■ V2R antagonist: no effect	[134]
Rat; 2 h tMCAO	SR49059 (6480, 720, 640 μL/h)	1 h before/after reperfusion	■ ↓ BWC in infarct area (720, 640 μL/h); ↓ sodium shift; ↓ Potassium loss prevention (only in"before reperfusion" timing condition)	[29]
Mouse; ICH; collagenase injection model	SR49059 (0.5 mg/kg, 2 mg/kg)	1 h after surgery	■ ↓ Cerebral edema at 24 h and 72 h post-ICH, ■ ↑ Neurobehavioral function at 72 h post-ICH; ■ ↓ BBB disruption at 72 h post-ICH; ↓ AQP4 expression	[135]
Mouse; 60 min MCAO	V1R antagonist (i.cv)V2R antagonist (i.cv)	5 min after MCAO	■ V1R antagonist- ↓ infarct volume, ↓ brain edema formation, ↓ BBB disruption, ↓ functional deficits ■ V2R antagonist: no effect	[180]
Rat; pMCAO	SR49059 (i.p. 30 mg/kg, 2 mg/kg)	Immediately, 1 h, 3 h, 6 h after MCAO	■ ↓ Infarct volume, ↓ Neurological deficits, ↓ Brain edema (in immediate and 1 h time conditions) ■ Higher dose (30 mg/kg) shows no advantage over 2 mg/kg (at 1 h after MCAO)	[30]
Rat; general cerebral hypoxia; bilateral common carotid ligation model	OPC-31260 (p.o 30 mg/kg)	Immediately after operation	■ ↑ Survival rate; ↓ increase in BWC and accumulation of Na^+; ↑ AVP level increase	[28]

Table 2. *Cont.*

Experimental Model/Type of Study	Antagonist (Dose)	Time of Administration	Results	Author
Rat; SAH; autologous blood administration on cerebral cortex	OPC-31260 (p.o 30 mg/kg, 10 mg/kg)	Immediately after operation or every 8 h following operation	▪ ↓ Water and Na+ accumulation; ↑ AVP level in plasma increase	[27]
Rat; MCAO	5% HS bolus +5% HS (HS) Conivaptan (Con) Conivaptan + 5% HS (Con+HS) Conivaptan + 5% HS bolus+ 5% HS (Con+HSb)	6 h after MCAO—early group 24 h after MCAO—late group	▪ ↓ Infarct volume, ↓ Brain edema in HS and Con treated early groups and HS treated late group ▪ ↓ Microglia/macrophage activation in Con treated early group	[186]
Rat; 30 min tCCAO	Conivaptan (i.v. 10 mg/mL, 20 mg/mL)	At reperfusion	▪ Conivaptan has better diuretic activity than mannitol ▪ ↑ AVP level increase, ↓ serum NSE, ↑ serum PGRN	[183]
Mouse; 60 min tMCAO	Conivaptan (i.v. 0.2 mg/day)	3, 5, and 20 h after reperfusion	▪ ↓ Edema, ↓ Neurological deficits improvement, ↓ BBB disruption (when administered 3 h after reperfusion) ▪ No effect on edema/BBB disruption (when administered 5 and 20 h after reperfusion)	[182]
Phase I clinical study, ICH at risk of developing PHE	Conivaptan (20 mg) + standard management	Every 12 h for 2 days	▪ Conivaptan is safe to use with ICH standard treatment	[184]
Case report; refractory brain edema following hemorrhagic stroke	Conivaptan (i.v 20 mg in 30 min + 20 mg over 24 h)	2 days after hemorrhage detection	▪ Rapid clinical and radiological improvement	[185]

tMCAO, Transient middle artery occlusion; BWC, Brain water content; BBB, Blood-brain barrier; ICH, Intracranial hemorrhage; AQP4, Aquaporin-4; V1R, Vasopressin V1 receptor; V2R, Vasopressin V2 receptor; MCAO, Middle cerebral artery occlusion; pMCAO, Permanent middle cerebral artery occlusion; SAH, Subarachnoid hemorrhage; HS, Hypertonic saline; tMCAO, transient middle cerebral artery occlusion; PGRN, progranulin; NSE, Neuron-specific enolase; PHE, Perihematomal edema. Down arrows (↓) indicate a decrease of the respective results cited; up arrows (↑) indicate an increase.

A limited number of studies have focused on the possibility of using AVP receptor antagonists in animal models of hemorrhagic stroke. Although ischemic stroke is epidemiologically more frequent than hemorrhagic stroke, we suggest that the efficacy of vasopressin antagonists in the latter should also be investigated [187]. Thus, we encourage researchers to include hemorrhagic stroke models in their studies.

In this review, we pointed out the involvement of AVP in some stroke-related pathophysiological events. The multiplicity of mechanisms involved in primary and secondary brain damage in stroke points out the still underexplored areas of possible AVP influence. We propose that further research should be undertaken in the following areas: (1) reperfusion injury, (2) neuroinflammation, (3) vasospasm, (4) oxidative stress, and (5) BBB disruption.

This review provides an update on understanding AVP involvement in ischemic stroke. Moreover, it highlights viable future research areas to explore. Our work would lend itself well to clinicians conducting clinical trials involving AVP receptor antagonists and researchers exploring the molecular mechanisms behind other neurological conditions. The ultimate goal of this review is to provide a springboard for further studies on the involvement of AVP in stroke settings and the implication thereof in modern medicine. In our view, the biggest challenge would be the time-consuming and cost-intensive process of clinical trials, particularly considering other competing research areas regarding stroke therapy, including neuroprotective drugs, safer/cheaper thrombolytic agents, and extended time windows protocols for thrombolysis/thrombectomy. Moreover, since the animal model for human ischemic stroke does not adequately reflect the human brain conditions in stroke, the discussed results of animal studies could be delusive, thus undermining the rationale for large-scale studies. To overcome these challenges, further experimental investigations are needed to establish the complete basis of the AVP mechanism of action in terms of stroke. For instance, the current animal stroke models should be improved by considering the age, metabolic state, and common comorbidities of stroke patients, as well as the anatomical and molecular differences between experimental animals and humans. Due to the significant burden of stroke in the world's rapidly growing and aging population, the scientific fields studying novel stroke therapeutics and stroke pathomechanism are one of the central issues currently being researched. The line of research regarding the involvement of AVP in stroke is especially significant, given that some drugs that could putatively diminish the detrimental effect of AVP are widely available for treatment in other medical scenarios, thus making them easily accessible in case of a positive translation process.

8. Conclusions

In conclusion, overlapping detrimental events occurring both in stroke and AVP hypersecretion models and the effectiveness of AVP receptor antagonists provide evidence to infer that AVP plays a vital role in stroke development. However, the indirect nature of the gathered data does not provide us with the essential details to determine the overall involvement of AVP-mediated mechanisms in stroke. Moreover, the differences between ischemic and hemorrhagic stroke regarding AVP release and the mechanism of action remain largely uncovered. Nevertheless, monitoring AVP levels and targeting AVP receptors are of paramount potential in clinical applications. Undoubtedly, molecular and prospective clinical studies are needed to discover the precise mechanisms behind AVP effects and integrate the measuring of copeptin or/and the use of AVP receptor antagonists in clinical practice.

Based on the data included in the sections of the manuscript, we can summarize that:

- A stroke is associated with the release of AVP into the blood in response to various endogenous stimuli;
- Overproduction of AVP is most likely detrimental to the brain, primarily due to the AVP-elicited brain edema;
- Copeptin is a surrogate marker of AVP and can be used as a prognostic biomarker of stroke outcome;

- AVP receptor antagonists can potentially be used as an add-on therapeutic intervention in stroke.

Author Contributions: Project administration: P.J.W., J.G., V.C.B., K.C. and M.O.; Conceptualization: M.O., K.C., J.G. and P.J.W.; Original manuscript preparation: K.C., M.O., J.G., J.R., A.V.A. and A.D.; Graphic and table preparation: K.C. and M.O.; Review and editing: P.J.W., V.C.B., J.G., K.C., M.O. and A.V.A. Supervision: P.J.W., J.G. and V.C.B. All authors have read and agreed to the published version of the manuscript.

Funding: This research received no external funding.

Institutional Review Board Statement: Not applicable.

Informed Consent Statement: Not applicable.

Data Availability Statement: Not applicable.

Conflicts of Interest: The authors declare no conflict of interest.

References

1. Global, regional, and national burden of stroke and its risk factors, 1990-2019: A systematic analysis for the Global Burden of Disease Study 2019. *Lancet. Neurol.* **2021**, *20*, 795–820. [CrossRef] [PubMed]
2. Matei, N.; Camara, J.; Zhang, J.H. The Next Step in the Treatment of Stroke. *Front. Neurol.* **2021**, *11*, 582605. [CrossRef] [PubMed]
3. Stuckey, S.M.; Ong, L.K.; Collins-Praino, L.E.; Turner, R.J. Neuroinflammation as a Key Driver of Secondary Neurodegeneration Following Stroke? *Int. J. Mol. Sci.* **2021**, *22*, 13101. [CrossRef] [PubMed]
4. Schrier, R.W.; Berl, T.; Anderson, R.J. Osmotic and nonosmotic control of vasopressin release. *Am. J. Physiol.* **1979**, *236*, F321–F332. [CrossRef]
5. Kozniewska, E.; Romaniuk, K. Vasopressin in vascular regulation and water homeostasis in the brain. *J. Physiol. Pharmacol. Off. J. Polish Physiol. Soc.* **2008**, *59* (Suppl 8), 109–116.
6. Szmydynger-Chodobska, J.; Chung, I.; Koźniewska, E.; Tran, B.; Harrington, F.; Duncan, J.; Chodobski, A. Increased Expression of Vasopressin V 1a Receptors after Traumatic Brain Injury. *J. Neurotrauma* **2004**, *21*, 1090–1102. [CrossRef]
7. Sakka, L.; Coll, G.; Chazal, J. Anatomy and physiology of cerebrospinal fluid. *Eur. Ann. Otorhinolaryngol. Head Neck Dis.* **2011**, *128*, 309–316. [CrossRef]
8. Bankir, L.; Bichet, D.G.; Morgenthaler, N.G. Vasopressin: Physiology, assessment and osmosensation. *J. Intern. Med.* **2017**, *282*, 284–297. [CrossRef]
9. Danziger, J.; Zeidel, M.L. Osmotic Homeostasis. *Clin. J. Am. Soc. Nephrol.* **2015**, *10*, 852–862. [CrossRef]
10. Toba, K.; Ohta, M.; Kimura, T.; Nagano, K.; Ito, S.; Ouchi, Y. Role of brain vasopressin in regulation of blood pressure. In *Advances in Brain Vasopressin*; Urban, I.J.A., Burbach, J.P.H., De Wed, D., Eds.; Progress in Brain Research; Elsevier: Amsterdam, The Netherlands, 1999; Volume 119, pp. 337–349.
11. Jia, S.W.; Liu, X.Y.; Wang, S.C.; Wang, Y.F. Vasopressin hypersecretion-associated brain edema formation in ischemic stroke: Underlying mechanisms. *J. Stroke Cerebrovasc. Dis.* **2016**, *25*, 1289–1300. [CrossRef]
12. Burton, T.M.; Bushnell, C.D. Reversible Cerebral Vasoconstriction Syndrome. *Stroke* **2019**, *50*, 2253–2258. [CrossRef]
13. Allen, C.L.; Bayraktutan, U. Oxidative stress and its role in the pathogenesis of ischaemic stroke. *Int. J. Stroke Off. J. Int. Stroke Soc.* **2009**, *4*, 461–470. [CrossRef]
14. Jiang, X.; Andjelkovic, A.V.; Zhu, L.; Yang, T.; Bennett, M.V.L.; Chen, J.; Keep, R.F.; Shi, Y. Blood-brain barrier dysfunction and recovery after ischemic stroke. *Prog. Neurobiol.* **2018**, *163–164*, 144–171. [CrossRef]
15. Martino, M.; Arnaldi, G. Copeptin and Stress. *Endocrines* **2021**, *2*, 384–404. [CrossRef]
16. Acher, R.; Chauvet, J.; Rouille, Y. Dynamic processing of neuropeptides: Sequential conformation shaping of neurohypophysial preprohormones during intraneuronal secretory transport. *J. Mol. Neurosci.* **2002**, *18*, 223–228. [CrossRef]
17. Morgenthaler, N.G.; Struck, J.; Alonso, C.; Bergmann, A. Assay for the measurement of copeptin, a stable peptide derived from the precursor of vasopressin. *Clin. Chem.* **2006**, *52*, 112–119. [CrossRef]
18. Christ-Crain, M. Vasopressin and Copeptin in health and disease. *Rev. Endocr. Metab. Disord.* **2019**, *20*, 283–294. [CrossRef]
19. Wendt, M.; Ebinger, M.; Kunz, A.; Rozanski, M.; Waldschmidt, C.; Weber, J.E.; Winter, B.; Koch, P.M.; Nolte, C.H.; Hertel, S.; et al. Copeptin Levels in Patients With Acute Ischemic Stroke and Stroke Mimics. *Stroke* **2015**, *46*, 2426–2431. [CrossRef]
20. De Marchis, G.M.; Katan, M.; Weck, A.; Fluri, F.; Foerch, C.; Findling, O.; Schuetz, P.; Buhl, D.; El-Koussy, M.; Gensicke, H.; et al. Copeptin adds prognostic information after ischemic stroke: Results from the CoRisk study. *Neurology* **2013**, *80*, 1278–1286. [CrossRef]
21. De Marchis, G.M.; Weck, A.; Audebert, H.; Benik, S.; Foerch, C.; Buhl, D.; Schuetz, P.; Jung, S.; Seiler, M.; Morgenthaler, N.G.; et al. Copeptin for the Prediction of Recurrent Cerebrovascular Events After Transient Ischemic Attack. *Stroke* **2014**, *45*, 2918–2923. [CrossRef]

22. Perovic, E.; Mrdjen, A.; Harapin, M.; Kuna, A.T.; Simundic, A.-M. Diagnostic and prognostic role of resistin and copeptin in acute ischemic stroke. *Top. Stroke Rehabil.* **2017**, *24*, 614–618. [CrossRef] [PubMed]

23. Purroy, F.; Suárez-Luis, I.; Cambray, S.; Farré, J.; Benabdelhak, I.; Mauri-Capdevila, G.; Sanahuja, J.; Quílez, A.; Begué, R.; Gil, M.I.; et al. The determination of copeptin levels helps management decisions among transient ischaemic attack patients. *Acta Neurol. Scand.* **2016**, *134*, 140–147. [CrossRef] [PubMed]

24. Xu, Q.; Tian, Y.; Peng, H.; Li, H. Copeptin as a biomarker for prediction of prognosis of acute ischemic stroke and transient ischemic attack: A meta-analysis. *Hypertens. Res.* **2017**, *40*, 465–471. [CrossRef] [PubMed]

25. Greisenegger, S.; Segal, H.C.; Burgess, A.I.; Poole, D.L.; Mehta, Z.; Rothwell, P.M. Copeptin and Long-Term Risk of Recurrent Vascular Events After Transient Ischemic Attack and Ischemic Stroke. *Stroke* **2015**, *46*, 3117–3123. [CrossRef] [PubMed]

26. Hertz, L.; Xu, J.; Chen, Y.; Gibbs, M.E.; Du, T.; Hertz, L.; Xu, J.; Chen, Y.; Gibbs, M.E.; Du, T. Antagonists of the Vasopressin V1 Receptor and of the β(1)-Adrenoceptor Inhibit Cytotoxic Brain Edema in Stroke by Effects on Astrocytes-but the Mechanisms Differ. *Curr. Neuropharmacol.* **2014**, *12*, 308–323. [CrossRef]

27. László, F.A.; Varga, C.; Nakamura, S. Vasopressin receptor antagonist OPC-31260 prevents cerebral oedema after subarachnoid haemorrhage. *Eur. J. Pharmacol.* **1999**, *364*, 115–122. [CrossRef]

28. Molnár, A.H.; Varga, C.; Berkó, A.; Rojik, I.; Párducz, A.; László, F.; László, F.A. Inhibitory effect of vasopressin receptor antagonist OPC-31260 on experimental brain oedema induced by global cerebral ischaemia. *Acta Neurochir.* **2008**, *150*, 265–271. [CrossRef]

29. Kleindienst, A.; Fazzina, G.; Dunbar, J.G.; Glisson, R.; Marmarou, A. Protective effect of the V1a receptor antagonist SR49059 on brain edema formation following middle cerebral artery occlusion in the rat. *Acta Neurochir. Suppl.* **2006**, *4*, 303–306. [CrossRef]

30. Shuaib, A.; Wang, C.X.; Yang, T.; Noor, R. Effects of nonpeptide V1 vasopressin receptor antagonist SR-49059 on infarction volume and recovery of function in a focal embolic stroke model. *Stroke* **2002**, *33*, 3033–3037. [CrossRef]

31. Marmarou, C.R.; Liang, X.; Abidi, N.H.; Parveen, S.; Taya, K.; Henderson, S.C.; Young, H.F.; Filippidis, A.S.; Baumgarten, C.M. Selective vasopressin-1a receptor antagonist prevents brain edema, reduces astrocytic cell swelling and GFAP, V1aR and AQP4 expression after focal traumatic brain injury. *Brain Res.* **2014**, *1581*, 89–102. [CrossRef]

32. Nakayama, S.; Amiry-Moghaddam, M.; Ottersen, O.P.; Bhardwaj, A. Conivaptan, a Selective Arginine Vasopressin V1a and V2 Receptor Antagonist Attenuates Global Cerebral Edema Following Experimental Cardiac Arrest via Perivascular Pool of Aquaporin-4. *Neurocrit. Care* **2016**, *24*, 273–282. [CrossRef]

33. Okuno, K.; Taya, K.; Marmarou, C.R.; Ozisik, P.; Fazzina, G.; Kleindienst, A.; Gulsen, S.; Marmarou, A. The modulation of aquaporin-4 by using PKC-activator (phorbol myristate acetate) and V1a receptor antagonist (SR49059) following middle cerebral artery occlusion/reperfusion in the rat. In *Proceedings of the Acta Neurochirurgica Supplements*; Steiger, H.-J., Ed.; Springer: Vienna, Austria, 2009; pp. 431–436.

34. Rajan, S.; Tosh, P.; Kadapamannil, D.; Srikumar, S.; Paul, J.; Kumar, L. Efficacy of vaptans for correction of postoperative hyponatremia: A comparison between single intravenous bolus conivaptan vs oral tolvaptan. *J. Anaesthesiol. Clin. Pharmacol.* **2018**, *34*, 193–197. [CrossRef]

35. Ameli, P.A.; Ameli, N.J.; Gubernick, D.M.; Ansari, S.; Mohan, S.; Satriotomo, I.; Buckley, A.K.; Maxwell, C.W.; Nayak, V.H.; Shushrutha Hedna, V. Role of vasopressin and its antagonism in stroke related edema. *J. Neurosci. Res.* **2014**, *92*, 1091–1099. [CrossRef]

36. Juul, K.V.; Bichet, D.G.; Nielsen, S.; Nørgaard, J.P. The physiological and pathophysiological functions of renal and extrarenal vasopressin V2 receptors. *Am. J. Physiol. Renal Physiol.* **2014**, *306*, F931–F940. [CrossRef]

37. García-Villalón, A.L.; Garcia, J.L.; Fernández, N.; Monge, L.; Gómez, B.; Diéguez, G. Regional differences in the arterial response to vasopressin: Role of endothelial nitric oxide. *Br. J. Pharmacol.* **1996**, *118*, 1848–1854. [CrossRef]

38. Aoyagi, T.; Koshimizu, T.A.; Tanoue, A. Vasopressin regulation of blood pressure and volume: Findings from V1a receptor-deficient mice. *Kidney Int.* **2009**, *76*, 1035–1039. [CrossRef]

39. Yamada, Y.; Yamamura, Y.; Chihara, T.; Onogawa, T.; Nakamura, S.; Yamashita, T.; Mori, T.; Tominaga, M.; Yabuuchi, Y. OPC-21268, a vasopressin V1 antagonist, produces hypotension in spontaneously hypertensive rats. *Hypertension* **1994**, *23*, 200–204. [CrossRef]

40. Friesenecker, B.E.; Tsai, A.G.; Martini, J.; Ulmer, H.; Wenzel, V.; Hasibeder, W.R.; Intaglietta, M.; Dünser, M.W. Arteriolar vasoconstrictive response: Comparing the effects of arginine vasopressin and norepinephrine. *Crit. Care* **2006**, *10*, R75. [CrossRef]

41. Sedhai, Y.R.; Shrestha, D.B.; Budhathoki, P.; Memon, W.; Acharya, R.; Gaire, S.; Pokharel, N.; Maharjan, S.; Jasaraj, R.; Sodhi, A.; et al. Vasopressin versus norepinephrine as the first-line vasopressor in septic shock: A systematic review and meta-analysis. *J. Clin. Transl. Res.* **2022**, *8*, 185–199.

42. Fassbender, K.; Schmidt, R.; Mössner, R.; Daffertshofer, M.; Hennerici, M. Pattern of activation of the hypothalamic-pituitary-adrenal axis in acute stroke. Relation to acute confusional state, extent of brain damage, and clinical outcome. *Stroke* **1994**, *25*, 1105–1108. [CrossRef]

43. Zelena, D.; Pintér, O.; Balázsfi, D.G.; Langnaese, K.; Richter, K.; Landgraf, R.; Makara, G.B.; Engelmann, M. Vasopressin signaling at brain level controls stress hormone release: The vasopressin-deficient Brattleboro rat as a model. *Amino Acids* **2015**, *47*, 2245–2253. [CrossRef] [PubMed]

44. Thibonnier, M.; Preston, J.A.; Dulin, N.; Wilkins, P.L.; Berti-Mattera, L.N.; Mattera, R. The human V3 pituitary vasopressin receptor: Ligand binding profile and density-dependent signaling pathways. *Endocrinology* **1997**, *138*, 4109–4122. [CrossRef] [PubMed]

45. Hannon, M.J.; Thompson, C.J. Chapter 18—Vasopressin, Diabetes Insipidus, and the Syndrome of Inappropriate Antidiuresis. In *Endocrinology: Adult and Pediatric*, 7th ed.; Jameson, J.L., De Groot, L.J., de Kretser, D.M., Giudice, L.C., Grossman, A.B., Melmed, S., Potts, J.T., Weir, G.C., Eds.; W.B. Saunders: Philadelphia, PA, USA, 2016; pp. 298–311.e4. ISBN 978-0-323-18907-1.

46. Perraudin, V.; Delarue, C.; Lefebvre, H.; Contesse, V.; Kuhn, J.M.; Vaudry, H. Vasopressin stimulates cortisol secretion from human adrenocortical tissue through activation of V1 receptors. *J. Clin. Endocrinol. Metab.* **1993**, *76*, 1522–1528. [CrossRef] [PubMed]

47. Niermann, H.; Amiry-Moghaddam, M.; Holthoff, K.; Witte, O.W.; Ottersen, O.P. A Novel Role of Vasopressin in the Brain: Modulation of Activity-Dependent Water Flux in the Neocortex. *J. Neurosci.* **2001**, *21*, 3045–3051. [CrossRef] [PubMed]

48. Tabarean, I. V Activation of Preoptic Arginine Vasopressin Neurons Induces Hyperthermia in Male Mice. *Endocrinology* **2020**, *162*, bqaa217. [CrossRef] [PubMed]

49. Buijs, R.M.; De Vries, G.J.; Van Leeuwen, F.W.; Swaab, D.F. Vasopressin and Oxytocin: Distribution and Putative Functions in the Brain. In *The Neurohypophysis: Structure, Function and Control*; Cross, B.A., Leng, G., Eds.; Progress in Brain Research; Elsevier: Amsterdam, The Netherlands, 1983; Volume 60, pp. 115–122.

50. Kasting, N.W.; Veale, W.L.; Cooper, K.E. Convulsive and hypothermic effects of vasopressin in the brain of the rat. *Can. J. Physiol. Pharmacol.* **1980**, *58*, 316–319. [CrossRef] [PubMed]

51. Kalsbeek, A.; van Heerikhuize, J.J.; Wortel, J.; Buijs, R.M. A Diurnal Rhythm of Stimulatory Input to the Hypothalamo{\textendash}Pituitary{\textendash}Adrenal System as Revealed by Timed Intrahypothalamic Administration of the Vasopressin V1Antagonist. *J. Neurosci.* **1996**, *16*, 5555–5565. [CrossRef] [PubMed]

52. Kalsbeek, A.; Fliers, E.; Hofman, M.A.; Swaab, D.F.; Buijs, R.M. Vasopressin and the output of the hypothalamic biological clock. *J. Neuroendocrinol.* **2010**, *22*, 362–372. [CrossRef]

53. Gomez, F.; Chapleur, M.; Fernette, B.; Burlet, C.; Nicolas, J.P.; Burlet, A. Arginine vasopressin (AVP) depletion in neurons of the suprachiasmatic nuclei affects the AVP content of the paraventricular neurons and stimulates adrenocorticotrophic hormone release. *J. Neurosci. Res.* **1997**, *50*, 565–574. [CrossRef]

54. Rohr, K.E.; Telega, A.; Savaglio, A.; Evans, J.A. Vasopressin regulates daily rhythms and circadian clock circuits in a manner influenced by sex. *Horm. Behav.* **2021**, *127*, 104888. [CrossRef]

55. Marcinkowska, A.B.; Biancardi, V.C.; Winklewski, P.J. Arginine-vasopressin, synaptic plasticity and brain networks. *Curr. Neuropharmacol.* **2022**, *20*, 2292–2302. [CrossRef]

56. Bosch, O.J.; Neumann, I.D. Brain vasopressin is an important regulator of maternal behavior independent of dams' trait anxiety. *Proc. Natl. Acad. Sci. USA* **2008**, *105*, 17139–17144. [CrossRef]

57. Neumann, I.D.; Landgraf, R. Balance of brain oxytocin and vasopressin: Implications for anxiety, depression, and social behaviors. *Trends Neurosci.* **2012**, *35*, 649–659. [CrossRef]

58. Boer, G.J.; Buijs, R.M.; Swaab, D.F.; De Vries, G.J. Vasopressin and the developing rat brain. *Peptides* **1980**, *1*, 203–209. [CrossRef]

59. Katan, M.; Christ-Crain, M. The stress hormone copeptin: A new prognostic biomarker in acute illness. *Swiss Med. Wkly.* **2010**, *140*, w13101. [CrossRef]

60. Morgenthaler, N.G.; Müller, B.; Struck, J.; Bergmann, A.; Redl, H.; Christ-Crain, M. Copeptin, a stable peptide of the arginine vasopressin precursor, is elevated in hemorrhagic and septic shock. *Shock* **2007**, *28*, 219–226. [CrossRef]

61. Liu, X.; Jin, Y.; Zheng, H.; Chen, G.; Tan, B.; Wu, B. Arginine vasopressin gene expression in supraoptic nucleus and paraventricular nucleus of hypothalamous following cerebral ischemia and reperfusion. *Chinese Med. Sci. J.* **2000**, *15*, 157–161.

62. Szmydynger-Chodobska, J.; Zink, B.J.; Chodobski, A. Multiple sites of vasopressin synthesis in the injured brain. *J. Cereb. blood flow Metab. Off. J. Int. Soc. Cereb. Blood Flow Metab.* **2011**, *31*, 47–51. [CrossRef]

63. Barreca, T.; Gandolfo, C.; Corsini, G.; Del Sette, M.; Cataldi, A.; Rolandi, E.; Franceschini, R. Evaluation of the secretory pattern of plasma arginine vasopressin in stroke patients. *Cerebrovasc. Dis.* **2001**, *11*, 113–118. [CrossRef]

64. Loesch, A. The effect of cerebral ischemia on the ultrastructure of the hypothalamo-neurohypophysial system of the mongolian gerbil. The neurohypophysial axons and pituicytes. *Aust. J. Exp. Biol. Med. Sci.* **1983**, *61*, 557–568. [CrossRef]

65. Hockel, K.; Schöller, K.; Trabold, R.; Nussberger, J.; Plesnila, N. Vasopressin V_{1a} Receptors Mediate Posthemorrhagic Systemic Hypertension Thereby Determining Rebleeding Rate and Outcome After Experimental Subarachnoid Hemorrhage. *Stroke* **2012**, *43*, 227–232. [CrossRef]

66. Chang, Y.; Chen, T.Y.; Chen, C.H.; Crain, B.J.; Toung, T.J.K.; Bhardwaj, A. Plasma arginine-vasopressin following experimental stroke: Effect of osmotherapy. *J. Appl. Physiol.* **2006**, *100*, 1445–1451. [CrossRef] [PubMed]

67. Arroja, M.M.C.; Reid, E.; McCabe, C. Therapeutic potential of the renin angiotensin system in ischaemic stroke. *Exp. Transl. Stroke Med.* **2016**, *8*, 8. [CrossRef] [PubMed]

68. Johanson, C.E.; Preston, J.E.; Chodobski, A.; Stopa, E.G.; Szmydynger-Chodobska, J.; McMillan, P.N. AVP V1 receptor-mediated decrease in Cl- efflux and increase in dark cell number in choroid plexus epithelium. *Am. J. Physiol.* **1999**, *276*, C82–C90. [CrossRef] [PubMed]

69. Xiong, L.; Tian, G.; Leung, H.; Soo, Y.O.Y.; Chen, X.; Ip, V.H.L.; Mok, V.C.T.; Chu, W.C.W.; Wong, K.S.; Leung, T.W.H. Autonomic Dysfunction Predicts Clinical Outcomes After Acute Ischemic Stroke: A Prospective Observational Study. *Stroke* **2018**, *49*, 215–218. [CrossRef]

70. Dütsch, M.; Burger, M.; Dörfler, C.; Schwab, S.; Hilz, M.J. Cardiovascular autonomic function in poststroke patients. *Neurology* **2007**, *69*, 2249–2255. [CrossRef]

71. Sykora, M.; Diedler, J.; Poli, S.; Rupp, A.; Turcani, P.; Steiner, T. Blood Pressure Course in Acute Stroke Relates to Baroreflex Dysfunction. *Cerebrovasc. Dis.* **2010**, *30*, 172–179. [CrossRef]

72. Sykora, M.; Diedler, J.; Turcani, P.; Hacke, W.; Steiner, T. Baroreflex: A New Therapeutic Target in Human Stroke? *Stroke* **2009**, *40*, e678–e682. [CrossRef]

73. Szczepanska-Sadowska, E.; Wsol, A.; Cudnoch-Jedrzejewska, A.; Czarzasta, K.; Żera, T. Multiple Aspects of Inappropriate Action of Renin-Angiotensin, Vasopressin, and Oxytocin Systems in Neuropsychiatric and Neurodegenerative Diseases. *J. Clin. Med.* **2022**, *11*, 908. [CrossRef]

74. Savić, B.; Murphy, D.; Japundžić-Žigon, N. The Paraventricular Nucleus of the Hypothalamus in Control of Blood Pressure and Blood Pressure Variability. *Front. Physiol.* **2022**, *13*, 858941. [CrossRef]

75. Saravia, F.E.; Gonzalez, S.L.; Roig, P.; Alves, V.; Homo-Delarche, F.; Nicola, A.F. De Diabetes Increases the Expression of Hypothalamic Neuropeptides in a Spontaneous Model of Type I Diabetes, the Nonobese Diabetic (NOD) Mouse. *Cell. Mol. Neurobiol.* **2001**, *21*, 15–27. [CrossRef]

76. Anai, H.; Ueta, Y.; Serino, R.; Nomura, M.; Kabashima, N.; Shibuya, I.; Takasugi, M.; Nakashima, Y.; Yamashita, H. Upregulation of the expression of vasopressin gene in the paraventricular and supraoptic nuclei of the lithium-induced diabetes insipidus rat. *Brain Res.* **1997**, *772*, 161–166. [CrossRef]

77. Keller, W.J.; Mullaj, E. Antidiuretic hormone release associated with increased intracranial pressure independent of plasma osmolality. *Brain Behav.* **2018**, *8*, e01005. [CrossRef]

78. Hacke, W.; Schwab, S.; Horn, M.; Spranger, M.; De Georgia, M.; von Kummer, R. "Malignant" middle cerebral artery territory infarction: Clinical course and prognostic signs. *Arch. Neurol.* **1996**, *53*, 309–315. [CrossRef]

79. Murtha, L.A.; McLeod, D.D.; Pepperall, D.; McCann, S.K.; Beard, D.J.; Tomkins, A.J.; Holmes, W.M.; McCabe, C.; Macrae, I.M.; Spratt, N.J. Intracranial pressure elevation after ischemic stroke in rats: Cerebral edema is not the only cause, and short-duration mild hypothermia is a highly effective preventive therapy. *J. Cereb. blood flow Metab. Off. J. Int. Soc. Cereb. Blood Flow Metab.* **2015**, *35*, 592–600. [CrossRef]

80. Zaghmi, A.; Dopico-López, A.; Pérez-Mato, M.; Iglesias-Rey, R.; Hervella, P.; Greschner, A.A.; Bugallo-Casal, A.; da Silva, A.; Gutiérrez-Fernández, M.; Castillo, J.; et al. Sustained blood glutamate scavenging enhances protection in ischemic stroke. *Commun. Biol.* **2020**, *3*, 729. [CrossRef]

81. Dávalos, A.; Shuaib, A.; Wahlgren, N.G. Neurotransmitters and pathophysiology of stroke: Evidence for the release of glutamate and other transmitters/mediators in animals and humans. *J. stroke Cerebrovasc. Dis. Off. J. Natl. Stroke Assoc.* **2000**, *9*, 2–8. [CrossRef]

82. Meng, X.; Li, N.; Guo, D.-Z.; Pan, S.-Y.; Li, H.; Yang, C. High plasma glutamate levels are associated with poor functional outcome in acute ischemic stroke. *Cell. Mol. Neurobiol.* **2015**, *35*, 159–165. [CrossRef]

83. Busnardo, C.; Crestani, C.C.; Resstel, L.B.M.; Tavares, R.F.; Antunes-Rodrigues, J.; Corrêa, F.M.A. Ionotropic glutamate receptors in hypothalamic paraventricular and supraoptic nuclei mediate vasopressin and oxytocin release in unanesthetized rats. *Endocrinology* **2012**, *153*, 2323–2331. [CrossRef]

84. Chen, Y.-J.; Wallace, B.K.; Yuen, N.; Jenkins, D.P.; Wulff, H.; O'Donnell, M.E. Blood-brain barrier KCa3.1 channels: Evidence for a role in brain Na uptake and edema in ischemic stroke. *Stroke* **2015**, *46*, 237–244. [CrossRef]

85. Sato-Numata, K.; Numata, T.; Ueta, Y.; Okada, Y. Vasopressin Neurons Respond to Hyperosmotic Stimulation with Regulatory Volume Increase and Secretory Volume Decrease by Activating Ion Transporters and Ca(2+) Channels. *Cell. Physiol. Biochem. Int. J. Exp. Cell. Physiol. Biochem. Pharmacol.* **2021**, *55*, 119–134. [CrossRef]

86. Wang, S.C.; Parpura, V.; Wang, Y.-F. Astroglial Regulation of Magnocellular Neuroendocrine Cell Activities in the Supraoptic Nucleus. *Neurochem. Res.* **2021**, *46*, 2586–2600. [CrossRef] [PubMed]

87. Wang, Y.-F.; Parpura, V. Central Role of Maladapted Astrocytic Plasticity in Ischemic Brain Edema Formation. *Front. Cell. Neurosci.* **2016**, *10*, 129. [CrossRef] [PubMed]

88. He, M.; Chen, M.; Wang, J.; Guo, G.; Zheng, Y.; Jiang, X.; Zhang, M. Relationship between glutamate in the limbic system and hypothalamus-pituitary-adrenal axis after middle cerebral artery occlusion in rats. *Chin. Med. J.* **2003**, *116*, 1492–1496. [PubMed]

89. Cui, D.; Jia, S.; Li, T.; Li, D.; Wang, X.; Liu, X.; Wang, Y.F.; Cui, D.; Jia, S.; Liu, X.; et al. Involvement of Supraoptic Astrocytes in Basilar Artery Occlusion-Evoked Differential Activation of Vasopressin Neurons and Vasopressin Secretion in Rats. *Neurochem. Res.* **2021**, *46*, 2651–2661. [CrossRef]

90. Cui, D.; Jia, S.; Li, T.; Li, D.; Wang, X.; Liu, X.; Wang, Y.F. Alleviation of brain injury by applying TGN-020 in the supraoptic nucleus via inhibiting vasopressin neurons in rats of focal ischemic stroke. *Life Sci.* **2021**, *264*, 118683. [CrossRef]

91. Jin, R.; Liu, L.; Zhang, S.; Nanda, A.; Li, G. Role of inflammation and its mediators in acute ischemic stroke. *J. Cardiovasc. Transl. Res.* **2013**, *6*, 834–851. [CrossRef]

92. Szmydynger-Chodobska, J.; Fox, L.M.; Lynch, K.M.; Zink, B.J.; Chodobski, A.; Cui, D.; Jia, S.; Li, T.; Li, D.; Wang, X.; et al. Vasopressin amplifies the production of proinflammatory mediators in traumatic brain injury. *J. Neurotrauma* **2010**, *27*, 1449–1461. [CrossRef]

93. Pascale, C.L.; Szmydynger-Chodobska, J.; Sarri, J.E.; Chodobski, A. Traumatic brain injury results in a concomitant increase in neocortical expression of vasopressin and its V1a receptor. *J. Physiol. Pharmacol. an Off. J. Polish Physiol. Soc.* **2006**, *57* (Suppl 1), 161–167.

94. Liamis, G.; Barkas, F.; Megapanou, E.; Christopoulou, E.; Makri, A.; Makaritsis, K.; Ntaios, G.; Elisaf, M.; Milionis, H. Hyponatremia in Acute Stroke Patients: Pathophysiology, Clinical Significance, and Management Options. *Eur. Neurol.* **2020**, *82*, 32–40. [CrossRef]

95. Robertson, G.L. Regulation of Arginine Vasopressin in the Syndrome of Inappropriate Antidiuresis. *Am. J. Med.* **2006**, *119*, 36–42. [CrossRef]

96. Hannon, M.J.; Thompson, C.J. The syndrome of inappropriate antidiuretic hormone: Prevalence, causes and consequences. *Eur. J. Endocrinol.* **2010**, *162*, 5–12. [CrossRef]

97. Giuliani, C.; Peri, A. Effects of hyponatremia on the brain. *J. Clin. Med.* **2014**, *3*, 1163–1177. [CrossRef]

98. Largeau, B.; Le Tilly, O.; Sautenet, B.; Salmon Gandonnière, C.; Barin-Le Guellec, C.; Ehrmann, S. Arginine Vasopressin and Posterior Reversible Encephalopathy Syndrome Pathophysiology: The Missing Link? *Mol. Neurobiol.* **2019**, *56*, 6792–6806. [CrossRef]

99. Huang, W.-Y.; Weng, W.-C.; Peng, T.-I.; Chien, Y.-Y.; Wu, C.-L.; Lee, M.; Hung, C.-C.; Chen, K.-H. Association of Hyponatremia in Acute Stroke Stage with Three-Year Mortality in Patients with First-Ever Ischemic Stroke. *Cerebrovasc. Dis.* **2012**, *34*, 55–62. [CrossRef]

100. Rodrigues, B.; Staff, I.; Fortunato, G.; McCullough, L.D. Hyponatremia in the prognosis of acute ischemic stroke. *J. Stroke Cerebrovasc. Dis. Off. J. Natl. Stroke Assoc.* **2014**, *23*, 850–854. [CrossRef]

101. Soiza, R.L.; Cumming, K.; Clark, A.B.; Bettencourt-Silva, J.H.; Metcalf, A.K.; Bowles, K.M.; Potter, J.F.; Myint, P.K. Hyponatremia predicts mortality after stroke. *Int. J. Stroke Off. J. Int. Stroke Soc.* **2015**, *10* (Suppl A), 50–55. [CrossRef]

102. Carcel, C.; Sato, S.; Zheng, D.; Heeley, E.; Arima, H.; Yang, J.; Wu, G.; Chen, G.; Zhang, S.; Delcourt, C.; et al. Prognostic Significance of Hyponatremia in Acute Intracerebral Hemorrhage: Pooled Analysis of the Intensive Blood Pressure Reduction in Acute Cerebral Hemorrhage Trial Studies. *Crit. Care Med.* **2016**, *44*, 1388–1394. [CrossRef]

103. Kalita, J.; Singh, R.K.; Misra, U.K. Cerebral Salt Wasting Is the Most Common Cause of Hyponatremia in Stroke. *J. Stroke Cerebrovasc. Dis. Off. J. Natl. Stroke Assoc.* **2017**, *26*, 1026–1032. [CrossRef]

104. Hannon, M.J.; Behan, L.A.; O'Brien, M.M.C.; Tormey, W.; Ball, S.G.; Javadpour, M.; Sherlock, M.; Thompson, C.J. Hyponatremia following mild/moderate subarachnoid hemorrhage is due to SIAD and glucocorticoid deficiency and not cerebral salt wasting. *J. Clin. Endocrinol. Metab.* **2014**, *99*, 291–298. [CrossRef]

105. Oikawa, R.; Hosoda, C.; Nasa, Y.; Daicho, T.; Tanoue, A.; Tsujimoto, G.; Takagi, N.; Tanonaka, K.; Takeo, S. Decreased susceptibility to salt-induced hypertension in subtotally nephrectomized mice lacking the vasopressin V1a receptor. *Cardiovasc. Res.* **2010**, *87*, 187–194. [CrossRef] [PubMed]

106. King, K.A.; Mackie, G.; Pang, C.C.; Wall, R.A. Central vasopressin in the modulation of catecholamine release in conscious rats. *Can. J. Physiol. Pharmacol.* **1985**, *63*, 1501–1505. [CrossRef] [PubMed]

107. Okada, S.; Murakami, Y.; Nakamura, K.; Yokotani, K. Vasopressin V(1) receptor-mediated activation of central sympatho-adrenomedullary outflow in rats. *Eur. J. Pharmacol.* **2002**, *457*, 29–35. [CrossRef] [PubMed]

108. Winklewski, P.J.; Radkowski, M.; Demkow, U. Cross-talk between the inflammatory response, sympathetic activation and pulmonary infection in the ischemic stroke. *J. Neuroinflamm.* **2014**, *11*, 213. [CrossRef] [PubMed]

109. Park, S.J.; Shin, J. Il Inflammation and hyponatremia: An underrecognized condition? *Korean J. Pediatr.* **2013**, *56*, 519–522. [CrossRef]

110. Chikanza, I.C.; Grossman, A.S. Hypothalamic-pituitary-mediated immunomodulation: Arginine vasopressin is a neuroendocrine immune mediator. *Br. J. Rheumatol.* **1998**, *37*, 131–136. [CrossRef]

111. Miller, A.H.; Spencer, R.L.; McEwen, B.S.; Stein, M. Depression, adrenal steroids, and the immune system. *Ann. Med.* **1993**, *25*, 481–487. [CrossRef]

112. De Winter, R.F.P.; van Hemert, A.M.; DeRijk, R.H.; Zwinderman, K.H.; Frankhuijzen-Sierevogel, A.C.; Wiegant, V.M.; Goekoop, J.G. Anxious–Retarded Depression: Relation with Plasma Vasopressin and Cortisol. *Neuropsychopharmacology* **2003**, *28*, 140–147. [CrossRef]

113. Roper, J.; O'Carroll, A.-M.; Young, W., 3rd; Lolait, S. The vasopressin Avpr1b receptor: Molecular and pharmacological studies. *Stress* **2011**, *14*, 98–115. [CrossRef]

114. Barugh, A.J.; Gray, P.; Shenkin, S.D.; MacLullich, A.M.J.; Mead, G.E. Cortisol levels and the severity and outcomes of acute stroke: A systematic review. *J. Neurol.* **2014**, *261*, 533–545. [CrossRef]

115. Geerlings, M.I.; Sigurdsson, S.; Eiriksdottir, G.; Garcia, M.E.; Harris, T.B.; Gudnason, V.; Launer, L.J. Salivary cortisol, brain volumes, and cognition in community-dwelling elderly without dementia. *Neurology* **2015**, *85*, 976–983. [CrossRef]

116. Lupien, S.J.; de Leon, M.; de Santi, S.; Convit, A.; Tarshish, C.; Nair, N.P.; Thakur, M.; McEwen, B.S.; Hauger, R.L.; Meaney, M.J. Cortisol levels during human aging predict hippocampal atrophy and memory deficits. *Nat. Neurosci.* **1998**, *1*, 69–73. [CrossRef]

117. Suri, D.; Vaidya, V.A. Glucocorticoid regulation of brain-derived neurotrophic factor: Relevance to hippocampal structural and functional plasticity. *Neuroscience* **2013**, *239*, 196–213. [CrossRef]

118. Green, K.N.; Billings, L.M.; Roozendaal, B.; McGaugh, J.L.; LaFerla, F.M. Glucocorticoids increase amyloid-beta and tau pathology in a mouse model of Alzheimer's disease. *J. Neurosci. Off. J. Soc. Neurosci.* **2006**, *26*, 9047–9056. [CrossRef]

119. Ferris, C.F.; Melloni Jr, R.H.; Koppel, G.; Perry, K.W.; Fuller, R.W.; Delville, Y. Vasopressin/Serotonin Interactions in the Anterior Hypothalamus Control Aggressive Behavior in Golden Hamsters. *J. Neurosci.* **1997**, *17*, 4331–4340. [CrossRef]

120. Coccaro, E.F.; Kavoussi, R.J.; Hauger, R.L.; Cooper, T.B.; Ferris, C.F. Cerebrospinal Fluid Vasopressin Levels: Correlates With Aggression and Serotonin Function in Personality-Disordered Subjects. *Arch. Gen. Psychiatry* **1998**, *55*, 708–714. [CrossRef]

121. Santos, C.O.; Caeiro, L.; Ferro, J.M.; Albuquerque, R.; Figueira, M.L. Denial in the first days of acute stroke. *J. Neurol.* **2006**, *253*, 1016–1023. [CrossRef]

122. Filep, J.; Rosenkranz, B. Mechanism of vasopressin-induced platelet aggregation. *Thromb Res.* **1987**, *45*, 7–15. [CrossRef]

123. Wun, T. Vasopressin and platelets: A concise review. *Platelets* **1997**, *8*, 15–22. [CrossRef]

124. Zheng, H.; Chen, C.; Zhang, J.; Hu, Z. Mechanism and therapy of brain edema after intracerebral hemorrhage. *Cerebrovasc. Dis.* **2016**, *42*, 155–169. [CrossRef]

125. Szczygielski, J.; Kopańska, M.; Wysocka, A.; Oertel, J. Cerebral Microcirculation, Perivascular Unit, and Glymphatic System: Role of Aquaporin-4 as the Gatekeeper for Water Homeostasis. *Front. Neurol.* **2021**, *12*, 767470. [CrossRef] [PubMed]

126. Nagelhus, E.A.; Ottersen, O.P. Physiological roles of aquaporin-4 in brain. *Physiol. Rev.* **2013**, *93*, 1543–1562. [CrossRef] [PubMed]

127. Chmelova, M.; Sucha, P.; Bochin, M.; Vorisek, I.; Pivonkova, H.; Hermanova, Z.; Anderova, M.; Vargova, L. The role of aquaporin-4 and transient receptor potential vaniloid isoform 4 channels in the development of cytotoxic edema and associated extracellular diffusion parameter changes. *Eur. J. Neurosci.* **2019**, *50*, 1685–1699. [CrossRef] [PubMed]

128. Hirt, L.; Fukuda, A.M.; Ambadipudi, K.; Rashid, F.; Binder, D.; Verkman, A.; Ashwal, S.; Obenaus, A.; Badaut, J. Improved long-term outcome after transient cerebral ischemia in aquaporin-4 knockout mice. *J. Cereb. blood flow Metab. Off. J. Int. Soc. Cereb. Blood Flow Metab.* **2017**, *37*, 277–290. [CrossRef]

129. Yao, X.; Derugin, N.; Manley, G.T.; Verkman, A.S. Reduced brain edema and infarct volume in aquaporin-4 deficient mice after transient focal cerebral ischemia. *Neurosci. Lett.* **2015**, *584*, 368–372. [CrossRef]

130. Chu, H.; Tang, Y.; Dong, Q. Protection of Vascular Endothelial Growth Factor to Brain Edema Following Intracerebral Hemorrhage and Its Involved Mechanisms: Effect of Aquaporin-4. *PLoS ONE* **2013**, *8*, e66051. [CrossRef]

131. Chiu, C.-D.; Chen, C.-C.V.; Shen, C.-C.; Chin, L.-T.; Ma, H.-I.; Chuang, H.-Y.; Cho, D.-Y.; Chu, C.-H.; Chang, C. Hyperglycemia exacerbates intracerebral hemorrhage via the downregulation of aquaporin-4: Temporal assessment with magnetic resonance imaging. *Stroke* **2013**, *44*, 1682–1689. [CrossRef]

132. Qiu, G.-P.; Xu, J.; Zhuo, F.; Sun, S.-Q.; Liu, H.; Yang, M.; Huang, J.; Lu, W.-T.; Huang, S.-Q. Loss of AQP4 polarized localization with loss of β-dystroglycan immunoreactivity may induce brain edema following intracerebral hemorrhage. *Neurosci. Lett.* **2015**, *588*, 42–48. [CrossRef]

133. Moeller, H.B.; Fenton, R.A.; Zeuthen, T.; MacAulay, N. Vasopressin-dependent short-term regulation of aquaporin 4 expressed in Xenopus oocytes. *Neuroscience* **2009**, *164*, 1674–1684. [CrossRef]

134. Liu, X.; Nakayama, S.; Amiry-Moghaddam, M.; Ottersen, O.P.; Bhardwaj, A. Arginine-vasopressin V1 but not V2 receptor antagonism modulates infarct volume, brain water content, and aquaporin-4 expression following experimental stroke. *Neurocrit. Care* **2010**, *12*, 124–131. [CrossRef]

135. Manaenko, A.; Fathali, N.; Khatibi, N.H.; Lekic, T.; Hasegawa, Y.; Martin, R.; Tang, J.; Zhang, J.H. Arginine-vasopressin V1a receptor inhibition improves neurologic outcomes following an intracerebral hemorrhagic brain injury. *Neurochem. Int.* **2011**, *58*, 542–548. [CrossRef]

136. Yuen, N.; Lam, T.I.; Wallace, B.K.; Klug, N.R.; Anderson, S.E.; O'Donnell, M.E. Ischemic factor-induced increases in cerebral microvascular endothelial cell Na/H exchange activity and abundance: Evidence for involvement of ERK1/2 MAP kinase. *Am. J. Physiol. Physiol.* **2014**, *306*, C931–C942. [CrossRef]

137. Wallace, B.K.; Foroutan, S.; O'Donnell, M.E. Ischemia-induced stimulation of Na-K-Cl cotransport in cerebral microvascular endothelial cells involves AMP kinase. *Am. J. Physiol. Physiol.* **2011**, *301*, C316–C326. [CrossRef]

138. Song, D.; Xu, J.; Du, T.; Yan, E.; Hertz, L.; Walz, W.; Peng, L. Inhibition of brain swelling after ischemia-reperfusion by β-adrenergic antagonists: Correlation with increased K+ and decreased Ca2+ concentrations in extracellular fluid. *Biomed Res. Int.* **2014**, *2014*, 873590. [CrossRef]

139. Yan, Y.; Dempsey, R.J.; Flemmer, A.; Forbush, B.; Sun, D. Inhibition of Na+–K+–Cl– cotransporter during focal cerebral ischemia decreases edema and neuronal damage. *Brain Res.* **2003**, *961*, 22–31. [CrossRef]

140. Faraco, G.; Wijasa, T.S.; Park, L.; Moore, J.; Anrather, J.; Iadecola, C. Water deprivation induces neurovascular and cognitive dysfunction through vasopressin-induced oxidative stress. *J. Cereb. Blood Flow Metab.* **2014**, *34*, 852–860. [CrossRef]

141. Cunningham, L.A.; Wetzel, M.; Rosenberg, G.A. Multiple roles for MMPs and TIMPs in cerebral ischemia. *Glia* **2005**, *50*, 329–339. [CrossRef]

142. Wang, X.; Jung, J.; Asahi, M.; Chwang, W.; Russo, L.; Moskowitz, M.A.; Dixon, C.E.; Fini, M.E.; Lo, E.H. Effects of matrix metalloproteinase-9 gene knock-out on morphological and motor outcomes after traumatic brain injury. *J. Neurosci. Off. J. Soc. Neurosci.* **2000**, *20*, 7037–7042. [CrossRef]

143. Harkness, K.A.; Adamson, P.; Sussman, J.D.; Davies-Jones, G.A.B.; Greenwood, J.; Woodroofe, M.N. Dexamethasone regulation of matrix metalloproteinase expression in CNS vascular endothelium. *Brain* **2000**, *123*, 698–709. [CrossRef]

144. Tahara, A.; Tsukada, J.; Tomura, Y.; Yatsu, T.; Shibasaki, M. Vasopressin regulates rat mesangial cell growth by inducing autocrine secretion of vascular endothelial growth factor. *J. Physiol. Sci.* **2011**, *61*, 115–122. [CrossRef]

145. Alonso, G.; Gallibert, E.; Lafont, C.; Guillon, G. Intrahypothalamic angiogenesis induced by osmotic stimuli correlates with local hypoxia: A potential role of confined vasoconstriction induced by dendritic secretion of vasopressin. *Endocrinology* **2008**, *149*, 4279–4288. [CrossRef] [PubMed]

146. Bele, S. Vasopressin Increases Cerebral Perfusion Pressure but not Cerebral Blood Flow in Neurosurgical Patients with Catecholamine-Refractory Hypotension: A Preliminary Evaluation Using the Non-Invasive Quantix ND in Comparison to the Literature. *J. Anesth. &Critical Care Open Access* **2014**, *1*, 17–22. [CrossRef]
147. Engelborghs, K.; Haseldonckx, M.; Van Reempts, J.; Van Rossem, K.; Wouters, L.; Borgers, M.; Verlooy, J. Impaired autoregulation of cerebral blood flow in an experimental model of traumatic brain injury. *J. Neurotrauma* **2000**, *17*, 667–677. [CrossRef] [PubMed]
148. Glushakova, O.Y.; Johnson, D.; Hayes, R.L. Delayed increases in microvascular pathology after experimental traumatic brain injury are associated with prolonged inflammation, blood-brain barrier disruption, and progressive white matter damage. *J. Neurotrauma* **2014**, *31*, 1180–1193. [CrossRef] [PubMed]
149. Aries, M.J.H.; Elting, J.W.; De Keyser, J.; Kremer, B.P.H.; Vroomen, P.C.A.J. Cerebral autoregulation in stroke: A review of transcranial doppler studies. *Stroke* **2010**, *41*, 2697–2704. [CrossRef]
150. Yamagata, K.; Sone, N.; Suguyama, S.; Nabika, T. Different effects of arginine vasopressin on high-mobility group box 1 expression in astrocytes isolated from stroke-prone spontaneously hypertensive rats and congenic SHRpch1_18 rats. *Int. J. Exp. Pathol.* **2016**, *97*, 97–106. [CrossRef]
151. Lei, C.; Lin, S.; Zhang, C.; Tao, W.; Dong, W.; Hao, Z.; Liu, M.; Wu, B. High-mobility group box1 protein promotes neuroinflammation after intracerebral hemorrhage in rats. *Neuroscience* **2013**, *228*, 190–199. [CrossRef]
152. Sun, S.Z.; Cao, H.; Yao, N.; Zhao, L.L.; Zhu, X.F.; Ni, E.A.; Zhu, Q.; Zhu, W. zhong β-Arrestin 2 mediates arginine vasopressin-induced IL-6 induction via the ERK1/2-NF-κB signal pathway in murine hearts. *Acta Pharmacol. Sin.* **2020**, *41*, 198–207. [CrossRef]
153. Kleindienst, A.; Dunbar, J.G.; Glisson, R.; Marmarou, A. The role of vasopressin V1A receptors in cytotoxic brain edema formation following brain injury. *Acta Neurochir.* **2013**, *155*, 151–164. [CrossRef]
154. Ribeiro, M.d.C.; Hirt, L.; Bogousslavsky, J.; Regli, L.; Badaut, J. Time course of aquaporin expression after transient focal cerebral ischemia in mice. *J. Neurosci. Res.* **2006**, *83*, 1231–1240. [CrossRef]
155. Rutkowsky, J.M.; Wallace, B.K.; Wise, P.M.; O'Donnell, M.E. Effects of estradiol on ischemic factor-induced astrocyte swelling and AQP4 protein abundance. *Am. J. Physiol. Cell Physiol.* **2011**, *301*, C204–C212. [CrossRef]
156. Frydenlund, D.S.; Bhardwaj, A.; Otsuka, T.; Mylonakou, M.N.; Yasumura, T.; Davidson, K.G.V.; Zeynalov, E.; Skare, O.; Laake, P.; Haug, F.-M.; et al. Temporary loss of perivascular aquaporin-4 in neocortex after transient middle cerebral artery occlusion in mice. *Proc. Natl. Acad. Sci. USA* **2006**, *103*, 13532–13536. [CrossRef]
157. Fernández, N.; Martínez, M.A.; Luis García-Villalón, A.L.; Monge, L.; Diéguez, G. Cerebral vasoconstriction produced by vasopressin in conscious goats: Role of vasopressin V1 and V2 receptors and nitric oxide. *Br. J. Pharmacol.* **2001**, *132*, 1837–1844. [CrossRef]
158. Van Haren, R.M.; Thorson, C.M.; Ogilvie, M.P.; Valle, E.J.; Guarch, G.A.; Jouria, J.A.; Busko, A.M.; Harris, L.T.; Bullock, M.R.; Jagid, J.R.; et al. Vasopressin for cerebral perfusion pressure management in patients with severe traumatic brain injury: Preliminary results of a randomized controlled trial. *J. Trauma Acute Care Surg.* **2013**, *75*, 1024–1030. [CrossRef]
159. Onoue, H.; Kaito, N.; Tomii, M.; Tokudome, S.; Nakajima, M.; Abe, T. Human basilar and middle cerebral arteries exhibit endothelium-dependent responses to peptides. *Am. J. Physiol.-Hear. Circ. Physiol.* **1994**, *267*, H880–H886. [CrossRef]
160. Takayasu, M.; Kajita, Y.; Suzuki, Y.; Shibuya, M.; Sugita, K.; Ishikawa, T.; Hidaka, H. Triphasic response of rat intracerebral arterioles to increasing concentrations of vasopressin in vitro. *J. Cereb. Blood Flow Metab.* **1993**, *13*, 304–309. [CrossRef]
161. Brillault, J.; Lam, T.I.; Rutkowsky, J.M.; Foroutan, S.; O'Donnell, M.E. Hypoxia effects on cell volume and ion uptake of cerebral microvascular endothelial cells. *Am. J. Physiol.-Cell Physiol.* **2008**, *294*, 88–96. [CrossRef]
162. O'Donnell, M.E.; Duong, V.; Suvatne, J.; Foroutan, S.; Johnson, D.M. Arginine vasopressin stimulation of cerebral microvascular endothelial cell Na-K-Cl cotransporter activity is V1 receptor and [Ca] dependent. *Am. J. Physiol.-Cell Physiol.* **2005**, *289*, 283–292. [CrossRef]
163. Lee, A.K.; Tse, F.W.; Tse, A. Arginine vasopressin potentiates the stimulatory action of CRH on pituitary corticotropes via a protein kinase C-dependent reduction of the background TREK-1 current. *Endocrinology* **2015**, *156*, 3661–3672. [CrossRef]
164. Johansson, A.; Olsson, T.; Carlberg, B.; Karlsson, K.; Fagerlund, M. Hypercortisolism after stroke–partly cytokine-mediated? *J. Neurol. Sci.* **1997**, *147*, 43–47. [CrossRef]
165. Mavani, G.P.; DeVita, M.V.; Michelis, M.F. A Review of the Nonpressor and Nonantidiuretic Actions of the Hormone Vasopressin. *Front. Med.* **2015**, *2*, 19. [CrossRef] [PubMed]
166. Katan, M.; Fluri, F.; Morgenthaler, N.G.; Schuetz, P.; Zweifel, C.; Bingisser, R.; Müller, K.; Meckel, S.; Gass, A.; Kappos, L.; et al. Copeptin: A novel, independent prognostic marker in patients with ischemic stroke. *Ann. Neurol.* **2009**, *66*, 799–808. [CrossRef] [PubMed]
167. De Marchis, G.M.; Dankowski, T.; König, I.R.; Fladt, J.; Fluri, F.; Gensicke, H.; Foerch, C.; Findling, O.; Kurmann, R.; Fischer, U.; et al. A novel biomarker-based prognostic score in acute ischemic stroke: The CoRisk score. *Neurology* **2019**, *92*, E1517–E1525. [CrossRef] [PubMed]
168. The CoRisk Score Online Calculator. Available online: https://stroke-outcome.github.io/the-corisk-score/ (accessed on 10 January 2022).
169. Katan, M.; Nigro, N.; Fluri, F.; Schuetz, P.; Morgenthaler, N.G.; Jax, F.; Meckel, S.; Gass, A.; Bingisser, R.; Steck, A.; et al. Stress hormones predict cerebrovascular re-events after transient ischemic attacks. *Neurology* **2011**, *76*, 563–566. [CrossRef]
170. Fluri, F.; Morgenthaler, N.G.; Mueller, B.; Christ-Crain, M.; Katan, M. Copeptin, Procalcitonin and Routine Inflammatory Markers–Predictors of Infection after Stroke. *PLoS ONE* **2012**, *7*, e48309. [CrossRef]

171. Hotter, B.; Hoffmann, S.; Ulm, L.; Montaner, J.; Bustamante, A.; Meisel, C.; Meisel, A. Inflammatory and stress markers predicting pneumonia, outcome, and etiology in patients with stroke: Biomarkers for predicting pneumonia, functional outcome, and death after stroke. *Neurol. Neuroimmunol. Neuroinflamm.* **2020**, *7*, e692. [CrossRef]

172. Hotter, B.; Hoffmann, S.; Ulm, L.; Meisel, C.; Bustamante, A.; Montaner, J.; Katan, M.; Smith, C.J.; Meisel, A. External Validation of Five Scores to Predict Stroke-Associated Pneumonia and the Role of Selected Blood Biomarkers. *Stroke* **2021**, *52*, 325–330. [CrossRef]

173. Deboevere, N.; Marjanovic, N.; Sierecki, M.; Marchetti, M.; Dubocage, M.; Magimel, E.; Mimoz, O.; Guenezan, J. Value of copeptin and the S-100b protein assay in ruling out the diagnosis of stroke-induced dizziness pattern in emergency departments. *Scand. J. Trauma. Resusc. Emerg. Med.* **2019**, *27*, 72. [CrossRef]

174. Zhang, N.; Zhang, L.; Wang, Q.; Zhao, J.; Liu, J.; Wang, G. Cerebrovascular risk factors associated with ischemic stroke in a young non-diabetic and non-hypertensive population: A retrospective case-control study. *BMC Neurol.* **2020**, *20*, 424. [CrossRef]

175. Biscetti, F.; Giovannini, S.; Straface, G.; Bertucci, F.; Angelini, F.; Porreca, C.; Landolfi, R.; Flex, A. RANK/RANKL/OPG pathway: Genetic association with history of ischemic stroke in Italian population. *Eur. Rev. Med. Pharmacol. Sci.* **2016**, *20*, 4574–4580.

176. Fuellen, G.; Walter, U.; Henze, L.; Böhmert, J.; Palmer, D.; Lee, S.; Schmitt, C.A.; Rudolf, H.; Kowald, A. Protein Biomarkers in Blood Reflect the Interrelationships Between Stroke Outcome, Inflammation, Coagulation, Adhesion, Senescence and Cancer. *Cell. Mol. Neurobiol.* **2022**, 1–12. [CrossRef]

177. Meisel, A.; Meisel, C.; Harms, H.; Hartmann, O.; Ulm, L. Predicting Post-Stroke Infections and Outcome with Blood-Based Immune and Stress Markers. *Cerebrovasc. Dis.* **2012**, *33*, 580–588. [CrossRef]

178. Manaenko, A.; Fathali, N.; Khatibi, N.H.; Lekic, T.; Shum, K.J.; Martin, R.; Zhang, J.H.; Tang, J. Post-treatment with SR49059 improves outcomes following an intracerebral hemorrhagic stroke in mice. *Acta Neurochir. Suppl.* **2011**, *111*, 191–196. [CrossRef]

179. Zeynalov, E.; Jones, S.M.; Seo, J.W.; Snell, L.D.; Elliott, J.P.; Ahmad, M. Arginine-vasopressin receptor blocker conivaptan reduces brain edema and blood-brain barrier disruption after experimental stroke in mice. *PLoS ONE* **2015**, *10*, e0136121. [CrossRef]

180. Vakili, A.; Kataoka, H.; Plesnila, N. Role of arginine vasopressin V1 and V2 receptors for brain damage after transient focal cerebral ischemia. *J. Cereb. Blood Flow Metab.* **2005**, *25*, 1012–1019. [CrossRef]

181. Li-Ng, M.; Verbalis, J.G. Conivaptan: Evidence supporting its therapeutic use in hyponatremia. *Core Evid.* **2009**, *4*, 83–92. [CrossRef]

182. Zeynalov, E.; Jones, S.M.; Elliott, J.P. Therapeutic time window for conivaptan treatment against stroke-evoked brain edema and blood-brain barrier disruption in mice. *PLoS ONE* **2017**, *12*, e0183985. [CrossRef]

183. Can, B.; Oz, S.; Sahinturk, V.; Musmul, A.; Alatas, İ.O. Effects of conivaptan versus mannitol on post-ischemic brain injury and edema. *Eurasian J. Med.* **2019**, *51*, 42–48. [CrossRef]

184. Corry, J.J.; Asaithambi, G.; Shaik, A.M.; Lassig, J.P.; Marino, E.H.; Ho, B.M.; Castle, A.L.; Banerji, N.; Tipps, M.E. Conivaptan for the Reduction of Cerebral Edema in Intracerebral Hemorrhage: A Safety and Tolerability Study. *Clin. Drug Investig.* **2020**, *40*, 503–509. [CrossRef]

185. Hedna, V.S.; Bidari, S.; Gubernick, D.; Ansari, S.; Satriotomo, I.; Khan, A.A.; Qureshi, A.I. Treatment of stroke related refractory brain edema using mixed vasopressin antagonism: A case report and review of the literature. *BMC Neurol.* **2014**, *14*, 213. [CrossRef]

186. Decker, D.; Collier, L.; Lau, T.; Olivera, R.; Roma, G.; Leonardo, C.; Seifert, H.; Rowe, D.; Pennypacker, K.R. The Effects of Clinically Relevant Hypertonic Saline and Conivaptan Administration on Ischemic Stroke. *Acta Neurochir. Suppl.* **2016**, *121*, 243–250. [CrossRef] [PubMed]

187. Tsao, C.W.; Aday, A.W.; Almarzooq, Z.I.; Alonso, A.; Beaton, A.Z.; Bittencourt, M.S.; Boehme, A.K.; Buxton, A.E.; Carson, A.P.; Commodore-Mensah, Y.; et al. Heart Disease and Stroke Statistics—2022 Update: A Report From the American Heart Association. *Circulation* **2022**, *145*, e153–e639. [CrossRef] [PubMed]

Review

Deep Brain Stimulation beyond the Clinic: Navigating the Future of Parkinson's and Alzheimer's Disease Therapy

Degiri Kalana Lasanga Senevirathne [1,†], Anns Mahboob [1,†], Kevin Zhai [1], Pradipta Paul [1], Alexandra Kammen [2], Darrin Jason Lee [2,3], Mohammad S. Yousef [1,*] and Ali Chaari [1,*]

[1] Weill Cornell Medicine-Qatar, Education City, Qatar Foundation, Doha 24144, Qatar
[2] Department of Neurosurgery, Keck School of Medicine, University of Southern California, Los Angeles, CA 90033, USA
[3] USC Neurorestoration Center, University of Southern California, Los Angeles, CA 90033, USA
* Correspondence: msy2001@qatar-med.cornell.edu (M.S.Y.); alc2033@qatar-med.cornell.edu (A.C.)
† These authors contributed equally to this work.

Abstract: Deep brain stimulation (DBS) is a surgical procedure that uses electrical neuromodulation to target specific regions of the brain, showing potential in the treatment of neurodegenerative disorders such as Parkinson's disease (PD) and Alzheimer's disease (AD). Despite similarities in disease pathology, DBS is currently only approved for use in PD patients, with limited literature on its effectiveness in AD. While DBS has shown promise in ameliorating brain circuits in PD, further research is needed to determine the optimal parameters for DBS and address any potential side effects. This review emphasizes the need for foundational and clinical research on DBS in different brain regions to treat AD and recommends the development of a classification system for adverse effects. Furthermore, this review suggests the use of either a low-frequency system (LFS) or high-frequency system (HFS) depending on the specific symptoms of the patient for both PD and AD.

Keywords: neurosurgery; neurology; geriatrics; dementia; dyskinesia; aging; deep brain stimulation; neuromodulation; Parkinson's disease; Alzheimer's disease

Citation: Senevirathne, D.K.L.; Mahboob, A.; Zhai, K.; Paul, P.; Kammen, A.; Lee, D.J.; Yousef, M.S.; Chaari, A. Deep Brain Stimulation beyond the Clinic: Navigating the Future of Parkinson's and Alzheimer's Disease Therapy. *Cells* **2023**, *12*, 1478. https://doi.org/10.3390/cells12111478

Academic Editors: Alexander E. Kalyuzhny, Lydia Giménez-Llort, Masaru Tanaka, Chong Chen, Simone Battaglia and Piril Hepsomali

Received: 5 March 2023
Revised: 30 April 2023
Accepted: 16 May 2023
Published: 25 May 2023

1. Introduction

Alzheimer's disease (AD) is a neurodegenerative disorder that affects millions of people worldwide, particularly the elderly. According to a WHO report in 2019, approximately 50 million people worldwide suffer from AD, with an incidence rate for dementia of 712 cases per 100,000 people globally in 2016 [1].

Most cases of AD are idiopathic, but studies have isolated certain risk factors that serve for the diagnosis of AD; these include mainly old age, genetic predisposition, and a healthy lifestyle (encapsulating dietary misbalance resulting in obesity and dyslipidemia, physical activity, marital status, and sleep) [2]. Moreover, other disease-related risk factors could include hypertension, cerebrovascular diseases, depression, and even microwave exposure [2,3]. The pathogenesis of AD is complex and involves several processes, including the hyperphosphorylation and oligomerization of tau protein, accumulation of amyloid beta (Aβ) plaques [4], and loss of cholinergic neurons and synapses [5]. These changes lead to progressive cognitive decline coupled with executive function depreciation as well as behavioral and psychiatric symptoms [4].

Parkinson's disease (PD) is another neurodegenerative disease characterized by amyloid aggregation and deposition. It involves degeneration of dopaminergic neurons in the substantia nigra (SN) and intraneuronal formation of protein aggregates called Lewy bodies and Lewy neurites [6]. PD is a burdensome disease, with an annual incidence of 5–30 cases per 100,000 people globally [7]. Whilst most cases of PD are idiopathic, risk factors have been identified. Namely, genetic factors which increase susceptibility alongside rural living are most important, as well as pesticide usage and exposure to the chemical MPTP, which

have been marked as significant risk factors [8,9]. More recently, other risk factors such as dairy consumption, cancer, usage of methamphetamine, traumatic brain injury, diabetes, cholesterol, alcohol, and hypertension have been linked to an increased risk of developing PD. Clinically, PD presents with progressive motor decline, including asymmetric resting tremor, cogwheel rigidity, and bradykinesia, as well as nonmotor symptoms such as anosmia, depression, and cognitive decline [10].

Despite their distinct clinical presentations, AD and PD share similar underlying pathophysiological mechanisms. Both are protein-misfolding disorders that result in the aggregation of misfolded amyloid proteins in neurons. In AD, tau (τ) and Aβ proteins are involved, while α-synuclein plays a major role in PD [11–13]. Additionally, both conditions are associated with dementia, and both PD and AD dementia have similar symptoms that may be relieved by deep brain stimulation (DBS) [12,14].

Current pharmacologic therapies for both disorders are primarily symptomatic and do not significantly impact the underlying disease processes [15,16]. These drugs can also have undesirable side effects in addition to being ineffective for large demographics of the patient population, especially younger patients and those who have worse clinical prognoses and disease progression. Therefore, there is a need for more effective disease-modifying therapies with a favorable side-effect profile. DBS is a potential treatment option that has shown promise in this regard.

This review will examine the development and current use of DBS in the treatment of PD with the aim of using the knowledge gained from studying PD to apply similar protocols in the treatment of AD patients. Currently, a large body of research points to similarities in the clinical pathologies of PD and AD, though studies have yet to draw a parallel between DBS use in PD and AD. As such, this review considers the potential of DBS as an effective treatment to both PD and AD patients, what has currently been found from the use of DBS as a licensed treatment for PD, and what translational qualities can be transferred to aid in the development of DBS for AD. Moreover, this review aims to highlight future developments that may improve the efficacy of the procedure and develop a series of recommendations for the use of DBS in the treatment of these two neurodegenerative diseases.

2. Search Strategy

The literature for this review was obtained from various sources including PubMed, Google Scholar, and Scopus, using search terms including "deep brain stimulation", "Parkinson's Disease", "Alzheimer's Disease", and combinations of these terms. The literature was included in a narrative manner to highlight the development of DBS, its current application, and its future in the treatment of patients. Human trials were obtained from Google Scholar, Embase, and the United States National Library of Medicine's clinical trials database. The information extracted included the region of implantation, laterality of the procedure, configuration parameters used for the DBS systems, duration of the study, number of patients involved, and current status of the study.

3. Deep Brain Stimulation

DBS is an invasive but established neurosurgical procedure that involves the implantation of one or more electrodes into a targeted brain area, an implantable pulse generator (IPG), and an extension connecting the electrode to the IPG [16]. The IPG contains the battery and circuitry, which generate the electrical signal that is delivered to the targeted brain structure. The DBS system allows for the delivery of electrical pulses to specific areas of the brain with minimal effects on nearby regions [17].

3.1. History of DBS

DBS is a relatively novel procedure, with most of the research conducted in the past 20 years. Prior to its development, surgical solutions were limited to ablation and the removal of affected brain regions, collectively known as craniotomy [18]. An example of such a procedure is thalamotomy, which entails the excision of the thalamus from the brain or sections of it [19]. While these procedures alleviated bradykinesia (slowness of movement) and dyskinesia (erratic excessive movement), they had major drawbacks including aphasia and dysphagia [20]. Even with the development of more precise surgical procedures such as stereo lesioning [21], the potential for adverse effects from the removal of brain tissue underscores the need for less invasive, better-tolerated therapies.

The use of DBS dates back to 1964, when Ohye et al. used a stereotactic procedure to stimulate the ventrolateral thalamic nucleus [22]. Since then, the volume of research on DBS has steadily increased (Figure 1). Initially, DBS was intended to treat PD and other movement disorders, and the FDA approved its use for PD and essential tremor in 1997 [23]. However, funding for DBS research declined following the development of levodopa in 1969, which may explain the lack of interest until the late 1990s [20]. The approval of thalamic DBS for the treatment of Parkinsonian symptoms in 1997 reignited interest in the field, and the potential applications of DBS in the treatment of other conditions such as AD, dystonia, epilepsy, and psychiatric disorders are currently being heavily researched [20]. As evident in Figure 1, while a significant amount of research has been conducted on the effectiveness of DBS treatment in PD, AD has received far less attention (Figure 1). Thus, further research on the potential use of DBS in AD is necessary.

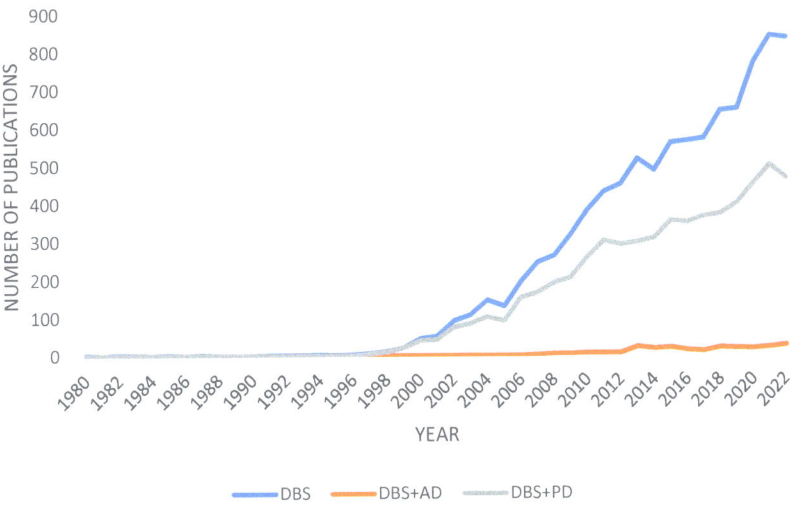

Figure 1. A graph of the number of research articles published on deep brain stimulation between 1980 and 2022. The queries performed were "DBS and deep brain stimulation", "DBS and deep brain stimulation and AD and Alzheimer's Disease", and "DBS and deep brain stimulation and PD and Parkinson's Disease". Search parameters where limited to articles, books, and book chapters. Data extracted from Scopus: https://www.scopus.com (Accessed on 15 December 2022).

3.2. Current Treatment Plans

DBS can be used to modulate almost all regions of the brain for numerous neurologic conditions. The treatment process begins with the implantation of wires that taper into electrodes in the brain using a precise stereo lesioning procedure [16,21]. Different brain regions can be stimulated to alleviate the effects of different conditions (Table 1).

Table 1. A selection of conditions in which DBS is being actively researched as a viable treatment option.

Condition	Area of Stimulation	First Author	Year of Publication	References
Parkinson's Disease	Subthalamic nucleus (STN)	Deuschl, Xie	2006, 2015	[24,25]
	Globus Pallidus internus (GPi)	Bloomstedt, Baker	2018, 2010	[26,27]
	Zona Incerta	Bloonstedt, Ossowska	2018, 2020	[26,28]
	Pedunculopontine nucleus	Thevathasan	2018	[29]
Essential Tremor	Ventral intermediate thalamus	Baizabal-Carvallo	2014	[30]
	Zona incerta	Fytagoridis	2012	[31]
Dystonia	GPi	Vidailhet	2013	[32]
Alzheimer's Disease	Fornix			[33,34]
	Ventromedial prefrontal cortex	Mao, Lozano	2018, 2016	[35,36]
	Hippocampus	Scharre, Chakravarty	2018, 2016	[37]
	Nucleus Basalis Meynert	Kuhn	2014	[38]
Huntington's Disease	GPi	Velez-Lago	2013	[39]
Obsessive Compulsive Disorder (OCD)	Anteromedial GPi	Nair	2014	[40]
	Nucleus Accumbens	Huff, Denys	2010, 2010	[41,42]
	Anterior limb of the internal capsule (ALIC)	Denys, Kammen	2020, 2022	[43,44]
	Ventral Capsule/Ventral Striatum (VC/VS)	Park, Kammen, Greenburg	2019, 2022, 2008	[44–46]
	STN	Kammen, Li, Chabardes	2022, 2020, 2013	[44,47,48]
	Inferior thalamic peduncle	Germann, Lee, Kammen	2022, 2019, 2022	[44,49,50]
	Bed nucleus of the stria terminalis	Luyten, Mosley, Raymaekers, Kammen	2015, 2021, 2016, 2022	[44,51–53]
Epilepsy	Anterior nucleus of thalamus	Salanova	2018	[54]
Depression	VC/VS	Malone	2009	[55]
	Nucleus Accumbens	Bewernick	2010	[56]

Once the electrodes are positioned into the desired area of the brain, a separate implantable pulse generator (IPG) device is implanted in the chest wall with leads running under the skin [16]. The IPG can be adjusted externally to deliver a range of frequencies, pulse widths, and amplitudes to the targeted brain region unilaterally and bilaterally [20]. Different frequencies have been used to target specific symptoms, with high frequency stimulation (HFS) defined as above 100 Hz and low frequency stimulation (LFS) as 60–80 Hz [57–59]. However, this is quite subjective to each disease, as some papers cite that the range of 10–70 Hz is generally avoided due to the risk of eliciting seizures for epilepsy studies [59]. Some studies have found that frequencies around 60 Hz have restorative effects on PD symptoms, such as reducing aspiration and improving gait [25,60]. In particular, it is noted by di Blaise et al. that HFS systems decrease the levodopa-responsive PD symptoms, whereas LFS systems decrease more axial symptoms of the disease [58]. Indeed, Ramdhani et al. found that PD patients who experienced worsening of their symptoms after receiving HFS DBS had improvements in freezing of gait and dyskinesia when they were switched to a 60 Hz LFS system [60]. Contrastingly, Wyckhuys et al. discovered that HFS systems of 130 Hz are more effective at increasing the threshold and latency of after-discharges in kindled rats, a common animal model of epilepsy [59]. These studies suggest that the use of HFS and LFS systems is dependent on the type of disease that is to be treated, as well as the specific symptoms to be alleviated. The next section details the exact surgical methods used to implant the DBS system in patients.

3.3. Patient Screening

Given the wide range of neurological disorders that can be treated by DBS, it is important to carefully select candidates for the treatment. One important factor to consider is the length of time since the patient's diagnosis. There is conflicting evidence whether DBS is more effective for patients who were diagnosed within the past 5 years or for those who were diagnosed earlier. For PD, some studies suggest that earlier treatment may be more beneficial [61], while others suggest that it is important to monitor patients over a longer period of time to identify atypical symptoms and assess treatment suitability [62]. Moreover, surgically invasive procedures are usually left as a last-resort treatment and thus are usually used on older patients who have been diagnosed for longer. However, this approach benefits from the ability to rule out other "Parkinson-plus" diseases which may not be clinically treatable through DBS, such as comorbidities of PD and other diseases with similar symptoms [61].

Patients being considered for device-assisted therapies must undergo evaluation by a specialized center's multi-disciplinary team, which typically includes a functional neurosurgeon, an anesthesiologist, a movement disorder neurologist, and a neuropsychologist, as well as representatives from departments such as radiology, psychiatry, physical therapy, nursing, and social work.

Another factor to consider when selecting candidates for DBS is the patient's responsiveness to levodopa, a common medication used to treat PD. Many studies have found that patients who are responsive to levodopa are more likely to also respond to DBS [62,63], while some others have found conflicting evidence [64,65]. The site of stimulation may affect the relationship between levodopa responsiveness and DBS efficacy, but the reason for this is not well understood. Lin et al. found a positive correlation between levodopa responsiveness and GPi-DBS efficacy (R^2 = 0.283, p = 0.016), though STN DBS was more efficacious than GPi-DBS for levodopa-resistant tremor control [66]. There is no current indication of how medication for AD may interact with DBS treatment, but co-therapy to mitigate the side effects is certainly something future clinical trials could investigate. In the future, online tools may be developed to help improve the selection process for DBS candidates and provide more uniformity in the selection process. One study found that the use of an online tool called Stimulus to screen 3128 patients led to a significantly higher acceptance rate than conventional multi-disciplinary screening [67].

3.4. Procedure and Mechanism of Action

The DBS hardware includes multiconductor intracranial quadripolar electrodes, a programmable single- or dual-channel internal pulse generator with battery unit, and an extension cable connecting the DBS electrodes to the pulse generator [20].

The procedure involves stereotactic placement of electrodes in the target area, either unilaterally or bilaterally (Figure 2). This is the first stage of the procedure and is often performed while the patient is awake. In the second stage, the electrode and extension cable are tunneled under the skin to the infraclavicular area, where they are connected to the battery-powered pulse generator [44].

When activated, the pulse generator delivers electric stimulation to the targeted area. The exact mechanism of action is not fully understood, but it is hypothesized that high-frequency DBS in PD suppresses GPi neuronal activity and efferent fiber pathways. This disrupts the flow of abnormal information through the cortico-basal ganglia circuits and downstream pathways, improving symptoms [68,69].

Overall, DBS as a procedure has come a long way since its initial development in the 1960s. Moreover, it is increasingly being trialed as an efficacious invasive procedure to alleviate the symptoms of many diseases. Additionally, more data are being generated on patient criteria which allows them to undergo ameliorative DBS treatment as well as the potential mechanisms by which DBS may affect different neurodegenerative diseases. Currently, the DBS only has regulatory approval in the treatment of PD, which will be explored in the next section.

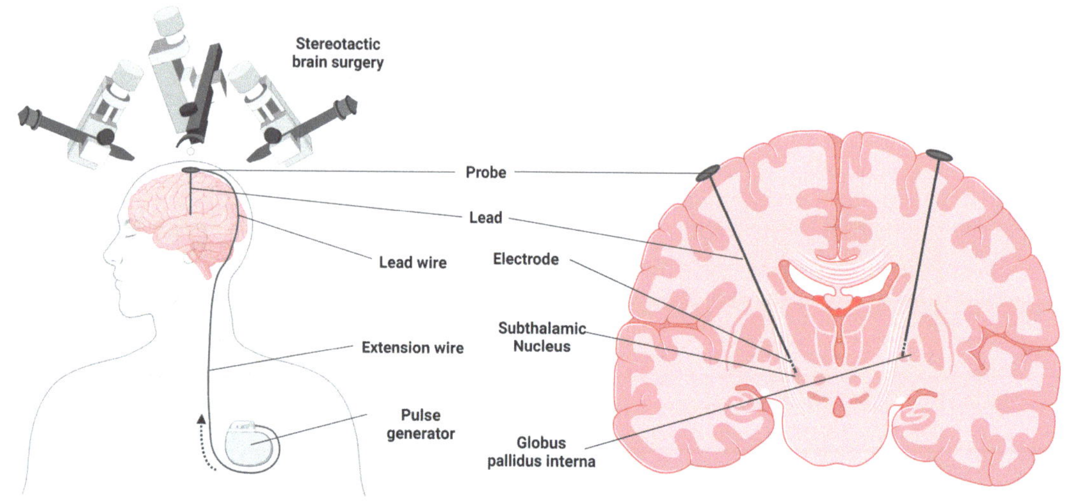

Figure 2. Electrode implantation for deep brain stimulation in Parkinson's disease. Created using https://BioRender.com (Accessed on 19 December 2022).

4. DBS in Parkinson's Disease

PD is a complex and progressive neurodegenerative disease that is primarily characterized by motor symptoms. These motor symptoms may include a slowly progressive asymmetric resting tremor, stiffness in the muscles (cogwheel rigidity), and difficulty with movement and coordination (bradykinesia). Other common nonmotor symptoms associated with PD include loss of sense of smell (anosmia), constipation, depression, and sleep disorders. Autonomic dysfunction, pain, and cognitive decline may occur in the later stages of disease [10]. This section explores our current understanding of PD pathophysiology and how DBS has been found to affect its mechanisms at a pathological and symptomatic level. Moreover, the parameters for use of DBS on patients and the effect of age are discussed, along with currently identified adverse effects of treatment.

4.1. Pathology of PD

In terms of neuropathology, PD involves the degeneration of dopaminergic neurons in the substantia nigra (SN) and the presence of intraneuronal protein aggregates called Lewy bodies and Lewy neurites [6]. Recent studies have suggested that loss of dopaminergic terminals in the striatum, rather than the SN, may be more responsible for the motor symptoms observed in PD [70]. Lewy pathology, which is the microscopic evidence of the misfolding of the α-synuclein protein [71], is a characteristic finding in most cases of PD (some rare genetic forms of PD involve the loss of striatal dopaminergic neurons without these protein aggregates [72]). α-synuclein protein is commonly found in synapses, where it plays a role in the function of synaptic vesicles; it is also found in non-neuronal cells such as hepatocytes, myocytes, lymphocytes, and erythrocytes, although its functions in these cells are not yet fully understood [73].

Braak and colleagues proposed a six-step pathological pathway for Lewy pathology in PD; the first stage is limited to the dorsal motor nucleus of the vagal nerve, while the proteins progressively spread to the rest of the brain in the subsequent stages [74]. These conclusions were based on post-mortem observations and do not have contemporary evidence to support them; indeed, recent observations suggest that clinically diagnosed PD patients may have a different pathology compared to what the Braak stages describe [75]. Nonetheless, Braak et al.'s model has gained popularity in recent years, and the earliest stages of Lewy pathology (before the aggregates reach the SN) are thought to be linked to the signs and symptoms of premotor PD [75].

Different α-synuclein assemblies can be secreted by neurons, and this process is upregulated if the lysosomal–autophagy system is inhibited. These assemblies can then be taken up by nearby neurons, where they can seed monomeric α-synuclein into Lewy-like aggregates. This prion-like property of α-synuclein assemblies may explain how Lewy pathology propagates between brain regions [76]. However, the cause of this aggregation is unknown. One theory is that environmental substances such as pesticides and pollutants, as well as pathogens, can enter α-synuclein-containing cells in the olfactory system and gastrointestinal tract. In a healthy individual, these aggregates would be mitigated by cellular proteostatic mechanisms and not lead to the spread of Lewy aggregates; however, in the presence of factors such as aging, genetic predisposition, and peripheral inflammation, it is proposed that α-synuclein may bypass the cell's normal clearance mechanism and cause Lewy aggregates to form in the brain [77].

4.2. Current Treatments for PD

While there is currently no truly disease-modifying treatment for PD, there are several drug therapies available for symptom management [78]. These include dopamine medication, the most common of which is levodopa. These dopamine agonists can be taken orally or through injection and are a preferred treatment for PD patients, especially in the later stages of the disease. However, levodopa loses effectiveness over time, requiring increasing dosage, and can cause side effects [78]. There are also monoamine oxidase inhibitors (MAOIs) and antagonists that can be taken orally and function to prevent the breakdown of levodopa and dopamine [79]. These drugs do not need to have their doses increased over time and have milder side effects compared to levodopa. However, these are not as effective in later stage PD and usually need to be supplemented with another group of drugs that have their own associated symptoms [78].

Another set of oral medications, anticholinergics, reduces acetylcholine activity at choline receptors and can be used as a monotherapy in the early stages, but there is reduced tolerance in elder patients and limited pharmacokinetic information available [80]. In addition to these medications, some surgical treatments, such as ablative surgery, are also used, although their side effects and symptoms may not be reversible [78].

4.3. Use of DBS in the Treatment of PD

The surgical treatments of PD have evolved over time. The early 1940s saw surgical procedures focused on lesioning the thalamus and GPi. These treatments, known as pallidotomy, became common before the use of levodopa became widespread. Later, surgical procedures saw a resurgence due to the side effects of levodopa. In the 1990s, DBS emerged as a replacement for lesioning treatments due to the adverse effects associated with bilateral lesions and the irreversible side effects of poorly placed lesions [81].

There have been many clinical trials that have studied the use of DBS in the treatment of PD (Figure 3). Some of these trials have compared DBS combined with medical therapy to medical therapy alone [82], while others have focused on the differences in effects seen when different brain regions are stimulated [83]. Moreover, several clinical trials have looked at how DBS treatment for PD may affect patients of different ages [84]. These studies combined seem to suggest that DBS can be an effective therapy for PD [85]. However, it is worth noting that many of these trials are not double-blind and have relied on open-label data from individual institutions [85]. Further clinical trials, with more objective rating scales and controls, could provide stronger evidence for the effectiveness of DBS in PD treatment.

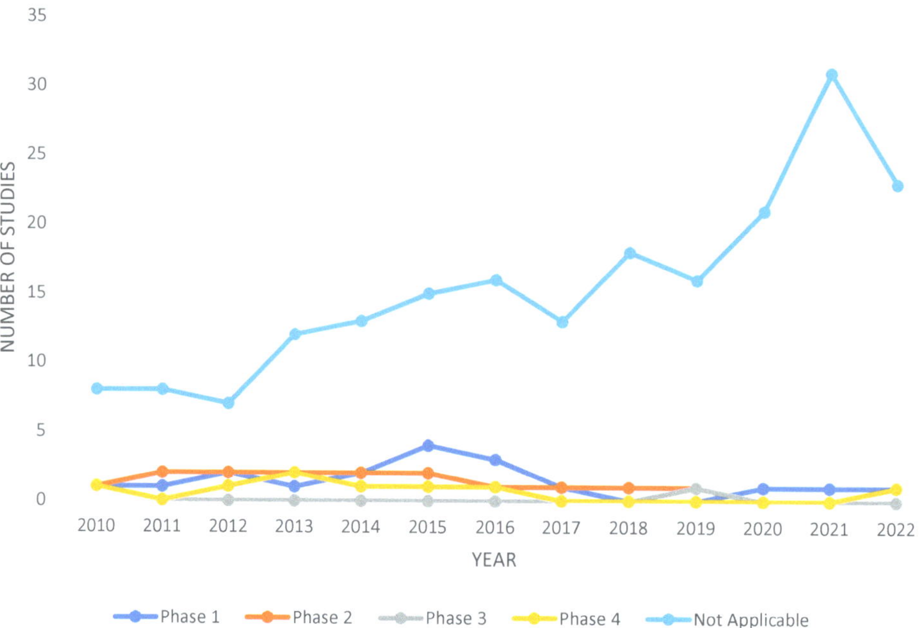

Figure 3. A summary of clinical trials that have been completed or are currently active on DBS and PD. **Phase I trials** focus on the safety of a procedure and involve a small number of healthy volunteers. They are designed to determine the most serious and common adverse side effects. **Phase II trials** are the first time that a procedure is tested on the relevant patient population. Adverse side effects continue to be evaluated. **Phase III trials** involve a larger population of patients and are used to evaluate the correct doses and criteria for successful use of a procedure for the relevant affected population. **Phase IV** trials are conducted after a treatment has been approved by the FDA and are used to continue to evaluate the efficacy, best dose, and safety of the approved treatment. **Not applicable**: these trials do not fit into the FDA-defined phases and may include trials of devices or behavioral interventions. All data were collected from https://www.clinicaltrials.gov/ (Accessed on 20 December 2022).

DBS can be used to stimulate different brain regions in PD patients, and each can have different effects on the symptoms of the disease. Stimulation of the GPi may reduce motor symptoms [86] and also alleviate painful dystonia and sensory symptoms, but it does not usually allow for a reduction in medication usage [87]. Stimulation of the STN can reduce the need for dopaminergic medication [88] and also improve gait, tremor, and bradykinesia [89,90]. However, stimulation of the STN may not address all of the major motor symptoms of PD.

4.4. Mechanisms of Action of DBS on PD

The exact mechanism by which DBS works in the treatment of PD is not fully understood and is a subject of ongoing research and debate [91]. Ongoing studies are attempting to understand the communication between brain structures at various levels (subcellular, neuronal, and fiber-pathway levels), how DBS modulates dysfunctional pathways in the brains of PD patients, and the plastic changes that occur in the brain due to DBS [92]. While there is still much to learn about the mechanisms of DBS in PD, some theories have been proposed [92]. One theory suggests that DBS works by inhibiting the activity of target neurons through a variety of means such as blocking the transmission of nerve impulses, disrupting the release of neurotransmitters such as glutamate, or increasing the release of inhibitory neurotransmitters such as GABA and adenosine [92–95]. Another theory proposes that DBS activates target neurons by increasing the levels of certain neurotransmitters such as glutamate [96,97]. A third theory suggests that DBS may both inhibit and activate target neurons by affecting the communication between the cell body and axons of neurons [70,71]. Additionally, DBS may disrupt abnormal oscillatory patterns by replacing irregularly firing cells with regularly firing cells, producing "jamming" signals that can lead to the release of neurotrophins and the generation of new neurons [98]. The varied and sometimes conflicting results of studies on the mechanism of DBS may be due to the many experimental variables that can affect research outcomes, such as the different approaches, types of stimulation, and time intervals at which the effects are observed.

4.5. Criteria for Successful Treatment of DBS in PD

The success of DBS for treating PD is highly dependent on careful selection of patients. Over 30% of DBS failed attempts are attributed to an inadequate selection process [99]. According to a 2007 survey, young patients with (1) good levodopa response, (2) no or few axial non-levodopa motor responsive symptoms, (3) few or mild cognitive impairments, and (4) absent (or well-controlled) psychiatric disease are best suited for DBS treatment for PD [100]. Admittedly, strict adherence to these criteria may exclude many patients who could still benefit from the surgery. The most reliable predictor of success with DBS is patients' response to levodopa [101]. A 30% improvement in the Unified Parkinson's Disease Rating Scale III (UPSDIII) score is commonly used as a marker of levodopa responsiveness, with severe tremor resistance to levodopa therapy being an exception [102]. Patients with atypical forms of parkinsonism and dementia tend to have less favorable outcomes with DBS [102,103]. Additionally, DBS surgery is delayed for patients with psychiatric symptoms, until their symptoms are properly managed [81]. Furthermore, the prior experience of the surgical team is also a key factor in determining the outcome of DBS surgeries. It is recommended that DBS programming is best performed by a highly trained clinician with expertise in both DBS technology and PD-related issues as well as pharmacological management [103].

4.6. Parameters for DBS use in PD

The optimal parameters for DBS in treating PD are not well understood and can be challenging to determine. Some studies suggest that a frequency around 60 Hz is best for use in PD [25,60], but there is still debate among experts in the field. Currently, the widely accepted range for treating parkinsonian symptoms is between 130 and 180 Hz [104,105], though emerging research has shown that long-term HFS treatment may not treat axial symptoms of the disease or even be deteriorative [106]. A consensus of experts has agreed that the relevant parameters of pulse width, frequency, voltage, and electrode configuration need to be optimized within 3 to 6 months during four to five programming sessions to ensure maximum benefit to patients [81]. However, finding the optimal parameters is considered as much of an "art" as it is a science [92]. There are several ways to improve the process of creating optimal parameters [92], including careful imaging to identify the exact electrode contacts in the brain, modeling the spread of electrical fields [98], identifying physiological changes associated with stimulation that show promise for long-term benefit, and automating the optimization process to reduce the risk of error and the time required [107].

4.7. Effect of Age on the Effectiveness of DBS on PD

The appropriateness of DBS as a treatment option for Parkinson's disease in elderly patients has been a topic of debate [81]. Historically, DBS has been considered a viable treatment option for patients under the age of 70, while those older than 70 have been excluded. However, Mathkour et al. evaluated the short- and long-term outcomes of DBS in patients with PD who were older than 70 years old and underwent DBS [108]. The study found a significant decrease in UPDRS III score (preoperative 31.8 to postoperative 15.6;

$p < 0.0001$) as well as a significant reduction in medication doses per day (preoperative 11.54 to postoperative 7.97; $p = 0.0112$) [108].

A separate study compared the long-term outcomes of DBS in Parkinson's disease patients that were young and old [109]. Both groups experienced a significant decrease in levodopa-equivalent dose daily (LEDD), with the elderly group seeing a more significant decrease [109]. Moreover, both groups had a reduction in UPDRS III score, which in both groups indicated significant improvements in motor function. Furthermore, elderly patients experienced a greater reduction in daily doses compared to the younger group. These results indicate that DBS can be beneficial for both younger and older patients [109]. A higher incidence of negative side effects or follow-up issues with older patients may be attributed to age-related comorbidities such as cognitive decline [102], a higher incidence of levodopa-resistant symptoms [84], and a higher overall risk of surgical complications [7].

4.8. Adverse Effects of DBS in PD

Currently, there exists no unified framework in which to describe the adverse effects of DBS, though some studies have tried to create such a system and put it into practice [110]. However, adverse effects can definitely be seen with the use of both STN and GPi DBS. Broadly, these issues can be categorized into procedure-related, hardware-related, and stimulation/disease-progression-related issues [110].

Procedurally, surgical implantation has its risks given the invasiveness of the procedure. Indeed, complications of the procedure may include intracranial hemorrhage (0–10%), stroke (0–2%), infection (0–15%), lead erosion without infection (1–2.5%), lead fracture (0–15%), lead migration (0–19%), and death (0–4.4%), according to Bronstein et al. [81]. Overall, STN-DBS has demonstrated a higher rate of surgical complications in the studies assessed by Videnovic et al., though this may be attributed to the larger sample size of patients who underwent STN-DBS treatment [110].

In terms of disease progression, patients with STN-DBS required less dopaminergic agents than those who received GPi-DBS. However, patients undergoing GPi-DBS showed an improvement in levels of depression post-operatively and decreased loss of visuomotor processing speed than STN-DBS patients in the long term [81,111]. Moreover, patients also reported a high incidence rate of weight gain (37.5%), speech disturbance (12.8%), eyelid opening apraxia (11.3%), and cognitive decline (5.8%) in GPi-DBS, though STN-DBS patients reported an uncharacteristically high gait ignition failure rate (17.6%) [110].

Long-term effects of DBS on PD include improvement in motor fluctuations and tremors for up to five years [112–114], but eventually patients may develop symptoms that are levodopa-resistant including freezing of gait, postural instability, and cognitive decline [81]. As such, it can be said that both forms of treatment have their individual advantages and drawbacks, though more research is required into GPi-DBS and its adverse effects to have a fair comparison between the two types of therapies.

In summary, DBS has shown versatility in the treatment of patients with PD. The GPi and STN regions have been identified as positive targets for DBS therapy. Moreover, there is a growing body of research which highlights the effect of DBS on the pathology of PD as well as marked improvements in the symptoms of patients who have undergone DBS treatment. While broadly applicable standards for settings to use in treatment and age eligibility criteria still remain elusive, the identification of possible side effects and their mitigation has allowed DBS to be established as a versatile technology in the treatment of PD. As such, this knowledge base provides the foundational basis to explore the use of DBS in AD treatment, which will be discussed in the next section.

5. DBS in Alzheimer's Disease

AD is a neurodegenerative disorder that affects memory and cognitive functions. It is the most prevalent type of dementia, accounting for the majority of dementia cases worldwide [115]. AD is characterized by a gradual decline in memory and cognitive abilities. Imaging studies, such as magnetic resonance imaging (MRI), can show shrinkage of the hippocampus, an area important for memory. Additionally, biomarkers such as decreased levels of the protein amyloid $\beta42$ in the cerebrospinal fluid and the presence of phosphorylated τ proteins detected by positron emission tomography (PET) can also indicate the presence of AD [116]. This section explores our current understanding of AD pathology, conventional methods of treatment, and what DBS has to offer. Moreover, selection criteria for patients and potential adverse effects found in the present literature are also reviewed.

5.1. Pathology of Alzheimer's Disease

AD and PD are similar in that both are caused by protein misfolding. In AD, the aggregation of extracellular $A\beta$ and intracellular hyperphosphorylation of τ protein leads to the formation of plaques and neurofibrillary tangles (NFTs) [13]. This results in the loss of cholinergic neurons and synapses [115].

The cause of $A\beta$ aggregation is not fully understood, but genetic factors such as missense mutations in the genes encoding amyloid precursor protein (APP), presenilin 1 (PSEN 1), presenilin 2 (PSEN 2), and apolipoprotein E (APOE) are known to increase the risk of AD [13,115]. In particular, early onset AD is often inherited and caused by mutations in the APP, PSEN 1, or PSEN 2 genes [13]. While there is ongoing debate as to what these genes code for, it is well established that these mutations can lead to γ-secretase forming more $A\beta$ fibrils [117]. Late-onset AD is associated with mutations in the APOE gene, which increases the risk of vascular dementia, Lewy body dementia, and others [115].

Regardless of the cause, the overproduction of $A\beta$ leads to its aggregation as plaques extracellularly, though the exact mechanism of its exocytosis remains unclear [118]. This leads to the hyperphosphorylation of τ protein by the dysregulation of kinases such as CDK-5 and GSK-3β by $A\beta$ fibrils [119]. This in turn leads to conformational changes in τ, resulting in the formation of NFTs inside the neuron [119]. $A\beta$ also triggers immune responses in the microglial cells through toll-like receptors, leading to inflammation, receptor-mediated phagocytosis, and cell clearance [120]. These mechanisms ultimately result in decreased brain weight and neuronal loss, particularly in the white matter and hippocampus [115]. DBS is being evaluated as a potential treatment to address these mechanisms.

5.2. Current Therapeutics in AD

The primary physiological feature finding of AD is the presence of NFTs and $A\beta$ fibril plaques. As a result, most currently FDA-approved drugs aim to prevent their formation, thereby reducing symptoms. Currently, the FDA has approved three acetylcholinesterase enzyme (AChE) inhibitors: donepezil, galantamine, and rivastigmine, as well as one N-methyl-D-aspartate (NMDA) receptor antagonist: memantine [121,122].

AChE inhibitors function by addressing cognitive dysfunction in AD caused by the loss of cholinergic nerves in the brain [123]. They perform this by inhibiting the enzyme acetylcholinesterase that breaks down acetylcholine, leading to increased levels of the neurotransmitter in cholinergic neurons [122]. This approach has been shown to be effective in clinical trials but may only halt or slow the rate of cognitive decline and not address underlying neuronal loss and brain atrophy [124,125]. NMDA receptor antagonists, on the other hand, prevent excitotoxicity and cell death by inhibiting the rapid influx of Ca^{2+} ions into neurons [121]. They can also mediate neurotoxicity induced by the presence of glutamate, thereby preventing neuronal death [122]. However, this treatment is typically used for mild to moderate cases of AD [121].

Recently, the FDA has also partially approved a drug called Aducanumab developed by Biogen, which has been somewhat controversial [126,127]. This drug involves injecting monoclonal antibodies that bind to $A\beta$ and its aggregates, allowing for their removal by the body [122]. This is the first disease-modifying therapy approved for AD patients [127]. However, some studies have pointed out that that this drug focuses heavily on the $A\beta$ aggregate pathology without taking into consideration τ protein aggregation in NFTs, and the effects on cognitive decline in phase III trials were limited [122,126]. Unfortunately, anti-τ therapies are mostly still in the early stages of clinical research, and it is uncertain how they will fare in further testing [122]. As such, an alternative approach to address AD has been found in DBS treatments.

5.3. Use of DBS Treatment in AD

DBS treatment for AD is not yet approved by the FDA or other regulatory bodies. However, several clinical studies are being conducted to better understand the effects of DBS on humans. These studies primarily focus on stimulating specific areas of the brain such as the fornix, hippocampus, and nucleus basalis of Meynert (NBM). The results of these studies are summarized in Table 2:

Overall, the clinical trials presently completed paint a positive outlook for the use of DBS in the treatment of AD. Specifically, phase I and II trials conducted on AD patients using bilateral fornix DBS have been shown to have positive outcomes on symptoms of AD [34,37]. Moreover, present clinical trials have also shown good tolerance and a lack of negative outcomes from the bilateral stimulation of the NBM region. Most importantly, DBS treatment seems to have halted or even reversed cognitive decline among patients undergoing treatment, highlighting a potentially bright future for the use of NBM-DBS in the treatment AD patients [38].

Table 2. Summary of all clinical trials run on DBS and AD research. The data were extracted from https://www.clinicaltrials.gov (Accessed on 19 December 2022) with 16 data points returned filtered to 14 research projects upon analysis. Abbreviations used include not disclosed (ND), year of publication (YoP), principal investigator (PI), and not published (NP).

Brain Region	Laterality	Stimulus Settings	Duration (Months)	Patients	Trial Status	YoP	Author/PI	Reference
-Fornix	Bilateral	3.9–7.5 mA, 90 µs, 130 Hz	24	1	Completed	2022	Barcia	[128]
-Fornix	ND	3.0–3.5 V, 90 µs, 130 Hz	12	6	Completed (phase I)	2010	Laxton	[37]
-Fornix	Bilateral	3.0–3.5 V, 90 µs, 130 Hz	12	42	Completed (phase II)	2016	Lozano	[34]
-Fornix	ND	ND	12	12	Active, not Recruiting	NP	Lozano	[129]
-ND	Bilateral	ND	23	3	Completed	NP	Rezai	[130]
-ND	ND	ND	12	10	Recruiting	NP	Luming	[131]
-Fornix	Bilateral	1–5 V, 90 ms, 130 Hz	12	6	Completed (phase I)	2018	Mao	[33]
-NBM	Bilateral	2.0–4.5 V, 90 µs, 20 Hz	12	6	Completed	2014	Kuhn	[38]
-NBM	ND	2.0–4.5 V, 60 µs, 20 Hz	12	30	Recruiting	NP	Chen	[132]
-NBM	Bilateral	ND	ND	6	Completed	NP	Sturm	[133]
-Hypothalamus-Fornix	Bilateral	2–3 V, 120 ms, 180 Hz	24	5	Recruiting	NP	Fontaine	[134]
-Fornix	Bilateral	ND	12	6	Completed	NP	Laxton	[135]
-Fornix	ND	ND	12	210	Recruiting	NP	ND	[136]
-Fornix-NBM	ND	ND	12	30	Recruiting	NP	ND	[137]

Additionally, only one trial has investigated the use of DBS to stimulate the hypothalamus, which has been found to be heavily affected by AD pathology [134]. There is also a lack of research using in vivo techniques in rat models to study the effectiveness of DBS in treating AD symptoms, although some studies have shown promising results for fornix DBS in transgenic mice models [138,139]. Research on the neuroprotective effects of DBS on the NBM has also been conducted in rat models with positive results [140,141]. Therefore, further research using in vivo techniques to identify suitable regions of the brain for DBS and progression of existing clinical trials for fornix DBS would be beneficial in obtaining approval for DBS as a treatment for AD. Additionally, understanding the underlying mechanism of how DBS affects AD pathology would help in identifying areas of the brain to target, which will be discussed in the next section.

Many of the completed studies above are relatively outdated and have small sample sizes, with the other clinical trials currently only in the recruiting phase. Moreover, many studies have yet to disclose certain vital information regarding the procedures conducted such as laterality and the specific settings used for stimulation therapy. As such, further transparency in the clinical studies in progress and further exploration of other potential sites for DBS stimulation would go a long way to advance the present paucity in the literature.

5.4. Mechanisms of Action of DBS in AD

DBS can affect the underlying pathology of AD in various ways, although the majority of these mechanisms are still being researched. Currently, it is known that DBS increases neuronal activity in the Papez circuit of the brain by activating neurons in the hippocampus, parahippocampal gyrus, and default mode network (precuneus, parietal, and temporal lobe) [142]. The most common target for DBS in AD is the fornix within the hippocampus, which has been shown to increase glucose metabolism and utilization in the cortico–thalamic and cortico–hippocampal networks [143]. This is associated with better clinical outcomes and also mitigates neuronal loss and synapse reduction [144], leading to greater hippocampal volume [142].

Stimulation of the NBM appears to have similar effects on fornix DBS. Cholinergic neuron degeneration in the medial forebrain is a characteristic sign of AD. Multiple phase I clinical trials have shown that stimulation of the NBM can result in a slowed increase their Alzheimer's Disease Assessment scale (ADAS) scores and stabilization of Mini-Mental State Examination (MMSE) or Clinical Dementia Rating (CDR), increased cortical glucose uptake, and decreased motor disability [38,145]. Additionally, it is hypothesized this may be the result of increased acetylcholine levels in the cortex, leading to improved cognitive functions [38].

A third potential treatment of DBS in AD is the stimulation of the ventral capsule/ventral striatum (VC/VS), but data on this are limited. Jordan et al. report on a phase 1 clinical trial that showed increased prefrontal glucose metabolism and decreased clinical decline as compared to patients without VC/VS DBS [146]. Unfortunately, no further trials have been conducted in this area. More research is needed to understand the underlying mechanism by which AD pathology is affected by DBS and how it can be modulated through treatments.

5.5. Criteria for Successful Applications of DBS in AD

The regulation of selecting viable candidates for DBS treatment of Alzheimer's disease is not well established. Studies have used a variety of selection criteria, such as ADAS scores, MMSE, and CDR.

The ADAS is a cognitive assessment used to quantify the extent of disease progression among AD patients [147]. Moreover, the practicality of ADAS in clinical assessment has also been noted in previous studies [148]. The MMSE is a similar but more wide-ranging test meant to assess the severity of cognitive impairment and serves as a quantifiable metric to judge cognitive decline [149]. Clinically, the ADAS scale has been used in conjunction with other tests such as the MMSE and shows significant correlation between the two tests [150]. The CDR scale is also comparable with the MMSE and ADAS, though it is a standardized scale which tests the extent of progression of dementia among afflicted patients [151].

Depending on the study, different selection criteria have been used to assess the suitability of candidates. The Advance study group led by Lozano et al. is the most recent group of researchers to conduct clinical trials into the effectiveness of fornix DBS on AD [34]. Their study selected patients with mild AD who had CDR scores of 0.5–1 and ADAS scores of 12–24 [34]. However, other studies used the MMSE to assess the extent of AD progression. The study led by Fontaine Denys et al. required the patient to have an MMSE score between 20 and 24, which is indicative of mild cognitive decline [134], yet other studies used none of the above metrics and instead chose to use the criteria of the National Institute of Neurological and Communicative Disorders and Stroke–Alzheimer's Disease and Related Disorders Association (NINCDS-ADRDA) [130]. Consequently, it is difficult to draw any definitive conclusions regarding the efficacy of DBS interventions in AD patients.

It is important to note that the metrics used to evaluate the effectiveness of AD diagnosis are outdated. Newer technology and standards may be more accurate [152]. A review of the current diagnostic tools and clearer criteria for selecting AD patients for DBS clinical trials is needed. Additionally, the age of the patient should also be considered.

5.6. The Effect of Age on the Effectiveness of DBS in AD

As with other areas of DBS research into AD, age and its correlation with DBS efficacy has not yet been well documented. Many of the studies listed in Table 2 focused on a narrow age range of patients. For instance, Fontaine et al.'s current study included patients between 50 and 65 years old, while Laxton et al.'s study required patients to be between 40 and 80 years old, but only had six patients between the ages of 51 and 68 [37,134].

The study conducted by Lozano et al. during the Advance phase II trials is the only research found to investigate the varying effects of DBS on different age groups. Their study, which involved fornix DBS, revealed that patients under 65 years of age experienced a deterioration in MMSE and CDR scores with the use of this technique [34].

Subsequent review by Lozano et al. and analysis by Aldehri et al. attributes the smaller age range used to a difference in the pathophysiology of AD between younger versus older populations, leading to a difference in clinical efficacy [34,153]. To this end, Lozano et al. suggest that the lack of ameliorative effects from DBS may be due to a more severe disease progression among younger patients, even though clinical symptoms may be similar to older patients. Indeed, a broader review by Schneider et al. showed that younger patients with AD are at greater risk of cognitive decline than older participants over a period of 12–24 months [154]. Conversely, this may also be due to different genetic predispositions which skew the results, given that only a small sample size of patients was used [34].

Consequently, the need for younger patients to receive DBS treatment in different regions of the brain is imperative. Moreover, further research into the reason behind the lack of DBS efficacy in younger patients should also be explored. In line with this finding, more recent research conducted by Lozano et al. has adopted larger age ranges of 45–85 years old [34].

5.7. Adverse Effects of DBS in AD

The present dearth in clinical trials with the use of DBS in AD makes it very difficult to discern possible adverse effects of the treatment. Certainly, none of the completed trials report any adverse effects related to hardware malfunction or manifestation of psychiatric complications.

However, the incidence of side effects post-operatively and changes to disease progression have been reported in the clinical trials. Laxton et al. reported that some patients experienced stimulation-induced autobiographical memory recall with a general warm, flushing feeling [37]. This may be due to stimulation of the fornix, which is not the primary target of treatment. Moreover, patients also increased heart rate and systolic blood pressure during stimulation [37]. In terms of clinical outcome, most patients were reported to have a decreased ADAS score in the short term, which were predicted to increase to pre-operative levels after 12 months post-operatively. Within this cohort, one patient did show improvement in their MMSE score but a deterioration in ADAS score [37]. Other studies such as Mao et al. and Kuhn et al. corroborate these findings with having patients in their cohorts that showed minimal amelioration of their ADAS and MMSE scores or even worsening of it post-operatively [33,38]. Mao et al. reported one of their patients also having worsening clinical presentation with a degeneration of his basic activities of daily living [33]. Moreover, Kuhn et al. also reported stimulation-induced inner restlessness in one of the patients of their study [38].

The outlook for using DBS to treat AD appears promising. Despite ongoing research into the precise mechanism of AD pathology, DBS may be a highly viable alternative to conventional pharmaceuticals, which often come with unappealing side effects. However, it is unfortunate that there is still some uncertainty or limitation surrounding its efficacy and safety. There is definitely a shortage of research into the use of DBS for AD in both animal models and humans. Only the fornix and NBM have been identified as potential targets in ongoing or completed human trials. Therefore, further efforts are needed to understand the specific criteria for selecting AD patients who could benefit from DBS treatment, the impact of DBS on AD pathology, and the potential adverse effects associated with DBS. These are crucial steps needed to fully integrate DBS as a frontline treatment for AD.

6. Discussion

Current research suggests a strong connection between AD and PD, as well as the effectiveness of DBS in treating the symptoms of both conditions. Both diseases result in neuronal loss through different mechanisms [6,115]. Moreover, many PD patients also develop AD within 20 years of diagnosis, highlighting the importance of treating co-occurring conditions simultaneously, such as AD in PD [155].

As noted in previous sections, the current application of DBS in PD has had positive responses. While we are still exploring the exact pathology of the disease, it is known that Lewy body aggregates are the primary cause [72]. Moreover, conventional therapies used in PD have been shown to have many adverse effects and a lack of efficacy at advanced stages of the disease [78]. Consequently, the use of DBS in treatment has allowed for better coverage and versality in treatment. Additionally, while current pathologies and mechanisms of action of DBS are listed below in Table 3, there is much to be learned in the action of multiple comorbidities on the human body.

Table 3. Summary of different mechanistic pathways by which DBS affects AD and PD, as well as their consequent physiological outcomes.

Disease	Action of DBS	Physiological Effects	Year	Author	References
AD	Neuronal activation through Fornix DBS.	Increased hippocampal volume.	2020, 2012, 2019, 2015	Jakobs, Smith, Aldehri, Sankar	[142–144,156]
	Increased acetylcholine levels using NBM DBS.	Increased glucose uptake in amygdalo-hippocampal, temporal, and superior lingual gyrus.	2021, 2014	Maltête, Kuhn	[38,145]
	Increased prefrontal glucose uptake.	Decreased clinical decline.	2021	Lam	[146]

Table 3. *Cont.*

Disease	Action of DBS	Physiological Effects	Year	Author	References
PD	Neuronal inhibition.	Depletion of glutamate and release of GABA and adenosine.	2012, 2001, 2001	Lozano, Contreras, Wu	[92–94]
	Neuronal activation.	Increased glutamate and dopamine levels.	2013, 2005, 1992	Lozano, Stefani, Benazzouz	[92,96,97]
	Neuronal activation and inhibition.	Decoupling of the soma and axons.	2013	Lozano	[92]
	Disrupt pathologic oscillatory patterns.	Neurotrophin release and generation of new neurons.	2013, 2010	Lozano, McIntyre	[92,98]

It is important to note that the proposed mechanisms of action outlined above in Table 3 pertain to known human trials, while animal studies have revealed additional mechanisms. For example, DBS treatment has been found to decrease amyloidosis, inflammation, neuronal loss, pathological τ-protein formation, and cholinergic nerve degeneration in animal models [138,157,158]. Other mechanisms such as synaptic plasticity, τ clearance, increased neurotropic factors, and increased tyrosine hydroxylase (TH) in substantia nigra pars compacta (SNpc) neurons have also been observed in animal models but have yet to be extensively tested in humans [142,159].

Studies on both AD and PD use a combination of quantitative and qualitative tests to determine eligibility for DBS treatment. A multi-disciplinary team is also involved in the selection process for both diseases. Age is also an important factor to consider, with younger AD patients being more at risk of cognitive decline [154] and older PD patients being more likely to experience adverse effects associated with age-related comorbidities [7].

DBS is used to treat PD and AD, but the specific targets vary between the two diseases (Figure 4A,B). Studies have shown that DBS of the STN and GPi has similar effects on motor function in PD, but one study found that STN DBS may lead to faster cognitive decline [24,69,86,160]. Moreover, it seems that fornix-DBS is effective at combating hippocampal and forniceal volume degeneration. Further research is needed to determine which target leads to improved clinical outcomes.

There is limited research on DBS for AD, but the most promising results have been found with fornix and NBM stimulation [153,161]. It is suggested that while DBS of the fornix is more effective at slowing the degeneration of the hippocampus, DBS of the NBM targets the loss of cholinergic neurons [162].

Both diseases require high-frequency DBS, but more research is needed on the effects of DBS at different frequencies, amplitudes, and pulse widths, as well as potential side effects. The currently used frequencies are more based on trial and error rather than observable data sets and would benefit from the use of more automated systems [107].

Overall, DBS is a promising treatment for PD and AD by targeting the underlying pathologies of the diseases. However, there is still room for advancement to minimize adverse effects and maximize the benefit of DBS. Suggestions for improvements include developing new DBS leads that improve performance, minimizing complications from errant electrode placement, and reducing programming time with AI-guided parameter selection [81]. The pulse generators could be made smaller, with longer battery life, allow for situation-based stimulation patterns, and be shielded from electromagnetic interference. It is recommended to systematically test the effects of stimulation for all electrodes during the initial programming session, gradually reduce anti-PD medications, and use lower-frequency stimulation and alternative electrode configurations if problems are not adequately treated.

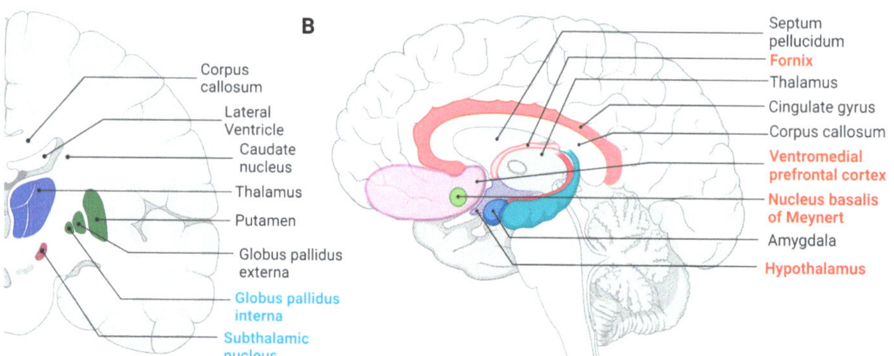

Figure 4. Anatomical representation of areas of the brain targeted by DBS treatment. (**A**) Frontal coronal cross-section of the brain depicting regions of the brain targeted for treatment of PD (highlighted in blue). (**B**) Frontal sagittal cross-section of the brain depicting regions targeted for treatment of AD (highlighted in red). Figure made and compiled using https://biorender.com/ (Accessed on 20 December 2022).

Advancements in the field also include the potential use of functional magnetic resonance imaging (fMRI) and machine learning (ML) in combination with DBS. A review of 83 studies found the fMRI and DBS combination to be a powerful tool for observing and manipulating neuronal circuits simultaneously [163]. ML has the potential to improve DBS outcomes [164], with a systematic review of 73 studies finding that ML can help process electrical and imaging data, though many challenges remain [165]. A recent trial showed the potential of ML and fMRI to complement each other, with fMRI serving as a biomarker of clinical response in DBS [166].

Future use of DBS may also have ethical implications that may impact its widespread application for PD and AD. These include the informed consent of patients with neuropsychiatric symptoms, informed consent given by guardians or caregivers, and guidelines to ensure safe and effective use across all centers and professionals [92,167]. Addressing issues of expertise and accessibility will ensure that DBS is accessible to all, regardless of their location or socioeconomic status [168]. Additionally, as DBS technology advances, policymakers must consider the potential impact on a person's personality [167] and whether it is ethical to use DBS on "healthy" individuals for desired effects [169].

Certainly, the relative success of using DBS in treating PD can offer guidance for utilizing this treatment in other neurodegenerative diseases, particularly in AD. Both diseases share protein misfolding pathologies, and clinical trials have demonstrated the effectiveness of similar DBS configurations for treating both PD and AD. Additionally, ongoing research has identified fundamental selection criteria and the underlying mechanistic actions of DBS treatment in PD, providing valuable lessons to be applied in advancing the literature on using DBS for AD.

There are undoubtedly limitations to our study approach. Notably, our paper relies on a narrative search strategy, which may limit the scope of the current research literature on DBS. Despite our efforts to explore all aspects of the topic comprehensively, some literature may have been inadvertently excluded. Furthermore, it is possible that our selection criteria may have exhibited bias towards research that supports the overall narrative of our paper. However, we have taken care to ensure that our coverage of all topics is transparent and accurate, and any conflicting evidence to the overall narrative has been reported with the utmost precision.

The main goal of this study is to provide a comprehensive overview of the field of DBS, including its development, current usage, and potential future advancements. Our findings indicate that DBS systems are highly effective in treating patients with PD, where a solid understanding of the disease progression, the mechanism by which DBS affects this pathology, selection criteria for treatment, and adverse effects and their mitigations has been established. As such, the current guidelines for selecting criteria and configuring DBS systems can also be used to evaluate their efficacy in AD patients. However, further in vivo and in vitro research is necessary to ensure the precise targeting of brain regions. It is recommended to use differential HFS/LFS systems of stimulus depending on disease progression and similar age criteria for the selection of DBS as a method of best practice. Nonetheless, ethical considerations, accessibility issues, and a shortage of current research literature present ongoing challenges. Despite these obstacles, DBS is a promising tool in the treatment of neurodegenerative diseases, and establishing a solid foundation of research literature and guidelines of practice will ensure its future efficacy, reliability, and adaptability to individual patient needs.

7. Conclusions

Both Parkinson's disease (PD) and Alzheimer's disease (AD) are global health burdens characterized by neurodegeneration and amyloid protein accumulation. Due to the lack of disease-modifying pharmaceutical therapies, non-pharmaceutical therapies such as DBS have gained interest. While initial clinical trials show promise for DBS in AD, further research is required to establish its efficacy. In contrast, DBS has been used in PD for decades and is already FDA-approved. However, further clinical trials are needed to clarify the effectiveness and side effects of DBS in a variety of patients with different prognoses. This can be achieved by developing a unified system of classification with tighter controls and determining the recommended age or time after disease diagnosis that is optimal for DBS treatment. Regarding AD, foundational in vivo work to identify the most susceptible brain regions to DBS in AD-equivalent animal models is required. Clinical trials of different brain regions using either HFS or LFS systems depending on the disease prognosis of the patient should also be conducted. Continual assessment of currently identified brain regions for DBS treatment should also be conducted, as well as reviews of the current diagnostic tools and clearer criteria for selecting AD patients before DBS becomes a reliable reality for AD therapy.

Author Contributions: Conceptualization, A.C. and M.S.Y.; literature review and resources, A.C., D.K.L.S. and A.M.; writing—original draft preparation, D.K.L.S. and A.M.; writing—review and editing, D.K.L.S., A.M., K.Z., P.P., A.K., D.J.L., M.S.Y. and A.C.; figure and table preparations and editing, A.C., D.K.L.S. and A.M.; visualization, A.C.; supervision, A.C. and M.S.Y.; project administration, A.C.; funding acquisition, A.C. All authors have read and agreed to the published version of the manuscript.

Funding: This research received no external funding.

Institutional Review Board Statement: Not applicable.

Informed Consent Statement: Not applicable.

Data Availability Statement: Not applicable.

Acknowledgments: We thank the Qatar National Library and the Health Sciences Library at Weill Cornell Medicine–Qatar for open-access funding.

Conflicts of Interest: The authors declare no conflict of interest.

References

1. Nichols, E.; Szoeke, C.E.I.; Vollset, S.E.; Abbasi, N.; Abd-Allah, F.; Abdela, J.; Aichour, M.T.E.; Akinyemi, R.O.; Alahdab, F.; Asgedom, S.W.; et al. Global, regional, and national burden of Alzheimer's disease and other dementias, 1990–2016: A systematic analysis for the Global Burden of Disease Study 2016. *Lancet Neurol.* **2019**, *18*, 88–106. [CrossRef] [PubMed]
2. Silva, M.V.F.; Loures, C.D.M.G.; Alves, L.C.V.; de Souza, L.C.; Borges, K.B.G.; Carvalho, M.D.G. Alzheimer's disease: Risk factors and potentially protective measures. *J. Biomed. Sci.* **2019**, *26*, 33. [CrossRef] [PubMed]
3. Mumtaz, S.; Rana, J.N.; Choi, E.H.; Han, I. Microwave Radiation and the Brain: Mechanisms, Current Status, and Future Prospects. *Int. J. Mol. Sci.* **2022**, *23*, 9288. [CrossRef] [PubMed]

4. Scheltens, P.; De Strooper, B.; Kivipelto, M.; Holstege, H.; Chételat, G.; Teunissen, C.E.; Cummings, J.; van der Flier, W.M. Alzheimer's disease. *Lancet* **2021**, *397*, 1577–1590. [CrossRef] [PubMed]
5. Nelson, P.T.; Head, E.; Schmitt, F.A.; Davis, P.R.; Neltner, J.H.; Jicha, G.A.; Abner, E.L.; Smith, C.D.; Van Eldik, L.J.; Kryscio, R.J.; et al. Alzheimer's disease is not "brain aging": Neuropathological, genetic, and epidemiological human studies. *Acta Neuropathol.* **2011**, *121*, 571–587. [CrossRef]
6. Poewe, W.; Seppi, K.; Tanner, C.M.; Halliday, G.M.; Brundin, P.; Volkmann, J.; Schrag, A.E.; Lang, A.E. Parkinson disease. *Nat. Rev. Dis. Primers* **2017**, *3*, 17013. [CrossRef]
7. Twelves, D.; Perkins, K.S.; Counsell, C. Systematic review of incidence studies of Parkinson's disease. *Mov. Disord.* **2003**, *18*, 19–31. [CrossRef]
8. Delamarre, A.; Meissner, W.G. Epidemiology, environmental risk factors and genetics of Parkinson's disease. *La Presse Médicale* **2017**, *46*, 175–181. [CrossRef]
9. Ascherio, A.; Schwarzschild, M.A. The epidemiology of Parkinson's disease: Risk factors and prevention. *Lancet Neurol.* **2016**, *15*, 1257–1272. [CrossRef]
10. Sung, V.W.; Nicholas, A.P. Nonmotor Symptoms in Parkinson's Disease: Expanding the View of Parkinson's Disease Beyond a Pure Motor, Pure Dopaminergic Problem. *Neurol. Clin.* **2013**, *31*, S1–S16. [CrossRef]
11. Lozano, A.M.; Lipsman, N.; Bergman, H.; Brown, P.; Chabardes, S.; Chang, J.W.; Matthews, K.; McIntyre, C.C.; Schlaepfer, T.E.; Schulder, M.; et al. Deep brain stimulation: Current challenges and future directions. *Nat. Rev. Neurol.* **2019**, *15*, 148–160. [CrossRef]
12. Lv, Q.; Du, A.; Wei, W.; Li, Y.; Liu, G.; Wang, X.P. Deep Brain Stimulation: A Potential Treatment for Dementia in Alzheimer's Disease (AD) and Parkinson's Disease Dementia (PDD). *Front. Neurosci.* **2018**, *12*, 360. [CrossRef]
13. Chouraki, V.; Seshadri, S. Genetics of Alzheimer's Disease. *Adv. Genet.* **2014**, *87*, 245–294. [CrossRef]
14. Cui, L.; Hou, N.-N.; Wu, H.-M.; Zuo, X.; Lian, Y.-Z.; Zhang, C.-N.; Wang, Z.-F.; Zhang, X.; Zhu, J.-H. Prevalence of Alzheimer's Disease and Parkinson's Disease in China: An Updated Systematical Analysis. *Front. Aging Neurosci.* **2020**, *12*, 603854. [CrossRef]
15. Simon, D.K.; Tanner, C.M.; Brundin, P. Parkinson Disease Epidemiology, Pathology, Genetics, and Pathophysiology. *Clin. Geriatr. Med.* **2019**, *36*, 1–12. [CrossRef]
16. Mandybur, G. Deep Brain Stimulation (DBS) for Parkinson's & Essential Tremor | Mayfield Brain & Spine, Cincinnati, (n.d.). Available online: https://mayfieldclinic.com/pe-dbs.htm (accessed on 13 December 2022).
17. Amon, A.; Alesch, F. Systems for deep brain stimulation: Review of technical features. *J. Neural Transm.* **2017**, *124*, 1083–1091. [CrossRef]
18. Parrent, A.G. History of Surgery for Movement Disorders. In *Textbook of Stereotactic and Functional Neurosurgery*; Springer: Berlin/Heidelberg, Germany, 2009; pp. 1467–1485. [CrossRef]
19. Burchiel, K.J. Thalamotomy for Movement Disorders. Neurosurg. *Clin. N. Am.* **1995**, *6*, 55–71. [CrossRef]
20. Miocinovic, S.; Somayajula, S.; Chitnis, S.; Vitek, J.L. History, Applications, and Mechanisms of Deep Brain Stimulation. *JAMA Neurol.* **2013**, *70*, 163–171. [CrossRef]
21. Spiegel, E.A. Stereoencephalotomy. *J. Am. Med. Assoc.* **1952**, *148*, 446–451. [CrossRef]
22. Ohye, C.; Kubota, K.; Hongo, T.; Nagao, T.; Narabayashi, H. Ventrolateral and Subventrolateral Thalamic Stimulation. *Arch. Neurol.* **1964**, *11*, 427–434. [CrossRef]
23. Premarket Approval (PMA), (n.d.). Available online: https://www.accessdata.fda.gov/scripts/cdrh/cfdocs/cfPMA/pma.cfm?id=P960009 (accessed on 23 December 2022).
24. Deuschl, G.; Schade-Brittinger, C.; Krack, P.; Volkmann, J.; Schäfer, H.; Bötzel, K.; Daniels, C.; Deutschländer, A.; Dillmann, U.; Eisner, W.; et al. A Randomized Trial of Deep-Brain Stimulation for Parkinson's Disease. *N. Engl. J. Med.* **2006**, *355*, 896–908. [CrossRef] [PubMed]
25. Xie, T.; Vigil, J.; MacCracken, E.; Gasparaitis, A.; Young, J.; Kang, W.; Bernard, J.; Warnke, P.; Kang, U.J. Low-frequency stimulation of STN-DBS reduces aspiration and freezing of gait in patients with PD. *Neurology* **2014**, *84*, 415–420. [CrossRef] [PubMed]
26. Blomstedt, P.; Persson, R.S.; Hariz, G.-M.; Linder, J.; Fredricks, A.; Häggström, B.; Philipsson, J.; Forsgren, L.; Hariz, M. Deep brain stimulation in the caudal zona incerta versus best medical treatment in patients with Parkinson's disease: A randomised blinded evaluation. *J. Neurol. Neurosurg. Psychiatry* **2018**, *89*, 710–716. [CrossRef] [PubMed]
27. Baker, K.B.; Lee, J.Y.; Mavinkurve, G.; Russo, G.S.; Walter, B.; DeLong, M.R.; Bakay, R.A.; Vitek, J.L. Somatotopic organization in the internal segment of the globus pallidus in Parkinson's disease. *Exp. Neurol.* **2010**, *222*, 219–225. [CrossRef] [PubMed]
28. Ossowska, K. Zona incerta as a therapeutic target in Parkinson's disease. *J. Neurol.* **2019**, *267*, 591–606. [CrossRef]
29. Thevathasan, W.; Debu, B.; Aziz, T.; Bloem, B.R.; Blahak, C.; Butson, C.; Czernecki, V.; Foltynie, T.; Fraix, V.; Grabli, D.; et al. Pedunculopontine nucleus deep brain stimulation in Parkinson's disease: A clinical review. *Mov. Disord.* **2017**, *33*, 10–20. [CrossRef]
30. Baizabal-Carvallo, J.F.; Kagnoff, M.N.; Jimenez-Shahed, J.; Fekete, R.; Jankovic, J. The safety and efficacy of thalamic deep brain stimulation in essential tremor: 10 years and beyond. *J. Neurol. Neurosurg. Psychiatry* **2013**, *85*, 567–572. [CrossRef]
31. Fytagoridis, A.; Sandvik, U.; Åström, M.; Bergenheim, T.; Blomstedt, P. Long term follow-up of deep brain stimulation of the caudal zona incerta for essential tremor. *J. Neurol. Neurosurg. Psychiatry* **2011**, *83*, 258–262. [CrossRef]
32. Vidailhet, M.; Jutras, M.-F.; Roze, E.; Grabli, D. Deep brain stimulation for dystonia. *Handb. Clin. Neurol.* **2013**, *116*, 167–187. [CrossRef]

33. Yu, X.-G.; Mao, Z.-Q.; Wang, X.; Xu, X.; Cui, Z.-Q.; Pan, L.-S.; Ning, X.-J.; Xu, B.-X.; Ma, L.; Ling, Z.-P.; et al. Partial improvement in performance of patients with severe Alzheimer's disease at an early stage of fornix deep brain stimulation. *Neural Regen. Res.* **2018**, *13*, 2164–2172. [CrossRef]

34. Lozano, A.M.; Fosdick, L.; Chakravarty, M.M.; Leoutsakos, J.-M.; Munro, C.; Oh, E.; Drake, K.E.; Lyman, C.H.; Rosenberg, P.B.; Anderson, W.S.; et al. A Phase II Study of Fornix Deep Brain Stimulation in Mild Alzheimer's Disease. *J. Alzheimer's Dis.* **2016**, *54*, 777–787. [CrossRef]

35. Scharre, D.W.; Weichart, E.; Nielson, D.; Zhang, J.; Agrawal, P.; Sederberg, P.B.; Knopp, M.V.; Rezai, A.R.; Initiative, F.T.A.D.N. Deep Brain Stimulation of Frontal Lobe Networks to Treat Alzheimer's Disease. *J. Alzheimer's Dis.* **2018**, *62*, 621–633. [CrossRef]

36. Chakravarty, M.M.; Hamani, C.; Martinez-Canabal, A.; Ellegood, J.; Laliberté, C.; Nobrega, J.N.; Sankar, T.; Lozano, A.M.; Frankland, P.W.; Lerch, J.P. Deep brain stimulation of the ventromedial prefrontal cortex causes reorganization of neuronal processes and vasculature. *Neuroimage* **2016**, *125*, 422–427. [CrossRef]

37. Laxton, A.W.; Tang-Wai, D.F.; McAndrews, M.P.; Zumsteg, D.; Wennberg, R.; Keren, R.; Wherrett, J.; Naglie, G.; Hamani, C.; Smith, G.S.; et al. A phase I trial of deep brain stimulation of memory circuits in Alzheimer's disease. *Ann. Neurol.* **2010**, *68*, 521–534. [CrossRef]

38. Kuhn, J.; Hardenacke, K.; Lenartz, D.; Gruendler, T.; Ullsperger, M.; Bartsch, C.; Mai, J.K.; Zilles, K.; Bauer, A.; Matusch, A.; et al. Deep brain stimulation of the nucleus basalis of Meynert in Alzheimer's dementia. *Mol. Psychiatry* **2014**, *20*, 353–360. [CrossRef]

39. Velez-Lago, F.M.; Thompson, A.; Oyama, G.; Hardwick, A.; Sporrer, J.M.; Zeilman, P.; Foote, K.D.; Bowers, D.; Ward, H.E.; Sanchez-Ramos, J.; et al. Differential and Better Response to Deep Brain Stimulation of Chorea Compared to Dystonia in Huntington's Disease. *Ster. Funct. Neurosurg.* **2013**, *91*, 129–133. [CrossRef]

40. Nair, G.; Evans, A.; Bear, R.E.; Velakoulis, D.; Bittar, R.G. The anteromedial GPi as a new target for deep brain stimulation in obsessive compulsive disorder. *J. Clin. Neurosci.* **2014**, *21*, 815–821. [CrossRef]

41. Huff, W.; Lenartz, D.; Schormann, M.; Lee, S.-H.; Kuhn, J.; Koulousakis, A.; Mai, J.; Daumann, J.; Maarouf, M.; Klosterkötter, J.; et al. Unilateral deep brain stimulation of the nucleus accumbens in patients with treatment-resistant obsessive-compulsive disorder: Outcomes after one year. *Clin. Neurol. Neurosurg.* **2010**, *112*, 137–143. [CrossRef]

42. Denys, D.; Mantione, M.; Figee, M.; Munckhof, P.V.D.; Koerselman, F.; Westenberg, H.; Bosch, A.; Schuurman, R. Deep Brain Stimulation of the Nucleus Accumbens for Treatment-Refractory Obsessive-Compulsive Disorder. *Arch. Gen. Psychiatry* **2010**, *67*, 1061–1068. [CrossRef]

43. Denys, D.; Graat, I.; Mocking, R.; de Koning, P.; Vulink, N.; Figee, M.; Ooms, P.; Mantione, M.; Munckhof, P.V.D.; Schuurman, R. Efficacy of Deep Brain Stimulation of the Ventral Anterior Limb of the Internal Capsule for Refractory Obsessive-Compulsive Disorder: A Clinical Cohort of 70 Patients. *Am. J. Psychiatry* **2020**, *177*, 265–271. [CrossRef]

44. Kammen, A.; Cavaleri, J.; Lam, J.; Frank, A.C.; Mason, X.; Choi, W.; Penn, M.; Brasfield, K.; Van Noppen, B.; Murray, S.B.; et al. Neuromodulation of OCD: A review of invasive and non-invasive methods. *Front. Neurol.* **2022**, *13*, 909264. [CrossRef] [PubMed]

45. Park, Y.-S.; Sammartino, F.; Young, N.A.; Corrigan, J.; Krishna, V.; Rezai, A.R. Anatomic Review of the Ventral Capsule/Ventral Striatum and the Nucleus Accumbens to Guide Target Selection for Deep Brain Stimulation for Obsessive-Compulsive Disorder. *World Neurosurg.* **2019**, *126*, 1–10. [CrossRef] [PubMed]

46. Greenberg, B.D.; A Gabriels, L.; A Malone, D.; Rezai, A.R.; Friehs, G.M.; Okun, M.; A Shapira, N.; Foote, K.; Cosyns, P.R.; Kubu, C.S.; et al. Deep brain stimulation of the ventral internal capsule/ventral striatum for obsessive-compulsive disorder: Worldwide experience. *Mol. Psychiatry* **2008**, *15*, 64–79. [CrossRef] [PubMed]

47. Li, N.; Baldermann, J.C.; Kibleur, A.; Treu, S.; Akram, H.; Elias, G.J.B.; Boutet, A.; Lozano, A.M.; Al-Fatly, B.; Strange, B.; et al. A unified connectomic target for deep brain stimulation in obsessive-compulsive disorder. *Nat. Commun.* **2020**, *11*, 3364. [CrossRef]

48. Chabardès, S.; Polosan, M.; Krack, P.; Bastin, J.; Krainik, A.; David, O.; Bougerol, T.; Benabid, A.L. Deep Brain Stimulation for Obsessive-Compulsive Disorder: Subthalamic Nucleus Target. *World Neurosurg.* **2012**, *80*, S31.e1–S31.e8. [CrossRef]

49. Germann, J.; Boutet, A.; Elias, G.J.; Gouveia, F.V.; Loh, A.; Giacobbe, P.; Bhat, V.; Kucharczyk, W.; Lozano, A.M. Brain Structures and Networks Underlying Treatment Response to Deep Brain Stimulation Targeting the Inferior Thalamic Peduncle in Obsessive-Compulsive Disorder. *Ster. Funct. Neurosurg.* **2022**, *100*, 236–243. [CrossRef]

50. Lee, D.J.; Dallapiazza, R.F.; De Vloo, P.; Elias, G.J.; Fomenko, A.; Boutet, A.; Giacobbe, P.; Lozano, A.M. Inferior thalamic peduncle deep brain stimulation for treatment-refractory obsessive-compulsive disorder: A phase 1 pilot trial. *Brain Stimul.* **2019**, *12*, 344–352. [CrossRef]

51. Luyten, L.; Hendrickx, S.; Raymaekers, S.; Gabriels, L.; Nuttin, B. Electrical stimulation in the bed nucleus of the stria terminalis alleviates severe obsessive-compulsive disorder. *Mol. Psychiatry* **2015**, *21*, 1272–1280. [CrossRef]

52. Mosley, P.E.; Windels, F.; Morris, J.; Coyne, T.; Marsh, R.; Giorni, A.; Mohan, A.; Sachdev, P.; O'leary, E.; Boschen, M.; et al. A randomised, double-blind, sham-controlled trial of deep brain stimulation of the bed nucleus of the stria terminalis for treatment-resistant obsessive-compulsive disorder. *Transl. Psychiatry* **2021**, *11*, 190. [CrossRef]

53. Raymaekers, S.; Vansteelandt, K.; Luyten, L.; Bervoets, C.; Demyttenaere, K.; Gabriels, L.; Nuttin, B. Long-term electrical stimulation of bed nucleus of stria terminalis for obsessive-compulsive disorder. *Mol. Psychiatry* **2016**, *22*, 931–934. [CrossRef]

54. Salanova, V. Deep brain stimulation for epilepsy. *Epilepsy Behav.* **2018**, *88*, 21–24. [CrossRef]

55. Malone, D.A., Jr.; Dougherty, D.D.; Rezai, A.R.; Carpenter, L.L.; Friehs, G.M.; Eskandar, E.N.; Rauch, S.L.; Rasmussen, S.A.; Machado, A.G.; Kubu, C.S.; et al. Deep Brain Stimulation of the Ventral Capsule/Ventral Striatum for Treatment-Resistant Depression. *Biol. Psychiatry* **2009**, *65*, 267–275. [CrossRef]

56. Bewernick, B.H.; Hurlemann, R.; Matusch, A.; Kayser, S.; Grubert, C.; Hadrysiewicz, B.; Axmacher, N.; Lemke, M.; Cooper-Mahkorn, D.; Cohen, M.X.; et al. Nucleus Accumbens Deep Brain Stimulation Decreases Ratings of Depression and Anxiety in Treatment-Resistant Depression. *Biol. Psychiatry* **2010**, *67*, 110–116. [CrossRef]
57. Baizabal-Carvallo, J.F.; Alonso-Juarez, M. Low-frequency deep brain stimulation for movement disorders. *Park. Relat. Disord.* **2016**, *31*, 14–22. [CrossRef]
58. di Biase, L.; Fasano, A. Low-frequency deep brain stimulation for Parkinson's disease: Great expectation or false hope? *Mov. Disord.* **2016**, *31*, 962–967. [CrossRef]
59. Wyckhuys, T.; Raedt, R.; Vonck, K.; Wadman, W.; Boon, P. Comparison of hippocampal Deep Brain Stimulation with high (130Hz) and low frequency (5Hz) on afterdischarges in kindled rats. *Epilepsy Res.* **2010**, *88*, 239–246. [CrossRef]
60. Ramdhani, R.A.; Patel, A.; Swope, D.; Kopell, B.H. Early Use of 60 Hz Frequency Subthalamic Stimulation in Parkinson's Disease: A Case Series and Review. *Neuromodulation Technol. Neural Interface* **2015**, *18*, 664–669. [CrossRef]
61. Okun, M.S.; Foote, K.D. Parkinson's disease DBS: What, when, who and why? The time has come to tailor DBS targets. *Expert Rev. Neurother.* **2010**, *10*, 1847–1857. [CrossRef]
62. Rodriguez, R.L.; Fernandez, H.H.; Haq, I.; Okun, M. Pearls in Patient Selection for Deep Brain Stimulation. *Neurologist* **2007**, *13*, 253–260. [CrossRef]
63. Ghika, J.; Villemure, J.-G.; Fankhauser, H.; Favre, J.; Assal, G.; Ghika-Schmid, F. Efficiency and safety of bilateral contemporaneous pallidal stimulation (deep brain stimulation) in levodopa-responsive patients with Parkinson's disease with severe motor fluctuations: A 2-year follow-up review. *J. Neurosurg.* **1998**, *89*, 713–718. [CrossRef]
64. Yamada, K.; Hamasaki, T.; Kuratsu, J.-I. Thalamic stimulation alleviates levodopa-resistant rigidity in a patient with non-Parkinson's disease parkinsonian syndrome. *J. Clin. Neurosci.* **2014**, *21*, 882–884. [CrossRef] [PubMed]
65. Chang, S.J.; Cajigas, I.; Guest, J.D.; Noga, B.R.; Widerström-Noga, E.; Haq, I.; Fisher, L.; Luca, C.C.; Jagid, J.R. MR Tractography-Based Targeting and Physiological Identification of the Cuneiform Nucleus for Directional DBS in a Parkinson's Disease Patient With Levodopa-Resistant Freezing of Gait. *Front. Hum. Neurosci.* **2021**, *15*, 676755. [CrossRef] [PubMed]
66. Lin, Z.; Zhang, X.; Wang, L.; Zhang, Y.; Zhou, H.; Sun, Q.; Sun, B.; Huang, P.; Li, D. Revisiting the L-Dopa Response as a Predictor of Motor Outcomes After Deep Brain Stimulation in Parkinson's Disease. *Front. Hum. Neurosci.* **2021**, *15*, 604433. [CrossRef] [PubMed]
67. Wächter, T.; Mínguez-Castellanos, A.; Valldeoriola, F.; Herzog, J.; Stoevelaar, H. A tool to improve pre-selection for deep brain stimulation in patients with Parkinson's disease. *J. Neurol.* **2010**, *258*, 641–646. [CrossRef]
68. Emamikhah, M.; Akhoundi, F.H.; Rohani, M. Mechanism of Deep Brain Stimulation. *Handb. Neuromodulation* **2022**, *1*, 245–264.
69. Okun, M.S. Deep-Brain Stimulation for Parkinson's Disease. *N. Engl. J. Med.* **2012**, *367*, 1529–1538. [CrossRef]
70. Marras, C.; Beck, J.C.; Bower, J.H.; Roberts, E.; Ritz, B.; Ross, G.W.; Abbott, R.D.; Savica, R.; Van Den Eeden, S.K.; Willis, A.W.; et al. Prevalence of Parkinson's disease across North America. *NPJ Parkinson's Dis.* **2018**, *4*, 21. [CrossRef]
71. Chartier, S.; Duyckaerts, C. Is Lewy pathology in the human nervous system chiefly an indicator of neuronal protection or of toxicity? *Cell Tissue Res.* **2018**, *373*, 149–160. [CrossRef]
72. Schneider, S.A.; Alcalay, R.N. Neuropathology of genetic synucleinopathies with parkinsonism: Review of the literature. *Mov. Disord.* **2017**, *32*, 1504–1523. [CrossRef]
73. Burré, J.; Sharma, M.; Südhof, T.C. Cell Biology and Pathophysiology of α-Synuclein. *Cold Spring Harb. Perspect. Med.* **2017**, *8*, a024091. [CrossRef]
74. Braak, H.; Del Tredici, K.; Rüb, U.; de Vos, R.A.; Steur, E.N.J.; Braak, E. Staging of brain pathology related to sporadic Parkinson's disease. *Neurobiol. Aging* **2003**, *24*, 197–211. [CrossRef]
75. Postuma, R.B.; Berg, D. Advances in markers of prodromal Parkinson disease. *Nat. Rev. Neurol.* **2016**, *12*, 622–634. [CrossRef]
76. Brundin, P.; Melki, R. Prying into the Prion Hypothesis for Parkinson's Disease. *J. Neurosci.* **2017**, *37*, 9808–9818. [CrossRef]
77. Johnson, M.E.; Stecher, B.; Labrie, V.; Brundin, L.; Brundin, P. Triggers, Facilitators, and Aggravators: Redefining Parkinson's Disease Pathogenesis. *Trends Neurosci.* **2018**, *42*, 4–13. [CrossRef]
78. Lee, T.K.; Yankee, E.L. A review on Parkinson's disease treatment. Neuroimmunol. *Neuroinflammation* **2022**, *8*, 222. [CrossRef]
79. Iarkov, A.; Barreto, G.E.; Grizzell, J.A.; Echeverria, V. Strategies for the Treatment of Parkinson's Disease: Beyond Dopamine. *Front. Aging Neurosci.* **2020**, *12*, 4. [CrossRef]
80. Brocks, D.R. Anticholinergic drugs used in Parkinson's disease: An overlooked class of drugs from a pharmacokinetic perspective. *J. Pharm. Pharm. Sci.* **1999**, *2*, 39–46. Available online: https://pubmed-ncbi-nlm-nih-gov.wcmq.idm.oclc.org/10952768/ (accessed on 27 February 2023).
81. Bronstein, J.M.; Tagliati, M.; Alterman, R.L.; Lozano, A.M.; Volkmann, J.; Stefani, A.; Horak, F.B.; Okun, M.S.; Foote, K.D.; Krack, P.; et al. Deep brain stimulation for Parkinson disease: An expert consensus and review of key issues. *Arch. Neurol.* **2011**, *68*, 165. [CrossRef]
82. Williams, A.; Gill, S.; Varma, T.; Jenkinson, C.; Quinn, N.; Mitchell, R.; Scott, R.; Ives, N.; Rick, C.; Daniels, J.; et al. Deep brain stimulation plus best medical therapy versus best medical therapy alone for advanced Parkinson's disease (PD SURG trial): A randomised, open-label trial. *Lancet Neurol.* **2010**, *9*, 581–591. [CrossRef]
83. Sun, L.; Xu, F.; Ma, W.; Huang, Y.; Qiu, Z. Deep brain stimulation of pallidal versus subthalamic for patients with Parkinson's disease: A meta-analysis of controlled clinical trials. *Neuropsychiatr. Dis. Treat.* **2016**, *12*, 1435–1444. [CrossRef]

84. Ory-Magne, F.; Brefel-Courbon, C.; Simonetta-Moreau, M.; Fabre, N.; Lotterie, J.A.; Chaynes, P.; Berry, I.; Lazorthes, Y.; Rascol, O. Does ageing influence deep brain stimulation outcomes in Parkinson's disease? *Mov. Disord.* **2007**, *22*, 1457–1463. [CrossRef] [PubMed]

85. Walter, B.L.; Vitek, J.L. Surgical Treatment for Parkinson's Disease. *Lancet Neurol.* **2004**, *3*, 719–728. [CrossRef] [PubMed]

86. Anderson, V.C.; Burchiel, K.J.; Hogarth, P.; Favre, J.; Hammerstad, J.P. Pallidal vs Subthalamic Nucleus Deep Brain Stimulation in Parkinson Disease. *Arch. Neurol.* **2005**, *62*, 554–560. [CrossRef] [PubMed]

87. Loher, T.J.; Burgunder, J.-M.; Weber, S.; Sommerhalder, R.; Krauss, J.K. Effect of chronic pallidal deep brain stimulation on off period dystonia and sensory symptoms in advanced Parkinson's disease. *J. Neurol. Neurosurg. Psychiatry* **2002**, *73*, 395–399. [CrossRef]

88. Rizzone, M.; Ferrarin, M.; Pedotti, A.; Bergamasco, B.; Bosticco, E.; Lanotte, M.; Perozzo, P.; Tavella, A.; Torre, E.; Recalcati, M.; et al. High-frequency electrical stimulation of the subthalamic nucleus in Parkinson's disease: Kinetic and kinematic gait analysis. *Neurol. Sci.* **2002**, *23*, s103–s104. [CrossRef]

89. Taha, J.M.; Janszen, M.A.; Favre, J. Thalamic deep brain stimulation for the treatment of head, voice, and bilateral limb tremor. *J. Neurosurg.* **1999**, *91*, 68–72. [CrossRef]

90. Ferrarin, M.; Rizzone, M.; Bergamasco, B.; Lanotte, M.; Recalcati, M.; Pedotti, A.; Lopiano, L. Effects of bilateral subthalamic stimulation on gait kinematics and kinetics in Parkinson?s disease. *Exp. Brain Res.* **2004**, *160*, 517–527. [CrossRef]

91. Kringelbach, M.L.; Green, A.L.; Owen, S.L.F.; Schweder, P.M.; Aziz, T.Z. Sing the mind electric—principles of deep brain stimulation. *Eur. J. Neurosci.* **2010**, *32*, 1070–1079. [CrossRef]

92. Lozano, A.M.; Lipsman, N. Probing and Regulating Dysfunctional Circuits Using Deep Brain Stimulation. *Neuron* **2013**, *77*, 406–424. [CrossRef]

93. Contreras, D.; Llinás, R. Voltage-Sensitive Dye Imaging of Neocortical Spatiotemporal Dynamics to Afferent Activation Frequency. *J. Neurosci.* **2001**, *21*, 9403–9413. [CrossRef]

94. Wu, Y.; Levy, R.; Ashby, P.; Tasker, R.; Dostrovsky, J. Does stimulation of the GPi control dyskinesia by activating inhibitory axons? *Mov. Disord.* **2001**, *16*, 208–216. [CrossRef]

95. Dostrovsky, J.O.; Levy, R.; Wu, J.P.; Hutchison, W.; Tasker, R.R.; Lozano, A. Microstimulation-Induced Inhibition of Neuronal Firing in Human Globus Pallidus. *J. Neurophysiol.* **2000**, *84*, 570–574. [CrossRef]

96. Benazzouz, A.; Gross, C.; Dupont, J.; Bioulac, B. MPTP induced hemiparkinsonism in monkeys: Behavioral, mechanographic, electromyographic and immunohistochemical studies. *Exp. Brain Res.* **1992**, *90*, 116–120. [CrossRef]

97. Stefani, A.; Fedele, E.; Galati, S.; Pepicelli, O.; Frasca, S.; Pierantozzi, M.; Peppe, A.; Brusa, L.; Orlacchio, A.; Hainsworth, A.H.; et al. Subthalamic stimulation activates internal pallidus: Evidence from cGMP microdialysis in PD patients. *Ann. Neurol.* **2005**, *57*, 448–452. [CrossRef]

98. McIntyre, C.C.; Hahn, P.J. Network perspectives on the mechanisms of deep brain stimulation. *Neurobiol. Dis.* **2010**, *38*, 329–337. [CrossRef]

99. Okun, M.S.; Tagliati, M.; Fernandez, H.H.; Rodriguez, R.L.; Alterman, R.L.; Foote, K. Management of Referred Deep Brain Stimulation Failures: A Retrospective Analysis From Two Movement Disorder Centers. *Neurosurgery* **2005**, *57*, 401. [CrossRef]

100. Morgante, L.; Morgante, F.; Moro, E.; Epifanio, A.; Girlanda, P.; Ragonese, P.; Antonini, A.; Barone, P.; Bonuccelli, U.; Contarino, M.F.; et al. How many parkinsonian patients are suitable candidates for deep brain stimulation of subthalamic nucleus? Results of a questionnaire. *Park. Relat. Disord.* **2007**, *13*, 528–531. [CrossRef]

101. Kleiner-Fisman, G.; Herzog, J.; Fisman, D.N.; Tamma, F.; Lyons, K.E.; Pahwa, R.; Lang, A.; Deuschl, G. Subthalamic nucleus deep brain stimulation: Summary and meta-analysis of outcomes. *Mov. Disord.* **2006**, *21*, S290–S304. [CrossRef]

102. Saint-Cyr, J.A.; Trépanier, L.L.; Kumar, R.; Lozano, A.M.; Lang, A.E. Neuropsychological consequences of chronic bilateral stimulation of the subthalamic nucleus in Parkinson's disease. *Brain* **2000**, *123*, 2091–2108. [CrossRef]

103. Shih, L.; Tarsy, D. Deep brain stimulation for the treatment of atypical parkinsonism. *Mov. Disord.* **2007**, *22*, 2149–2155. [CrossRef]

104. Moreau, C.; Defebvre, L.; Destee, A.; Bleuse, S.; Clement, F.; Blatt, J.L.; Krystkowiak, P.; Devos, D. STN-DBS frequency effects on freezing of gait in advanced Parkinson disease. *Neurology* **2008**, *71*, 80–84. [CrossRef] [PubMed]

105. Benabid, A.L.; Chabardes, S.; Mitrofanis, J.; Pollak, P. Deep brain stimulation of the subthalamic nucleus for the treatment of Parkinson's disease. *Lancet Neurol.* **2009**, *8*, 67–81. [CrossRef]

106. Jia, F.; Shukla, A.W.; Hu, W.; Almeida, L.; Holanda, V.; Zhang, J.; Meng, F.; Okun, M.S.; Li, L. Deep Brain Stimulation at Variable Frequency to Improve Motor Outcomes in Parkinson's Disease. *Mov. Disord. Clin. Pr.* **2018**, *5*, 538–541. [CrossRef] [PubMed]

107. Roediger, J.; A Dembek, T.; Achtzehn, J.; Busch, J.L.; Krämer, A.-P.; Faust, K.; Schneider, G.-H.; Krause, P.; Horn, A.; A Kühn, A. Automated deep brain stimulation programming based on electrode location: A randomised, crossover trial using a data-driven algorithm. *Lancet Digit. Heal.* **2022**, *5*, e59–e70. [CrossRef] [PubMed]

108. Mathkour, M.; Garces, J.; Scullen, T.; Hanna, J.; Valle-Giler, E.; Kahn, L.; Arrington, T.; Houghton, D.; Lea, G.; Biro, E.; et al. Short- and Long-Term Outcomes of Deep Brain Stimulation in Patients 70 Years and Older with Parkinson Disease. *World Neurosurg.* **2016**, *97*, 247–252. [CrossRef]

109. Hanna, J.A.; Scullen, T.; Kahn, L.; Mathkour, M.; Gouveia, E.E.; Garces, J.; Evans, L.M.; Lea, G.; Houghton, D.J.; Biro, E.; et al. Comparison of elderly and young patient populations treated with deep brain stimulation for Parkinson's disease: Long-term outcomes with up to 7 years of follow-up. *J. Neurosurg.* **2019**, *131*, 807–812. [CrossRef]

110. Videnovic, A.; Metman, L.V. Deep brain stimulation for Parkinson's disease: Prevalence of adverse events and need for standardized reporting. *Mov. Disord.* **2007**, *23*, 343–349. [CrossRef]
111. Follett, K.A.; Weaver, F.M.; Stern, M.; Hur, K.; Harris, C.L.; Luo, P.; Marks, W.J.; Rothlind, J.; Sagher, O.; Moy, C.; et al. Pallidal versus Subthalamic Deep-Brain Stimulation for Parkinson's Disease. *N. Engl. J. Med.* **2010**, *362*, 2077–2091. [CrossRef]
112. Volkmann, J.; Albanese, A.; Kulisevsky, J.; Tornqvist, A.-L.; Houeto, J.-L.; Pidoux, B.; Bonnet, A.-M.; Mendes, A.; Benabid, A.-L.; Fraix, V.; et al. Long-term effects of pallidal or subthalamic deep brain stimulation on quality of life in Parkinson's disease. *Mov. Disord.* **2009**, *24*, 1154–1161. [CrossRef]
113. Limousin, P.; Speelman, J.D.; Gielen, F.; Janssens, M. Multicentre European study collaborators Multicentre European study of thalamic stimulation in parkinsonian and essential tremor. *J. Neurol. Neurosurg. Psychiatry* **1999**, *66*, 289–296. [CrossRef]
114. Gervais-Bernard, H.; Xie-Brustolin, J.; Mertens, P.; Polo, G.; Klinger, H.; Adamec, D.; Broussolle, E.; Thobois, S. Bilateral subthalamic nucleus stimulation in advanced Parkinson's disease: Five year follow-up. *J. Neurol.* **2009**, *256*, 225–233. [CrossRef]
115. De Ture, M.A.; Dickson, D.W. The neuropathological diagnosis of Alzheimer's disease. *Mol. Neurodegener.* **2019**, *14*, 32. [CrossRef]
116. Joe, E.; Ringman, J.M. Cognitive symptoms of Alzheimer's disease: Clinical management and prevention. *BMJ* **2019**, *367*, l6217. [CrossRef]
117. Crews, L.; Masliah, E. Molecular mechanisms of neurodegeneration in Alzheimer's disease. *Hum. Mol. Genet.* **2010**, *19*, R12–R20. [CrossRef]
118. O'Brien, R.J.; Wong, P.C. Amyloid Precursor Protein Processing and Alzheimer's Disease. *Annu. Rev. Neurosci.* **2011**, *34*, 185–204. [CrossRef]
119. Zhang, H.; Wei, W.; Zhao, M.; Ma, L.; Jiang, X.; Pei, H.; Cao, Y.; Li, H. Interaction between Aβ and Tau in the Pathogenesis of Alzheimer's Disease. *Int. J. Biol. Sci.* **2021**, *17*, 2181–2192. [CrossRef]
120. Yu, Y.; Ye, R.D. Microglial Aβ Receptors in Alzheimer's Disease. *Cell. Mol. Neurobiol.* **2015**, *35*, 71–83. [CrossRef]
121. Fish, P.V.; Steadman, D.; Bayle, E.D.; Whiting, P. New approaches for the treatment of Alzheimer's disease. *Bioorganic Med. Chem. Lett.* **2018**, *29*, 125–133. [CrossRef]
122. Vaz, M.; Silvestre, S. Alzheimer's disease: Recent treatment strategies. *Eur. J. Pharmacol.* **2020**, *887*, 173554. [CrossRef]
123. Bartus, R.T.; Dean, R.L., III; Beer, B.; Lippa, A.S. The cholinergic hypothesis of geriatric memory dysfunction. *Science* **1982**, *217*, 408–414. [CrossRef]
124. Birks, J. Cholinesterase inhibitors for Alzheimer's disease. In *Cochrane Database of Systematic Reviews*; The Cochrane Collaboration: Oxford, UK, 2006. [CrossRef]
125. Bullock, R.; Dengiz, A. Cognitive performance in patients with Alzheimer's disease receiving cholinesterase inhibitors for up to 5 years. *Int. J. Clin. Pract.* **2005**, *59*, 817–822. [CrossRef] [PubMed]
126. Tagliavini, F.; Tiraboschi, P.; Federico, A. Alzheimer's disease: The controversial approval of Aducanumab. *Neurol. Sci.* **2021**, *42*, 3069–3070. [CrossRef] [PubMed]
127. Nisticò, R.; Borg, J.J. Aducanumab for Alzheimer's disease: A regulatory perspective. *Pharmacol. Res.* **2021**, *171*, 105754. [CrossRef] [PubMed]
128. Barcia, J.A.; Viloria, M.A.; Yubero, R.; Sanchez-Sanchez-Rojas, L.; López, A.; Strange, B.A.; Cabrera, M.; Canuet, L.; Gil, P.; Nombela, C. Directional DBS of the Fornix in Alzheimer's Disease Achieves Long-Term Benefits: A Case Report. *Front. Aging Neurosci.* **2022**, *14*. [CrossRef] [PubMed]
129. Deep Brain Stimulation in Alzheimer's Disease: Biomarkers and Dose Optimization—Tabular View—ClinicalTrials.gov, (n.d.). Available online: https://clinicaltrials.gov/ct2/show/record/NCT04856072?term=Deep+Brain+Stimulation&cond=Alzheimer+Disease&draw=2&rank=2 (accessed on 19 December 2022).
130. Deep Brain Stimulation for the Treatment of Alzheimer's Disease—Tabular View—ClinicalTrials.gov, (n.d.). Available online: https://clinicaltrials.gov/ct2/show/record/NCT01559220?term=Deep+Brain+Stimulation&cond=Alzheimer+Disease&draw=2&rank=3 (accessed on 19 December 2022).
131. The Safety and Efficacy of Long-term Treatment of PINS Stimulator System for Patients with Alzheimer's Disease—Tabular View—ClinicalTrials.gov, (n.d.). Available online: https://clinicaltrials.gov/ct2/show/record/NCT02253043?term=Deep+Brain+Stimulation&cond=Alzheimer+Disease&draw=2&rank=4 (accessed on 19 December 2022).
132. Deep Brain Stimulation for Alzheimer's Disease—Tabular View—ClinicalTrials.gov, (n.d.). Available online: https://clinicaltrials.gov/ct2/show/record/NCT03959124?term=Deep+Brain+Stimulation&cond=Alzheimer+Disease&draw=2&rank=6 (accessed on 19 December 2022).
133. Deep Brain Stimulation (DBS) of the Nucleus Basalis Meynert (NBM) to Treat Cognitive Deficits in Light to Moderate Alzheimer's Disease—Tabular View—ClinicalTrials.gov, (n.d.). Available online: https://clinicaltrials.gov/ct2/show/record/NCT01094145?term=Deep+Brain+Stimulation&cond=Alzheimer+Disease&draw=2&rank=7 (accessed on 19 December 2022).
134. Study of the Brain Stimulation Effect on Memory Impairment in Alzheimer Disease—Tabular View—ClinicalTrials.gov, (n.d.). Available online: https://clinicaltrials.gov/ct2/show/record/NCT00947934?term=Deep+Brain+Stimulation&cond=Alzheimer+Disease&draw=2&rank=9 (accessed on 19 December 2022).
135. Deep Brain Stimulation (DBS) for Alzheimer's Disease—Tabular View—ClinicalTrials.gov, (n.d.). Available online: https://clinicaltrials.gov/ct2/show/record/NCT00658125?term=Deep+Brain+Stimulation&cond=Alzheimer+Disease&draw=2&rank=10 (accessed on 19 December 2022).

136. ADvance II Study: DBS-f in Patients With Mild Alzheimer's Disease—Tabular View—ClinicalTrials.gov, (n.d.). Available online: https://clinicaltrials.gov/ct2/show/record/NCT03622905?term=Deep+Brain+Stimulation&cond=Alzheimer+Disease&draw=2&rank=11 (accessed on 19 December 2022).

137. Fornix and NbM as Targets of Stimulation In Alzheimer's Disease—Tabular View—ClinicalTrials.gov, (n.d.). Available online: https://clinicaltrials.gov/ct2/show/record/NCT03352739?term=Deep+Brain+Stimulation&cond=Alzheimer+Disease&draw=2&rank=14 (accessed on 19 December 2022).

138. Leplus, A.; Lauritzen, I.; Melon, C.; Goff, L.K.-L.; Fontaine, D.; Checler, F. Chronic fornix deep brain stimulation in a transgenic Alzheimer's rat model reduces amyloid burden, inflammation, and neuronal loss. *Brain Struct Funct.* **2019**, *224*, 363–372. [CrossRef]

139. Li, R.; Zhang, C.; Rao, Y.; Yuan, T.-F. Deep brain stimulation of fornix for memory improvement in Alzheimer's disease: A critical review. *Ageing Res. Rev.* **2022**, *79*, 101668. [CrossRef]

140. Koulousakis, P.; Hove, D.V.D.; Visser-Vandewalle, V.; Sesia, T. Cognitive Improvements After Intermittent Deep Brain Stimulation of the Nucleus Basalis of Meynert in a Transgenic Rat Model for Alzheimer's Disease: A Preliminary Approach. *J. Alzheimers Dis.* **2020**, *73*, 461–466. [CrossRef]

141. Huang, C.; Chu, H.; Ma, Y.; Zhou, Z.; Dai, C.; Huang, X.; Fang, L.; Ao, Q.; Huang, D. The neuroprotective effect of deep brain stimulation at nucleus basalis of Meynert in transgenic mice with Alzheimer's disease. *Brain Stimul.* **2018**, *12*, 161–174. [CrossRef]

142. Jakobs, M.; Lee, D.J.; Lozano, A.M. Modifying the progression of Alzheimer's and Parkinson's disease with deep brain stimulation. *Neuropharmacology* **2019**, *171*, 107860. [CrossRef]

143. Smith, G.S.; Laxton, A.W.; Tang-Wai, D.F.; McAndrews, M.P.; Diaconescu, A.O.; Workman, C.; Lozano, A. Increased Cerebral Metabolism After 1 Year of Deep Brain Stimulation in Alzheimer Disease. *Arch. Neurol.* **2012**, *69*, 1141–1148. [CrossRef]

144. Aldehri, M.; Temel, Y.; Jahanshahi, A.; Hescham, S. Fornix deep brain stimulation induces reduction of hippocampal synaptophysin levels. *J. Chem. Neuroanat.* **2018**, *96*, 34–40. [CrossRef]

145. Maltête, D.; Wallon, D.; Bourilhon, J.; Lefaucheur, R.; Danaila, T.; Thobois, S.; Defebvre, L.; Dujardin, K.; Houeto, J.-L.; Godefroy, O.; et al. Nucleus Basalis of Meynert Stimulation for Lewy Body Dementia. *Neurology* **2020**, *96*, e684–e697. [CrossRef]

146. Lam, J.; Lee, J.; Liu, C.Y.; Lozano, A.M.; Lee, D.J. Deep Brain Stimulation for Alzheimer's Disease: Tackling Circuit Dysfunction. *Neuromodulation Technol. Neural Interface* **2020**, *24*, 171–186. [CrossRef]

147. Mohs, R.C.; Knopman, D.; Petersen, R.C.; Ferris, S.H.; Ernesto, C.; Grundman, M.; Sano, M.; Beiliauskas, L.; Geldmacher, D.; Clark, C.; et al. Development of cognitive instruments for use in clinical trials of antidementia drugs: Additions to the Alzheimer's Disease Assessment Scale that broaden its scope. *Alzheimer Dis. Assoc. Disord.* **1997**, *11*, 13–21. Available online: https://journals.lww.com/alzheimerjournal/pages/articleviewer.aspx?year=1997&issue=00112&article=00003&type=Abstract&casa_token=T3Ve16P7d0QAAAAA:l-UlJenZW3fORcJ6SK6RyU0Pc1RsMlD4XfRQ-7QkgtX4Rn4XDeW9bM0ggRFLQzF83VUjBBkcmm4kh7qqiVjmNJkGQQ (accessed on 19 December 2022). [CrossRef]

148. Peña-Casanova, J. Alzheimer's Disease Assessment Scale–Cognitive in Clinical Practice. *Int. Psychogeriatr.* **1997**, *9*, 105–114. [CrossRef]

149. Tombaugh, T.N.; McIntyre, N.J. The Mini-Mental State Examination: A Comprehensive Review. *J. Am. Geriatr. Soc.* **1992**, *40*, 922–935. [CrossRef]

150. Doraiswamy, P.M.; Bieber, F.; Kaiser, L.; Krishnan, K.R.; Reuning-Scherer, J.; Gulanski, B. The Alzheimer's disease assessment scale. *Neurology* **1997**, *48*, 1511–1517. [CrossRef]

151. Morris, J.C. Clinical Dementia Rating: A Reliable and Valid Diagnostic and Staging Measure for Dementia of the Alzheimer Type. *Int. Psychogeriatr.* **1997**, *9*, 173–176. [CrossRef]

152. Carnero-Pardo, C. Should the Mini-Mental State Examination be retired? *Neurología* **2014**, *29*, 473–481. [CrossRef]

153. Hescham, S.; Aldehri, M.; Temel, Y.; Alnaami, I.; Jahanshahi, A. Deep brain stimulation for Alzheimer's Disease: An update. *Surg. Neurol. Int.* **2018**, *9*, 58. [CrossRef]

154. Schneider, L.S.; Kennedy, R.E.; Wang, G.; Cutter, G.R. Differences in Alzheimer disease clinical trial outcomes based on age of the participants. *Neurology* **2015**, *84*, 1121–1127. [CrossRef]

155. Hely, M.A.; Reid, W.G.J.; Adena, M.A.; Halliday, G.M.; Morris, J.G.L. The Sydney multicenter study of Parkinson's disease: The inevitability of dementia at 20 years. *Mov. Disord.* **2008**, *23*, 837–844. [CrossRef]

156. Sankar, T.; Chakravarty, M.M.; Bescos, A.; Lara, M.; Obuchi, T.; Laxton, A.W.; McAndrews, M.P.; Tang-Wai, D.F.; Workman, C.I.; Smith, G.S.; et al. Deep Brain Stimulation Influences Brain Structure in Alzheimer's Disease. *Brain Stimul.* **2014**, *8*, 645–654. [CrossRef]

157. Akwa, Y.; Gondard, E.; Mann, A.; Capetillo-Zarate, E.; Alberdi, E.; Matute, C.; Marty, S.; Vaccari, T.; Lozano, A.M.; E Baulieu, E.; et al. Synaptic activity protects against AD and FTD-like pathology via autophagic-lysosomal degradation. *Mol. Psychiatry* **2017**, *23*, 1530–1540. [CrossRef] [PubMed]

158. Chen, J.; Hofman, K.; Koprich, J.; Brotchie, J.; Volkmann, J.; Ip, C. P 15 Long-term subthalamic deep brain stimulation modulates pathological beta oscillations in the AAV-A53T-Synuclein Parkinson's disease rat model. *Clin. Neurophysiol.* **2022**, *137*, e23–e24. [CrossRef]

159. Gondard, E.; Chau, H.N.; Mann, A.; Tierney, T.S.; Hamani, C.; Kalia, S.K.; Lozano, A.M. Rapid Modulation of Protein Expression in the Rat Hippocampus Following Deep Brain Stimulation of the Fornix. *Brain Stimul.* **2015**, *8*, 1058–1064. [CrossRef] [PubMed]

160. Okun, M.; Fernandez, H.H.; Wu, S.S.; Kirsch-Darrow, L.; Bowers, D.; Bova, F.J.; Bs, M.S.; Jacobson, C.E.; Wang, X.; Gordon, C.W.; et al. Cognition and mood in Parkinson's disease in subthalamic nucleus versus globus pallidus interna deep brain stimulation: The COMPARE Trial. *Ann. Neurol.* **2009**, *65*, 586–595. [CrossRef] [PubMed]

161. Laxton, A.W.; Lozano, A.M. Deep Brain Stimulation for the Treatment of Alzheimer Disease and Dementias. *World Neurosurg.* **2013**, *80*, S28.e1–S28.e8. [CrossRef]

162. Yu, D.; Yan, H.; Zhou, J.; Yang, X.; Lu, Y.; Han, Y. A circuit view of deep brain stimulation in Alzheimer's disease and the possible mechanisms. *Mol. Neurodegener.* **2019**, *14*, 1–12. [CrossRef]

163. Loh, A.; Gwun, D.; Chow, C.T.; Boutet, A.; Tasserie, J.; Germann, J.; Santyr, B.; Elias, G.; Yamamoto, K.; Sarica, C.; et al. Probing responses to deep brain stimulation with functional magnetic resonance imaging. *Brain Stimul.* **2022**, *15*, 683–694. [CrossRef]

164. Krause, K.J.; Phibbs, F.; Davis, T.; Fabbri, D. Predicting Motor Responsiveness to Deep Brain Stimulation with Machine Learning. *AMIA Annu. Symp. Proc.* **2021**, *2021*, 651.

165. Peralta, M.; Jannin, P.; Baxter, J.S. Machine learning in deep brain stimulation: A systematic review. *Artif. Intell. Med.* **2021**, *122*, 102198. [CrossRef]

166. Boutet, A.; Madhavan, R.; Elias, G.J.B.; Joel, S.E.; Gramer, R.; Ranjan, M.; Paramanandam, V.; Xu, D.; Germann, J.; Loh, A.; et al. Predicting optimal deep brain stimulation parameters for Parkinson's disease using functional MRI and machine learning. *Nat. Commun.* **2021**, *12*, 1–13. [CrossRef]

167. Lipsman, N.; Glannon, W. Brain, mind and machine: What are the implications of deep brain stimulation for perceptions of personal identity, agency and free will? *Bioethics* **2012**, *27*, 465–470. [CrossRef]

168. Bell, E.; Maxwell, B.; McAndrews, M.P.; Sadikot, A.; Racine, E. Deep Brain Stimulation and Ethics: Perspectives from a Multisite Qualitative Study of Canadian Neurosurgical Centers. *World Neurosurg.* **2011**, *76*, 537–547. [CrossRef]

169. Mendelsohn, D.; Lipsman, N.; Bernstein, M. Neurosurgeons' perspectives on psychosurgery and neuroenhancement: A qualitative study at one center. *J. Neurosurg.* **2010**, *113*, 1212–1218. [CrossRef]

International Journal of
Molecular Sciences

Article

Safety of Special Waveform of Transcranial Electrical Stimulation (TES): In Vivo Assessment

Muhammad Adeel [1,2,†], Chun-Ching Chen [3,†], Bor-Shing Lin [4], Hung-Chou Chen [5,6], Jian-Chiun Liou [1], Yu-Ting Li [7] and Chih-Wei Peng [1,2,8,*]

[1] School of Biomedical Engineering, College of Biomedical Engineering, Taipei Medical University, Taipei 110, Taiwan; d845108004@tmu.edu.tw (M.A.); jcliou@tmu.edu.tw (J.-C.L.)
[2] International PhD Program in Biomedical Engineering, College of Biomedical Engineering, Taipei Medical University, Taipei 110, Taiwan
[3] Department of Interaction Design, College of Design, National Taipei University of Technology, Taipei 106, Taiwan; cceugene@ntut.edu.tw
[4] Department of Computer Science and Information Engineering, National Taipei University, New Taipei City 237, Taiwan; bslin@mail.ntpu.edu.tw
[5] Department of Physical Medicine and Rehabilitation, School of Medicine, College of Medicine, Taipei Medical University, Taipei 110, Taiwan; 10462@s.tmu.edu.tw
[6] Department of Physical Medicine and Rehabilitation, Shuang Ho Hospital, Taipei Medical University, New Taipei City 235, Taiwan
[7] Taiwan Instrument Research Institute, National Applied Research Laboratories, Hsinchu 30261, Taiwan; ytl@tiri.narl.org.tw
[8] School of Gerontology Health Management, College of Nursing, Taipei Medical University, Taipei 110, Taiwan
* Correspondence: cwpeng@tmu.edu.tw
† These authors contributed equally to this study.

Citation: Adeel, M.; Chen, C.-C.; Lin, B.-S.; Chen, H.-C.; Liou, J.-C.; Li, Y.-T.; Peng, C.-W. Safety of Special Waveform of Transcranial Electrical Stimulation (TES): In Vivo Assessment. *Int. J. Mol. Sci.* **2022**, *23*, 6850. https://doi.org/10.3390/ ijms23126850

Academic Editors: Masaru Tanaka, Simone Battaglia, Lydia Giménez-Llort, Chong Chen and Piril Hepsomali

Received: 22 April 2022
Accepted: 17 June 2022
Published: 20 June 2022

Publisher's Note: MDPI stays neutral with regard to jurisdictional claims in published maps and institutional affiliations.

Abstract: Intermittent theta burst (iTBS) powered by direct current stimulation (DCS) can safely be applied transcranially to induce neuroplasticity in the human and animal brain cortex. tDCS-iTBS is a special waveform that is used by very few studies, and its safety needs to be confirmed. Therefore, we aimed to evaluate the safety of tDCS-iTBS in an animal model after brain stimulations for 1 h and 4 weeks. Thirty-one Sprague Dawley rats were divided into two groups: (1) short-term stimulation for 1 h/session (sham, low, and high) and (2) long-term for 30 min, 3 sessions/week for 4 weeks (sham and high). The anodal stimulation applied over the primary motor cortex ranged from 2.5 to 4.5 mA/cm^2. The brain biomarkers and scalp tissues were assessed using ELISA and histological analysis (H&E staining) after stimulations. The caspase-3 activity, cortical myelin basic protein (MBP) expression, and cortical interleukin (IL-6) levels increased slightly in both groups compared to sham. The serum MBP, cortical neuron-specific enolase (NSE), and serum IL-6 slightly changed from sham after stimulations. There was no obvious edema or cell necrosis seen in cortical histology after the intervention. The short- and long-term stimulations did not induce significant adverse effects on brain and scalp tissues upon assessing biomarkers and conducting histological analysis.

Keywords: intermittent theta burst stimulation (iTBS); transcranial direct current stimulation (tDCS); electrical stimulation; safety parameters; primary cortex; in vivo; current density; duration; frequency; scalp

1. Introduction

Among the non-invasive neuromodulation approaches, transcranial magnetic (TMS) and direct current stimulations (DCS) are used to regulate cortical excitability [1,2]. Clinically, both stimulation techniques can induce neuroplasticity in improving motor and memory functions in patients [3–6]. Transcranial magnetic stimulation (TMS) can be applied in the form of continuous, repetitive, or burst waveforms through a magnetic coil over the head that works by generating an electric field in the brain via electromagnetism [7].

A special form of rTMS consists of a series of short bursts at a high inner frequency interspersed with brief periods of no stimulation, called theta-burst stimulation (TBS). It consists of short bursts of 50 Hz rTMS which are repeated at a rate in the theta range (5 Hz) as a continuous (cTBS) or intermittent (iTBS) train, and is the most widely utilized method [8,9]. One commonly used transcranial electrical stimulation (TES) is a transcranial DCS (tDCS), which utilizes large (25–35 cm^2) electrodes to provide a small amount of direct current (1–2 mA) to the scalp [10,11]. The stimulation alters brain function by depolarizing or hyperpolarizing the resting membrane potential of the cell. The current used in anodal tDCS depolarizes the resting membrane potential, increasing neuronal firing and producing long-term potentiation (LTP)-like effects. Cathodal tDCS induces a long-term depression (LTD)-like effect, which causes the resting membrane potential to become hyperpolarized. Because of reduced impulsive cell firing, this LTD-like effect reduces neuronal excitability [2,12].

According to Lisman and Idiart, theta-frequency oscillations regulate high-frequency gamma oscillations, which are related to cognitive processing in human recognition memory [13]. Furthermore, iTBS raises the amplitude of subsequent I-waves and generates LTP-like alterations at synaptic junctions in the motor cortex by altering the intrinsic circuitry of the motor cortex [8,14]. Furthermore, tDCS generates plastic after-effects via membrane polarization and NMDA receptor-mediated glutamatergic synaptic transmission, according to several human and animal investigations [15,16]. Previous studies reported that the iTBS-like anodal DC stimulation elicited the two mentioned neuroplastic processes of iTBS and tDCS at the same time, resulting in a cumulative impact on neuroplasticity. In comparison to traditional tDCS, it was found that iTBS-like anodal DC stimulation caused a larger amplitude of motor evoked potential (MEP) [7].

A previous study reported the use of 10 Hz and iTBS dorsomedial prefrontal cortex (DMPFC) rTMS as a safe and tolerable stimulation intervention for major depression. The outcomes were compared for 6 min iTBS and 30 min 10 Hz protocols in a clinical population [17]. In recent years, TES—specifically tDCS—has received more attention as a potential therapy for neurological and psychiatric illnesses. Several attractive clinical properties, including safety, tolerability, the convenience of use, affordability, and portability, have piqued interest in employing TES for therapeutic purposes [10]. It has been hypothesized that daily anodal tDCS across the left dorsolateral prefrontal cortex (DLPFC) relieves depression by reversing the hypoactivity found in major depressive disorder (MDD) [18]. One study proposed that depression is linked to lower levels of brain-derived neurotrophic factor (BDNF), a neurotrophin important for synaptic strengthening and neuronal endurance [19], and suggested that the upregulation of BDNF levels, as a critical neurobiological mechanism for depression alleviation, might be involved in antidepressant actions.

Based on simulated and animal studies, only a percentage of the electric current penetrates the cortex, resulting in neuronal polarization and excitability in the cortex [20–23] and hippocampus [20,24]. tDCS has been studied for many clinical and assistive applications, including depression [25,26], motor rehabilitation [27], speech rehabilitation [28–30], pain control [31–33], and working memory [34]. Regarding safety, tDCS was used in previous studies without any adverse effects when appropriate standards were implemented [10,35–39]. Currently, the use of tDCS across many applications is widespread, so more studies are required to standardize the safe tDCS dosing parameters [40–42]. Few studies have evaluated the effects of the special waveform of tDCS, i.e., iTBS. Conventionally, iTBS is outputted from TMS to induce neuroplasticity in nerve tissue [43,44]. However, recently, two studies by our search team reported the use of iTBS waveform delivered through electrical stimulations in animals in vivo experiments [7] and stroke patients [45] for its efficacy. In an animal study, it was reported that tDCS-iTBS (0.25 mA tDCS and 50 Hz iTBS for 190 s) is a feasible therapeutic approach for enhancing neuromodulatory effects [7], while the clinical study implied that the stroke patients had significantly improved upper limb functioning from using tDCS-iTBS (1 mA anodal tDCS and 1.5 mA iTBS for 20 min for

3 days/week for 6 weeks) stimulation along with rehabilitation compared to rehabilitation alone [45].

To assess the safety of tDCS on microglial cells in rodents, one study reported the activation of these cells by 0.5 mA anodal or cathodal stimulation for 15 min [46]. Another study showed that anodal tDCS at a current intensity of 0.4 mA (31.8 A/m^2 electrode current density) induced microglial alterations in neurodegeneration-related morphology [47]. The histological analysis of one study mentioned that anodal tDCS can induce brain lesions in rats at a stimulation intensity of 0.5 mA using 25 mm^2 electrodes (20 A/m^2) [48]. Liebetanz et al. reported the brain lesion threshold intensity for cathodal tDCS as 0.001–1 mA for 15–270 min (143 A/m^2) [49]. The safety effects of tDCS on the human brain were studied by using neuron-specific enolase (NSE), magnetic resonance imaging (MRI), and electroencephalography (EEG) data in previous studies [15,50,51]. There is a scarcity of research on the safety of special waveform of tDCS, i.e., iTBS, in animal cortex using biomarkers and immunohistological assays. One of the studies examined the effects of deep brain stimulation (DBS) on apoptosis in the hippocampal pedunculopontine tegmental nucleus (PPTg). It reported a decreased level of caspase-3 for apoptotic activity and myelin basic protein (MBP) for brain injury biomarkers, without any change in cytokine interleukin-6 (IL-6) for inflammation after the application of DBS [52]. According to the analysis of cortical activity markers, the expressions of GABA-synthesizing enzyme GAD67 (67 kD isoform of glutamate decarboxylase), calcium-binding proteins parvalbumin (PV), and calbindin (CB) were lowered by iTBS but not by cTBS one day after the last session. Because the magnetic stimulation did not target a specific cortical location, these alterations were visible in many cortical areas in all animals, regardless of whether they completed the task. However, the frontal and barrel cortex, which are engaged in the learning process, exhibited a considerably smaller reduction in PV and CB expression than the visual cortex, which was not involved in the activity [53].

Until the present research, no study has reported the neuroprotective effects of the special waveform of tDCS, i.e., iTBS, on a rat brain after short- and long-term stimulations using a low current density. Therefore, the rationale of this study was to evaluate the safety of the iTBS waveform delivered through electrical current power by the previously designed and implemented novel transcranial burst electrostimulation device, which confirmed the feasibility and neuroplastic effects of tDCS-iTBS waveform for neurostimulation on the brain [7] in vivo experiments. Thus, in our current study, we conducted brain and scalp tissue analysis to evaluate the safety of the tDCS-iTBS in the animal model after stimulations for 1 h and 4 weeks.

2. Results

The results of the present study demonstrate a non-significant change in biomarkers (caspase-3 activity, cortical and serum MBP levels, NSE expression, and cortical and serum IL-6 levels) and histology of cortical tissues after short- and long-term stimulations to highlight the safety of tDCS-iTBS waveform intervention (Table 1).

2.1. Effect of tDCS-iTBS on Caspase-3 Activity

In this study, we found that caspase-3 activity increased in low (2.5 mA/cm^2) and high current density (4.5 mA/cm^2) interventions in short- and long-term stimulations compared to sham. The high-density stimulation resulted in slightly more increased activity of caspase-3 in the rat cortex. The difference between sham, low, and high current density for short- and long-term stimulations did not reach a significance level (p values > 0.05, Figure 1).

Table 1. Descriptive statistics of biomarkers during short- and long-term stimulation ($n = 31$).

Subgroups		Sham		Low		High	
	Parameters	Mean ± SD	Min–Max	Mean ± SD	Min–Max	Mean ± SD	Min–Max
Groups							
	Caspase-3 (units)	0.39 ± 0.10	0.24–0.50	0.47 ± 0.11	0.34–0.56	0.48 ± 0.08	0.34–0.56
	Cortical MBP (ng/mg)	2.57 ± 0.57	2.10–3.55	2.62 ± 0.44	2.10–3.30	2.71 ± 0.27	2.51–3.18
Short-term stimulation ($n = 15$)	Serum MBP (ng/mL)	0.01 ± 0.02	0.00–0.05	0.02 ± 0.02	0.01–0.05	0.01 ± 0.02	0.00–0.05
	Cortical NSE (ng/mg)	5.96 ± 0.69	4.81–6.60	5.86 ± 1.16	3.93–6.80	6.26 ± 0.80	5.48–7.58
	Cortical IL-6 (pg/mg)	1748.81 ± 122.30	1577.53–1902.30	1871.44 ± 332.47	1435.70–2300.69	1985.92 ± 511.68	1275.53–2542.05
	Serum IL-6 (pg/mg)	848.71 ± 368.95	492.57–1436.97	830.49 ± 263.27	535.99–1200.00	888.29 ± 347.84	283.70–1138.80
	Caspase-3 (units)	0.47 ± 0.14	0.24–0.72	–	–	0.51 ± 0.21	0.23–0.87
	Cortical MBP (ng/mg)	2.71 ± 0.27	2.34–3.10	–	–	2.87 ± 0.25	2.50–3.20
Long-term stimulation ($n = 16$)	Serum MBP (ng/mL)	0.03 ± 0.01	0.01–0.04	–	–	0.02 ± 0.01	0.01–0.04
	Cortical NSE (ng/mg)	6.48 ± 0.72	5.40–7.30	–	–	6.33 ± 1.21	3.90–7.70
	Cortical IL-6 (pg/mg)	2046.38 ± 175.93	1833.00–2432.00	–	–	1930.00 ± 263.71	1576.00–2455.00
	Serum IL-6 (pg/mg)	801.63 ± 213.89	490.00–1200.00	–	–	865.75 ± 502.56	145.00–1600.00

Mean ± SD, average and standard deviation; min, minimum value; max, maximum value; short-term stimulation for 1 h/session ($n = 15$: sham = 5, low = 5, high = 5), long-term stimulation for 3 sessions/week for total 4 weeks ($n = 16$: sham = 8, high = 8), subgroups (low = 2.5 mA/cm^2, high = 4.5 mA/cm^2); n, number of animals; ng/mg, nanogram per milligram; ng/mL, nanogram per milliliter; pg/mg, picogram per milligram; –, study included three subgroups (sham, low, & high) for short-term stimulation while long-term stimulation has only two subgroups (sham & high).

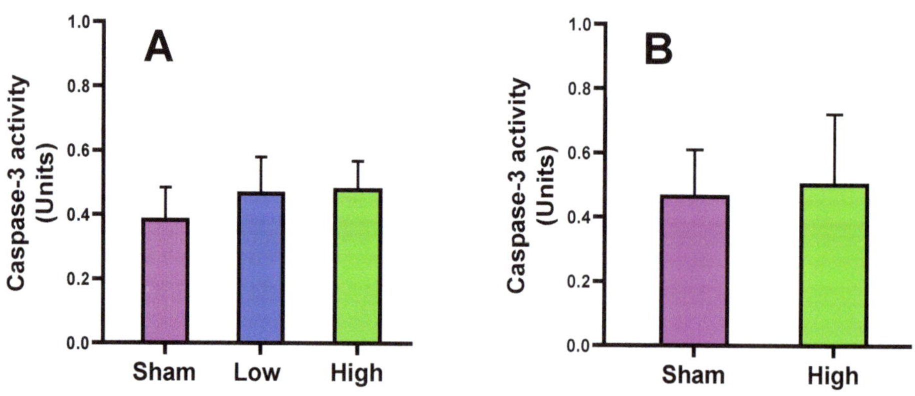

Figure 1. Effect of tDCS-iTBS on caspase-3 activity: (**A**) short-term ($n = 15$) and (**B**) long-term stimulations ($n = 16$).

2.2. *Effect of tDCS-iTBS on Brain and Skin/Serum Biomarkers*

2.2.1. Cortical MBP Expression

The myelin basic protein (MBP) is the most abundant protein of myelin in oligodendrocytes in the central nervous system (CNS) and Schwan cells in the peripheral nervous system (PNS). It is responsible for the myelination of neurons. In this study, the cortical expression of MBP (ng/mg) was increased in both short- and long-term stimulation groups on high current density interventions. None of the groups attained the level of significance (p values > 0.05, Figure 2).

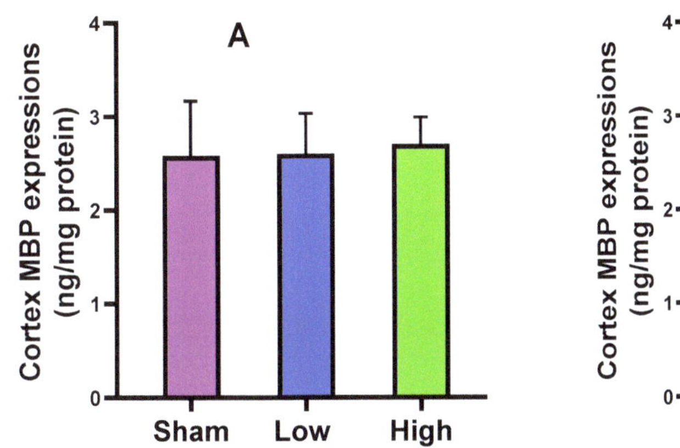
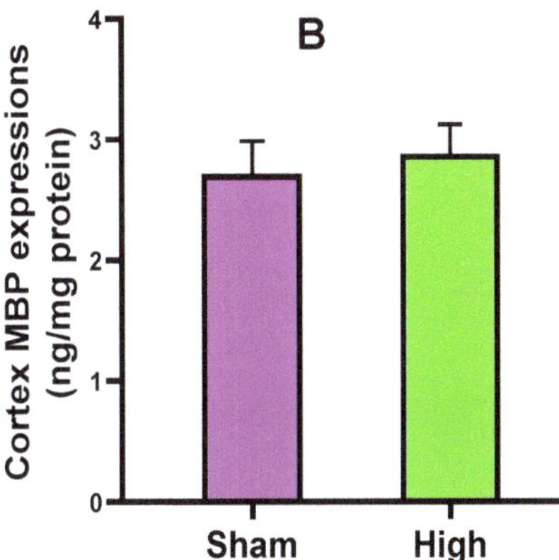

Figure 2. Effect of tDCS-iTBS on cortical MBP expressions (ng/mg protein): (**A**) short-term ($n = 15$) and (**B**) long-term stimulations ($n = 16$).

2.2.2. Serum MBP Expression

The serum MBP expression after short-term stimulation was higher in low current density intervention but lower in high current density intervention than sham for both short- and long-term stimulation groups without reaching a significant level (Figure 3).

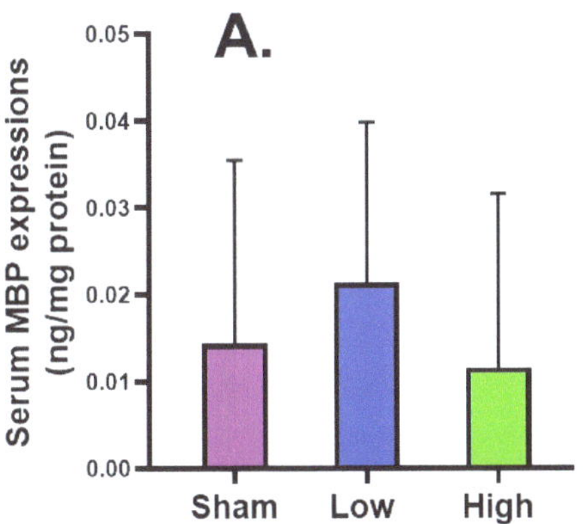
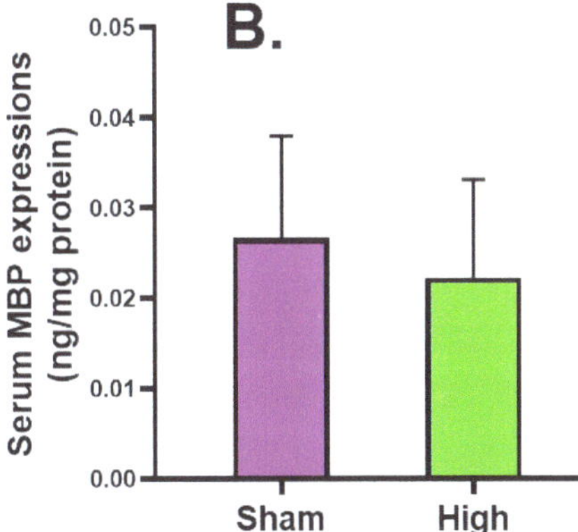

Figure 3. Effect of tDCS-iTBS on serum MBP expressions (ng/mg protein): (**A**) short-term ($n = 15$) and (**B**) long-term stimulations ($n = 16$).

2.2.3. Cortical NSE Expression

The neuron-specific enolase (NSE) is a prominent biomarker of ischemic brain injury. By the application of short-term stimulation, NSE expression was higher in high current intervention but did not change in low current intervention as compared to the sham. For long-term stimulation, no change in NSE expression was observed from the sham. For both groups, no intervention attained the level of significance (Figure 4).

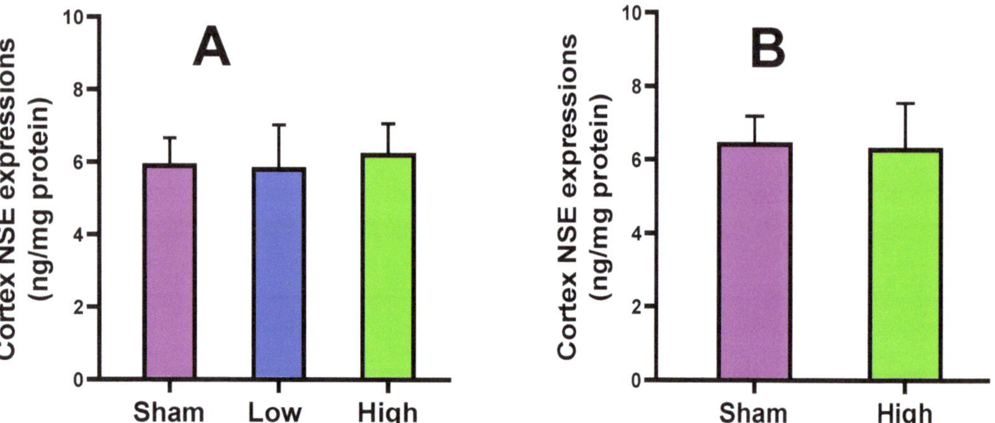

Figure 4. Effect of tDCS-iTBS on cortex NSE expressions (ng/mg protein): (**A**) short-term (n = 15) and (**B**) long-term stimulations (n = 16).

2.2.4. Cortical IL-6 Levels

The cortical IL-6 was increased for high current density rather than low current density intervention during short-term stimulation, and was decreased for high current density compared to sham for long-term stimulation, without any significant difference (Figure 5).

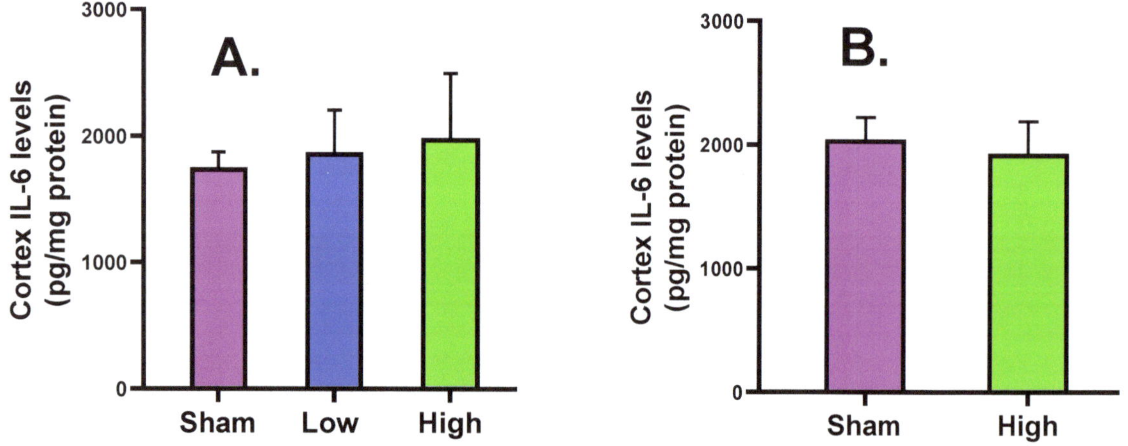

Figure 5. Effect of tDCS-iTBS on cortical IL-6 levels (pg/mg protein): (**A**) short-term (n = 15) and (**B**) long-term stimulations (n = 16).

2.2.5. Serum Tissue IL-6 Levels

The scalp skin tissue pro-inflammatory biomarker IL-6 level showed that high current intervention increased the IL-6 levels for both short- and long-term stimulation groups, but low current intervention decreased for short-term stimulation. There is no statistically significant difference between the three interventions, suggesting that high current density can affect the scalp tissue under-stimulation more than low current intervention (Figure 6).

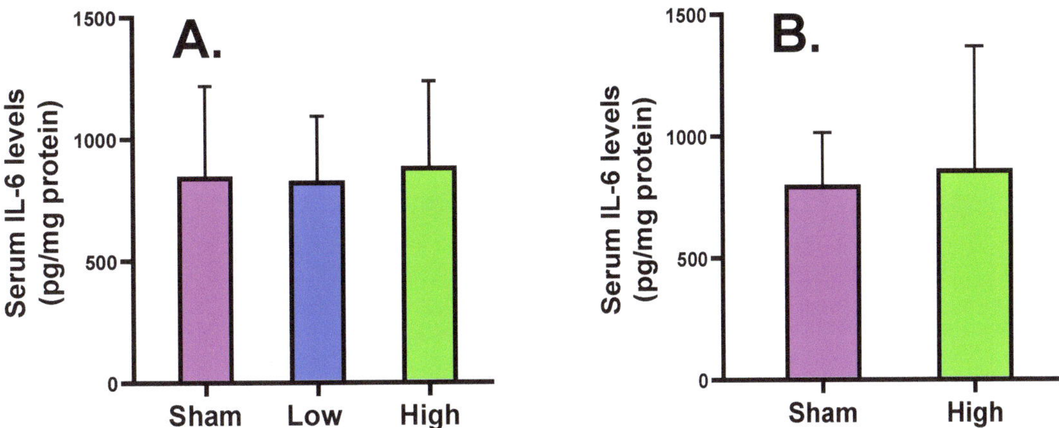

Figure 6. Effect of tDCS-iTBS on serum IL-6 levels (pg/mg protein): (**A**) short-term (n = 15) and (**B**) long-term stimulations (n = 16).

2.3. *Effect of tDCS-iTBS on Brain Tissue Morphology (H&E Staining)*

2.3.1. Short-Term Stimulation H&E Staining

Figure 7a–c presents cortical tissue histology using H&E staining. After applying short-term stimulation, only high current density intervention (Figure 7c) showed mild edema without obvious cell shrinkage, necrosis, or microglial changes. Low current density intervention (Figure 7b) did not show any change from the sham intervention (Figure 7a). These results reflect the safety of short-term stimulations on rat cortical tissue because there was no obvious difference in tissue morphology after either intervention.

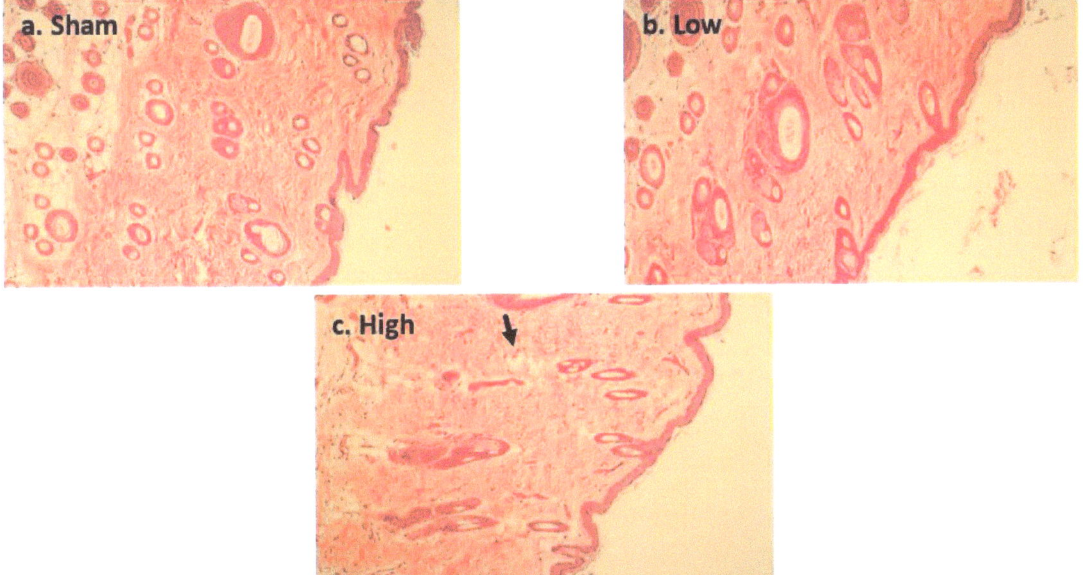

Figure 7. Effect of tDCS-iTBS on cortical tissue morphology (H&E staining) during short-term stimulation (n = 15); (**a**) sham stimulation, (**b**) low-intensity stimulation (2.5 mA/cm^2), and (**c**) high-intensity stimulation (4.5 mA/cm^2).

2.3.2. Long-Term Stimulation Staining

During long-term stimulation, no obvious change was observed in H&E staining, except for a mild decrease in the number of neuron cells in high current density intervention. There was no clear difference from sham intervention, and no edema, shrinkage, or necrosis was observed (Figure 8a–d).

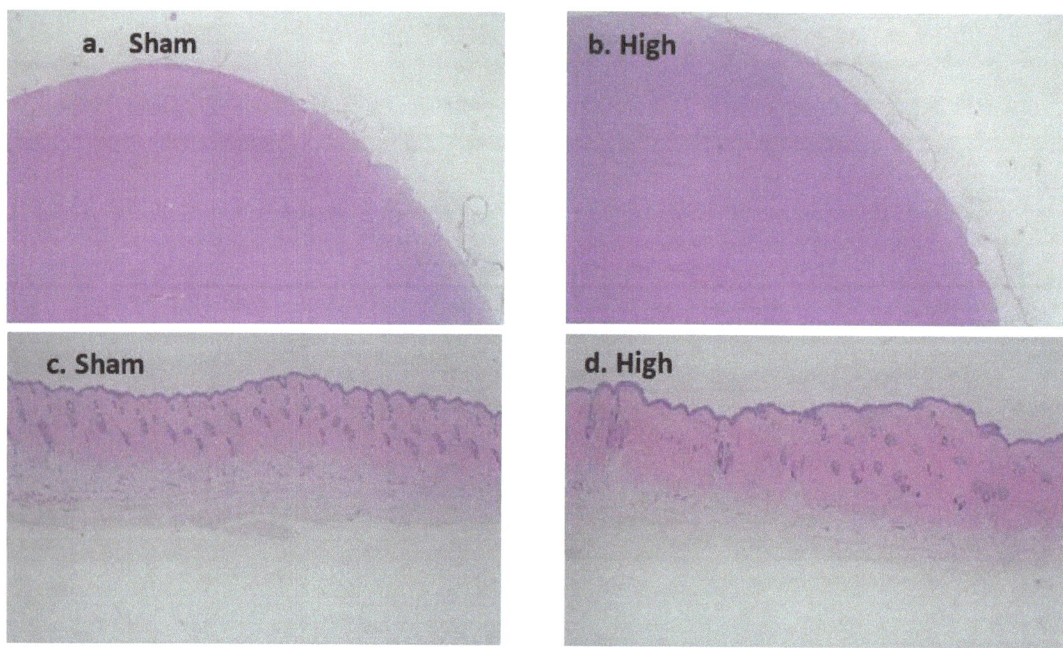

Figure 8. Effect of tDCS-iTBS on cortical tissue morphology (H&E staining) during long-term stimulation (n = 16); (**a**) sham and (**b**) high-intensity stimulation (4.5 mA/cm^2) subgroups 40 μm coronal tissue section, while (**c**) sham and (**d**) high-intensity stimulation (4.5 mA/cm^2) magnified tissue section to represent stimulation efffect.

3. Discussion

In the present study, we report the effects of the special waveform of tDCS, i.e., iTBS, which is electrically powered as opposed to conventional magnetic stimulation. For the two groups including short- and long-term stimulations, we did not find any significant effect of the low and high current density on the rat cortex and scalp skin tissues. Based on the biomarker analysis and histological staining, it is thus confirmed that tDCS-iTBS in the range of 2.5 to 4.5 mA/cm^2 did not induce any brain and skin damage because the current density is below the lesion threshold of 14.2 mA/cm^2 [49], ensuring the safety of tDCS-iTBS waveform for use in animals to induce neuroplasticity.

The safety of the tDCS-iTBS waveform was evaluated after short- and long-term stimulations. The caspase-3 activity of the rat cortex was not significantly changed in either group. For short-term stimulation, caspase-3 activity did not obviously increase (sham = 0.39 ± 0.10, low = 0.47 ± 0.11, and high = 0.48 ± 0.08), and the same effect was noted for long-term stimulation (sham = 0.47 ± 0.14 and high = 0.51 ± 0.21). One previous study reported the effect of tDCS on caspase-3 activity after middle cerebral artery occlusion in ischemic stroke rats. They observed that cathodal tDCS can significantly reduce caspase-3 expression (p = 0.000 *), which is raised after ischemic injury, and inhibit apoptosis at the injury site [54]. Based on our study results, after applying the different intensity tDCS-iTBS stimulations, the protein caspase-3 activity did not significantly differ from the sham.

The cortical MBP level was slightly increased for low (2.62 ± 0.44) and high (2.71 ± 0.27) current density compared to sham (2.57 ± 0.57) in the short-term stimulation group, and also slightly increased for high (2.87 ± 0.25) current density compared to sham (2.71 ± 0.27) in the long-term stimulation, without a significant difference. The serum MBP level slightly increased for only low (0.02 ± 0.02) current density compared to sham (0.01 ± 0.02) in the short-term stimulation, but a decreasing trend was noted for high (0.02 ± 0.01) current density compared to sham (0.03 ± 0.01) in the long-term stimulation. The NSE expression slightly decreased for low current density (5.86 ± 1.16) and increased for high current density (6.26 ± 0.80) compared to sham (5.96 ± 0.69) in the short-term stimulation, while it decreased for high current density (6.33 ± 1.21) compared to sham (6.48 ± 0.72) in the long-term stimulation. The cortical IL-6 level was increased for low (1871.44 ± 332.47) and high (1985.92 ± 511.68) current density compared to sham (1748.81 ± 122.30) in the short-term stimulation group but slightly decreased for high (1930.00 ± 263.71) current density compared to sham (2046.38 ± 175.93) in the long-term stimulation, without a significant change. The serum IL-6 level slightly decreased for low current density (830.49 ± 263.27) and increased for high current density (888.29 ± 347.84) compared to sham (848.71 ± 368.95) in the short-term stimulation, but an increase was noted for high (865.75 ± 502.56) current density compared to sham (801.63 ± 213.89) in the long-term stimulation. The histology of brain tissues of both groups reported no significant difference from sham in terms of edema, cell shrinkage, and necrosis, confirming the safety of the tDCS-iTBS waveform in the present study.

Based on recent research, tDCS protocols can be tolerable and safe if well-defined electrodes, stimulus durations, and intensities are employed [10]. The tDCS application is determined by its tolerability and safety. Tolerability means the presence of unpleasant and unexpected effects (such as tingling and itching sensation below the electrodes), while safety implies harmful effects. Overall, the presently used tDCS profiles are tolerable [55,56]. tDCS may cause erythema under the electrodes due to vasodilation, which is not a safety concern [57]. The safety is determined strictly by structural brain tissue damage, as reported by previous research [15,58]. Skin damage is occasionally reported in other studies [59] but mostly occurs due to the drying of contact media under the electrodes. The use of tap water which can induce skin burns or electrode gel is not appropriate, so electrode cream is recommended between the skin and electrode surface [60,61].

The stimulation current density used in this study ranged from 2.5 to 4.5 mA/cm^2 (tDCS: 1–3 mA; iTBS: 1.5 mA) applied for 1 h during short-term stimulations and 30 min during long-term stimulations. Previously, a study by Bikson et al. (2016) reported the safety of tDCS for 33,000 sessions on 1000 patients who received tDCS. Their report mentioned that there was no serious adverse effect on the brain by using conventional tDCS in clinical trials (≤40 min, ≤4 mA, and ≤7.2 C) [41]. The parameters for stimulation in our study spanned were 30 min–1 h, 1–3 mA tDCS combined with 1.5 mA iTBS for each subgroup, charge of 8.1–16.2 C, and current density of 2.5–4.5 mA/cm^2. Three research groups mentioned the brain injury threshold for tDCS application: (1) Liebetanz et al., 14.2 mA/cm^2 (used 500 µA current with 3.5 × 3.5 mm electrodes for 10 min) [49]; (2) Fritsch, applied 600 µA through 4 mm diameter electrode for 20 min; and (3) Jankord, through 500 µA using 5 × 5 mm electrodes for 60 min [41].

To assess the safety of the tDCS-iTBS waveform, there are only a few studies available. One study utilized anodal tDCS-iTBS waveform on rat cortex and found significant neuro-plastic effects (35% and 100% increase in MEP amplitude to induce LTP-like effects) than traditional tDCS waveform. They used iTBS in the range of 0 to 1.5 mA (0.25 mA/step) combined with tDCS of 0.25 mA for 190 s and did not report any adverse effects [7]. Another study applied 1 mA tDCS combined with 1.5 mA iTBS for 20 min on the primary motor cortex of stroke patients and found a significant improvement in upper limb recovery without adverse events [45]. Based on the available research, one study reported a scaling factor to determine the threshold damage of tDCS current application from rats to humans as 173 mA from Fritsch, 120 mA from Liebetanz, and 67 mA from Jankord [41].

Regarding the limitations of this study, (1) the safety of the tDCS-iTBS waveform was tested on only animals. Future studies should be conducted to test the safety of the tDCS-iTBS waveform on humans to start the clinical application to treat neurological disorders such as depression, stroke, spinal cord injury, or Alzheimer's disease. (2) A limited number of animals were tested in the present study; a large sample size in different subgroups should be considered for the reproducibility of results in future research. (3) Safety on the brain and scalp should be determined through EEG or MRI-based studies using the same waveform in future research.

4. Materials and Methods

4.1. Animal Handling and Tissue Extraction

4.1.1. Selection and Care of Animals

The study included 31 (n = 31) adult male Sprague Dawley rats with weights ranging from 200 to 260 g from BioLASCO Taiwan, Yilan, Taiwan. According to ethical guidelines, all of the rats were kept in a well-equipped animal facility in a sterile, temperature- and humidity-controlled setting. All of the rats were housed on a 12:12 h light:dark cycle with unlimited access to pellet nutrition and water ad libitum. The animals were accustomed for 7 to 10 days before being used in the study, and the experimental procedures and animal use/methods were authorized by Taipei Medical University's Institutional Animal Care and Use Committee (IACUC-TMU approval no. LAC 2016-0316, 2017-01-01).

4.1.2. Grouping and Experiment Schedule of Animals

All the study animals were categorized into two groups: (1) short-term stimulation (n = 15) and (2) long-term stimulation (n = 16). The animals in the short-term stimulation group were divided into three subgroups, each containing 5 animals: sham, low, and high current intensity tDCS-iTBS. The animals in the long-term stimulation group were divided into two subgroups, with each having 8 animals: sham and high current intensity tDCS-iTBS. By applying tDCS-iTBS stimulation, the rat brain and scalp skin tissues were extracted after 24 h for group 1 and after 4 weeks for group 2.

4.1.3. tDCS-iTBS Brain Stimulation Protocol

The tDCS-iTBS intervention was applied through an electrostimulation device which produces varied current intensity ranging from 0 to 3 mA tDCS and from 0 to 1.5 mA iTBS (0.25 mA/step) [7]. Before stimulation, each rat was given urethane anesthesia (1.2 g/kg). The two disposable electrodes (Medihightec Medical Company, Keelung, Taiwan) were placed at the same location for both groups [62]: an active anode (1 cm × 1 cm) over the right skull area on the primary motor cortex of the right forelimb concerning functional brain mapping already carried out in rat model [63], and a reference electrode (5 cm × 3 cm) over an abdomen.

The applied iTBS waveform output from tDCS (electric power) consisted of a 2 s train repeated after 10 s for 20 cycles (190 s, total 600 pulses). The frequency of each burst was 50 Hz, with three pulses of each 5 Hz repeated every 200 ms [9,64]. For the short-term stimulation, subgroups received (1) sham 0 mA tDCS-iTBS, (2) low 1 mA tDCS + 1.5 mA iTBS (2.5 mA/cm^2), and (3) high 3 mA tDCS + 1.5 mA iTBS (4.5 mA/cm^2) for 1 h for a single session. Long-term stimulation subgroups received (1) sham 0 mA tDCS-iTBS and (2) high 3 mA tDCS + 1.5 mA iTBS (4.5 mA/cm^2) for 30 min per day, 3 days a week for a total of 4 weeks (Table 2).

Table 2. Stimulation parameters for short- and long-term stimulations (tDCS-iTBS) ($n = 31$).

Parameters	Short-Term Stimulation ($n = 15$)			Long-Term Stimulation ($n = 16$)	
	Sham	Low	High	Sham	High
Current (mA)	0	2.5	4.5	0	4.5
Current density (mA/cm^2)	0	2.5	4.5	0	4.5
Charge (C)	0	9	16.2	0	8.1
Charge density (C/m^2)	0	90,000	162,000	0	81,000
Duration	1 h	1 h	1 h	30 min	30 min
Number of sessions	1 session	1 session	1 session	12 sessions	12 sessions

Short-term ($n = 15$: sham = 5, low = 5, high = 5) and long-term ($n = 16$: sham = 8, high = 8) stimulation consisted of iTBS output from tDCS; iTBS, 1.5 mA; tDCS, 1–3 mA (low =1 mA, high =3 mA); n, number of animals; 12 sessions, 3 sessions/week for 4 total weeks.

4.1.4. Sacrifice, Removal of Brain Samples, and Brain and Skin Tissue Collection

Animals were sacrificed using an overdose of urethane (4 g/kg) by cardiac puncture to extract the brain tissue. After being sacrificed, animals were beheaded using a scissor, and their brains were gently extracted. Then, the brain and skin tissues were rinsed in cold isotonic saline before being blotted on filter paper [65]. The brain tissue from the cortex and skin tissue were collected and rinsed again with saline. Excess saline was dried with absorbent paper, and the brain and skin tissues were frozen at $-80\ ^\circ$C. The tissues were then homogenized in saline (10% w/v) at 4 $^\circ$C. The homogenous specimens were spun at 3000 rpm for 10 min in a refrigerating centrifuge, and the supernatant was collected for examination [52]. Cortical apoptotic activity through caspase-3, cortical and serum MBP expression, cortical NSE, and cortical and serum IL-6 levels were analyzed.

4.2. Biochemical Analysis

4.2.1. Evaluation of Caspase-3 Activity

A caspase-3 colorimetric assay kit (Biovisiom, Milpitas, CA, USA) was utilized to measure caspase-3 activity. First, 50 μL of brain tissue specimen was added to a 96-well plate, followed by 50 μL of 2× reaction buffer (containing 10 mM DTT) and 5 μL of the 4 mM DEVD-pNA substrate. The plates were then put into an incubator at 37 $^\circ$C for 1–2 h, and the absorbance was determined in an ELISA reader at 405 nm [52].

4.2.2. Enzyme-Linked Immunosorbent Assay (ELISA)

The levels of MBP, NSE, and IL-6 were calculated through ELISA kits (Duo-Set; R&D Systems Inc., Minneapolis, MN, USA). The tissue specimens were placed into an incubator for 2 h with a biotinylated rabbit antibody, followed by the addition of streptavidin-conjugated horseradish peroxidase for 20 min. 3,3′,5,5′-tetramethylbenzidine/H$_2$O$_2$ (R&D Systems Inc., Minneapolis, MN, USA) was added for 30 min, which started the peroxidase reaction, and 0.5 M H$_2$SO$_4$ stopped it. The absorbance was recorded at 450 nm [52].

4.3. Histological Analysis

The 40 μm thick coronal section of brain tissues was processed for hematoxylin and eosin (H&E) staining. The histology of tissues was assessed through light microscopy for pathological changes such as edema, necrosis, and hematoma, as described in the literature on electrical neurotrauma [49,66] (Figure 9).

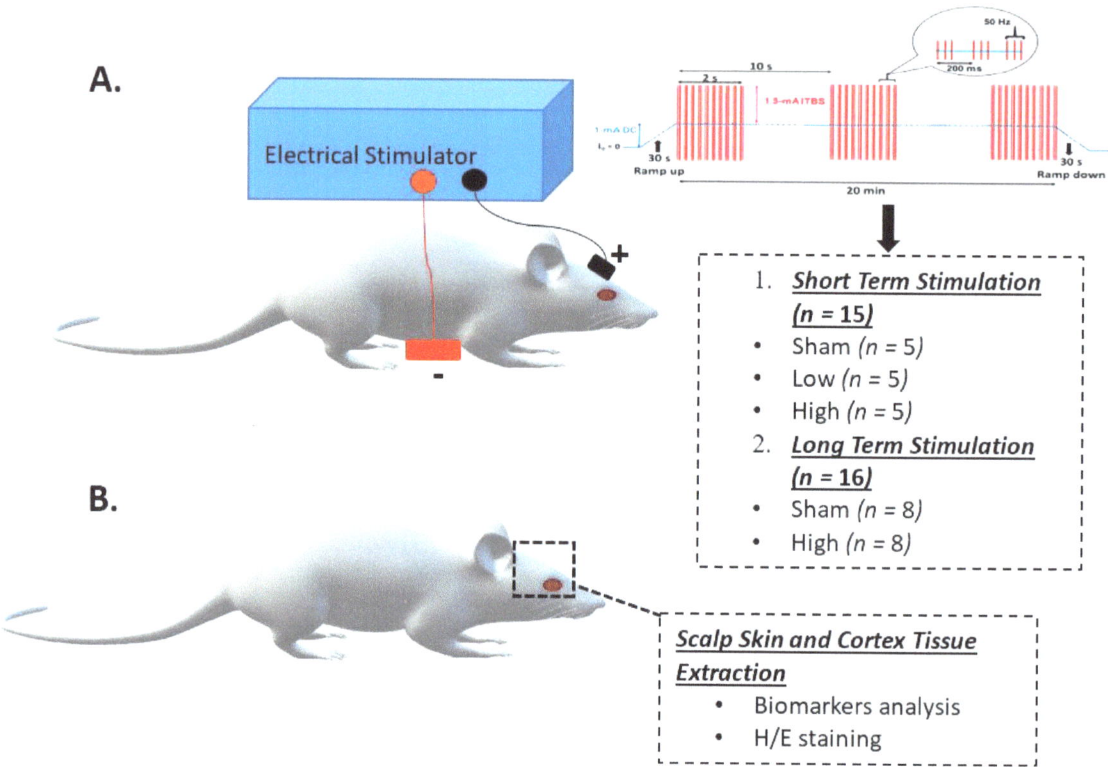

Figure 9. Schematic representation of (**A**) electrical stimulation of rat primary cortex; (**B**) scalp skin and cortex tissue extraction after short- and long-term stimulation interventions.

4.4. Statistical Analysis

The study data are reported using mean ± standard deviation (SD) and one-way analysis of variance (ANOVA) for comparing inter-group intervention for cortical caspase-3 activity, cortical and serum MBP expression, cortical NSE expression, and cortical and serum IL-6 levels. The level of significance was set to a *p*-value < 0.05. GraphPad Prism 6 software (GraphPad Software, San Diego, CA, USA) was used for statistical analysis.

5. Conclusions

The safety of tDCS-iTBS can be ensured using electrode placement (animal: active electrode on brain and reference electrode on abdomen; human: active electrode on head and reference on scalp), electrode size (animal: 1 cm × 1 cm; human: 5 cm × 5 cm), electric current penetration into the brain (animal: 1.5 mm thick skull; human: 6–7 mm thick skull), and current density (animal: cathodal tDCS = 142.9 A/m^2, anodal tDCS = 20 A/m^2 (our study = 25–45 A/m^2); human tDCS: 0.23–0.32 A/m^2). Other factors for safety include stimulation duration, polarity, and electrode shape.

In our current study, the non-significant results suggest no obvious difference from the sham group and reflect that tDCS-iTBS waveform in the range of 2.5 to 4.5 mA/cm^2 is safe for use in animals. The rationale behind this study was to determine the safety of our newly developed electrical current waveform for brain stimulation. We have reported the same waveform neuroplastic effects in previous research in rats and stroke patients using MEP recording and upper limb physical activity questionnaires, and time to expand tDCS-iTBS waveform outcomes to animal and clinical populations. According to the FDA Medwatch program database, tDCS caused no adverse effects in clinical trials. From this study, we report non-significant results after short- and long-term stimulations in an animal model. These outcomes open new avenues for other researchers to know the safety parameters and translate this stimulation protocol for large-scale animal testing and clinical settings by adjusting dosing parameters without adverse effects.

Based on the results, we conclude that short-term and long-term stimulation did not induce significant adverse effects on brain and skin tissues assessed through biomarkers and histological analysis. The tDCS-iTBS waveform from 2.5 to 4.5 mA/cm^2 can be used safely for therapeutic purposes in animals to attain stronger neuroplastic effects. In the present study, the current density was lower than the threshold reported in previous studies. It is worth noting that in this study, with 2.5–4.5 mA/cm^2 intensity, the effect on brain tissue biomarkers and skin/serum markers did not show a significant change from the sham, which proves that these stimulation interventions are safe to use for animal studies without causing brain tissue injury. This stimulation waveform can have applications in clinical use for various neurological diseases to generate neuroplastic effects without brain and scalp tissue lesions.

Future Research Direction

In the current study, we tried to determine the safety of the tDCS-iTBS waveform in an animal model and employed brain (caspase-3, MBP, and NSE) and scalp tissue injury biomarkers (IL-6) and brain tissue morphology (H&E staining). The genetic biomarkers of localized brain tissue stimulation, joule heating, and polarity of cathodal or anodal stimulation should be researched in future studies.

Author Contributions: Conceptualization, M.A. and C.-C.C.; methodology, B.-S.L. and C.-W.P.; software, M.A. and H.-C.C.; validation, J.-C.L., Y.-T.L., and C.-W.P.; formal analysis, M.A. and C.-C.C.; investigation, M.A. and C.-C.C.; resources, C.-W.P.; data curation, B.-S.L.; writing—original draft preparation, M.A.; writing—review and editing, C.-C.C. and C.-W.P.; visualization, H.-C.C.; supervision, Y.-T.L. and C.-W.P.; project administration, J.-C.L.; funding acquisition, C.-W.P. All authors have read and agreed to the published version of the manuscript.

Funding: The present study was generously funded by the Ministry of Science and Technology (110-2314-B-038-001, 110-2811-E-038-500-MY3, 110-2314-B-305-001, 109-2314-B-305-001, 109-2221-E-305-001-MY2, 109-2221-E-038-005-MY3, and 109-2314-B-038-132) and the University System of Taipei Joint Research Program USTP-NTUT-TMU-111-03 of Taiwan.

Institutional Review Board Statement: The study was approved by the experimental procedures and animal use/methods were authorized by Taipei Medical University's Institutional Animal Care and Use Committee (IACUC-TMU approval no. LAC 2016-0316, 2017-01-01).

Informed Consent Statement: Not applicable.

Data Availability Statement: Not applicable.

Acknowledgments: We would like to acknowledge our very dedicated research assistant and the animal care center of Taipei Medical University for helping with animal care, data collection, processing, etc.

Conflicts of Interest: The authors declare no conflict of interest.

References

1. Rossi, S.; Hallett, M.; Rossini, P.M.; Pascual-Leone, A. Safety, ethical considerations, and application guidelines for the use of transcranial magnetic stimulation in clinical practice and research. *Clin. Neurophysiol.* **2009**, *120*, 2008–2039. [CrossRef] [PubMed]
2. Nitsche, M.A.; Paulus, W. Excitability changes induced in the human motor cortex by weak transcranial direct current stimulation. *Brain Res.* **2000**, *527*, 633. [CrossRef] [PubMed]
3. Boroojerdi, B.; Prager, A.; Muellbacher, W.; Cohen, L.G. Reduction of human visual cortex excitability using 1-Hz transcranial magnetic stimulation. *Neurology* **2000**, *54*, 1529–1531. [CrossRef] [PubMed]
4. Chen, R.; Classen, J.; Gerloff, C.; Celnik, P.; Wassermann, E.M.; Hallett, M.; Cohen, L.G. Depression of motor cortex excitability by low-frequency transcranial magnetic stimulation. *Neurology* **1997**, *48*, 1398–1403. [CrossRef] [PubMed]
5. Pascual-Leone, A.; Valls-Solé, J.; Wassermann, E.M.; Hallett, M. Responses to rapid-rate transcranial magnetic stimulation of the human motor cortex. *Brain* **1994**, *117*, 847–858. [CrossRef]

6. Cantello, R.; Gianelli, M.; Bettucci, D.; Civardi, C.; De Angelis, M.S.; Mutani, R. Parkinson's disease rigidity: Magnetic motor evoked potentials in a small hand muscle. *Neurology* **1991**, *41*, 1449–1456. [CrossRef]
7. Li, Y.-T.; Chen, S.-C.; Yang, L.-Y.; Hsieh, T.-H.; Peng, C.-W. Designing and implementing a novel transcranial electrostimulation system for neuroplastic applications: A preliminary study. *IEEE Trans. Neural Syst. Rehabil. Eng.* **2019**, *27*, 805–813. [CrossRef]
8. Di Lazzaro, V.; Pilato, F.; Dileone, M.; Profice, P.; Oliviero, A.; Mazzone, P.; Insola, A.; Ranieri, F.; Meglio, M.; Tonali, P.A.; et al. The physiological basis of the effects of intermittent theta burst stimulation of the human motor cortex. *J. Physiol.* **2008**, *586*, 3871–3879. [CrossRef]
9. Huang, Y.-Z.; Edwards, M.J.; Rounis, E.; Bhatia, K.P.; Rothwell, J.C. Theta burst stimulation of the human motor cortex. *Neuron* **2005**, *45*, 201–206. [CrossRef]
10. Woods, A.J.; Antal, A.; Bikson, M.; Boggio, P.S.; Brunoni, A.R.; Celnik, P.; Cohen, L.G.; Fregni, F.; Herrmann, C.S.; Kappenman, E.S. A technical guide to tDCS, and related non-invasive brain stimulation tools. *Clin. Neurophysiol.* **2016**, *127*, 1031–1048. [CrossRef]
11. Brunoni, A.R.; Nitsche, M.A.; Bolognini, N.; Bikson, M.; Wagner, T.; Merabet, L.; Edwards, D.J.; Valero-Cabre, A.; Rotenberg, A.; Pascual-Leone, A. Clinical research with transcranial direct current stimulation (tDCS): Challenges and future directions. *Brain Stimul.* **2012**, *5*, 175–195. [CrossRef] [PubMed]
12. Nitsche, M.A.; Cohen, L.G.; Wassermann, E.M.; Priori, A.; Lang, N.; Antal, A.; Paulus, W.; Hummel, F.; Boggio, P.S.; Fregni, F. Transcranial direct current stimulation: State of the art 2008. *Brain Stimul.* **2008**, *1*, 206–223. [CrossRef] [PubMed]
13. Lisman, J.E.; Idiart, M.A. Storage of 7 +/- 2 short-term memories in oscillatory subcycles. *Science* **1995**, *267*, 1512–1515. [CrossRef] [PubMed]
14. Di Lazzaro, V.; Pilato, F.; Saturno, E.; Oliviero, A.; Dileone, M.; Mazzone, P.; Insola, A.; Tonali, P.A.; Ranieri, F.; Huang, Y.Z.; et al. Theta-burst repetitive transcranial magnetic stimulation suppresses specific excitatory circuits in the human motor cortex. *J. Physiol.* **2005**, *565*, 945–950. [CrossRef] [PubMed]
15. Nitsche, M.A.; Nitsche, M.S.; Klein, C.C.; Tergau, F.; Rothwell, J.C.; Paulus, W. Level of action of cathodal DC polarisation induced inhibition of the human motor cortex. *Clin. Neurophysiol.* **2003**, *114*, 600–604. [CrossRef]
16. Stagg, C.J.; Nitsche, M.A. Physiological basis of transcranial direct current stimulation. *Neurosci. A Rev. J. Bringing Neurobiol. Neurol. Psychiatry* **2011**, *17*, 37–53. [CrossRef]
17. Bakker, N.; Shahab, S.; Giacobbe, P.; Blumberger, D.M.; Daskalakis, Z.J.; Kennedy, S.H.; Downar, J. rTMS of the dorsomedial prefrontal cortex for major depression: Safety, tolerability, effectiveness, and outcome predictors for 10 Hz versus intermittent theta-burst stimulation. *Brain Stimul.* **2015**, *8*, 208–215. [CrossRef]
18. Mayberg, H.S.; Brannan, S.K.; Tekell, J.L.; Silva, J.A.; Mahurin, R.K.; McGinnis, S.; Jerabek, P.A. Regional metabolic effects of fluoxetine in major depression: Serial changes and relationship to clinical response. *Biol. Psychiatry* **2000**, *48*, 830–843. [CrossRef]
19. Duman, R.S.; Monteggia, L.M. A neurotrophic model for stress-related mood disorders. *Biol. Psychiatry* **2006**, *59*, 1116–1127. [CrossRef]
20. Rohan, J.G.; Carhuatanta, K.A.; McInturf, S.M.; Miklasevich, M.K.; Jankord, R. Modulating hippocampal plasticity with in vivo brain stimulation. *J. Neurosci.* **2015**, *35*, 12824–12832. [CrossRef]
21. Rahman, A.; Reato, D.; Arlotti, M.; Gasca, F.; Datta, A.; Parra, L.C.; Bikson, M. Cellular effects of acute direct current stimulation: Somatic and synaptic terminal effects. *J. Physiol.* **2013**, *591*, 2563–2578. [CrossRef] [PubMed]
22. Márquez-Ruiz, J.; Leal-Campanario, R.; Sánchez-Campusano, R.; Molaee-Ardekani, B.; Wendling, F.; Miranda, P.C.; Ruffini, G.; Gruart, A.; Delgado-García, J.M. Transcranial direct-current stimulation modulates synaptic mechanisms involved in associative learning in behaving rabbits. *Proc. Natl. Acad. Sci. USA* **2012**, *109*, 6710–6715. [CrossRef] [PubMed]
23. Datta, A.; Bansal, V.; Diaz, J.; Patel, J.; Reato, D.; Bikson, M. Gyri-precise head model of transcranial direct current stimulation: Improved spatial focality using a ring electrode versus conventional rectangular pad. *Brain Stim.* **2009**, *2*, 201–207. [CrossRef] [PubMed]
24. Kronberg, G.; Bridi, M.; Abel, T.; Bikson, M.; Parra, L.C. Direct current stimulation modulates LTP and LTD: Activity dependence and dendritic effects. *Brain Stim.* **2017**, *10*, 51–58. [CrossRef] [PubMed]
25. Loo, C.K.; Alonzo, A.; Martin, D.; Mitchell, P.B.; Galvez, V.; Sachdev, P. Transcranial direct current stimulation for depression: 3-week, randomised, sham-controlled trial. *Br. J. Psychiatry* **2012**, *200*, 52–59. [CrossRef] [PubMed]
26. Brunoni, A.; Ferrucci, R.; Bortolomasi, M.; Vergari, M.; Tadini, L.; Boggio, P.; Giacopuzzi, M.; Barbieri, S.; Priori, A. Transcranial direct current stimulation (tDCS) in unipolar vs. bipolar depressive disorder. *Prog. Neuro-Psychopharmacol. Biol. Psych.* **2011**, *35*, 96–101. [CrossRef]
27. Edwards, D.; Krebs, H.; Rykman, A.; Zipse, J.; Thickbroom, G.; Mastaglia, F.; Pascual-Leone, A.; Volpe, B.T. Raised corticomotor excitability of M1 forearm area following anodal tDCS is sustained during robotic wrist therapy in chronic stroke. *Restor. Neurol. Neurosci.* **2009**, *27*, 199–207. [CrossRef] [PubMed]
28. Galletta, E.E.; Cancelli, A.; Cottone, C.; Simonelli, I.; Tecchio, F.; Bikson, M.; Marangolo, P. Use of computational modeling to inform tDCS electrode montages for the promotion of language recovery in post-stroke aphasia. *Brain Stim.* **2015**, *8*, 1108–1115. [CrossRef]
29. Fridriksson, J.; Richardson, J.D.; Baker, J.M.; Rorden, C. Transcranial direct current stimulation improves naming reaction time in fluent aphasia: A double-blind, sham-controlled study. *Stroke* **2011**, *42*, 819–821. [CrossRef]
30. Baker, J.M.; Rorden, C.; Fridriksson, J. Using transcranial direct-current stimulation to treat stroke patients with aphasia. *Stroke* **2010**, *41*, 1229–1236. [CrossRef]

31. Castillo-Saavedra, L.; Gebodh, N.; Bikson, M.; Diaz-Cruz, C.; Brandao, R.; Coutinho, L.; Truong, D.; Datta, A.; Shani-Hershkovich, R.; Weiss, M.; et al. Clinically effective treatment of fibromyalgia pain with high-definition transcranial direct current stimulation: Phase II open-label dose optimization. *J. Pain* **2016**, *17*, 14–26. [CrossRef] [PubMed]

32. DaSilva, A.F.; Mendonca, M.E.; Zaghi, S.; Lopes, M.; Dossantos, M.F.; Spierings, E.L.; Bajwa, Z.; Datta, A.; Bikson, M.; Fregni, F. tDCS-induced analgesia and electrical fields in pain-related neural networks in chronic migraine. *Headache J. Head Face Pain* **2012**, *52*, 1283–1295. [CrossRef] [PubMed]

33. Fregni, F.; Boggio, P.S.; Lima, M.C.; Ferreira, M.J.; Wagner, T.; Rigonatti, S.P.; Castro, A.W.; Souza, D.R.; Riberto, M.; Freedman, S.D.; et al. A sham-controlled, phase II trial of transcranial direct current stimulation for the treatment of central pain in traumatic spinal cord injury. *Pain* **2006**, *122*, 197–209. [CrossRef] [PubMed]

34. Brunoni, A.R.; Vanderhasselt, M.-A. Working memory improvement with non-invasive brain stimulation of the dorsolateral prefrontal cortex: A systematic review and meta-analysis. *Brain Cogn.* **2014**, *86*, 1–9. [CrossRef] [PubMed]

35. Palm, U.; Segmiller, F.M.; Epple, A.N.; Freisleder, F.-J.; Koutsouleris, N.; Schulte-Körne, G.; Padberg, F. Transcranial direct current stimulation in children and adolescents: A comprehensive review. *J. Neural Trans.* **2016**, *123*, 1219–1234. [CrossRef]

36. Gbadeyan, O.; Steinhauser, M.; McMahon, K.; Meinzer, M. Safety, tolerability, blinding efficacy and behavioural effects of a novel MRI-compatible, high-definition tDCS set-up. *Brain Stim.* **2016**, *9*, 545–552. [CrossRef]

37. Nitsche, M.A.; Paulus, W. Vascular safety of brain plasticity induction via transcranial direct currents. *Neurology* **2015**, *84*, 556–557. [CrossRef]

38. Kasschau, M.; Sherman, K.; Haider, L.; Frontario, A.; Shaw, M.; Datta, A.; Bikson, M.; Charvet, L. A protocol for the use of remotely-supervised transcranial direct current stimulation (tDCS) in multiple sclerosis (MS). *J. Vis. Exp. JoVE* **2015**, *106*, e53542. [CrossRef]

39. Bikson, M.; Datta, A.; Elwassif, M. Establishing safety limits for transcranial direct current stimulation. *Clin. Neurophysiol.* **2009**, *120*, 1033. [CrossRef]

40. Jackson, M.P.; Rahman, A.; Lafon, B.; Kronberg, G.; Ling, D.; Parra, L.C.; Bikson, M. Animal models of transcranial direct current stimulation: Methods and mechanisms. *Clin. Neurophysiol.* **2016**, *127*, 3425–3454. [CrossRef]

41. Bikson, M.; Grossman, P.; Thomas, C.; Zannou, A.L.; Jiang, J.; Adnan, T.; Mourdoukoutas, A.P.; Kronberg, G.; Truong, D.; Boggio, P.; et al. Safety of transcranial direct current stimulation: Evidence based update 2016. *Brain Stim.* **2016**, *9*, 641–661. [CrossRef] [PubMed]

42. Peterchev, A.V.; Wagner, T.A.; Miranda, P.C.; Nitsche, M.A.; Paulus, W.; Lisanby, S.H.; Pascual-Leone, A.; Bikson, M. Fundamentals of transcranial electric and magnetic stimulation dose: Definition, selection, and reporting practices. *Brain Stim.* **2012**, *5*, 435–453. [CrossRef] [PubMed]

43. Koc, G.; Gokcil, Z.; Bek, S.; Kasikci, T.; Eroglu, E.; Odabasi, Z. Effects of continuous theta burst transcranial magnetic stimulation on cortical excitability in patients with idiopathic generalized epilepsy. *Epilepsy* **2017**, *77*, 26–29. [CrossRef] [PubMed]

44. Kim, D.H.; Shin, J.C.; Jung, S.; Jung, T.-M.; Kim, D.Y. Effects of intermittent theta burst stimulation on spasticity after stroke. *Neuroreport* **2015**, *26*, 561. [CrossRef] [PubMed]

45. Chen, S.-C.; Yang, L.-Y.; Adeel, M.; Lai, C.-H.; Peng, C.-W. Transcranial electrostimulation with special waveforms enhances upper-limb motor function in patients with chronic stroke: A pilot randomized controlled trial. *J. NeuroEng. Rehab.* **2021**, *18*, 106. [CrossRef]

46. Rueger, M.A.; Keuters, M.H.; Walberer, M.; Braun, R.; Klein, R.; Sparing, R.; Fink, G.R.; Graf, R.; Schroeter, M. Multi-session transcranial direct current stimulation (tDCS) elicits inflammatory and regenerative processes in the rat brain. *PLoS ONE* **2012**, e43776. [CrossRef]

47. Gellner, A.-K.; Reis, J.; Fritsch, B. Glia: A neglected player in non-invasive direct current brain stimulation. *Front. Cell. Neurosci.* **2016**, *10*, 188. [CrossRef]

48. Jackson, M.P.; Truong, D.; Brownlow, M.L.; Wagner, J.A.; McKinley, R.A.; Bikson, M.; Jankord, R. Safety parameter considerations of anodal transcranial direct current stimulation in rats. *Brain Behav. Immun.* **2017**, *64*, 152–161. [CrossRef]

49. Liebetanz, D.; Koch, R.; Mayenfels, S.; König, F.; Paulus, W.; Nitsche, M.A. Safety limits of cathodal transcranial direct current stimulation in rats. *Clin. Neurophysiol.* **2009**, *120*, 1161–1167. [CrossRef]

50. Iyer, M.; Mattu, U.; Grafman, J.; Lomarev, M.; Sato, S.; Wassermann, E. Safety and cognitive effect of frontal DC brain polarization in healthy individuals. *Neurology* **2005**, *64*, 872–875. [CrossRef]

51. Nitsche, M.; Niehaus, L.; Hoffmann, K.; Hengst, S.; Liebetanz, D.; Paulus, W.; Meyer, B.-U. MRI study of human brain exposed to weak direct current stimulation of the frontal cortex. *Clin. Neurophysiol.* **2004**, *115*, 2419–2423. [CrossRef] [PubMed]

52. Praveen Rajneesh, C.; Hsieh, T.-H.; Chen, S.-C.; Lai, C.-H.; Yang, L.-Y.; Chin, H.-Y.; Peng, C.W. Deep brain stimulation of the pedunculopontine tegmental nucleus renders neuroprotection through the suppression of hippocampal apoptosis: An experimental animal study. *Brain Sci.* **2020**, *10*, 25. [CrossRef] [PubMed]

53. Mix, A.; Benali, A.; Eysel, U.T.; Funke, K. Continuous and intermittent transcranial magnetic theta burst stimulation modify tactile learning performance and cortical protein expression in the rat differently. *Eur. J. Neurosci.* **2010**, *32*, 1575–1586. [CrossRef] [PubMed]

54. Zhang, K.-Y.; Rui, G.; Zhang, J.-P.; Guo, L.; An, G.-Z.; Lin, J.-J.; He, W.; Ding, G.-R. Cathodal tDCS exerts neuroprotective effect in rat brain after acute ischemic stroke. *BMC Neurosci.* **2020**, *21*, 1–13. [CrossRef] [PubMed]

55. Fertonani, A.; Ferrari, C.; Miniussi, C. What do you feel if I apply transcranial electric stimulation? Safety, sensations and secondary induced effects. *Clin. Neurophysiol.* **2015**, *126*, 2181–2188. [CrossRef] [PubMed]

56. Poreisz, C.; Boros, K.; Antal, A.; Paulus, W. Safety aspects of transcranial direct current stimulation concerning healthy subjects and patients. *Brain Res. Bull.* **2007**, *72*, 208–214. [CrossRef]

57. Durand, S.; Fromy, B.; Bouyé, P.; Saumet, J.; Abraham, P. Vasodilatation in response to repeated anodal current application in the human skin relies on aspirin-sensitive mechanisms. *J. Physiol.* **2002**, *540*, 261–269. [CrossRef]

58. Nitsche, M.A.; Paulus, W. Sustained excitability elevations induced by transcranial DC motor cortex stimulation in humans. *Neurology* **2001**, *57*, 1899–1901. [CrossRef]

59. Palm, U.; Keeser, D.; Schiller, C.; Fintescu, Z.; Nitsche, M.; Reisinger, E.; Padberg, F.; Nitsche, M. Skin lesions after treatment with transcranial direct current stimulation (tDCS). *Brain Stimul.* **2008**, *1*, 386–387. [CrossRef]

60. Voss, U.; Holzmann, R.; Hobson, A.; Paulus, W.; Koppehele-Gossel, J.; Klimke, A.; Nitsche, M.A. Induction of self awareness in dreams through frontal low current stimulation of gamma activity. *Nature Neurosci.* **2014**, *17*, 810–812. [CrossRef]

61. Nitsche, M.A.; Jakoubkova, M.; Thirugnanasambandam, N.; Schmalfuss, L.; Hullemann, S.; Sonka, K.; Paulus, W.; Trenkwalder, C.; Happe, S. Contribution of the premotor cortex to consolidation of motor sequence learning in humans during sleep. *J. Neurophysiol.* **2010**, *104*, 2603–2614. [CrossRef] [PubMed]

62. Hsieh, T.-H.; Huang, Y.-Z.; Chen, J.-J.J.; Rotenberg, A.; Chiang, Y.-H.; Chien, W.-S.C.; Chang, H.-S.; Wang, J.-Y.; Peng, C.-W. Novel use of theta burst cortical electrical stimulation for modulating motor plasticity in rats. *J. Med. Biol. Eng.* **2015**, *35*, 62–68. [CrossRef]

63. Fonoff, E.T.; Pereira, J.F., Jr.; Camargo, L.V.; Dale, C.S.; Pagano, R.L.; Ballester, G.; Teixeira, M.J. Functional mapping of the motor cortex of the rat using transdural electrical stimulation. *Behav. Brain Res.* **2009**, *202*, 138–141. [CrossRef] [PubMed]

64. Gamboa, O.L.; Antal, A.; Laczo, B.; Moliadze, V.; Nitsche, M.A.; Paulus, W. Impact of repetitive theta burst stimulation on motor cortex excitability. *Brain Stimul.* **2011**, *4*, 145–151. [CrossRef] [PubMed]

65. Cenedella, R.; Galli, C.; Paoletti, R. Brain free fatty acid levels in rats sacrificed by decapitation versus focused microwave irradiation. *Lipids* **1975**, *10*, 290–293. [CrossRef] [PubMed]

66. Oehmichen, M.; Auer, R.N.; König, H.G. Electrical trauma. In *Forensic Neuropathology and Associated Neurology*; Springer: Berlin/Heidelberg, Germany, 2006. [CrossRef]

Article

Hepatic Oxi-Inflammation and Neophobia as Potential Liver–Brain Axis Targets for Alzheimer's Disease and Aging, with Strong Sensitivity to Sex, Isolation, and Obesity

Juan Fraile-Ramos [1,2], Anna Garrit [3], Josep Reig-Vilallonga [3] and Lydia Giménez-Llort [1,2,*]

1   Institut de Neurociències, Universitat Autònoma de Barcelona, 08193 Barcelona, Spain
2   Department of Psychiatry and Forensic Medicine, School of Medicine, Universitat Autònoma de Barcelona, 08193 Barcelona, Spain
3   Department of Anatomy, School of Medicine, Universitat Autònoma de Barcelona, 08193 Barcelona, Spain
*   Correspondence: lidia.gimenez@uab.cat

Abstract: Research on Alzheimer's disease (AD) has classically focused on alterations that occur in the brain and their intra- and extracellular neuropathological hallmarks. However, the oxi-inflammation hypothesis of aging may also play a role in neuroimmunoendocrine dysregulation and the disease's pathophysiology, where the liver emerges as a target organ due to its implication in regulating metabolism and supporting the immune system. In the present work, we demonstrate organ (hepatomegaly), tissue (histopathological amyloidosis), and cellular oxidative stress (decreased glutathione peroxidase and increased glutathione reductase enzymatic activities) and inflammation (increased IL-6 and TNFα) as hallmarks of hepatic dysfunction in 16-month-old male and female 3xTg-AD mice at advanced stages of the disease, and as compared to age- and sex-matched non-transgenic (NTg) counterparts. Moreover, liver–brain axis alterations were found through behavioral (increased neophobia) and HPA axis correlations that were enhanced under forced isolation. In all cases, sex (male) and isolation (naturalistic and forced) were determinants of worse hepatomegaly, oxidative stress, and inflammation progression. In addition, obesity in old male NTg mice was translated into a worse steatosis grade. Further research is underway determine whether these alterations could correlate with a worse disease prognosis and to establish potential integrative system targets for AD research.

Keywords: Alzheimer's disease; 3xTg-AD; liver–brain axis; obesity; HPA axis; corticosterone; oxidative stress; social isolation; amyloidosis; steatosis

Citation: Fraile-Ramos, J.; Garrit, A.; Reig-Vilallonga, J.; Giménez-Llort, L. Hepatic Oxi-Inflammation and Neophobia as Potential Liver–Brain Axis Targets for Alzheimer's Disease and Aging, with Strong Sensitivity to Sex, Isolation, and Obesity. *Cells* **2023**, *12*, 1517. https://doi.org/10.3390/cells12111517

Academic Editors: Antonella Caccamo, Illana Gozes and Jacques P. Tremblay

Received: 27 March 2023
Revised: 6 May 2023
Accepted: 29 May 2023
Published: 30 May 2023

1. Introduction

Alzheimer's disease (AD), the most prevalent form of dementia among older adults, is expected to increasingly affect the global population [1–3]. To date, research on AD has classically focused on alterations in the brain based on the original findings of the main pathological hallmarks of the disease: extracellular accumulation of β-amyloid (Aβ) plaques and intraneuronal neurofibrillary tangles of hyperphosphorylated tau protein [4,5]. New clinical evidence also suggests the influence of peripheral organs on the pathophysiological development of the disease, which can be analyzed using classical oxidative stress and inflammation theories of aging [6,7]. Biomarkers of oxidative stress identified in the urine of patients with AD and peripheral inflammation biomarkers that correlate with the risk of dementia suggest physiological effects not only in the brain but also in peripheral tissues [8,9]. In support of these emergent areas in AD research, animal models have confirmed crosstalk between central and peripheral inflammation [10–12].

The liver is considered a hotspot for oxidative stress because it regulates metabolism and inflammation by supporting the immune system [13]. Moreover, its role in the peripheral metabolism of Aβ has been described [14]. Oxidative stress and inflammation are risk factors for chronic hepatic diseases and negatively affect each other [15]. Recent findings have implicated this organ in the pathophysiology of AD, as an association between peripheral markers of liver function and central markers associated with AD has been described in people with the disease [16]. Moreover, liver dysfunction and impaired Aβ clearance have been suggested as early events in the pathophysiology of AD and are decisive in the progression of the disease [17,18]. Preliminary results from our laboratory for 3xTg-AD mice also suggested hepatic oxidative stress dysfunction at the onset of the disease [19,20]. Therefore, biochemical communication between these two organs via the so-called liver–brain axis has begun to attract attention.

In addition to this new perspective on the involvement of peripheral organs in AD, the classical cognitive conceptualization of this disease, centered on memory impairment at the clinical level, now embraces the disease's broad and heterogeneous spectrum of neuropsychiatric and behavioral manifestations [21]. It has been well documented that mouse AD models have substantial defects in memory and cognition. In our extensive work with 3xTg-AD mice created by LaFerla's laboratory [22], we described that this animal model for AD exhibits a conspicuous neuropsychiatric (NPS)-like phenotype [23,24]. However, it is not yet known how these cognitive changes are related to peripheral abnormalities. Among these, psychological stress and anxiety are commonly present in patients with AD, underlined by dysregulation of the hypothalamic–pituitary–adrenal (HPA) axis and consequent negative feedback of increased glucocorticoid levels in the limbic system that worsens AD progression [25]. Moreover, in mammals, glucocorticoids regulate the pro- and anti-inflammatory balance in the immune response. Our research has also demonstrated dysregulation of the HPA axis, neuroimmune communication derangement with increased peripheral oxidative stress, and inflammation, as observed in spleen and peritoneal cells [23,26–29]. This neuroimmune–endocrine effect has also been reproduced in other animal models of AD [30]. Studies in chickens have confirmed that chronic corticosterone administration leads to increased lipid deposition in the liver, which may be accompanied by oxidative stress and inflammation [31]. This reaffirms the involvement of peripheral organs in the disease and how they are influenced by neuropsychiatric and behavioral manifestations.

Extrinsic environmental factors are decisive in maintaining neuroimmune–endocrine homeostasis [26,29] and many physiological and psychological processes [32]. We have recently demonstrated short naturalistic isolation effects due to the phenotype of survivors [10,33,34]. In contrast, social isolation and poor responses to stress and anxiety, male sex, menopause, and obesity are models of premature aging in rodents and humans. In animals, premature aging can be monitored through behavioral tests and is characterized by peripheral immunosenescence, oxidative and inflammatory stress, and nervous system derangement [35]. Therefore, the extrinsic factors considered were standard (grouped), naturalistic (natural death of congeners), and forced (experimental) social isolation housing conditions.

Taking advantage of the conspicuous NPS-like phenotype of 3xTg-AD mice [24], in the present study, we investigated liver dysfunction at different levels of study, using cellular oxi-inflammation, histological analysis, and functional correlates with the HPA axis (corticosterone) and the liver–brain axis (NPS-like phenotype). Male and female 3xTg-AD mice were studied in a longitudinal design starting at 13 months and ending at 16 months, corresponding to advanced stages of the disease [36,37]. However, their NTg counterparts, the gold standard C57BL/6 mice, are prone to exhibit a sex-dependent obese profile at this age. The extrinsic factors considered were standard (grouped), naturalistic (natural death of congeners), and forced (experimental) social isolation housing conditions.

2. Materials and Methods

2.1. Animals

Genetically engineered homozygous triple-transgenic 3xTg-AD mice carrying the human PS1/M146V, APP$_{Swe}$, and tau$_{P301L}$ transgenes were produced at the University of

California, Irvine [22]. Briefly, two independent transgenes encoding human APP_{Swe} and tau_{P301L} (both regulated by mouse Thy1.2) were co-injected into single-cell embryos from PS1KI (PS1M146V knock-in) mice.

A cohort of sixty-six male and female mice from Spanish colonies of C57BL/6 wild-type mice (from now referred to as NTg, non-transgenic mice) ($n = 29$) and homozygous 3xTg-AD mice ($n = 37$) from litters of a breeding program established at Universitat Autònoma de Barcelona after embryonic transfer to C57BL/6 strain background were used [38]. All animals were housed in groups of 3–4 individuals of the same sex and genotype until the experiment. The longitudinal design studied the animals from 13 to 16 months of age. They were maintained in cages (Makrolon, $35 \times 15 \times 15$ cm^3) under standard laboratory conditions (12 h light/dark, cycle starting at 8:00 a.m., food and water ad libitum, $22 \pm 2\,°C$, 50–60% humidity). Biochemical and histopathological analyses were performed in a counterbalanced manner, blinded to the experiment.

All procedures were performed according to the Spanish legislation on the "Protection of Animals Used for Experimental and Other Scientific Purposes" and the EU Council directive (2010/63/EU). The protocol CEEAH 3588/DMAH 9452 was approved on the 8th of March 2019 by Departament de Medi Ambient i Habitatge, Generalitat de Catalunya. ARRIVE guidelines developed by NC3Rs were followed, to reduce the number of animals used in the experimental procedures [39].

2.2. Extrinsic Factors: Naturalistic and Forced Social Isolation

Naturalistic isolation (nISO) resulting from increased mortality in 3xTg-AD mice [23] elicited a group of male 3xTg-AD mice that were naturally isolated for 2–3 months after the death of their cage mates. To mirror this scenario in both genotypes, a forced isolation (fISO) paradigm was conducted; at 13 months of age, a group of animals per each sex and genotype was individualized in cages and kept isolated for 3 months until they were euthanized. The rest of the animals remained in group-housed conditions. Namely, the following groups were established: Males NTg grouped ($n = 6$), Males NTg fISO ($n = 8$), Males 3xTg-AD grouped ($n = 10$), Males 3xTg-AD fISO ($n = 10$), Females NTg grouped ($n = 7$), Females NTg fISO ($n = 8$), Females 3xTg-AD grouped ($n = 10$), Females 3xTg-AD fISO ($n = 7$) and, additionally in Table 1, Males 3xTg-AD nISO ($n = 5$). In all cases, the animals were still socially connected olfactorily and auditorily, and the nesting conditions were the same as for the grouped animals.

Table 1. Effect of different isolation scenarios, naturalistic (nISO) and forced (fISO), in 3xTg-AD males with AD-pathological aging. Number of visited corners; number of rearings; latency of rearing; body weight; liver weight; liver index; GSH levels; GR activity; GPx activity; IL-6 levels; TNFα levels and IL-1β levels. N = 5–10 mice per group, mean ± SEM. Isolation: # $p \leq 0.05$.

	3xTg-AD, Male Grouped ($n = 8$) Mean ± SEM	3xTg-AD, Male ISO ($n = 15$) Mean ± SEM	3xTg-AD, Male fISO ($n = 10$) Mean ± SEM	3xTg-AD, Male nISO ($n = 5$) Mean ± SEM
Visited corners	3.38 ± 1.07	3.00 ± 0.59	2.70 ± 0.82	3.60 ± 0.75
Rearings	0.50 ± 0.27	1.13 ± 0.34	0.90 ± 0.38	1.60 ± 0.68
Lat. Rearing (s)	22.75 ± 3.70	16.93 ± 2.83	18.20 ± 3.31	14.40 ± 5.70
Body weight (g)	33.60 ± 1.08	31.27 ± 0.49 #	30.97 ± 0.51 #	31.88 ± 1.12
Liver weight (g)	1.59 ± 0.08	1.43 ± 0.05	1.48 ± 0.07	1.33 ± 0.09
Liver index (%)	0.05 ± 0.002	0.05 ± 0.001	0.05 ± 0.002	0.4 ± 0.002
GSH (nmol/mg prot.)	66.13 ± 23.19	55.19 ± 3.46	57.12 ± 4.06	51.34 ± 6.76
GR (mU/mg prot.)	23.19 ± 0.86	22.85 ± 0.51	22.49 ± 0.46	23.57 ± 1.24
GPx (mU/mg prot.)	291.46 ± 17.49	300.90 ± 26.93	303.58 ± 35.02	295.55 ± 45.67
IL-6 (pg/ mg prot.)	8742.12 ± 847.56	7125.19 ± 696.54	6735.39 ± 689.15	7904.79 ± 1652.04
TNFα (pg/ mg prot.)	1930.90 ± 169.66	1769.08 ± 101.78	1727.44 ± 79.16	1852.35 ± 279.25
IL-1β (pg/ mg prot.)	19.63 ± 1.20	18.92 ± 1.22	18.98 ± 1.65	18.81 ± 1.81

2.3. Longitudinal Behavioral Assessment

All mice were evaluated following the same procedure in a pre- and post-longitudinal design at 13 months and again at 16 months. The corner test was used to evaluate the immediate neophobic response of the animals in a new standard home cage with a clear bed of wood save. Once the animal was placed in the new standard home cage, the following information was recorded immediately during the first 30 s: horizontal activity (number of visited corners) and vertical activity (number of rearings and latency of rearing). Behavioral assessment was performed from 9:00 a.m. to 1:00 p.m., counterbalanced and blinded to the experimenter.

2.4. Physical Status: Body Weight and Liver Index

Animals were weighed at the time of sacrifice, and their livers were dissected and stored at –80 °C until the protein extract was processed for biochemical analysis. The liver index or relative weight of the liver was calculated according to the following equation: Liver index = Weight of liver (g)/Bodyweight (g).

2.5. Histopathological Assessment

After euthanasia, a part of the liver was dissected, fixed with 10% formalin for 24 h, and embedded in paraffin for further histopathological analysis. Slices of 5 and 10 μm thickness were de-paraffinized and re-hydrated by several passes through ethanol at different concentrations.

2.5.1. Morphological and Structural Analysis with H&E

Standard staining with hematoxylin and eosin (H&E) was performed, briefly, a 5-min incubation with Harris hematoxylin and a 2-min incubation with Eosin Y, with a dedifferentiation phase in between. Finally, the slices were manually mounted with DPX mounting medium after dehydration. Images were taken with a ×20 objective on a Nikon Eclipse 80i microscope, using a digital camera running on the control software ACT-1 (ver2.70).

2.5.2. Congo Red Staining for Amyloidosis

Amyloidosis was assessed using an Amyloid Stain, Congo Red Kit (Sigma-Aldrich, Saint Louis, MO, USA). The staining procedure was performed according to the kit indications, with incubation in Mayer's hematoxylin and then in Congo red solution. Finally, the slices were manually mounted with DPX mounting medium after dehydration. Images were taken with a ×20 objective on a Nikon Eclipse 80i microscope under normal light (bright field) and fluorescence light (fluorescent filter G-2A), using a digital camera running on the control software ACT-1 (ver2.70).

2.6. Biochemical Analysis

2.6.1. HPA Axis: Glucocorticoids Measurements

Blood samples were collected from each animal during euthanasia. After centrifugation at 3500 rpm × 14 min, serum was preserved at −20 °C. Corticosterone levels (Ng/mL) were measured using a commercial kit (Corticosterone EIA Immunodiagnostic Systems Ltd., Boldon, UK). Absorbance was read at 450 nm using a Varioskan LUX ESW 1.00.38 (Thermo Fisher Scientific, Waltham, MA, USA).

2.6.2. Hepatic Oxidative Stress

Antioxidant capacity was studied by evaluating the levels of total glutathione (GSH), the main non-enzymatic reducing agent of the organism, and glutathione peroxidase (GPx) and reductase (GR) enzymatic activity in liver homogenates, benefiting from the enzymatic recycling method previously described [40].

Total GSH was measured by absorbance at 412 nm and adapted to 96-well plates with slight modifications [41].

The enzymatic activity of glutathione reductase was assessed following the method described by Massey and Williams [42], with slight modifications. Monitoring the oxidation of NADPH spectrophotometrically was performed at 340 nm for 300 s.

The enzymatic activity of glutathione peroxidase was measured using a technique that was modified after a previous description [43,44]. The reaction was monitored spectrophotometrically by the decreasing absorbance at 340 nm for 300 s.

The results were expressed as milliunits (mU) of enzymatic activity per milligram (mg) of organ protein.

2.6.3. Hepatic Inflammation

Liver homogenates were produced by lysing frozen liver samples in cold lysis buffer containing protease and phosphatase inhibitors (Sigma-Aldrich, Saint Louis, MO, USA). The protein content was quantified using the BCA Protein Assay Kit (Thermo Fisher Scientific, Waltham, MA, USA). TNFα and the cytokines IL-6 and IL-1β were quantified by sandwich ELISA using commercial kits (Biospes, Antibodies, Cambridge, UK). Absorbance was read at 450 nm using a Varioskan LUX ESW 1.00.38. (Thermo Fisher Scientific, Waltham, MA, USA).

2.7. Functional Correlates

The correlations between the functional variables, namely, behavioral phenotype, physical status, HPA axis, liver oxidative stress, and liver inflammation, were determined using Pearson's test, considering the two intrinsic (genotype and sex) and extrinsic (isolation) factors. The data are shown as a heat map with correlation coefficients. The most important relationships with biological meaning are shown in scatter plots.

2.8. Statistics

Results are expressed as the mean $\pm$ SEM. GraphPad Prism 8.0 and R 4.1.2 software were used. Three-way and two-way analyses for multiple variables (ANOVA) and nested one-way (ANOVA) followed by Tukey's post hoc test for multiple comparisons were used. Student's t-test was used to assess the significance of differences between two independent groups. Functional correlations were analyzed using RStudio 1.4.1103 software. In all tests, statistical significance was set at $p \leq 0.05$.

3. Results

3.1. Longitudinal Behavioral Assessment

The pre- and post-isolation assessment of the horizontal activity (number of visited corners) and vertical activity (number of rearings and latency of rearing) of the neophobia response were longitudinally assessed. At the age of 13 months, genotype differences were already observed among the females (Figure 1A; F $(1, 28)$ = 6.285, p = 0.0183), with an increased number of visited corners in 3xTg-AD females as compared to their NTg counterparts. At 16 months, genotype effects were found for both sexes in horizontal and vertical counts (Figure 1D–F; all F's $(1, 22) \geq 6.784$, p = 0.0152). A genotype $\times$ isolation interaction effect (Figure 1E; F $(1, 28)$ = 4.241, p = 0.0489) was present in the number of rearings in males. Further statistical analysis revealed that these genotype effects were attributable to a strong reduction in the number of visited corners and rearings in all 3xTg-AD males. In contrast, only grouped females showed a statistically significant reduction in horizontal activity, while isolated 3xTg-AD females showed this reduction in vertical activity. The isolation effect was shown by the number of rearings (Figure 1E; F $(1, 50)$ = 5.1, p = 0.0283), with fISO groups exhibiting decreased vertical activity. (Figure 1G–I; all F's $(3, 12) \geq 9.008$, p = 0.002) showed pre- and post-differences that were clearly manifested in the 3xTg-AD animals. No sex differences were found (all F's $(3, 12) \geq 9.008$, p = 0.0021).

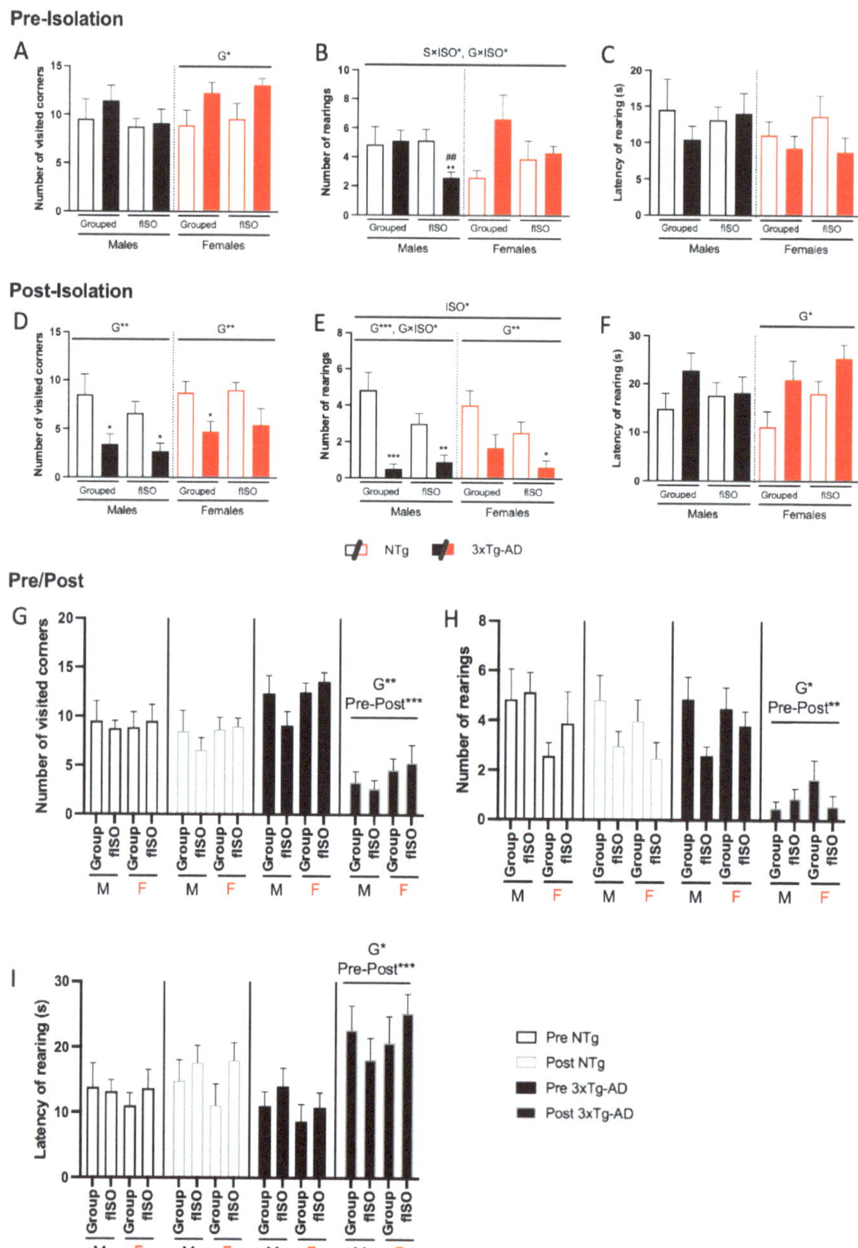

Figure 1. Pre- and post-isolation longitudinal behavioral assessment in the corner test for neophobia. Pre-isolation test: horizontal activity (**A**) and vertical activity (**B**,**C**). Post-isolation test: horizontal activity (**D**) and vertical activity (**E**,**F**). Comparison between pre- and post-test in horizontal activity (**G**) and vertical activity (**H**,**I**). N = 7–10 mice per group, mean ± SEM. Genotype: G*, $p \leq 0.05$; G**, $p \leq 0.01$; G***, $p \leq 0.001$. Forced isolation: ISO*, $p \leq 0.05$. Interaction sex and isolation: S × ISO* $p \leq 0.05$. Interaction genotype and isolation: G × ISO* $p \leq 0.05$. Pre- and post-differences: Pre-Post**, $p \leq 0.01$; Pre-Post***, $p \leq 0.001$. Differences between two counterparts, genotype: * $p \leq 0.05$, ** $p \leq 0.01$, *** $p \leq 0.001$ vs. NTg counterpart; isolation: ## $p \leq 0.01$ vs. grouped counterpart.

3.2. Physical Status: Body Weight and Liver Index

Sex and genotype effects were found in all the variables studied (Figure 2; S, all F's $(1, 58) \geq 5.309$, $p = 0.0248$; G, all F's $(1, 58) \geq 7.140$, $p = 0.0098$), and interaction between these factors occurred in the body weight of the animal and the liver index (Figure 2A,C; all F's $(1, 58) = 4.721$, $p = 0.0339$). Body weight measurements (Figure 1A) showed sexual dimorphism. The genotype effect, with a higher body weight recorded in NTg mice, was mainly due to males, whereas genotype differences in females were due to reduced weight in isolated animals. Similar differences were observed in the liver weight (Figure 2B). Thus, genotype differences in liver weight were enhanced, reaching statistical significance, in both male and female isolated animals compared to their NTg isolated counterparts. The relative weight of the livers revealed hepatomegaly in the 3xTg-AD mice, and this effect was primarily due to male sex (Figure 2C), in addition to sexual dimorphism. The effect of isolation was found in body weight, which was due to the reduction observed in the 3xTg-AD mice.

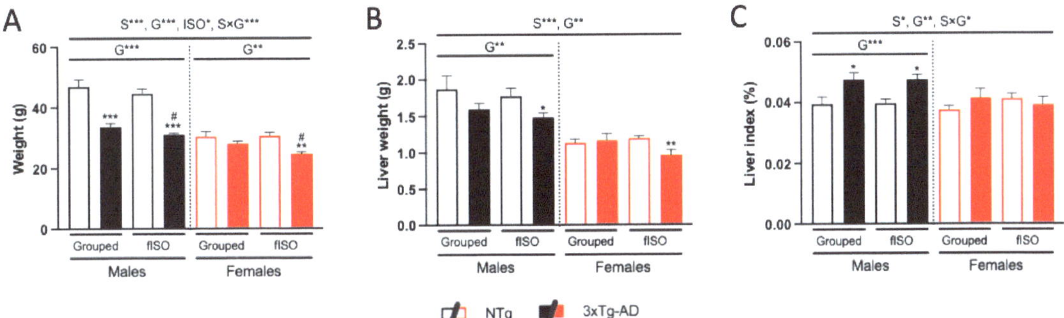

Figure 2. Physical status: Body (**A**) and liver (**B**) weights and liver index (**C**). N = 7–10 mice per group, mean ± SEM. Sex: S*, $p \leq 0.05$; S***, $p \leq 0.001$. Genotype: G**, $p \leq 0.01$; G***, $p \leq 0.001$. Forced isolation: ISO*, $p \leq 0.05$. Interaction sex and genotype: S × G* $p \leq 0.05$; S × G*** $p \leq 0.001$. Differences between two counterparts, genotype: * $p \leq 0.05$, ** $p \leq 0.01$, *** $p \leq 0.001$ vs. NTg counterpart; isolation: # $p \leq 0.05$ vs. grouped counterpart.

3.3. Histopathological Assessment: H&E and Congo Red Staining

Histopathological sections were prepared and stained for microscopic analysis to determine morphological and physiological alterations in the liver. Two genotype-dependent morphological alterations were observed (Figures 3 and 4; all F's $(3, 56) = 3.840$, $p = 0.0143$). In NTg mice, the main alteration observed was accumulation of lipids in the form of fat droplets, indicating a state of steatosis and possibly an early stage of non-alcoholic fatty liver disease (NAFLD). Steatosis was present in all NTg groups to varying degrees, tending to affect more males than females. In contrast, in 3xTg-AD mice, excessive thickening of the blood vessel parenchyma, losing its typical circular structure due to amyloidosis, tended to be the main alteration, reaching statistical significance in isolated males. Isolation in males caused the change in these alterations by genotype to be more pronounced, while in females, the only effect was the attenuation of amyloidosis in 3xTg-AD females. Infiltration of immune cells was observed in all groups, suggesting a possible inflammatory state in the liver. Additionally, cell damage in the form of "ballooning" was observed in all animals regardless of sex and genotype, although an isolation effect was found in 3xTg-AD females, reducing this damage.

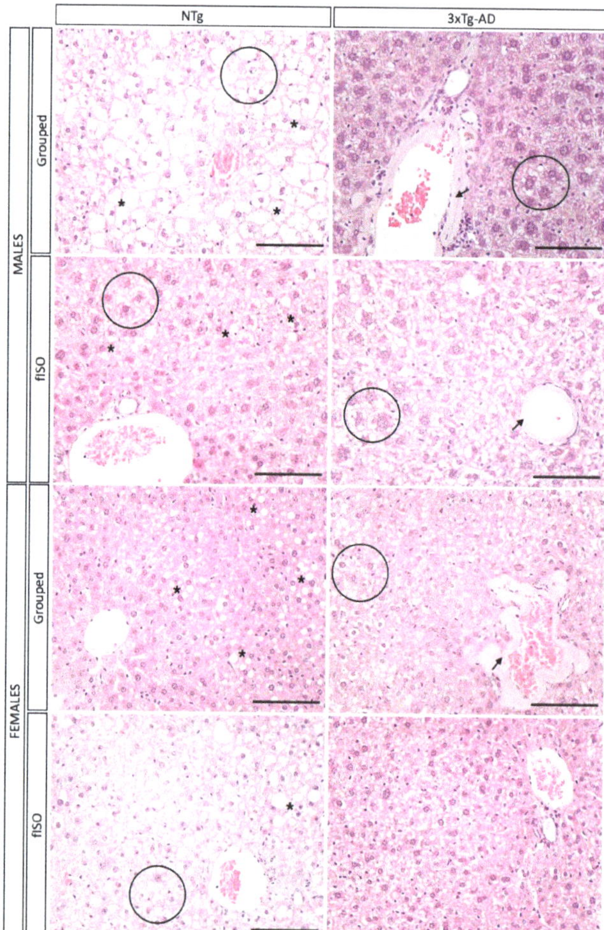

Figure 3. H&E staining of liver tissue. Morphological and structural comparison of liver tissue in NTg and 3xTg-AD males and females grouped and those submitted to forced isolation. H&E stained. Symbols indicate the morphological features as follows: *—steatosis; black arrow—amyloidosis; white arrow—inflammatory infiltration; black circle—ballooning. The images were taken with ×20 objective lens.

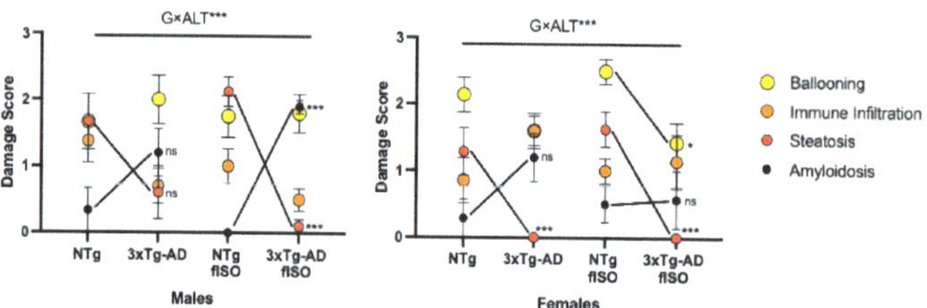

Figure 4. Histopathological damage score of liver tissue. Comparison between NTg and 3xTg-AD males and females and those submitted to forced isolation. The score was determined according to the severity presented in four morphological alterations: ballooning, immune infiltration, steatosis, and amyloidosis, each receiving a score between 0 and 4 for each animal. N = 7–10 mice per group, mean ± SEM. Genotype (G) and morphological alterations (ALT) interaction effect: G × ALT*** $p \leq 0.001$. Differences between two counterparts: * $p \leq 0.05$, *** $p \leq 0.001$ vs. NTg counterparts.

Congo red staining was performed to detect amyloidosis in the liver. As shown in Figure 5, in the bright field images, NTg animals did not show stained areas, whereas 3xTg-AD animals showed parenchyma of the blood vessels with the characteristic reddish tone of this staining. In addition, when the sections were viewed under fluorescent light, the areas marked by the staining showed intense red fluorescence.

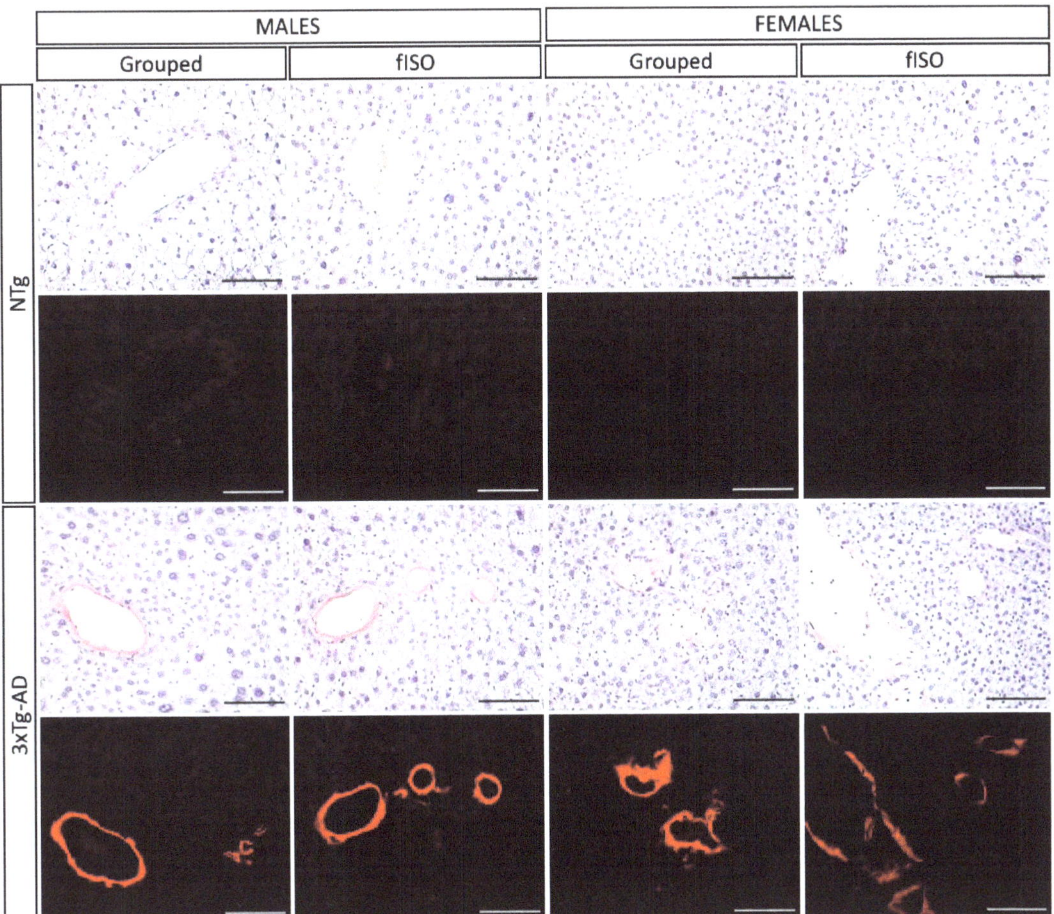

Figure 5. Congo red staining of liver tissue. Differences in vascular parenchyma of liver tissue in NTg and 3xTg-AD males and females grouped and those submitted to forced isolation. Congo red stained. The images were taken with ×20 objective lens; magnification bar = 100 μm.

3.4. Biochemical Status of the Liver–Brain and HPA Axis

The corticosterone levels of NTg and 3xTg-AD mice living in standard housing conditions did not differ. However, genotype differences emerged in the groups subjected to isolation, resulting in an interaction between genotype and isolation, with increased levels in both male and female 3xTg-AD mice compared to their NTg counterparts (Figure 6A; $F_{(1, 58)} = 4.352$, $p = 0.0414$). Sexual dimorphism, with higher levels in females than in males, was also observed (Figure 6A; $F_{(1, 58)} = 5.525$, $p = 0.0222$).

Liver oxidative stress was assessed by measuring GSH levels and GR and GPx enzymatic activity. Sexual dimorphism was observed in GSH levels and GPx activity (Figure 6B,C; all F's $(1, 58) \geq 12.4$, $p = 0.0009$), with females having lower levels of GSH and higher GPx activity. An isolation effect was also observed in the GSH levels

(Figure 6B; F(1, 57) = 6.178, p = 0.0159), which were reduced in the isolated mice. Genotype differences were found for both GPx and GR enzymatic activities (Figure 6C,D; all F's (1, 28) = 5.056, p = 0.0326) were more marked in GPx activity. The 3xTg-AD animals had reduced GPx activity and increased GR activity compared to NTg animals. Therefore, the AD genotype affects liver metabolism in terms of oxidative stress, altering the functioning of antioxidant defenses.

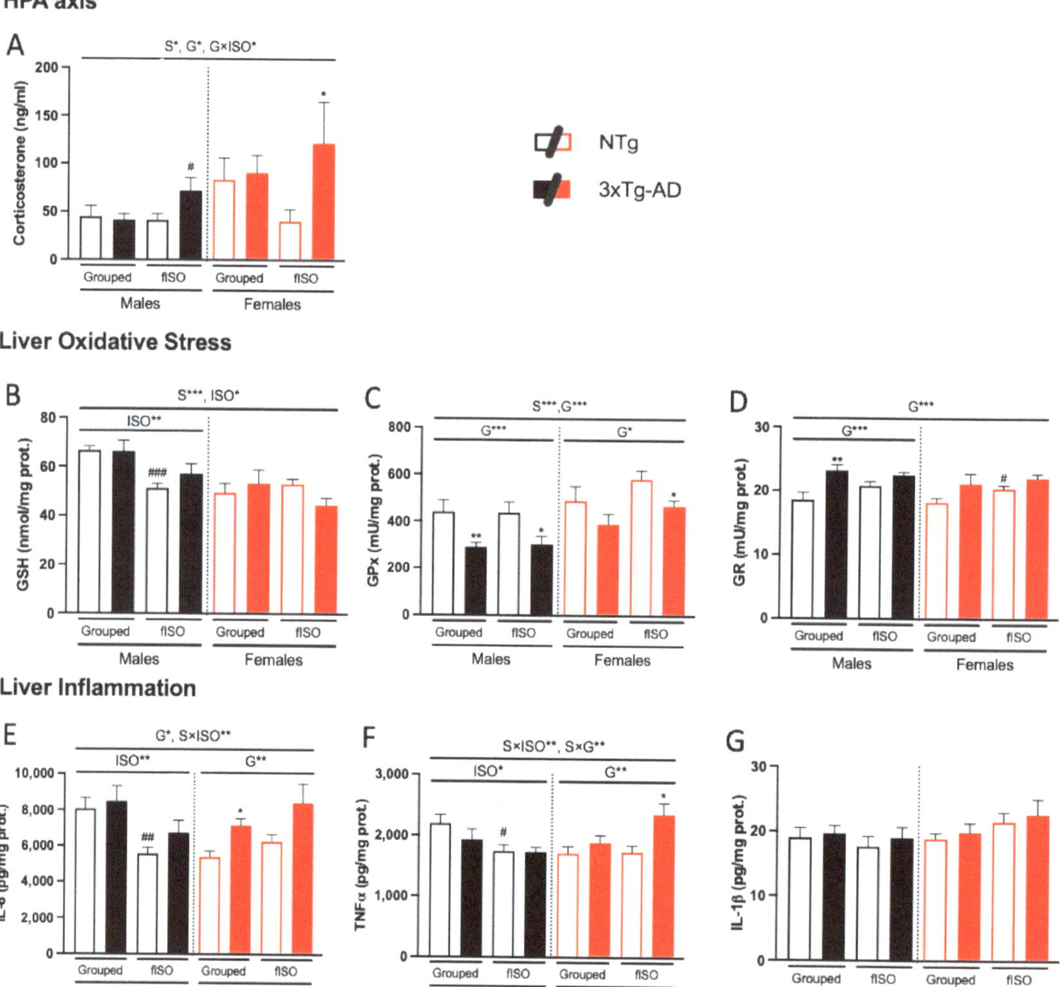

Figure 6. Biochemical analysis of the liver–brain and HPA axis. Corticosterone levels in serum (**A**). Total levels of GSH in the liver (**B**). Enzymatic activity of GR and GPx (**C**,**D**, respectively). Pro-inflammatory cytokines levels of IL-6 (**E**), TNFα (**F**), and IL-1β (**G**). N= 7–10 mice per group, mean ± SEM. Sex: S*, $p \leq 0.05$; S***, $p \leq 0.001$. Genotype: G*, $p \leq 0.05$; G**, $p \leq 0.01$; G***, $p \leq 0.001$. Forced isolation: ISO*, $p \leq 0.05$; ISO**, $p \leq 0.01$. Interaction effects: sex and genotype: S × G**, $p \leq 0.01$; sex and isolation: S × ISO**, $p \leq 0.01$; sex and genotype: S × G**, $p \leq 0.01$. Differences between two counterparts, genotype: * $p \leq 0.05$, ** $p \leq 0.01$ vs. NTg counterpart; isolation: # $p \leq 0.05$, ## $p \leq 0.01$, ### $p \leq 0.001$ vs. grouped counterpart. G × ISO*, $p \leq 0.05$.

Liver inflammation, on the other hand, was assessed through the levels of IL-6, TNFα, and IL-1β. Genotype and isolation had differential effects according to sex (Figure 6E,F; S $\times$ ISO F (1, 56) = 8.407, p = 0.0053; S $\times$ G F (1, 56) = 7.326, p = 0.009). Genotype had an effect on females (Figure 6E,F; all F's (1, 55) $\geq$ 6.465, p = 0.0138), resulting in higher levels of IL-6 and TNFα in 3xTg-AD animals. Isolation had an effect on males (Figure 6E,F; all F's (1, 29) $\geq$ 5.927, p = 0.0213), showing lower levels of IL-6 and TNFα in isolated males compared to grouped males. IL-1β levels were not affected. In males, the levels were very similar in all groups, whereas in females, there was a slight tendency to increase, although the difference was not statistically significant.

Additionally, due to the death of their cage mates, a group of transgenic males was isolated (Males 3xTg-AD nISO) and added to the study to determine the effect of two different levels of isolation (Table 1). The results of the variables analyzed for this group followed the trend observed with the fISO. Only forced isolation achieved statistical significance in comparison to its grouped counterpart in body weight (Table 1; t (18) = 2.201, p = 0.041), consisting in a reduction.

3.5. Meaningful Functional Correlates between Behavioral, Physical Metrics, Liver Oxidative Stress, and Inflammation Variables

Statistical correlation tests were performed between all variables, separated by the different factors considered in the study, to determine the possible relationships (Figure 7). Next, relationships were established between the functional variables with biological significance and how they behaved according to the genotype (Figure 8).

These correlations varied according to sex, genotype, and isolation (Figure 7). In all cases, it can be seen how relationships changed in intensity and how some changed from positive to negative, or vice versa. Females established more and stronger relationships between the variables than males. This was more pronounced in the 3xTg-AD animals, where males presented 26.92% of null relationships (correlation equal to 0) and females presented 7.69%. The genotype effect observed differed depending on sex. Thus, in males, the range of null relations increased from 10.26% in NTg mice to 20.51% in 3xTg-AD mice, while in females the opposite occurred, going from 12.82% in NTg to 7.69% in 3xTg-AD females. Regarding for the effect of isolation on the relationships between variables, no remarkable changes were observed, although sexual dimorphism was preserved. In this case, if strong positive and negative relationships (with coefficients between 0.5 and 1) are considered, isolated males accounted for 7.69% and 5.13%, respectively, while isolated females accounted for 19.23% and 15.38%, respectively.

In view of this and our focus on the possible relationships between behavior, physical status, oxidative stress, and inflammation in physiological and AD aging, the most relevant correlations were further analyzed. Starting with the relationship between the brain and liver, NTg animals with physiological aging present a connection between the variables studied with the corner test and oxidative stress variables. Specifically, the number of rearings and rearing latency were related to the enzymatic activity of antioxidant enzymes (Figure 8A,B). In the 3xTg-AD animals, in the other hand, a relationship was established using the liver index (Figure 8C,D). Regarding the HPA axis, in NTg animals, it was related to rearing latency but not to oxidative stress or inflammation. In 3xTg-AD animals, the opposite occurred, relating to GSH levels but not to behavior variables (Figure 8E,F). If we focused on oxidative stress and inflammation in the liver, it was observed in NTg animals that the liver index was related to GR activity (Figure 8G), which was also related to TNFα levels (Figure 8I). Finally, it should be noted that a mismatch between the activity of both anti-inflammatory enzymes was observed. In NTg animals, both enzymes that are part of GSH antioxidant cycle had a positive relationship (Figure 8H). However, in 3xTg-AD animals this relationship was lost, indicating that the activity of both enzymes was unbalanced in the 3xTg-AD genotype (Figure 8J).

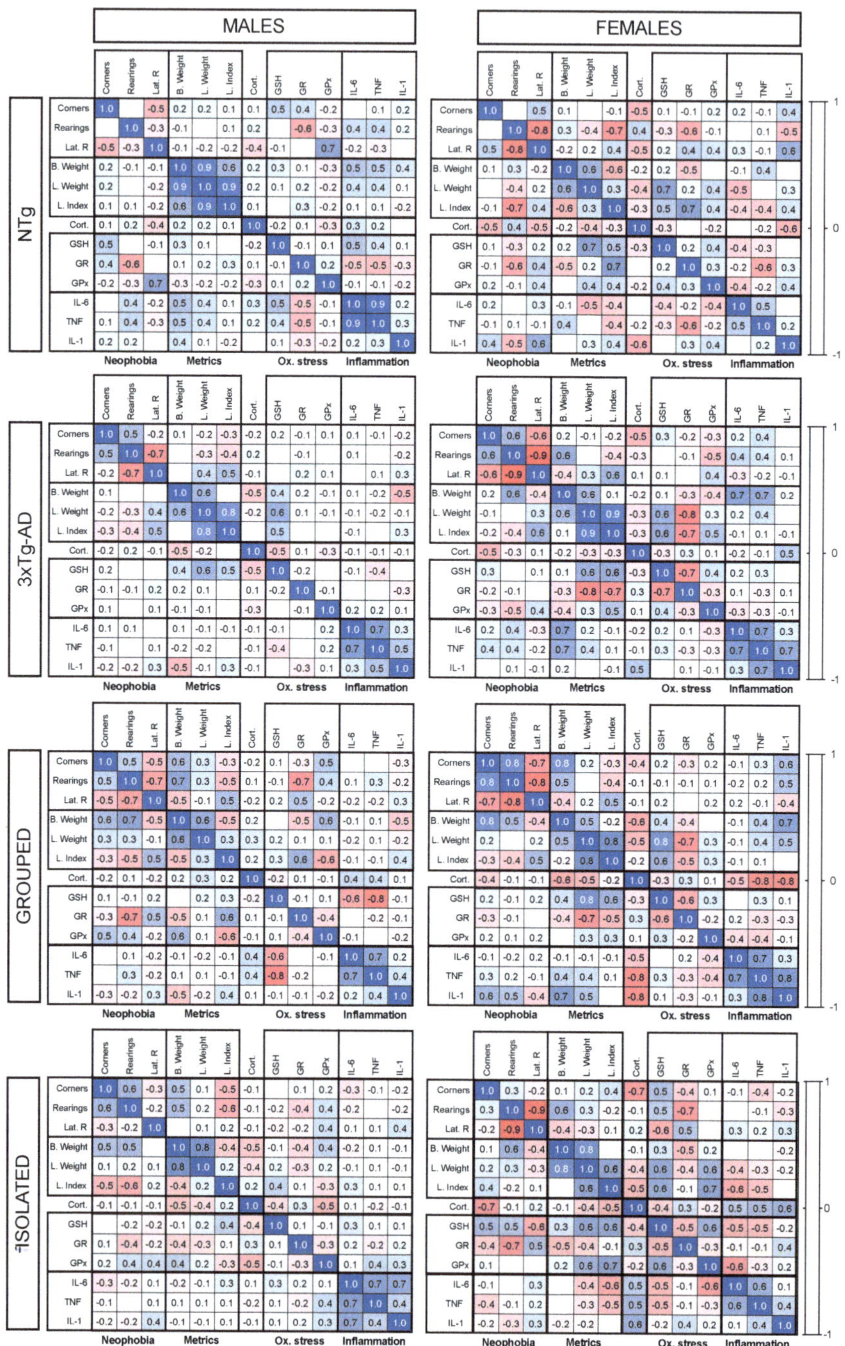

Figure 7. Heat map of correlations between the studied variables. Neophobia: number of visited corners, number of rearings, and latency of rearing; physical metrics: body weight, liver weight, and liver index; corticosterone; oxidative stress: GSH levels, GR activity, and GPx activity; inflammation: IL-6, TNFα, and IL-1β levels. The color scale shows the correlation intensity between 1 and -1, with positive correlations in blue and negative correlations in red. The statistical analysis was based on Pearson's test.

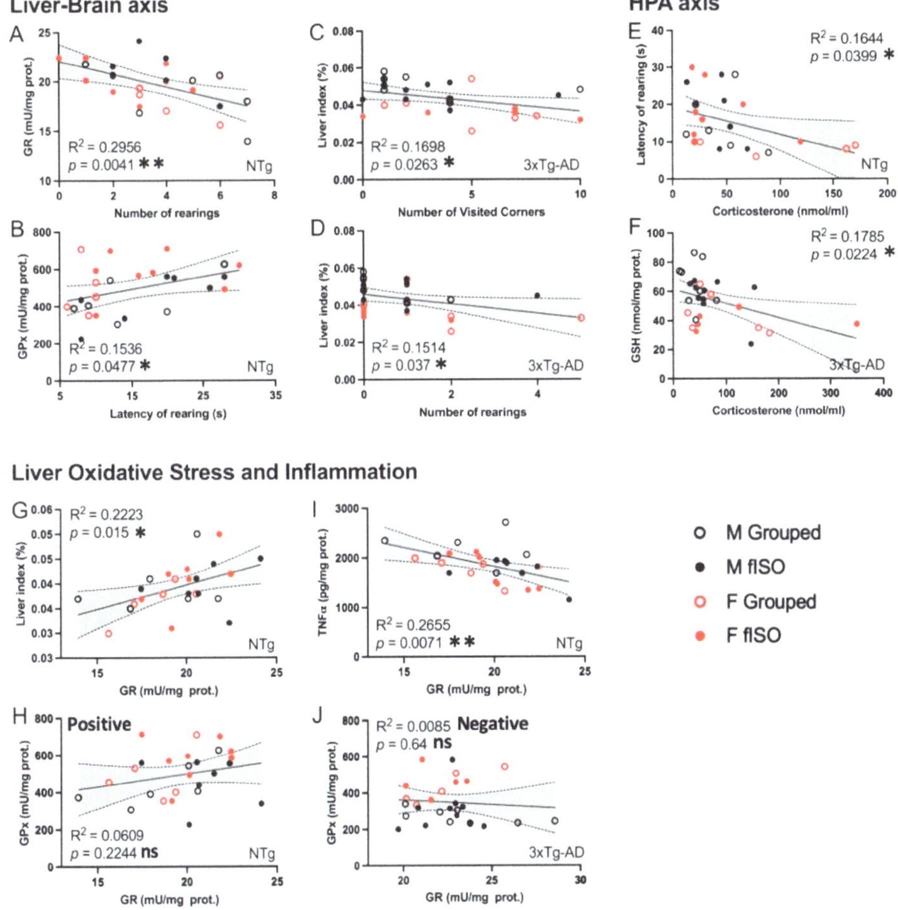

Figure 8. Scatter plot representing the most biologically relevant correlations: GR activity and number of rearings by NTg animals (**A**); GPx activity and latency of rearings by NTg animals (**B**); liver index and number of corners visited by 3xTg-AD animals (**C**); liver index and number of rearings by 3xTg-AD animals (**D**); latency of rearing and corticosterone levels in NTg animals (**E**); GSH levels and corticosterone leves in 3xTg-AD animals (**F**); liver index and GR activity in NTg animals (**G**); TNFα levels and GR activity in NTg animals (**H**); GPx activity and GR activity in NTg animals (**I**); GPx activity and GR activity in 3xTg-AD animals (**J**). N = 6–10 mice per group, each linear regression includes the correlation index (R^2) and the level of significance of the correlation (*p*), * $p \leq 0.05$, ** $p \leq 0.01$.

4. Discussion

AD is one of the most important neurodegenerative diseases, with extensive literature on brain changes, in contrast to research on the effects or involvement of peripheral organs. The liver, the main organ responsible for metabolism, is one of the most susceptible organs to AD [45]. Previous studies by our group have shown peripheral alterations at prodromal (4 months of age) and early stages of disease (6 months of age) in oxidative stress in 3xTg-AD mice, suggesting accelerated aging. However, these genotype differences were lost at very late ages (18 months), when physiological AD neuropathological aging seems to converge [19,20,27,46]. This study assessed the histopathological profile of mice at 16 months of age, which is a very advanced stage of the disease. Furthermore, oxidative stress status and inflammation in the liver of 3xTg-AD mice were assessed to determine the relationship between the AD genotype and this organ. The impact of sex and isolation as intrinsic (biological) and extrinsic (environmental) factors, respectively, have also been studied. In addition, old male C57BL/6 mice have been used as animal models of aging [47].

4.1. Hepatic Pathology Changes according to Genotype

The physical status and histopathology of the liver indicated that this organ's state was disturbed in both 3xTg-AD and NTg mice, but in a different manner. NTg mice, the gold-standard C57BL/6 mice, as the genetic background of our 3xTg-AD mice colony, showed the highest body and liver weights, indicating an obese phenotype. This genetic background has been widely used to develop mouse models predisposed to obesity and other liver disorders [48,49], which is consistent with our results considering the old age of the animals. Massive accumulation of lipids was present in their livers, an event commonly seen in various liver diseases that have been associated with aging and obesity, such as NAFLD. This disease is characterized by lipid deposition, leading to steatosis, which causes oxidative stress and inflammation, and ultimately leads to NASH [50,51]. Hepatic steatosis increases liver size and weight, which is commonly referred to as hepatomegaly [52].

These effects were expected to be more marked in the 3xTg-AD animals, where hepatomegaly was more marked. Surprisingly, the outcome was the opposite, since 3xTg-AD animals presented lower grades of steatosis, and instead, they presented with amyloidosis, a group of diseases caused by the extracellular deposition of insoluble amyloid fibrils. Congo red has been shown to be a reliable stain for the detection of amyloid plaques [53] and diagnosis of amyloidosis [53,54]. In this study, most amyloid accumulation was observed in the portal fields of 3xTg-AD mice in the form of vascular and interstitial deposits. From the clinical manifestation of amyloidosis, it is known that it can develop in a systemic way affecting several organs, such as the liver, leading to hepatocellular atrophy and progressive organ failure [55,56]. The liver could be one of the organs most affected, since one of its primary functions is to detoxify noxious compounds that are toxic to the organism [57]. Moreover, hepatocytes are involved in the process promoting Aβ clearance by degradation or by excretion with the bile [58]. The accumulation observed in the portal fields denotes a failure in this clearance of Aβ by the liver. The 3xTg-AD mouse model carries two transgenes for β-amyloid pathology [22], and there is evidence to suggest that Aβ is not only restricted to the brain but can also develop in other peripheral organs, such as the liver. In fact, Aβ from the brain can be transported to these peripheral organs via the brain–blood barrier (BBB) or the blood–CFS barrier, which we demonstrated to be disrupted in 3xTg-AD mice at this age [15,59]. 3xTg-AD males also had hepatomegaly, as shown by their liver index, which is considered a manifestation of amyloidosis in this mouse model [60]. The present results establish a point of connection between both organs in this mouse model and for both sexes, providing further evidence that reinforce the relevance of the liver–brain axis in AD.

Hepatocellular damage, characterized by swelling of the hepatocytes and clearing of the cytoplasm is called "ballooning", and occurs when the liver suffers severe damage and the cells begin to degenerate [61]. Interestingly, ballooning was observed in all animals and therefore considered not genotype-dependent. This is an alteration associated with NASH [62] and could explain the affectation of NTg animals.

4.2. Hepatic Oxi-Inflammation Metabolic Disruption

GSH is one of the most important antioxidant peptides in the liver to manage oxidative stress [63,64], and the GSSG/GSH ratio can be used as a rutinary biomarker of oxidative stress in hepatocytes [65]. In the present study, the possible increase in oxidative stress was evaluated by the state of the cycle of GSH. Under normal conditions, this peptide is almost completely reduced; however, under stress conditions, it is oxidized by the action of GPx and reduced by the action of GR, forming a mechanism that aims to reduce reactive species within the cell [66]. GSH also promotes the expression of antioxidant genes [67]. Although sex differences and isolation effects were observed in GSH levels, the lack of genotype changes suggests the same level of protection from oxidative stress in both NTg and 3xTg-AD animals. Interestingly, GPx activity, which is responsible for the detoxification of reactive species due to the oxidation of GSH [68], was reduced in 3xTg-AD animals, indicating a loss in antioxidant capacity. Our group determined peripheral alterations in the oxidative stress status of male and female 3xTg-AD mice occurring at the prodromal (4 months of age) and early stages of the disease (6 and 9 months of age). These alterations consisted of a reduction in GSH levels as well as a reduction in GPx and GR enzyme activity in the liver, spleen, and kidney. In addition, the levels of xanthine oxidase also increased [19,20,27]. This is consistent with the results of this study, where the reduction in GR activity was preserved at 16 months of age at a much more advanced stage of the disease. However, the differences in GSH levels were lost, and GPx enzymatic activity increased rather than decreased. This indicated that the liver continued to exhibit persistent oxidative stress. All differences in oxidative stress were lost by the age of 18 months, suggesting that at this age NTg animals reached the level of oxidative stress in 3xTg-AD animals [46].

Another important aspect to consider is inflammation, as it has been widely reported to have a close relationship with oxidative stress in this organ [15,69]. Indeed, we found that the levels of these cytokines were increased in 3xTg-AD females, indicating that this could be related to the levels of corticosterone and oxidative stress in the liver.

4.3. HPA Axis Disruption, Oxidative Stress, and Inflammation in the Liver

The HPA axis is responsible for psychological stress responses; alterations in this axis may be reflected in cognitive impairment, oxidative stress, and inflammation. Anxiety and psychological factors are among the most relevant psychological factors in patients with AD. Psychological stress also disrupts the HPA axis [70,71]. These psychological effects must be considered when evaluating the relationship between AD and liver function. One advantage of using animal models is that they can reproduce an important number of characteristics observed in humans [38,72]. In particular, the 3xTg-AD animals exhibited marked neophobia at 16 months of age, and their performance indicated worsening of long-term memory in such an experimental scenario, considering that the animals were already tested at 13 months. Moreover, anxiety, depression, and stress-like behavior are also present [24,72].

3xTg-AD female mice at 4 months of age at the initial stages of AD already show HPA axis hyperactivity and overproduction of corticosterone [11,73]. This study also found this difference in 3xTg-AD females, with a tendency toward higher corticosterone levels. Moreover, this is consistent with previous findings from our group, where old females presented higher levels of corticosterone than males, suggesting that these sex differences in corticosterone levels may be triggered by estrogens [23]. Corticosterone also influences oxidative stress and inflammation. Chronically increased corticosterone levels alter lipid metabolism and oxidative phosphorylation in liver mitochondria, leading to oxidative stress [74]. Isolated perfused livers from rats exposed to psychological stress showed an increase in tumor necrosis factor (TNFα) and interleukin-6 (IL-6) [75]. This highlights the relationship between corticosterone and oxidative stress and inflammation, which would agree with the results obtained from the biochemical analysis of 3xTg-AD females.

4.4. Meaningful Functional Correlates between Behavioral, Physical Metrics, Liver Oxidative Stress, and Inflammation Variables

The statistical correlations shown in the results help us understand how the variables behave with each other and can provide information on the possible relationship between AD genotype and alterations observed in the liver.

It can be seen that there is a relationship between behavior and the physiological processes that are happening in the liver. The involvement of cognitive processes in the physiology of peripheral organs is beginning to be studied. In humans, the connection between the central nervous system and periphery through fear has already been determined [76,77]. In this study, we observed that neophobia was directly related to the activity of antioxidant enzymes. In addition, 3xTg-AD animals showed high neophobia, which was correlated with the liver index, establishing a link between behavior and hepatomegaly, possibly through amyloidosis in the liver. On the other hand, another axis of connection could also be established through corticosterone. In 3xTg-AD animals, a negative correlation was observed between corticosterone and GSH levels, indicating a relationship between corticosterone levels and the oxidative status of the liver. In rat models with a chronic glucocorticoid-enriched endogenous environment, it has already been shown that oxidative stress in the liver is increased by a reduction in GSH levels [78]. Furthermore, this is consistent with what we observed for the relationship between GR and GPx enzymatic activity, since in NTg animals, the activity of both enzymes is timed for the cycle to work, but in 3xTg-AD animals this relationship is lost. This imbalance in the activity of these two enzymes caused by the AD genotype indicates that the reduction in reactive species is not efficient and therefore increases the oxidative stress generated in the liver. This imbalance in the activity of these two enzymes caused by the AD genotype indicates that the reduction in reactive species is not efficient and therefore increases the oxidative stress generated in the liver. This has already been observed in the brains of mice following severe ethanol administration. This causes the overproduction of reactive oxygen species that impairs the antioxidant system by reducing GPx activity and increasing GR activity, which is considered a compensatory mechanism to preserve the cell oxidative stress status [79]. Finally, excess accumulation of lipids in the liver, known as steatosis, impairs antioxidant defenses that eventually cause liver damage due to oxidative stress and simultaneously activate the entire pro-inflammatory machinery. This damage leads to the development of chronic liver diseases, particularly NAFLD [80,81]. This is consistent with our histopathological and biochemical results for NTg animals and becomes evident with the correlations between the liver index, enzymatic activity of antioxidant enzymes, and levels of pro-inflammatory factors. Despite not presenting with this liver disease, 3xTg-AD animals were found to be affected by a group of diseases known as amyloidosis. Based on these results, oxidative stress and inflammation were found to be exacerbated.

The paradigm comes when we focus on the effect of social isolation as an environmental factor in the model. Many patients with AD face situations of social isolation that imply major psychological stress in addition to the intrinsic stress generated by the disease [82]. In our experimental scenario, the isolated 3xTg-AD mice showed a similar pattern of neophobia as the grouped mice. This contrasts with the results obtained in the group's previous studies, where isolated mice exhibited considerable changes in the cognitive tests performed, indicating that they suffered hyperactivity, bizarre behaviors, anxiogenic patterns, and stress-coping responses that shifted to flight behavior [33]. Histopathological and biochemical analysis of the liver showed a main effect increasing the genotype difference. The different isolation scenarios were compared, and no differences were found between naturalistic and forced isolation in the study variables, except in body weight. Weight reduction was observed in forced-isolated animals, which was not as marked in naturally isolated animals.

5. Conclusions

This longitudinal study confirms the existence of the liver–brain axis alterations in Alzheimer's disease and identifies hepatic oxi-inflammation and neophobia as potential integrative system targets that are under the modulation of intrinsic (genotype and sex) and extrinsic (social conditions) factors. In particular:

Hepatic dysfunction was determined at the organ, tissue, and cellular levels in 16-month-old animals. The hallmarks of 3xTg-AD liver dysfunction included hepatomegaly, acute amyloidosis, and ballooning. In addition, increased oxidative stress and inflammation were shown as the dysregulation of antioxidant enzymes and increased levels of pro-inflammatory cytokines, with a sexual dimorphism. In contrast, liver steatosis was present in their male and female C57BL/6 wild-type counterparts and, in the case of males, was co-morbid with obesity.

Dysregulation of the HPA axis may also be involved in liver oxidative stress and inflammation, as shown by the higher corticosterone levels in 3xTg-AD females, which correlated with GSH levels.

Neophobia has previously been shown to exert a clear psychological effect by increasing hyperactivity, anxiogenic patterns, and bizarre/flight behavior. In this study, neophobia correlated with oxidative stress variables in the liver.

In forced isolated animals, the genotype differences on hepatic histopathology (liver damage score) and oxi-inflammation mechanisms (reduced GSH levels and increased IL-6 and TNFα) were enhanced or appear for the first time.

The liver–brain axis is an emerging research area in AD that demands further efforts to unveil the interplay of the pathways involved the systemic component of this disease.

Author Contributions: Conceptualization, L.G.-L.; experimental design performance, L.G.-L.; harvesting, biochemical analysis, and statistics, J.F.-R.; pathology: J.F.-R. and A.G.; laboratory resources, J.R.-V.; illustrations: J.F.-R.; writing—original draft preparation, J.F.-R.; writing—review and editing, J.F.-R. and L.G.-L.; funding acquisition, L.G.-L. All authors have read and agreed to the published version of the manuscript.

Funding: This work was funded by AGAUR, Generalitat de Catalunya, 2017-SGR-1468 and Universitat Autònoma de Barcelona, UAB-GE-260408 to L.G.-L. The colony of 3xTg-AD mice is sustained by ArrestAD H2020, Fet-OPEN-1-2016-2017-737390, European Union's Horizon 2020 research and innovation program under grant agreement No 737390 to L.G.-L. It also received financial support from Memorial Mercedes Llort Sender, 2021/80/09241941.8.

Institutional Review Board Statement: The study was conducted according to the guidelines of the Declaration of Helsinki and approved by the Ethics Committee of Departament de Medi Ambient i Habitatge, Generalitat de Catalunya (CEEAH 3588/DMAH 9452) on the 8 March 2019.

Informed Consent Statement: Not applicable.

Data Availability Statement: The data presented in this study are available on request to the corresponding author.

Acknowledgments: We thank Frank M LaFerla, Institute for Memory Impairments and Neurological Disorders, Department of Neurobiology and Behavior, University of California, Irvine, USA, for kindly providing the progenitors of the Spanish colonies of 3xTg-AD and NTg mice. We also thank BioRender.com for providing graphical tools for scientific illustration.

Conflicts of Interest: The authors declare no conflict of interest. The funders had no role in the study's design, in the collection, analyses, or interpretation of data; in the writing of the manuscript, or in the decision to publish the results.

References

1. García-Mesa, Y.; Colie, S.; Corpas, R.; Cristòfol, R.; Comellas, F.; Nebreda, A.R.; Giménez-Llort, L.; Sanfeliu, C. Oxidative Stress Is a Central Target for Physical Exercise Neuroprotection Against Pathological Brain Aging. *J. Gerontol. A Biol. Sci. Med. Sci.* **2016**, *71*, 40–49. [CrossRef] [PubMed]

2. Tiwari, S.; Atluri, V.; Kaushik, A.; Yndart, A.; Nair, M. Alzheimer's Disease: Pathogenesis, Diagnostics, and Therapeutics. *Int. J. Nanomed.* **2019**, *14*, 5541–5554. [CrossRef] [PubMed]

3. Scheltens, P.; De Strooper, B.; Kivipelto, M.; Holstege, H.; Chételat, G.; Teunissen, C.E.; Cummings, J.; van der Flier, W.M. Alzheimer's Disease. *Lancet* **2021**, *397*, 1577–1590. [CrossRef]

4. Alzheimer's Association. 2021 Alzheimer's Disease Facts and Figures. *Alzheimer's Dement.* **2021**, *17*, 327–406. [CrossRef] [PubMed]

5. Hasegawa, M. Molecular Mechanisms in the Pathogenesis of Alzheimer's Disease and Tauopathies-Prion-Like Seeded Aggregation and Phosphorylation. *Biomolecules* **2016**, *6*, 24. [CrossRef]

6. Vida, C.; Martinez de Toda, I.; Garrido, A.; Carro, E.; Molina, J.A.; De la Fuente, M. Impairment of Several Immune Functions and Redox State in Blood Cells of Alzheimer's Disease Patients. Relevant Role of Neutrophils in Oxidative Stress. *Front. Immunol.* **2018**, *8*, 1974. [CrossRef]

7. Morris, G.; Berk, M.; Maes, M.; Puri, B.K. Could Alzheimer's Disease Originate in the Periphery and If So How So? *Mol. Neurobiol.* **2019**, *56*, 406–434. [CrossRef]

8. Gibson, G.E.; Huang, H.-M. Oxidative Stress in Alzheimer's Disease. *Neurobiol. Aging* **2005**, *26*, 575–578. [CrossRef]

9. Pugazhenthi, S.; Qin, L.; Reddy, P.H. Common Neurodegenerative Pathways in Obesity, Diabetes, and Alzheimer's Disease. *Biochim. Biophys. Acta (BBA)—Mol. Basis Dis.* **2017**, *1863*, 1037–1045. [CrossRef]

10. Muntsant, A.; Giménez-Llort, L. Crosstalk of Alzheimer's Disease-Phenotype, HPA Axis, Splenic Oxidative Stress and Frailty in Late-Stages of Dementia, with Special Concerns on the Effects of Social Isolation: A Translational Neuroscience Approach. *Front. Aging Neurosci.* **2022**, *14*, 969381. [CrossRef]

11. Baeta-Corral, R.; De la Fuente, M.; Giménez-Llort, L. Sex-Dependent Worsening of NMDA-Induced Responses, Anxiety, Hypercortisolemia, and Organometry of Early Peripheral Immunoendocrine Impairment in Adult 3xTg-AD Mice and Their Long-Lasting Ontogenic Modulation by Neonatal Handling. *Behav. Brain Res.* **2023**, *438*, 114189. [CrossRef]

12. Angiulli, F.; Conti, E.; Zoia, C.P.; Da Re, F.; Appollonio, I.; Ferrarese, C.; Tremolizzo, L. Blood-Based Biomarkers of Neuroinflammation in Alzheimer's Disease: A Central Role for Periphery? *Diagnostics* **2021**, *11*, 1525. [CrossRef] [PubMed]

13. Racanelli, V.; Rehermann, B. The Liver as an Immunological Organ. *Hepatology* **2006**, *43*, S54–S62. [CrossRef] [PubMed]

14. Sarkar, S.R.; Dutta, R. Role of Liver in Alzheimer's Disease. *ARC J. Neurosci.* **2020**, *4*, 20–25. [CrossRef]

15. Li, S.; Hong, M.; Tan, H.-Y.; Wang, N.; Feng, Y. Insights into the Role and Interdependence of Oxidative Stress and Inflammation in Liver Diseases. *Oxidative Med. Cell. Longev.* **2016**, *2016*, 4234061. [CrossRef]

16. Nho, K.; Kueider-Paisley, A.; Ahmad, S.; MahmoudianDehkordi, S.; Arnold, M.; Risacher, S.L.; Louie, G.; Blach, C.; Baillie, R.; Han, X.; et al. Association of Altered Liver Enzymes With Alzheimer Disease Diagnosis, Cognition, Neuroimaging Measures, and Cerebrospinal Fluid Biomarkers. *JAMA Netw. Open* **2019**, *2*, e197978. [CrossRef] [PubMed]

17. Huang, Z.; Lin, H.W.; Zhang, Q.; Zong, X. Targeting Alzheimer's Disease: The Critical Crosstalk between the Liver and Brain. *Nutrients* **2022**, *14*, 4298. [CrossRef]

18. Estrada, L.D.; Ahumada, P.; Cabrera, D.; Arab, J.P. Liver Dysfunction as a Novel Player in Alzheimer's Progression: Looking Outside the Brain. *Front. Aging Neurosci.* **2019**, *11*, 174. [CrossRef]

19. Gimenez-Llort, L.; De Las Casas-Engel, M.; Laferla, F.M.; De La Fuente, M. P4-245: Decrease of the Antioxidant Defenses in 3xTg-AD Triple-transgenic Adult Mice for Alzheimer's Disease. *Alzheimer's Dement.* **2009**, *5*, P502. [CrossRef]

20. Hernanz, A.; Vida, C.; Manassra, R.; Gimenez-Llort, L.; De la Fuente, M. P2-063: Early Peripheral Oxidative Stress Status In Male And Female Triple-Transgenic Mice For Alzheimer's Disease. *Alzheimer's Dement.* **2014**, *10*, P492–P493. [CrossRef]

21. Cummings, J.L. The Neuropsychiatric Inventory: Assessing Psychopathology in Dementia Patients. *Neurology* **1997**, *48*, 10S–16S. [CrossRef] [PubMed]

22. Oddo, S.; Caccamo, A.; Shepherd, J.D.; Murphy, M.P.; Golde, T.E.; Kayed, R.; Metherate, R.; Mattson, M.P.; Akbari, Y.; LaFerla, F.M. Triple-Transgenic Model of Alzheimer's Disease with Plaques and Tangles. *Neuron* **2003**, *39*, 409–421. [CrossRef] [PubMed]

23. Giménez-Llort, L.; Arranz, L.; Maté, I.; De la Fuente, M. Gender-Specific Neuroimmunoendocrine Aging in a Triple-Transgenic 3×Tg-AD Mouse Model for Alzheimer's Disease and Its Relation with Longevity. *Neuroimmunomodulation* **2008**, *15*, 331–343. [CrossRef] [PubMed]

24. Giménez-Llort, L.; Blázquez, G.; Cañete, T.; Johansson, B.; Oddo, S.; Tobeña, A.; LaFerla, F.M.; Fernández-Teruel, A. Modeling Behavioral and Neuronal Symptoms of Alzheimer's Disease in Mice: A Role for Intraneuronal Amyloid. *Neurosci. Biobehav. Rev.* **2007**, *31*, 125–147. [CrossRef]

25. Ahmad, M.H.; Fatima, M.; Mondal, A.C. Role of Hypothalamic-Pituitary-Adrenal Axis, Hypothalamic-Pituitary-Gonadal Axis and Insulin Signaling in the Pathophysiology of Alzheimer's Disease. *Neuropsychobiology* **2019**, *77*, 197–205. [CrossRef] [PubMed]

26. Giménez-Llort, L.; Maté, I.; Manassra, R.; Vida, C.; De la Fuente, M. Peripheral Immune System and Neuroimmune Communication Impairment in a Mouse Model of Alzheimer's Disease. *Ann. N. Y. Acad. Sci.* **2012**, *1262*, 74–84. [CrossRef]

27. De la Fuente, M.; Manassra, R.; Mate, I.; Vida, C.; Hernanz, A.; Gimenez-Llort, L. Early Oxidation and Inflammation State of the Immune System in Male and Female Triple Transgenic Mice for Alzheimer's Disease (3xTgAD). *Free Radic. Biol. Med.* **2012**, *53*, S165. [CrossRef]

28. Maté, I.; Cruces, J.; Giménez-Llort, L.; De la Fuente, M. Function and Redox State of Peritoneal Leukocytes as Preclinical and Prodromic Markers in a Longitudinal Study of Triple-Transgenic Mice for Alzheimer's Disease. *J. Alzheimer's Dis.* **2014**, *43*, 213–226. [CrossRef]

29. Blázquez, G.; Cañete, T.; Tobeña, A.; Giménez-Llort, L.; Fernández-Teruel, A. Cognitive and Emotional Profiles of Aged Alzheimer's Disease (3×TgAD) Mice: Effects of Environmental Enrichment and Sexual Dimorphism. *Behav. Brain Res.* **2014**, *268*, 185–201. [CrossRef]

30. Esmaeili, M.H.; Bahari, B.; Salari, A.-A. ATP-Sensitive Potassium-Channel Inhibitor Glibenclamide Attenuates HPA Axis Hyperactivity, Depression- and Anxiety-Related Symptoms in a Rat Model of Alzheimer's Disease. *Brain Res. Bull.* **2018**, *137*, 265–276. [CrossRef]

31. Feng, Y.; Hu, Y.; Hou, Z.; Sun, Q.; Jia, Y.; Zhao, R. Chronic Corticosterone Exposure Induces Liver Inflammation and Fibrosis in Association with M6A-Linked Post-Transcriptional Suppression of Heat Shock Proteins in Chicken. *Cell Stress Chaperones* **2020**, *25*, 47–56. [CrossRef] [PubMed]

32. Berkman, L.F.; Glass, T.; Brissette, I.; Seeman, T.E. From Social Integration to Health: Durkheim in the New Millennium. *Soc. Sci. Med.* **2000**, *51*, 843–857. [CrossRef] [PubMed]

33. Muntsant, A.; Giménez-Llort, L. Impact of Social Isolation on the Behavioral, Functional Profiles, and Hippocampal Atrophy Asymmetry in Dementia in Times of Coronavirus Pandemic (COVID-19): A Translational Neuroscience Approach. *Front. Psychiatry* **2020**, *11*, 572583. [CrossRef]

34. Gimenez-Llort, L.; Alveal-Mellado, D. Digging Signatures in 13-Month-Old 3xTg-AD Mice for Alzheimer's Disease and Its Disruption by Isolation Despite Social Life Since They Were Born. *Front. Behav. Neurosci.* **2021**, *14*, 611384. [CrossRef]

35. Baeza, I.; De Castro, N.M.; Giménez-Llort, L.; De la Fuente, M. Ovariectomy, a Model of Menopause in Rodents, Causes a Premature Aging of the Nervous and Immune Systems. *J. Neuroimmunol.* **2010**, *219*, 90–99. [CrossRef] [PubMed]

36. Torres-Lista, V.; De la Fuente, M.; Giménez-Llort, L. Survival Curves and Behavioral Profiles of Female 3xTg-AD Mice Surviving to 18-Months of Age as Compared to Mice with Normal Aging. *J. Alzheimers Dis. Rep.* **2017**, *1*, 47–57. [CrossRef]

37. Jack, C.R.; Lowe, V.J.; Weigand, S.D.; Wiste, H.J.; Senjem, M.L.; Knopman, D.S.; Shiung, M.M.; Gunter, J.L.; Boeve, B.F.; Kemp, B.J.; et al. Serial PIB and MRI in Normal, Mild Cognitive Impairment and Alzheimer's Disease: Implications for Sequence of Pathological Events in Alzheimer's Disease. *Brain* **2009**, *132*, 1355–1365. [CrossRef] [PubMed]

38. Baeta-Corral, R.; Giménez-Llort, L. Persistent Hyperactivity and Distinctive Strategy Features in the Morris Water Maze in 3xTg-AD Mice at Advanced Stages of Disease. *Behav. Neurosci.* **2015**, *129*, 129–137. [CrossRef]

39. Kilkenny, C.; Browne, W.J.; Cuthill, I.C.; Emerson, M.; Altman, D.G. Improving Bioscience Research Reporting: The ARRIVE Guidelines for Reporting Animal Research. *PLoS Biol.* **2010**, *8*, e1000412. [CrossRef]

40. Tietze, F. Enzymic Method for Quantitative Determination of Nanogram Amounts of Total and Oxidized Glutathione: Applications to Mammalian Blood and Other Tissues. *Anal. Biochem.* **1969**, *27*, 502–522. [CrossRef]

41. Rahman, I.; Kode, A.; Biswas, S.K. Assay for Quantitative Determination of Glutathione and Glutathione Disulfide Levels Using Enzymatic Recycling Method. *Nat. Protoc.* **2006**, *1*, 3159–3165. [CrossRef] [PubMed]

42. Massey, V.; Williams, C.H. On the Reaction Mechanism of Yeast Glutathione Reductase. *J. Biol. Chem.* **1965**, *240*, 4470–4480. [CrossRef]

43. Lawrence, R.A.; Burk, R.F. Glutathione Peroxidase Activity in Selenium-Deficient Rat Liver. *Biochem. Biophys. Res. Commun.* **1976**, *71*, 952–958. [CrossRef]

44. Alvarado, C.; Álvarez, P.; Jiménez, L.; De la Fuente, M. Oxidative Stress in Leukocytes from Young Prematurely Aging Mice Is Reversed by Supplementation with Biscuits Rich in Antioxidants. *Dev. Comp. Immunol.* **2006**, *30*, 1168–1180. [CrossRef]

45. Trefts, E.; Gannon, M.; Wasserman, D.H. The Liver. *Curr. Biol.* **2017**, *27*, R1147–R1151. [CrossRef]

46. Muntsant, A.; Giménez-Llort, L. Genotype Load Modulates Amyloid Burden and Anxiety-Like Patterns in Male 3xTg-AD Survivors despite Similar Neuro-Immunoendocrine, Synaptic and Cognitive Impairments. *Biomedicines* **2021**, *9*, 715. [CrossRef]

47. De la Fuente, M.; Gimenez-Llort, L. Models of Aging of Neuroimmunomodulation: Strategies for Its Improvement. *Neuroimmunomodulation* **2010**, *17*, 213–216. [CrossRef]

48. Siersbæk, M.S.; Ditzel, N.; Hejbøl, E.K.; Præstholm, S.M.; Markussen, L.K.; Avolio, F.; Li, L.; Lehtonen, L.; Hansen, A.K.; Schrøder, H.D.; et al. C57BL/6J Substrain Differences in Response to High-Fat Diet Intervention. *Sci. Rep.* **2020**, *10*, 14052. [CrossRef] [PubMed]

49. Li, J.; Wu, H.; Liu, Y.; Yang, L. High Fat Diet Induced Obesity Model Using Four Strains of Mice: Kunming, C57BL/6, BALB/c and ICR. *Exp. Anim.* **2020**, *69*, 326–335. [CrossRef] [PubMed]

50. Friedman, S.L.; Neuschwander-Tetri, B.A.; Rinella, M.; Sanyal, A.J. Mechanisms of NAFLD Development and Therapeutic Strategies. *Nat. Med.* **2018**, *24*, 908–922. [CrossRef]

51. Ipsen, D.H.; Lykkesfeldt, J.; Tveden-Nyborg, P. Molecular Mechanisms of Hepatic Lipid Accumulation in Non-Alcoholic Fatty Liver Disease. *Cell. Mol. Life Sci.* **2018**, *75*, 3313–3327. [CrossRef]

52. Simon, G.; Heckmann, V.; Tóth, D.; Pauka, D.; Petrus, K.; Molnár, T.F. The Effect of Hepatic Steatosis and Fibrosis on Liver Weight and Dimensions. *Leg. Med.* **2020**, *47*, 101781. [CrossRef]

53. Sarkar, S.; Raymick, J.; Cuevas, E.; Rosas-Hernandez, H.; Hanig, J. Modification of Methods to Use Congo-Red Stain to Simultaneously Visualize Amyloid Plaques and Tangles in Human and Rodent Brain Tissue Sections. *Metab. Brain Dis.* **2020**, *35*, 1371–1383. [CrossRef] [PubMed]

54. Clement, C.G.; Truong, L.D. An Evaluation of Congo Red Fluorescence for the Diagnosis of Amyloidosis. *Hum. Pathol.* **2014**, *45*, 1766–1772. [CrossRef] [PubMed]

55. Gioeva, Z.; Kieninger, B.; Röcken, C. Amyloidose in Leberbiopsien. *Pathologe* **2009**, *30*, 240–245. [CrossRef]

56. Muchtar, E.; Dispenzieri, A.; Magen, H.; Grogan, M.; Mauermann, M.; McPhail, E.D.; Kurtin, P.J.; Leung, N.; Buadi, F.K.; Dingli, D.; et al. Systemic Amyloidosis from A (AA) to T (ATTR): A Review. *J. Intern. Med.* **2021**, *289*, 268–292. [CrossRef]

57. Więckowska-Gacek, A.; Mietelska-Porowska, A.; Chutorański, D.; Wydrych, M.; Długosz, J.; Wojda, U. Western Diet Induces Impairment of Liver-Brain Axis Accelerating Neuroinflammation and Amyloid Pathology in Alzheimer's Disease. *Front. Aging Neurosci.* **2021**, *13*, 654509. [CrossRef] [PubMed]

58. Cheng, Y.; Tian, D.-Y.; Wang, Y.-J. Peripheral Clearance of Brain-Derived Aβ in Alzheimer's Disease: Pathophysiology and Therapeutic Perspectives. *Transl. Neurodegener.* **2020**, *9*, 16. [CrossRef]

59. González-Marrero, I.; Giménez-Llort, L.; Johanson, C.E.; Carmona-Calero, E.M.; Castañeyra-Ruiz, L.; Brito-Armas, J.M.; Castañeyra-Perdomo, A.; Castro-Fuentes, R. Choroid Plexus Dysfunction Impairs Beta-Amyloid Clearance in a Triple Transgenic Mouse Model of Alzheimer's Disease. *Front. Cell. Neurosci.* **2015**, *9*, 17. [CrossRef]

60. Marchese, M.; Cowan, D.; Head, E.; Ma, D.; Karimi, K.; Ashthorpe, V.; Kapadia, M.; Zhao, H.; Davis, P.; Sakic, B. Autoimmune Manifestations in the 3xTg-AD Model of Alzheimer's Disease. *J. Alzheimer's Dis.* **2014**, *39*, 191–210. [CrossRef] [PubMed]
61. Machado, M.V.; Cortez-Pinto, H. Cell Death and Nonalcoholic Steatohepatitis: Where Is Ballooning Relevant? *Expert Rev. Gastroenterol. Hepatol.* **2011**, *5*, 213–222. [CrossRef] [PubMed]
62. Lackner, C. Hepatocellular Ballooning in Nonalcoholic Steatohepatitis: The Pathologist's Perspective. *Expert Rev. Gastroenterol. Hepatol.* **2011**, *5*, 223–231. [CrossRef] [PubMed]
63. Presnell, E.C.; Bhatti, G.; Numan, L.S.; Lerche, M.; Alkhateeb, S.K.; Ghalib, M.; Shammaa, M.; Kavdia, M. Computational Insights into the Role of Glutathione in Oxidative Stress. *Curr. Neurovascular Res.* **2013**, *10*, 185–194. [CrossRef]
64. Wu, G.; Lupton, J.R.; Turner, N.D.; Fang, Y.-Z.; Yang, S. Glutathione Metabolism and Its Implications for Health. *J. Nutr.* **2004**, *134*, 489–492. [CrossRef] [PubMed]
65. Sentellas, S.; Morales-Ibanez, O.; Zanuy, M.; Albertí, J.J. GSSG/GSH Ratios in Cryopreserved Rat and Human Hepatocytes as a Biomarker for Drug Induced Oxidative Stress. *Toxicol. Vitr.* **2014**, *28*, 1006–1015. [CrossRef] [PubMed]
66. Zitka, O.; Skalickova, S.; Gumulec, J.; Masarik, M.; Adam, V.; Hubalek, J.; Trnkova, L.; Kruseova, J.; Eckschlager, T.; Kizek, R. Redox Status Expressed as GSH:GSSG Ratio as a Marker for Oxidative Stress in Paediatric Tumour Patients. *Oncol. Lett.* **2012**, *4*, 1247–1253. [CrossRef]
67. Han, D.; Hanawa, N.; Saberi, B.; Kaplowitz, N. Mechanisms of Liver Injury. III. Role of Glutathione Redox Status in Liver Injury. *Am. J. Physiol.-Gastrointest. Liver Physiol.* **2006**, *291*, G1–G7. [CrossRef] [PubMed]
68. Chen, J.; Bhandar, B.; Kavdia, M. Interaction of ROS and RNS with GSH and GSH/GPX Systems. *FASEB J.* **2015**, *29*, 636.7. [CrossRef]
69. Ferre, N.; Claria, J. New Insights into the Regulation of Liver Inflammation and Oxidative Stress. *Mini-Rev. Med. Chem.* **2006**, *6*, 1321–1330. [CrossRef] [PubMed]
70. Lyketsos, C.G.; Carrillo, M.C.; Ryan, J.M.; Khachaturian, A.S.; Trzepacz, P.; Amatniek, J.; Cedarbaum, J.; Brashear, R.; Miller, D.S. Neuropsychiatric Symptoms in Alzheimer's Disease. *Alzheimer's Dement.* **2011**, *7*, 532–539. [CrossRef]
71. Canet, G.; Hernandez, C.; Zussy, C.; Chevallier, N.; Desrumaux, C.; Givalois, L. Is AD a Stress-Related Disorder? Focus on the HPA Axis and Its Promising Therapeutic Targets. *Front. Aging Neurosci.* **2019**, *11*, 269. [CrossRef]
72. Kosel, F.; Pelley, J.M.S.; Franklin, T.B. Behavioural and Psychological Symptoms of Dementia in Mouse Models of Alzheimer's Disease-Related Pathology. *Neurosci. Biobehav. Rev.* **2020**, *112*, 634–647. [CrossRef]
73. Nguyen, E.T.; Selmanovic, D.; Maltry, M.; Morano, R.; Franco-Villanueva, A.; Estrada, C.M.; Solomon, M.B. Endocrine Stress Responsivity and Social Memory in 3xTg-AD Female and Male Mice: A Tale of Two Experiments. *Horm. Behav.* **2020**, *126*, 104852. [CrossRef]
74. Xie, X.; Shen, Q.; Yu, C.; Xiao, Q.; Zhou, J.; Xiong, Z.; Li, Z.; Fu, Z. Depression-like Behaviors Are Accompanied by Disrupted Mitochondrial Energy Metabolism in Chronic Corticosterone-Induced Mice. *J. Steroid Biochem. Mol. Biol.* **2020**, *200*, 105607. [CrossRef] [PubMed]
75. Liao, J.; Keiser, J.A.; Scales, W.E.; Kunkel, S.L.; Kluger, M.J. Role of Corticosterone in TNF and IL-6 Production in Isolated Perfused Rat Liver. *Am. J. Physiol.-Regul. Integr. Comp. Physiol.* **1995**, *268*, R699–R706. [CrossRef]
76. Battaglia, S.; Thayer, J.F. Functional Interplay between Central and Autonomic Nervous Systems in Human Fear Conditioning. *Trends Neurosci.* **2022**, *45*, 504–506. [CrossRef] [PubMed]
77. Battaglia, S.; Orsolini, S.; Borgomaneri, S.; Barbieri, R.; Diciotti, S.; di Pellegrino, G. Characterizing Cardiac Autonomic Dynamics of Fear Learning in Humans. *Psychophysiology* **2022**, *59*, e14122. [CrossRef]
78. Villagarcía, H.G.; Sabugo, V.; Castro, M.C.; Schinella, G.; Castrogiovanni, D.; Spinedi, E.; Massa, M.L.; Francini, F. Chronic Glucocorticoid-Rich Milieu and Liver Dysfunction. *Int. J. Endocrinol.* **2016**, *2016*, 7838290. [CrossRef] [PubMed]
79. Baliño, P.; Romero-Cano, R.; Sánchez-Andrés, J.V.; Valls, V.; Aragón, C.G.; Muriach, M. Effects of Acute Ethanol Administration on Brain Oxidative Status: The Role of Acetaldehyde. *Alcohol. Clin. Exp. Res.* **2019**, *43*, 1672–1681. [CrossRef]
80. Tao, W.; Sun, W.; Liu, L.; Wang, G.; Xiao, Z.; Pei, X.; Wang, M. Chitosan Oligosaccharide Attenuates Nonalcoholic Fatty Liver Disease Induced by High Fat Diet through Reducing Lipid Accumulation, Inflammation and Oxidative Stress in C57BL/6 Mice. *Mar. Drugs* **2019**, *17*, 645. [CrossRef] [PubMed]
81. Ke, Z.; Zhao, Y.; Tan, S.; Chen, H.; Li, Y.; Zhou, Z.; Huang, C. Citrus Reticulata Blanco Peel Extract Ameliorates Hepatic Steatosis, Oxidative Stress and Inflammation in HF and MCD Diet-Induced NASH C57BL/6 J Mice. *J. Nutr. Biochem.* **2020**, *83*, 108426. [CrossRef] [PubMed]
82. Shen, L.-X.; Yang, Y.-X.; Kuo, K.; Li, H.-Q.; Chen, S.-D.; Chen, K.-L.; Dong, Q.; Tan, L.; Yu, J.-T. Social Isolation, Social Interaction, and Alzheimer's Disease: A Mendelian Randomization Study. *J. Alzheimer's Dis.* **2021**, *80*, 665–672. [CrossRef] [PubMed]

International Journal of
Molecular Sciences

MDPI

Article

Glomerular Hypertrophy and Splenic Red Pulp Degeneration Concurrent with Oxidative Stress in 3xTg-AD Mice Model for Alzheimer's Disease and Its Exacerbation with Sex and Social Isolation

Juan Fraile-Ramos [1,2], Josep Reig-Vilallonga [3] and Lydia Giménez-Llort [1,2,*]

[1]   Institut de Neurociències, Universitat Autònoma de Barcelona, 08193 Barcelona, Spain; juan.fraile@uab.cat
[2]   Department of Psychiatry and Forensic Medicine, School of Medicine, Universitat Autònoma de Barcelona, 08193 Barcelona, Spain
[3]   Department of Anatomy, School of Medicine, Universitat Autònoma de Barcelona, 08193 Barcelona, Spain; josep.reig@uab.cat
*   Correspondence: Correspondence: lidia.gimenez@uab.cat

Abstract: The continuously expanding field of Alzheimer's disease (AD) research is now beginning to defocus the brain to take a more systemic approach to the disease, as alterations in the peripheral organs could be related to disease progression. One emerging hypothesis is organ involvement in the process of Aβ clearance. In the present work, we aimed to examine the status and involvement of the kidney as a key organ for waste elimination and the spleen, which is in charge of filtering the blood and producing lymphocytes, and their influence on AD. The results showed morphological and structural changes due to acute amyloidosis in the kidney (glomeruli area) and spleen (red pulp area and red/white pulp ratio) together with reduced antioxidant defense activity (GPx) in 16-month-old male and female 3xTg-AD mice when compared to their age- and sex-matched non-transgenic (NTg) counterparts. All these alterations correlated with the anxious-like behavioral phenotype of this mouse model. In addition, forced isolation, a cause of psychological stress, had a negative effect by intensifying genotype differences and causing differences to appear in NTg animals. This study further supports the relevance of a more integrative view of the complex interplay between systems in aging, especially at advanced stages of Alzheimer's disease.

Keywords: Alzheimer's disease; peripheric organs; kidney; spleen; amyloidosis; oxidative stress

Citation: Fraile-Ramos, J.; Reig-Vilallonga, J.; Giménez-Llort, L. Glomerular Hypertrophy and Splenic Red Pulp Degeneration Concurrent with Oxidative Stress in 3xTg-AD Mice Model for Alzheimer's Disease and Its Exacerbation with Sex and Social Isolation. *Int. J. Mol. Sci.* **2024**, *25*, 6112. https://doi.org/10.3390/ijms25116112

Academic Editor: Anna Atlante

Received: 5 April 2024
Revised: 28 May 2024
Accepted: 30 May 2024
Published: 1 June 2024

1. Introduction

The field of Alzheimer's disease (AD) has continuously expanded since its inception. Since the first neurophysiological findings, specifically the accumulation of the amyloid-beta (Aβ) protein in plaques and tau neurofilaments, exhaustive efforts have been made to characterize the changes in the brain and their causes. New approaches have also considered additional alterations involved in this disease, such as oxidative stress, reactive glia, and changes in microglia, as important factors involved in the generation and progression of this disease [1]. However, one of the lines of research yet to be expanded in this field is the communication between the brain and the rest of the body during the progression of the disease and its effects. Thus, new investigations challenge the classical view with a more systemic approach to AD. The involvement of the peripheral immune system in neuroinflammation, which is a central feature of AD, has already been discussed, and crosstalk has been established [2–4]. In addition, peripheral tissues and organs have the capacity to clear Aβ derived from the brain [5]. This indicates that the peripheral organs are involved in the development and progression of AD, suggesting a new possible valid therapeutic approach for AD by enhancing peripheral Aβ clearance.

The main connections established so far pose the neuro-immunoendocrine and other homeostatic systems as major AD-vulnerable peripheral systems. Previous studies by our group have already determined an altered status of the liver at advanced stages of the disease. Amyloidosis in the hepatic blood vessels and the presence of oxidative stress and inflammation support the hypothesis of the involvement of the liver–brain axis in this neurodegenerative disease [6–8]. In addition to the liver, which is capable of degrading Aβ and excreting it in bile [9], other organs such as the kidney and the spleen have been seen to have amyloid depositions in systemic amyloidosis and may be affected during disease progression [10]. Renal function has been linked to the risk of dementia on many occasions. Patients with chronic kidney disease (CKD) are likely to present cognitive disorders and dementia [11,12]. Moreover, the central role of the kidney in the clearance of Aβ by excretion in the urine has strongly supported the use of dialysis as a possible therapeutic intervention [13]. A reduction in Aβ plasma levels in humans and the attenuation of AD phenotypes in the APP/PS1 model has been demonstrated after peritoneal dialysis [14]. On the other hand, the spleen has also been related to AD pathology. The spleen can affect the brain's function through immune modulation, supporting a communication pathway between these two organs [15]. Moreover, alterations in inflammatory cytokines associated with splenic dysfunction in the APP/PS1/Tau model of AD also support the involvement of the spleen in AD [16].

Chronic stress has been classically described as the hyperactivation of the hypothalamic–pituitary–adrenal (HPA) axis and cellular inflammatory oxidative stress responses [17,18]. But chronic stress is also able to accelerate the neurodegeneration process of AD, affecting some of its features, such as the accumulation of β-amyloid, oxidative stress, inflammation, and the impairment of mitochondrial function [19]. In this context, social isolation, as an extrinsic environmental factor associated with psychological stress, can affect cognitive status and the development and prognosis of dementia. This fact has gained attention since the COVID-19 pandemic when imposed social isolation exerted a negative impact [20]. We recently demonstrated that male 3xTg-AD mice were strongly affected by naturalistic isolation (derived from the death of home cage mates), as shown by disrupted behavior and the liver, spleen, and kidney organometrics [21]. Thus, further study of behavioral and peripheral interplay under socially stressful scenarios can bring a better understanding of intrinsic and extrinsic vulnerability in aging and disease.

The 3xTg-AD mouse model has been extensively used for AD research; this model offers both neuropathological hallmarks of the disease, Aβ plaques, and tau tangles in an age-related manner [22]. Moreover, these mice also exhibit the prominent neuropsychiatric-like profile of the disease, which can already be recorded from the age of 4–6 months [23–25]. Taking advantage of these characteristics, our precedent work demonstrates hepatic oxi-inflammation and neophobia as potential targets for AD, with a strong sensitivity to sex and isolation [6]. The present work aims to further study the contribution of other peripheral-brain axes, namely the kidney and spleen axis, at advanced stages of the disease and how social isolation and stressful scenarios can impact this.

2. Results

2.1. Physical Metrics of Kidney and Spleen

The first measurement was the weight of both organs. In the case of the kidney, sex, and genotype effects were found both in the weight of the organ and the index (Figure 1A,B, S, all F's $(1,56) > 22.61$, $p < 0.0001$; G, all F's $(1,56) > 15.79$, $p = 0.0002$). In addition, the genotype had an interaction effect with sex (Figure 1A,B, all F's $(1,56) > 23.61$, $p < 0.0001$) and isolation (Figure 1A,B, all F's $(1,56) > 4.062$, $p = 0.0487$). Males had larger kidneys than females, and 3xTg-AD males had larger kidneys than NTg males. These differences were more marked in the index, where 3xTg-AD males showed a larger index, and females showed slight significance. Moreover, an isolation-dependent increase was observed in the kidney index of the NTg animals.

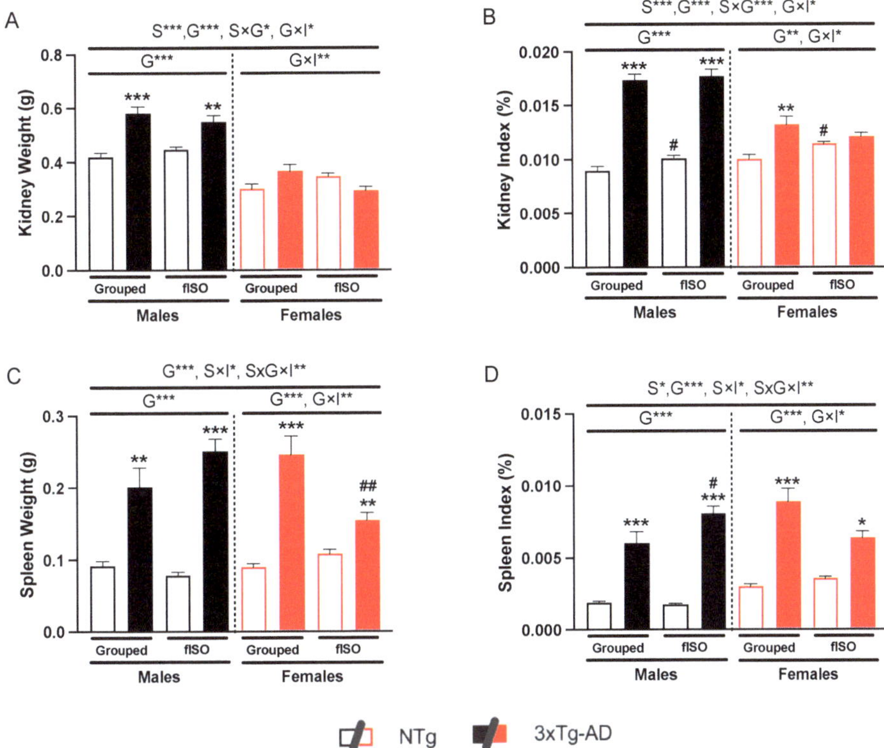

Figure 1. Physical status: kidney weights (**A**), kidney index (**B**), spleen weight (**C**), and spleen index (**D**). n = 5–10 mice per group, mean ± SEM. Sex: S*, $p < 0.05$; S***, $p < 0.001$. Genotype: G**, $p < 0.01$; G***, $p < 0.001$. Forced isolation: I***, $p < 0.001$. Interaction between sex and genotype: S × G*, $p < 0.05$; S × G***, $p < 0.001$. Interaction between genotype and isolation: G × I*, $p < 0.05$; G × I***, $p < 0.001$. Interaction between sex and isolation: S × I*, $p < 0.05$; S × I**, $p < 0.01$. Interaction between sex, genotype and isolation: S × G × I*, $p < 0.05$; S × G × I**, $p < 0.01$. Differences between two counterparts for genotype: * $p < 0.05$, ** $p < 0.01$, *** $p < 0.001$ vs. NTg counterpart; isolation: # $p < 0.05$, ## $p < 0.01$ vs. grouped counterpart.

As for the spleen, the most marked effect was observed for the genotype (Figure 1C,D, all F's (1,53) > 78.48, $p = 0.0004$). 3xTg-AD mice had twice the weight compared to NTg mice. Interaction effects between all factors and between sex and isolation (Figure 1C, S × I F (1,53) = 8.827, $p = 0.0062$; S × G × I F (1,53) = 10.02, $p = 0.0026$) were also present because of the reduction in the spleen's weight in the 3xTg-AD isolated females. This effect was also present in the spleen index, although it did not reach statistical significance. The index showed a sex effect (Figure 1D, F (1,53) = 4.978, $p = 0.0299$) and an interaction effect (Figure 1D, S × I F (1,53) = 4.852, $p = 0.032$; S × G × I F (1,53) = 8.835, $p = 0.0044$). All 3xTg-AD animals had a higher spleen index than NTg animals, and isolated 3xTg-AD males had a higher index than non-isolated males. These results indicate that 3xTg-AD animals have renomegaly and splenomegaly.

2.2. Histopathological Assessment: H&E and Congo Red Staining

The kidney and spleen sections were prepared and stained for microscopic analysis to determine morphological and physiological alterations (Figure 2). The kidney showed a marked alteration of the glomeruli in the 3xTg-AD animals. In NTg animals, the glomeruli were round and small, but in 3xTg-AD, their size was considerably higher, and their round form was lost. Quantitative analysis of this alteration measuring the glomerular area showed a sex and genotype effect (Figure 3A; S, F (1,57) = 4.31, p = 0.042; G, F (1,57) = 28.9, p < 0.0001). 3xTg-AD animals had higher glomerular areas than NTg animals, and this effect was more marked in males than females. Another aspect taken into account was the immune infiltration. which had genotype differences (Figure 3B, F (1,57) = 19.16, p < 0.0001) that were significantly higher in 3xTg-AD animals. On the other hand, the spleen also presented structural changes in the red pulp; the typical microarchitecture of the spleen was lost when the red pulp enlarged and disintegrated, making it look clear and almost disappear. As for the spleen, the quantification of the red pulp area showed a clear genotype effect (Figure 3C; F (1,53) = 240.1, p < 0.001), with all 3xTg-AD animals showing wider red pulp areas. The ratio of the red pulp and white pulp in the spleen also showed this enlargement of the red pulp according to genotype (Figure 3D; F (1,53) = 59.12, p < 0.001); in the ratio, a tendency was also observed in isolated males to have higher RP/WP indexes, whereas isolated females showed a lower RP/WP index.

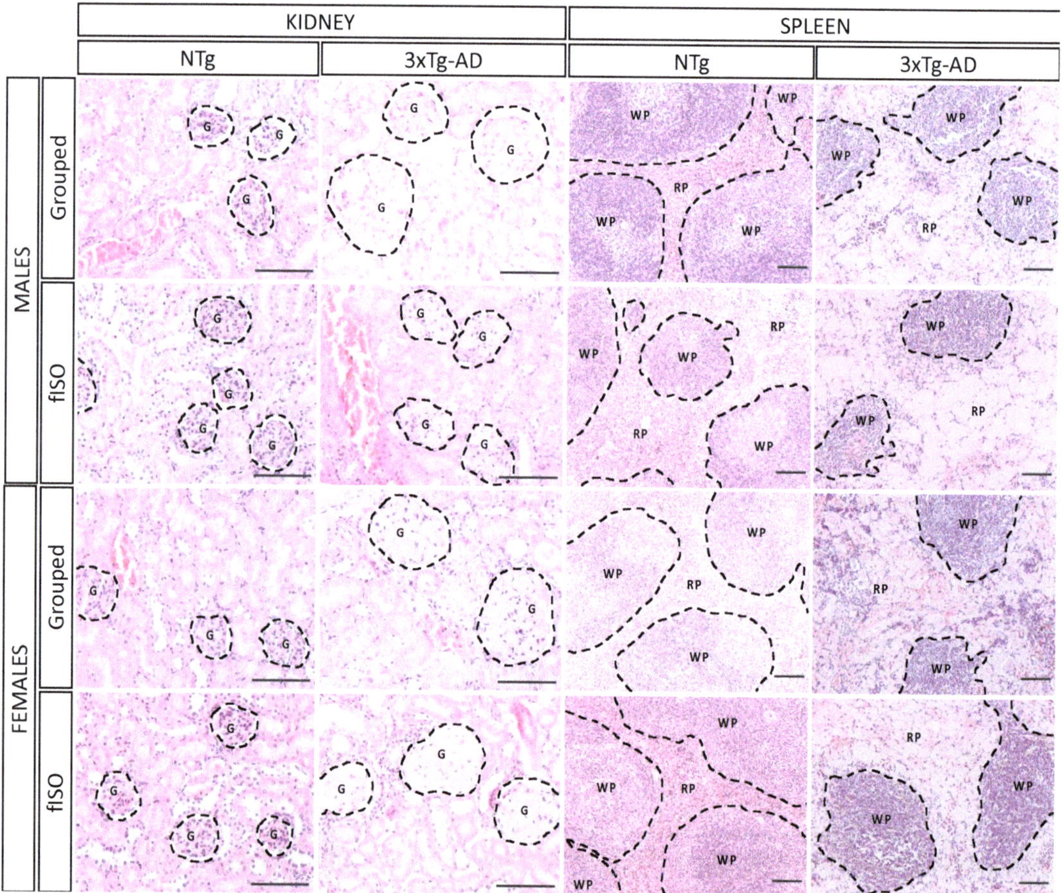

Figure 2. Histopathological assessment. Morphological comparison of kidney and spleen tissue in NTg and 3xTg-AD males and females for mice grouped and those submitted to forced isolation. H&E stained. Dashed lines indicate altered morphological features as follows: G, glomeruli; WP, white pulp; and RP, red pulp. The images were taken with ×20 and ×10 objective lenses; magnification bar = 100 μm.

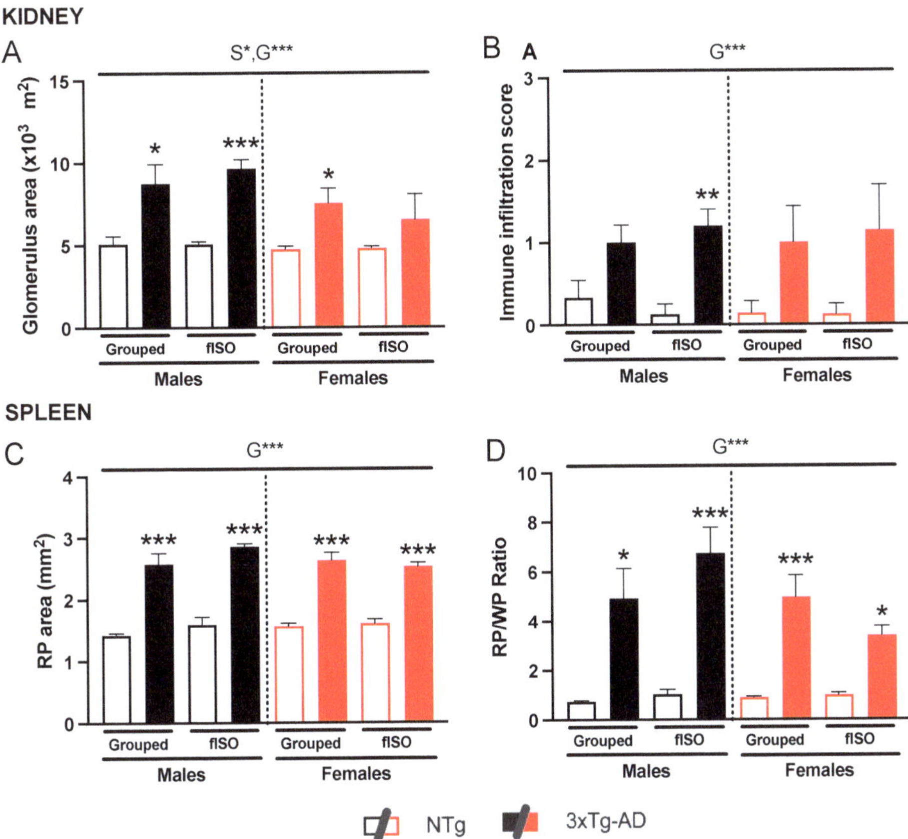

Figure 3. Quantification of different histopathological aspects. (**A**) Quantification of the glomerulus area and at least 20 glomeruli from each mouse were quantified. (**B**) Immune infiltration score of the kidney comprising the grades 0, none; 1, mild; 2, moderate; and 3, severe. (**C**) Quantification of the red pulp area and at least three different areas from each spleen were quantified. (**D**) The ratio between the red pulp area and white pulp area; at least three different areas from each spleen were quantified. n = 6–10 mice per group, mean ± SEM. Sex: S*, $p < 0.05$. Genotype: G***, $p < 0.001$. Differences between two counterparts for genotype: * $p < 0.05$,** $p < 0.01$, *** $p < 0.001$ vs. NTg counterpart.

Deepening the histopathological analysis of the kidney and spleen, staining with Congo red was performed to determine amyloidosis (Figure 4). In the case of the kidney, the staining marked the kidneys of 3xTg-AD animals exclusively with that characteristic reddish color. In addition, fluorescence images showed a more highly stained area in males than in females; in any case, the affected area was always the glomeruli. In the spleen, the result was similar; only 3xTg-AD animals presented Congo red-stained areas, and in accordance with the H&E staining, the area affected was the red pulp. It can be seen in the bright field image that the red pulp has a reddish tone, and focusing on the fluorescent images, the red pulp clearly shows red fluorescence. Thus, 3xTg-AD animals presented amyloidosis in the kidney (glomeruli) and the spleen (red pulp).

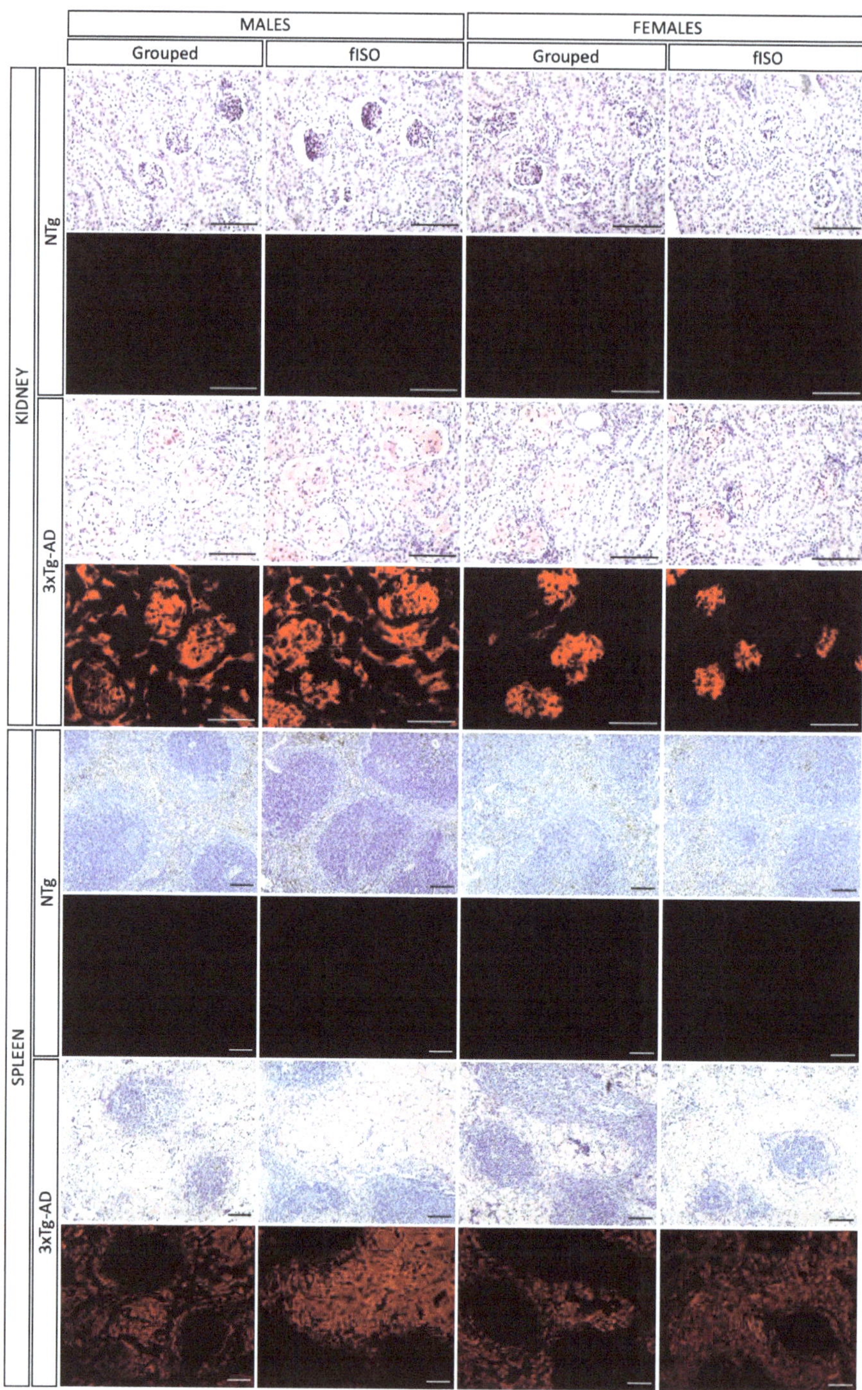

Figure 4. Amyloidosis assessment. Differences in amyloidosis in kidney and spleen tissue in NTg and 3xTg-AD males and females both grouped and those submitted to forced isolation. Congo red-stained. The images were taken with a ×20 and x10 objective lens; magnification bar = 100 μm.

2.3. Oxidative Stress Status

The oxidative status of the kidney and spleen was determined based on the GSH cycle, considering the levels of GSH and the activity of the antioxidant enzymes involved in the process (GPx and GR).

Focusing first on the kidney, GSH levels presented sex and genotype differences (Figure 5A; S, $F_{(1,57)} = 22.86$, $p < 0.0001$; G, $F_{(1,57)} = 5.105$, $p < 0.028$), with females having higher levels than males, and 3xTg-AD animals having higher levels than NTg. The activity of the antioxidant enzymes also showed changes according to genotype (Figure 5B,C, all F's $(1,51) > 11.45$, $p = 0.0014$). These differences consisted of a reduction in GPx activity and an increase in GR activity in the 3xTg-AD animals. In the case of GPx, sex differences were also present (Figure 5B, $F_{(1,55)} = 9.56$, $p = 0.003$), and the enzyme activity was higher in females, while only males achieved statistical significance by a reduction in this enzyme's activity. In the kidney, the only effect of the isolation was the interaction effect with the genotype (Figure 5C, $F_{(1,51)} = 4.11$, $p = 0.048$) shown by the Ntg isolated males.

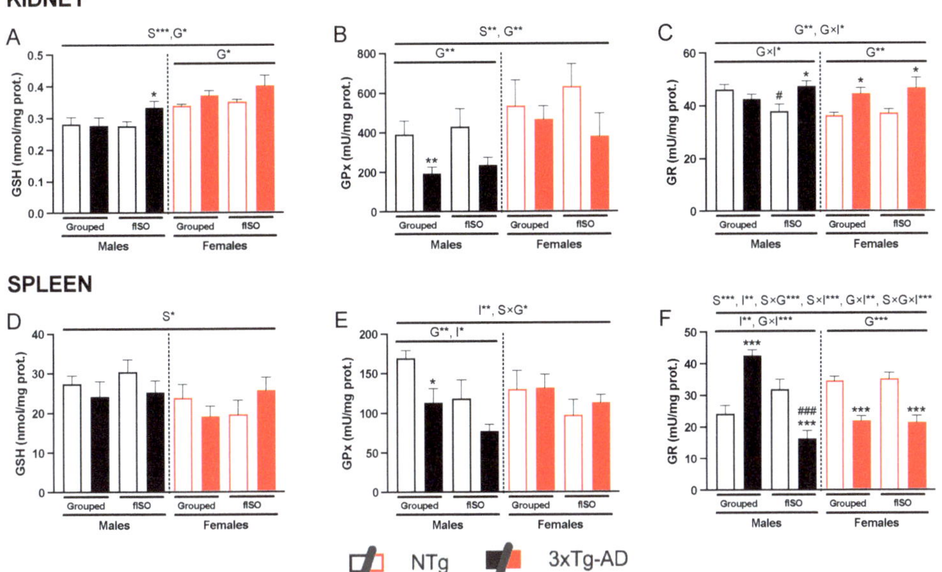

Figure 5. Oxidative stress in peripheral organs: kidney GSH levels (**A**), GPx activity (**B**), GR activity (**C**), spleen GSH levels (**D**), GPx activity (**E**), and GR activity (**F**). n = 5–10 mice per group, mean ± SEM. Sex: S*, $p < 0.05$; S**, $p < 0.01$; S***, $p < 0.001$. Genotype: G*, $p < 0.05$; G**, $p < 0.01$; G***, $p < 0.001$. Forced isolation: I*, $p < 0.05$; I**, $p < 0.01$. Interaction between sex and genotype: S × G*, $p < 0.05$; S × G***, $p < 0.001$. Interaction between genotype and isolation: G × I*, $p < 0.05$; G × I**, $p < 0.01$; G × I***, $p < 0.001$. Interaction between sex and isolation: S × I***, $p < 0.001$; S × I**, $p < 0.01$. Interaction between sex, genotype, and isolation: S × G × I***, $p < 0.001$. Differences between two counterparts for genotype: * $p < 0.05$, ** $p < 0.01$, *** $p < 0.001$ vs. NTg counterpart; isolation: # $p < 0.05$, ### $p < 0.001$ vs. grouped counterpart.

As for the spleen, GSH levels only showed sex differences (Figure 5D, $F_{(1,50)} = 4.47$, $p = 0.039$); in this case, males tended to have higher levels than females. GPx activity presented isolation differences (Figure 5E, $F_{(1,53)} = 7.338$, $p = 0.009$), with all isolated animals showing a slight reduction in GPx activity. In addition, males also presented genotype differences (Figure 5E, $F_{(1,27)} = 8.06$, $p = 0.0085$), as 3xTg-AD males presented lower activity levels than NTg males. GR activity showed a clear genotype effect, which differed according to sex (Figure 5F; S × G, $F_{(1,55)} = 18.56$, $p < 0.001$). In males, GR activity acted differently depending on isolation (Figure 5F; G × I, F (1,30) = 41.61, $p < 0.001$); in grouped males, 3xTg-AD had higher levels of activity than NTg, whereas, in isolation, 3xTg-AD males had lower levels of activity than NTg males. In females, genotype differences were the same regardless of the social conditions (Figure 5F; G, F (1,25) = 43.91, $p < 0.001$). 3xTg-AD females all presented a reduction in the activity of GR when compared to NTg females.

2.4. Statistical Correlation between the Studied Variables

The statistical analysis of the correlations of the variables was also included in this study for each organ, with the behavioral and corticosterone data measured for the same mice in a previous study [6]. In Figure 6A–G, we can see the correlations from 3xTg-AD animals established between behavioral variables, corticosterone levels, metrics variables, histopathological variables, and oxidative stress variables. These correlations were not present in NTg animals.

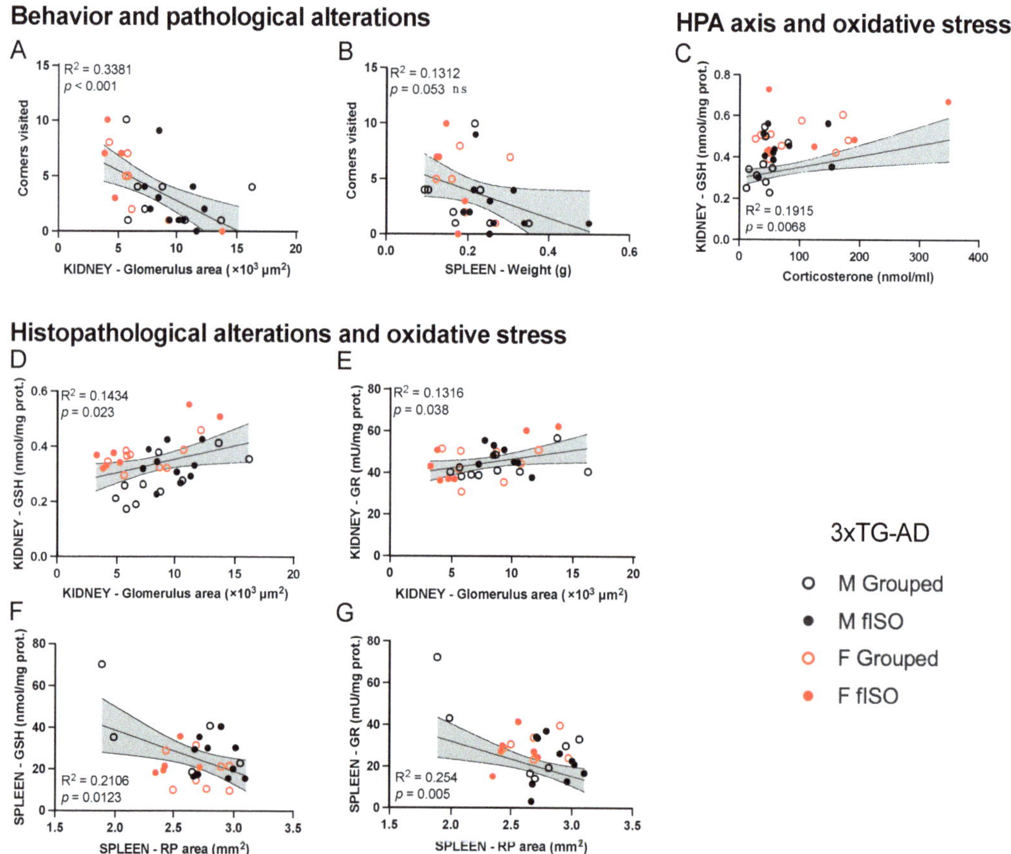

Figure 6. Correlation between the studied variables of the kidney and the spleen in 3xTg-AD animals. Correlations from behavior and histopathological alterations corresponding to the number of visited corners and glomerulus area (**A**) in the kidney and spleen weight (**B**) In the spleen, correlations between the HPA axis and oxidative stress corresponding to the levels of GSH in the kidney and serum levels of corticosterone (**C**). Correlations between the histopathological alterations and the oxidative stress corresponding to the glomerulus area in the kidney, the levels of GSH (**D**) and GR activity (**E**) and the red pulp (RP) area in the spleen, and levels of GSH (**F**) and GR activity (**G**).

Neophobia behavior shown by the number of visited corners correlated with histopathological (glomerulus area) in the kidney (Figure 6A), indicating a lower exploration rate as the glomerular area is larger. In the spleen, the correlation between behavioral variables and physical disturbances did not reach statistical significance, but they did show the same tendency (Figure 6B). The HPA axis showed a relationship with the oxidative stress status as the levels of corticosterone were related to the levels of GSH in the kidney (Figure 6C). Finally, all these histopathological alterations also correlated with oxidative stress status. In the case of the kidney, the enlargement of the glomeruli positively correlated with GSH levels (Figure 6D) and GR activity (Figure 6E), indicating higher oxidative stress defense with a higher glomerulus area. In the case of the spleen, however, the red pulp enlargement negatively correlated with GSH levels (Figure 6F) and GR activity (Figure 6G), indicating a reduction in antioxidant defenses with higher RP areas.

3. Discussion

AD is gaining relevance within an aging world as the most important neurodegenerative disease and form of dementia. This disease has a significant impact on the organism beyond cognitive impairment. However, most of the research has focused on changes in the brain, whereas the involvement of peripheral organs has been scarcely investigated until now. Recent studies suggest that Alzheimer's disease also has a systemic component, producing metabolic dysfunction and different pathological processes in peripheral organs [26]. We have recently provided evidence that the liver is affected in 3xTg-AD mice, and we propose hepatic amyloidosis and the dysregulation of its oxi-inflammatory defenses as potential targets for AD [6]. In this systemic approach to AD, here, we further investigate the histopathological profile and the oxidative stress status of the kidney and the spleen of these animals. Current findings show how morphological, structural, and oxidative stress alterations in these two organs further support the relevance of a peripheral organ–brain axis in the integrative system understanding of AD. In addition to the genotype differences, compared to the same age gold standard of C57BL/6 mice with physiological aging, the impact of sex and forced isolation as intrinsic (biological) and extrinsic (social conditions) factors, respectively, were also demonstrated.

3.1. Physical and Histopathological Alterations of Kidney and Spleen

Abnormalities were found in physical measurements and the histopathological analysis of the kidney and spleen of 3xTg-AD mice compared to NTg mice. Their weight and index (relative weight) were considerably high, denoting the hypertrophy of both organs, mostly in the spleen, in both sexes of 3xTg-AD mice. The kidney, with an important function in maintaining the homeostasis of the organism as it is currently, is also considered to have a role in the peripheral clearance of β-amyloid [13,27]. In the case of the spleen, the splenomegaly found can be important since it has been usually related to serious systemic diseases [28,29]. In previous works in mouse models for AD, splenomegaly has become a common feature and is related to inflammation and autoimmune processes [16,30]. Considering that the spleen also acts as a blood filter and is attributed to the physiological cleansing of Aβ [31], this splenomegaly supports the idea that the organs related to this function are the most affected in AD-pathogenic conditions.

In the present work, histopathological analysis showed physical abnormalities in both organs. In the case of the kidney, the increase in the glomeruli is almost twice its normal size, and the loss of its typical round morphology were significant features associated with the AD phenotype. On the other hand, the spleen presented structural destruction regarding the red pulp. This was consistent with previous results from 19-month-old 3xTg-AD males [7], and it is now seen that these alterations occur in both sexes, with higher severity in males than in females at advanced stages of the disease.

Congo red staining revealed that these histopathological changes were due to the accumulation of Aβ in the glomeruli of the kidney and the red pulp of the spleen. Therefore, these observations mimic the morphological alterations observed in patients with systemic amyloidosis as Aβ infiltration was seen in the glomeruli from kidney biopsies [32]. Moreover, unilateral nephrectomy has been shown to increase the deposition of Aβ in the brain, aggravating AD pathologies in the brain [33]. In the case of the spleen, the accumulation of Aβ has been seen in the mouse models of AD [34], and its potential to physiologically clear Aβ from the blood suggests that splenectomy could aggravate AD pathogenesis [31]. However, this effect has not yet been constated in humans. Moreover, in other studies performed with the 3xTg-AD mouse model, severe splenomegaly and structural destruction were reported [16]. This could indicate systemic amyloidosis in 3xTg-AD animals and pose the kidney and the spleen as main targets for its clearance. In fact, it is known that β-amyloid efflux from the brain to the periphery via the Blood–brain barrier (BBB) occurs during the disease for peripheral clearance, contributing to the brain's mechanisms for the removal of Aβ [5]. In advanced stages of the disease, the BBB of 3xTg-AD mice was disrupted, and its permeability increased, allowing even more Aβ to reach the blood flow and the periphery [35,36]. The results of the present study further demonstrate the disruption or saturation of these clearing systems, supporting the clearing of brain Aβ and leading to its accumulation in the peripheral organs at the age of 16 months.

3.2. Kidney and Spleen Oxidative Stress

GSH metabolism has been widely used as a biomarker of oxidative stress along with the activity of both GR and GPx activity. The imbalance in the levels of GSH and the activity of these enzymes is considered to be a sign of oxidative stress in the cell [37,38]. Therefore, molecularly deepening in both organs, the level of oxidative stress was determined, and the results showed that the previous observations in the liver were reproduced in the kidney and spleen [6]. GSH levels remained fairly constant, although, in the kidney, they experienced a slight increase with regard to sex, with females having higher levels and genotypes, with 3xTg-AD animals tending to have more GSH. High levels of intracellular GSH are considered to be an adaptive response to oxidative stress [39].

In all organs, a decrease in GPx activity was observed in 3xTg-AD animals in both sexes, coinciding with studies in humans and animals where the decrease in the activity of this antioxidant enzyme is related to oxidative stress in this disease [40]. This oxidative stress was sex-dependent as it affected males more significantly. GR activity, despite having the most pronounced genotype differences, had opposite effects depending on the organ. In the case of the kidney, activity increased, possibly to compensate for the reduction in GPx activity. However, in the spleen, the activity of GR was also reduced, so in this organ, both enzymes showed lower activity. It should be noted that in the case of males, GR activity showed differences by social isolation, these differences pointing to greater oxidative stress in both organs. To better understand these systems and their interplay, the correlations between the different variables of organometrics, histopathology, cellular oxidative stress, neophobia, and HPA axis stress (corticosterone) were analyzed.

3.3. Correlations between Behavior, Histopathology, HPA Axis and Peripheral Oxidative Stress

In both organs, several significant correlations were found for the 3xTg-AD genotype but were not present in NTg mice. Anatomical and histopathological alterations correlated with neophobia, which is a fearful response to novelty, typically described as part of the neuropsychiatric-like phenotype of the 3xTg-AD mouse model [41–43]. The correlation with neophobia shows that both behavioral and peripheral changes happen at the same time in the advanced stages of the disease and agrees with previous results from the liver [6]. Moreover, these alterations were also related to oxidative stress defense changes, as shown by GSH levels and GR activity. Further correlations between histopathological and oxidative stress suggest that affectation due to Aβ in the glomeruli contributed to this oxidative stress and renal health. This is important since oxidative stress can affect kidney function and lead to renal damage [44,45]. Additionally, impaired kidney function has been associated with cognitive decline in older adults [46]. In the case of the spleen, despite splenomegaly and the destruction of the spleen's microarchitecture being detected in the

3xTg-AD model on numerous occasions [7,16,26,47], this is the first time that red pulp is identified as a splenic structural target in AD. Noteworthy, the red pulp area correlated with GSH levels and GR activity in the same way as the glomerular area from the kidney.

Psychological stress during the development of this disease [48] can also affect the oxidative stress status of peripheral organs [49,50]. It has been found that chronic kidney disease can disturb cortisol regulation, leading to subclinical hypercortisolism and reduced cortisol clearance [51]. High cortisol levels may worsen cognitive status and become relevant to the generation and progression of AD [52]. In the present work, 3xTg-AD mice, but not NTg mice, showed a correlation between corticosterone and GSH levels, suggesting that corticosterone levels may be related to their increased oxidative stress. In fact, high levels of corticosterone have already been shown to produce oxidative stress in rats [53], and prolonged isolation stress has already been shown to accelerate the onset of AD in the 5xFAD mouse model [54].

Overall, these correlations, at different levels of study, indicate the active interplay between these organs and the brain. Further research about the dynamics of this communication under the AD pathological condition can allow for a broader but integrative vision of this neurodegenerative disease.

3.4. Study Limitations

Different limitations to this basic translational research study were found. Starting with the scarce knowledge of the non-classical roles of the peripheral organs in the involvement of AD limits the conceptual framework and discussion and creates demands for primary screening. Further biomarkers of organ functioning should have been included to confirm the histological and oxidative stress alterations. It would have also been interesting to evaluate other distinctive behavioral hallmarks and their relationship with peripheral changes in addition to examining the neurological changes happening in the brain. Only one model was used in this study, but other genetic and/or sporadic models of AD should be assessed, as well as different stages of the disease. Social isolation was assessed as an extrinsic factor, but other factors may also be affecting the modulation of these changes and should also be considered.

3.5. Future Perspective of Peripheral Organs in AD

Apart from the classical brain, structural and functional alterations in the onset of the disease [55], a more integrated vision of AD has already allowed for consideration of other hallmarks, such as neuroinflammation [56,57] and impaired brain bioenergetics [58]. In addition, clinical and basic research is now providing evidence of multiple connections between neuroinflammation and systemic inflammation [59], opening a third-generation approach to the disease that considers peripheric implications in AD. For instance, we have previously demonstrated the interplay of the immune system and behavior in the 3xTg-AD model [4,60]. The present and the previous work [6,8,61] show, at different levels of study, the implication of the liver, the kidney, and the spleen in Aβ pathology in AD. A clearly augmented relative weight in these organs, the histopathological alterations localized in the hepatic blood vessels, the renal glomeruli, and the splenic red pulp, as well as the reduction in oxidative stress defenses, stood out as the main alterations. However, the involvement of peripheral organs in AD is still a novel and little-explored idea. Therefore, identifying and quantifying these pathological peripheral signatures of AD can lead to new translational findings that may unveil their significance in the clinical field. In fact, blood and urine levels of Aβ have been considered as potential biomarkers for AD [62–64]. Similarly to classical visualization methods in organometry and other peripheral organ parameters used in the diagnosis of different diseases [61], the present results suggest that the relative weight of the liver, the kidney, and the spleen could be of use in AD diagnosis and non-invasive tracking. Also, new technologies of imaging, such as magnetic resonance imaging (MRI), could be useful not only to see the structural and functional organization of the brain [65] but also the peripheral organs [66] and their possible hypertrophy. Further research endeavors are needed to clarify the complex bidirectional communication between these brain–peripheral axes, posing a new and promising area of research. Similarly, under integrative system analysis, whether other systems can also take part in the development and progression of AD and be affected as a consequence of disease progression is yet to be determined.

4. Materials and Methods

4.1. Animals

Homozygous triple-transgenic 3xTg-AD mice carrying human PS1/M146V, APPSwe, and tauP301L transgenes were produced at the University of California, Irvine [67]. Briefly, two independent transgenes encoding human APPSwe and tauP301L, both regulated by the mouse Thy1.2 transcription factor, were co-injected into single-cell embryos from PS1KI (PS1M146V knock-in) mice. A cohort of sixty-six 16-month-old male and female mice from the Spanish colonies of homozygous 3xTg-AD mice (n = 37) and from litters of a breeding program established at Universitat Autònoma de Barcelona after embryonic transfer to a C57BL/6 strain background [68], from now on referred to as "3xTg-AD", and C57BL/6 wildtype mice (n = 29), from now on referred to as "NTg", non-transgenic mice, were used. All animals were housed in groups of 3–4 individuals of the same sex and genotype until the experiment. They were maintained in cages (Makrolon, 35 × 15 × 15 cm3) under standard laboratory conditions (12 h light/dark, cycle starting at 8:00 a.m., food and water ad libitum, 22 ± 2 °C, 50–60% humidity). Biochemical and histopathological analyses were performed blind to the experiment in a counterbalanced manner.

All procedures were performed according to the Spanish legislation on the 'Protection of Animals Used for Experimental and Other Scientific Purposes' and the EU Council directive (2010/63/EU) on this matter. The protocol CEEAH 3588/DMAH 9452 was approved on the 8th of March 2019 by the Departament de Medi Ambient i Habitatge, Generalitat de Catalunya. The ARRIVE guidelines developed by the NC3Rs were followed with the objective of reducing the number of animals used in experimental procedures [69].

4.2. Experimental Design

To mirror the translational psychological stressful scenario that was lived in during the COVID-19 pandemic, mice from both genotypes and sexes were subjected to a forced isolation (fISO) paradigm; at 13 months of age, the animals were separated and individualized in cages and were kept for 3 months until they were euthanized. In all the cases, even though isolated animals were in their cages and could not physically interact with others, they were still connected through olfaction and audition. Nesting conditions were the same as for the grouped animals.

4.3. Physical Status

Animals were weighed at the time of sacrifice, and the organs were dissected, weighed, and stored at −80 °C until protein extraction for biochemical analysis. The index of the organ or relative weight was calculated according to the following equation: Organ index = Organ weight (g) / Body weight (g).

4.4. Behavioral Phenotype

The corner test was used to evaluate the immediate neophobia response of the animals in a new standard home cage with a clean bed of wood save. The animals were placed in the center of the cage and observed for 30 s. The following information was recorded: the number of visited corners, the latency of the first rearing, and the number of rearings. Behavioral assessment was performed from 9:00 a.m. to 1:00 p.m., counterbalanced and blind to the experimenter.

4.5. Histopathological Assessment

After euthanasia, one kidney and half of the spleen were dissected, fixed with 10% formalin for 24 h, and embedded in paraffin for further histopathological analysis. Slices of 5 μm and 10 μm thickness were deparaffinized and re-hydrated by several passes through ethanol at different concentrations.

4.5.1. Morphological and Structural Analysis with H&E

Standard staining with hematoxylin and eosin (H&E) (Sigma-Aldrich, San Luis, MO, USA) was performed. Briefly, a 5 min incubation with Harris hematoxylin and a 2 min incubation with Eosin Y was performed, with a dedifferentiation phase in between. Finally, the slices were manually mounted with a DPX mounting medium after dehydration. The images were taken with a ×20 objective on a Nikon Eclipse 80i microscope, using a digital camera running on the control software ACT-1 (ver2.70). For histopathological measurements in the kidney, the glomerular area of at least 20 glomeruli of each mouse was measured. In the spleen, three random areas across the spleen of each mouse were selected, and the area of red pulp and white pulp were quantified. In both cases, measurements were performed using Fiji software version 2.14.0/1.54f for image processing.

4.5.2. Congo Red Staining for Amyloidosis

Amyloidosis was assessed using a Congo red kit (Sigma-Aldrich, San Luis, MO, USA) for amyloid staining. The procedure was followed as the kit indicated, with incubation in Mayer's hematoxylin and then in Congo red solution. Finally, the slices were manually mounted with the DPX mounting medium after dehydration. The images were taken with a ×20 objective on a Nikon Eclipse 80i microscope under normal light (bright field) and fluorescence light (fluorescent filter G-2A), using a digital camera running on the control software ACT-1 (ver2.70).

4.6. Oxidative Stress Analysis

The antioxidant capacity of the organs was evaluated from homogenates of the kidney and spleen tissue by the levels of total glutathione (GSH), the main non-enzymatic reducing agent of the organism, and the enzymatic activity of glutathione peroxidase (GPx) and reductase (GR). This benefited from the enzymatic recycling method previously described [70]. All reagents from Sigma-Aldrich, San Luis, MO, USA.

Total GSH was determined by measuring the absorbance at 412 nm, which was adapted to 96-well plates with slight modifications [71].

The enzymatic activity of glutathione reductase was assessed following the method described by Massey and Williams [72] with slight modifications. The oxidation of NADPH was monitored spectrophotometrically and performed at 340 nm for 300 s.

The enzymatic activity of glutathione peroxidase was measured using the modified previously described technique [73,74]. The reaction was followed spectrophotometrically by a decrease in the absorbance at 340 nm for 300 s.

The results were expressed as milliunits (mU) of enzymatic activity per milligram (mg) of the organ protein.

4.7. Functional Correlates

Correlations between the variables measured in this study and the behavioral phenotype and HPA axis variables from the same animals [6] were carried out using Pearson's test. The data are shown as scattered plots with the total correlation coefficient and the significance, as well as separated coefficients by genotype (NTg and 3xTg-AD).

4.8. Statistics

Results are expressed as mean ± SEM. GraphPad Prism 8.0. Three-way and two-way analyses for multiple variables (ANOVA) followed by Tukey's post hoc test for multiple comparisons were used. Student's *t*-test was used to assess the significance between two independent groups. In all the tests, statistical significance was considered at $p < 0.05$.

5. Conclusions

Taking advantage of the 3xTg-AD genotype and considering sex as an intrinsic factor and isolation as an extrinsic factor, alterations in peripheral organs were studied in the translational advanced stage scenario of the disease. The present work provides evidence of renal and splenic organometric and histopathological alterations accompanied by oxidative stress in the 3xTg-AD mouse model in a sex- (worse in males) and social condition- (worse in forced isolation) dependent manner. Renal glomeruli and the red pulp of the spleen were determined as key morphological and structural alterations accumulating Aβ, confirming that both organs were compromised in AD. The concurrent enhancement of oxidative stress was evidenced by a decrease in the activity of the antioxidant enzyme GPx in both organs. Furthermore, these alterations correlated with the 3xTg-AD animals' behavioral phenotype. This work further highlights the relevance of the peripheral organ–brain axis in aging, especially in AD. It brings the research closer to taking a more comprehensive view of integrative systems where intrinsic (sex) and extrinsic (social conditions) factors play a role.

Author Contributions: Conceptualization, L.G.-L.; experimental design performance, L.G.-L.; harvesting, biochemical analysis, and statistics, J.F.-R.; pathology: J.F.-R. and J.R-V; laboratory resources, J.R.-V.; illustrations: J.F.-R.; writing—original draft preparation, J.F.-R.; writing—review and editing, J.F.-R. and L.G.-L.; funding acquisition, L.G.-L. All authors have read and agreed to the published version of the manuscript.

Funding: This work was funded by AGAUR, Generalitat de Catalunya, 2017-SGR-1468 and Universitat Autònoma de Barcelona, UAB-GE-260408 to L.G.-L. The colony of 3xTg-AD mice was obtained by ArrestAD H2020, Fet-OPEN-1-2016-2017-737390, European Union's Horizon 2020 research and innovation program under grant agreement No 737390 to L.G.-L. This study also received financial support from Memorial Mercedes Llort Sender, 2021/80/09241941.8.

Institutional Review Board Statement: This study was conducted according to the guidelines of the Declaration of Helsinki and approved by the Ethics Committee of Departament de Medi Ambient I Habitatge, Generalitat de Catalunya (CEEAH 3588/DMAH 9452) on the 8 March 2019.

Informed Consent Statement: Not applicable.

Data Availability Statement: The data presented in this study are available on request to the corresponding author.

Acknowledgments: We thank Frank M LaFerla, Institute for Memory Impairments and Neurological Disorders, Department of Neurobiology and Behavior, University of California, Irvine, USA, for kindly providing the progenitors of the Spanish colonies of 3xTg-AD and NTg mice. We thank Anna Garrit, lab technician in the Department of Anatomy, School of Medicine, Universitat Autònoma de Barcelona, for helping with histology techniques and methodology. We also thank BioRender.com for providing graphical tools for scientific illustration.

Conflicts of Interest: The authors declare no conflicts of interest. The funders had no role in the study's design; in the collection, analyses, or interpretation of data; in the writing of the manuscript; or in the decision to publish the results.

References

1. Agostinho, P.; Cunha, R.A.; Oliveira, C. Neuroinflammation, Oxidative Stress and the Pathogenesis of Alzheimer's Disease. *Curr. Pharm. Des.* **2010**, *16*, 2766–2778. [CrossRef]
2. Angiulli, F.; Conti, E.; Zoia, C.P.; Da Re, F.; Appollonio, I.; Ferrarese, C.; Tremolizzo, L. Blood-Based Biomarkers of Neuroinflammation in Alzheimer's Disease: A Central Role for Periphery? *Diagnostics* **2021**, *11*, 1525. [CrossRef] [PubMed]
3. Cao, W.; Zheng, H. Peripheral Immune System in Aging and Alzheimer's Disease. *Mol. Neurodegener.* **2018**, *13*.
4. Giménez-Llort, L.; Maté, I.; Manassra, R.; Vida, C.; De la Fuente, M. Peripheral Immune System and Neuroimmune Communication Impairment in a Mouse Model of Alzheimer's Disease. *Ann. N. Y. Acad. Sci.* **2012**, *1262*, 74–84. [CrossRef] [PubMed]

5.   Cheng, Y.; Tian, D.Y.; Wang, Y.J. Peripheral Clearance of Brain-Derived Aβ in Alzheimer's Disease: Pathophysiology and Therapeutic Perspectives. *Transl. Neurodegener.* **2020**, *9*. [CrossRef] [PubMed]

6.   Fraile-Ramos, J.; Garrit, A.; Reig-Vilallonga, J.; Giménez-Llort, L. Hepatic Oxi-Inflammation and Neophobia as Potential Liver–Brain Axis Targets for Alzheimer's Disease and Aging, with Strong Sensitivity to Sex, Isolation, and Obesity. *Cells* **2023**, *12*, 1517. [CrossRef]

7.   Muntsant, A.; Giménez-Llort, L. Genotype Load Modulates Amyloid Burden and Anxiety-Like Patterns in Male 3xTg-AD Survivors despite Similar Neuro-Immunoendocrine, Synaptic and Cognitive Impairments. *Biomedicines* **2021**, *9*, 715. [CrossRef]

8.   Muntsant, A.; Giménez-Llort, L. Crosstalk of Alzheimer's Disease-Phenotype, HPA Axis, Splenic Oxidative Stress and Frailty in Late-Stages of Dementia, with Special Concerns on the Effects of Social Isolation: A Translational Neuroscience Approach. *Front. Aging. Neurosci.* **2022**, *14*. [CrossRef]

9.   Estrada, L.D.; Ahumada, P.; Cabrera, D.; Arab, J.P. Liver Dysfunction as a Novel Player in Alzheimer's Progression: Looking Outside the Brain. *Front. Aging Neurosci.* **2019**, *11*. [CrossRef]

10.   Joo Kim, M.; Baek, D.; Truong, L.; Ro, Y.J. Pathologic Findings of Amyloidosis: Recent Advances. In *Amyloid Diseases*; IntechOpen: Rijeka, Croatia, 2019.

11.   Etgen, T.; Chonchol, M.; Frstl, H.; Sander, D. Chronic Kidney Disease and Cognitive Impairment: A Systematic Review and Meta-Analysis. *Am. J. Nephrol.* **2012**, *35*, 474–482. [CrossRef]

12.   Bugnicourt, J.M.; Godefroy, O.; Chillon, J.M.; Choukroun, G.; Massy, Z.A. Cognitive Disorders and Dementia in CKD: The Neglected Kidney-Brain Axis. *J. Am. Soc. Nephrol.* **2013**, *24*, 353–363. [CrossRef] [PubMed]

13.   Liu, Y.H.; Xiang, Y.; Wang, Y.R.; Jiao, S.S.; Wang, Q.H.; Bu, X.L.; Zhu, C.; Yao, X.Q.; Giunta, B.; Tan, J.; et al. Association Between Serum Amyloid-Beta and Renal Functions: Implications for Roles of Kidney in Amyloid-Beta Clearance. *Mol. Neurobiol.* **2015**, *52*, 115–119. [CrossRef] [PubMed]

14.   Jin, W.S.; Shen, L.L.; Bu, X.L.; Zhang, W.W.; Chen, S.H.; Huang, Z.L.; Xiong, J.X.; Gao, C.Y.; Dong, Z.; He, Y.N.; et al. Peritoneal Dialysis Reduces Amyloid-Beta Plasma Levels in Humans and Attenuates Alzheimer-Associated Phenotypes in an APP/PS1 Mouse Model. *Acta Neuropathol.* **2017**, *134*, 207–220. [CrossRef] [PubMed]

15.   Wei, Y.; Wang, T.; Liao, L.; Fan, X.; Chang, L.; Hashimoto, K. Brain-Spleen Axis in Health and Diseases: A Review and Future Perspective. *Brain Res. Bull* **2022**, *182*, 130–140. [CrossRef] [PubMed]

16.   Yang, S.-H.; Kim, J.; Lee, M.J.; Kim, Y. Abnormalities of Plasma Cytokines and Spleen in Senile APP/PS1/Tau Transgenic Mouse Model. *Sci. Rep.* **2015**, *5*, 15703. [CrossRef] [PubMed]

17.   Hayashi, T. Conversion of Psychological Stress into Cellular Stress Response: Roles of the Sigma-1 Receptor in the Process. *Psychiatry Clin. Neurosci.* **2015**, *69*, 179–191. [CrossRef] [PubMed]

18.   Guilliams, T.G.; Edwards, L. *The Stress Response System Chronic Stress and the HPA Axis: Clinical Assessment and Therapeutic Considerations*; The Point Institute of Nutraceutical Research: Stevens Point, WI, USA, 2010.

19.   Machado, A.; Herrera, A.J.; de Pablos, R.M.; Espinosa-Oliva, A.M.; Sarmiento, M.; Ayala, A.; Venero, J.L.; Santiago, M.; Villarán, R.F.; Delgado-Cortés, M.J.; et al. Chronic Stress as a Risk Factor for Alzheimer's Disease. *Rev. Neurosci.* **2014**, *25*. [CrossRef]

20.   Drinkwater, E.; Davies, C.; Spires-Jones, T.L. Potential Neurobiological Links between Social Isolation and Alzheimer's Disease Risk. *Eur. J. Neurosci.* **2022**, *56*, 5397–5412. [CrossRef] [PubMed]

21.   Muntsant, A.; Giménez-Llort, L. Impact of Social Isolation on the Behavioral, Functional Profiles, and Hippocampal Atrophy Asymmetry in Dementia in Times of Coronavirus Pandemic (COVID-19): A Translational Neuroscience Approach. *Front. Psychiatry* **2020**, *11*. [CrossRef]

22.   Javonillo, D.I.; Tran, K.M.; Phan, J.; Hingco, E.; Kramár, E.A.; da Cunha, C.; Forner, S.; Kawauchi, S.; Milinkeviciute, G.; Gomez-Arboledas, A.; et al. Systematic Phenotyping and Characterization of the 3xTg-AD Mouse Model of Alzheimer's Disease. *Front. Neurosci.* **2022**, *15*. [CrossRef]

23.   Giménez-Llort, L.; Blázquez, G.; Cañete, T.; Johansson, B.; Oddo, S.; Tobeña, A.; LaFerla, F.M.; Fernández-Teruel, A. Modeling Behavioral and Neuronal Symptoms of Alzheimer's Disease in Mice: A Role for Intraneuronal Amyloid. *Neurosci. Biobehav. Rev.* **2007**, *31*, 125–147. [CrossRef]

24.   Giménez-Llort, L.; Blázquez, G.; Cañete, T.; Rosa, R.; Vivó, M.; Oddo, S.; Navarro, X.; LaFerla, F.M.; Johansson, B.; Tobeña, A. Modeling Neuropsychiatric Symptoms of Alzheimer's Disease Dementia in 3xTg-AD Mice. In *Alzheimer's Disease: New Advances*; Medimond SRL: Pianoro, Italy, 2006; pp. 513–516.

25.   Baeta-Corral, R.; Giménez-Llort, L. Bizarre Behaviors and Risk Assessment in 3xTg-AD Mice at Early Stages of the Disease. *Behav. Brain Res.* **2014**, *258*, 97–105. [CrossRef]

26.   Manna, J.; Dunbar, G.L.; Maiti, P. Curcugreen Treatment Prevented Splenomegaly and Other Peripheral Organ Abnormalities in 3xTg and 5xFAD Mouse Models of Alzheimer's Disease. *Antioxidants* **2021**, *10*, 899. [CrossRef]

27.   Tian, D.-Y.; Cheng, Y.; Zhuang, Z.-Q.; He, C.-Y.; Pan, Q.-G.; Tang, M.-Z.; Hu, X.-L.; Shen, Y.-Y.; Wang, Y.-R.; Chen, S.-H.; et al. Physiological Clearance of Amyloid-Beta by the Kidney and Its Therapeutic Potential for Alzheimer's Disease. *Mol. Psychiatry* **2021**, *26*, 6074–6082. [CrossRef]

28.   Tangde, A.; Sonwane, B.; Bindu, R. A Clinicohematological Profile of Splenomegaly. *Int. J. Res. Med. Sci.* **2019**, *7*, 1934. [CrossRef]

29.   Gibson, P.R.; Gibson, R.N.; Ditchfield, M.R.; Donlan, J.D. Splenomegaly-an Insensitive Sign of Portal Hypertension. *Aust. New Zealand J. Med.* **1990**, *20*, 771–774. [CrossRef]

30. Marchese, M.; Cowan, D.; Head, E.; Ma, D.; Karimi, K.; Ashthorpe, V.; Kapadia, M.; Zhao, H.; Davis, P.; Sakic, B. Autoimmune Manifestations in the 3xTg-AD Model of Alzheimer's Disease. *JJ. Alzheimer's Dis.* **2014**, *39*, 191–210. [CrossRef]
31. Yu, Z.; Chen, D.; Tan, C.; Zeng, G.; He, C.; Wang, J.; Bu, X.; Wang, Y. Physiological Clearance of Aβ by Spleen and Splenectomy Aggravates Alzheimer-type Pathogenesis. *Aging Cell* **2022**, *21*. [CrossRef]
32. Hawkins, P.N. Hereditary Systemic Amyloidosis with Renal Involvement. *J. Nephrol.* **2003**, *3*, 443–448.
33. Xiang, Y.; Bu, X.L.; Liu, Y.H.; Zhu, C.; Shen, L.L.; Jiao, S.S.; Zhu, X.Y.; Giunta, B.; Tan, J.; Song, W.H.; et al. Physiological Amyloid-Beta Clearance in the Periphery and Its Therapeutic Potential for Alzheimer's Disease. *Acta Neuropathol.* **2015**, *130*, 487–499. [CrossRef]
34. Di Benedetto, G.; Burgaletto, C.; Carta, A.R.; Saccone, S.; Lempereur, L.; Mulas, G.; Loreto, C.; Bernardini, R.; Cantarella, G. Beneficial Effects of Curtailing Immune Susceptibility in an Alzheimer's Disease Model. *J. Neuroinflammation* **2019**, *16*, 166. [CrossRef]
35. Gonzalez-Marrero, I.; Gimenez-Llort, L.; Johanson, C.E.; Carmona-Calero, E.M.; Castañera-Ruiz, L.; Brito-Armas, J.M.; Castañeyra-Perdomo, A.; Castro-Fuentes, R. Choroid Plexus Dysfunction Impairs Beta-Amyloid Clearance in a Triple Transgenic Mouse Model of Alzheimerâ€™s Disease. *Front. Cell Neurosci.* **2015**, *9*. [CrossRef]
36. Sweeney, M.D.; Sagare, A.P.; Zlokovic, B.V. Blood–Brain Barrier Breakdown in Alzheimer Disease and Other Neurodegenerative Disorders. *Nat. Rev. Neurol.* **2018**, *14*, 133–150. [CrossRef]
37. Teskey, G.; Abrahem, R.; Cao, R.; Gyurjian, K.; Islamoglu, H.; Lucero, M.; Martinez, A.; Paredes, E.; Salaiz, O.; Robinson, B.; et al. Glutathione as a Marker for Human Disease. *Adv. Clin. Chem.* **2018**, *87*, 141–159.
38. Presnell, C.E.; Bhatti, G.; Numan, L.S.; Lerche, M.; Alkhateeb, S.K.; Ghalib, M.; Shammaa, M.; Kavdia, M. Computational Insights into the Role of Glutathione in Oxidative Stress. *Curr. Neurovasc. Res.* **2013**, *10*, 185–194. [CrossRef]
39. Dorval, J. Role of Glutathione Redox Cycle and Catalase in Defense against Oxidative Stress Induced by Endosulfan in Adreno-cortical Cells of Rainbow Trout (Oncorhynchus Mykiss). *Toxicol. Appl. Pharmacol.* **2003**, *192*, 191–200. [CrossRef]
40. Iane Garlet, Q.; Vaitsa Losh Haskel, M.; Picada Pereira, R.; Cláudio Francisco Nunes Da Silva, W.; Batista Teixeira Da Rocha, J.; Sirlene Oliveira, C.; Sartori Bonini, J. Δ-Aminolevulinate Dehydratase and Glutathione Peroxidase Activity in Alzheimer's Disease: A Case-Control Study. *EXCLI J.* **2019**, *18*, 866–875. [CrossRef]
41. Torres-Lista, V.; Parrado-Fernández, C.; Alvarez-Montón, I.; Frontiñán-Rubio, J.; Durán-Prado, M.; Peinado, J.R.; Johansson, B.; Alcaín, F.J.; Giménez-Llort, L. Neophobia, NQO1 and SIRT1 as Premorbid and Prodromal Indicators of AD in 3xTg-AD Mice. *Behav. Brain Res.* **2014**, *271*, 140–146. [CrossRef]
42. España, J.; Giménez-Llort, L.; Valero, J.; Miñano, A.; Rábano, A.; Rodriguez-Alvarez, J.; LaFerla, F.M.; Saura, C.A. Intraneuronal β-Amyloid Accumulation in the Amygdala Enhances Fear and Anxiety in Alzheimer's Disease Transgenic Mice. *Biol. Psychiatry* **2010**, *67*, 513–521. [CrossRef]
43. Roda, A.R.; Esquerda-Canals, G.; Martí-Clúa, J.; Villegas, S. Cognitive Impairment in the 3xTg-AD Mouse Model of Alzheimer's Disease Is Affected by Aβ-ImmunoTherapy and Cognitive Stimulation. *Pharmaceutics* **2020**, *12*, 944. [CrossRef]
44. Ozbek, E. Induction of Oxidative Stress in Kidney. *Int. J. Nephrol.* **2012**, *2012*, 1–9. [CrossRef]
45. Tamay-Cach, F.; Quintana-Pérez, J.C.; Trujillo-Ferrara, J.G.; Cuevas-Hernández, R.I.; Del Valle-Mondragón, L.; García-Trejo, E.M.; Arellano-Mendoza, M.G. A Review of the Impact of Oxidative Stress and Some Antioxidant Therapies on Renal Damage. *Ren. Fail* **2016**, *38*, 171–175. [CrossRef]
46. Guerville, F.; De Souto Barreto, P.; Coley, N.; Andrieu, S.; Mangin, J.; Chupin, M.; Payoux, P.; Ousset, P.; Rolland, Y.; Vellas, B. Kidney Function and Cognitive Decline in Older Adults: Examining the Role of Neurodegeneration. *J. Am. Geriatr. Soc.* **2021**, *69*, 651–659. [CrossRef]
47. Nava Catorce, M.; Acero, G.; Gevorkian, G. Age- and Sex-Dependent Alterations in the Peripheral Immune System in the 3xTg-AD Mouse Model of Alzheimer's Disease: Increased Proportion of CD3+CD4-CD8- Double-Negative T Cells in the Blood. *J. Neuroimmunol.* **2021**, *360*, 577720. [CrossRef]
48. Justice, N.J. The Relationship between Stress and Alzheimer's Disease. *Neurobiol. Stress* **2018**, *8*, 127–133. [CrossRef]
49. Møller, P.; Wallin, H.; Knudsen, L.E. Oxidative Stress Associated with Exercise, Psychological Stress and Life-Style Factors. *Chem. Biol. Interact.* **1996**, *102*, 17–36. [CrossRef]
50. Ghezzi, P.; Floridi, L.; Boraschi, D.; Cuadrado, A.; Manda, G.; Levic, S.; D'Acquisto, F.; Hamilton, A.; Athersuch, T.J.; Selley, L. Oxidative Stress and Inflammation Induced by Environmental and Psychological Stressors: A Biomarker Perspective. *Antioxid. Redox Signal.* **2018**, *28*, 852–872. [CrossRef]
51. Sagmeister, M.S.; Harper, L.; Hardy, R.S. Cortisol Excess in Chronic Kidney Disease – A Review of Changes and Impact on Mortality. *Front. Endocrinol.* **2023**, *13*. [CrossRef]
52. Ouanes, S.; Popp, J. High Cortisol and the Risk of Dementia and Alzheimer's Disease: A Review of the Literature. *Front. Aging Neurosci.* **2019**, *11*. [CrossRef]
53. Zafir, A.; Banu, N. Modulation of in Vivo Oxidative Status by Exogenous Corticosterone and Restraint Stress in Rats. *Stress* **2009**, *12*, 167–177. [CrossRef]
54. Peterman, J.L.; White, J.D.; Calcagno, A.; Hagen, C.; Quiring, M.; Paulhus, K.; Gurney, T.; Eimerbrink, M.J.; Curtis, M.; Boehm, G.W.; et al. Prolonged Isolation Stress Accelerates the Onset of Alzheimer's Disease-Related Pathology in 5xFAD Mice despite Running Wheels and Environmental Enrichment. *Behav. Brain Res.* **2020**, *379*, 112366. [CrossRef]

55. Ávila-Villanueva, M.; Marcos Dolado, A.; Gómez-Ramírez, J.; Fernández-Blázquez, M. Brain Structural and Functional Changes in Cognitive Impairment Due to Alzheimer's Disease. *Front. Psychol.* **2022**, *13*. [CrossRef]
56. Heneka, M.T.; Carson, M.J.; Khoury, J.E.; Landreth, G.E.; Brosseron, F.; Feinstein, D.L.; Jacobs, A.H.; Wyss-Coray, T.; Vitorica, J.; Ransohoff, R.M.; et al. Neuroinflammation in Alzheimer's Disease. *Lancet Neurol.* **2015**, *14*, 388–405. [CrossRef]
57. Onyango, I.G.; Jauregui, G.V.; Čarná, M.; Bennett, J.P.; Stokin, G.B. Neuroinflammation in Alzheimer's Disease. *Biomedicines* **2021**, *9*, 524. [CrossRef]
58. Ardanaz, C.G.; Ramírez, M.J.; Solas, M. Brain Metabolic Alterations in Alzheimer's Disease. *Int. J. Mol. Sci.* **2022**, *23*, 3785. [CrossRef]
59. Walker, K.A.; Ficek, B.N.; Westbrook, R. Understanding the Role of Systemic Inflammation in Alzheimer's Disease. *ACS Chem. Neurosci.* **2019**, *10*, 3340–3342. [CrossRef]
60. Gimenez-Llort, L.; Torres-Lista, V.; Fuente, M. Crosstalk between Behavior and Immune System During the Prodromal Stages of Alzheimer's Disease. *Curr. Pharm. Des.* **2014**, *20*, 4723–4732. [CrossRef]
61. Baeta-Corral, R.; De la Fuente, M.; Giménez-Llort, L. Sex-Dependent Worsening of NMDA-Induced Responses, Anxiety, Hypercortisolemia, and Organometry of Early Peripheral Immunoendocrine Impairment in Adult 3xTg-AD Mice and Their Long-Lasting Ontogenic Modulation by Neonatal Handling. *Behav. Brain Res.* **2023**, *438*, 114189. [CrossRef]
62. Takata, M.; Nakashima, M.; Takehara, T.; Baba, H.; Machida, K.; Akitake, Y.; Ono, K.; Hosokawa, M.; Takahashi, M. Detection of Amyloid β Protein in the Urine of Alzheimer's Disease Patients and Healthy Individuals. *Neurosci. Lett.* **2008**, *435*, 126–130. [CrossRef]
63. Wang, M.J.; Yi, S.; Han, J.; Park, S.Y.; Jang, J.-W.; Chun, I.K.; Kim, S.E.; Lee, B.S.; Kim, G.J.; Yu, J.S.; et al. Oligomeric Forms of Amyloid-β Protein in Plasma as a Potential Blood-Based Biomarker for Alzheimer's Disease. *Alzheimers Res. Ther.* **2017**, *9*, 98. [CrossRef]
64. Nabers, A.; Perna, L.; Lange, J.; Mons, U.; Schartner, J.; Güldenhaupt, J.; Saum, K.; Janelidze, S.; Holleczek, B.; Rujescu, D.; et al. Amyloid Blood Biomarker Detects Alzheimer's Disease. *EMBO Mol. Med.* **2018**, *10*. [CrossRef]
65. Tanaka, M.; Diano, M.; Battaglia, S. Editorial: Insights into Structural and Functional Organization of the Brain: Evidence from Neuroimaging and Non-Invasive Brain Stimulation Techniques. *Front. Psychiatry* **2023**, *14*. [CrossRef]
66. Khan, T.K.; Alkon, D.L. Peripheral Biomarkers of Alzheimer's Disease. *J. Alzheimer's Dis.* **2015**, *44*, 729–744. [CrossRef]
67. Oddo, S.; Caccamo, A.; Shepherd, J.D.; Murphy, M.P.; Golde, T.E.; Kayed, R.; Metherate, R.; Mattson, M.P.; Akbari, Y.; Laferla, F.M. Triple-Transgenic Model of Alzheimer's Disease with Plaques and Tangles: Intracellular A and Synaptic Dysfunction. *Neuron* **2003**, *39*, 409–421. [CrossRef] [PubMed]
68. Baeta-Corral, R.; Giménez-Llort, L. Persistent Hyperactivity and Distinctive Strategy Features in the Morris Water Maze in 3xTg-AD Mice at Advanced Stages of Disease. *Behav. Neurosci.* **2015**, *129*, 129–137. [CrossRef]
69. Kilkenny, C.; Browne, W.J.; Cuthill, I.C.; Emerson, M.; Altman, D.G. Improving Bioscience Research Reporting: The Arrive Guidelines for Reporting Animal Research. *PLoS Biol.* **2010**, *8*. [CrossRef] [PubMed]
70. Tietze, F. Enzymic Method for Quantitative Determination of Nanogram Amounts of Total and Oxidized Glutathione: Applications to Mammalian Blood and Other Tissues. *Anal. Biochem.* **1969**, *27*, 502–522. [CrossRef]
71. Rahman, I.; Kode, A.; Biswas, S.K. Assay for Quantitative Determination of Glutathione and Glutathione Disulfide Levels Using Enzymatic Recycling Method. *Nat. Protoc.* **2007**, *1*, 3159–3165. [CrossRef] [PubMed]
72. Massey, V.; Williams, C.H. On the Reaction Mechanism of Yeast Glutathione Reductase. *J. Biol. Chem.* **1965**, *240*, 4470–4480. [CrossRef]
73. Lawrence, R.A.; Burk, R.F. Glutathione Peroxidase Activity in Selenium-Deficient Rat Liver. *Biochem. Biophys. Res. Commun.* **1976**, *71*, 952–958. [CrossRef]
74. Alvarado, C.; Álvarez, P.; Jiménez, L.; De la Fuente, M. Oxidative Stress in Leukocytes from Young Prematurely Aging Mice Is Reversed by Supplementation with Biscuits Rich in Antioxidants. *Dev. Comp. Immunol.* **2006**, *30*, 1168–1180. [CrossRef]

pharmaceuticals

Article

Unveiling the Enigma: Exploring Risk Factors and Mechanisms for Psychotic Symptoms in Alzheimer's Disease through Electronic Medical Records with Deep Learning Models

Peihao Fan [1], Oshin Miranda [1], Xiguang Qi [1], Julia Kofler [2], Robert A. Sweet [3,4] and Lirong Wang [1,*]

[1] Computational Chemical Genomics Screening Center, Department of Pharmaceutical Sciences, School of Pharmacy, University of Pittsburgh, Pittsburgh, PA 15213, USA; pef14@pitt.edu (P.F.); osm7@pitt.edu (O.M.); xiq24@pitt.edu (X.Q.)

[2] Department of Pathology, Division of Neuropathology, UPMC Presbyterian Hospital, Pittsburgh, PA 15213, USA; koflerjk@upmc.edu

[3] Department of Psychiatry, School of Medicine, University of Pittsburgh, Pittsburgh, PA 15213, USA; sweetra@upmc.edu

[4] Department of Neurology, School of Medicine, University of Pittsburgh, Pittsburgh, PA 15213, USA

* Correspondence: liw30@pitt.edu

Abstract: Around 50% of patients with Alzheimer's disease (AD) may experience psychotic symptoms after onset, resulting in a subtype of AD known as psychosis in AD (AD + P). This subtype is characterized by more rapid cognitive decline compared to AD patients without psychosis. Therefore, there is a great need to identify risk factors for the development of AD + P and explore potential treatment options. In this study, we enhanced our deep learning model, DeepBiomarker, to predict the onset of psychosis in AD utilizing data from electronic medical records (EMRs). The model demonstrated superior predictive capacity with an AUC (area under curve) of 0.907, significantly surpassing conventional risk prediction models. Utilizing a perturbation-based method, we identified key features from multiple medications, comorbidities, and abnormal laboratory tests, which notably influenced the prediction outcomes. Our findings demonstrated substantial agreement with existing studies, underscoring the vital role of metabolic syndrome, inflammation, and liver function pathways in AD + P. Importantly, the DeepBiomarker model not only offers a precise prediction of AD + P onset but also provides mechanistic understanding, potentially informing the development of innovative treatments. With additional validation, this approach could significantly contribute to early detection and prevention strategies for AD + P, thereby improving patient outcomes and quality of life.

Keywords: Alzheimer's disease; dementia; psychosis; deep learning; electronic medical records; comorbidity

Citation: Fan, P.; Miranda, O.; Qi, X.; Kofler, J.; Sweet, R.A.; Wang, L. Unveiling the Enigma: Exploring Risk Factors and Mechanisms for Psychotic Symptoms in Alzheimer's Disease through Electronic Medical Records with Deep Learning Models. *Pharmaceuticals* **2023**, *16*, 911. https://doi.org/10.3390/ph16070911

Academic Editors: Masaru Tanaka, Lydia Giménez-Llort, Simone Battaglia, Chong Chen and Piril Hepsomali

Received: 3 May 2023
Revised: 14 June 2023
Accepted: 16 June 2023
Published: 21 June 2023

1. Introduction

Alzheimer's Disease (AD) is the most common neurodegenerative disease affecting 50 million people worldwide [1]. Its presence is associated with a substantial decline in the quality of life [2]. It is estimated that the cost of AD is $604 billion worldwide per year and will triple by the year 2050 [3]. Psychosis, defined by the occurrence of delusions and/or hallucinations, is observed as a common complication of AD. Approximately 50% of patients are likely to suffer from psychotic symptoms after the onset of AD (AD with psychosis, or AD + P) [4]. AD + P patients have more severe cognitive impairments and a quicker cognitive decline than AD patients without psychosis (AD-P) [5,6]. AD + P is also associated with higher rates of co-occurring agitation, aggression, and depression compared to AD-P [5]. These non-cognitive symptoms create burdens not only for people with AD or other dementias but also for their caregivers and are associated with poor outcomes in terms of function, quality of life, disease course, mortality, and economic cost [7–9].

The occurrence of psychosis in AD has been shown to be familial, with an estimated heritability of 61%. This suggests that it arises from distinctive biology that could be effectively targeted pharmacologically [10]. More recent studies have helped to elucidate some of the underlying biologic risks of psychosis. A small proportion of the risk is conferred by a greater burden of hyperphosphorylated tau, one of the hallmark pathologies of Alzheimer's disease. Other pathologies that are often comorbid in AD, such as Lewy bodies, TDP-43 inclusions, and vascular lesions, show less consistent and/or smaller contributions to the risk of psychosis in AD patients [11]. More recent data have highlighted a role for the excess vulnerability of excitatory neurons and synapses in psychosis risk in AD [11–13].

The use of pharmacotherapy-based treatment options for AD + P has been limited [14–16]. Currently, the Food and Drug Administration (FDA) has not approved any specific medication for AD + P. Second-generation antipsychotics (SGAs) are frequently employed and endorsed by geriatric specialists for managing AD + P. However, their application is significantly constrained due to an elevated risk of adverse events and the potential for co-existing health conditions. This led the FDA to issue a "black-box" warning in 2005 to emphasize the heightened risk of mortality for dementia patients receiving SGAs [17]. Simultaneously, antipsychotics have shown only moderate effectiveness in managing psychosis, aggression, and agitation in individuals with dementia. Furthermore, the accelerated cognitive decline present in AD + P (relative to AD without psychosis) is present years before the onset of AD and of psychosis [18], suggesting that there is a window of opportunity to intervene if AD + P can be accurately predicted and if preventative treatments can be identified. This emphasizes the urgent need to discover and develop more effective and safer therapeutic alternatives for AD + P and or agents that may prevent its development [19].

Drawing from our previous research, the limited effectiveness of antipsychotics in treating AD + P can be attributed to their failure to effectively target the underlying biology of the condition [20]. Another study performed a large genome-wide association meta-analysis on 12,317 AD subjects with or without psychosis [21]. The authors reported that AD + P is not significantly genetically correlated with schizophrenia, but it is negatively correlated with bipolar disorder and positively correlated with depression. These findings serve as a reminder to broaden our horizons in search of better treatment options for AD + P.

Deep Neural Networks (DNNs) represent the pinnacle of current machine learning and big data analytics, with a wide array of applications that span from defense and surveillance to human–computer interaction and question-answering systems [22]. DNN architectures can be further categorized into more subtypes for different tasks and data types, including Feed-forward Neural Networks, Convolution Neural Networks (CNNs), and Recurrent Neural Networks (RNNs) [23]. Recently, self-attention-based DNN architecture, transformers, and their variations have been reported to have better model performance than other DNNs [24]. The integration of deep learning in healthcare offers physicians precise disease analysis, leading to improved treatment strategies and, consequently, enhanced medical decision-making. By incorporating deep learning technologies into hospital management information systems, multiple benefits can be achieved: reduction in costs, minimization of hospital stays and their duration, prevention of insurance fraud, detection of changes in disease patterns, improvement in healthcare quality, and more efficient allocation of medical resources [25].

Deep learning/data mining algorithms can translate data into information for hypothesis generation through deep hierarchical feature construction to capture long-range dependencies in EMR data. Recently, a variety of deep learning techniques and frameworks have been applied to information extraction, representation learning, outcome prediction, phenotyping, and de-identification [26–30] and yielded better performance than traditional methods and required less time-consuming preprocessing and feature engineering. Specifically, deep learning techniques learn optimal features directly from the data itself, without any human guidance, allowing for the automatic discovery of latent data relationships that might otherwise be unknown or hidden [31]. The proposed research aims to identify potential therapeutic drugs for preventing, delaying, or treating AD + P through an examination of the EMR of AD patients. We predict that drugs capable of minimizing the risk of psychosis development would be beneficial for AD + P management. The deep learning model's capacity to discern intricate patterns among a multitude of longitudinal features provides us with a comprehensive view of how these variables intersect with AD + P [32]. This, in turn, facilitates the identification of possible novel therapeutic options.

In our previous study, we built a deep-learning-based model, DeepBiomarker, through modification of an established deep-learning framework, Pytorch_EHR [33,34]. In DeepBiomarker, we used diagnosis, medication use, and lab tests as the input, implemented data augmentation technologies to improve the model performance, and also integrated a perturbation-based approach [35] for risk factor identification. In this study, we showed the application of DeepBiomarker on the prediction of AD + P in AD patients for risk/beneficial factor identifications. Compared to the previous version, we have updated the calculation of relative contribution for feature importance.

2. Results

2.1. The Performance of DeepBiomarker in AD + P Patients

We have identified 16,294 AD patients from the University of Pittsburgh Medical Center (UPMC) EMR data. We further identified 3535 cases of patients who developed AD + P and 3535 controls who did not develop AD + P within 3 months after the index dates, and all of whom had more than 1 year of EMR data before the index dates. Among the 3535 cases, the average age is 81.9 years old with a standard deviation of 9.06, and the average for the control group is 83.9 years old with a standard deviation of 8.26. There are 2368 female and 1167 male patients in the case cohort and 2333 female/1202 male in the control cohort. The index date for each individual was the date of any encounter after the diagnosis of AD but before the diagnosis of AD + P. Those samples were split into an 8:1:1 ratio for training, validation, and test sets. The validation datasets were used to optimize the parameters in the DeepBiomarker model, and the test datasets were used to evaluate the performance of the established models. The performance of the DeepBiomarker can be found in Table 1. The higher AUC (area under the receiver operating characteristic curve) indicates that the model demonstrates greater accuracy in making accurate predictions. We can see that the two deep-learning-based models implemented in DeepBiomarker, TLSTM, and RETAIN, both achieved high accuracy with an AUC above 0.90 in the validation dataset and test dataset, while the traditional machine learning approach, such as logistic regression, only yielded prediction accuracy with an AUC of 0.837 and 0.822 in the validation dataset and the test dataset, respectively.

Table 1. Performance of three AD + P predicting models.

	Validation AUC	Test AUC	Validation AUC Std.	Test AUC Std.
T-LSTM	0.921	0.903	0.006	0.005
RETAIN	0.935	0.907	0.004	0.002
LR	0.837	0.822	0.009	0.012

AUC: Area under the receiver operating characteristic curve. Std: Standard deviation. T-LSTM: Temporal information enhancing long short-term memory. RETAIN: Reverse time attention model. LR: Logistic regression.

2.2. Risk Factors Identified by the DeepBiomarker Model with Significant Contributions

As previously mentioned, we used a perturbation-based estimation to determine the relative contribution (RC) of each feature in predicting AD + P. Out of the 65 features that demonstrated significant effects in our results, 23 were diagnoses; 36 were drugs, and 6 were lab tests. To enhance the reliability of our findings, we only included diagnoses that appeared in more than 10% of the entire population.

To provide a more comprehensive understanding of the significant features, we have provided three tables (Tables 2–4) that detail their specific effects. The RC values in these tables indicate the association of each feature with AD + P. An RC above 1 indicates that the feature contributes more to cases than controls, suggesting a hazardous effect, while an RC below 1 suggests a protective effect. These results shed light on the critical features that contribute to the development of AD + P and can be used to improve the accuracy of our predictive model. By identifying the specific risk factors associated with AD + P, we can design more targeted treatment and prevention strategies to improve patient outcomes.

Table 2. Relative contributions of diagnosis features that showed significant association (Q < 0.05) with AD + P, listed in order of increasing RC value.

Feature	RC	CI95up	CI95down	Q Value *
Hypoxemia	0.718	0.85	0.606	0.002
Arthropathy, unspecified, site unspecified [1]	0.747	0.831	0.672	<0.001
Pain in joint, shoulder region	0.764	0.898	0.651	0.01
Activities involving walking, marching, and hiking	0.782	0.876	0.699	0.001
Acute kidney failure, unspecified [1]	0.86	0.945	0.783	0.014
Unspecified osteoarthritis, unspecified site [1]	0.877	0.944	0.814	0.006
Esophageal reflux	1.112	1.174	1.054	0.002
Depressive disorder, not elsewhere classified	1.117	1.195	1.045	0.01
Hypothyroidism, unspecified [1]	1.136	1.234	1.045	0.02
Disorientation, unspecified [1]	1.148	1.268	1.039	0.039
Atherosclerotic heart disease of native coronary artery without angina pectoris	1.154	1.228	1.085	<0.001
Abnormality of gait	1.171	1.292	1.061	0.014
Type 2 diabetes mellitus without complications	1.191	1.261	1.125	<0.001
Obstructive sleep apnea	1.207	1.354	1.076	0.012
Central pain syndrome	1.221	1.397	1.067	0.025
Diabetes mellitus without mention of complication, type II or unspecified type, not stated as uncontrolled [1]	1.234	1.328	1.147	<0.001
Aortic valve disorders	1.274	1.506	1.077	0.03
Atrial fibrillation	1.358	1.477	1.249	<0.001
Dependence on renal dialysis	1.361	1.602	1.156	0.003
Hypocalcemia	1.417	1.689	1.189	0.002
Long term (current) use of insulin	1.582	1.722	1.454	<0.001
Primary hypercoagulable state	1.582	1.722	1.454	<0.001
Acute venous embolism and thrombosis of unspecified deep vessels of lower extremity	1.687	2.325	1.223	0.012

* FDR-adjusted *p*-value. [1] The term "unspecified" is used in ICD9 and ICD10 codes when the available medical record information is not sufficient to assign a more specific code. RC: relative contribution; CI: confidential interval.

Table 2 highlights several common comorbid diseases among AD patients, such as Diabetes, Esophageal Reflux, Atrial Fibrillation, Depression, and Primary Hypercoagulable State. These diseases have RC values greater than 1, indicating that they are associated with an increased risk of AD + P. Conversely, a few of the comorbidities, such as hypoxemia, pain in the joint/shoulder region, arthropathy, and osteoarthritis, have RC values less than 1, indicating that they are associated with a lower risk of AD + P.

Table 3. Relative contributions of medication features that showed significant association (Q < 0.05) with AD + P, listed in order of increasing RC value.

Feature	Indication/Drug Class	RC	CI95up	CI95down	Q Value *
Glucosamine–Chondroitin	Osteoarthritis, reduce joint pain and inflammation	0.359	0.7	0.184	0.02
Dextromethorphan–Guaifenesin	Cough suppressant	0.439	0.757	0.255	0.022
Fish Oil	Dietary supplement	0.456	0.73	0.285	0.01
Sucralfate	Gastrointestinal ulcers	0.472	0.716	0.311	0.005
Midodrine	Alpha-adrenergic agonist for hypotension	0.477	0.65	0.35	<0.001
Irbesartan	Angiotensin receptor blocker for hypertension	0.509	0.769	0.338	0.012
Esomeprazole Magnesium	Proton pump inhibitor	0.538	0.698	0.414	<0.001
Cyclobenzaprine	Skeletal muscle relaxant	0.572	0.734	0.445	<0.001
Budesonide–Formoterol	Corticosteroid/beta2-adrenergic receptor agonist	0.578	0.763	0.437	0.002
Lactulose	Constipation and portal systemic encephalopathy	0.599	0.835	0.429	0.019
Duloxetine	Antidepressant	0.606	0.757	0.485	<0.001
Ezetimibe	Hyperlipidemia	0.624	0.854	0.457	0.022
Magnesium Hydroxide	Laxative and antacid	0.66	0.823	0.529	0.003
Famotidine	Histamine H2 receptor antagonist	0.672	0.823	0.548	0.002
Nitroglycerin	Nitrate vasodilator	0.679	0.84	0.549	0.005
Alprazolam	Anxiety disorders and panic disorders	0.692	0.857	0.56	0.008
Isosorbide Mononitrate	Prevent and treat angina in coronary artery disease	0.719	0.89	0.582	0.018
Quetiapine	Antipsychotics	0.726	0.875	0.603	0.008
Glipizide	Type 2 diabetes	0.732	0.923	0.581	0.046
Memantine	Alzheimer's disease	0.749	0.83	0.676	<0.001
Triamcinolone Acetonide	Corticosteroid	0.751	0.923	0.612	0.038
Losartan	Angiotensin receptor blocker for hypertension	0.766	0.869	0.676	0.001
Clopidogrel	Antiplatelet	0.777	0.918	0.658	0.022
Docusate Sodium	Stool softener	0.78	0.907	0.671	0.011
Calcium Carbonate-Vitamin D3	Calcium supplement/osteoporosis	0.781	0.902	0.676	0.008
Cephalexin	Antibiotics	0.793	0.917	0.686	0.014
Tramadol	Opioid agonist and serotonin/norepinephrine reuptake inhibitor	0.795	0.937	0.674	0.037

Table 3. *Cont.*

Feature	Indication/Drug Class	RC	CI95up	CI95down	Q Value *
Aspirin	Antiplatelet/Non-steroidal anti-inflammatory drugs	0.808	0.904	0.721	0.003
Pantoprazole	Proton pump inhibitor	0.835	0.95	0.734	0.036
Warfarin	Anticoagulated/vitamin K antagonist	1.289	1.478	1.124	0.004
Fluconazole	Antifungal medication	1.58	2.175	1.147	0.032
Allopurinol	Xanthine oxidase inhibitor/reduce uric acid concentrations	1.639	2.157	1.245	0.005
Cholestyramine–Aspartame	Lower cholesterol levels	1.642	2.233	1.207	0.013
Terazosin	Alpha-1 adrenergic antagonist/hypertension	1.842	2.633	1.288	0.009
Metoclopramide	Antiemetic agent and dopamine D2 antagonist	1.879	2.697	1.309	0.007
Clobetasol	High potency corticosteroid topical medication	2.054	3.043	1.387	0.004

* FDR-adjusted p-value. RC: relative contribution; CI: confidential interval.

Table 4. Relative contributions of lab test features that showed significant association ($Q < 0.05$) with AD + P.

Feature	RC	CI95up	CI95down	Q Value *
Aspartate Aminotransferase (AST) Test	0.704	0.864	0.574	0.008
Alkaline Phosphatase (ALP) Test	0.826	0.924	0.739	0.009
Urea Nitrogen	0.84	0.909	0.776	0.001
Anion Gap	0.863	0.944	0.789	0.012
Glucose	0.871	0.941	0.806	0.006
Chloride (Cl)	0.886	0.952	0.825	0.01

* FDR-adjusted p-value. RC: relative contribution; CI: confidential interval.

Table 3 reveals several medications with RC values of less than 1 that are associated with reduced rates of onset of AD + P. These include hypertension medications, such as losartan and irbesartan, the hypotension drug midodrine, antidiabetic drug glipizide, anti-gastroesophageal reflux drugs pantoprazole, famotidine, sucralfate, and esomeprazole, laxatives docusate and lactulose, pain medications, such as aspirin, tramadol, and glucosamine-chondroitin, medications for chest pain, nitroglycerin and isosorbide mononitrate, cholesterol-lowering medication ezetimibe, asthma or chronic obstructive pulmonary disease drug budesonide-formoterol, antidepressants, duloxetine and cyclobenzaprine, vitamin D and fish oil, along with the antipsychotic quetiapine.

In contrast, Table 3 also identifies medications with RC values greater than 1 that are associated with a higher risk of developing AD + P. These include metoclopramide, an antiemetic and gut motility stimulator, allopurinol, uric acid reducer, warfarin, an anticoagulant, fluconazole, an antifungal drug, cholestyramine–aspartame, a cholesterol-lowering medication, and terazosin, an antihypertensive drug and urinary retention medication.

Interestingly, Table 4 reveals that all six lab tests with $Q < 0.05$, such as Chloride (Cl), Glucose, Urea Nitrogen, Anion Gap, Alkaline Phosphatase (ALP) Test, and Aspartate Aminotransferase (AST) Test, have RC values of less than 1 indicating that they are associated with a reduced risk of developing AD + P.

In general, this study built a cutting-edge deep-learning-based predictive model that can accurately predict the risk of an AD patient developing into an AD + P patient. In addition, this interpretable model allowed us to identify a set of risk factors and beneficial elements that may play important roles in the development of AD + P. These findings can further lead to novel therapeutics, mechanism explorations, and better treatment options.

3. Discussion

This study presents a significant stride forward in our understanding of risk factors and protective measures related to AD + P. By leveraging EMRs and deep learning models, we uncovered novel clinical features and potential pharmacological interventions that deserve further exploration. Notably, our work validates and extends previous research conducted at the University of Pittsburgh Alzheimer's Disease Research Center (ADRC), reinforcing the protective role of vitamin D and uncovering the potential benefits of quetiapine, duloxetine, and memantine in reducing AD + P risk.

Our results have highlighted the significant beneficial effects of vitamin D, quetiapine, duloxetine, and memantine in reducing the risk of developing AD + P. In our previous study on research data collected by ADRC and network analysis, the use of vitamin D was associated with a lower risk of developing AD + P [36,37]. The strong beneficial effect of vitamin D in this observational study using EMR data further supports the potential beneficial effect of vitamin D in preventing AD + P. For quetiapine, though there are studies suggesting no significant difference in treating behavioral symptoms and cognitive decline, their effect in managing psychotic symptoms in AD + P remains unclear [38,39]. In an 8-week, double-blind, placebo-controlled trial, duloxetine exhibited a strong effect in improving cognition, depression, and some pain measures and was safe and well-tolerated in elderly patients [40]. As for memantine, mixed results have been reported for its potential effect in treating psychotic symptoms associated with dementia; some studies found that there is no significant beneficial effect between memantine and placebo in terms of their Positive and Negative Syndrome Scale (PANSS) score [41,42]. While some studies suggested that the use of memantine in addition to atypical antipsychotics can be beneficial in managing negative symptoms, several randomized, double-blind, placebo-controlled studies have reported that memantine alone showed a significant beneficial effect by decreasing the PANSS score [43–45].

In a deep learning model, the relationship between features can be complex and not entirely independent [46]. When estimating the effect of these features using a perturbation-based approach, the model's ability to capture intricate patterns and dependencies between input features is taken into account. Perturbation-based methods involve making small changes to the input features and observing their impact on the model's predictions [35]. By doing so, these methods can help disentangle the complex relationships and interactions among multiple features, providing insights into their individual and joint contributions to the model's output.

Another point we should bear in mind when interpreting the results for the model is that compared to traditional statistical analysis (such as Logistic Regression or Survival Analysis), our deep learning model can take advantage of the features that showed high collinearity, such as the diagnosis of a disease and their treatments. Deep learning models are designed to capture complex relationships and interactions between input features [47]. During the training process, the model learns to recognize patterns and dependencies in the data, including correlations between different features. The model's neurons can learn to capture interactions between features by combining them in various ways. For example, the model may learn to recognize that a specific medication is prescribed for a particular disease diagnosis. This information can be represented in the weights and biases of the neurons.

When we consider the combined effects of comorbidities and their treatments, we observe some interesting findings. For instance, joint disorders, such as arthropathy, osteoarthritis, and pain in the joint or shoulder region, are associated with inflammation. These tend to reduce the risk of developing AD + P. While this might seem surprising, it is plausible that patients with these conditions are more likely to take calcium/vitamin D supplements, which can target both peripheral and brain-specific inflammatory processes, ultimately reducing the risk of AD + P.

A key area of our investigation focused on glucose metabolism. Our findings highlight the pivotal role it plays in the development and severity of AD + P, a relationship that is intricate and warrants further exploration. Our study, for instance, identified diabetes as a risk factor for AD + P but found that the effects of treatments for diabetes on AD + P risk were mixed. Glipizide, an oral glucose-lowering drug, was associated with reduced AD + P risk, while the use of insulin was linked with an increased risk. These findings underline the potential for targeted therapies in this domain and invite a more comprehensive exploration of glucose metabolism's role in the AD + P development [48].

Our research also shed light on the significant connection between cardiovascular disease and AD + P. Hypertension, a major risk for cerebrovascular disease, was not associated with AD + P risk, potentially due to the protective effects of hypertension medications, such as irbesartan and losartan. However, the complex pattern of cerebrovascular risk and its treatments, such as atrial fibrillation, aortic valve disorders, and their associated medications (clopidogrel and warfarin), call for an in-depth investigation into the underlying pharmacological mechanisms. A more complex pattern is highlighted by atrial fibrillation and aortic valve disorders, which both increase cerebrovascular risk and are associated with an increased risk of AD + P. However, the effects of their treatments differed. Clopidogrel was associated with reduced AD + P risk, suggesting that this therapy effectively mitigated the risks associated with their indications. In contrast, warfarin, which is also prescribed for these indications, was associated with an increased risk of AD + P. Examining the mechanisms and targets of medications that are associated with AD + P (Table 5) sheds further light on the reasons why these drugs with the same indications might have opposite effects on AD + P risk. Warfarin is an anticoagulant medication that inhibits the production of vitamin K-dependent clotting factors in the liver, whereas clopidogrel is an antiplatelet medication that inhibits the activation and aggregation of platelets. As warfarin and clopidogrel can penetrate the blood–brain barrier, they may exert central nervous system effects that are independent of their anticoagulant and antiplatelet functions. For example, it has been reported that clopidogrel can delay the closure of compromised blood–brain barriers by inhibiting the purinergic receptor P2RY12 [49]. More molecular mechanistic studies are needed to fully understand the underlying mechanisms for the beneficial effects of the medications identified in our study. Nonetheless, our findings provide important clues for the development of new medications that can target specific pathways and mechanisms involved in the development of AD + P. This has led us to key targets, such as adrenergic receptors and angiotensin II receptors, which are influenced by certain medications. The fact that cardiovascular-related drugs and conditions emerged as key factors in our analysis points to potential targets and mechanisms in this domain, such as adrenergic receptors and angiotensin II receptors [50]. The diverse effects of different antihypertensive medications suggest that the protective influence of certain drugs may not be solely attributed to their blood pressure-lowering effect, warranting further investigation. Intriguingly, not all antihypertensive medications demonstrated similar protective effects, as seen with Irbesartan and Losartan. This suggests that the beneficial effects of these drugs could be attributed to mechanisms beyond their blood-pressure-lowering capacity. The absence of similar effects in the case of diuretics and calcium channel blockers strengthens this assertion. Another salient point is the association between the impairment of mitochondrial functions and a range of neurological diseases, including AD, depression, anxiety, and psychiatric symptoms [51,52]. These findings underscore the need for further exploration to unravel the intricate mechanisms that contribute to AD + P's development. This will aid in the identification of innovative therapeutic targets to effectively prevent or manage this disease subtype.

Table 5. Mechanism actions, targets, and blood–brain barrier (BBB) penetration ability for the medications identified associated with the development of AD + P.

Drugs	Mechanism of Action	Targets *	Ability to Penetrate Blood–Brain Barrier	Predicted Effects against AD + P
Glucosamine–Chondroitin	Maintain healthy cartilage by providing the building blocks for its synthesis and supporting its repair	UDP-glucose 2-epimerase/ManNAc kinase (GNE) gene	No	Beneficial
Dextromethorphan–Guaifenesin	Cough suppressant that works by acting on the cough center in the brain/thinning and loosening mucus in the airways	GRIN1 GRIN2A GRIN2B SIGMAR1 HTR3A HTR3B	Yes	Beneficial
Fish Oil	Incorporating into cell membranes and modulating the production of eicosanoids		Yes	Beneficial
Sucralfate	Forming a protective barrier over the ulcer or damaged area, which helps to prevent further damage and promote healing		No	Beneficial
Midodrine	Selective alpha-1 adrenergic agonist, which increases peripheral vascular resistance and blood pressure	ADRA1A	No	Beneficial
Irbesartan	Selectively blocking the angiotensin II receptor type 1 (AT1) in the renin–angiotensin–aldosterone system, which leads to vasodilation and a decrease in blood pressure	AT1 AGTR1	No	Beneficial
Esomeprazole Magnesium	Inhibiting the proton pump (H+/K+ ATPase) in the stomach	the proton pump (H+/K+ ATPase)	Yes	Beneficial
Cyclobenzaprine	A centrally-acting muscle relaxant, which reduces muscle tone and spasm by blocking the activity of alpha motor neurons in the spinal cord	alpha motor neurons in the spinal cord	Yes	Beneficial
Budesonide–Formoterol	Binding to glucocorticoid receptors in the lungs, leading to the suppression of inflammation and immune responses	Glucocorticoid receptors Beta-2 adrenergic receptors	Yes	Beneficial
Lactulose				Beneficial
Duloxetine	Inhibition of the reuptake of two neurotransmitters in the brain, serotonin and norepinephrine	SLC6A2 SLC6A4	Yes	Beneficial
Ezetimibe	Increasing the osmotic pressure in the colon, which draws water into the colon and softens the stool	NPC1L1 SOAT1	No	Beneficial
Magnesium Oxide	Providing magnesium ions to the body, which are essential for many biological processes		Yes	Beneficial
Famotidine	Inhibiting the activity of histamine h2 receptors in the stomach	histamine H2 receptor	Yes	Beneficial

Table 5. *Cont.*

Drugs	Mechanism of Action	Targets *	Ability to Penetrate Blood–Brain Barrier	Predicted Effects against AD + P
Nitroglycerin	A potent vasodilator by releasing nitric oxide in the smooth muscle of blood vessels, leading to relaxation of vascular smooth muscle and vasodilation	NPR1	Yes	Beneficial
Alprazolam	Enhancing the activity of gamma-aminobutyric acid (GABA) in the brain	GABA-A receptor benzodiazepine receptor	Yes	Beneficial
Isosorbide Mononitrate	It acts as a vasodilator by releasing nitric oxide in the smooth muscle of blood vessels, leading to relaxation of vascular smooth muscle and vasodilation	NPR1	No	Beneficial
Quetiapine	Antagonist of several neurotransmitter receptors in the brain, including dopamine, serotonin, and histamine receptors	DRD2 HTR1A HTR2A HRH1	Yes	Beneficial
Glipizide	Stimulating the release of insulin from the beta cells of the pancreas.	ATP-sensitive potassium channels in pancreatic beta cells SUR1	No	Beneficial
Memantine	Blocking of the activity of the NMDA (n-methyl-d-aspartate) subtype of glutamate receptors in the brain.	NMDA subtype of glutamate receptors	Yes	Beneficial
Triamcinolone Acetonide	A synthetic glucocorticoid, which reduces inflammation and swelling by inhibiting the production and release of inflammatory mediators	Inflammatory mediators and their signaling pathways. NR3C1	No	Beneficial
Losartan	Angiotensin II receptor antagonist, blocking the binding of angiotensin II to specific receptors in the body, which inhibits its vasoconstrictive and pro-inflammatory effects	angiotensin II receptor	Yes	Beneficial
Clopidogrel	Irreversibly inhibits the P2Y12 receptor, which is found on the surface of platelets. Reduces the activation and aggregation of platelets.	P2Y12 receptor	Yes	Beneficial
Docusate Sodium	Increasing the amount of water and fat in the stool		No	Beneficial
Vitamin D	Binding to vitamin d receptors (VDR) in cells, leading to changes in gene expression and protein synthesis	Vitamin D receptor (VDR)	Yes	Beneficial
Cephalexin	Inhibiting bacterial cell wall synthesis by binding to penicillin-binding proteins (PBPS)	bacterial PBPs	No	Beneficial

Table 5. *Cont.*

Drugs	Mechanism of Action	Targets *	Ability to Penetrate Blood–Brain Barrier	Predicted Effects against AD + P
Tramadol	An opioid agonist, which means it binds to and activates opioid receptors in the brain, inhibits the reuptake of serotonin and norepinephrine, which are neurotransmitters involved in pain processing, further enhancing its analgesic effect	OPRM1 SLC6A2 SLC6A4 SCN2A NMDA receptors ADORA1	Yes	Beneficial
Aspirin	Irreversibly inhibit the cyclooxygenase (COX) enzyme	PTGS1 PTGS2 AKR1C1 EDNRA TP53 HSPA5 RPS6KA3 NFKBIA	Yes	Beneficial
Pantoprazole	Irreversibly blocking the H+/K+-atpase enzyme in the parietal cells of the stomach	ATP4A ATP4B	No	Beneficial
Warfarin	Inhibiting the synthesis of vitamin K-dependent clotting factors in the liver, specifically factors II, VII, IX, and X	VKORC1 NR1I2	Yes	Hazardous
Fluconazole	Inhibiting fungal cytochrome P450-dependent enzymes	fungal cytochrome P450-dependent enzymes	Yes	Hazardous
Allopurinol	Inhibiting the xanthine oxidase enzyme, which is involved in the metabolism of purines	xanthine oxidase enzyme	Yes	Hazardous
Cholestyramine–Aspartame	Binding to bile acids in the intestine and preventing their reabsorption	bile acids	No	Hazardous
Terazosin	Blocking the alpha-1 adrenergic receptors in smooth muscle tissue, including the prostate and blood vessels	ADRA1A ADRA1B ADRA1D	No	Hazardous
Metoclopramide	Blocking dopamine receptors and stimulating 5-HT4 serotonin receptors in the gastrointestinal tract	DRD1 DRD2 DRD3 DRD4 DRD5 HTR4	Yes	Hazardous
Clobetasol	Binding to and activating glucocorticoid receptors in skin cells	NR3C1	No	Hazardous

* Some of the medications have unclear mechanism of action/targets.

The significant association between the results of Aspartate Aminotransferase (AST) and Alkaline Phosphatase (ALP) tests and AD + P that we found in our study calls for an intriguing interpretation. Both AST and ALP are enzymes that, among other roles, are key indicators of liver and kidney function [53,54]. Their elevated levels often indicate some level of dysfunction or damage in these organs. The connection we've identified suggests that there could be an underlying link between kidney-related indicators and brain health, possibly through metabolic or vascular pathways. The metabolic waste products that the kidneys filter from the blood, for instance, might have a more intricate relationship with the brain's health and function than previously understood [53,54]. If these waste products are not efficiently removed, they might indirectly impact brain health, potentially contributing to the development or exacerbation of AD + P. Likewise, kidney diseases are often associated with vascular problems, such as hypertension, which has been linked to cognitive decline and dementia [55]. This adds another layer of complexity to the relationship between kidney health and brain function. With these interpretations in mind, it is plausible that monitoring kidney health could be of great importance in the early detection, prevention, and management of AD + P. Health strategies could involve regular screenings for kidney function, especially in populations at risk of AD + P. Such measures could help to identify potential issues early, allowing for timely intervention. Furthermore, treatments aimed at improving kidney function or managing kidney diseases might also prove beneficial for AD + P patients.

The DeepBiomarker 1.5 model identified several important features that were strongly associated with AD + P, including inflammation-, glucose-metabolism-, cardiovascular-, and kidney-related biomarkers/mechanisms. Though the beneficial effect of glucose-lowering medications (glipizide, insulin) [56,57], cardiovascular medications (midodrine, irbesartan, losartan, clopidogrel, warfarin, terazosin, losartan) [58–60], antibiotics (cephalexin), topical corticosteroids (clobetasol), dietary supplements (glucosamine–chondroitin), and anti-inflammatory medications (triamcinolone acetonide, aspirin) [61] in AD have already been reported by multiple studies, there is no clear evidence supporting their correlation with AD + P. It is possible that these medications exert their protective effects against AD + P by treating comorbidities in AD. However, our findings also provide some mechanistic insight into the association between AD, AD + P, and these comorbidities.

Our research has limitations that need to be taken into account when interpreting the results. First, because the patients had a limited number of medications, we could not assess the effects of medications with few to no users. Additionally, there may be inconsistencies in patients' biochemical test results due to enrollment bias, and some laboratory tests may be underrepresented in our database, limiting the analysis's ability to detect their effects. Furthermore, although we investigated the influence of biomarkers, comorbidities had a greater impact since the diagnosis considers the patients' past status, while biomarkers only take into account their current status.

We would also like to point out that the population used in this study partially overlapped with the population in one of our previous studies that reported the beneficial effect of Vitamin D against AD + P [36]. We were unable to match and exclude the overlapping patients because of the deidentification process conducted by the data management team at the UPMC. However, with a total of 502 subjects included in the previous study, the overlapping sample size is too small to cause a significant impact.

Despite the potential limitations inherent to our observational study and the complexity of the identified dependencies among various health conditions and treatments, our findings pave the way for future research. They underline the need for a comprehensive approach that considers these interdependencies and advances toward more targeted preventive measures and therapeutic strategies for AD + P. Our study ultimately emphasizes the potential of big data and machine learning in this domain while underlining the need for an integrated approach that considers not only the primary disease symptoms but also co-existing conditions and their treatments. Our analysis has highlighted a novel angle in drug development for AD + P by revealing many drugs that have CNS penetration and impact varying protein targets present in the brain, and thus, may not impact AD + P through their effects on their original indications, but rather through other underlying mechanisms that are directly involved in the development of AD + P. These findings have important implications for drug development in AD + P and suggest that novel therapeutic targets should be explored to effectively prevent or treat this subtype. For example, looking at the overlap of the gene networks of these drugs' targets within the

CNS and that of AD + P may prove fruitful for the identification of new mechanisms for the prevention of AD + P [20,36]. Moving forward, our research emphasizes the necessity of a holistic approach to disease management. Future research should be directed toward disentangling the intricate web of dependencies among various health conditions and treatments. Ideally, the goal is to develop effective prevention and treatment strategies that account for these multifaceted interactions. To reach this goal, it is crucial to expand upon our findings with more detailed mechanistic studies and randomized clinical trials. In the grand scheme of the field, this research further underscores the immense potential of big data and machine learning in advancing our understanding of such complex diseases as AD + P. At the same time, it also emphasizes the critical role of a comprehensive and integrative approach, considering not only primary disease symptoms but also co-existing conditions and their treatments.

4. Materials and Methods

The application of deep learning models to predict clinical outcomes using electronic medical records (EMR) data has gained significant attention recently [62,63]. EMR data, which typically represents a patient's history as a sequence of visits with multiple events per visit, is well-suited for such sequence models as RNNs [64,65]. Recent studies indicated that simple-gated RNN models, such as Gated Recurrent Units (GRUs) and Long Short-Term Memory (LSTMs), when finely tuned using Bayesian Optimization, often deliver competitive outcomes [33]. Due to the limited sample size, we did not use Transformer-based models, which require a large amount of data for training.

4.1. Data Source

We examined the data from January 2004 to October 2019 from the Neptune system at the UPMC, which manages the use of patient EMRs from the UPMC health system for research purposes (rio.pitt.edu/services, accessed on 2 May 2023) [66]. The database includes demographic information, diagnoses, encounters, medication prescriptions, prescription fill history, and laboratory tests. AD patients and psychosis patients were identified using a series of diagnosis terms in the EMR systems (Supplementary Lists S1 and S2). In addition, to avoid the possible misdiagnosis of psychosis by short-term delirium symptoms, a psychosis diagnosis that co-occurred with a delirium diagnosis (Supplementary List S3) was excluded.

4.2. Inclusion/Exclusion Criteria and Data Preparation

For each AD patient, we would like to predict whether the patient will have psychosis within the next 3 months, given the history of EMRs. The inclusion criteria for cases and controls are described as follows. To be included in this study, an AD patient had to have at least a one-year EMR record prior to the first AD diagnosis, and the patient had no previous history of psychosis diagnosis or at least a one-year washout period before AD. This is to make sure that (1) we have collected enough comorbidity information for this patient and (2) the psychosis onset was new to this patient after the AD diagnosis. At any encounter, an AD patient who had a record of psychosis within the following 3 months is defined as a case, while no record of psychosis within the following 3 months is defined as a control. For a patient with multiple encounters satisfying the criteria of control, only the last encounter was included to mimic the latest status of these patients. We also require no records of psychosis during this period to the index date to make sure that this is a new onset of AD + P. Moreover, we used data augmentation to increase the number of cases (see below). The date of this encounter will be the index date. We used the medication, diagnosis, and lab tests 1 year preceding the index date as the input. For lab tests, we only included those abnormal ones in our modeling by searching those RESULT_FLAG labeled as "ABNORMAL", "HIGH", or "LOW". We also excluded those lab tests with low frequency and kept the 89 top frequently tested ones. The diagnosis was coded in ICD9 before the year 2015 and ICD10 after the year 2015. As such, we used a lookup table from https://www.cms.gov/Medicare/Coding/ICD10/2018-ICD-10-CM-and-GEMs (accessed on 8 December 2022) to convert ICD9 to ICD10 codes. The first three characters of the ICD10, which designate the category of the diagnosis, were extracted, yielding 1614 diagnosis groups. Medication names were converted to DrugBank IDs by name matching, and 1407 unique DrugBank IDs were mapped. Finally, for each encounter, the associated medications, diagnosis, and abnormal lab test results were packed into a sequence with the indices of DrugBank IDs, categories of the diagnosis, and lab test IDs, respectively.

4.3. Data Augementation

Data augmentation is a technology used to increase the data size and reduce overfitting. At any encounter, the chance of having psychosis within the next 3 months was much lower than that of having no psychosis, even within these AD patients with high risk. We included all encounters nearby the psychosis, which satisfied the inclusion criteria for positive cases while under-sampling the encounters which satisfied the inclusion criteria for controls (Figure 1). The purpose of data augmentation is to enhance the influence of factors nearby the events while reducing the effects of factors far from the events.

The dataset was split with a ratio of 8:1:1, and 8 of 10 subsets were used as the training dataset, while 1 of 10 subsets was used as the validation dataset to find the optimal parameters, and 1 subset was used as the test set to evaluate the generalization of our model.

4.4. Model Construction and Assessment

We adopted the Pytorch_EHR framework established by ZhiGroup, where Deep learning models with Vanilla RNN, GRU, LSTM, Bidirectional RNN, Bidirectional GRU, Bidirectional LSTM, Dilated RNN, Dilated GRU, Dilated LSTM, QRNN, and T-LSTM were used to analyze and predict clinical outcomes [33]. We further modified the framework, as highlighted in Figure 1, by (A) data augmenting to improve the model performance, (B) including individual lab tests and medications along with the diagnosis groups as the input so that we could assess the effects of each lab tests and medications, and (C) integrating contribution analysis [35] module for the importance estimation of key factors (see below for more details). The structure we used here was the LSTM model, which stores previous illness history, infers current illness states, and predicts future medical outcomes. The memory cell is gated to moderate the information flow to or from the cell. LSTMs have been adapted to many applications, such as machine translation, handwriting recognition, and speech recognition. In this study, the following parameters are used: embed dimension: 128; hidden size: 128; dropout rate: 0.2; the number of layers: 2; input size: 30,000; patience: 3. The calculations were repeated ten times for each deep learning algorithm to estimate the standard deviations of the accuracy.

To further investigate the importance of those factors on the prediction of psychosis, we calculated the relative contribution (*RC*) of each feature on the psychosis [35]. The *RC* of a feature was calculated as the average contribution of the feature to events divided by the average contributions of this feature to no events. The contributions were estimated by a perturbation-based approach. Such an approach has been used in a recent study on the important features of the heart failure incidence prediction [67]. The equation is shown as follows, where *FC* represents the feature contribution:

$$RC = \frac{mean\ FC\ in\ patients\ with\ event}{mean\ FC\ in\ patients\ without\ event} \tag{1}$$

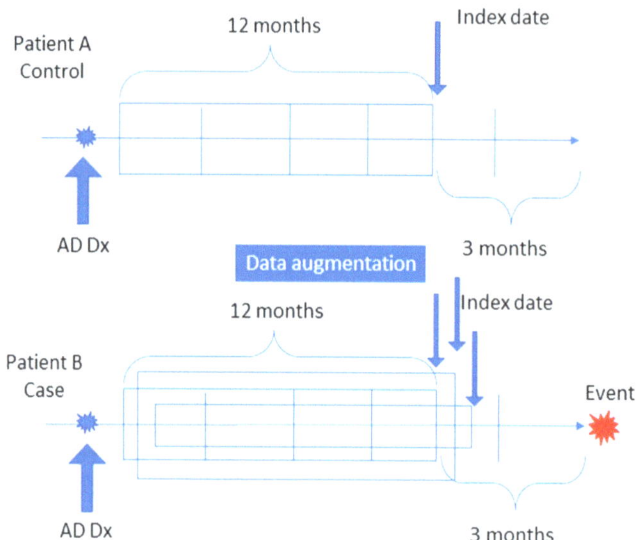

A. Data sampling from electronic medical records

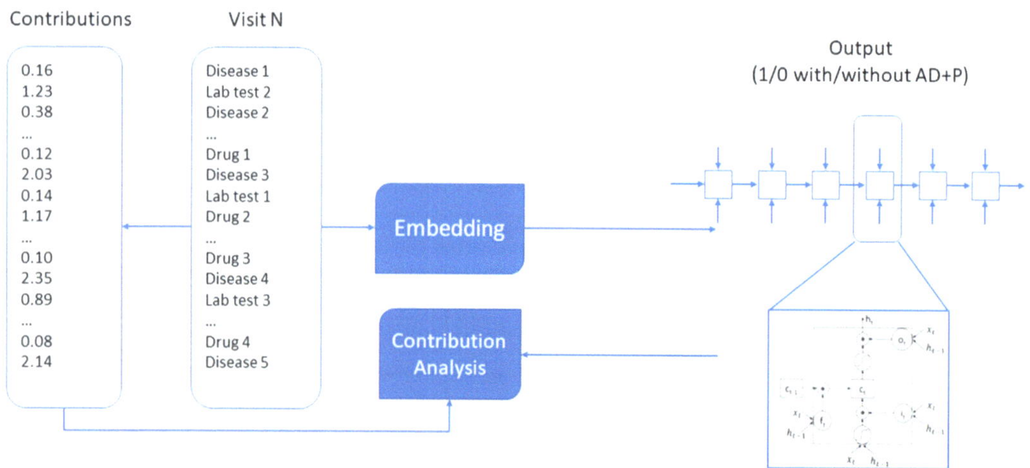

B. Data embedding & Contribution analysis   **C. Prediction by neural network**

Figure 1. Overview of DeepBiomarker. (**A**) Data sampling process from EMRs to create case and control cohorts; data augmentation was applied to oversample AD + P patients to create balanced datasets. (**B**) Data embedding is a process of representing higher dimensional data in a lower-dimensional space while preserving the relevant properties of the original data. (**C**) Prediction by neural network with LSTM as the basic prediction unit. Perturbation-based contribution analysis was used to identify important features.

FC value was the total value of the feature within the same patient if the feature appeared more than once in that patient. The natural logarithm form variance for RC was calculated as follows:

$$Variance(\ln(RC)) = \frac{\left(\frac{sd\ of\ FC\ of\ patients\ with\ event}{mean\ of\ FC\ of\ patients\ with\ event}\right)^2}{number\ of\ patients\ with\ event} + \frac{\left(\frac{sd\ of\ FC\ of\ patients\ without\ event}{mean\ of\ FC\ of\ patients\ without\ event}\right)^2}{number\ of\ patients\ without\ event} \qquad (2)$$

Thus, the 95% confidence interval (CI) of RC was given by

$$95\%\text{CI} = e^{(\ln(RC))\pm 1.96\sqrt{Variance(\ln(RC))}} \qquad (3)$$

The p-value was under the assumption of z distribution [68]. Bonferroni correction [69] was used to reduce the type I error caused by multiple comparisons.

In this version of DeepBiomarker (V1.5), we refined our RC calculation by normalizing the FC value and scaling the RC values for all our features. The improved FC formula is the ratio of the summary of the contribution of a feature and the summary of the contribution of all the features. Next, we performed scaling, where the RC for an AD diagnosis was scaled to 1, and the factor generated was multiplied to get the real RC value for each of the other features. The rationale of normalization is to consider the different occurrences of a feature in patients because of differences in the number of encounters and consider the influence of other features. The rationale for scaling the RC of AD to 1 is that all the patients had AD, and AD will have similar effects on the risk of psychosis.

Model performance was evaluated by the area under the ROC curve (AUROC).

5. Conclusions

In this study, we presented a cutting-edge deep learning model, DeepBiomarker, that is capable of accurately predicting the onset the AD + P and identifying risk factors and potential treatment options for AD + P. The results generated by this study not only provided a powerful tool for clinical care but also provided insights for mechanism studies related to AD + P and can further facilitate the development of effective treatment options for AD + P.

Our research emphasizes the remarkable potential of big data and deep learning in unraveling the multifaceted influences on the development of AD+P. Through our work, we spotlight the intricate interplay of comorbid conditions and their treatments in AD+P pathogenesis, highlighting glucose metabolism, cardiovascular factors, and kidney/liver functions as key areas of interest. This provides a valuable springboard for targeted therapies that extend beyond conventional symptom management. However, the complexity of these interdependencies necessitates refined methodologies for accurate interpretation, emphasizing a need for comprehensive, holistic approaches in both research and clinical practice. Future research should aim to disentangle these relationships further, fostering the development of innovative preventive measures and therapeutic strategies for AD + P. This study reaffirms the pivotal role of translational research, bridging the gap between theory and clinical application as we continue our pursuit to alleviate the burden of Alzheimer's disease.

Supplementary Materials: The following supporting information can be downloaded at: https://www.mdpi.com/article/10.3390/ph16070911/s1. Supplementary material contains: Supplementary List S1: diagnosis terms used to identify AD patients. Supplementary List S2: diagnosis terms used to identify psychosis patients. Supplementary List S3: diagnosis terms used to identify delirium patients.

Author Contributions: Conceptualization, P.F. and L.W.; methodology, P.F., X.Q. and L.W.; software, P.F. and L.W.; validation, X.Q., O.M. and L.W.; formal analysis, P.F.; data curation, X.Q.; writing—original draft preparation, P.F.; writing—review and editing, O.M., J.K., R.A.S. and L.W.; supervision, J.K., R.A.S. and L.W. All authors have read and agreed to the published version of the manuscript.

Funding: This study was funded by the following National Institutes of Health grants: S10 OD028483/OD/NIH HHS/United States; P50 AG005133/AG/NIA NIH HHS/United States; R01 AG027224/AG/NIA NIH HHS/United States; R01 MH116046/MH/NIMH NIH HHS/United States; P30 AG066468/AG/NIA NIH HHS/United States; UL1 TR001857/TR/NCATS NIH HHS/United States.

Institutional Review Board Statement: Not applicable.

Informed Consent Statement: Not applicable.

Data Availability Statement: Data are available upon request from the authors.

Conflicts of Interest: The authors declare no conflict of interest.

References

1. Alzheimer's Association. 2016 Alzheimer's disease facts and figures. *Alzheimers Dement.* **2016**, *12*, 459–509. [CrossRef]
2. Thorgrimsen, L.; Selwood, A.; Spector, A.; Royan, L.; de Madariaga Lopez, M.; Woods, R.; Orrell, M. Whose quality of life is it anyway?: The validity and reliability of the Quality of Life-Alzheimer's Disease (QoL-AD) scale. *Alzheimer Dis. Assoc. Disord.* **2003**, *17*, 201–208. [CrossRef] [PubMed]
3. Wimo, A.; Prince, M. *World Alzheimer Report. The Global Economic Impact of Dementia*; Alzheimer's Disease International: London, UK, 2010.
4. Murray, P.S.; Kumar, S.; Demichele-Sweet, M.A.; Sweet, R.A. Psychosis in Alzheimer's disease. *Biol. Psychiatry* **2014**, *75*, 542–552. [CrossRef]
5. Ropacki, S.A.; Jeste, D.V. Epidemiology of and risk factors for psychosis of Alzheimer's disease: A review of 55 studies published from 1990 to 2003. *Am. J. Psychiatry* **2005**, *162*, 2022–2030. [CrossRef]
6. Battaglia, S.; Nazzi, C.; Thayer, J.F. Fear-induced bradycardia in mental disorders: Foundations, current advances, future perspectives. *Neurosci. Biobehav. Rev.* **2023**, *149*, 105163. [CrossRef]
7. Lanctôt, K.L.; Amatniek, J.; Ancoli-Israel, S.; Arnold, S.E.; Ballard, C.; Cohen-Mansfield, J.; Ismail, Z.; Lyketsos, C.; Miller, D.S.; Musiek, E.; et al. Neuropsychiatric signs and symptoms of Alzheimer's disease: New treatment paradigms. *Alzheimers Dement.* **2017**, *3*, 440–449. [CrossRef] [PubMed]
8. Lyketsos, C.G.; Carrillo, M.C.; Ryan, J.M.; Khachaturian, A.S.; Trzepacz, P.; Amatniek, J.; Cedarbaum, J.; Brashear, R.; Miller, D.S. Neuropsychiatric symptoms in Alzheimer's disease. *Alzheimers Dement.* **2011**, *7*, 532–539. [CrossRef] [PubMed]
9. Battaglia, S.; Di Fazio, C.; Vicario, C.M.; Avenanti, A. Neuropharmacological Modulation of N-methyl-D-aspartate, Noradrenaline and Endocannabinoid Receptors in Fear Extinction Learning: Synaptic Transmission and Plasticity. *Int. J. Mol. Sci.* **2023**, *24*, 5926. [CrossRef] [PubMed]
10. Sweet, R.A.; Bennett, D.A.; Graff-Radford, N.R.; Mayeux, R. Assessment and familial aggregation of psychosis in Alzheimer's disease from the National Institute on Aging Late Onset Alzheimer's Disease Family Study. *Brain J. Neurol.* **2010**, *133*, 1155–1162. [CrossRef]
11. Krivinko, J.M.; Erickson, S.L.; Ding, Y.; Sun, Z.; Penzes, P.; MacDonald, M.L.; Yates, N.A.; Ikonomovic, M.D.; Lopez, O.L.; Sweet, R.A.; et al. Synaptic Proteome Compensation and Resilience to Psychosis in Alzheimer's Disease. *Am. J. Psychiatry* **2018**, *175*, 999–1009. [CrossRef]
12. DeChellis-Marks, M.R.; Wei, Y.; Ding, Y.; Wolfe, C.M.; Krivinko, J.M.; MacDonald, M.L.; Lopez, O.L.; Sweet, R.A.; Kofler, J. Psychosis in Alzheimer's Disease Is Associated With Increased Excitatory Neuron Vulnerability and Post-transcriptional Mechanisms Altering Synaptic Protein Levels. *Front. Neurol.* **2022**, *13*, 778419. [CrossRef] [PubMed]
13. Krivinko, J.; DeChellis-Marks, M.; Zeng, L.; Fan, P.; Lopez, O.; Ding, Y.; Wang, L.; Kofler, J.; MacDonald, M.; Sweet, R. Targeting the Post-Synaptic Proteome in Alzheimer Disease with Psychosis. *Commun. Biol.* **2023**, *6*, 538. [CrossRef] [PubMed]
14. Madhusoodanan, S.; Shah, P. Management of psychosis in patients with Alzheimer's disease: Focus on aripiprazole. *Clin. Interv. Aging* **2008**, *3*, 491–501. [CrossRef]
15. Alexopoulos, G.S.; Streim, J.; Carpenter, D.; Docherty, J.P. Expert consensus guidelines for using antipsychotic agents in older patients. *J. Clin. Psychiatry* **2004**, *65*, 5–99. [PubMed]
16. Burke, A.D.; Burke, W.J. Antipsychotics FOR patients WITH dementia: The road less traveled: Second-generation agents have an important but limited role in treating behavioral and psychological symptoms. *Curr. Psychiatry* **2018**, *17*, 26–36.
17. Dorsey, E.R.; Rabbani, A.; Gallagher, S.A.; Conti, R.M.; Alexander, G.C. Impact of FDA black box advisory on antipsychotic medication use. *Arch. Intern. Med.* **2010**, *170*, 96–103. [CrossRef] [PubMed]
18. Emanuel, J.E.; Lopez, O.L.; Houck, P.R.; Becker, J.T.; Weamer, E.A.; Demichele-Sweet, M.A.; Kuller, L.; Sweet, R.A. Trajectory of cognitive decline as a predictor of psychosis in early Alzheimer disease in the cardiovascular health study. *Am. J. Geriatr. Psychiatry Off. J. Am. Assoc. Geriatr. Psychiatry* **2011**, *19*, 160–168. [CrossRef]
19. Tampi, R.R.; Tampi, D.J.; Balachandran, S.; Srinivasan, S. Antipsychotic use in dementia: A systematic review of benefits and risks from meta-analyses. *Ther. Adv. Chronic Dis.* **2016**, *7*, 229–245. [CrossRef]
20. Fan, P.; Kofler, J.; Ding, Y.; Marks, M.; Sweet, R.A.; Wang, L. Efficacy difference of antipsychotics in Alzheimer's disease and schizophrenia: Explained with network efficiency and pathway analysis methods. *Brief. Bioinform.* **2022**, *23*, bbac394. [CrossRef]

21. DeMichele-Sweet, M.A.A.; Klei, L.; Creese, B.; Harwood, J.C.; Weamer, E.A.; McClain, L.; Sims, R.; Hernandez, I.; Moreno-Grau, S.; Tarraga, L.; et al. Genome-wide association identifies the first risk loci for psychosis in Alzheimer disease. *Mol. Psychiatry* **2021**, *26*, 5797–5811. [CrossRef]

22. Bote-Curiel, L.; Muñoz-Romero, S.; Gerrero-Curieses, A.; Rojo-Álvarez, J.L. Deep Learning and Big Data in Healthcare: A Double Review for Critical Beginners. *Appl. Sci.* **2019**, *9*, 2331. [CrossRef]

23. Salehinejad, H.; Sankar, S.; Barfett, J.; Colak, E.; Valaee, S.J. Recent advances in recurrent neural networks. *arXiv* **2017**, arXiv:1801.01078.

24. Vaswani, A.; Shazeer, N.; Parmar, N.; Uszkoreit, J.; Jones, L.; Gomez, A.N.; Kaiser, Ł.; Polosukhin, I. Attention is all you need. *Adv. Neural Inf. Process. Syst.* **2017**, *30*.

25. Supriya, M.; Deepa, A.J. Machine learning approach on healthcare big data: A review. *Big Data Inf. Anal.* **2020**, *5*, 58–75. [CrossRef]

26. Pham, T.; Tran, T.; Phung, D.; Venkatesh, S. Deepcare: A deep dynamic memory model for predictive medicine. In Proceedings of the Pacific-Asia Conference on Knowledge Discovery and Data Mining, Auckland, New Zealand, 19–22 April 2016; pp. 30–41.

27. Dernoncourt, F.; Lee, J.Y.; Uzuner, O.; Szolovits, P. De-identification of patient notes with recurrent neural networks. *J. Am. Med. Inform. Assoc.* **2017**, *24*, 596–606. [CrossRef] [PubMed]

28. Che, Z.; Purushotham, S.; Khemani, R.; Liu, Y. Distilling knowledge from deep networks with applications to healthcare domain. *arXiv* **2015**, arXiv:1512.03542.

29. Cheng, Y.; Wang, F.; Zhang, P.; Hu, J. Risk prediction with electronic health records: A deep learning approach. In Proceedings of the SIAM International Conference on Data Mining, Miami, FL, USA, 5–7 May 2016; pp. 432–440.

30. Choi, Y.; Chiu, C.Y.; Sontag, D. Learning Low-Dimensional Representations of Medical Concepts. *AMIA Summits Transl. Sci. Proc.* **2016**, *2016*, 41–50.

31. Shickel, B.; Tighe, P.J.; Bihorac, A.; Rashidi, P. Deep EHR: A Survey of Recent Advances in Deep Learning Techniques for Electronic Health Record (EHR) Analysis. *IEEE J. Biomed. Health Inform.* **2018**, *22*, 1589–1604. [CrossRef]

32. Wiemken, T.L.; Kelley, R.R. Machine Learning in Epidemiology and Health Outcomes Research. *Annu. Rev. Public. Health* **2020**, *41*, 21–36. [CrossRef]

33. Rasmy, L.; Zhu, J.; Li, Z.; Hao, X.; Tran, H.T.; Zhou, Y.; Tiryaki, F.; Xiang, Y.; Xu, H.; Zhi, D. Simple Recurrent Neural Networks is all we need for clinical events predictions using EHR data. *arXiv* **2021**, arXiv:.00998.

34. Miranda, O.; Fan, P.; Qi, X.; Yu, Z.; Ying, J.; Wang, H.; Brent, D.A.; Silverstein, J.C.; Chen, Y.; Wang, L. DeepBiomarker: Identifying Important Lab Tests from Electronic Medical Records for the Prediction of Suicide-Related Events among PTSD Patients. *J. Pers. Med.* **2022**, *12*, 524. [CrossRef] [PubMed]

35. Guan, C.; Wang, X.; Zhang, Q.; Chen, R.; He, D.; Xie, X. Towards a deep and unified understanding of deep neural models in nlp. In Proceedings of the International Conference on Machine Learning, Long Beach, CA, USA, 9–15 June 2019; pp. 2454–2463.

36. Wang, L.; Ying, J.; Fan, P.; Weamer, E.A.; DeMichele-Sweet, M.A.A.; Lopez, O.L.; Kofler, J.K.; Sweet, R.A. Effects of Vitamin D Use on Outcomes of Psychotic Symptoms in Alzheimer Disease Patients. *Am. J. Geriatr. Psychiatry Off. J. Am. Assoc. Geriatr. Psychiatry* **2019**, *27*, 908–917. [CrossRef]

37. Fan, P.; Qi, X.; Sweet, R.A.; Wang, L. Network systems pharmacology-based mechanism study on the beneficial effects of vitamin d against psychosis in Alzheimer's disease. *Sci. Rep.* **2020**, *10*, 6136. [CrossRef] [PubMed]

38. Calsolaro, V.; Antognoli, R.; Okoye, C.; Monzani, F. The Use of Antipsychotic Drugs for Treating Behavioral Symptoms in Alzheimer's Disease. *Front. Pharmacol.* **2019**, *10*, 1465. [CrossRef]

39. Vigen, C.L.; Mack, W.J.; Keefe, R.S.; Sano, M.; Sultzer, D.L.; Stroup, T.S.; Dagerman, K.S.; Hsiao, J.K.; Lebowitz, B.D.; Lyketsos, C.G.; et al. Cognitive effects of atypical antipsychotic medications in patients with Alzheimer's disease: Outcomes from CATIE-AD. *Am. J. Psychiatry* **2011**, *168*, 831–839. [CrossRef] [PubMed]

40. Raskin, J.; Wiltse, C.G.; Siegal, A.; Sheikh, J.; Xu, J.; Dinkel, J.J.; Rotz, B.T.; Mohs, R.C. Efficacy of duloxetine on cognition, depression, and pain in elderly patients with major depressive disorder: An 8-week, double-blind, placebo-controlled trial. *Am. J. Psychiatry* **2007**, *164*, 900–909. [CrossRef] [PubMed]

41. Lee, J.G.; Lee, S.W.; Lee, B.J.; Park, S.W.; Kim, G.M.; Kim, Y.H. Adjunctive memantine therapy for cognitive impairment in chronic schizophrenia: A placebo-controlled pilot study. *Psychiatry Investig.* **2012**, *9*, 166–173. [CrossRef]

42. Lieberman, J.A.; Papadakis, K.; Csernansky, J.; Litman, R.; Volavka, J.; Jia, X.D.; Gage, A.; Group, M.-M.-S. A randomized, placebo-controlled study of memantine as adjunctive treatment in patients with schizophrenia. *Neuropsychopharmacol. Off. Publ. Am. Coll. Neuropsychopharmacol.* **2009**, *34*, 1322–1329. [CrossRef]

43. Veerman, S.R.; Schulte, P.F.; Smith, J.D.; de Haan, L. Memantine augmentation in clozapine-refractory schizophrenia: A randomized, double-blind, placebo-controlled crossover study. *Psychol. Med.* **2016**, *46*, 1909–1921. [CrossRef]

44. Krivoy, A.; Weizman, A.; Laor, L.; Hellinger, N.; Zemishlany, Z.; Fischel, T. Addition of memantine to antipsychotic treatment in schizophrenia inpatients with residual symptoms: A preliminary study. *Eur. Neuropsychopharmacol. J. Eur. Coll. Neuropsychopharmacol.* **2008**, *18*, 117–121. [CrossRef]

45. Omranifard, V.; Rajabi, F.; Mohammadian-Sichani, M.; Maracy, M. The effect of add-on memantine on global function and quality of life in schizophrenia: A randomized, double-blind, controlled, clinical trial. *Adv. Biomed. Res.* **2015**, *4*, 211.

46. Kotsiantis, S.B.; Zaharakis, I.; Pintelas, P.J. Supervised machine learning: A review of classification techniques. *Emerg. Artif. Intell. Appl. Comput. Eng.* **2007**, *160*, 3–24.

47. LeCun, Y.; Bengio, Y.; Hinton, G. Deep learning. *Nature* **2015**, *521*, 436–444. [CrossRef] [PubMed]

48. Kinney, J.W.; Bemiller, S.M.; Murtishaw, A.S.; Leisgang, A.M.; Salazar, A.M.; Lamb, B.T. Inflammation as a central mechanism in Alzheimer's disease. *Alzheimers Dement.* **2018**, *4*, 575–590. [CrossRef]

49. Lou, N.; Takano, T.; Pei, Y.; Xavier, A.L.; Goldman, S.A.; Nedergaard, M. Purinergic receptor P2RY12-dependent microglial closure of the injured blood-brain barrier. *Proc. Natl. Acad. Sci. USA* **2016**, *113*, 1074–1079. [CrossRef] [PubMed]

50. Huang, Y. Mechanisms linking apolipoprotein E isoforms with cardiovascular and neurological diseases. *Curr. Opin. Lipidol.* **2010**, *21*, 337–345. [CrossRef] [PubMed]

51. Tanaka, M.; Szabó, Á.; Vécsei, L. Preclinical modeling in depression and anxiety: Current challenges and future research directions. *Adv. Clin. Exp. Med.* **2023**, *32*, 505–509. [CrossRef]

52. Tanaka, M.; Szabó, Á.; Spekker, E.; Polyák, H.; Tóth, F.; Vécsei, L. Mitochondrial Impairment: A Common Motif in Neuropsychiatric Presentation? The Link to the Tryptophan—Kynurenine Metabolic System. *Cells* **2022**, *11*, 2607. [CrossRef] [PubMed]

53. Shi, Y.; Liu, Z.; Shen, Y.; Zhu, H. A novel perspective linkage between kidney function and alzheimer's disease. *Front. Cell. Neurosci.* **2018**, *12*, 384. [CrossRef]

54. Han, S.W.; Park, Y.H.; Jang, E.S.; Nho, K.; Kim, S. Implications of liver enzymes in the pathogenesis of alzheimer's disease. *J. Alzheimers Dis.* **2022**, *88*, 1371–1376. [CrossRef]

55. Harciarek, M.; Williamson, J.B.; Biedunkiewicz, B.; Lichodziejewska-Niemierko, M.; Dębska-Ślizień, A.; Rutkowski, B. Risk factors for selective cognitive decline in dialyzed patients with end-stage renal disease: Evidence from verbal fluency analysis. *J. Int. Neuropsychol. Soc.* **2012**, *18*, 162–167. [CrossRef]

56. Liao, W.; Xu, J.; Li, B.; Ruan, Y.; Li, T.; Liu, J. Deciphering the Roles of Metformin in Alzheimer's Disease: A Snapshot. *Front. Pharmacol.* **2021**, *12*, 728315. [CrossRef] [PubMed]

57. Mishra, S.; Prusty, S.K.; Sahu, P.K.; Das, D.J. Irbesartan protects against aluminium chloride induced amyloidogenesis and cognitive impairment. *J. Krishna Inst. Med. Sci.* **2022**, *11*, 18–30.

58. Sushko, V.V.; Sushko, V.V. Use Acetazolamide in the Complex Therapy of Alzheimer's Disease. *Alzheimer's Dement.* **2022**, *18*, e059982. [CrossRef]

59. Sberna, G.; Saez-Valero, J.; Beyreuther, K.; Masters, C.L.; Small, D.H. The amyloid beta-protein of Alzheimer's disease increases acetylcholinesterase expression by increasing intracellular calcium in embryonal carcinoma P19 cells. *J. Neurochem.* **1997**, *69*, 1177–1184. [CrossRef] [PubMed]

60. Umukoro, S.; Bakre, T.O.; Onwuchekwa, C. Anti-psychotic and sedative effect of calcium channel blockers in mice. *Afr. J. Med. Med. Sci.* **2010**, *39*, 61–66.

61. Stuve, O.; Weideman, R.A.; McMahan, D.M.; Jacob, D.A.; Little, B.B. Diclofenac reduces the risk of Alzheimer's disease: A pilot analysis of NSAIDs in two US veteran populations. *Ther. Adv. Neurol. Disord.* **2020**, *13*, 1756286420935676. [CrossRef]

62. Yin, C.; Zhao, R.; Qian, B.; Lv, X.; Zhang, P. Domain knowledge guided deep learning with electronic health records. In Proceedings of the IEEE International Conference on Data Mining (ICDM), Beijing, China, 8–11 November 2019; pp. 738–747.

63. Su, C.; Xu, Z.; Pathak, J.; Wang, F. Deep learning in mental health outcome research: A scoping review. *Transl. Psychiatry* **2020**, *10*, 116. [CrossRef]

64. Zehraoui, F.; Sendi, N.; Abchiche-Mimouni, N. MS-LSTMEA: Predicting clinical events for Hypertension using Multi-Sources LSTM Explainable Approach. *SSRN 4123459* **2022**. [CrossRef]

65. Zhang, J. Representation Learning of Longitudinal Electronic Health Record Data for Patient Characterization and Prediction of Health Outcomes. Ph.D. Thesis, University of Virginia, Charlottesville, VA, USA, 2017.

66. Visweswaran, S.; McLay, B.; Cappella, N.; Morris, M.; Milnes, J.T.; Reis, S.E.; Silverstein, J.C.; Becich, M.J. An atomic approach to the design and implementation of a research data warehouse. *J. Am. Med. Inform. Assoc.* **2022**, *29*, 601–608. [CrossRef]

67. Rao, S.; Li, Y.; Ramakrishnan, R.; Hassaine, A.; Canoy, D.; Zhu, Y.; Salimi-Khorshidi, G.; Rahimi, K. BEHRT-HF: An interpretable transformer-based, deep learning model for prediction of incident heart failure. *Eur. Heart J.* **2020**, *41*, ehaa946.3553. [CrossRef]

68. Altman, D.G.; Bland, J.M. How to obtain the P value from a confidence interval. *BMJ* **2011**, *343*, d2304. [CrossRef] [PubMed]

69. Bonferroni, C. *Teoria Statistica delle Classi e Calcolo delle Probabilita*; Pubblicazioni del R. Istituto Superiore di Scienze Economiche e Commericiali di Firenze: Florence, Italy, 1936; Volume 8, pp. 3–62.

International Journal of
Molecular Sciences

Article

The STIM1/2-Regulated Calcium Homeostasis Is Impaired in Hippocampal Neurons of the 5xFAD Mouse Model of Alzheimer's Disease

Ksenia Skobeleva [1], Alexey Shalygin [1], Elena Mikhaylova [1], Irina Guzhova [1], Maria Ryazantseva [2,*] and Elena Kaznacheyeva [1,*]

[1] Institute of Cytology of the Russian Academy of Sciences (RAS), 194064 St. Petersburg, Russia
[2] Neuroscience Center, University of Helsinki, HiLIFE, P.O. Box 63, 00014 Helsinki, Finland
* Correspondence: mariaandreevnar@gmail.com (M.R.); evkazn@incras.ru (E.K.)

Abstract: Alzheimer's disease (AD) is the most common cause of age-related dementia. Neuronal calcium homeostasis impairment may contribute to AD. Here we demonstrated that voltage-gated calcium (VGC) entry and store-operated calcium (SOC) entry regulated by calcium sensors of intracellular calcium stores STIM proteins are affected in hippocampal neurons of the 5xFAD transgenic mouse model. We observed excessive SOC entry in 5xFAD mouse neurons, mediated by STIM1 and STIM2 proteins with increased STIM1 contribution. There were no significant changes in cytoplasmic calcium level, endoplasmic reticulum (ER) bulk calcium levels, or expression levels of STIM1 or STIM2 proteins. The potent inhibitor BTP-2 and the FDA-approved drug leflunomide reduced SOC entry in 5xFAD neurons. In turn, excessive voltage-gated calcium entry was sensitive to the inhibitor of L-type calcium channels nifedipine but not to the T-type channels inhibitor ML218. Interestingly, the depolarization-induced calcium entry mediated by VGC channels in 5xFAD neurons was dependent on STIM2 but not STIM1 protein in cells with replete Ca^{2+} stores. The result gives new evidence on the VGC channel modulation by STIM2. Overall, the data demonstrate the changes in calcium signaling of hippocampal neurons of the AD mouse model, which precede amyloid plaque accumulation or other signs of pathology manifestation.

Keywords: Alzheimer's disease; 5xFAD; hippocampal cultures; store-operated calcium entry; voltage-gated calcium entry; STIM1; STIM2; leflunomide; nifedipine; BTP-2; calcium hypothesis

Citation: Skobeleva, K.; Shalygin, A.; Mikhaylova, E.; Guzhova, I.; Ryazantseva, M.; Kaznacheyeva, E. The STIM1/2-Regulated Calcium Homeostasis Is Impaired in Hippocampal Neurons of the 5xFAD Mouse Model of Alzheimer's Disease. *Int. J. Mol. Sci.* **2022**, *23*, 14810. https://doi.org/10.3390/ijms232314810

Academic Editors: Lydia Giménez-Llort, Masaru Tanaka, Chong Chen, Simone Battaglia and Piril Hepsomali

Received: 20 October 2022
Accepted: 23 November 2022
Published: 26 November 2022

Publisher's Note: MDPI stays neutral with regard to jurisdictional claims in published maps and institutional affiliations.

1. Introduction

Alzheimer's disease (AD) is the most common cause of age-related dementia. AD is a neurodegenerative disease that primarily affects hippocampal and cortical neurons, leading to memory loss and cognitive impairment. A familial form of AD (FAD), in most cases, is caused by mutations in *APP*, *PSEN1*, and *PSEN2* genes, manifests early and exhibits a severe phenotype (reviewed in [1]). Mouse models of FAD usually express mutant human *APP* and *PSEN1* genes associated with the disease. The 5xFAD transgenic mouse model is a double transgenic model with five FAD-associated mutations: Swedish (K670N, M671L), Florida (I716V), and London (V717I) mutations in the amyloid precursor protein (APP) gene, and M146L and L286V mutations in the presenilin-1 (*PSEN1*) gene under the control of the neuron-specific promoter Thy1 [2]. It has a phenotype with rapid amyloidogenesis, neuronal loss, behavior abnormalities, and memory deficits [2,3]. The amyloid plaques in the brain of 5xFAD mice manifest between 2 and 4 months of age. However, emerging evidence supports the idea that pathological changes related to FAD may occur earlier than amyloid plaque deposition [4,5]. Here, we use hippocampal neurons of 5xFAD mice cultured for 10–13 days in vitro. At this stage neurons do not exhibit severe amyloid pathology [6] but express human *APP* and *PSEN1* genes with mutations associated with familial AD [7].

A growing number of studies demonstrate deregulation of calcium homeostasis in modeling FAD and other neurodegenerative diseases in neurons of animals [3,8–11] (reviewed in [12–15]). Calcium homeostasis plays a critical role in neuron survival, synaptic transmission, and neural plasticity, as well as in the mitochondria function and cell metabolism (reviewed in [16,17]). Therefore, impairments in calcium homeostasis can affect many processes in neurons leading to neurodegeneration, cognitive impairment, and memory loss. There are two ways to raise calcium concentration within the neuron: by calcium entry through ionotropic receptors and ion channels in the plasmatic membrane (PM) and by calcium release from intracellular stores. Several publications have demonstrated changes in the endoplasmic reticulum (ER) calcium storage and release in neurons that appear earlier than the AD-associated plaques [18,19]. Calcium stores of the endoplasmic reticulum (ER) could rapidly release calcium to the cytosol through ryanodine- or IP3-calcium channels. Stromal interactional molecules (STIM) sense calcium concentrations within the ER and produce an intracellular signal in response to a drop in ER calcium [20]. STIMs are a central link between intracellular calcium stores and calcium channels in the plasma membrane (PM). They play a notable role in the regulation of calcium homeostasis in health and disease (reviewed in [21,22]). There are two homologs of STIMs, which differ in their role in calcium homeostasis and intracellular signaling. Both STIMs are known to activate store-operated calcium (SOC) entry by activating Orai and transient receptor potential canonical (TRPC) channels in the PM [23–25]. SOC entry disturbances were reported for 3xTg, PS1-M146V knock-in and APPKI transgenic mouse models of AD [26–29], and other neuronal models of AD and aging [26,27,30–33]. The STIM2 homolog has a lower affinity for calcium than STIM1 and, therefore, is more sensitive to small changes in ER luminal calcium concentrations. Tonic activation of STIM2 regulates basal calcium entry and refilling of ER stores, while STIM1 is regarded as a signal transducer of the more robust receptors-induced calcium release from ER [34,35]. STIM1 interacts with L-type and T-type voltage-gated calcium (VGC) channels and inhibits their activity [36–38]. Although STIM2 was described as a weaker CaV1.2 inhibitor in vascular smooth muscle cells, its role in VGC channel regulation has not been investigated further [37]. Moreover, despite the prominent STIM2 role in neuronal calcium homeostasis, it is unknown whether STIM2 regulates VGC channels in neurons. Considering the importance of SOC and VGC channels in neurons for processes such as synaptic plasticity and neurotransmission [39], the malfunction of STIMs may lead to drastic consequences for the brain. STIM1 expression was shown to be attenuated in the brains of patients with AD, possibly leading to hyperactivation of L-type VGC channels [40]. STIM2 functional deficiency was reported in the dendrites of neurons in mouse AD models [26,41]. Taking into account the role of STIM proteins in the modulation of VGC and SOC channels as well as ER store refilling, we aimed to address STIMs' contribution to calcium homeostasis impairment in the hippocampal neurons of 5xFAD transgenic mice.

Here for the first time, we investigated functional roles of the STIMs in the disruption of calcium homeostasis in the cultured hippocampal neurons of the newborn 5xFAD transgenic mice. In particular, we addressed the contribution of STIMs to SOC and VGC entry by combining in vitro calcium imaging and viral-induced gene knock-down. In addition, we analyzed SOC and VGC entry sensitivity to the specific pharmacological inhibitors and FDA-approved drugs in vitro.

2. Results

2.1. Store-Operated Calcium Entry Is Enhanced in 5xFAD Hippocampal Neurons

Primary cultures of the hippocampi of 5xFAD mice and their non-transgenic littermates (WT) were probed with Fura2-AM to study store-operated calcium (SOC) entry. SOC entry in 5xFAD mice has not been addressed via calcium imaging before. We used thapsigargin (Tg), a sarco-/endoplasmic reticulum ATPase inhibitor, to deplete internal calcium stores. All solutions were supplemented with 1 μM TTX and 1 μM nifedipine to prevent neuronal activity. We measured SOC entry as calcium influx in 2 mM Ca^{2+} aCSF, which followed the internal calcium stores depletion with 1 µM Tg in a Ca-free solution. Calcium imaging revealed a 42% elevation of SOC entry in the 5xFAD hippocampal neurons compared to WT neurons (Figure 1A). Normalized SOC entry was 1.00 ± 0.07 for WT neurons and 1.42 ± 0.05 for 5xFAD neurons ($p < 0.001$, Mann–Whitney test). To study SOC entry without affecting store depletion/repletion machinery, we used chelator TPEN followed by Ca^{2+} add-back. TPEN chelates the calcium within internal stores, leading to SOC channel activation and calcium entry [42]. This approach confirmed that SOC calcium entry was elevated by 131%. Normalized calcium entry after TPEN treatment was 1.00 ± 0.09 for WT neurons and 2.31 ± 0.22 for 5xFAD neurons ($p < 0.001$, Mann–Whitney test, Figure 1B).

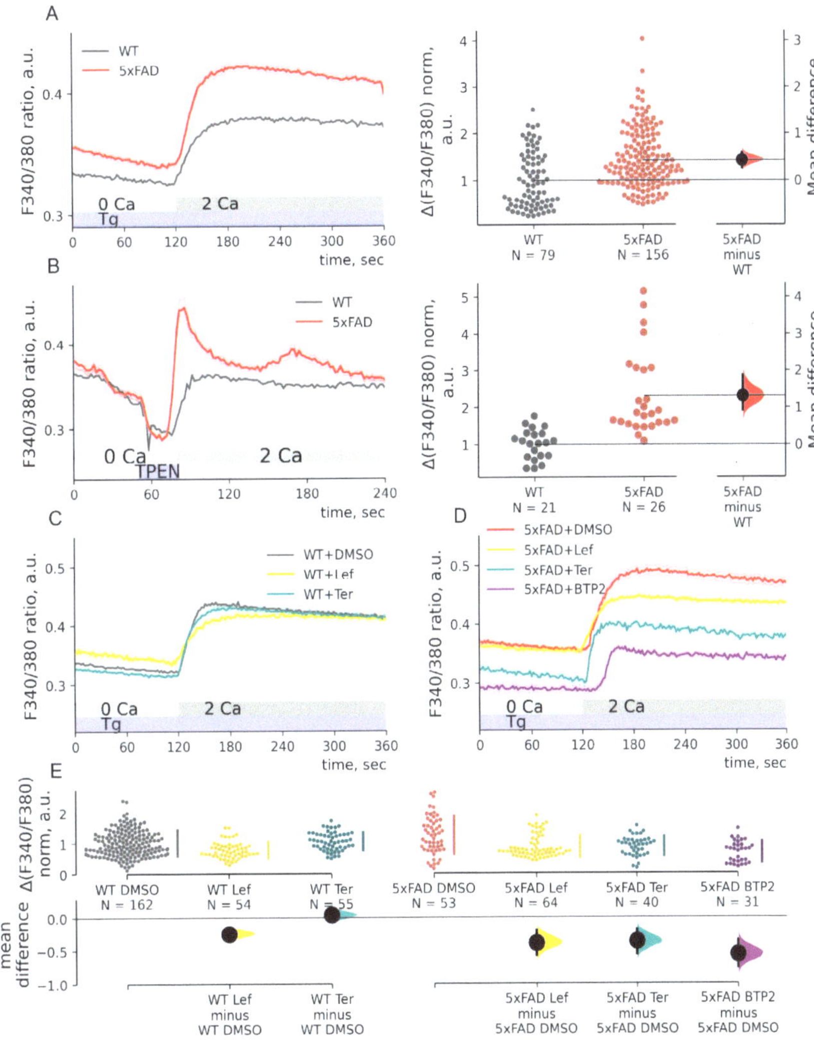

Figure 1. Effect of inhibitors Leflunomide, Teriflunomide, and BTP2 on elevated store-operated calcium entry in hippocampal neurons from 5XFAD mice. (**A**,**B**) Calcium imaging with Fura2 of neurons of primary hippocampal cultures of wild-type (WT, gray) or 5xFAD (red) mice. Mean traces of F340/380 ratio are represented as mean $\pm$ SEM. Calcium entry (ΔF340/380) was calculated as the F340/F380 peak of each trace after Ca^{2+} add-back, and the baseline was subtracted. ΔF340/F380 data were normalized by the mean WT calcium entry and shown in the estimation plots. WT and 5xFAD

calcium entry data are plotted on the left axes; the mean difference is plotted on floating axes on the right as a bootstrap sampling distribution. (**A**) Thapsigargin (Tg) was added to deplete internal calcium stores. The unpaired mean difference between WT and 5xFAD Tg-induced calcium entry is 0.42, $p = 0.2 \times 10^{-6}$, Mann–Whitney test. (**B**) Treatment with TPEN for 20 s led to Ca^{2+}-binding and mimicking depleted ER stores. Ca^{2+} add-back resulted in calcium entry. The unpaired mean difference between WT and 5xFAD data is 1.31; $p = 4.2 \times 10^{-7}$, Mann–Whitney test. (**C–E**) Calcium imaging with Fura2 of primary hippocampal cell cultures of WT (**C**) or 5xFAD (**D**) mice treated for 20 h with 1 μM of leflunomide (Lef), teriflunomide (Ter), BTP2 or DMSO. 1 μM thapsigargin (Tg) was added to deplete internal calcium stores. Mean traces of F340/380 ratio are represented as mean ± SEM. Calcium entry (ΔF340/F380) was calculated as before and normalized by median calcium entry in WT neurons treated with DMSO and shown in the estimation plot. ΔF340/F380 data are plotted on the upper axes; mean differences are plotted on floating axes on the bottom as bootstrap sampling distributions. The Kruskal–Wallis ANOVA ($p < 0.001$) with post-hoc Dunn's test was implemented; pairwise p-values were calculated. Mean differences between WT DMSO-treated neurons and WT neurons treated with Lef or Ter are −0.25 and 0.04; $p = 0.002$ and 1.0. Mean differences between 5xFAD DMSO-treated neurons and 5xFAD neurons treated with Lef, Ter or BTP2 are −0.40, −0.37 and −0.57; $p = 2.4 \times 10^{-4}$, 0.07 and 1.8×10^{-6}.

2.2. FDA-Approved Drugs Reduce SOC Entry in 5xFAD Mouse Neurons

Inhibiting abnormal SOC entry could be a therapeutic opportunity to normalize calcium homeostasis. Screening of FDA-approved drug libraries yielded leflunomide and teriflunomide as possible SOC channel inhibitors [43]. The drugs have been approved for the treatment of inflammatory diseases such as arthritis and multiple sclerosis. Teriflunomide is a leflunomide metabolite that is converted in the plasma and the gastrointestinal tract [44]; both share structural similarities with a potent SOC channel inhibitor BTP2 [43]. Leflunomide and teriflunomide, as well as BTP2, demonstrated inhibition of SOC entry at clinically relevant doses. A 24 h pretreatment of rat basophilic leukemia cells with a low dose (1 μM) of a test drug reduced SOC entry [43]. Here we pretreated primary hippocampal cultures for 20 h with 1 μM leflunomide, teriflunomide, or 0.1% DMSO. The SOC entry was measured as a peak after Ca^{2+} add-back after 1 μM Tg treatment in Ca^{2+}-free aCSF. Leflunomide reduced the SOC entry in both WT and 5xFAD neurons to 74% and 68% of the baseline compared to SOC entry in WT and 5xFAD neurons treated with DMSO ($p < 0.001$, Dunn's test, Figure 1C–E). Normalized calcium entry for WT and 5xFAD neurons treated with DMSO were 1.00 ± 0.03 and 1.22 ± 0.08 and for WT and 5xFAD neurons treated with leflunomide, 0.74 ± 0.04 and 0.83 ± 0.05. Teriflunomide did not affect store-operated calcium entry in either WT or 5xFAD neurons: 1.04 ± 0.05 for WT and 0.85 ± 0.05 for 5xFAD ($p > 0.5$, Dunn's test). BTP2 treatment reduced the store-operated calcium entry to 54% of the 5xFAD neurons baseline (0.66 ± 0.06; $p < 0.001$ compared to SOC entry in 5xFAD neurons treated with DMSO, Dunn' test, Figure 1D,E). Therefore, SOC entry in 5xFAD hippocampal neurons was sensitive to inhibitors BTP2 and its FDA-approved analog leflunomide.

2.3. SOC Entry Enhancement in 5xFAD Mouse Neurons Is Associated with Changes in the Passive Leak but Not with Expression Levels of STIMs

The enhanced SOC entry in 5xFAD could be associated with ER store calcium levels, SOC machinery protein expression level, or SOC entry regulation. We addressed those possibilities to study the origin of the SOC entry increase.

The resting calcium levels measured in 2 mM Ca^{2+} aCSF at the beginning of experiments were similar between 5xFAD and WT neurons (Figure 2B). On the other hand, Tg-induced calcium release was significantly lower in the 5xFAD mouse neurons in respect to both amplitude and total calcium release assessed as an area under the curve (AUC). The reduction was estimated at about 25% (−0.24 for amplitude peaks and −0.25 for their

AUC; $p = 0.04$ and 0.03, Mann–Whitney test). Moreover, we found a difference in the stores' sensitivity to Tg as 5xFAD mouse neurons more frequently responded with calcium release. Tg failed to induce calcium release in 65% ($n = 51/79$) and 39% ($n = 61/156$) in WT and 5xFAD neurons correspondingly; Fisher's exact conditional p-value = 0.3E-4. Application of 5 µM ionomycin in Ca^{2+}-free aCSF led to calcium release in both 5xFAD and WT neurons (Figure 2D), showing no statistically significant difference in total intracellular Ca^{2+} store content assessed as an area under the curve (AUC); $p = 0.84$, Mann–Whitney test. Caffeine-induced calcium release, driven by RyRs activation, was similar in its amplitude, AUC, and slope in 5xFAD and WT neurons (Figure 2E), showing the absence of impact on the caffeine-sensitive/RyR-regulated stores; $p = 0.28$ for amplitude and 0.84 for AUC, Mann–Whitney test. We analyzed protein levels of key SOC entry molecular players in hippocampal neurons: channel subunits Orai1, Orai2, TRPC1, and calcium sensors STIM1 and STIM2 [45–47]. We found no difference in the protein levels in total lysates of WT and 5xFAD hippocampal cultures ($p > 0.05$, t-test, Figure 2A). Immunofluorescence of neurons stained with STIM1 or STIM2 antibodies did not reveal any difference in STIM1 or STIM2 protein levels between WT and 5xFAD neurons (Supplementary Figure S1).

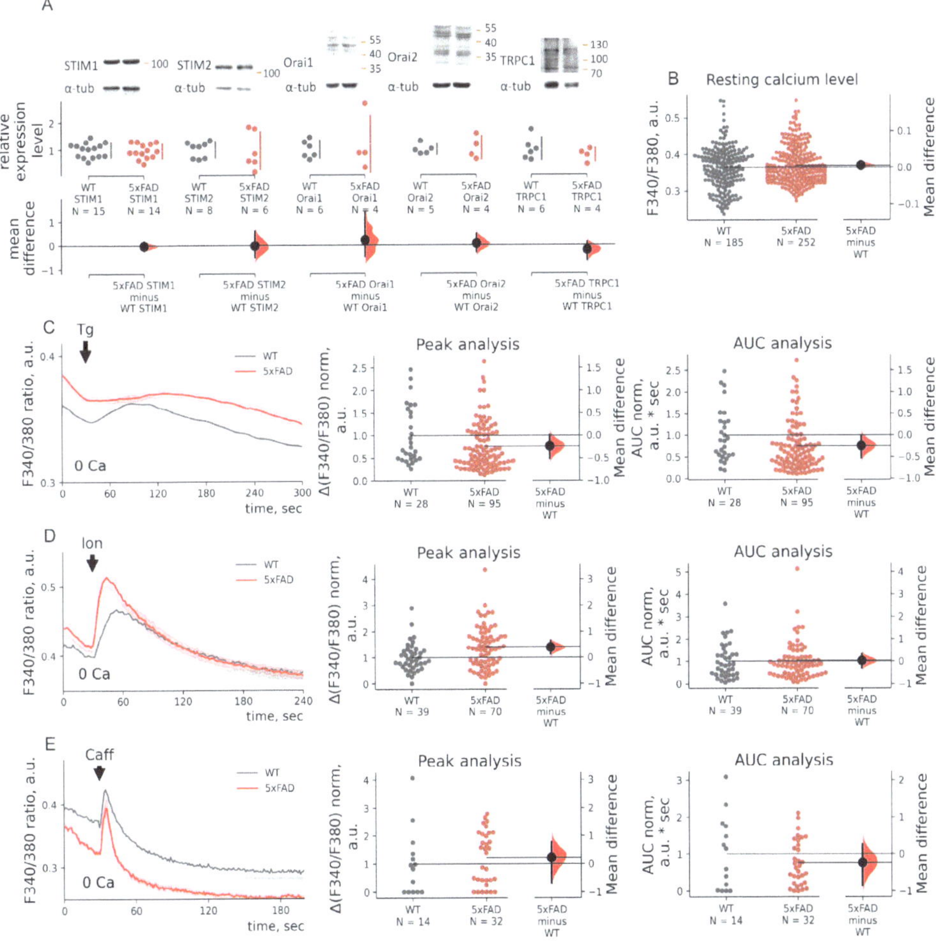

Figure 2. Intracellular basal calcium concentration or calcium store content and STIM, Orai, and TRPC1 protein expression levels are not affected in 5xFAD hippocampal neurons. (**A**) Western blots

with neuronal culture lysates of wild-type (WT) and 5xFAD mice. Relative expression levels were calculated as densitometry of values for the target protein and normalized to α-tubulin values. Data are plotted on the upper axes; WT values were taken as 1; mean differences are plotted on floating axes on the bottom as bootstrap sampling distributions. Normalized data for WT and 5xFAD lysates were compared using unpaired t-test. Mean differences for STIM1 is -0.03 ($p = 0.74$), for STIM2 is ($p = 0.90$), for Orai1 is 0.20 ($p = 0.67$), for Orai2 is 0.05 ($p = 0.80$), and for TRPC1 is -0.21 ($p = 0.38$). (**B**) Resting calcium levels were measured as F340/F380 in neurons in 2Ca aCSF solution before any drugs application. Resting calcium levels in WT and 5xFAD neurons are shown in the estimation plot. Raw F340/F380 data are plotted on the left axes; the mean difference is plotted on floating axes on the right as a bootstrap sampling distribution. The mean difference between basal calcium entry of WT and 5xFAD neurons is 0.006; $p = 0.75$, Mann–Whitney test. (**C–E**) Calcium imaging with Fura2 of neurons of primary hippocampal cultures of wild-type (WT, gray) or 5xFAD (red) mice. Mean traces of F340/380 ratio are represented as mean $\pm$ SEM. Areas under the curve (AUC) are for response peaks after stimulation Ca^{2+} release. ΔF340/F380 and AUC values are normalized by the mean WT values and shown in estimation plots. ΔF340/F380 and AUC values are plotted on the left axes; the mean difference is plotted on floating axes on the right as a bootstrap sampling distribution. (**C**) 1 μM Thapsigargin (Tg) was added to deplete internal calcium stores. We excluded from analysis neurons that did not have Tg-induced calcium release. The normalized mean difference between Tg-induced Ca^{2+}-release peaks of WT and 5xFAD neurons is -0.24; $p = 0.04$, Mann–Whitney test. The normalized mean difference between Tg-induced Ca^{2+}-release AUC values of WT and 5xFAD neurons is -0.25; $p = 0.03$, Mann–Whitney test. (**D**) Five μM ionomycin (Ion) was added to deplete internal calcium stores. The normalized mean difference between Tg-induced Ca^{2+}-release peaks of WT and 5xFAD neurons is 0.39; $p = 0.009$, Mann–Whitney test. The normalized mean difference between Tg-induced Ca^{2+}-release AUC values of WT and 5xFAD neurons was 0.02; $p = 0.08$, Mann–Whitney test. (**E**) An amount of 25 mM caffeine (Caff) was added to evoke caffeine-induced calcium release driven by RyRs. The normalized mean difference between Tg-induced Ca^{2+}-release peaks of WT and 5xFAD neurons is 0.22; $p = 0.28$, Mann–Whitney test. The normalized mean difference between Tg-induced Ca^{2+}-release AUC values of WT and 5xFAD neurons is -0.24; $p = 0.84$, Mann–Whitney test.

2.4. STIM1 Contribution to SOC Entry Is Increased in 5xFAD Mouse Neurons

To determine the role of regulation by STIMs in the excessive SOC entry, we performed a knock-down of STIM1 and STIM2 with shRNAs (STIM1 KD and STIM2 KD). STIM1 and STIM2 protein levels were reduced by the knock-down approximately to 40–50% of the baseline (mock shRNA) (Figure 3B,C). Calcium imaging revealed that neurons of wild-type mice with STIM1 KD had the same level of SOC entry as mock shRNA controls (Figure 3A). Normalized Tg-induced calcium entry amplitudes (ΔF340/F380) were 1.00 $\pm$ 0.07 for WT neurons and 1.15 $\pm$ 0.10 for WT STIM1 KD neurons ($p = 1$, Dunns' test). On the other hand, the STIM1 KD in 5xFAD neurons decreased SOC entry by 66%, leading to a lower SOC entry than in WT neurons. Normalized Tg-induced calcium entry amplitudes (ΔF340/F380) were 1.42 $\pm$ 0.05 for 5xFAD neurons and 0.49 $\pm$ 0.09 for 5xFAD STIM1 KD neurons ($p <$ 0.001, Dunns' test, Figure 3A). Meanwhile, STIM2 KD decreased SOC entry in both WT and 5xFAD neurons to comparable levels. ΔF340/F380 normalized data for WT STIM2 and 5xFAD STIM2 were 0.48 $\pm$ 0.08 and 0.82 $\pm$ 0.09, and p-values of the Dunns' test were <0.001 compared to WT and 5xFAD neurons (Figure 3A). The data indicates a dominant role of STIM1 contribution to SOC entry in 5xFAD hippocampal neurons.

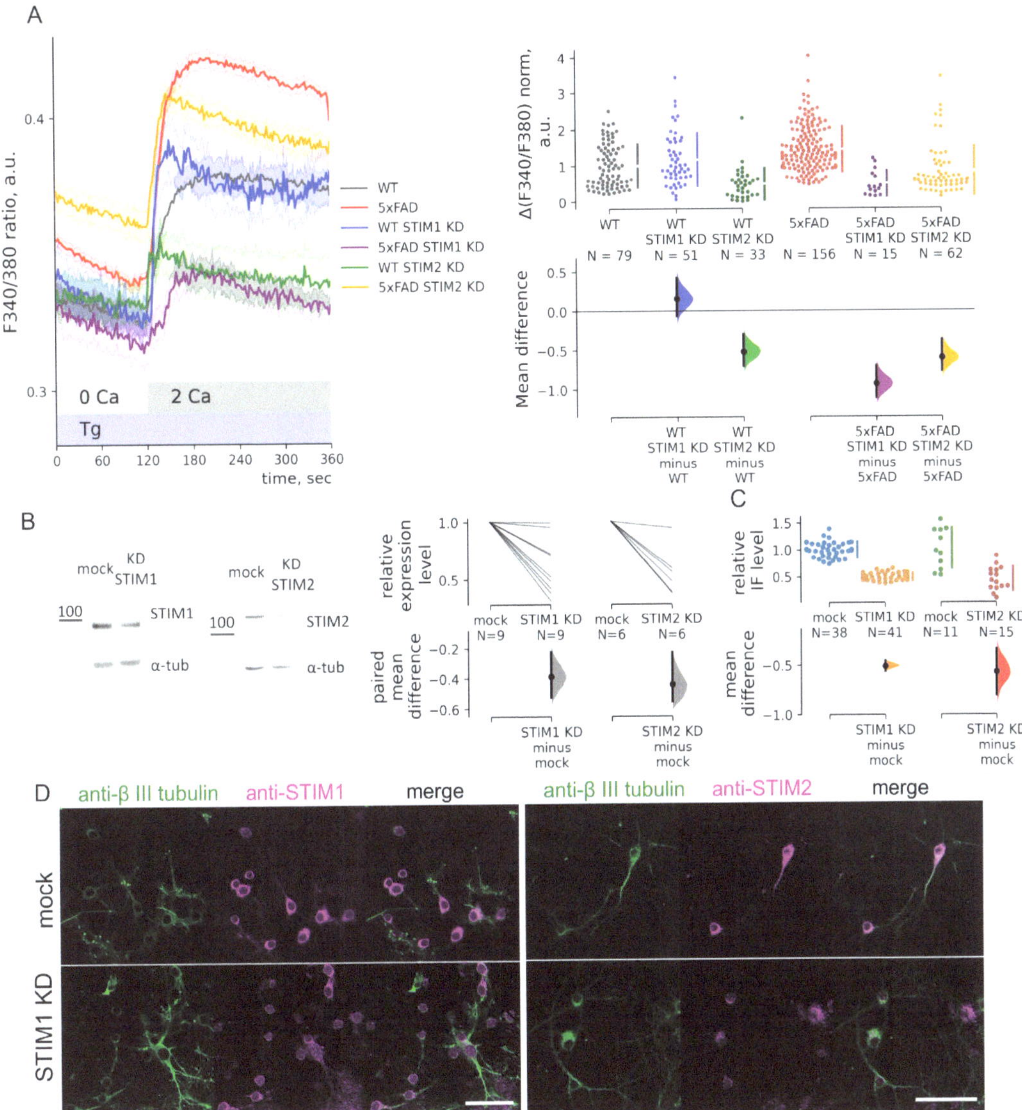

Figure 3. Store-operated calcium entry regulation by STIM1/2 proteins is changed in 5xFAD hippocampal neurons. (**A**) Calcium imaging with Fura2 of neurons of primary hippocampal cultures of wild-type (WT, gray) or 5xFAD (red) mice and treated with STIM1/STIM2 shRNA lentiviruses (STIM1 KD and STIM2 KD) or mock shRNA lentivirus. Mean traces of F340/380 ratio are represented as mean ± SEM. Thapsigargin (Tg) was added to deplete internal calcium stores. Calcium entry (ΔF340/F380) was calculated as the F340/F380 peak of each trace after Ca^{2+} add-back, and the baseline was subtracted. ΔF340/F380 data are normalized by the mean WT calcium entry and shown in the estimation plot. ΔF340/F380 data are plotted on the upper axes; mean differences are plotted on floating axes on the bottom as bootstrap sampling distributions. The Kruskal–Wallis ANOVA ($p < 0.0001$) with post-hoc Dunn's test was implemented; pairwise p-values were calculated. Mean differences between calcium entry in WT and WT STIM1 KD or WT STIM2 KD neurons are 0.15 and -0.52; $p = 1.0$ and 2.3×10^{-4} compared to WT neurons. The mean difference between calcium entry in 5xFAD and 5xFAD STIM1 KD or 5xFAD STIM2 KD neurons are -0.94 and -0.61; $p = 1.8 \times 10^{-7}$

and 6.4×10^{-11} compared to 5xFAD. (**B**) Western blots with neuronal culture lysates. Neurons were treated with STIM1/STIM2 shRNA lentiviruses (STIM1 KD and STIM2 KD) or mock shRNA lentivirus (mock). Relative expression levels were computed as STIM1/2 densitometry values and normalized to α-tubulin values. Data pairs are plotted on the upper axes; mock values were taken as 1; mean differences are plotted on floating axes on the bottom as a bootstrap sampling distribution. The paired mean difference between mock and STIM1 KD or STIM2 KD lysates are -0.39 and -0.44 ($p = 0.001$ and 0.003, paired t-test). (**C,D**) Confocal immunofluorescence images of neuronal cultures probed with anti-STIM1/2 (magenta, anti-STIM1 and anti-STIM2) and β-III neuro-tubulin antibodies (green, anti-β III neuro-tubulin). Scale bar is 50 μm. Integrated immunofluorescence for anti-STIM1/2 staining was calculated for every neuron counterstained with anti-β III neuro-tubulin. Datasets were normalized to mock shRNA lentivirus values (mock) to determine relative immunofluorescence (IF) levels. Normalized data are plotted on the upper axes; mean differences are plotted on floating axes on the bottom as a bootstrap sampling distribution. Mean differences in relative immunofluorescence levels between mock and STIM1 KD or STIM2 KD neurons are -0.51 and -0.52 ($p < 0.001$, t-test).

2.5. Depolarization-Induced Calcium Entry Is Enhanced in 5xFAD Neurons

To address the effect of endogenous STIMs on VGC entry in 5xFAD mice hippocampal neurons, we induced depolarization of the cells by applying high K+ aCSF (50 mM KCl). Depolarization-induced calcium entry was elevated by 20% compared to WT (1.00 ± 0.04 and 1.19 ± 0.05, $p < 0.001$ Mann–Whitney test, Figure 4A). Calcium stores depletion with 1 mM chelator TPEN reduced amplitudes of KCl-induced calcium entry to the same level in WT and 5xFAD neurons. The store depletion reduced VGC entry in both groups of neurons, with the 5xFAD mouse neurons decreasing the AUC of the response peak by 41% compared to the replete state (0.93 ± 0.04, $p < 0.001$, Dunn's test), while WT mice show a reduction by 27% compared to WT replete state (0.73 ± 0.04, $p = 0.036$, Dunn's test). These data pointed out that the increase in VGC entry in 5xFAD neurons is due to the store-sensitive VGC component. At the same time, the total calcium entry after the stores depletion, assessed as AUC, remained significantly elevated in 5xFAD neurons (1.00 ± 0.02 for WT and 1.28 ± 0.05 for 5xFAD; $p = 0.006$, Mann–Whitney test, Figure 4B). Therefore, the part of the difference in VGC entry was not dependent on ER calcium release or ER calcium content and persisted in the store-depleted condition.

The excessive VGC entry in 5xFAD neurons was mainly sensitive to the L-type VGC channel inhibitor nifedipine (10 μM) but not to the T-type channel inhibitor ML218 (Figure 4C). Nifedipine reduced amplitude by 37% in 5xFAD neurons (0.73 ± 0.06 $p < 0.001$, Dunn's test). KCl-induced calcium entry in control neurons was sensitive to both L- and T-type VGC channel inhibitors. Nifedipine reduced calcium entry amplitude by 26% (0.74 ± 0.04, $p < 0.001$, Dunn's test), and ML218 by 15% (0.85 ± 0.04, $p = 0.01$, Dunn's test).

STIM1 KD did not impact the KCl-induced calcium entry peak amplitudes in hippocampal neurons from WT or 5xFAD animals. Normalized calcium entry amplitudes were 1.09 ± 0.08 for WT STIM1 KD neurons and 1.28 ± 0.08 for 5xFAD STIM1 KD neurons, $p > 0.5$, Dunn's test (Figure 5A,B). However, STIM2 KD enhanced calcium entry in both WT and 5xFAD neurons (Figure 5A,B). Furthermore, in neurons from 5xFAD mice, STIM2 KD induced both higher peaks and long-lasting elevation of intracellular calcium increase after depolarization. Normalized calcium entry amplitude was 1.38 ± 0.05 ($p = 0.005$, Dunn's test), and AUC value was 1.44 ± 0.05 ($p < 0.001$, Dunn's test). The data imply that STIM2 regulates depolarization-induced VGC entry in hippocampal neurons with replete stores.

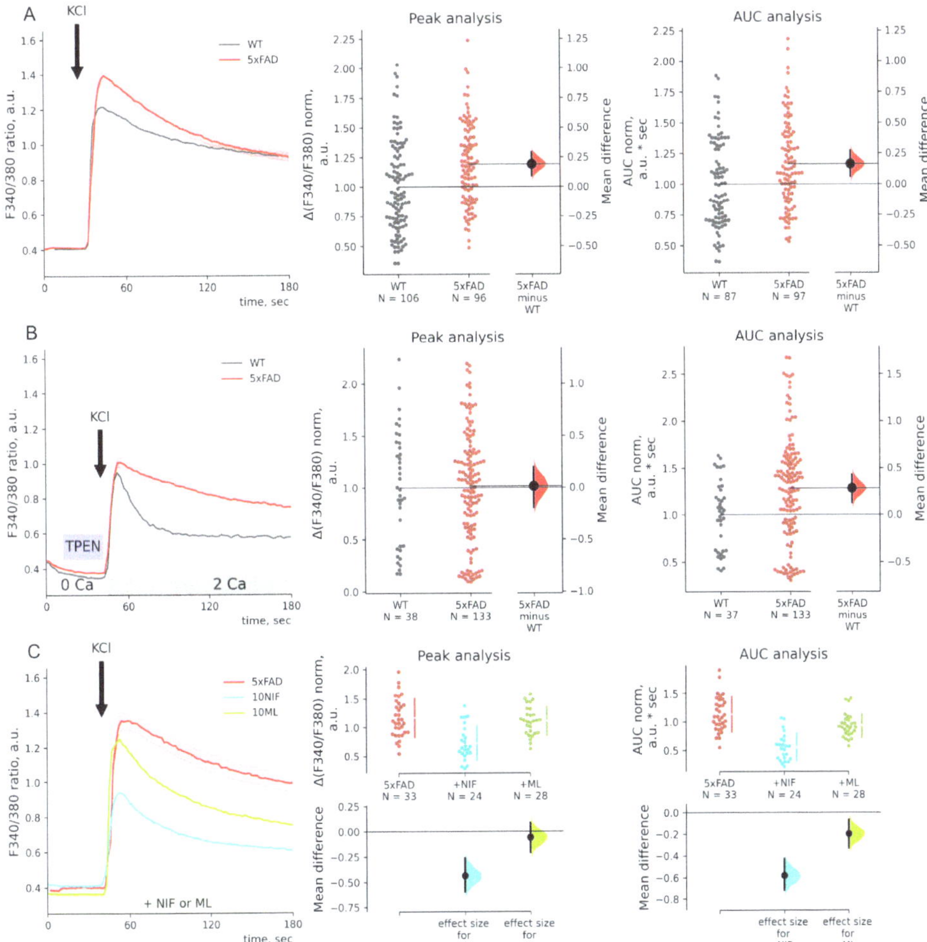

Figure 4. In hippocampal neurons, voltage-gated calcium entry is sensitive to intracellular calcium stores. (**A–C**) Calcium imaging with Fura2 of neurons of primary hippocampal cultures of wild-type (WT, gray) and 5xFAD mice. Mean traces of F340/380 ratio are represented as mean ± SEM. 50 mM KCl (KCl) was added to induce depolarization. Calcium entry (ΔF340/F380) was calculated as the F340/F380 peak of each trace after KCl application, and the baseline was subtracted. Areas under the curve (AUC) were calculated for 40 s after KCl applications. ΔF340/F380 and AUC data are normalized by the mean WT neuron values and shown in separate estimation plots. The estimation plots have normalized data plotted on the upper axes and mean differences plotted on floating axes on the bottom as a bootstrap sampling distribution. (**A**) The mean difference between KCl-induced calcium entry in WT and 5xFAD neurons is 0.19 for amplitudes and 0.17 for AUC values; $p = 1.6 \times 10^{-4}$ and 0.002, Mann–Whitney test. (**B**) Calcium stores were depleted with 1 mM chelator TPEN for 20 s before KCl application. The mean difference between KCl-induced calcium entry in WT and 5xFAD neurons is 0.01 for amplitudes and 0.28 for AUC values; $p = 0.98$ and 0.006, Mann–Whitney test. (**C**) Inhibitory analysis of depolarization-induced calcium entry in 5xFAD neurons was done with the Kruskal–Wallis ANOVA ($p < 0.001$ for peaks and AUC values) with post hoc Dunn's test was implemented; pairwise p-values were calculated for untreated 5xFAD neurons. Mean differences between KCl-induced calcium entry in 5xFAD neurons and 10 μM nifedipine or ML218-treated 5xFAD neurons are -0.44 and -0.06 ($p = 3.2 \times 10^{-5}$ and 1.0) for amplitudes and -0.58 and -0.20 ($p = 1.9 \times 10^{-9}$ and 0.10) for AUC values.

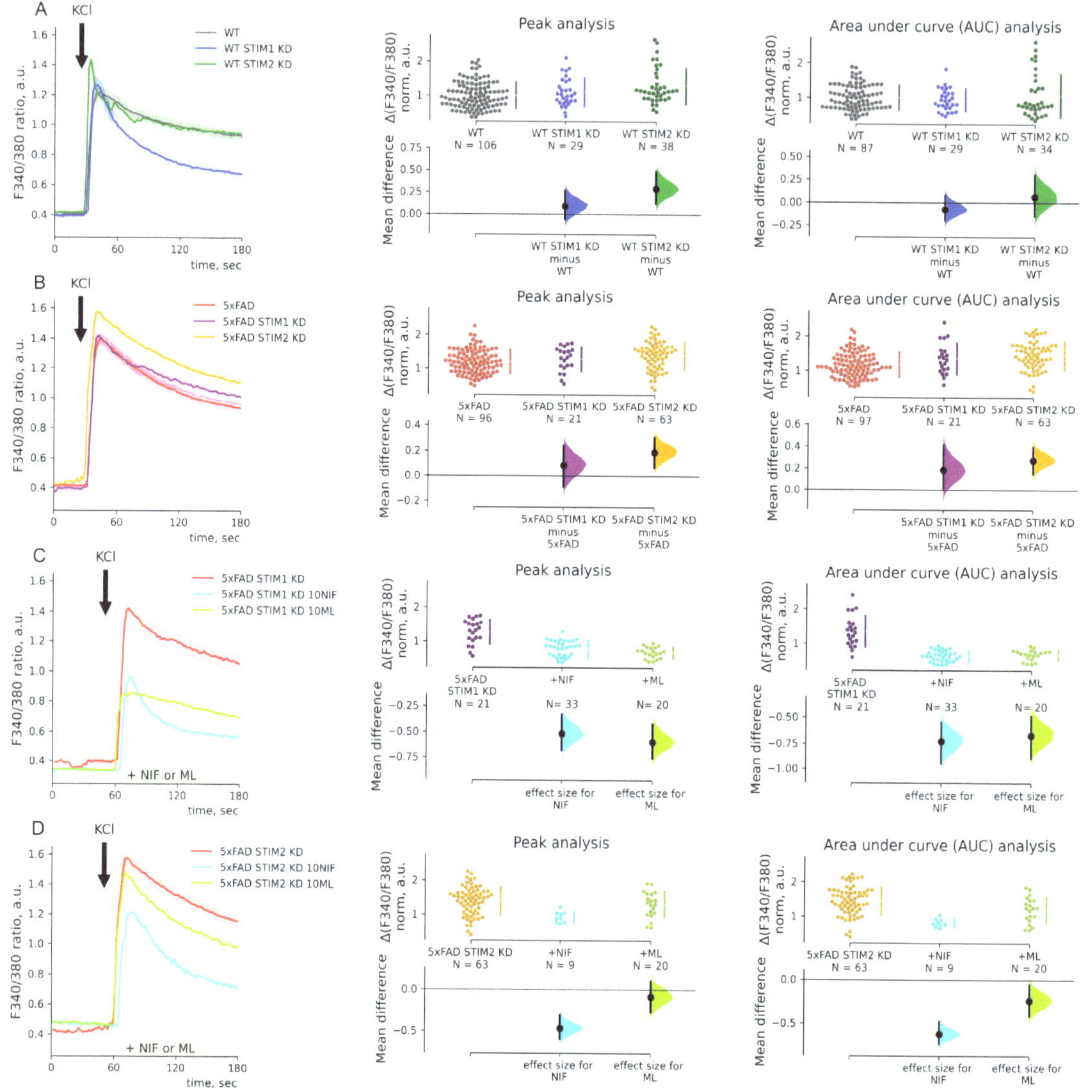

Figure 5. In hippocampal neurons, voltage-gated calcium entry is regulated by STIM2 sensors. (**A–D**) Calcium imaging with Fura2 of neurons of primary hippocampal cultures of wild-type (WT, gray) and 5xFAD mice and treated with STIM1/STIM2 shRNA lentiviruses (STIM1 KD and STIM2 KD) or mock shRNA lentivirus. Inhibitory analysis has been performed with 10 μM nifedipine (NIF), an inhibitor of L-type calcium channels, and 10 μM ML218 (ML), an inhibitor of T-type calcium channels. The inhibitors were introduced 1 min before KCl application. Mean traces of F340/380 ratio are represented as mean ± SEM. 50 mM KCl (KCl) was added to induce calcium entry. Calcium entry (ΔF340/F380) was calculated as the F340/F380 peak of each trace after KCl application, and the baseline was subtracted. Areas under the curve (AUC) were calculated for 40 s after KCl applications. ΔF340/F380 and AUC data are normalized by the mean WT neuron values and shown in separate estimation plots. The estimation plots have normalized ΔF340/F380 data plotted on the upper axes and mean differences plotted on floating axes on the bottom as bootstrap sampling distributions. (**A**) The Kruskal–Wallis ANOVA ($p < 0.01$ for peaks and AUC values) with post hoc Dunn's test was implemented; pairwise p-values were calculated for WT values comparison. Mean differences between KCl-induced calcium entry in WT and WT STIM1 KD or WT STIM2 KD neurons are 0.09

and 0.29 with $p = 0.91$ and 0.005 for amplitudes and -0.07 and 0.06 with both $p = 1.0$ for AUC values. (**B**) The Kruskal–Wallis ANOVA ($p < 0.001$ for peaks and AUC values) with post hoc Dunn's test was implemented; pairwise p-values were calculated for 5XFAD values comparison. Mean differences between KCl-induced calcium entry in 5xFAD and 5xFAD STIM1 KD or 5xFAD STIM2 KD neurons are 0.08 and 0.19 ($p = 0.78$ and 0.005) for amplitudes and 0.18 and 0.28 ($p = 0.26$ and 4.8×10^{-5}) for AUC. (**C**) The Kruskal–Wallis ANOVA ($p < 0.001$ for peaks and AUC values) with post hoc Dunn's test was implemented; pairwise p-values were calculated for 5xFAD neurons with STIM1 KD comparison. Mean differences between KCl-induced calcium entry in 5xFAD STIM1 KD neurons and 10 µM nifedipine or ML218 treated 5xFAD STIM1 KD neurons are -0.52 and -0.60 ($p = 4.4 \times 10^{-5}$ and 0.3×10^{-5}) for amplitudes and -0.73 and -0.67 ($p = 6.6 \times 10^{-9}$ and 2.1×10^{-5}) for AUC values. (**D**) The Kruskal–Wallis ANOVA ($p < 0.004$ for peaks and AUC values) with post hoc Dunn's test was implemented; pairwise p-values were calculated for 5xFAD neurons with STIM2 KD comparison. Mean differences between KCl-induced calcium entry in 5xFAD STIM2 KD neurons and 10 µM nifedipine or ML218 treated 5xFAD STIM2 KD neurons are -0.47 and -0.08 with $p = 0.002$ and 1.0 for amplitudes and -0.62 and -0.24 with $p = 8.2 \times 10^{-5}$ and 0.09 for AUC values.

Experiments with pharmacological inhibitors of the VGC channels demonstrated that around 50% of KCl-induced calcium entry amplitude was sensitive to an L-type calcium channel inhibitor (10 µM nifedipine) in 5xFAD neurons with STIM1 or STIM2 KD, which was not different from neurons with mock shRNA or untreated 5xFAD neurons (the Kruskal–Wallis ANOVA is $p > 0.05$). The nifedipine reduced the calcium entry amplitudes by 41% in 5xFAD with STIM1 KD neurons (0.76 ± 0.05, $p < 0.001$, Dunn's test, Figure 5C) and by 33% in 5xFAD neurons with STIM2 KD (0.92 ± 0.05, $p < 0.001$, Dunn's test, Figure 5D). Surprisingly, the STIM1 KD in 5xFAD neurons affected the sensitivity of VGC entry to the inhibitor of T-type calcium channels (10 µM ML218), leading to VGC entry being blocked by the inhibitor by 46% compared to 5xFAD mock-treated neurons (0.68 ± 0.04, $p < 0.001$, Dunn's test). The STIM2 KD or mock shRNA did not affect the sensitivity of VGC entry to ML-218; amplitudes were 1.10 ± 0.09 for mock shRNA and 1.30 ± 0.08 for STIM2 KD 5xFAD neurons; $p > 0.05$, Dunn's test, Figure 5C,D.

3. Discussion

3.1. STIM1 Drives Increased SOC Entry in Hippocampal Neurons of the 5xFAD Mouse Model

Several studies have demonstrated diverse effects on SOC entry in neurons of AD animal models, AD patient cells, and other cellular models (reviewed in [12,48]). A variety of mechanisms have been proposed to explain those effects: changes in expression levels of STIM proteins [26,40,49], changes in ER calcium storage and/or release [8,10,50,51], and enzymatic cleavage of STIM1 by gamma-secretase [52]. We studied early cultures (10–13 day in vitro) of the hippocampus of 5xFAD mice with human *APP* and *PSEN1* transgenes driven by a neuron-specific Thy1 promoter. Thy1 drives expression of transgenes in the hippocampus of E17.5 mouse embryos [53] and in primary cultures of neurons as early as 4–5 days in vitro. Cultured hippocampal 5xFAD neurons at 10–13 days in vitro exhibit neither severe amyloid pathology, due to limited time to accumulate human APP and PS1 transgenes, nor detectable level of amyloids [6]. Meanwhile, changes in calcium signaling were demonstrated already at this stage in cultured neurons of mouse transgenic AD models such as 3xTg and PS2(N141I) [27,54]. Here we observed for the first time elevated SOC entry in the neurons of 10–13 day in vitro primary cultures from the hippocampus of 5xFAD newborn mice. Both STIM homologs were involved in the elevated SOC entry. At the same time, the contribution of STIM1 to the somatic SOC entry in 5xFAD hippocampal neurons was more pronounced compared to neurons of WT mice. The effect was not associated with changes in protein levels of STIM1, STIM2, Orai1, Orai2, and TRPC1. The ER store content, estimated as ionomycin (by AUC) or caffeine-induced calcium release, was not different between 5xFAD and WT hippocampal neurons. At the same time, Tg-induced calcium release was decreased in the 5xFAD neurons, while neurons were more readily responsive to the Tg application. Therefore, the difference in SOC entry was not caused by changes in the bulk calcium in the ER stores or by changes in the levels of key SOC entry proteins. We speculate that ER stores in the 5xFAD mouse hippocampus have a higher sensitivity for Tg-induced calcium release, which increases the contribution of STIM1 in SOC entry regulation. This is in line with previous studies [31,55–57] where the passive leak, enhanced calcium store release, or STIM1 contributed to the disruption of calcium homeostasis in the neurons of AD models. The contribution of STIM1 and STIM2 splice isoforms is yet to be investigated [58,59].

Previously, we have demonstrated elevation of somatic STIM1-induced SOC entry in mouse hippocampal neurons expressing human presenilin-1 ΔE9 associated with FAD [31]. Later the enhancement was confirmed in dendritic spines of presenilin-1 ΔE9 expressing neurons [33]. However, several other AD models, including 3xTg, PS1-M146V knock-in and APP knock-in transgenic mice, demonstrated diminished SOC entry in soma and dendrites of neurons [26,27,29,41]. The difference in SOC entry between AD models could be an effect of different AD pathological mutations and variations in somatic/dendritic measurements. Pharmacological modulation of the SOC entry has been proposed as a strategy for treating neurodegenerative diseases (reviewed in [12,48]). For example, SOC entry inhibitor EVP4593 was shown to restore mushroom spines in mouse hippocampal neurons expressing presenilin-1 ΔE9 [33]. A less specific SOC entry inhibitor 2-APB was able to restore short-term memory in the Drosophila model of FAD [31] and restore LTP in the hippocampus of the FAD mouse model expressing presenilin-1 ΔE9 [60]. Here we demonstrated that elevated SOC entry in 5xFAD mouse hippocampal neurons was sensitive to the FDA-approved immunomodulatory drug leflunomide. The drug has been studied as an inhibitor of Orai1 and TRPC-dependent SOC entry at clinically relevant doses but has never been tested in neuronal cultures [43]. Pharmacological data can be useful for the future understanding of the drug's action and side effects.

3.2. STIM2 Regulates VGC Entry in 5xFAD Neurons with Replete ER Calcium Store

Several studies suggested the contribution of VGC channels in the pathology of AD based on post-mortem samples and animal models [61–65]. Here we observed the elevation of VGC entry in hippocampal neurons of 5xFAD mice in response to depolarization with 50 mM KCl. Increased VGC entry was sensitive to L-type channels inhibitor nifedipine but not to T-type channels inhibitor ML-218. The data is in line with previously observed elevated VGC entry in 3xTG and cellular AD models [40,63]. Both store-sensitive and insensitive VGC channels (defined by the sensitivity to store depletion with TPEN) contributed to the elevated calcium entry after depolarization.

It has been reported earlier that STIM1 acts as a negative regulator of endogenous or overexpressed L-type calcium channels upon ER calcium store depletion in arterial smooth muscle cells (A7r5 VSMCs), cortical neurons, Jurkat T cells, HEK293 cells, and neuroblastoma Neuro2a [36,37]. The same effect upon SR store depletion has been reported for the endogenous T-type calcium channels in HL-1 cells [38]. The purported mechanism involves a reciprocal regulation of SOC and VGC channels by STIM1 upon store depletion and receptor-induced calcium release [36,37,66]. Although the STIM1 KD did not affect the VGC entry in VSMCs at a steady state (without store depletion), the STIM1-KO was reported to increase VGC entry in differentiated human neuroblastoma SH-SY5Y under the same conditions [40,67]. In contrast to STIM1, its ortholog STIM2 was described as a weaker negative regulator of voltage-gated calcium channels when it was overexpressed in VSMCs [37]. STIM2 KD in a VSMC line and human iPSC-derived GABAergic neurons did not affect the VGC entry with replete ER stores [67,68]. Here we studied endogenous VGC channel activation in hippocampal neurons without internal calcium stores pre-depletion. Our data demonstrated an absence of the effect of STIM1 KD on the VGC entry induced by depolarization with 50 mM KCl. At the same time, we observed an increase in VGC entry due to STIM2 KD. STIM2 is known to regulate steady-state calcium entry and could be active in resting neurons, as it was shown for hippocampal and cortical neuronal cultures [34,35,69]. The influence of overexpressed STIM1 on VGC entry was increased by STIM1 association with the plasma membrane. STIM2 proteins have a more robust association with the plasma membrane under resting conditions [70], which probably facilitates VGC channel regulation by STIM2. Interestingly, depolarization-induced calcium entry in neurons with STIM1 KD is more sensitive to ML-218 inhibitors of T-type VGC channels. Further investigation is required to reveal the mechanisms of VGC channel regulation by STIMs.

Overall, our data demonstrate the changes in calcium signaling of hippocampal neurons of the AD mouse model, which precede amyloid plaque accumulation or other signs of pathology manifestation. Activity of both store-operated and voltage-gated calcium channels were affected in 5xFAD mouse neurons. Excessive store-operated calcium entry in 5xFAD is mediated by STIM1 and is sensitive to the FDA-approved drug leflunomide. Moreover, for the first time, we showed that STIM2 regulates voltage-gated calcium entry in hippocampal neurons. The data obtained could help to develop a strategy for treatment at the early stage of AD.

4. Materials and Methods

4.1. Animals

Hemizygous 5xFAD mice, their non-transgenic littermates, and C3HA mice of both sexes were used. Non-transgenic littermates were taken as controls and referred to as wild-type (WT). The 5xFAD transgenic mouse model is a double transgenic model with five FAD-associated mutations: Swedish (K670N, M671L), Florida (I716V), and London (V717I) mutations in the *APP* gene, and M146L and L286V mutations in the *PSEN1* gene under the neuron-specific Thy1 promoter [2]. Transgenic mice were acquired from the Jackson Laboratory and maintained on a mixed SJL/C57Bl6. All animal research was carried out under the guidelines for the welfare of animals of the ethical committee of the Institute of Cytology, Russian Academy of Sciences (No. F18-00380, approved on 12 October 2017). The animals were maintained in their home cages in a climate-controlled room (21–23 °C) with a 12:12 h light–dark cycle and had free access to water and food. 5xFAD mice were genotyped according to Jackson Laboratory protocol (№31769) with polymerase chain reaction (PCR) analysis of biopsies: ear for adult animals or cortical brain fragments for pups. The primers were: CGG GCC TCT TCG CTA TTA C (Chr3 Mutant Reverse), ACC CCC ATG TCA GAG TTC CT (Chr3 Common), TAT ACA ACC TTG GGG GAT GG (Chr3 WT Reverse). PCR from hemizygous animals produced two bands. Heterozygous and wild-type yielded a single band at the size of 129 and 216 bp, respectively. Homozygotes were excluded from the experiments.

4.2. Cell Cultures

Cell cultures were maintained at 37 °C with 5% CO_2 in a humidified incubator. Human embryonic kidney 293T (HEK293T) cells were cultured in Dulbecco's Modified Eagle Medium (DMEM) with 10% fetal bovine serum (FBS) and penicillin/streptomycin. Mouse primary cultures of hippocampal neurons were prepared as described earlier [71]. Shortly, the hippocampi of newborn mice were dissected, digested with trypsin, and seeded onto poly-L-lysine treated coverslips or Petri's dishes in Neurobasal-A medium with 3% FBS and 3% B27 supplement. We took a maximum of 8 pups from one litter for parallel hippocampal cultures. Tissues of a pup yielded one hippocampal culture in a Petri dish with 10–14 small (4 mm × 4 mm) coverslips for calcium imaging or 4–5 coverslips for immunostaining. If required, coverslips were split to different dishes for lentiviral infections or drug treatments. Experiments were performed on in vitro days 10–13.

4.3. Lentiviral Infection

Some cultures were infected with lentiviruses five days before experiments. Lentiviruses were generated by transfecting HEK293T with polyethyleneimine and a mixture of three plasmids: envelope vector pMD2.G, packaging vector psPAX2 (AddGene #12259 and

#12260 were gifts from Didier Trono), pGFP-C-shLenti vector (OriGene, Rockville, MD, USA, TR30021, referred to as mock shRNA), shRNA STIM1 or shRNA STIM2 vectors (a kind gift from prof. I. Bezprozvanny, Southwestern. Medical Center, Dallas, TX, USA). The supernatant with viruses was harvested 48 and 72 h after transfection, filtered, and concentrated by centrifugation (47,000 *g*, 2 h, 4 °C). Pellets were resuspended in NeuroBasal-A, aliquoted, and stored at −80 °C. Transduction efficacy was confirmed by western blotting and immunostaining of primary hippocampal neuron cultures of C3Ha mice.

4.4. Western Blotting

Total protein lysates were separated by 8–10% SDS-PAGE and transferred onto the nitrocellulose membrane. Membranes were blocked with 5–10% bovine serum albumin (BSA) and probed with antibodies. Antibodies raised against α-tubulin (Sigma, St. Louis, MO, USA, T9026, 1:1000), STIM1 (Cell Signaling, Danvers, MA, USA, №5668, 1:1000), STIM2 (Cell Signaling №4917, 1:1000), Orai1 (Sigma SAB4200273, 1:500), Orai2 (ProSci, Poway, CA, USA, №4111, 1:300), TRPC1 (Alomone, Jerusalem, Israel, ACC-010, 1:200) were used as primary antibodies; anti-mouse and anti-rabbit horseradish peroxidase-labeled antibodies were used as secondary antibodies (Sigma A0168 and A0545, 1:30,000). Signal was detected with SuperSignal West Femto Maximum Sensitivity Substrates and imaged with ChemiDoc Imaging System. Densitometry values were determined using Image Lab software. Target protein values were normalized to α-tubulin values.

4.5. Immunostaining

The modified AbCam protocol for immunostaining (www.abcam.com, accessed on 22 July 2019) was used. Cells on coverslips were fixed with 4% formaldehyde for 10 min, permeabilized with 0.25% TritonX-100 for 10 min, blocked with 10% FBS and 0.3M glycine in PBST (PBS with 0.1% Tween20) for 1 h, and subsequently incubated with antibodies in 1% BSA in PBST for 1 h. Primary antigen-specific antibodies included: STIM1 (Cell Signaling №5668, 1:800), STIM2 (Cell Signaling №4917, 1:50), and neuron-specific β-tubulin III (Sigma T8578, 1:500). Secondary antibodies were Alexa Fluor 555 anti-rabbit (Thermo Fisher Scientific, Waltham, USA, A-21428, 1:500) and Alexa Fluor 488 anti-mouse (Thermo Fisher Scientific A-11001, 1:2000). Nuclei were counterstained with DAPI in the mounting medium. Images were obtained with an IX83 microscope and Olympus FV3000 confocal laser system using an Olympus 40x objective (oil, 1.3 NA); the pinhole was one Airy disk. Fluorescence was excited by 405, 488, and 561 nm laser lines and detected. Representative z-stack images were captured using Olympus FluoView software (v2.1.1.98) and analyzed with FiJi software (v1.53f51) [72].

4.6. Calcium Imaging

Cells were loaded with 5 μM Fura-2 AM in 0.2 mM probenecid) in artificial cerebrospinal fluid (aCSF) for 55 min at 37 °C. aCSF contained (in mM): 140 NaCl, 5 KCl, 2 $CaCl_2$, 1 $MgCl_2$, 10 HEPES, 10 glucose; pH 7.3. The cytosolic Ca^{2+} concentration change was determined as the ratio of Fura-2 fluorescence intensities at 340 and 380 nm excitation wavelengths (F340/F380) with emission at 510 nm. Cells were imaged by an Andor's Zyla 4.2 sCMOS camera driven by MicroManager Software [73]. Images were analyzed with FiJi [72]. F340/F380 curves are represented as the means with a standard error of the mean (SEM). Resting calcium levels were measured in 2 mM Ca^{2+} aCSF solution before any treatment. Calcium entry was calculated as the F340/F380 peak of each trace, and the baseline was subtracted (ΔF340/F380). Before internal calcium store depletion, cells were kept in Ca^{2+}-free aCSF for 30 sec. Internal calcium stores were depleted in Ca^{2+}-free aCSF solution with an addition of 1 μM thapsigargin (Tg; Sigma, T9033), or 5 μM ionomycin (Ion; Sigma I0634) or 25 μM caffeine (Sigma, C0750), supplemented with 1 μM nifedipine (Sigma, N7634) and 1 μM tetrodotoxin (TTX, Tocris, 1069). SOC entry was measured after Ca^{2+}-readdition. Ca^{2+}-free aCSF had 100 nM of free Ca^{2+} (MaxChelator calculation) and contained (in mM) 47.4 nM CaCl2 and 100 μM EGTA in aCSF. Some cells were pretreated for 20 h with 1 µM teriflunomide (Sigma SML0936) or leflunomide (Sigma, L5025). VGC entry was activated with a depolarizing solution containing (in mM) 50 KCl and 97.5 NaCl in aCSF (50 KCl aCSF). One mM of TPEN (Sigma, P4413) chelator in Ca^{2+}-free aCSF was used to mimic store depletion. Calcium entry through L-type and T-type VGC channels was blocked by the application of 10 µM nifedipine or 10 µM ML218 (Sigma, SML0385) 1 min before depolarization with 50 KCl. A calcium imaging experiment yielded 3 to 30 individual neuron records from one coverslip. All the experiments were done blind to the genotype, and genotyping was done after gathering the data. Therefore, all the data sets represent all the data points obtained based on the genotyping results. No data points were removed. At least 3 animals per group were used in data sets.

4.7. Statistics

Statistical analysis was performed with OriginPro8.1 and Python. We analyzed datasets with estimation plots [74]. Estimation plots were created with a package for Data Analysis using Bootstrap-Coupled ESTimation (DABEST). Each estimation plot contains a scatter plot with raw data normalized by the control group and descriptive statistics for mean differences with effect sizes, sampling distributions, 95% confidence intervals (the vertical error bars). A total of 5000 bootstrap samples were taken from each group to draw sampling distributions. The Shapiro–Wilk test was applied to determine the data distribution to choose between parametric or non-parametric tests for further analysis. Unpaired and paired *t*-tests were applied for the comparison of Western blotting and immunofluorescence data. The Mann–Whitney test or non-parametric ANOVA (Kruskal–Wallis test) with post hoc Dunn's test with Bonferroni correction was applied for calcium data analysis. Both Mann–Whitney and Dunn's were used as non-parametric tests that do not assume normal distributions. Dunn's test was used for non-parametric pairwise multiple comparisons [75]. Fisher's exact test was used to assess frequency difference. p values below 0.05 were considered significant. Mean, SEM, and statistical data are presented in Supplementary Tables S1–S14.

Supplementary Materials: The following supporting information can be downloaded at: https://www.mdpi.com/article/10.3390/ijms232314810/s1.

Author Contributions: Conceptualization, M.R., K.S., and E.K.; Methodology, K.S. and M.R.; Validation, K.S. and M.R.; Formal Analysis, K.S.; Investigation, K.S. and E.M.; Resources, E.K.; Data Curation, K.S.; Writing—Original Draft Preparation, K.S., A.S., and M.R.; Writing—Review and Editing, K.S., A.S., and M.R.; Visualization, K.S.; Supervision, M.R. and E.K.; Project Administration, E.K. and I.G.; Funding Acquisition, E.K and I.G. All authors have read and agreed to the published version of the manuscript.

Funding: This research was funded by the Grant from the Ministry of Science and Higher Education of Russia—Research Project N 075-15-2020-795, local identifier 13.1902.21.0027.

Institutional Review Board Statement: The study was conducted according to the guidelines for the welfare of animals of the ethical committee of the Institute of Cytology, Russian Academy of Sciences (No. F18-00380, approved on 12 October 2017).

Informed Consent Statement: Not applicable.

Data Availability Statement: The raw and analyzed datasets are available on request to the corresponding author.

Conflicts of Interest: The authors declare no conflict of interest.

References

1. Bekris, L.M.; Yu, C.-E.; Bird, T.D.; Tsuang, D.W. Genetics of Alzheimer Disease. *J. Geriatr. Psychiatry Neurol.* **2010**, *23*, 213–227. [CrossRef]
2. Oakley, H.; Cole, S.L.; Logan, S.; Maus, E.; Shao, P.; Craft, J.; Guillozet-Bongaarts, A.; Ohno, M.; Disterhoft, J.; Eldik, L.V.; et al. Intraneuronal Beta-Amyloid Aggregates, Neurodegeneration, and Neuron Loss in Transgenic Mice with Five Familial Alzheimer's Disease Mutations: Potential Factors in Amyloid Plaque Formation. *J. Neurosci.* **2006**, *26*, 10129–10140. [CrossRef]

3. Ghoweri, A.O.; Ouillette, L.; Frazier, H.N.; Anderson, K.L.; Lin, R.-L.; Gant, J.C.; Parent, R.; Moore, S.; Murphy, G.G.; Thibault, O. Electrophysiological and Imaging Calcium Biomarkers of Aging in Male and Female 5×FAD Mice. *J. Alzheimers Dis.* **2020**, *78*, 1419–1438. [CrossRef]

4. Reitz, C. Alzheimer's Disease and the Amyloid Cascade Hypothesis: A Critical Review. *J. Alzheimers Dis.* **2012**, *2012*, 369808. [CrossRef] [PubMed]

5. Guan, P.-P.; Cao, L.-L.; Wang, P. Elevating the Levels of Calcium Ions Exacerbate Alzheimer's Disease via Inducing the Production and Aggregation of β-Amyloid Protein and Phosphorylated Tau. *Int. J. Mol. Sci* **2021**, *22*, 5900. [CrossRef]

6. Shim, H.S.; Horner, J.W.; Wu, C.-J.; Li, J.; Lan, Z.D.; Jiang, S.; Xu, X.; Hsu, W.-H.; Zal, T.; Flores, I.I.; et al. Telomerase Reverse Transcriptase Preserves Neuron Survival and Cognition in Alzheimer's Disease Models. *Nat. Aging* **2021**, *1*, 1162–1174. [CrossRef]

7. Pilat, D.; Paumier, J.-M.; García-González, L.; Louis, L.; Stephan, D.; Manrique, C.; Khrestchatisky, M.; Di Pasquale, E.; Baranger, K.; Rivera, S. MT5-MMP Promotes Neuroinflammation, Neuronal Excitability and Aβ Production in Primary Neuron/Astrocyte Cultures from the 5xFAD Mouse Model of Alzheimer's Disease. *J. Neuroinflammation* **2022**, *19*, 65. [CrossRef]

8. Stutzmann, G.E.; Caccamo, A.; LaFerla, F.M.; Parker, I. Dysregulated IP3 Signaling in Cortical Neurons of Knock-in Mice Expressing an Alzheimer's-Linked Mutation in Presenilin1 Results in Exaggerated Ca^{2+} Signals and Altered Membrane Excitability. *J. Neurosci.* **2004**, *24*, 508–513. [CrossRef] [PubMed]

9. Smith, I.F.; Hitt, B.; Green, K.N.; Oddo, S.; LaFerla, F.M. Enhanced Caffeine-Induced Ca^{2+} Release in the 3xTg-AD Mouse Model of Alzheimer's Disease. *J. Neurochem.* **2005**, *94*, 1711–1718. [CrossRef] [PubMed]

10. Lerdkrai, C.; Asavapanumas, N.; Brawek, B.; Kovalchuk, Y.; Mojtahedi, N.; Moral, M.O.D.; Garaschuk, O. Intracellular Ca^{2+} Stores Control in Vivo Neuronal Hyperactivity in a Mouse Model of Alzheimer's Disease. *Proc. Natl. Acad. Sci. USA* **2018**, *115*, E1279–E1288. [CrossRef] [PubMed]

11. Rigotto, G.; Zentilin, L.; Pozzan, T.; Basso, E. Effects of Mild Excitotoxic Stimulus on Mitochondria Ca^{2+} Handling in Hippocampal Cultures of a Mouse Model of Alzheimer's Disease. *Cells* **2021**, *10*, 2046. [CrossRef]

12. Popugaeva, E.; Bezprozvanny, I.; Chernyuk, D. Reversal of Calcium Dysregulation as Potential Approach for Treating Alzheimer's Disease. *Curr. Alzheimer Res.* **2020**, *17*, 344–354. [CrossRef] [PubMed]

13. Galla, L.; Redolfi, N.; Pozzan, T.; Pizzo, P.; Greotti, E. Intracellular Calcium Dysregulation by the Alzheimer's Disease-Linked Protein Presenilin 2. *Int. J. Mol. Sci.* **2020**, *21*, 770. [CrossRef] [PubMed]

14. O'Day, D.H. Calmodulin Binding Proteins and Alzheimer's Disease: Biomarkers, Regulatory Enzymes and Receptors That Are Regulated by Calmodulin. *Int. J. Mol. Sci.* **2020**, *21*, E7344. [CrossRef]

15. Grekhnev, D.A.; Kaznacheyeva, E.V.; Vigont, V.A. Patient-Specific IPSCs-Based Models of Neurodegenerative Diseases: Focus on Aberrant Calcium Signaling. *Int. J. Mol. Sci.* **2022**, *23*, 624. [CrossRef] [PubMed]

16. Ryan, K.C.; Ashkavand, Z.; Norman, K.R. The Role of Mitochondrial Calcium Homeostasis in Alzheimer's and Related Diseases. *Int. J. Mol. Sci.* **2020**, *21*, 9153. [CrossRef]

17. Kumar, A. Calcium Signaling During Brain Aging and Its Influence on the Hippocampal Synaptic Plasticity. *Adv. Exp. Med. Biol.* **2020**, *1131*, 985–1012. [CrossRef]

18. Etcheberrigaray, R.; Hirashima, N.; Nee, L.; Prince, J.; Govoni, S.; Racchi, M.; Tanzi, R.E.; Alkon, D.L. Calcium Responses in Fibroblasts from Asymptomatic Members of Alzheimer's Disease Families. *Neurobiol. Dis.* **1998**, *5*, 37–45. [CrossRef]

19. LaFerla, F.M. Calcium Dyshomeostasis and Intracellular Signalling in Alzheimer's Disease. *Nat. Rev. Neurosci.* **2002**, *3*, 862–872. [CrossRef]

20. Prakriya, M.; Lewis, R.S. Store-Operated Calcium Channels. *Physiol. Rev.* **2015**, *95*, 1383–1436. [CrossRef]

21. Wang, X.; Zheng, W. Ca^{2+} Homeostasis Dysregulation in Alzheimer's Disease: A Focus on Plasma Membrane and Cell Organelles. *FASEB J.* **2019**, *33*, 6697–6712. [CrossRef] [PubMed]

22. Serwach, K.; Gruszczynska-Biegala, J. Target Molecules of STIM Proteins in the Central Nervous System. *Front. Mol. Neurobiol.* **2020**, *13*, 617422. [CrossRef] [PubMed]

23. Lewis, R.S. Store-Operated Calcium Channels: From Function to Structure and Back Again. *Cold Spring Harb. Perspect. Biol.* **2020**, *12*, a035055. [CrossRef] [PubMed]

24. Lopez, J.J.; Jardin, I.; Sanchez-Collado, J.; Salido, G.M.; Smani, T.; Rosado, J.A. TRPC Channels in the SOCE Scenario. *Cells* **2020**, *9*, 126. [CrossRef] [PubMed]

25. Shalygin, A.; Kolesnikov, D.; Glushankova, L.; Gusev, K.; Skopin, A.; Skobeleva, K.; Kaznacheyeva, E.V. Role of STIM2 and Orai Proteins in Regulating TRPC1 Channel Activity upon Calcium Store Depletion. *Cell Calcium* **2021**, *97*, 102432. [CrossRef] [PubMed]

26. Sun, S.; Zhang, H.; Liu, J.; Popugaeva, E.; Xu, N.-J.; Feske, S.; White, C.L.; Bezprozvanny, I. Reduced Synaptic STIM2 Expression and Impaired Store-Operated Calcium Entry Cause Destabilization of Mature Spines in Mutant Presenilin Mice. *Neuron* **2014**, *82*, 79–93. [CrossRef] [PubMed]

27. Zhang, H.; Sun, S.; Herreman, A.A.; Strooper, B.D.; Bezprozvanny, I. Role of Presenilins in Neuronal Calcium Homeostasis. *J. Neurosci.* **2010**, *30*, 8566–8580. [CrossRef] [PubMed]

28. Zhang, H.; Wu, L.; Pchitskaya, E.; Zakharova, O.; Saito, T.; Saido, T.; Bezprozvanny, I. Neuronal Store-Operated Calcium Entry and Mushroom Spine Loss in Amyloid Precursor Protein Knock-In Mouse Model of Alzheimer's Disease. *J. Neurosci.* **2015**, *35*, 13275–13286. [CrossRef] [PubMed]

29. Zhang, H.; Sun, S.; Wu, L.; Pchitskaya, E.; Zakharova, O.; Tacer, K.F.; Bezprozvanny, I. Store-Operated Calcium Channel Complex in Postsynaptic Spines: A New Therapeutic Target for Alzheimer's Disease Treatment. *J. Neurosci.* **2016**, *36*, 11837–11850. [CrossRef] [PubMed]

30. Calvo-Rodriguez, M.; Hernando-Pérez, E.; López-Vázquez, S.; Núñez, J.; Villalobos, C.; Núñez, L. Remodeling of Intracellular Ca^{2+} Homeostasis in Rat Hippocampal Neurons Aged In Vitro. *Int. J. Mol. Sci.* **2020**, *21*, 1549. [CrossRef] [PubMed]

31. Ryazantseva, M.; Goncharova, A.; Skobeleva, K.; Erokhin, M.; Methner, A.; Georgiev, P.; Kaznacheyeva, E. Presenilin-1 Delta E9 Mutant Induces STIM1-Driven Store-Operated Calcium Channel Hyperactivation in Hippocampal Neurons. *Mol. Neurobiol.* **2018**, *55*, 4667–4680. [CrossRef]

32. Ryazantseva, M.; Skobeleva, K.; Kaznacheyeva, E. Familial Alzheimer's Disease-Linked Presenilin-1 Mutation M146V Affects Store-Operated Calcium Entry: Does Gain Look like Loss? *Biochimie* **2013**, *95*, 1506–1509. [CrossRef]

33. Chernyuk, D.; Zernov, N.; Kabirova, M.; Bezprozvanny, I.; Popugaeva, E. Antagonist of Neuronal Store-Operated Calcium Entry Exerts Beneficial Effects in Neurons Expressing PSEN1ΔE9 Mutant Linked to Familial Alzheimer Disease. *Neuroscience* **2019**, *410*, 118–127. [CrossRef]

34. Brandman, O.; Liou, J.; Park, W.S.; Meyer, T. STIM2 Is a Feedback Regulator That Stabilizes Basal Cytosolic and Endoplasmic Reticulum Ca^{2+} Levels. *Cell* **2007**, *131*, 1327–1339. [CrossRef] [PubMed]

35. Gruszczynska-Biegala, J.; Pomorski, P.; Wisniewska, M.B.; Kuznicki, J. Differential Roles for STIM1 and STIM2 in Store-Operated Calcium Entry in Rat Neurons. *PLoS ONE* **2011**, *6*, e19285. [CrossRef] [PubMed]

36. Park, C.Y.; Shcheglovitov, A.; Dolmetsch, R. The CRAC Channel Activator STIM1 Binds and Inhibits L-Type Voltage-Gated Calcium Channels. *Science* **2010**, *330*, 101–105. [CrossRef]

37. Wang, Y.; Deng, X.; Mancarella, S.; Hendron, E.; Eguchi, S.; Soboloff, J.; Tang, X.D.; Gill, D.L. The Calcium Store Sensor, STIM1, Reciprocally Controls Orai and CaV1.2 Channels. *Science* **2010**, *330*, 105–109. [CrossRef] [PubMed]

38. Nguyen, N.; Biet, M.; Simard, É.; Béliveau, É.; Francoeur, N.; Guillemette, G.; Dumaine, R.; Grandbois, M.; Boulay, G. STIM1 Participates in the Contractile Rhythmicity of HL-1 Cells by Moderating T-Type Ca^{2+} Channel Activity. *Biochim. Biophys. Acta -Mol. Cell Res.* **2013**, *1833*, 1294–1303. [CrossRef]

39. Majewski, L.; Kuznicki, J. SOCE in Neurons: Signaling or Just Refilling? *Biochim. Biophys. Acta-Mol. Cell Res.* **2015**, *1853*, 1940–1952. [CrossRef] [PubMed]

40. Pascual-Caro, C.; Berrocal, M.; Lopez-Guerrero, A.M.; Alvarez-Barrientos, A.; Pozo-Guisado, E.; Gutierrez-Merino, C.; Mata, A.M.; Martin-Romero, F.J. STIM1 Deficiency Is Linked to Alzheimer's Disease and Triggers Cell Death in SH-SY5Y Cells by Upregulation of L-Type Voltage-Operated Ca^{2+} Entry. *J. Mol. Med.* **2018**, *96*, 1061–1079. [CrossRef] [PubMed]

41. Popugaeva, E.; Pchitskaya, E.; Speshilova, A.; Alexandrov, S.; Zhang, H.; Vlasova, O.; Bezprozvanny, I. STIM2 Protects Hippocampal Mushroom Spines from Amyloid Synaptotoxicity. *Mol. Neurodegener.* **2015**, *10*, 37. [CrossRef] [PubMed]

42. Hofer, A.M.; Fasolato, C.; Pozzan, T. Capacitative Ca^{2+} entry is closely linked to the filling state of internal Ca^{2+} stores: A study using simultaneous measurements of I_{CRAC} and intraluminal [Ca^{2+}]. *J. Cell Biol.* **1998**, *140*, 325–334. [CrossRef] [PubMed]

43. Rahman, S.; Rahman, T. Unveiling Some FDA-Approved Drugs as Inhibitors of the Store-Operated Ca^{2+} Entry Pathway. *Sci. Rep.* **2017**, *7*, 12881. [CrossRef] [PubMed]

44. Breedveld, F.C.; Dayer, J.-M. Leflunomide: Mode of Action in the Treatment of Rheumatoid Arthritis. *Ann. Rheum Dis.* **2000**, *59*, 841–849. [CrossRef] [PubMed]

45. Skibinska-Kijek, A.; Wisniewska, M.B.; Gruszczynska-Biegala, J.; Methner, A.; Kuznicki, J. Immunolocalization of STIM1 in the Mouse Brain. *Acta Neurobiol. Exp.* **2009**, *69*, 413–428.

46. Chen-Engerer, H.J.; Hartmann, J.; Karl, R.M.; Yang, J.; Feske, S.; Konnerth, A. Two Types of Functionally Distinct Ca^{2+} Stores in Hippocampal Neurons. *Nat. Commun.* **2019**, *10*, 3223. [CrossRef] [PubMed]

47. Gualdani, R.; Gailly, P. How TRPC Channels Modulate Hippocampal Function. *Int. J. Mol. Sci.* **2020**, *21*, 3915. [CrossRef]

48. Secondo, A.; Bagetta, G.; Amantea, D. On the Role of Store-Operated Calcium Entry in Acute and Chronic Neurodegenerative Diseases. *Front. Mol. Neurosci.* **2018**, *11*, 87. [CrossRef] [PubMed]

49. Bojarski, L.; Pomorski, P.; Szybinska, A.; Drab, M.; Skibinska-Kijek, A.; Gruszczynska-Biegala, J.; Kuznicki, J. Presenilin-Dependent Expression of STIM Proteins and Dysregulation of Capacitative Ca^{2+} Entry in Familial Alzheimer's Disease. *Biochim. Biophys. Acta* **2009**, *1793*, 1050–1057. [CrossRef] [PubMed]

50. Tu, H.; Nelson, O.; Bezprozvanny, A.; Wang, Z.; Lee, S.-F.; Hao, Y.-H.; Serneels, L.; De Strooper, B.; Yu, G.; Bezprozvanny, I. Presenilins Form ER Ca^{2+} Leak Channels, a Function Disrupted by Familial Alzheimer's Disease-Linked Mutations. *Cell* **2006**, *126*, 981–993. [CrossRef] [PubMed]

51. Zatti, G.; Burgo, A.; Giacomello, M.; Barbiero, L.; Ghidoni, R.; Sinigaglia, G.; Florean, C.; Bagnoli, S.; Binetti, G.; Sorbi, S.; et al. Presenilin Mutations Linked to Familial Alzheimer's Disease Reduce Endoplasmic Reticulum and Golgi Apparatus Calcium Levels. *Cell Calcium* **2006**, *39*, 539–550. [CrossRef]

52. Tong, B.C.-K.; Lee, C.S.-K.; Cheng, W.-H.; Lai, K.-O.; Foskett, J.K.; Cheung, K.-H. Familial Alzheimer's Disease–Associated Presenilin 1 Mutants Promote γ-Secretase Cleavage of STIM1 to Impair Store-Operated Ca^{2+} Entry. *Sci. Signal.* **2016**, *9*, ra89. [CrossRef] [PubMed]

53. Alić, I.; Kosi, N.; Kapuralin, K.; Gorup, D.; Gajović, S.; Pochet, R.; Mitrečić, D. Neural Stem Cells from Mouse Strain Thy1 YFP-16 Are a Valuable Tool to Monitor and Evaluate Neuronal Differentiation and Morphology. *Neurosci. Lett.* **2016**, *634*, 32–41. [CrossRef] [PubMed]

54. Kipanyula, M.J.; Contreras, L.; Zampese, E.; Lazzari, C.; Wong, A.K.C.; Pizzo, P.; Fasolato, C.; Pozzan, T. Ca^{2+} Dysregulation in Neurons from Transgenic Mice Expressing Mutant Presenilin 2. *Aging Cell* **2012**, *11*, 885–893. [CrossRef] [PubMed]

55. Nelson, O.; Tu, H.; Lei, T.; Bentahir, M.; de Strooper, B.; Bezprozvanny, I. Familial Alzheimer Disease–Linked Mutations Specifically Disrupt Ca^{2+} Leak Function of Presenilin 1. *J. Clin. Investig.* **2007**, *117*, 1230–1239. [CrossRef]

56. Cheung, K.-H.; Mei, L.; Mak, D.-O.D.; Hayashi, I.; Iwatsubo, T.; Kang, D.E.; Foskett, J.K. Gain-of-Function Enhancement of IP3 Receptor Modal Gating by Familial Alzheimer's Disease-Linked Presenilin Mutants in Human Cells and Mouse Neurons. *Sci. Signal.* **2010**, *3*, ra22. [CrossRef] [PubMed]

57. Greotti, E.; Capitanio, P.; Wong, A.; Pozzan, T.; Pizzo, P.; Pendin, D. Familial Alzheimer's Disease-Linked Presenilin Mutants and Intracellular Ca^{2+} Handling: A Single-Organelle, FRET-Based Analysis. *Cell Calcium* **2019**, *79*, 44–56. [CrossRef] [PubMed]

58. Rana, A.; Yen, M.; Sadaghiani, A.M.; Malmersjö, S.; Park, C.Y.; Dolmetsch, R.E.; Lewis, R.S. Alternative Splicing Converts STIM2 from an Activator to an Inhibitor of Store-Operated Calcium Channels. *J. Cell Biol.* **2015**, *209*, 653–670. [CrossRef] [PubMed]

59. Ramesh, G.; Jarzembowski, L.; Schwarz, Y.; Poth, V.; Konrad, M.; Knapp, M.L.; Schwär, G.; Lauer, A.A.; Grimm, M.O.W.; Alansary, D.; et al. A Short Isoform of STIM1 Confers Frequency-Dependent Synaptic Enhancement. *Cell Rep.* **2021**, *34*, 108844. [CrossRef] [PubMed]

60. Hu, W.-Y.; He, Z.-Y.; Yang, L.-J.; Zhang, M.; Xing, D.; Xiao, Z.-C. The Ca^{2+} Channel Inhibitor 2-APB Reverses β-Amyloid-Induced LTP Deficit in Hippocampus by Blocking BAX and Caspase-3 Hyperactivation. *Br. J. Pharmacol.* **2015**, *172*, 2273–2285. [CrossRef] [PubMed]

61. Coon, A.L.; Wallace, D.R.; Mactutus, C.F.; Booze, R.M. L-Type Calcium Channels in the Hippocampus and Cerebellum of Alzheimer's Disease Brain Tissue. *Neurobiol. Aging* **1999**, *20*, 597–603. [CrossRef] [PubMed]

62. Thibault, O.; Pancani, T.; Landfield, P.W.; Norris, C.M. Reduction in Neuronal L-Type Calcium Channel Activity in a Double Knock-in Mouse Model of Alzheimer's Disease. *Biochim. Biophys. Acta-Mol. Basis Dis.* **2012**, *1822*, 546–549. [CrossRef] [PubMed]

63. Wang, Y.; Mattson, M.P. L-Type Ca^{2+} Currents at CA1 Synapses, but Not CA3 or Dentate Granule Neuron Synapses, Are Increased in 3xTgAD Mice in an Age-Dependent Manner. *Neurobiol. Aging* **2014**, *35*, 88–95. [CrossRef] [PubMed]

64. Ishii, M.; Hiller, A.J.; Pham, L.; McGuire, M.J.; Iadecola, C.; Wang, G. Amyloid-Beta Modulates Low-Threshold Activated Voltage-Gated L-Type Calcium Channels of Arcuate Neuropeptide Y Neurons Leading to Calcium Dysregulation and Hypothalamic Dysfunction. *J. Neurosci.* **2019**, *39*, 8816–8825. [CrossRef] [PubMed]

65. Li, T.; Jiao, J.-J.; Su, Q.; Hölscher, C.; Zhang, J.; Yan, X.-D.; Zhao, H.-M.; Cai, H.-Y.; Qi, J.-S. A GLP-1/GIP/Gcg Receptor Triagonist Improves Memory Behavior, as Well as Synaptic Transmission, Neuronal Excitability and Ca^{2+} Homeostasis in 3xTg-AD Mice. *Neuropharmacology* **2020**, *170*, 108042. [CrossRef]

66. Harraz, O.F.; Altier, C. STIM1-Mediated Bidirectional Regulation of Ca(2+) Entry through Voltage-Gated Calcium Channels (VGCC) and Calcium-Release Activated Channels (CRAC). *Front. Cell Neurosci.* **2014**, *8*, 43. [CrossRef] [PubMed]

67. Lu, W.; Wang, J.; Peng, G.; Shimoda, L.A.; Sylvester, J.T. Knockdown of Stromal Interaction Molecule 1 Attenuates Store-Operated Ca^{2+} Entry and Ca^{2+} Responses to Acute Hypoxia in Pulmonary Arterial Smooth Muscle. *Am. J. Physiol. Lung Cell Mol. Physiol.* **2009**, *297*, L17–L25. [CrossRef] [PubMed]

68. Vigont, V.A.; Grekhnev, D.A.; Lebedeva, O.S.; Gusev, K.O.; Volovikov, E.A.; Skopin, A.Y.; Bogomazova, A.N.; Shuvalova, L.D.; Zubkova, O.A.; Khomyakova, E.A.; et al. STIM2 Mediates Excessive Store-Operated Calcium Entry in Patient-Specific IPSC-Derived Neurons Modeling a Juvenile Form of Huntington's Disease. *Front. Cell Dev. Biol.* **2021**, *9*, 625231. [CrossRef] [PubMed]

69. Berna-Erro, A.; Braun, A.; Kraft, R.; Kleinschnitz, C.; Schuhmann, M.K.; Stegner, D.; Wultsch, T.; Eilers, J.; Meuth, S.G.; Stoll, G.; et al. STIM2 Regulates Capacitive Ca^{2+} Entry in Neurons and Plays a Key Role in Hypoxic Neuronal Cell Death. *Sci. Signal.* **2009**, *2*, ra67. [CrossRef] [PubMed]

70. Ercan, E.; Momburg, F.; Engel, U.; Temmerman, K.; Nickel, W.; Seedorf, M. A Conserved, Lipid-Mediated Sorting Mechanism of Yeast Ist2 and Mammalian STIM Proteins to the Peripheral ER. *Traffic* **2009**, *10*, 1802–1818. [CrossRef] [PubMed]

71. Wu, J.; Shih, H.-P.; Vigont, V.; Hrdlicka, L.; Diggins, L.; Singh, C.; Mahoney, M.; Chesworth, R.; Shapiro, G.; Zimina, O.; et al. Neuronal Store-Operated Calcium Entry Pathway as a Novel Therapeutic Target for Huntington's Disease Treatment. *Chem. Biol.* **2011**, *18*, 777–793. [CrossRef] [PubMed]

72. Schindelin, J.; Arganda-Carreras, I.; Frise, E.; Kaynig, V.; Longair, M.; Pietzsch, T.; Preibisch, S.; Rueden, C.; Saalfeld, S.; Schmid, B.; et al. Fiji: An Open-Source Platform for Biological-Image Analysis. *Nat. Methods* **2012**, *9*, 676–682. [CrossRef] [PubMed]

73. Edelstein, A.D.; Tsuchida, M.A.; Amodaj, N.; Pinkard, H.; Vale, R.D.; Stuurman, N. Advanced Methods of Microscope Control Using MManager Software. *J. Biol. Methods* **2014**, *1*, e10. [CrossRef] [PubMed]

74. Ho, J.; Tumkaya, T.; Aryal, S.; Choi, H.; Claridge-Chang, A. Moving beyond P Values: Data Analysis with Estimation Graphics. *Nat. Methods* **2019**, *16*, 565–566. [CrossRef] [PubMed]

75. Lee, S.; Lee, D.K. What is the proper way to apply the multiple comparison test? *Korean J. Anesthesiol.* **2018**, *71*, 353–360. [CrossRef] [PubMed]

Article

Intraoperative Hypothermia Induces Vascular Dysfunction in the CA1 Region of Rat Hippocampus

Tianjia Li, Guangyan Xu, Jie Yi *and Yuguang Huang

Department of Anesthesiology, Peking Union Medical College Hospital, Chinese Academy of Medical Sciences and Peking Union Medical College, Beijing 100730, China; litianjia@pumch.cn (T.L.); bonnie_xgy08@163.com (G.X.); garypumch@163.com (Y.H.)
* Correspondence: neyil@pumch.cn

Abstract: Intraoperative hypothermia is very common and leads to memory decline. The hippocampus is responsible for memory formation. As a functional core area, the cornu ammonis 1 (CA1) region of the hippocampus contains abundant blood vessels and is susceptible to ischemia. The aim of the study was to explore vascular function and neuronal state in the CA1 region of rats undergoing intraoperative hypothermia. The neuronal morphological change and activity-regulated cytoskeleton-associated protein (Arc) expression were evaluated by haematoxylin-eosin staining and immunofluorescence respectively. Histology and immunohistochemistry were used to assess vascular function. Results showed that intraoperative hypothermia inhibited the expression of vascular endothelial growth factor and endothelial nitric oxide synthase, and caused reactive oxygen species accumulation. Additionally, the phenotype of vascular smooth muscle cells was transformed from contractile to synthetic, showing a decrease in smooth muscle myosin heavy chain and an increase in osteopontin. Ultimately, vascular dysfunction caused neuronal pyknosis in the CA1 region and reduced memory-related Arc expression. In conclusion, neuronal disorder in the CA1 region was caused by intraoperative hypothermia-related vascular dysfunction. This study could provide a novel understanding of the effect of intraoperative hypothermia in the hippocampus, which might identify a new research target and treatment strategy.

Keywords: intraoperative hypothermia; hippocampus; CA1 region; vascular endothelial cells; endothelial dysfunction; vascular smooth muscle cells; phenotypic transformation; activity-regulated cytoskeleton-associated protein; reactive oxygen species; endothelial nitric oxide synthase

Citation: Li, T.; Xu, G.; Yi, J.; Huang, Y. Intraoperative Hypothermia Induces Vascular Dysfunction in the CA1 Region of Rat Hippocampus. *Brain Sci.* **2022**, *12*, 692. https://doi.org/10.3390/brainsci12060692

Academic Editors: Masaru Tanaka, Lydia Giménez-Llort, Simone Battaglia, Chong Chen, Piril Hepsomali and Simona Lattanzi

Received: 7 April 2022
Accepted: 25 May 2022
Published: 27 May 2022

Publisher's Note: MDPI stays neutral with regard to jurisdictional claims in published maps and institutional affiliations.

1. Introduction

With an increase in longevity and a growing older population, many people may undergo operations at least once in their lifetime. Intraoperative hypothermia, defined as a core temperature <36 °C, is a common complication, and its incidence can be up to 25–70% [1,2]. Previous studies and our epidemiological investigation found that intraoperative hypothermia postponed memory recovery, and increased dependency on families and society [3,4]. However, the underlying mechanism of this phenomenon remained unclear. Recently, studies showed that there were no obvious brain abnormalities related to synaptic plasticity, neurotransmitters, or other physiological function [5–10].

Brain tissue is sensitive to ischemia, as cerebral function is critically dependent on adequate blood supply [11]. Despite representing only 2% of total body weight, the brain receives over 15% of the cardiac output and consumes more than 20% of the glucose and oxygen sources [12]. The vascular contributions to cognitive and memory decline should be one of the priority research areas for potential breakthroughs [13]. Studies on Alzheimer's and other memory disorders show that the hippocampus is vulnerable to hypoperfusion and vascular density, diameter, and branch angle are reduced in the hippocampus [13,14]. Moreover, as a crucial structure of neural circuits in contextual retrieval and emotional learning, the hippocampus functionally binds with the prefrontal cortex, suggesting that it is preferentially vulnerable [15].

The hippocampus is a crucial structure for connecting with other brain structures and storing memories which are embedded deep into the temporal lobe [16]. This hippocampal region needs a large blood supply to function properly supplied by the posterior cerebral artery and the anterior choroidal artery [17], which branch into numerous arterioles passing through the hippocampus [18]. In memory decline diseases, vascular dysfunction is often the early features [12,19]. Patients with vascular diseases, such as atherosclerosis, are likely more susceptible to memory impairment [20]. Thus, vascular dysfunction may be the mechanism of the intraoperative hypothermia-related hippocampal damage.

Blood vessels are composed of vascular endothelial cells (VECs) and vascular smooth muscle cells (VSMCs). For VECs, endothelial nitric oxide synthase (eNOS) is the key functional molecule that catalyzes nitric oxide to dilate blood vessels and ensure blood supply [21]. Vascular endothelial growth factor (VEGF) is responsible for the survival of VECs, which is required for angiogenesis and the maintenance of endothelium integrity [22,23]. Functionally, VSMCs can be divided into contractile and synthetic phenotypes according to different characteristics [24,25]. Their phenotypic transformation is involved in the pathophysiology. Physiologically, marked by the presence of smooth muscle myosin heavy chain (SMMHC), VSMCs are in the state of contractile phenotype, which regulates vascular tension to control blood supply. In response to pathological stimulation, VSMCs are transformed to the synthetic phenotype with the expression of osteopontin. The phenotypic transformation induces VSMCs proliferation and migration, eventually leading to the overexpression of retinol-binding protein (RBP) and stenosis of the vascular lumen [26,27].

VECs, VSMCs, cerebrovascular basement membrane, pericytes, astrocytes and neurons are collectively referred to as the neurovascular unit (NVU) [28]. Physiologically, the interaction between vascular function and neuronal activity is known as neurovascular coupling [12,29,30]. This mechanism ensures the delivery of oxygen and nutrients in the vascular network of arteries, arterioles, and capillaries. And then oxygen and nutrients are transport across the blood-brain barrier [31], that composes of endothelial cells, pericytes, and astrocytic end feet and forms a protective sheath around brain capillaries [32]. Regarding the NVU and the hypothesis that neurovascular coupling, VSMC contractility, and endothelial-related arteriolar dilation can become dysfunctional in memory decline [12], our study focused on the arterioles connecting neurovascular dysfunction and neurodegeneration in the hippocampus.

Cytoarchitecturally, the rat hippocampus is divided into cornu ammonis 1 (CA1), CA3, and dentate gyrus (DG) regions. The head of the hippocampus is mainly composed of the CA1 region with a complete signal output loop [33,34]. As a functional core area, the CA1 region is particularly sensitive to ischemia [35,36]. Moreover, enduring forms of memory rely on the structural connectivity of synapses that typically require activity-dependent expression of the activity-regulated cytoskeleton-associated protein (Arc). In response to external information, Arc is the immediate-early protein that converts synaptic input into neuronal changes and forms memory [37,38]. Arc knockout mice fail to form long-lasting memories. Moreover, reducing basal Arc expression by less than 60% can significantly attenuate spatial memory [39]. Because of these strong correlations it has been widely assumed that Arc plays a role in memory formation and indicates hippocampal function.

The inadvertent intraoperative hypothermia has a high incidence, and the memory decline is one of its important complications. Few studies addressed the effect of intraoperative hypothermia in the memory decline. As the functional core area of memory, the hippocampal CA1 region is rich in blood vessels and susceptible to ischemia. Therefore, the mechanism of ischemic injury of the CA1 region might provide a new research target and the development of a prevention strategy for intraoperative hypothermia-induced memory decline. The purpose of this study was to explore the vascular function in the CA1 region and to analyze the memory-related neuron status.

2. Materials and Methods

2.1. Ethics Statement

This study complied with the Guide for the Care and Use of Laboratory Animals by the National Institutes of Health and the Animal Management Rule of the Chinese Ministry of Health, and was approved by the Animal Care Committee of Peking Union Medical College (XHDW-2019-003).

2.2. Animals and Experimental Design

Adult male Sprague Dawley rats (300–400 g, 12-weeks-old) were purchased from HFK Bioscience Co., Ltd. (Beijing, China). All rats were housed in a specific pathogen-free room (temperature $23 \pm 1\,°C$, 12 h light/dark cycle) with free access to food and water. These rats were randomly divided into three groups ($n = 10$ per group). The intraoperative hypothermia group received exploratory laparotomies on cryogenic pads (rectal temperature $33 \pm 0.5\,°C$). The intraoperative normothermia group received exploratory laparotomies on temperature blankets (rectal temperature $37 \pm 0.5\,°C$). The surgical procedure lasted for 60 minutes. The control group was fed normally without operation (rectal temperature $37 \pm 0.5\,°C$). After 12-days, the rats were anesthetized by the injection of pentobarbital sodium to collect hippocampus for following experiments. Each harvested hippocampus was evenly divided into two parts, and was fixed in paraffin and frozen in liquid nitrogen respectively.

During operation, an abdominal median incision of approximately 3 cm was made to allow penetration of the peritoneal cavity, and 5–10 cm of the small intestine was exteriorised and left in air for 60 min. The intestine was then placed inside the peritoneal cavity, and the wound was sutured with 4-0 non-absorbable sutures in three layers consisting of the peritoneal lining, abdominal muscles, and skin. Core body temperatures were measured using a rectal temperature probe. After the wounds were sutured, the rats were immediately intraperitoneally injected with penicillin (60,000 U) to prevent postoperative infections, and the body temperatures of those in intraoperative hypothermia group were raised to $37 \pm 0.5\,°C$ using a heating pad.

2.3. Histology and Immunohistochemistry

The harvested hippocampi ($n = 10$ per group) were fixed in 4% paraformaldehyde, embedded in paraffin. As for the orientation and distribution of vasculature in hippocampus, the vascular network exhibited a rake-like distribution of equally spaced transverse hippocampal arterioles arising from longitudinal hippocampal arteriole. The capillaries mainly formed the microvascular network in the distal end. And the hippocampal arterioles entered the hippocampus perpendicularly to the coronal plane [13]. In our study, the hippocampal paraffin block was sectioned along the coronal plane (5-μm thick) to ensure the hippocampal vasculature was properly outlined [29,40,41]. Additionally, quantitative data on the vasculature of the hippocampus of mice showed the maximum vascular diameter (30.02 ± 0.90 μm) [13]. For the function of NVU, our study focused on the arterioles of hippocampus, and the mean diameter of arterioles in the study was 21.16 ± 0.87 μm (Carl Zeiss Jena, Oberkochen, Germany).

The slides were stained with hematoxylin-eosin (H&E), and observed by microscopy (Ti-S, Olympus FluoView software, Olympus, Japan). The total number of damaged neurons in each image was counted and analysed by Image-Pro Plus 6.0 Software. Immunohistochemical (IHC) staining was also performed to assess the expression of vascular functional molecules. After blocking in phosphate-buffered saline containing 5% goat serum for 30 min at room temperature ($20–25\,°C$), the sections were incubated overnight at $4\,°C$ with primary antibodies. The following primary antibodies were used: anti-CD31 (1:2000, #ab182981, Abcam, Cambridge, UK), anti-eNOS (1:2000, #ab76198, Abcam), anti-VEGF (1:500, #19003-1-AP, Proteintech Group, Chicago, IL, USA), anti-SMMHC (1:200, #21404-1-AP, Proteintech Group), anti-osteopontin (1:200, #22952-1-AP, Proteintech Group), and anti-RBP (1:200, #22683-1-AP, Proteintech Group). The following day, signal amplification was performed with horseradish peroxidase-conjugated goat anti-rabbit or anti-mouse IgG secondary antibody (1:500; ZhongshanJinqiao Biotechnology Co., Ltd., Beijing, China). Cytokines expression levels were quantified using integrated optical density (IOD) values generated by Image-Pro Plus 6.0 software.

2.4. Detection of Reactive Oxygen Species (ROS)

Dihydroethidium was used to evaluate ROS formation (n = 10 per group) in the frozen sections of hippocampus (12 μm thick). The sections were incubated with dihydroethidium (10 μmol/L; Sigma-Aldrich, St Louis, MO, USA) for 30 min at 37 °C in the dark, and then washed with phosphate-buffered saline three times. The nuclei were stained with 4′,6-diamidino-2-phenylindole (DAPI). The level of ROS was assessed by fluorescence microscopy. Confocal images of hippocampal sections were sequentially acquired with Zeiss AIM software on a Zeiss LSM 700 confocal microscope system (Carl Zeiss Jena, Oberkochen, Germany). In fluorescence intensity analysis, the whole section was scanned to one image and stored as the "MView MRXS File". And then we used the CaseViewer 2.4 software (3DHISTECH Ltd., Budapest, Hungary) to get multiple images from different regions.

2.5. Immunofluorescence

Immunofluorescence staining was performed on the sectioned hippocampus (n = 10 per group) to assess the expression of Arc. Hippocampus sections were blocked with 5% goat serum for 30 min at room temperature, then incubated with primary anti-Arc (sc-17839, 1:100, Santa Cruz, CA, USA) and anti-NeuN (1:100, ab12763, Abcam) overnight at 4 °C. The secondary antibodies used were anti-mouse and anti-rabbit respectively (1:1000; Invitrogen, Carlsbad, CA, USA). Sections were incubated with the secondary antibodies in the dark for 2 h at room temperature. The nuclei were then stained with DAPI. The anti-NeuN was used as neuronal marker. The expression level of Arc was assessed by fluorescence microscopy. Confocal images of sections were sequentially acquired with Zeiss AIM software on a Zeiss LSM 700 confocal microscope system. The analysis method is the same as ROS mentioned above.

2.6. Statistics

All data were expressed as mean ± SEM and were compared by ANOVA and Tukey post hoc test (using SPSS26.0). Statistical significance was determined at $p < 0.05$.

3. Results

3.1. The Abundance of Blood Vessels in the CA1 Region of the Hippocampus and Its Relationship with Hippocampal Function

CD31 is the endothelial marker of vascular tissue. There are abundant blood vessels in the hippocampus (Figure 1A), indicating a large demand for blood supply. Compared with other regions, there are more blood vessels in the CA1 region (Figure 1B), which demonstrates its susceptibility to ischemia. Additionally, the schematic diagram displays the development from vascular dysfunction to neuronal damage (Figure 1C). First, vascular dysfunction consists of VECs dysfunction and VSMCs overproliferation. Second, the endothelial dysfunction is characterized by the low expression of VEGF and eNOS, while the excessive proliferation of VSMCs is induced by the phenotypic transformation. VEGF is the required growth factor for the survival of VECs and the maintenance of endothelium integrity. eNOS is the functional molecule that synthesizes nitric oxide and maintains vascular diastolic function. For VSMCs, the phenotypic transformation from physiological contraction to pathological synthesis indicated an excessive proliferation and subsequent vascular stenosis, marked by osteopontin overexpression. Moreover, the low-expression of eNOS can cause ROS accumulation, further promoting the phenotypic transformation. All these vascular abnormalities eventually led to insufficient blood supply, causing hippocampal neuronal damage, as indicated by neuronal pyknosis and the low-expression of Arc. As hippocampus is the key region for memory, vascular dysfunction ultimately causes memory decline. Briefly, the schematic diagram in the Figure 1 provides a general description of the research.

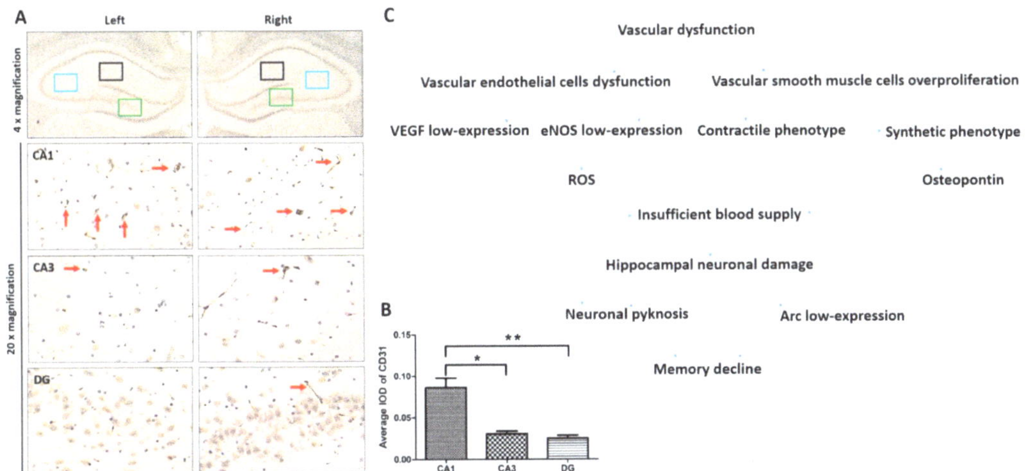

Figure 1. Analysis of hippocampal blood vessels and the relationship between vascular dysfunction and neuronal damage. (**A**) CD31 immunohistochemical staining of blood vessels (red arrows) in the rat bilateral hippocampus. In 4× magnification images, the circled rectangles in the different regions of the hippocampus were magnified by 20×. Black rectangles represent CA1, blue rectangles represent CA3, and green rectangles represent DG. (**B**) Integrated optical density (IOD) values of CD31. * p = 0.0027 vs. CA3, ** p = 0.0019 vs. DG. (**C**) The schematic diagram demonstrates the transition from vascular dysfunction to neuronal damage in the study. Statistics data were presented in Table 1.

Table 1. ANOVA analysis of factors associated with vascular and neuronal function.

Effect	R-Squared	F	p
CD31	0.9478	36.30	0.0027
VEGF	0.8281	14.45	0.0051
eNOS	0.9524	59.98	0.0001
ROS	0.8021	12.16	0.0077
SMMHC	0.9100	30.33	0.0007
Osteopontin	0.9632	78.55	<0.0001
RBP	0.9472	53.85	0.0001
Neuronal pyknosis	0.8866	23.45	0.0015
Fluorescence intensity	0.8617	18.69	0.0026

VEGF, vascular endothelial growth factor; eNOS, endothelial nitric oxide synthase; ROS, reactive oxygen species; SMMHC, smooth muscle myosin heavy chain; RBP, retinol binding protein.

3.2. Intraoperative Hypothermia Caused Dysfunction of VECs and Phenotypic Transformation of VSMCs

In representative cross-sections of the hippocampus, black rectangles in the CA1 region were magnified to 20× to show blood vessels (Figure 2). For VECs, intraoperative hypothermia significantly inhibited the expression of VEGF and eNOS (Figure 2A–D), while intraoperative normothermia preserved those expressions. As VEGF and eNOS function to preserve endothelial integrity and reduce oxidative damage, respectively, ROS was accumulated in the intraoperative hypothermia group (Figure 2E,F), indicating the presence of hippocampal oxidative damage. For VSMCs, intraoperative hypothermia led to the phenotypic transformation, in which the expression of contractile phenotype marker

SMMHC was decreased and the synthetic phenotype marker osteopontin was increased (Figure 2G–J). Inversely, intraoperative normothermia maintained the physiological contractile phenotype. The dysfunction of VECs and phenotypic transformation of VSMCs could induce atherosclerosis, which displayed as the overexpression of the proliferation regulator RBP in the intraoperative hypothermia group (Figure 2K,L). These results demonstrated that intraoperative hypothermia disturbed vascular function of the CA1 region, leading to subsequent neuronal damage.

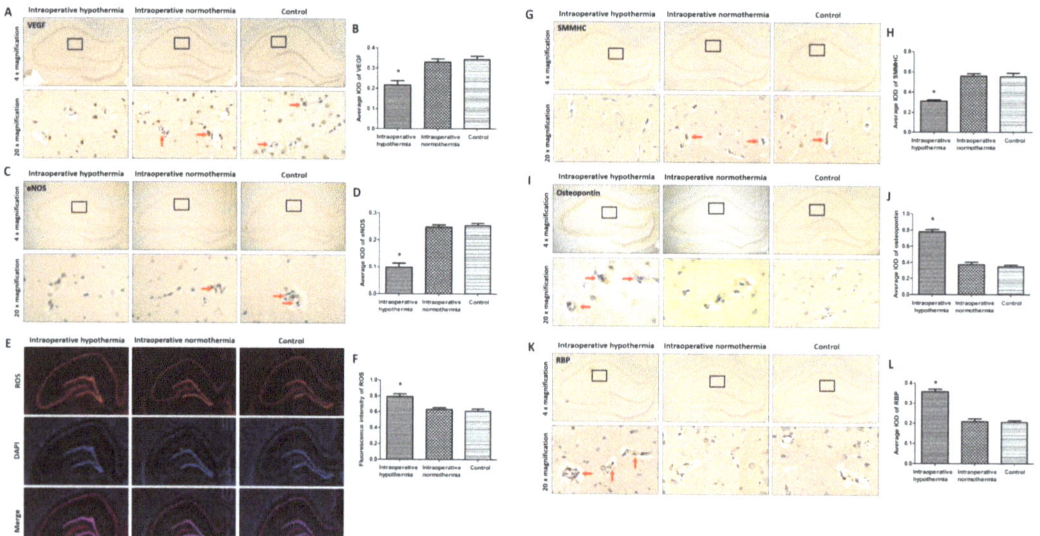

Figure 2. Intraoperative hypothermia leads to endothelial dysfunction and the phenotypic transformation of vascular smooth muscle cells. (**A–D**) Immunohistochemical staining of vascular endothelial growth factor (VEGF) and endothelial nitric oxide synthase (eNOS) in rat hippocampus and summary integrated optical density (IOD) values. * $p = 0.0051$ vs. Control, in quantification of VEGF; * $p = 0.0001$ vs. Control, in quantification of eNOS; * $p = 0.0077$ vs. Control, in quantification of ROS. (**E,F**) Reactive oxygen species (ROS) fluorescence staining of rat hippocampus and fluorescence intensity (4× magnification). (**G–L**) Immunohistochemical staining of smooth muscle myosin heavy chain (SMMHC), osteopontin, and retinol binding protein (RBP) in rat hippocampus and summary IOD values. * $p = 0.0007$ vs. Control, in quantification of SMMHC; * $p < 0.0001$ vs. Control, in quantification of Osteopontin; * $p = 0.0001$ vs. Control, in quantification of RBP. In 4× magnification images, black rectangles in the CA1 region are selected to perform 20× magnification. Intraoperative hypothermia: rats experienced hypothermia during surgery. Intraoperative normothermia: rats maintained normal body temperature during surgery. Control: rats were fed normally without operation. Statistics data were presented in Table 1.

3.3. The Overall Impact of Intraoperative Hypothermia on the Blood Vessel

The schematic diagram shows pathological changes in blood vessels induced by intraoperative hypothermia (Figure 3). For VECs, the dysfunction was characterized by disturbed organelles with decreased expression of VEGF and eNOS. In endothelium, increased vascular permeability could lead to infiltration of inflammatory cells and thrombosis. Additionally, ROS accumulated in the blood vessels, compromising vascular integrity. For VSMCs, actin is the main intracellular protein in the physiological contractile phenotype with vasomotor function. After the phenotypic transformation induced by intraoperative hypothermia, VSMCs deformed from shuttle to oval in shape. In microstructure, VSMCs could synthesize more protein and possessed more organelles, such as endoplasmic reticulum, Golgi apparatus, etc. Ultimately, the synthetic phenotype of highly proliferative VSMCs caused vascular stenosis and hypoperfusion.

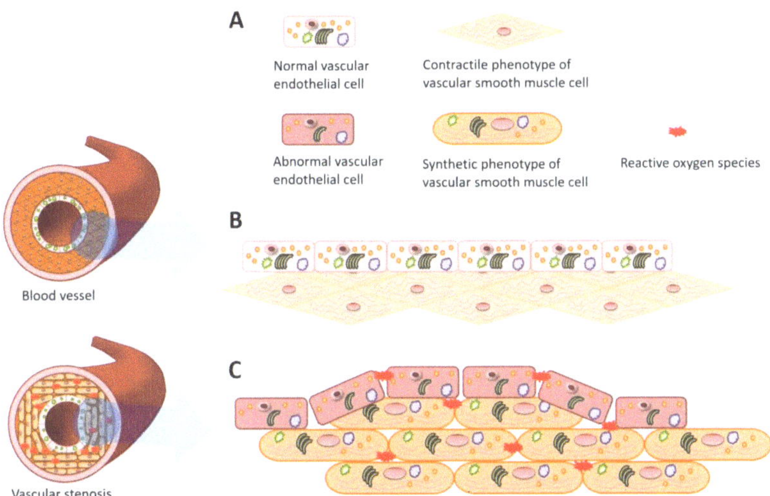

Figure 3. Schematic representation of the effects of intraoperative hypothermia on vascular endothelial cells (VECs) and vascular smooth muscle cells (VSMCs). (**A**) Key legends for the schematic. (**B**) Normal VECs and VSMCs with contractile phenotype make up the wall of blood vessels, which preserve the blood supply. (**C**) Dysfunctional VECs increase the vascular permeability while VSMCs with synthetic phenotype proliferate excessively, causing stenosis of the vascular lumen.

3.4. Intraoperative Hypothermia Damaged Hippocampal Neurons in the CA1 Region

Figure 4A illustrates three regions of the rat hippocampus: CA1, CA2, and DG. The CA1 region is most susceptible to ischemia. H&E staining was used to analyze the state of hippocampal neurons (Figure 4B,C). The CA1 region was framed in the $4\times$ magnification image. In $20\times$ magnification images, there were neuronal pyknosis and necrosis in the intraoperative hypothermia group. In contrast, in control and intraoperative normothermia groups, neurons were round or oval, with lightly stained cytoplasm. These morphological changes of neurons in the CA1 region demonstrated that intraoperative hypothermia caused neuronal damage. Besides, the neuronal activity was marked by Arc expression, which plays a critical role in memory storage. The hippocampal immunofluorescence analysis showed that the expression of Arc in the CA1 region was significantly decreased by intraoperative hypothermia (Figure 4D,E). Together with previous results, the study indicated intraoperative hypothermia-related vascular dysfunction induced hippocampal neuronal damage, especially in the CA1 region with high susceptibility to ischemia.

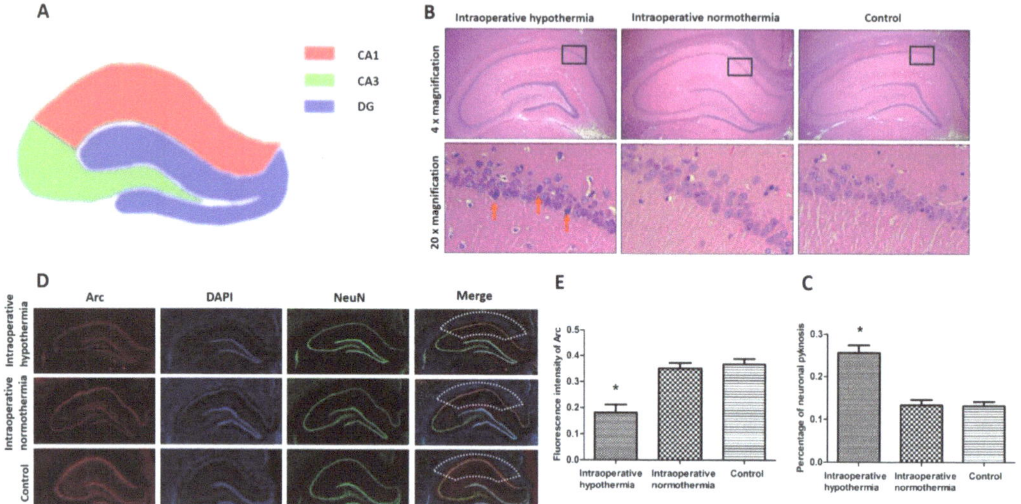

Figure 4. Histopathology of the hippocampus caused by intraoperative hypothermia. (**A**) Cross-sectional diagram of rat unilateral hippocampus with three regions encoded by different colors. (**B,C**) Neuronal pyknosis (red arrows) detected by H&E staining and summary results. Black rectangles in the CA1 region are selected to perform 20× magnification. * $p = 0.0015$ vs. Control. (**D,E**) Representative immunofluorescence (4× magnification) staining for Arc in the hippocampus. The red fluorescence shows Arc expression. The blue fluorescence represents nuclei stained with 4′,6-diamidino-2-phenylindole (DAPI). The green fluorescence shows neurons labeled with NeuN. In the merged figure, the CA1 region is framed in a dotted line. * $p = 0.0026$ vs. Control. Intraoperative hypothermia: rats experienced hypothermia during surgery. Intraoperative normothermia: rats maintained normal body temperature during surgery. Control: rats were fed normally without operation. Statistics data were presented in Table 1.

4. Discussion

Clinical investigations and rat maze experiments both demonstrated that intraoperative hypothermia could lead to memory decline [4,42]. As memory is generated in the hippocampus, intraoperative hypothermia may negatively impact the hippocampus. While blood supply is critically required for neuronal function, hippocampal blood vessels are very thin and lack capillary anastomosis (Figure 1). In the high-resolution mapping of brain vasculature, the low vascular diameter, length density (total vascular length per tissue volume) and volume fraction (total vascular volume relative to the total tissue volume) of the hippocampus also suggest that the hippocampus is more vulnerable to cerebrovascular dysfunction [13]. Therefore, the hippocampus is particularly sensitive to ischemia. Studies of Alzheimer's disease also revealed ischemic damage in the hippocampus [14]. Thus, the vascular function could be a key point to explore the effect of intraoperative hypothermia on hippocampus. The hippocampus in a rat is divided into CA1, CA3, and DG regions. The CA1 region, with widespread projections, is the key signal output node of the hippocampal memory circuit [43]. Due to its functional features, the CA1 region needs more blood supply and is highly susceptible to ischemia.

Intraoperative hypothermia induced vascular dysfunction in the CA1 region of rat hippocampus. The vascular wall consists of VECs and VSMCs, which are important in preserving vascular function. VECs line the vascular lumen and contact with blood. Their dysfunction is the pathological basis of thrombosis and stenosis. Among them, VEGF is required for the survival of VECs and angiogenesis. Intraoperative hypothermia significantly reduced VEGF expression (Figure 2A). The decreased VEGF could result in shedding of VECs and changes in vascular permeability that attenuated blood supply. VEGF also has a neurotrophic effect [22], especially in the ischemia-sensitive CA1 region. Additionally, eNOS induces nitric oxide that preserves vascular relaxation and eliminates ROS. In vivo measurements, nitric oxide increased the local cerebral blood flow in hippocampus, while ROS compromised nitric oxide mediated neurovascular response during memory decline [44]. eNOS expression was significantly decreased in the intraoperative hypothermia group in comparison to that of the control group (Figure 2C). The decrease in eNOS indicated the presence of endothelial impairment, which could reduce hippocampal blood supply. Furthermore, the decrease in antioxidant molecule eNOS and subsequent reduction in blood supply resulted in the accumulation of ROS, which induced neuronal oxidative damage in the intraoperative hypothermia group (Figure 2E). In clinical practice, the factors contributing to the inadvertent intraoperative hypothermia might also lead to the vascular dysfunction and the accumulation of ROS, such as impairment of thermoregulation by general anesthesia and undergoing major-plus surgery [45,46]. Collectively, these results showed that intraoperative hypothermia disturbed the function of VECs, subsequently reducing the blood supply of the hippocampal CA1 region.

VSMCs constitute the media of the vascular wall and control the vasomotor to regulate blood supply. Normal vascular structure and function depend on the contractile phenotype of VSMCs with the presence of SMMHC that is an actin-based motor protein essential to VSMCs motility. Intraoperative hypothermia caused the SMMHC suppression and osteopontin overexpression (Figure 2G–J). The phenotypic transformation leads to wall thickening and decreased compliance. Osteopontin mediates cell-cell and cell-extracellular matrix interactions, which is associated with VSMCs adhesion and vascular remodeling [47]. Overall, these pathological changes can lead to the hypoperfusion in hippocampus. With the regard to the NVU, neuron also can induce nitric oxide and regulate VSMCs relaxation [12,48]. Further neural signal pathway studies shown that activation of neuronal N-methyl D-aspartate receptor could lead to concentration dependent nitric oxide-mediated dilatation of arterioles [49,50]. Therefore, the intraoperative hypothermia-induced vascular dysfunction might also involve the N-methyl D-aspartate receptor signal pathway which will be detected in our following study. Additionally, more organelles, such as the endoplasmic reticulum and Golgi complex, could be observed in synthetic VSMCs (Figure 3), indicating the enhancement in synthesis and secretory functions. Besides, the phenotypic transformation could lead to angiosclerosis [51]. Overexpression of RBP was shown in the intraoperative hypothermia group (Figure 2K), which is implicated in lipid deposition and mineralization. The series of pathological changes in the contractile phenotype exacerbates luminal stenosis and hippocampus hypoperfusion. Within the NVU, the cellular composition differs along the vascular tree [12]. Pericytes in the terminal capillary level are the ones with contractile properties and control blood flow, as shown for VSMCs in the upper arteriole segment of NVU [52]. Memory decline diseases also demonstrate pericyte dysfunction in the hippocampus [31,53]. Collectively, VSMCs were transformed into the synthetic phenotype by intraoperative hypothermia. The phenotypic transformation could result in vasomotor dysfunction and vascular stenosis, ultimately attenuating the blood supply of the CA1 region.

As two major components of blood vessels, VECs and VSMCs have complementary functions. The intact VECs act as a barrier to protect VSMCs from blood flow stimulation, which maintains the normal contractile function. In addition, endothelial release of nitric oxide leads to VSMC relaxation and arteriolar dilation. Following endothelial damage, the subsequent excessive proliferation of VSMCs could lead to atherosclerosis [54,55]. Moreover, the accumulation of ROS also induced the development of the synthetic phenotype in VSMCs [56]. Likewise, the synthetic phenotype VSMCs could excessively proliferate and destroy the integrity of the endodermis, which further exacerbates the progression of vascular disorder. Eventually, these series of events resulted in lumen narrowing and hippocampus hypoperfusion.

Hypoperfusion could damage structural integrity, such as gray matter atrophy and white matter microstructure disorder [57]. In the vulnerable CA1 region, neurons are more sensitive to ischemia and prone to pathological changes. There is clear neuronal pyknosis in the CA1 region of the intraoperative hypothermia group (Figure 4B). In terms of structure and function, the CA1 region is divided into the pyramidal, polymorphic, and molecular layers, of which individual values of neuronal density are relatively heterogeneous. The bodies of neurons are mainly located in the pyramidal layer, while axons and dendrites are in the polymorphic and molecular layers. Not all neuronal layers in the CA1 region have the same sensitivity to ischemia. We observed the neuronal pyknosis mainly in the pyramidal layer (Figure 4B). This finding confirms that the CA1 pyramidal neurons were sensitive to intraoperative hypothermia-related hypoperfusion. Additionally, the vascular dysfunction contributed to the activation of astrocytes and the ensuing inflammatory response [32], and neuronal damage could disturb specific neurochemical content. Particularly, the low expression of Arc further demonstrated neuronal damage (Figure 4D).

Arc is an active neuronal marker of hippocampal function. As the immediate-early protein, Arc accumulates in synapse and drives the formation of memory. Moreover, during the recall of memory, Arc has an anatomical distribution in the CA1 region. In this study, the expression of Arc in the CA1 region was decreased by intraoperative hypothermia (Figure 4). In addition, our maze experiment showed that reducing Arc expression in the CA1 region could attenuate the spatial memory of rats [42]. The low expression of Arc is associated with memory decline in different neurological diseases. Alzheimer's disease shows a depletion of Arc in the hippocampus [5,58]. Arc knockout mice also failed to form long-term memories and would forget what they learned 24 h earlier [39]. In electrophysiology, decreased expression of Arc resulted in depolarization through the inhibitory neurotransmitter GABA and intracellular Cl- [59], which might be the underlying neurotransmitter mechanism of hippocampal neuronal damage in response to intraoperative hypothermia. The Arc reduction observed in this study further revealed the neuronal damage in the CA1 region.

The regulation of cerebral blood flow is essential for normal brain function. Specifically, the neurovascular coupling is modulated by NVU that interacts with local neuron activation [15,60–62]. Disrupted NVU and neurovascular uncoupling were found in the early stages of neurological disorders [12,63,64]. Our study showed NVU dysfuntion that VSMC contractility and endothelial-related arteriolar repairment became dysfunctional in the intraoperative hypothermia-induced memory decline.

As the anatomical and methodological difficulties [15,18], our study has a limitation in measuring the coordinated interaction between neurons and vasculature. Probing dynamically the hippocampal blood flow might allow characterization of vascular function during intraoperative hypothermia, which is what needs to be achieved in our future research. Besides, neuronal pyknosis (Figure 4b) implies that neurodegeneration can reflect the impairment and dysfunction of the mnemonic and cognitive nervous systems [65]. Except for the dynamic blood flow detection, magnetic resonance imaging and positron emission tomography may help identify the affected region of hippocampal neuron pyknosis [15,31,66]. Furthermore, during the progression of neurodegeneration, such as Alzheimer's disease and Parkinson's disease, neuroinflammation is the common prelude [67]. Vascular dysfunction induces hypoperfusion, increase of ROS, and oligodendrocyte malfunction, leading to the release of inflammatory cytokines, such as TNF-α, IL-1β, IL-6, and IL-10, which amplifies the risk of vascular dysfunction [67]. The alterations of inflammatory factors levels were observed in the cerebrospinal fluid of neurodegeneration diseases [68]. The inflammatory factors detection in cerebrospinal fluid may further reveal the mystery. Additionally, studies show that the therapeutic hypothermia is a neuroprotective strategy, and the mechanisms involved are complex, such as reducing oxygen requirement and cytotoxic edema, interrupting anoxic injury, preventing free-radical injury, protecting the blood-brain barrier integrity, and etc. [69–72]. In other words, the precise protective mechanisms are not yet fully understood. As for our study, the memory decline was observed after the inadvertent intraoperative hypothermia both in clinical practice and animal experiment [4,42,45,46]. In regard to cerebral blood flow, the therapeutic hypothermia also induced reduction in post-ischemic hyperperfusion and sustained hypoperfusion [72]. Anyway, there are differences between therapeutic hypothermia and inadvertent intraoperative hypothermia in their neurological effects. Still, given the magnitude and complexity of the brain function, further studies of the effect of intraoperative hypothermia in hippocampus-related memory should be given a high priority.

5. Conclusions

The CA1 region contains abundant blood vessels and is susceptible to ischemia. Intraoperative hypothermia disturbed the blood supply by causing the dysfunction of VECs and the phenotypic transformation of VSMCs. Ultimately, neurons in the CA1 region displayed morphological pyknosis and biological suppression of Arc. The findings from this study suggest future research directions. Further measurement of regional blood flow in the hippocampus might allow characterization of vascular function during intraoperative hypothermia. Aside from neuronal morphological pyknosis and reduced Arc expression, subcellular synaptic plasticity might also be an important neuroelectrophysiological change. Lastly, but importantly, intraoperative hypothermia is very common, but usually neglected. While intraoperative hypothermia can be prevented by thermal insulation, it cannot be completely eliminated. Notably, some operations require hypothermia. The recovery of memory function is important to determine the quality of life after such operations. This study could provide a novel understanding of the effect of intraoperative hypothermia in the hippocampus. As such, it might identify a new research target and treatment strategy for memory decline induced by intraoperative hypothermia.

Author Contributions: T.L. and G.X. conducted the experiments, analysed the results and drafted the manuscript; J.Y. and Y.H. conceived and designed the study and helped to edit the manuscript. All authors have read and agreed to the published version of the manuscript.

Funding: This work was supported by the Post-doctoral Science Foundation of China [No. 2019M660550], the National Natural Science Youth Foundation of China [No. 81900433], the National Natural Science Foundation of China [No. 82071252], and the Capital's Funds for Health Improvement and Research [No. 2018-2-4013].

Institutional Review Board Statement: The study was conducted according to the guidelines of the Declaration of Helsinki, and approved by the Ethical Committee for Animal Experimentation of the Peking Union Medical College Hospital (XHDW-2019-003).

Informed Consent Statement: Not applicable.

Data Availability Statement: All data included in this study are available upon request from the corresponding authors.

Acknowledgments: We thank Institute of Laboratory Animal Sciences, Chinese Academy of Medical Sciences and Peking Union Medical College for providing behavioral devices and guidance.

Conflicts of Interest: The authors declare no conflict of interest.

References

1. Long, K.C.; Tanner, E.J.; Frey, M.; Leitao, M.M.; Levine, D.A.; Gardner, G.J.; Sonoda, Y.; Abu-Rustum, N.R.; Barakat, R.R.; Chi, D.S. Intraoperative hypothermia during primary surgical cytoreduction for advanced ovarian cancer: Risk factors and associations with postoperative morbidity. *Gynecol. Oncol.* **2013**, *131*, 525–530. [CrossRef]
2. Sari, S.; Aksoy, S.M.; But, A. The incidence of inadvertent perioperative hypothermia in patients undergoing general anesthesia and an examination of risk factors. *Int. J. Clin. Pract.* **2021**, *75*, e14103. [CrossRef]
3. Scott, A.V.; Stonemetz, J.L.; Wasey, J.O.; Johnson, D.J.; Rivers, R.J.; Koch, C.G.; Frank, S.M. Compliance with Surgical Care Improvement Project for Body Temperature Management (SCIP Inf-10) Is Associated with Improved Clinical Outcomes. *Anesthesiology* **2015**, *123*, 116–125. [CrossRef]
4. Yi, J.; Lei, Y.; Xu, S.; Si, Y.; Li, S.; Xia, Z.; Shi, Y.; Gu, X.; Yu, J.; Xu, G.; et al. Intraoperative hypothermia and its clinical outcomes in patients undergoing general anesthesia: National study in China. *PLoS ONE* **2017**, *12*, e0177221. [CrossRef] [PubMed]
5. Zhang, H.; Bramham, C.R. Arc/Arg3.1 function in long-term synaptic plasticity: Emerging mechanisms and unresolved issues. *Eur. J. Neurosci.* **2021**, *54*, 6696–6712. [CrossRef] [PubMed]

6. Wang, D.; Cui, Z.; Zeng, Q.; Kuang, H.; Wang, L.P.; Tsien, J.Z.; Cao, X. Genetic Enhancement of Memory and Long-Term Potentiation but Not CA1 Long-Term Depression in NR2B Transgenic Rats. *PLoS ONE* **2009**, *4*, e7486. [CrossRef] [PubMed]

7. Calabresi, P.; Gubellini, P.; Centonze, D.; Picconi, B.; Bernardi, G.; Chergui, K.; Svenningsson, P.; Fienberg, A.A.; Greengard, P. Dopamine and cAMP-regulated phosphoprotein 32 kDa controls both striatal long-term depression and long-term potentiation, opposing forms of synaptic plasticity. *J. Neurosci.* **2000**, *20*, 8443–8451. [CrossRef] [PubMed]

8. Nikolaienko, O.; Patil, S.; Eriksen, M.S.; Bramham, C.R. Arc protein: A flexible hub for synaptic plasticity and cognition. *Semin. Cell Dev. Biol.* **2018**, *77*, 33–42. [CrossRef] [PubMed]

9. Verhage, M.; Maia, A.S.; Plomp, J.J.; Brussaard, A.B.; Heeroma, J.H.; Vermeer, H.; Toonen, R.F.; Hammer, R.E.; van Den, T.K.; Berg, T.V.D.; et al. Synaptic Assembly of the Brain in the Absence of Neurotransmitter Secretion. *Science* **2000**, *287*, 864–869. [CrossRef] [PubMed]

10. Chu, M.; Teng, J.; Guo, L.; Wang, Y.; Zhang, L.; Gao, J.; Liu, L. Mild hyperhomocysteinemia induces blood–brain barrier dysfunction but not neuroinflammation in the cerebral cortex and hippocampus of wild-type mice. *Can. J. Physiol. Pharmacol.* **2021**, *99*, 847–856. [CrossRef]

11. Han, X.-J.; Shi, Z.-S.; Xia, L.-X.; Zhu, L.-H.; Zeng, L.; Nie, J.-H.; Xu, Z.-C.; Ruan, Y.-W. Changes in synaptic plasticity and expression of glutamate receptor subunits in the CA1 and CA3 areas of the hippocampus after transient global ischemia. *Neuroscience* **2016**, *327*, 64–78. [CrossRef]

12. Kisler, K.; Nelson, A.R.; Montagne, A.; Zlokovic, B.V. Cerebral blood flow regulation and neurovascular dysfunction in Alzheimer disease. *Nat. Rev. Neurosci.* **2017**, *18*, 419–434. [CrossRef]

13. Zhang, X.; Yin, X.; Zhang, J.; Li, A.; Gong, H.; Luo, Q.; Zhang, H.; Gao, Z.; Jiang, H. High-resolution mapping of brain vasculature and its impairment in the hippocampus of Alzheimer's disease mice. *Natl. Sci. Rev.* **2019**, *6*, 1223–1238. [CrossRef]

14. Pluta, R.; Bogucka-Kocka, A.; Ułamek-Kozioł, M.; Bogucki, J.; Januszewski, S.; Kocki, J.; Czuczwar, S.J. Ischemic tau protein gene induction as an additional key factor driving development of Alzheimer's phenotype changes in CA1 area of hippocampus in an ischemic model of Alzheimer's disease. *Pharmacol. Rep.* **2018**, *70*, 881–884. [CrossRef]

15. Battaglia, S.; Garofalo, S.; Di Pellegrino, G. Context-dependent extinction of threat memories: Influences of healthy aging. *Sci. Rep.* **2018**, *8*, 12592. [CrossRef]

16. Zhang, H.; Roman, R.J.; Fan, F. Hippocampus is more susceptible to hypoxic injury: Has the Rosetta Stone of regional variation in neurovascular coupling been deciphered? *Geroscience* **2022**, *44*, 127–130. [CrossRef]

17. Estefan, D.P.; Sánchez-Fibla, M.; Duff, A.; Principe, A.; Rocamora, R.; Zhang, H.; Axmacher, N.; Verschure, P.F.M.J. Coordinated representational reinstatement in the human hippocampus and lateral temporal cortex during episodic memory retrieval. *Nat. Commun.* **2019**, *10*, 2255. [CrossRef]

18. Rusinek, H.; Brys, M.; Glodzik, L.; Switalski, R.; Tsui, W.-H.; Haas, F.; Mcgorty, K.; Chen, Q.; de Leon, M.J. Hippocampal blood flow in normal aging measured with arterial spin labeling at 3T. *Magn. Reson. Med.* **2011**, *65*, 128–137. [CrossRef]

19. Coughlan, G.; Laczó, J.; Hort, J.; Minihane, A.-M.; Hornberger, M. Spatial navigation deficits—Overlooked cognitive marker for preclinical Alzheimer disease? *Nat. Rev. Neurol.* **2018**, *14*, 496–506. [CrossRef]

20. Hort, J.; Vališ, M.; Kuča, K.; Angelucci, F. Vascular Cognitive Impairment: Information from Animal Models on the Pathogenic Mechanisms of Cognitive Deficits. *Int. J. Mol. Sci.* **2019**, *20*, 2405. [CrossRef]

21. Peleli, M.; Zollbrecht, C.; Montenegro, M.F.; Hezel, M.; Zhong, J.; Persson, E.G.; Holmdahl, R.; Weitzberg, E.; Lundberg, J.O.; Carlström, M. Enhanced XOR activity in eNOS-deficient mice: Effects on the nitrate-nitrite-NO pathway and ROS homeostasis. *Free Radic. Biol. Med.* **2016**, *99*, 472–484. [CrossRef] [PubMed]

22. Apte, R.S.; Chen, D.S.; Ferrara, N. VEGF in Signaling and Disease: Beyond Discovery and Development. *Cell* **2019**, *176*, 1248–1264. [CrossRef] [PubMed]

23. Karaman, S.; Leppänen, V.-M.; Alitalo, K. Vascular endothelial growth factor signaling in development and disease. *Development* **2018**, *145*, dev151019. [CrossRef] [PubMed]

24. Wanjare, M.; Kuo, F.; Gerecht, S. Derivation and maturation of synthetic and contractile vascular smooth muscle cells from human pluripotent stem cells. *Cardiovasc. Res.* **2013**, *97*, 321–330. [CrossRef] [PubMed]

25. Tosun, Z.; McFetridge, P.S. Variation in Cardiac Pulse Frequencies Modulates vSMC Phenotype Switching During Vascular Remodeling. *Cardiovasc. Eng. Technol.* **2015**, *6*, 59–70. [CrossRef] [PubMed]

26. Bennick, R.A.; Nagengast, A.A.; DiAngelo, J.R. The SR proteins SF2 and RBP1 regulate triglyceride storage in the fat body of Drosophila. *Biochem. Biophys. Res. Commun.* **2019**, *516*, 928–933. [CrossRef] [PubMed]

27. Monroe, D.G.; Hawse, J.R.; Subramaniam, M.; Spelsberg, T.C. Retinoblastoma binding protein-1 (RBP1) is a Runx2 coactivator and promotes osteoblastic differentiation. *BMC Musculoskelet. Disord.* **2010**, *11*, 104. [CrossRef]

28. Iadecola, C. The Neurovascular Unit Coming of Age: A Journey through Neurovascular Coupling in Health and Disease. *Neuron* **2017**, *96*, 17–42. [CrossRef]

29. Shaw, K.; Bell, L.; Boyd, K.; Grijseels, D.M.; Clarke, D.; Bonnar, O.; Crombag, H.S.; Hall, C.N. Neurovascular coupling and oxygenation are decreased in hippocampus compared to neocortex because of microvascular differences. *Nat. Commun.* **2021**, *12*, 3190. [CrossRef]

30. Iadecola, C. Neurovascular regulation in the normal brain and in Alzheimer's disease. *Nat. Rev. Neurosci.* **2004**, *5*, 347–360. [CrossRef]

31. Montagne, A.; Barnes, S.R.; Sweeney, M.D.; Halliday, M.R.; Sagare, A.P.; Zhao, Z.; Toga, A.W.; Jacobs, R.E.; Liu, C.Y.; Amezcua, L.; et al. Blood-Brain Barrier Breakdown in the Aging Human Hippocampus. *Neuron* **2015**, *85*, 296–302. [CrossRef]
32. Senatorov, V.V.; Friedman, A.R.; Milikovsky, D.Z.; Ofer, J.; Saar-Ashkenazy, R.; Charbash, A.; Jahan, N.; Chin, G.; Mihaly, E.; Lin, J.M.; et al. Blood-brain barrier dysfunction in aging induces hyperactivation of TGFβ signaling and chronic yet reversible neural dysfunction. *Sci. Transl. Med.* **2019**, *11*, eaaw8283. [CrossRef]
33. Ginsberg, S.D.; Malek-Ahmadi, M.H.; Alldred, M.J.; Che, S.; Elarova, I.; Chen, Y.; Jeanneteau, F.; Kranz, T.M.; Chao, M.V.; Counts, S.E.; et al. Selective decline of neurotrophin and neurotrophin receptor genes within CA1 pyramidal neurons and hippocampus proper: Correlation with cognitive performance and neuropathology in mild cognitive impairment and Alzheimer's disease. *Hippocampus* **2019**, *29*, 422–439. [CrossRef]
34. Chen, S.-Y.; Lin, M.-C.; Tsai, J.-S.; He, P.-L.; Luo, W.-T.; Chiu, I.-M.; Herschman, H.R.; Li, H.-J. Exosomal 2′,3′-CNP from mesenchymal stem cells promotes hippocampus CA1 neurogenesis/neuritogenesis and contributes to rescue of cognition/learning deficiencies of damaged brain. *Stem Cells Transl. Med.* **2020**, *9*, 499–517. [CrossRef]
35. Rasouli, B.; Ghahari, L.; Safari, M.; Shahroozian, E.; Naeimi, S. Combination therapy of the granulocyte colony stimulating factor and intravenous lipid emulsion protect the hippocampus after global ischemia in rat: Focusing on CA1 region. *Metab. Brain Dis.* **2020**, *35*, 991–997. [CrossRef]
36. Lei, S.; Zhang, P.; Li, W.; Gao, M.; He, X.; Zheng, J.; Li, X.; Wang, X.; Wang, N.; Zhang, J.; et al. Pre- and Posttreatment With Edaravone Protects CA1 Hippocampus and Enhances Neurogenesis in the Subgranular Zone of Dentate Gyrus After Transient Global Cerebral Ischemia in Rats. *ASN Neuro* **2014**, *6*, 1759091414558417. [CrossRef]
37. Wall, M.; Collins, D.R.; Chery, S.L.; Allen, Z.D.; Pastuzyn, E.D.; George, A.J.; Nikolova, V.D.; Moy, S.S.; Philpot, B.D.; Shepherd, J.D.; et al. The Temporal Dynamics of Arc Expression Regulate Cognitive Flexibility. *Neuron* **2018**, *98*, 1124–1132.e7. [CrossRef]
38. Shepherd, J.D.; Bear, M.F. New views of Arc, a master regulator of synaptic plasticity. *Nat. Neurosci.* **2011**, *14*, 279–284. [CrossRef]
39. Plath, N.; Ohana, O.; Dammermann, B.; Errington, M.L.; Schmitz, D.; Gross, C.; Mao, X.; Engelsberg, A.; Mahlke, C.; Welzl, H.; et al. Arc/Arg3.1 Is Essential for the Consolidation of Synaptic Plasticity and Memories. *Neuron* **2006**, *52*, 437–444. [CrossRef]
40. He, Z.; He, B.; Behrle, B.L.; Fejleh, M.P.C.; Cui, L.; Paule, M.G.; Greenfield, L.J. Ischemia-Induced Increase in Microvascular Phosphodiesterase 4D Expression in Rat Hippocampus Associated with Blood Brain Barrier Permeability: Effect of Age. *ACS Chem. Neurosci.* **2012**, *3*, 428–432. [CrossRef]
41. Yang, A.C.; Vest, R.T.; Kern, F.; Lee, D.P.; Agam, M.; Maat, C.A.; Losada, P.M.; Chen, M.B.; Schaum, N.; Khoury, N.; et al. A human brain vascular atlas reveals diverse mediators of Alzheimer's risk. *Nature* **2022**, *603*, 885–892. [CrossRef] [PubMed]
42. Xu, G.; Li, T.; Huang, Y. The Effects of Intraoperative Hypothermia on Postoperative Cognitive Function in the Rat Hippocampus and Its Possible Mechanisms. *Brain Sci.* **2022**, *12*, 96. [CrossRef]
43. Soltesz, I.; Losonczy, A. CA1 pyramidal cell diversity enabling parallel information processing in the hippocampus. *Nat. Neurosci.* **2018**, *21*, 484–493. [CrossRef] [PubMed]
44. Lourenço, C.F.; Santos, R.M.; Barbosa, R.M.; Cadenas, E.; Radi, R.; Laranjinha, J. Neurovascular coupling in hippocampus is mediated via diffusion by neuronal-derived nitric oxide. *Free Radic. Biol. Med.* **2014**, *73*, 421–429. [CrossRef] [PubMed]
45. Yi, J.; Xiang, Z.; Deng, X.; Fan, T.; Fu, R.; Geng, W.; Guo, R.; He, N.; Li, C.; Li, L.; et al. Incidence of Inadvertent Intraoperative Hypothermia and Its Risk Factors in Patients Undergoing General Anesthesia in Beijing: A Prospective Regional Survey. *PLoS ONE* **2015**, *10*, e0136136. [CrossRef] [PubMed]
46. Yi, J.; Zhan, L.; Lei, Y.; Xu, S.; Si, Y.; Li, S.; Xia, Z.; Shi, Y.; Gu, X.; Yu, J.; et al. Establishment and Validation of a Prediction Equation to Estimate Risk of Intraoperative Hypothermia in Patients Receiving General Anesthesia. *Sci. Rep.* **2017**, *7*, 13927. [CrossRef] [PubMed]
47. Jiang, H.; Lun, Y.; Wu, X.; Xia, Q.; Zhang, X.; Xin, S.; Zhang, J. Association between the Hypomethylation of Osteopontin and Integrin β3 Promoters and Vascular Smooth Muscle Cell Phenotype Switching in Great Saphenous Varicose Veins. *Int. J. Mol. Sci.* **2014**, *15*, 18747–18761. [CrossRef]
48. Faraci, F.M.; Breese, K.R. Nitric oxide mediates vasodilatation in response to activation of N-methyl-D-aspartate receptors in brain. *Circ. Res.* **1993**, *72*, 476–480. [CrossRef]
49. Bhardwaj, A.; Northington, F.J.; Carhuapoma, J.R.; Falck, J.R.; Harder, D.R.; Traystman, R.J.; Koehler, R.C. P-450 epoxygenase and NO synthase inhibitors reduce cerebral blood flow response to N-methyl-d-aspartate. *Am. J. Physiol. Heart Circ. Physiol.* **2000**, *279*, H1616–H1624. [CrossRef]
50. Buerk, D.G.; Ances, B.M.; Greenbergbc, J.H.; Detrebc, J.A. Temporal Dynamics of Brain Tissue Nitric Oxide during Functional Forepaw Stimulation in Rats. *NeuroImage* **2003**, *18*, 1–9. [CrossRef]
51. Ren, X.-S.; Tong, Y.; Ling, L.; Chen, D.; Sun, H.-J.; Zhou, H.; Qi, X.-H.; Chen, Q.; Li, Y.-H.; Kang, Y.-M.; et al. NLRP3 Gene Deletion Attenuates Angiotensin II-Induced Phenotypic Transformation of Vascular Smooth Muscle Cells and Vascular Remodeling. *Cell. Physiol. Biochem.* **2017**, *44*, 2269–2280. [CrossRef]
52. Cauli, B. Revisiting the role of neurons in neurovascular coupling. *Front. Neuroenergetics* **2010**, *2*, 9. [CrossRef]
53. Nation, D.A.; Sweeney, M.D.; Montagne, A.; Sagare, A.P.; D'Orazio, L.M.; Pachicano, M.; Sepehrband, F.; Nelson, A.R.; Buennagel, D.P.; Harrington, M.G.; et al. Blood–brain barrier breakdown is an early biomarker of human cognitive dysfunction. *Nat. Med.* **2019**, *25*, 270–276. [CrossRef]
54. Li, T.; Ni, L.; Liu, X.; Wang, Z.; Liu, C. High glucose induces the expression of osteopontin in blood vessels in vitro and in vivo. *Biochem. Biophys. Res. Commun.* **2016**, *480*, 201–207. [CrossRef]

55. Li, T.; Liu, X.; Ni, L.; Wang, Z.; Wang, W.; Shi, T.; Liu, X.; Liu, C. Perivascular adipose tissue alleviates inflammatory factors and stenosis in diabetic blood vessels. *Biochem. Biophys. Res. Commun.* **2016**, *480*, 147–152. [CrossRef]
56. Yang, M.; Fang, J.; Liu, Q.; Wang, Y.; Zhang, Z. Role of ROS-TRPM7-ERK1/2 axis in high concentration glucose-mediated proliferation and phenotype switching of rat aortic vascular smooth muscle cells. *Biochem. Biophys. Res. Commun.* **2017**, *494*, 526–533. [CrossRef]
57. Perosa, V.; Priester, A.; Ziegler, G.; Cardenas-Blanco, A.; Dobisch, L.; Spallazzi, M.; Assmann, A.; Maass, A.; Speck, O.; Oltmer, J.; et al. Hippocampal vascular reserve associated with cognitive performance and hippocampal volume. *Brain* **2020**, *143*, 622–634. [CrossRef]
58. Morin, J.-P.; Cerón-Solano, G.; Velázquez-Campos, G.; Pacheco-López, G.; Bermúdez-Rattoni, F.; Díaz-Cintra, S. Spatial Memory Impairment is Associated with Intraneural Amyloid-β Immunoreactivity and Dysfunctional Arc Expression in the Hippocampal-CA3 Region of a Transgenic Mouse Model of Alzheimer's Disease. *J. Alzheimer's Dis.* **2016**, *51*, 69–79. [CrossRef]
59. Fuentes, J.G.; Serrano, R.I.; Sánchez, S.C.; Donate, M.J.L.; Infante, E.R.; Jiménez, M.D.M.A.; Rabal, M.D.P.M. Neuropeptides in the developing human hippocampus under hypoxic–ischemic conditions. *J. Anat.* **2021**, *239*, 856–868. [CrossRef]
60. Ungvari, Z.; Tarantini, S.; Donato, A.J.; Galvan, V.; Csiszar, A. Mechanisms of Vascular Aging. *Circ. Res.* **2018**, *123*, 849–867. [CrossRef]
61. Fan, F.; Booz, G.W.; Roman, R.J. Aging diabetes, deconstructing the cerebrovascular wall. *Aging* **2021**, *13*, 9158–9159. [CrossRef]
62. Mughal, A.; Harraz, O.F.; Gonzales, A.L.; Hill-Eubanks, D.; Nelson, M.T. PIP2 Improves Cerebral Blood Flow in a Mouse Model of Alzheimer's Disease. *Function* **2021**, *2*, zqab010. [CrossRef]
63. Zlokovic, B.V. Neurovascular pathways to neurodegeneration in Alzheimer's disease and other disorders. *Nat. Rev. Neurosci.* **2011**, *12*, 723–738. [CrossRef]
64. Iadecola, C. The Pathobiology of Vascular Dementia. *Neuron* **2013**, *80*, 844–866. [CrossRef] [PubMed]
65. Gerez, J.A.; Riek, R. Neurodegenerative diseases distinguished through protein-structure analysis. *Nature* **2020**, *578*, 223–224. [CrossRef] [PubMed]
66. Shimizu, S.; Hirose, D.; Hatanaka, H.; Takenoshita, N.; Kaneko, Y.; Ogawa, Y.; Sakurai, H.; Hanyu, H. Role of Neuroimaging as a Biomarker for Neurodegenerative Diseases. *Front. Neurol.* **2018**, *9*, 265. [CrossRef] [PubMed]
67. Tanaka, M.; Toldi, J.; Vécsei, L. Exploring the Etiological Links behind Neurodegenerative Diseases: Inflammatory Cytokines and Bioactive Kynurenines. *Int. J. Mol. Sci.* **2020**, *21*, 2431. [CrossRef]
68. Brosseron, F.; Krauthausen, M.; Kummer, M.; Heneka, M.T. Body Fluid Cytokine Levels in Mild Cognitive Impairment and Alzheimer's Disease: A Comparative Overview. *Mol. Neurobiol.* **2014**, *50*, 534–544. [CrossRef]
69. Zhang, M.; Wang, H.; Zhao, J.; Chen, C.; Leak, R.K.; Xu, Y.; Vosler, P.; Chen, J.; Gao, Y.; Zhang, F. Drug-Induced Hypothermia in Stroke Models: Does it Always Protect? *CNS Neurol. Disord. -Drug Targets* **2013**, *12*, 371–380. [CrossRef]
70. Curfman, G.D. Hypothermia to Protect the Brain. *N. Engl. J. Med.* **2002**, *346*, 546. [CrossRef]
71. Olsen, T.S.; Weber, U.J.; Kammersgaard, L.P. Therapeutic hypothermia for acute stroke. *Lancet Neurol.* **2003**, *2*, 410–416. [CrossRef]
72. Yenari, M.A.; Han, H.S. Neuroprotective mechanisms of hypothermia in brain ischaemia. *Nat. Rev. Neurosci.* **2012**, *13*, 267–278. [CrossRef]

brain sciences

Review

Somatic Cell Reprogramming for Nervous System Diseases: Techniques, Mechanisms, Potential Applications, and Challenges

Jiafeng Chen [1], Lijuan Huang [1], Yue Yang [1], Wei Xu [1], Qingchun Qin [1], Rongxing Qin [1], Xiaojun Liang [1], Xinyu Lai [2], Xiaoying Huang [1], Minshan Xie [1] and Li Chen [1,2,*]

1   Department of Neurology, the First Affiliated Hospital of Guangxi Medical University, Nanning 530021, China
2   Key Laboratory of Longevity and Aging-Related Diseases of Chinese Ministry of Education, Nanning 530021, China
*   Correspondence: chenli@gxmu.edu.cn

Abstract: Nervous system diseases present significant challenges to the neuroscience community due to ethical and practical constraints that limit access to appropriate research materials. Somatic cell reprogramming has been proposed as a novel way to obtain neurons. Various emerging techniques have been used to reprogram mature and differentiated cells into neurons. This review provides an overview of somatic cell reprogramming for neurological research and therapy, focusing on neural reprogramming and generating different neural cell types. We examine the mechanisms involved in reprogramming and the challenges that arise. We herein summarize cell reprogramming strategies to generate neurons, including transcription factors, small molecules, and microRNAs, with a focus on different types of cells.. While reprogramming somatic cells into neurons holds the potential for understanding neurological diseases and developing therapeutic applications, its limitations and risks must be carefully considered. Here, we highlight the potential benefits of somatic cell reprogramming for neurological disease research and therapy. This review contributes to the field by providing a comprehensive overview of the various techniques used to generate neurons by cellular reprogramming and discussing their potential applications.

Keywords: nervous system diseases; neuroscience; somatic cell; reprogramming; neurons; mechanisms; transcription factors; therapeutic; microRNA; molecules

Citation: Chen, J.; Huang, L.; Yang, Y.; Xu, W.; Qin, Q.; Qin, R.; Liang, X.; Lai, X.; Huang, X.; Xie, M.; et al. Somatic Cell Reprogramming for Nervous System Diseases: Techniques, Mechanisms, Potential Applications, and Challenges. *Brain Sci.* **2023**, *13*, 524. https://doi.org/10.3390/brainsci13030524

Academic Editors: Masaru Tanaka and Manuel Sánchez Malmierca

Received: 4 February 2023
Revised: 14 March 2023
Accepted: 20 March 2023
Published: 22 March 2023

1. Introduction

The world's population is aging quickly, and this is linked to several systemic diseases such as neurological disorders [1]. Notably, the number of people with neurological diseases increases as the population ages [2]. As of 2019, in the elderly, stroke and Alzheimer's disease are among the top 10 causes of death [3]. This growing patient population significantly burdens healthcare systems, society, and the economy [2]. Unfortunately, effective treatments for degenerating or damaged neurons have yet to be developed in neuroscience [4].

Obtaining neural tissue or neurons for research is often challenging due to limitations such as ethics or hardships in obtaining the tissue. The human brain and spinal cord are protected by the skull and spinal canal, and obtaining tissue often implies traumatic surgery. The lack of human in vitro models to investigate pathological alterations in neural cell function that affect the entire neurological disease process is a severe limitation in studying mechanisms and potential therapies to protect brain health and target age-related neurodegenerative diseases.

Finding new ways to study diseases and treat neurological disorders is crucial in this urgent situation. Somatic cell reprogramming, a new way to obtain neurons, is one way that scientists are trying to do this. Through somatic cell reprogramming, somatic cells can be transformed into neurons, thereby accelerating research progress and advancing the development of effective neural repair and replacement solutions [5]. Somatic reprogramming involves transforming one somatic cell type into induced pluripotent cells, which are then matured into the desired cell type. In contrast, transdifferentiation refers to the direct transformation of one somatic cell type into another without first reprogramming into pluripotent cells and then differentiating into functional somatic cells [6].

In general, mature neurons that are lost in adulthood are not replaced. The brain does contain some pockets of neural stem cells, including the subventricular zone across the lateral ventricles and the subgranular zone of the dentate gyrus in the hippocampus [7,8]. However, these neural stem cells are relatively sparse and cannot restore the large numbers of neurons lost during stroke, major trauma, or neurodegenerative disease [8]. Thus, scientists have focused on utilizing somatic cells to acquire neurons to address the scarcity of resources for studying human neurological diseases [9].

Early research on somatic cell reprogramming can be traced back to the middle of the last century, when scientists attempted to perform nucleus transplantation, and the transplanted cells were able to develop into intact individuals. For example, Gurdon et al. transferred a nucleus from an embryonic cell into an enucleated and unfertilized egg of the same species; this cell eventually developed into an entire sexually mature individual of *Xenopus laevis* [10]. Another example is Wilmut et al.'s method of cloning sheep by transplanting adult nuclei into unfertilized egg cells, a study known worldwide as the "birth of Dolly, the sheep" [11]. These two experiments showed that the nuclei of adult cells could be reprogrammed so that they exhibited cellular characteristics like those of fertilized eggs. In a later study, Japanese scientist Tomo Nakayama transplanted the nuclei of adult rat skin cells into embryonic stem cells. This study showed that the nuclei of adult cells could be reprogrammed into embryonic stem cells. These stem cells could differentiate into different types of cells [12].

Subsequent progress was that different reprogramming methods, such as transcription factors, were developed. In 2004, researchers dove into the mechanisms of nuclear reprogramming of adult cells. They found that specific transcription factors could reprogram somatic cells into pluripotent stem cells (capable of differentiating into many different types of cells). In 2006, scientist Shinya Yamanaka successfully transformed adult mouse skin cells into pluripotent stem cells capable of differentiating into multiple cell types by subjecting them to genetic transformations and cultures [13], which led to his winning the Nobel Prize. This achievement was considered a revolutionary advancement in cell biology and medicine and laid the foundation for subsequent research. In subsequent studies, scientists discovered that it was possible to obtain mature neurons by converting somatic cells into stem cells and then differentiating them into neural precursor cells with a dozen transcription factors [14–17].

Indeed, ectopic overexpression of specific neuronal determinants can reprogram non-neuronal cells directly into fully functional neurons in vivo and in vitro [18–20]. Since this pioneering work, researchers have explored how to exploit transcription factors, microRNAs (miRNAs), and gene silencing to reprogram highly differentiated, mature cells into neurons for therapeutic purposes. Generally, the rationale is to suppress the expression of genes subserving the current cell phenotype while activating genes that give rise to the target cell phenotype [21,22].

The complete reprogramming of somatic cells has evolved step by step, from early studies of nuclear transplantation of adult cells, to the discovery that pluripotency was induced in somatic cells overexpressing certain transcription factors, to the ability of somatic cells to transdifferentiate directly into neurons without going through the pluripotency stage. Sophisticated techniques and tools are now available to reprogram somatic cells into neurons. The next phase of research should be to dig deeper into the mechanisms of reprogramming, promote the efficiency gain of cell reprogramming, limit the risks of reprogramming, improve the safety of reprogramming in vivo, and, most importantly, develop therapeutic modalities that can be applied to clinical patients to truly bring the benefits of somatic cell reprogramming technology to patients with neurological diseases.

This review analyzes the history of efforts to reprogram adult somatic cells into neurons. First, we focus on somatic cell reprogramming and adult cell types that can be reprogrammed into neurons. We explore transdifferentiation strategies and factors and the molecular pathways involved. Subsequently, we explore the therapeutic potential of such somatic cell reprogramming as a treatment against neurological injury and illness and highlight the challenges that must be overcome. Researchers interested in reprogramming can use this review to obtain a quick overview of the technical means of reprogramming somatic cells into neurons and the related application studies. This review can also reveal future trends in somatic cell reprogramming into neurons.

2. Cell Types That Have Been Researched for Transdifferentiation into Neurons

Multiple types of somatic cells have been investigated in neural reprogramming. Reprogramming is easier and more effective when the somatic cell phenotype is similar to the target neuron lineage. Astrocytes are currently thought to be ideal candidates for neural repair. Human fibrocytes are also other ideal sources of reprogramming in this field because they are relatively easy to obtain.

Researchers have investigated the reprogramming of non-neuronal somatic cells into neurons using different cell types from other organs and embryonic cortices, with most studies focusing on ectodermal cell types. These ectodermal cells include fibroblasts [13,23–25], keratinocytes, oligodendrocytes [24], astrocytes [26], pericytes [27], and neuronal cells with the same neuronal identity [28]. T cells [29] and monocytes [30] from peripheral blood and hepatocytes [31] from visceral organs can also be reprogrammed into neurons. The success of these attempts has provided multiple cellular resources for neural reprogramming.

The more similar the initial adult cell phenotype is to the target neuronal phenotype, the more straightforward and effective the transdifferentiation procedure is. This is because embryonic stem and progenitor cells can more easily differentiate into cells with similar genealogical origins. The neuronal ectoderm produces fibroblasts, astrocytes, and pericytes [32]. Most early studies tried to generate neurons from fibroblasts, but astrocytes may be the ideal candidates for neuroregenerative reprogramming [33]. The main cell types used for brain regeneration and reprogramming are shown in Table 1.

Table 1. Cell types that can be reprogrammed into neurons, and the neuronal subtypes after reprogramming.

Original Cell Type	Type of Reprogrammed Neuron	References
Fibroblasts	Neurons	[24]
Infrapatellar fat pad stem cells	Neurons	[34]
Astrocytes	Neurons	[26,35]
Dental stem cell	Neurons	[36]
Hematopoietic cells	Induced neuronal cells	[37]
Urine-derived cells	Induced neuronal cells	[38]
Olfactory ensheathing cells	Neuronal cells	[39]
Spermatogonial stem cells	Dopaminergic neurons	[40]
Glioma cells	Neurons	[41]
Microglia	Neurons	[42]
Striatal neurons/post-mitotic callosal neurons	Neurons	[43]
Peripheral blood T cells	Neurons	[29]
Peripheral blood mononuclear cells	Neurons	[30]
Spiral ganglion non-neuronal cells	Cochlear hair cells and cochlear nucleus neurons	[44]
Pericytes	Cholinergic neurons	[27]
Pluripotent stem cell-derived cardiomyocytes	Neurons	[45]
Oligodendrocytes	Functional neurons	[46]

Table 1. *Cont.*

Original Cell Type	Type of Reprogrammed Neuron	References
NG2 cells	Neurons	[47]
Mesenchymal stem cells	Neural precursors	[48]
Hair follicle keratinocytes	Dopaminergic neurons	[32]
Müller glia	Neurogenic retinal progenitors	[49]
Adipose-derived stem cells	Neural stem cells and functional GABAergic neurons	[50]
Hepatocytes	Functional induced neuronal cells	[31]
Oligodendrocyte precursor cells	Neurons	[51]
Interfollicular keratinocytes	Neurons	[52]
Bone marrow-derived mesenchymal stem cells	GABAergic neurons	[53]

The somatic cell types presented in the table above are the somatic cells that are currently used for reprogramming into neurons. Some of these cells, such as fibroblasts and astrocytes, are used more frequently. Despite their abundance, most types of cells are not formed into sustainable cell lines, and the need to continuously isolate cultures from the body's limited tissues is a significant limitation.

3. Mature Somatic Cells Reprogram into Neurons through Different Pathways

Somatic cell reprogramming into neurons can occur by inducing pluripotency, stimulating neural stem cells, or directly reprogramming cells into specific neuronal subtypes. Induced pluripotent stem cells (iPSCs) can differentiate into different types of neurons, whereas neural stem cells only differentiate in specific environments. Adult neurons can be generated using direct reprogramming without the use of pluripotent or neural stem cells.

First, somatic cells can be forced to become pluripotent. Then, these stem cells differentiate into different kinds of neurons. For example, induced pluripotent stem cells (iPSCs) such as peripheral blood mononuclear cells display pluripotency after transdifferentiation; they can be converted into motor neurons using differentiation medium containing brain-derived neurotrophic factor (BDNF) or glial cell line-derived neurotrophic factor (GDNF) [30,54]. The resulting neurons express mature neuronal markers such as Tuj1, Map2, and synaptophysin, and they show spontaneous action potentials in patch clamp assays [14–17].

Second, somatic cells are stimulated to produce neural stem cells, which then differentiate into neurons in a particular environment, such as in differentiation medium. For example, treating oligodendrocyte precursors with bone morphogenetic protein generates neural stem cells that express stem cell markers and that can be incubated with exogenous factors to differentiate into oligodendrocytes, astrocytes, and neurons [51,55–57].

Third, somatic cells do not undergo the intermediate step of pluripotent stem cells or neural stem cells but direct reprogramming to specific neuronal subtypes. Overexpressing proneural transcription factors in astrocytes or applying exogenous factors can generate adult neurons without the need to induce pluripotent or neural stem cells [58–63]. During direct reprogramming, cells enter a transient pluripotent stage in which several neurogenesis-related genes are upregulated to complete the conversion into neurons [64]. Figure 1 depicts how induction can change non-neurons or neurons into target neurons.

Above all, somatic cells can be reprogrammed into neurons in three different ways. The reprogramming pathway differs for different types of cells, based primarily on the cells' characteristics and the method of induction used. The more general approach of direct reprogramming without going through the induced pluripotent stem cell phase may reduce the risk of tumorigenesis.

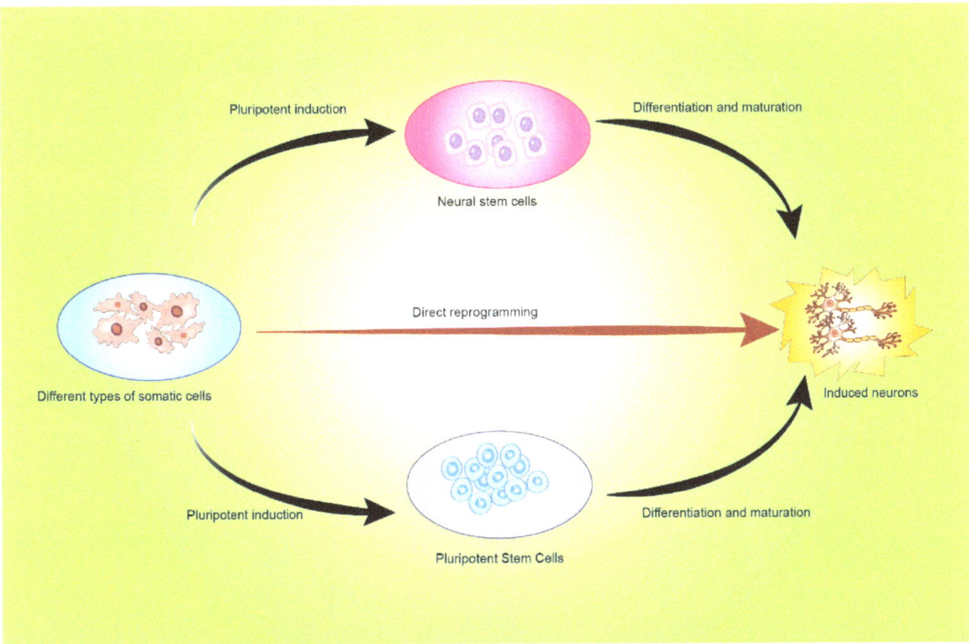

Figure 1. Pathways of neural reprogramming. First, some cells can be treated with induction factors to become pluripotent stem cells or neural stem cells, which then mature into neurons. Some somatic cells can differentiate directly into neurons.

4. Transcription Factors, Small Molecules, and miRNAs That Induce Transdifferentiation of Somatic Cells into Neurons

Scientists have used different methods to convert somatic cells into neurons, such as transcription factors, small chemical molecules, and microRNAs. Both transcription factors and microRNAs are essential in early neural development. The small molecules are generally involved in critical cellular pathways, and in turn, all three can be reshaped into somatic cell phenotypes, toward a neuronal phenotype.

The observation that bone morphogenetic factors can stimulate oligodendrocyte progenitors to differentiate into oligodendrocytes, astrocytes, and neurons [57] prompted researchers to examine new techniques for transforming somatic cells into neurons by using transcription factors, small molecules, and miRNAs (Tables 2 and 3).

Embryonic nervous system development involves numerous proneural transcription factors, including Ascl1, Mash1, Neurog1–3, Math1, KLF4, MYC, POU5F1, NeuroD1, Pax6, and Sox2. Ascl1 and Neurog2 induce growing stem cells to become an intermediate neuronal subtype [65]. Ascl1 regulates progenitor maintenance, neuronal differentiation, and neurite development in the central and peripheral nervous systems [66]. Based on their critical role in neurodevelopment, Ascl1 and Neurog2 cause somatic reprogramming by ectopic overexpression in somatic cells [48,49,67]. Pax6 is a member of the paired-box transcription factor family widely expressed in the developing central nervous system [68], where it regulates cortical progenitor cell proliferation, neurogenesis, migration, and forebrain axonal connections [69]. It also guides the differentiation of glial cells into neurons during embryonic mouse brain formation. Table 2 illustrates transcription factor families and their reprogramming roles.

Table 2. Transcription factors commonly used for reprogramming somatic cells into neurons.

Transcription Factor	Family Affiliation	Role in Neurogenesis, Differentiation, or Reprogramming	References
ASCL1	bHLH family	Determination of neuronal subtypes during neural development.	[70]
ATOH1	bHLH family	Unknown.	[71]
BCL11B	COUP TF1-interacting protein 2 (also known as Ctip2) and zinc finger-containing transcriptional repressors	Central to differentiation of medium spiny neurons and development of the striatum.	[72]
BCL2	Anti-apoptotic factor	Promotes DNA damage, genetic instability, and cell proliferation.	[33,73,74]
BRCA1	Tumor suppressor protein	Involved in BMP-2-mediated reactivation of Sox2.	[51]
BRN2a	Brain-specific homeobox/POU domain protein 2	Associated with neuroendocrine function.	[6,75,76]
CEND1	Neurogenic protein	Pathways involved in neuronal differentiation by CEND1 through activation by NEUROGENIN 1 and 2.	[77]
EBF1	Zinc finger	Acting downstream of Ngn, EBF-1 can promote ectopic neurogenesis.	[78,79]
FEZF2	Zinc finger transcriptional repressor	Fezf2 manipulates the origin, specific differentiation, and synaptic connectivity of corticospinal motor neurons by regulating neural progenitor cell lineage-directed differentiation signals.	[80]
FOXA2	Forkhead	Expressed in the ventral hindbrain's serotonergic progenitor regions and in midbrain dopaminergic neurons.	[28,81]
FOXG1	Forkhead	Involved in the primitive (anterior) neuroectoderm during development of embryonic stem cells.	[82,83]
GATA3	Zinc finger	Associated with noradrenergic phenotype and development of the sympathetic nervous system.	[84,85]
GATA4	GATA	GATA4 can drive embryonic Sertoli-like cell differentiation.	[86]
HAND2	bHLH family	Required for the acquisition of noradrenergic phenotype.	[84,85,87]
ISL2	LIM homeodomain-containing	Vital to the development and differentiation of visceral motor neurons in the spinal cord.	[78,88]
KLF4	Zinc finger	Directly represses p53.	[13]
LIN28	RNA binding protein	Lin-28 can shuttle between the nucleus and cytoplasm and regulate other genes that control the cell cycle.	[89]
LMO2	Key hematopoietic transcriptional regulator	Creates a regulatory complex that mediates transcription of multiple genes in hematopoietic progenitor cells; it is associated with the transcriptional control of stem/progenitor cells.	[90,91]
c-Myc	Myc	Alters expression of many proteins to enhance proliferation and transformation.	[13]
MYT1L	Neural zinc finger	Exemplifies a class of neural sequence-specific transcription factors that actively recruit histone deacetylases to selected genes during central nervous system development.	[92,93]
NANOG	Divergent homeodomain protein	NANOG sustains the identity of embryonic stem cells (ESCs).	[94]

Table 2. *Cont.*

Transcription Factor	Family Affiliation	Role in Neurogenesis, Differentiation, or Reprogramming	References
NEUROD1	bHLH family	Neural differentiation factor essential for late-stage neurogenesis and important in the development of the central nervous system, as well as in the auditory and vestibular systems.	[35,95,96]
NEUROD2	bHLH family	Essential for the maturation and survival of neurons in the central nervous system.	[97–99]
NEUROG2	bHLH family	NEUROG2 is a key contributor to early neurogenesis.	[67]
NURR1	Nr4a2 (ligand-independent nuclear receptors)	Essential for the differentiation, maturation, and maintenance of midbrain dopaminergic neurons.	[100,101]
OCT/4	POU5F1, a member of the POU class of homeodomain proteins	Central to the transcriptional regulatory hierarchy that specifies embryonic stem cell identity during early development.	[102]
OLIG2	Basic helix–loop–helix	Mediates self-renewal in the expansion of neurosphere cultures and promotes the generation of neurons and oligodendrocytes under differentiation conditions.	[103]
PAX6	Paired-box family	Critically important in multiple cell types and at several stages of forebrain development.	[104]
PHOX2A	Paired homeodomain	Selectively expressed and required for the specification of ventral motor neurons in the hindbrain and in the oculomotor nucleus, located laterally to dopaminergic neurons in the ventral midbrain.	[105]
Phox2B	Paired homeodomain	Same as Phox2a.	[105]
PTF1A	Basic helix–loop–helix	Mostly expressed in post-mitotic cells, and it specifies terminal cell fate in neural tissues.	[106]
SOX2	HMG-box	Central to the transcriptional regulatory hierarchy that specifies embryonic stem cell identity during early development.	[102]
SOX4	SoxC	Controls the survival of neural precursors and their differentiated progeny, in redundancy with SOX11.	[107]
SOX11	SoxC	Same as SOX4.	[107]
SV40LT	SV40 large T gene	Unknown.	[108]
TLX3	Tlx-class homeobox genes	Tlx3 functions as a post-mitotic selection gene in the embryonic spinal cord, determining the fate of dorsal glutamatergic neuronal cells.	[78,109]
ZEB1	Zinc finger E-box-binding transcription factor	During individual development, Zeb1 plays a crucial role in the nervous system. It is upregulated in growing neurons throughout the central nervous system and is required for the survival of spinal cord neural stem cells.	[110]

The basic helix–loop–helix protein family of transcription factors includes "proneuronal factors" [111], which play an indispensable role in neuronal commitment and in identifying neural progenitors. One such proneuronal factor is neurogenin2 (Ngn2), which controls neuronal development and identity [112]. Ngn2 regulates differentiation into the glutamatergic neuron phenotype, preventing the formation of γ-aminobutyric acid (GABA)-ergic neurons [70]. Ngn2 reprograms reactive astrocytes into deep cortical vertebral neurons that extend to the striatum, thalamus, and spinal cord [33,113,114].

Some of these transcription factors, alone or in combination, can induce non-neuronal somatic cells to reprogram into neurons. In a mouse model of cerebral infarction, over-expressed Olig2 and Pax6 reprogrammed glial cells into neurons in situ, as confirmed by the expression of the neuron-specific marker doublecortin and electrophysiological assays [115]. In another study, exogenous Brn2a, Myt1l, and ASCL1 induced mouse embryonic fibroblasts to differentiate into neurons that expressed microtubulin III (TUJ1) and microtubule-associated protein 2 (MAP2) [75]. POU5F1, SOX2, KLF4, and MYC can stimulate the development of non-neuronal somatic cells into neurons. One study overexpressed the transcription factors Oct4, Sox2, Klf4, and c-Myc in murine fibroblasts to obtain functional neural progenitor cells, which could differentiate into neurons and glial cells [116].

Hair follicle keratinocytes have been reprogrammed into iPSCs, which differentiate into neural precursors and, subsequently, into dopaminergic or glutamatergic neurons [32]. Overexpression of OCT3/4, Sox2, Klf4, and Myc in astrocytes led to iPSCs that expressed stem cell markers and showed potential to differentiate into neurons [117]. These various transcription factors can work as a complex or alone to reprogram non-neuronal mature somatic cells into neurons [49,62,118].

MicroRNAs also play an important role in the reprogramming of somatic cells into neurons. MicroRNAs (miRNAs) are key regulators of gene expression that control numerous cellular and developmental processes in eukaryotes [119]. Within the central nervous system (CNS), miRNAs play a critical role in the regulation of gene expression patterns during development [120], and are actively involved in the regulation of neurogenesis at each stage [121]. Moreover, miRNAs play a fundamental role in the establishment of specific neuronal phenotypes [122], such as miR-124, which is capable of inducing a neuronal phenotype when overexpressed in embryonic stem cells [123]. Embryonic stem cells are enriched in miRNA-124, miR-9/9*, and microRNAs 302-367, which can suppress the neural-gene-specific repressor REST complex, which is essential for the expression of genes related to neuronal function [26,98].

Researchers have successfully induced the reprogramming of fibroblasts into neurons using miR-9/9* and miR-124 [72,124], and active astrocytes into neurons in normal and Alzheimer's disease models using microRNA-302/367 [26]. However, microRNA-302/367 acts as a mediator and enhances the induction effect of miR-9/9* and miR-124 on fibroblasts, and may not be sufficient by itself for fibroblast reprogramming [25]. Furthermore, miRNAs can be used in combination with other methods, such as the combination of miR-9/9* and miR-124 with transcription factors to induce the reprogramming of fibroblasts from bipolar disorder patients into neurons [97]. Additionally, miR-124-9-9* has been shown to increase the efficiency of Ascl1-induced reprogramming of Müller glia into neurons [125]. Taken together, miRNAs are important players in the reprogramming of somatic cells into neurons, providing exciting prospects for regenerative medicine research.

Dozens of small compounds have been used as activators or inhibitors of critical signaling pathways to regulate reprogramming, preserve cellular stability during transdifferentiation, and avoid cell death (Table 3). In 2013, a study revealed that exogenous small molecules were sufficient to convert mouse somatic cells into pluripotent stem cells [126]. Since then, various research teams have explored hundreds of small molecules for somatic cell reprogramming, including valproic acid, CHIR99021, repsox, forskolin, SP600125, GO6983, and Y-27632 [54]. Combinations of small molecules such as these can induce the transdifferentiation of fibroblasts, astrocytes, and glioblastomas into neurons [54,127,128].

Table 3. List of small molecules commonly used for reprogramming somatic cells into neurons and differentiating them during their maturation.

Small Molecule	Description	References
616452	Repsox, an ALK5 inhibitor.	[129]
17-allylaminogeldanamycin	GSK3 inhibitor.	[129]
A83-01	ALK4/5/7 inhibitor.	[38,130,131]
All-trans retinoic acid	Physiologically active metabolite of vitamin A.	[53,132,133]
AM580	Retinoic acid agonist, stable benzoic derivative of retinoic acid.	[129,134]

Table 3. *Cont.*

Small Molecule	Description	References
Apicidin	HDAC inhibitor.	[51]
Azacytidine	Nucleic acid synthesis inhibitor.	[135]
Blebbistatin	NMII inhibitor.	[136]
Bradykinin	Bradykinin plays a role in neural fate determination and facilitates neurogenesis and migration.	[137]
CH55	Synthetic stable analog of retinoic acid.	[138]
CHIR99021	GSK3 inhibitor.	[129]
CpdE	Notch signaling pathway inhibitor.	[139]
DAPT	Inhibits γ-secretase and Notch signaling.	[53,140]
DMH1	BMP type I receptor inhibitor.	[139]
Dorsomorphin	Inhibitor of AMP-activated protein kinase and bone morphogenetic protein type 1 receptor.	[67]
DZNep	Histone methylation inhibitor.	[141]
EPZ004777	Dot1l inhibitor.	[138,142]
Forskolin	cAMP agonist.	[129]
GSK3β inhibitor	Glycogen synthase kinase-3β boosts the production of neuroprotective and neurotrophic factors in the context of spinal cord injury.	[143]
GO6983	PKC inhibitor.	[54,144]
Hh-Ag1.5	Unknown.	[131]
I-BET 151	BET family bromodomain inhibitor.	[145]
Insulin–transferrin–selenium	Insulin, transferring, and sodium selenium compound.	[135]
Isoxazole	Isoxazole is able to upregulate proneural marker genes and exhibit regulation of stem cells.	[41]
ISX9	Induces neuronal differentiation through myocyte enhancer factor 2 (Mef2), which is a vital pathway for neural differentiation and maturation.	[128]
Kenpaullone	GSK-3β inhibitor.	[146]
LDN193189	Inhibitor of bone morphogenetic protein type I receptors ALK2 and ALK3, used to suppress specification of mesoderm and endoderm.	[147]
LIF	Leukemia inhibitory factor.	[148,149]
LM-22A4	Growth factor.	[150]
Mercaptoethanol	Unknown.	[148]
MS-275	Benzamide.	[51]
Niclosamide	Wnt signaling inhibitor.	[34]
Noggin	SMAD inhibitor.	[151]
NT3	Unknown.	[150]
P7C3-A20	May stimulate NAMPT-relevant pathways to exert neurogenesis.	[24]
Parnate	Lysine-specific demethylase 1 inhibitor.	[24]
PD0325901	Mitogen-activated protein kinase inhibitor.	[24]
PS48	PDK1 activator.	[152]
Purmorphamine	Activator of the Shh signaling pathway.	[153]
Quercetin	PI3K signaling inhibitor.	[34]
QVD-OPH	Caspase inhibitor.	[144]
Repsox	Transforming growth factor-β inhibitor.	[154]
Retinoic acid	Induces neurogenesis and neuronal differentiation by activating retinoic acid receptors.	[38]
RG108	DNA methyltransferase inhibitor that is less toxic to cells than parnate.	[24]
Ruxolitinib	Selective JAK1/2 inhibitor.	[155]
SB203580	P38 MARK inhibitor.	[145,155]
SB43152	Unknown.	[153]
SB431542	Inhibits TGF-β type I receptors ALK4, ALK5, and ALK7.	[30,147]
SB4352	Transforming growth factor-beta inhibitor	[36]

Table 3. *Cont.*

Small Molecule	Description	References
SMER28	SMER28 shows neurotrophic and neuroprotective effects by inducing neurite growth and protecting against excitotoxin-induced axonal degeneration.	[156,157]
Smoothened agonist	Alone or in concert with other molecules, smoothened agonist stimulates proliferation of primary neuronal precursor cells.	[158,159]
Sodium butyrate	HDAC inhibitor, causes hyperacetylation of histones.	[38]
Sonic hedgehog (SHH)	Required for the development of dopaminergic neurons in multiple locations along the anterior neural tube.	[160,161]
SP600125	1,9-pyrazoloanthrone, JNK inhibitor.	[162]
SP600625	JNK inhibitor.	[144]
TD114-2	GSK3-beta inhibitor, preferred over CHIR99021.	[138]
Transforming growth factor beta 3	Required for the induction, differentiation, and survival of midbrain dopaminergic neurons.	[163]
Thiazovivin	Unknown.	[164]
Tranylcypromine	Lysine-specific histone demethylase LSD1 inhibitor.	[141]
Trichostatin A	Histone deacetylase inhibitor.	[165]
TTNPB	Agonist of retinoic acid receptors, which play an important role in neural differentiation.	[164]
Valproic acid	Inhibits histone deacetylase activity.	[166]
Vitamin C	Prevents cell death.	[144]
Y-27632	Rho-associated protein kinase inhibitor.	[162]

Abbreviations: ALK: Anaplastic lymphoma kinase; GSK: glycogen synthase kinase; NMII: Nonmuscle myosin II; BMP: Bone morphogenetic protein; AMP: Adenosine monophosphate; Dot1l: Known as KMT4L; cAMP: Cyclic adenosine monophosphate; PKC: Protein Kinase C; BET family: Bromodomain and extra-terminal domain family; SMAD: small mother against decapentaplegic; NAMPT: Nicotinamide phosphoribosyltransferase; PDK1: 3-phosphoinositide-dependent kinase 1; PI3K: Phosphoinositide 3-kinase; JAK1/2: Janus kinases 1/2; MARK: Microtubule-affinity regulating kinases; TGF: Transforming growth factor; HDAC: Histone deacetylase; JNK: c-Jun N-terminal kinase; LSD1: Lysine-specific histone demethylase 1.

Transcription factors, microRNAs, and small molecules can be used individually or in combination to induce cellular reprogramming. When used together, they complement each other and increase efficiency of the reprogramming process. This is partly due to the ability of transcription factors to bind microRNAs [72] and small molecules [67]. For example, cells can undergo early stages of reprogramming by overexpressing transcription factors or adding microRNAs. The reprogrammed cells can then be induced to mature through the addition of valproic acid, forskolin, vitamin C, and BDNF [16,30]. Small molecules can also enhance reprogramming by modifying DNA or histone structure [164].

We have listed the technical approaches to reprogram somatic cells into neurons above, but these induction factors still need to be ideal. What we require is a method that is simpler, more efficient, safe, and clinically applicable. The current approach may still be in its infancy.

5. Molecular Mechanisms of Somatic Cell Transdifferentiation into Neurons

Cells are significantly transformed in all aspects of the reprogramming process, including changes in DNA plasticity, transcriptome, and energy usage. DNA plasticity refers to how inducing factors cause chromatin and gene expression to change over time. During reprogramming, there is also a dramatic change in the transcriptome. Genes related to the original somatic cell phenotype are downregulated, while genes related to the neural type are upregulated. Moreover, the energy process changes from glycolysis to aerobic oxidation, in which an induced neuron can exercise normal neurological function. A deeper understanding of these mechanisms can help us improve the efficiency of somatic reprogramming and prevent potential harm to the cells. A growing body of literature has explored the roles of changes in chromatin, transcription, translation, and metabolism in somatic reprogramming. Analysis of cell-specific markers and phenotypes has revealed three levels of change that may explain cellular transdifferentiation [67].

The first level is DNA plasticity. When cells are adequately triggered, such as in microglia overexpressing the transcription factor NeuroD1, DNA is dynamic [167], with changes in the accessibility of chromatin [42]. The expression of genes linked to neurons is upregulated, while motifs linked to the initial somatic cell phenotype are downregulated, as revealed by extensive RNA sequencing [147,168].

Even in mature, highly differentiated somatic cells, chromatin is not immutable, as demonstrated by the cloning of Dolly the sheep, macaques, and salamanders [10,169,170]. Mature somatic cell chromatin can be activated and altered to express specific genes. When somatic cells are reprogrammed, chromatin undergoes DNA methylation [171] and demethylation [172], which upregulates neurogenesis-related gene expression. One study found that demethylation of H3K9me3 in chromatin improved somatic cell nuclear transfer and promoted cellular reprogramming in mice and humans [173]. Direct fibroblast-to-neuron reprogramming involves DNA methylation remodeling; changes in DNA methylation inhibit fibroblasts' myogenic program and promote activation of neuronal genes [171]. Genetic epistasis studies with H3K9 methyltransferase suggest that this chromatin change restricts plasticity by acting downstream of the end selector. Terminal selectors activate identity-specific genes and reduce the accessibility of non-identity-defining genes, balancing identity specification with cellular plasticity [174].

Secondly, emerging technologies such as RNA and single-cell sequencing support the idea that somatic cell reprogramming alters the transcriptome, such as in the case of glial cells [147] and fibroblasts [175]. Overall, genes related to maintaining the original somatic phenotype are downregulated during cell reprogramming, while genes associated with pluripotent stem cells, neuronal progenitors, or neuronal phenotypes are upregulated. For example, ASCL1 expression alters Müller cell chromatin and temporarily activates progenitor gene expression while suppressing glial cell gene expression, inducing Müller glia-derived progenitor cells to become neurons [49]. Ectopic expression of the neuronal transcription factor NeuroD1 causes rapid transcriptome changes during the astrocyte-to-neuron transition [176]. The shift toward a neuronal identity has been linked to changes in signaling pathways involving Notch and p21/p53 pathways [147,177]; indeed, inhibition of p53 or p21 increases production of induced adult neuroblasts from glial cells [177].

Third, during transdifferentiation, energy metabolism switches from glycolysis to aerobic oxidation. Other somatic cells, such as astrocytes, fibroblasts, brain stem cells, or progenitor cells, depend more on β-oxidation and anaerobic glycolysis for energy than neurons, which depend primarily on oxidative metabolism [178,179]. Furthermore, neuritogenesis can be enhanced by increasing the mitochondrial membrane potential, polarizing the mitochondria, and decreasing the reliance on glycolysis [180]. Because ATP generation is required for synaptic activity, neuronal activity strongly depends on normal functioning of neuronal mitochondria and energy metabolism [181,182]. As a result, one of the most visible aspects of the transdifferentiation process is the abrupt change in metabolism [33]. Cells surviving the severe energy change during reprogramming are candidates for becoming neurons [33]. At the same time, a dramatic energy shift and a significant accumulation of reactive oxygen species can harm cells and cause ferroptosis [33]. Therefore, a more in-depth study of metabolic shifts in successfully reprogramed somatic cells and exploration of ways to promote successful energy conversion and inhibit ferroptosis can improve reprogramming efficiency [33].

The central mechanism for reprogramming somatic cells into neurons is a phenotypic shift involving intracellular changes that should, in theory, be extensive. In addition to what we have shown above, other unknown mechanisms require more profound and advanced exploration methods.

6. Somatic Cell Transdifferentiation Provides New Possibilities for Research and Treatment of Neurological Disease

By reprogramming the somatic cells involved in neurological diseases, scientists can learn more about diseases and develop new treatment methods. This entails transforming somatic cells from patients with various genetic or age-related neurological diseases into neurons that retain the patient's chromosomal mutations and display disease symptoms. In vitro disease modeling with induced neurons assists researchers in determining disease causes. In addition, researchers are investigating the possibility of reprogramming to reconstruct damaged neural networks and enhance the neural function of animals. Transforming induced neural stem cells into the mouse brain improved axonal regeneration, motor function, and electrophysiological activity. It has significant implications for the treatment of numerous diseases.

Somatic cell reprogramming for neurological disease research allows access to elusive tissues, facilitating investigations into disease mechanisms and novel interventions. For instance, somatic cells procured from patients afflicted with a range of genetic and age-related neurological disorders can be induced to differentiate into neurons. These induced neurons, which retain the patient's chromosomal mutations [183] or display characteristics such as epilepsy-like hyperactivity [184] and defective neural networks in patients with autism [185,186], enable investigations into the underlying mechanisms of the diseases. Furthermore, motor neurons induced directly from fibroblasts taken from elderly patients exhibit age-related characteristics [187], making them an exceptional model for investigating late-stage motor neuron diseases. In vitro modeling is valuable for examining disease mechanisms, drug screening, and toxicity testing in patients with neurological disorders.

The induction of neurons holds potential for the treatment of neurological diseases. Research teams are using methods such as cell transplantation or in vivo reprogramming to reconstruct damaged neural networks and improve animal neural function. Corti et al. transplanted neural stem cells obtained from in vitro reprogramming of somatic cells into the mouse brain, which successfully integrated into the local brain cortex [94]. The transplanted induced neural stem cells could differentiate into all neural lineages and restore axonal regeneration in a spinal cord injury (SCI) model [188]. It was discovered that iNSC transplantation could promote motor function and electrophysiological activity recovery, as confirmed by functional evaluation. The use of somatic cell reprogramming has significant implications for treating a range of diseases, including neurodegenerative diseases, neurological disorders, and psychiatric conditions [18,20,35,95]. We have summarized the research on somatic cell reprogramming to become neurons for neurodegenerative diseases, neurological diseases, and psychiatric diseases in Table 4.

Table 4. List of neurodegenerative, neurological and psychiatric diseases where models based on reprogramming somatic cells into neurons have been used.

Disease	Applications/Results of Neuronal Reprogramming	Reference
Dravet syndrome	Fibroblasts derived from controls and patients were differentiated into neurons. Epilepsy-specific iPSC-derived neurons are helpful for modeling epilepsy-like hyperactivity.	[184]
MT-ATP6	Skin fibroblast reprogramming and iPSCs can model disease caused by the MT-ATP6 mutation.	[189]
Alzheimer's disease	Small molecules induce the reprogramming of patient fibroblasts into neurons for personalized modeling of neurological disease.	[54]
Fragile X syndrome	Fibroblasts from patients can be induced into iPSC lines to enable in vitro modeling of the human disease.	[190]

Table 4. *Cont.*

Disease	Applications/Results of Neuronal Reprogramming	Reference
Multiple sclerosis	iPSCs from peripheral blood mononuclear cells can be used to model multiple sclerosis.	[191]
Glioblastoma multiforme	Isoxazole acts as a stem cell modulator to trigger neuronal gene expression and block tumor cell proliferation, which may guide research into reprogramming as an antitumor strategy.	[41]
Frontotemporal dementia, amyotrophic lateral sclerosis	Fibroblasts were isolated from patients' skin to generate induced pluripotent stem cells to investigate the pathological mechanisms underlying frontotemporal dementia or amyotrophic lateral sclerosis.	[192]
Huntington's disease	Stable HD-iPS cell lines have been established to investigate disease mechanisms.	[193]
Schizophrenia	The authors directly reprogrammed fibroblasts/hair follicle-derived cells from schizophrenia patients into iPSCs, which they differentiated into neurons. These neurons were then studied for disease pathology.	[32,194]
Spinal cord injury	NOTCH1 signaling regulates the latent neurogenic program in adult reactive astrocytes after spinal cord injury.	[195]
Amyotrophic lateral sclerosis	Peripheral blood cells from an ALS patient carrying the TARDBP p.A382T mutation were reprogrammed into iPSCs.	[30]
Stroke	Overexpression of Ascl1 can convert astrocytes from the subventricular zone into neurons in vivo after stroke.	[8]
Parkinson's disease	Fibroblasts were taken from the pathology biopsies of Parkinson's disease patients and encouraged to develop into dopaminergic neurons, which can be used for future studies into the mechanistic underpinnings of the disease.	[151]
Rett syndrome	Overexpressing reprogramming factors in Rett syndrome fibroblasts generated iPSCs, which differentiated into neurons with a neuronal maturation phenotype similar to that of the clinical syndrome.	[196]
Neurodevelopmental disorders	Human hair follicle-derived iPSCs can be differentiated into various neural lineages. This experimental system provides an in vitro model to study normal and pathological neural development without the need for skin biopsies.	[197]
Ageing	Directly converted astrocytes retain the ageing features of the donor fibroblasts and clarify the astrocytic contribution to human CNS health and disease.	[198]
Bipolar disorder	Human fibroblasts can be reprogrammed into induced neurons or iPSCs, then differentiated into neurons for mechanistic studies of the disease.	[97]
Pain	Transcription factors can transform mouse and human fibroblasts into noxious-stimulus-detecting (injury receptor) neurons, which displayed TrpV1-mediated sensitization to inflammation.	[78]
Demyelinating diseases	Exposing mouse embryonic fibroblasts to chemical conditions could induce their differentiation into OPC-like cells, which may serve as a therapeutic strategy for treating demyelinating diseases.	[131]
Mitochondrial DNA mutations	This work generated stem cells from patients carrying the most common human disease mutation in mitochondrial DNA, m.3243A>G (MELAS).	[183]

Table 4. *Cont.*

Disease	Applications/Results of Neuronal Reprogramming	Reference
Autism spectrum disorders	Fibroblasts from patients can be reprogrammed into neurons with fewer excitatory synapses, a faulty neural network phenotype, or a synaptic phenotype comparable to that induced by autism-associated neuroligand protein-3 mutations, confirming the use of induced neuronal cells for disease modeling.	[185,186]

Abbreviations: iPSC: induced pluripotent stem cell; MT-ATP6: mitochondrial ATP synthase subunit 6 gene; HD: Huntington's disease; ALS: amyotrophic lateral sclerosis; TARDBP: Transactive response DNA binding protein; CNS: Central Nervous System; TrpV1: transient receptor potential vanilloid-1 channel; OPC: oligodendrocyte progenitor cells; MELAS: mitochondrial encephalomyopathy, lactic acidosis, and stroke-like episodes (MELAS) syndrome; iN: induced neuronal cells.

The reprogramming of somatic cells into neurons holds significant potential for the treatment of neurological diseases, with focus on disease processes and therapeutic applications. This approach can advance our understanding of disease development and and facilitate relevant drug screening and testing. There will still be gaps in treatment due to unresolved risks and safety issues. As a result, treatment is still in the animal model stage.

7. Limitations of Neural Reprogramming

Reprogramming somatic cells into neurons has the potential to be beneficial. However, limitations and risks cannot be overlooked, such as low efficiency, immune rejection, genetic mutations, tumorigenicity, and potentially unknown risks. The reprogramming process varies in efficiency, with reported rates ranging from as low as 0.01% [193] to as high as 90% [199]. The attainment of sufficient neuron numbers requires careful consideration of multiple factors, including cell type and reprogramming technique. Another limitation and danger of somatic cell reprogramming is the potential for immunological rejection, as reprogrammed cells may be viewed as foreign by the immune system. Even induced pluripotent stem cells (iPSCs) derived from host mice can elicit immunological rejection and cause teratomas when transplanted into animals [200]. The immunogenicity of iPSCs is attributed to genetic and epigenetic defects, which warrants caution.

The long-term stability of generated neurons is also a concern, as cells must maintain normal physiological function after reprogramming. However, concerns remain about the quality and durability of generated neurons, limiting the potential of somatic cell reprogramming.

Safety is the most critical issue in somatic cell reprogramming due to the potential risks. During reprogramming, genetic mutations may occur as a result of random integration of viral gene segments carried by retroviruses or lentiviruses into host cells, leading to alterations of essential host cell genes that may cause deleterious consequences or even cell death [201–203]. Uncontrolled proliferation and differentiation during reprogramming may result in tumor growth, while unanticipated genomic changes during gene editing can lead to the formation of malignant tumors [204]. The retroviruses used by Yamanaka carry a potential cancer risk, as they can fuse with the DNA of host cells. Furthermore, the Myc gene, one of Yamanaka's four reprogramming factors, is an oncogene [205,206].

The safety risks associated with therapeutic somatic cell reprogramming persist, requiring further research to bridge the gap between laboratory research and clinical application. Improving the safety and reliability of the technology should be a primary focus of future research to protect human health while facilitating practical use. Given the significant obstacles, it is clear that there are substantial hurdles to overcome before this approach can be employed at the bedside.

8. Discussion

Reprogramming somatic cells into neurons has emerged as a prominent field of research over the past 20 years. This breakthrough has not only highlighted the plasticity of somatic cells but also opened up new possibilities for research and therapy in regenerative medicine. A wide range of somatic cells, including fibroblasts [24], astrocytes [26], pericytes [27], peripheral blood mononuclear cells [30], T lymphocytes [29], hepatocytes [31], and other cells, can now be transformed into neurons. With the refinement and standardization of techniques, the means to achieve this transformation are becoming more sophisticated and standardized. Several factors, such as transcription factors [13], microRNA [72,124], and a combination of different small chemical molecules [54], have been shown to induce the reprogramming of somatic cells into neurons. These methods, when combined, can further enhance reprogramming efficiency and stabilize induced neuronal cell stability.

During the reprogramming process, some somatic cells reprogrammed into neurons may undergo cellular pluripotency, reversing their differentiation from somatic cells to induced pluripotent stem cells or neural stem cells, followed by induced differentiation and maturation. This process gradually allows induced pluripotent stem cells or neural stem cells to differentiate and mature, acquiring neuron-associated phenotypes, neuron-like morphology, electrophysiological functions, and neuron-associated markers [51,57]. On the other hand, some somatic cells do not undergo pluripotency in response to inducing factors, instead undergoing direct reprogramming into another phenotypically distinct neuronal cell [64]. This process is also known as transdifferentiation. The reprogramming of somatic cells into neurons is a promising area of research that has demonstrated the potential of somatic cells in regenerative medicine. With the refinement and standardization of reprogramming techniques, this field holds great promise for future advances in the field of neuroscience research and regenerative medicine.

Many studies have primarily focused on the reprogramming of adult cells into neurons, yet the physiological and biochemical mechanisms involved in this process remain largely unknown. In this review, we aimed to integrate several studies in the literature and to provide a comprehensive summary of the mechanisms involved in reprogramming at three levels of physiological and biochemical processes. Firstly, at the chromatin level, the reprogramming process involves changes to the epigenetic landscape of somatic cells. These alterations affect the accessibility of reprogramming techniques [67], leading to changes in gene expression patterns. Secondly, the overall transcriptome level of somatic cells shifts towards a neuronal identity [204], characterized by the upregulation of genes associated with the neuronal phenotype and the downregulation of genes related to the original cell phenotype. Finally, at the metabolic level, the reprogramming process requires significant changes to cellular metabolism to meet the high energy demands of neuronal cells [33]. The reprogramming of somatic cells into neurons is a complex process that involves physiological and biochemical processes at multiple levels. Our review sheds light on the fundamental mechanisms underlying this process and may pave the way for future advancements in regenerative medicine and neuroscience.

The technique of reprogramming somatic cells into neurons is now a well-established and feasible method. This process allows us to have a more sustainable source of research material for studying neurological diseases. Moreover, several studies have been conducted on the in vitro reprogramming of neurons to establish disease models for various neurological disorders, such as ALS [181]. In animal models, cell transplantation therapy has been shown to effectively restore damaged neurological function [207].

While the field of somatic cell reprogramming for neurons is rapidly advancing, it is essential to acknowledge that this process also poses risks and challenges. Issues such as tumor formation, immunological rejection, cellular stability, the highly variable efficiency of reprogramming, and the ability of reprogrammed neuronal cells to integrate into local networks and function correctly are all critical factors that require our attention [5].

As previously mentioned, reprogramming adult cells into neurons involves the overexpression of a neural transcription factor in adult cells using a viral vector carrying the target gene, which is then expressed after viral infection of the host cell. The target gene is integrated into the chromatin of the host cell, and random mutations in the chromatin of the somatic cell can occur [201–203]. Sometimes, the cell may even die outright if the mutation occurs at a binding site. The induction of chemical molecules in a manner that does not involve gene insertion can also have toxic effects [37].

Reducing the risks and concerns associated with somatic cell reprogramming is necessary. To achieve this, it is essential to continue to improve and standardize the reprogramming process to make it as safe and efficient as possible. Additionally, further research is needed to better understand the underlying mechanisms of reprogramming and the factors that influence its effectiveness.

While reprogramming somatic cells into neurons holds great promise for advancing our understanding and treatment of neurological diseases, we must remain vigilant of the potential risks and challenges associated with this technique. By continuing to refine our methods and understanding the underlying mechanisms involved, we can ensure this technology's safe and practical application in the clinic. Therefore, non-integrating approaches for somatic cell reprogramming have been proposed to address these issues. Various methods have been suggested, such as using a plasmid to construct the overexpressed gene and introducing it into the somatic cell via electroporation to achieve overexpression. Gotz's research team has made significant progress in somatic cell reprogramming by using the CRISPR-Cas9 system to facilitate gene overexpression while avoiding the introduction of random mutations during cell reprogramming [208].

Furthermore, inducing pluripotency in somatic cells before differentiation into neurons may lead to the overexpression of genes associated with tumor formation and uncontrolled cellular proliferation. Additionally, the transcription factor c-MYC, commonly used in pluripotency induction, is inherently tumorigenic and closely linked to tumor formation and invasion [205,206]. Hence, it is essential to carefully select reprogramming factors to ensure successful and efficient induction and develop more advanced monitoring and control tools to establish the basis for future clinical applications of somatic reprogramming. In addition, further animal models and mechanistic studies are necessary to apply this tool in the clinical setting. Several research teams are currently using non-human primates to establish studies related to somatic reprogramming [209]. Non-human primates are the animals closest to humans regarding physiology, biochemistry, and metabolism, and strengthening animal studies is necessary to translate research into the clinic.

Reprogramming somatic cells into neurons has opened up an innovative and promising avenue for developing cell-based therapies for neurological diseases. This method could overcome the limitations of conventional treatments and provide a more effective and targeted approach to treating neurological disorders. Furthermore, the ability to generate patient-specific neurons through reprogramming could improve our understanding of the underlying mechanisms of neurological diseases and facilitate the development of personalized treatments. Reprogramming somatic cells into neurons is valuable for investigating and treating neurological diseases. With continued advancements in cellular reprogramming, we hope for significant progress in developing novel treatments and therapies for various neurological disorders.

Therefore, current research confirms that applying this technology to clinical treatment still has a long way to go. We need to develop safer, more stable, more effective, and more precise reprogramming tools and more sensitive detection and control of reprogramming techniques. The ultimate goal of reprogramming somatic cells into neurons is to obtain enough neurons to replace the neurons lost in the nervous system for various reasons. Obtaining sufficient neurons and ensuring newborn neurons correctly integrate into the local neural network after reprogramming to function appropriately for an extended period through precise and controlled reprogramming are the critical factors in applying neural reprogramming techniques in the clinical treatment of patients.

9. Conclusions

Somatic reprogramming has demonstrated promise in neuroregeneration and the study of neurological disorders. Small molecules and transcription factors have the ability to transform cells into neurons, which can then be used to model diseases, test drugs, and rebuild neural networks. Nonetheless, the complex environment in vivo can result in a variety of outcomes, and there is insufficient information about the safety of gene vectors in animals or humans. These are issues that must be addressed. More research is needed to improve how reprogramming for neuronal regeneration works and what it can be used for. Millions of patients around the world could benefit from more intelligent and better therapies if the many paradoxes in somatic cell reprogramming can be resolved.

Author Contributions: J.C. developed the topic, integrated the literature, drew figures, and wrote the first draft. L.H. screened the literature and created tables. Y.Y., W.X. and Q.Q. edited and revised the manuscript. R.Q., X.L. (Xiaojun Liang), X.L. (Xinyu Lai), X.H. and M.X. did the initial screening of the extensive literature and the acquisition of basic information. L.C. obtained funding and provided scientific suggestions for the article. All authors reviewed the final manuscript. All authors have read and agreed to the published version of the manuscript.

Funding: This study was supported by grants from the Natural Science Foundation of China (82271371 and 82260367), Health Appropriate Technology Development and Application Project (S2021107), and the Clinical Research "Climbing" Program of the First Affiliated Hospital of Guangxi Medical University (YYZS2021002).

Conflicts of Interest: The authors declare no financial or non-financial interest or relationships that could be perceived as potentially influencing the research reported in this article.

References

1. Prabhakaran, D.; Anand, S.; Watkins, D.; Gaziano, T.; Wu, Y.; Mbanya, J.C.; Nugent, R. Cardiovascular, respiratory, and related disorders: Key messages from Disease Control Priorities, 3rd edition. *Lancet* **2018**, *391*, 1224–1236. [CrossRef] [PubMed]
2. Sharma, A.; Sharma, R.; Zhang, Z.; Liaw, K.; Kambhampati, S.P.; Porterfield, J.E.; Lin, K.C.; DeRidder, L.B.; Kannan, S.; Kannan, R.M. Dense hydroxyl polyethylene glycol dendrimer targets activated glia in multiple CNS disorders. *Sci. Adv.* **2020**, *6*, eaay8514. [CrossRef] [PubMed]
3. Collaborators, G.D.a.I. Global burden of 369 diseases and injuries in 204 countries and territories, 1990–2019: A systematic analysis for the Global Burden of Disease Study 2019. *Lancet* **2020**, *396*, 1204–1222. [CrossRef]
4. Tasoglu, S.; Gurkan, U.A.; Wang, S.; Demirci, U. Manipulating biological agents and cells in micro-scale volumes for applications in medicine. *Chem. Soc. Rev.* **2013**, *42*, 5788–5808. [CrossRef]
5. Grade, S.; Götz, M. Neuronal replacement therapy: Previous achievements and challenges ahead. *NPJ Regen. Med.* **2017**, *2*, 29. [CrossRef]
6. Torper, O.; Pfisterer, U.; Wolf, D.A.; Pereira, M.; Lau, S.; Jakobsson, J.; Björklund, A.; Grealish, S.; Parmar, M. Generation of induced neurons via direct conversion in vivo. *Proc. Natl. Acad. Sci. USA* **2013**, *110*, 7038–7043. [CrossRef]
7. Denoth-Lippuner, A.; Jessberger, S. Formation and integration of new neurons in the adult hippocampus. *Nat. Rev. Neurosci.* **2021**, *22*, 223–236. [CrossRef]
8. Faiz, M.; Sachewsky, N.; Gascón, S.; Bang, K.W.; Morshead, C.M.; Nagy, A. Adult Neural Stem Cells from the Subventricular Zone Give Rise to Reactive Astrocytes in the Cortex after Stroke. *Cell Stem Cell* **2015**, *17*, 624–634. [CrossRef] [PubMed]
9. Srivastava, D.; DeWitt, N. In Vivo Cellular Reprogramming: The Next Generation. *Cell* **2016**, *166*, 1386–1396. [CrossRef]
10. Gurdon, J.B.; Elsdale, T.R.; Fischberg, M. Sexually mature individuals of Xenopus laevis from the transplantation of single somatic nuclei. *Nature* **1958**, *182*, 64–65. [CrossRef]
11. Wilmut, I.; Schnieke, A.E.; McWhir, J.; Kind, A.J.; Campbell, K.H. Viable offspring derived from fetal and adult mammalian cells. *Nature* **1997**, *385*, 810–813. [CrossRef]
12. Wakayama, T.; Perry, A.C.; Zuccotti, M.; Johnson, K.R.; Yanagimachi, R. Full-term development of mice from enucleated oocytes injected with cumulus cell nuclei. *Nature* **1998**, *394*, 369–374. [CrossRef] [PubMed]
13. Takahashi, K.; Yamanaka, S. Induction of pluripotent stem cells from mouse embryonic and adult fibroblast cultures by defined factors. *Cell* **2006**, *126*, 663–676. [CrossRef]
14. de Leeuw, V.C.; van Oostrom, C.T.M.; Imholz, S.; Piersma, A.H.; Hessel, E.V.S.; Dollé, M.E.T. Going Back and Forth: Episomal Vector Reprogramming of Peripheral Blood Mononuclear Cells to Induced Pluripotent Stem Cells and Subsequent Differentiation into Cardiomyocytes and Neuron-Astrocyte Co-cultures. *Cell. Reprogram.* **2020**, *22*, 300–310. [CrossRef]
15. Takahashi, K.; Tanabe, K.; Ohnuki, M.; Narita, M.; Ichisaka, T.; Tomoda, K.; Yamanaka, S. Induction of pluripotent stem cells from adult human fibroblasts by defined factors. *Cell* **2007**, *131*, 861–872. [CrossRef] [PubMed]
16. Habekost, M.; Jorgensen, A.L.; Qvist, P.; Denham, M. MicroRNAs and Ascl1 facilitate direct conversion of porcine fibroblasts into induced neurons. *Stem Cell Res.* **2020**, *48*, 101984. [CrossRef]
17. Onorati, M.; Camnasio, S.; Binetti, M.; Jung, C.B.; Moretti, A.; Cattaneo, E. Neuropotent self-renewing neural stem (NS) cells derived from mouse induced pluripotent stem (iPS) cells. *Mol. Cell. Neurosci.* **2010**, *43*, 287–295. [CrossRef]

18. Wu, Z.; Parry, M.; Hou, X.Y.; Liu, M.H.; Wang, H.; Cain, R.; Pei, Z.F.; Chen, Y.C.; Guo, Z.Y.; Abhijeet, S.; et al. Gene therapy conversion of striatal astrocytes into GABAergic neurons in mouse models of Huntington's disease. *Nat. Commun.* **2020**, *11*, 1105. [CrossRef] [PubMed]

19. Rivetti di Val Cervo, P.; Romanov, R.A.; Spigolon, G.; Masini, D.; Martín-Montañez, E.; Toledo, E.M.; La Manno, G.; Feyder, M.; Pifl, C.; Ng, Y.H.; et al. Induction of functional dopamine neurons from human astrocytes in vitro and mouse astrocytes in a Parkinson's disease model. *Nat. Biotechnol.* **2017**, *35*, 444–452. [CrossRef]

20. Chen, Y.C.; Ma, N.X.; Pei, Z.F.; Wu, Z.; Do-Monte, F.H.; Keefe, S.; Yellin, E.; Chen, M.S.; Yin, J.C.; Lee, G.; et al. A NeuroD1 AAV-Based Gene Therapy for Functional Brain Repair after Ischemic Injury through In Vivo Astrocyte-to-Neuron Conversion. *Mol. Ther.* **2020**, *28*, 217–234. [CrossRef]

21. Zhou, H.; Su, J.; Hu, X.; Zhou, C.; Li, H.; Chen, Z.; Xiao, Q.; Wang, B.; Wu, W.; Sun, Y.; et al. Glia-to-Neuron Conversion by CRISPR-CasRx Alleviates Symptoms of Neurological Disease in Mice. *Cell* **2020**, *181*, 590–603.e516. [CrossRef]

22. Zaret, K.S. Pioneer Transcription Factors Initiating Gene Network Changes. *Annu. Rev. Genet.* **2020**, *54*, 367–385. [CrossRef] [PubMed]

23. Löhle, M.; Hermann, A.; Glass, H.; Kempe, A.; Schwarz, S.C.; Kim, J.B.; Poulet, C.; Ravens, U.; Schwarz, J.; Schöler, H.R.; et al. Differentiation efficiency of induced pluripotent stem cells depends on the number of reprogramming factors. *Stem. Cells* **2012**, *30*, 570–579. [CrossRef] [PubMed]

24. Yang, Y.; Chen, R.; Wu, X.; Zhao, Y.; Fan, Y.; Xiao, Z.; Han, J.; Sun, L.; Wang, X.; Dai, J. Rapid and Efficient Conversion of Human Fibroblasts into Functional Neurons by Small Molecules. *Stem Cell Rep.* **2019**, *13*, 862–876. [CrossRef]

25. Zhou, C.; Gu, H.; Fan, R.; Wang, B.; Lou, J. MicroRNA 302/367 Cluster Effectively Facilitates Direct Reprogramming from Human Fibroblasts into Functional Neurons. *Stem Cells Dev.* **2015**, *24*, 2746–2755. [CrossRef] [PubMed]

26. Ghasemi-Kasman, M.; Shojaei, A.; Gol, M.; Moghadamnia, A.A.; Baharvand, H.; Javan, M. miR-302/367-induced neurons reduce behavioral impairment in an experimental model of Alzheimer's disease. *Mol. Cell. Neurosci.* **2018**, *86*, 50–57. [CrossRef]

27. Liang, X.G.; Tan, C.; Wang, C.K.; Tao, R.R.; Huang, Y.J.; Ma, K.F.; Fukunaga, K.; Huang, M.Z.; Han, F. Myt1l induced direct reprogramming of pericytes into cholinergic neurons. *CNS Neurosci. Ther.* **2018**, *24*, 801–809. [CrossRef] [PubMed]

28. Niu, W.; Zang, T.; Wang, L.L.; Zou, Y.; Zhang, C.L. Phenotypic Reprogramming of Striatal Neurons into Dopaminergic Neuron-like Cells in the Adult Mouse Brain. *Stem Cell Rep.* **2018**, *11*, 1156–1170. [CrossRef]

29. Tanabe, K.; Ang, C.E.; Chanda, S.; Olmos, V.H.; Haag, D.; Levinson, D.F.; Sudhof, T.C.; Wernig, M. Transdifferentiation of human adult peripheral blood T cells into neurons. *Proc. Natl. Acad. Sci. USA* **2018**, *115*, 6470–6475. [CrossRef]

30. Bossolasco, P.; Sassone, F.; Gumina, V.; Peverelli, S.; Garzo, M.; Silani, V. Motor neuron differentiation of iPSCs obtained from peripheral blood of a mutant TARDBP ALS patient. *Stem Cell Res.* **2018**, *30*, 61–68. [CrossRef]

31. Marro, S.; Pang, Z.P.; Yang, N.; Tsai, M.C.; Qu, K.; Chang, H.Y.; Sudhof, T.C.; Wernig, M. Direct lineage conversion of terminally differentiated hepatocytes to functional neurons. *Cell Stem Cell* **2011**, *9*, 374–382. [CrossRef] [PubMed]

32. Robicsek, O.; Karry, R.; Petit, I.; Salman-Kesner, N.; Müller, F.J.; Klein, E.; Aberdam, D.; Ben-Shachar, D. Abnormal neuronal differentiation and mitochondrial dysfunction in hair follicle-derived induced pluripotent stem cells of schizophrenia patients. *Mol. Psychiatry* **2013**, *18*, 1067–1076. [CrossRef] [PubMed]

33. Gascon, S.; Murenu, E.; Masserdotti, G.; Ortega, F.; Russo, G.L.; Petrik, D.; Deshpande, A.; Heinrich, C.; Karow, M.; Robertson, S.P.; et al. Identification and Successful Negotiation of a Metabolic Checkpoint in Direct Neuronal Reprogramming. *Cell Stem Cell* **2016**, *18*, 396–409. [CrossRef]

34. Radhakrishnan, S.; Martin, C.A.; Dhayanithy, G.; Reddy, M.S.; Rela, M.; Kalkura, S.N.; Sellathamby, S. Hypoxic Preconditioning Induces Neuronal Differentiation of Infrapatellar Fat Pad Stem Cells through Epigenetic Alteration. *ACS Chem. Neurosci.* **2021**, *12*, 704–718. [CrossRef] [PubMed]

35. Puls, B.; Ding, Y.; Zhang, F.; Pan, M.; Lei, Z.; Pei, Z.; Jiang, M.; Bai, Y.; Forsyth, C.; Metzger, M.; et al. Regeneration of Functional Neurons After Spinal Cord Injury via in situ NeuroD1-Mediated Astrocyte-to-Neuron Conversion. *Front. Cell Dev. Biol.* **2020**, *8*, 591883. [CrossRef]

36. El Ayachi, I.; Zhang, J.; Zou, X.Y.; Li, D.; Yu, Z.; Wei, W.; O'Connell, K.M.S.; Huang, G.T. Human dental stem cell derived transgene-free iPSCs generate functional neurons via embryoid body-mediated and direct induction methods. *J. Tissue Eng. Regen. Med.* **2018**, *12*, e1836–e1851. [CrossRef]

37. Castaño, J.; Menendez, P.; Bruzos-Cidon, C.; Straccia, M.; Sousa, A.; Zabaleta, L.; Vazquez, N.; Zubiarrain, A.; Sonntag, K.C.; Ugedo, L.; et al. Fast and efficient neural conversion of human hematopoietic cells. *Stem Cell Rep.* **2014**, *3*, 1118–1131. [CrossRef]

38. Xu, G.; Wu, F.; Gu, X.; Zhang, J.; You, K.; Chen, Y.; Getachew, A.; Zhuang, Y.; Zhong, X.; Lin, Z.; et al. Direct Conversion of Human Urine Cells to Neurons by Small Molecules. *Sci. Rep.* **2019**, *9*, 16707. [CrossRef]

39. Sun, X.; Tan, Z.; Huang, X.; Cheng, X.; Yuan, Y.; Qin, S.; Wang, D.; Hu, X.; Gu, Y.; Qian, W.J.; et al. Direct neuronal reprogramming of olfactory ensheathing cells for CNS repair. *Cell Death Dis.* **2019**, *10*, 646. [CrossRef]

40. Yang, H.; Hao, D.; Liu, C.; Huang, D.; Chen, B.; Fan, H.; Liu, C.; Zhang, L.; Zhang, Q.; An, J.; et al. Generation of functional dopaminergic neurons from human spermatogonial stem cells to rescue parkinsonian phenotypes. *Stem Cell Res. Ther.* **2019**, *10*, 195. [CrossRef] [PubMed]

41. Zhang, L.; Li, P.; Hsu, T.; Aguilar, H.R.; Frantz, D.E.; Schneider, J.W.; Bachoo, R.M.; Hsieh, J. Small-molecule blocks malignant astrocyte proliferation and induces neuronal gene expression. *Differ. Res. Biol. Divers.* **2011**, *81*, 233–242. [CrossRef] [PubMed]

42. Matsuda, T.; Irie, T.; Katsurabayashi, S.; Hayashi, Y.; Nagai, T.; Hamazaki, N.; Adefuin, A.M.D.; Miura, F.; Ito, T.; Kimura, H.; et al. Pioneer Factor NeuroD1 Rearranges Transcriptional and Epigenetic Profiles to Execute Microglia-Neuron Conversion. *Neuron* **2019**, *101*, 472–485.e477. [CrossRef]
43. Rouaux, C.; Arlotta, P. Direct lineage reprogramming of post-mitotic callosal neurons into corticofugal neurons in vivo. *Nat. Cell Biol.* **2013**, *15*, 214–221. [CrossRef] [PubMed]
44. Noda, T.; Meas, S.J.; Nogami, J.; Amemiya, Y.; Uchi, R.; Ohkawa, Y.; Nishimura, K.; Dabdoub, A. Direct Reprogramming of Spiral Ganglion Non-neuronal Cells into Neurons: Toward Ameliorating Sensorineural Hearing Loss by Gene Therapy. *Front. Cell Dev. Biol.* **2018**, *6*, 16. [CrossRef] [PubMed]
45. Chuang, W.; Sharma, A.; Shukla, P.; Li, G.; Mall, M.; Rajarajan, K.; Abilez, O.J.; Hamaguchi, R.; Wu, J.C.; Wernig, M.; et al. Partial Reprogramming of Pluripotent Stem Cell-Derived Cardiomyocytes into Neurons. *Sci. Rep.* **2017**, *7*, 44840. [CrossRef]
46. Weinberg, M.S.; Criswell, H.E.; Powell, S.K.; Bhatt, A.P.; McCown, T.J. Viral Vector Reprogramming of Adult Resident Striatal Oligodendrocytes into Functional Neurons. *Mol. Ther.* **2017**, *25*, 928–934. [CrossRef]
47. Torper, O.; Ottosson, D.R.; Pereira, M.; Lau, S.; Cardoso, T.; Grealish, S.; Parmar, M. In Vivo Reprogramming of Striatal NG2 Glia into Functional Neurons that Integrate into Local Host Circuitry. *Cell Rep.* **2015**, *12*, 474–481. [CrossRef]
48. Cheng, F.; Lu, X.C.; Hao, H.Y.; Dai, X.L.; Da Qian, T.; Huang, B.S.; Tang, L.J.; Yu, W.; Li, L.X. Neurogenin 2 Converts Mesenchymal Stem Cells into a Neural Precursor Fate and Improves Functional Recovery after Experimental Stroke. *Cell. Physiol. Biochem.* **2014**, *33*, 847–858. [CrossRef]
49. Pollak, J.; Wilken, M.S.; Ueki, Y.; Cox, K.E.; Sullivan, J.M.; Taylor, R.J.; Levine, E.M.; Reh, T.A. ASCL1 reprograms mouse Muller glia into neurogenic retinal progenitors. *Development* **2013**, *140*, 2619–2631. [CrossRef]
50. Giorgetti, A.; Marchetto, M.C.; Li, M.; Yu, D.; Fazzina, R.; Mu, Y.; Adamo, A.; Paramonov, I.; Cardoso, J.C.; Monasterio, M.B.; et al. Cord blood-derived neuronal cells by ectopic expression of Sox2 and c-Myc. *Proc. Natl. Acad. Sci. USA* **2012**, *109*, 12556–12561. [CrossRef]
51. Lyssiotis, C.A.; Walker, J.; Wu, C.; Kondo, T.; Schultz, P.G.; Wu, X. Inhibition of histone deacetylase activity induces developmental plasticity in oligodendrocyte precursor cells. *Proc. Natl. Acad. Sci. USA* **2007**, *104*, 14982–14987. [CrossRef]
52. Grinnell, K.L.; Yang, B.; Eckert, R.L.; Bickenbach, J.R. De-differentiation of mouse interfollicular keratinocytes by the embryonic transcription factor Oct-4. *J. Investig. Dermatol.* **2007**, *127*, 372–380. [CrossRef]
53. Long, Q.; Luo, Q.; Wang, K.; Bates, A.; Shetty, A.K. Mash1-dependent Notch Signaling Pathway Regulates GABAergic Neuron-Like Differentiation from Bone Marrow-Derived Mesenchymal Stem Cells. *Aging Dis.* **2017**, *8*, 301–313. [CrossRef] [PubMed]
54. Hu, W.; Qiu, B.; Guan, W.; Wang, Q.; Wang, M.; Li, W.; Gao, L.; Shen, L.; Huang, Y.; Xie, G.; et al. Direct Conversion of Normal and Alzheimer's Disease Human Fibroblasts into Neuronal Cells by Small Molecules. *Cell Stem Cell* **2015**, *17*, 204–212. [CrossRef] [PubMed]
55. Dewald, L.E.; Rodriguez, J.P.; Levine, J.M. The RE1 binding protein REST regulates oligodendrocyte differentiation. *J. Neurosci.* **2011**, *31*, 3470–3483. [CrossRef] [PubMed]
56. Boshans, L.L.; Soh, H.; Wood, W.M.; Nolan, T.M.; Mandoiu, I.I.; Yanagawa, Y.; Tzingounis, A.V.; Nishiyama, A. Direct reprogramming of oligodendrocyte precursor cells into GABAergic inhibitory neurons by a single homeodomain transcription factor Dlx2. *Sci. Rep.* **2021**, *11*, 3552. [CrossRef]
57. Kondo, T.; Raff, M. Oligodendrocyte precursor cells reprogrammed to become multipotential CNS stem cells. *Science* **2000**, *289*, 1754–1757. [CrossRef]
58. Griffiths, B.B.; Bhutani, A.; Stary, C.M. Adult neurogenesis from reprogrammed astrocytes. *Neural Regen. Res.* **2020**, *15*, 973–979. [CrossRef]
59. Peng, Z.; Lu, H.; Yang, Q.; Xie, Q. Astrocyte Reprogramming in Stroke: Opportunities and Challenges. *Front. Aging Neurosci.* **2022**, *14*, 885707. [CrossRef]
60. Chen, C.; Zhong, X.; Smith, D.K.; Tai, W.; Yang, J.; Zou, Y.; Wang, L.L.; Sun, J.; Qin, S.; Zhang, C.L. Astrocyte-Specific Deletion of Sox2 Promotes Functional Recovery After Traumatic Brain Injury. *Cereb. Cortex* **2019**, *29*, 54–69. [CrossRef]
61. Aravantinou-Fatorou, K.; Vejdani, S.; Thomaidou, D. Cend1 and Neurog2 efficiently reprogram human cortical astrocytes to neural precursor cells and induced-neurons. *Int. J. Dev. Biol.* **2021**, *66*, 199–209. [CrossRef]
62. Liu, M.-H.; Li, W.; Zheng, J.-J.; Xu, Y.-G.; He, Q.; Chen, G. Differential neuronal reprogramming induced by NeuroD1 from astrocytes in grey matter versus white matter. *Neural Regen. Res.* **2020**, *15*, 342–351. [CrossRef]
63. Fujii, Y.; Arima, M.; Shimokawa, S.; Murakami, Y.; Sonoda, K.-h. The direct reprogramming of retinal astrocytes into neurons with small-molecule compounds. *Investig. Ophthalmol. Vis. Sci.* **2019**, *60*, 2.
64. Bocchi, R.; Masserdotti, G.; Götz, M. Direct neuronal reprogramming: Fast forward from new concepts toward therapeutic approaches. *Neuron* **2022**, *110*, 366–393. [CrossRef]
65. VandenBosch, L.S.; Wohl, S.G.; Wilken, M.S.; Hooper, M.; Finkbeiner, C.; Cox, K.; Chipman, L.; Reh, T.A. Developmental changes in the accessible chromatin, transcriptome and Ascl1-binding correlate with the loss in Müller Glial regenerative potential. *Sci. Rep.* **2020**, *10*, 13615. [CrossRef] [PubMed]
66. Ali, F.R.; Cheng, K.; Kirwan, P.; Metcalfe, S.; Livesey, F.J.; Barker, R.A.; Philpott, A. The phosphorylation status of Ascl1 is a key determinant of neuronal differentiation and maturation in vivo and in vitro. *Development* **2014**, *141*, 2216–2224. [CrossRef]
67. Smith, D.K.; Yang, J.; Liu, M.L.; Zhang, C.L. Small Molecules Modulate Chromatin Accessibility to Promote NEUROG2-Mediated Fibroblast-to-Neuron Reprogramming. *Stem Cell Rep.* **2016**, *7*, 955–969. [CrossRef]

68. Walther, C.; Gruss, P. Pax-6, a murine paired box gene, is expressed in the developing CNS. *Development* **1991**, *113*, 1435–1449. [CrossRef] [PubMed]

69. Georgala, P.A.; Carr, C.B.; Price, D.J. The role of Pax6 in forebrain development. *Dev. Neurobiol.* **2011**, *71*, 690–709. [CrossRef] [PubMed]

70. Chouchane, M.; Costa, M.R. Instructing neuronal identity during CNS development and astroglial-lineage reprogramming: Roles of NEUROG2 and ASCL1. *Brain Res.* **2019**, *1705*, 66–74. [CrossRef]

71. Todd, L.; Hooper, M.J.; Haugan, A.K.; Finkbeiner, C.; Jorstad, N.; Radulovich, N.; Wong, C.K.; Donaldson, P.C.; Jenkins, W.; Chen, Q.; et al. Efficient stimulation of retinal regeneration from Muller glia in adult mice using combinations of proneural bHLH transcription factors. *Cell Rep.* **2021**, *37*, 109857. [CrossRef]

72. Victor, M.B.; Richner, M.; Hermanstyne, T.O.; Ransdell, J.L.; Sobieski, C.; Deng, P.Y.; Klyachko, V.A.; Nerbonne, J.M.; Yoo, A.S. Generation of human striatal neurons by microRNA-dependent direct conversion of fibroblasts. *Neuron* **2014**, *84*, 311–323. [CrossRef]

73. Jin, Z.; May, W.S.; Gao, F.; Flagg, T.; Deng, X. Bcl2 suppresses DNA repair by enhancing c-Myc transcriptional activity. *J. Biol. Chem.* **2006**, *281*, 14446–14456. [CrossRef] [PubMed]

74. Gresita, A.; Glavan, D.; Udristoiu, I.; Catalin, B.; Hermann, D.M.; Popa-Wagner, A. Very Low Efficiency of Direct Reprogramming of Astrocytes into Neurons in the Brains of Young and Aged Mice After Cerebral Ischemia. *Front. Aging Neurosci.* **2019**, *11*, 334. [CrossRef]

75. Mattiassi, S.; Rizwan, M.; Grigsby, C.L.; Zaw, A.M.; Leong, K.W.; Yim, E.K.F. Enhanced efficiency of nonviral direct neuronal reprogramming on topographical patterns. *Biomater. Sci.* **2021**, *9*, 5175–5191. [CrossRef]

76. Michaud, J.L.; Rosenquist, T.; May, N.R.; Fan, C.M. Development of neuroendocrine lineages requires the bHLH-PAS transcription factor SIM1. *Genes Dev.* **1998**, *12*, 3264–3275. [CrossRef]

77. Aravantinou-Fatorou, K.; Thomaidou, D. In Vitro Direct Reprogramming of Mouse and Human Astrocytes to Induced Neurons. *Methods Mol. Biol.* **2020**, *2155*, 41–61. [CrossRef]

78. Wainger, B.J.; Buttermore, E.D.; Oliveira, J.T.; Mellin, C.; Lee, S.; Saber, W.A.; Wang, A.J.; Ichida, J.K.; Chiu, I.M.; Barrett, L.; et al. Modeling pain in vitro using nociceptor neurons reprogrammed from fibroblasts. *Nat. Neurosci.* **2015**, *18*, 17–24. [CrossRef]

79. Dubois, L.; Bally-Cuif, L.; Crozatier, M.; Moreau, J.; Paquereau, L.; Vincent, A. XCoe2, a transcription factor of the Col/Olf-1/EBF family involved in the specification of primary neurons in Xenopus. *Curr. Biol.* **1998**, *8*, 199–209. [CrossRef] [PubMed]

80. Arlotta, P.; Molyneaux, B.J.; Chen, J.; Inoue, J.; Kominami, R.; Macklis, J.D. Neuronal subtype-specific genes that control corticospinal motor neuron development in vivo. *Neuron* **2005**, *45*, 207–221. [CrossRef] [PubMed]

81. Aydin, B.; Sierk, M.; Moreno-Estelles, M.; Tejavibulya, L.; Kumar, N.; Flames, N.; Mahony, S.; Mazzoni, E.O. Foxa2 and Pet1 Direct and Indirect Synergy Drive Serotonergic Neuronal Differentiation. *Front. Neurosci.* **2022**, *16*, 903881. [CrossRef] [PubMed]

82. Lujan, E.; Chanda, S.; Ahlenius, H.; Sudhof, T.C.; Wernig, M. Direct conversion of mouse fibroblasts to self-renewing, tripotent neural precursor cells. *Proc. Natl. Acad. Sci. USA* **2012**, *109*, 2527–2532. [CrossRef] [PubMed]

83. Raciti, M.; Granzotto, M.; Duc, M.D.; Fimiani, C.; Cellot, G.; Cherubini, E.; Mallamaci, A. Reprogramming fibroblasts to neural-precursor-like cells by structured overexpression of pallial patterning genes. *Mol. Cell. Neurosci.* **2013**, *57*, 42–53. [CrossRef]

84. Li, S.; Shi, Y.; Yao, X.; Wang, X.; Shen, L.; Rao, Z.; Yuan, J.; Liu, Y.; Zhou, Z.; Zhang, Z.; et al. Conversion of Astrocytes and Fibroblasts into Functional Noradrenergic Neurons. *Cell Rep.* **2019**, *28*, 682–697.e687. [CrossRef]

85. Apostolova, G.; Dechant, G. Development of neurotransmitter phenotypes in sympathetic neurons. *Auton. Neurosci.* **2009**, *151*, 30–38. [CrossRef] [PubMed]

86. Buganim, Y.; Itskovich, E.; Hu, Y.C.; Cheng, A.W.; Ganz, K.; Sarkar, S.; Fu, D.; Welstead, G.G.; Page, D.C.; Jaenisch, R. Direct reprogramming of fibroblasts into embryonic Sertoli-like cells by defined factors. *Cell Stem Cell* **2012**, *11*, 373–386. [CrossRef]

87. Schmidt, M.; Lin, S.; Pape, M.; Ernsberger, U.; Stanke, M.; Kobayashi, K.; Howard, M.J.; Rohrer, H. The bHLH transcription factor Hand2 is essential for the maintenance of noradrenergic properties in differentiated sympathetic neurons. *Dev. Biol.* **2009**, *329*, 191–200. [CrossRef]

88. Triplett, J.W.; Wei, W.; Gonzalez, C.; Sweeney, N.T.; Huberman, A.D.; Feller, M.B.; Feldheim, D.A. Dendritic and axonal targeting patterns of a genetically-specified class of retinal ganglion cells that participate in image-forming circuits. *Neural Dev.* **2014**, *9*, 2. [CrossRef] [PubMed]

89. Ramachandran, R.; Fausett, B.V.; Goldman, D. Ascl1a regulates Müller glia dedifferentiation and retinal regeneration through a Lin-28-dependent, let-7 microRNA signalling pathway. *Nat. Cell Biol.* **2010**, *12*, 1101–1107. [CrossRef]

90. Batta, K.; Florkowska, M.; Kouskoff, V.; Lacaud, G. Direct reprogramming of murine fibroblasts to hematopoietic progenitor cells. *Cell Rep.* **2014**, *9*, 1871–1884. [CrossRef]

91. Wilson, N.K.; Foster, S.D.; Wang, X.; Knezevic, K.; Schütte, J.; Kaimakis, P.; Chilarska, P.M.; Kinston, S.; Ouwehand, W.H.; Dzierzak, E.; et al. Combinatorial transcriptional control in blood stem/progenitor cells: Genome-wide analysis of ten major transcriptional regulators. *Cell Stem Cell* **2010**, *7*, 532–544. [CrossRef]

92. Baek, S.; Oh, J.; Song, J.; Choi, H.; Yoo, J.; Park, G.Y.; Han, J.; Chang, Y.; Park, H.; Kim, H.; et al. Generation of Integration-Free Induced Neurons Using Graphene Oxide-Polyethylenimine. *Small* **2017**, *13*, 201601993. [CrossRef] [PubMed]

93. Romm, E.; Nielsen, J.A.; Kim, J.G.; Hudson, L.D. Myt1 family recruits histone deacetylase to regulate neural transcription. *J. Neurochem.* **2005**, *93*, 1444–1453. [CrossRef] [PubMed]

94. Corti, S.; Nizzardo, M.; Simone, C.; Falcone, M.; Donadoni, C.; Salani, S.; Rizzo, F.; Nardini, M.; Riboldi, G.; Magri, F.; et al. Direct reprogramming of human astrocytes into neural stem cells and neurons. *Exp. Cell Res.* **2012**, *318*, 1528–1541. [CrossRef]
95. Guo, Z.; Zhang, L.; Wu, Z.; Chen, Y.; Wang, F.; Chen, G. In vivo direct reprogramming of reactive glial cells into functional neurons after brain injury and in an Alzheimer's disease model. *Cell Stem Cell* **2014**, *14*, 188–202. [CrossRef]
96. Pennesi, M.E.; Cho, J.H.; Yang, Z.; Wu, S.H.; Zhang, J.; Wu, S.M.; Tsai, M.J. BETA2/NeuroD1 null mice: A new model for transcription factor-dependent photoreceptor degeneration. *J. Neurosci.* **2003**, *23*, 453–461. [CrossRef]
97. Bavamian, S.; Mellios, N.; Lalonde, J.; Fass, D.M.; Wang, J.; Sheridan, S.D.; Madison, J.M.; Zhou, F.; Rueckert, E.H.; Barker, D.; et al. Dysregulation of miR-34a links neuronal development to genetic risk factors for bipolar disorder. *Mol. Psychiatry* **2015**, *20*, 573–584. [CrossRef]
98. Yoo, A.S.; Sun, A.X.; Li, L.; Shcheglovitov, A.; Portmann, T.; Li, Y.; Lee-Messer, C.; Dolmetsch, R.E.; Tsien, R.W.; Crabtree, G.R. MicroRNA-mediated conversion of human fibroblasts to neurons. *Nature* **2011**, *476*, 228–231. [CrossRef]
99. Olson, J.M.; Asakura, A.; Snider, L.; Hawkes, R.; Strand, A.; Stoeck, J.; Hallahan, A.; Pritchard, J.; Tapscott, S.J. NeuroD2 is necessary for development and survival of central nervous system neurons. *Dev. Biol.* **2001**, *234*, 174–187. [CrossRef]
100. Rodríguez-Traver, E.; Solís, O.; Díaz-Guerra, E.; Ortiz, Ó.; Vergaño-Vera, E.; Méndez-Gómez, H.R.; García-Sanz, P.; Moratalla, R.; Vicario-Abejón, C. Role of Nurr1 in the Generation and Differentiation of Dopaminergic Neurons from Stem Cells. *Neurotox. Res.* **2016**, *30*, 14–31. [CrossRef] [PubMed]
101. Perlmann, T.; Wallén-Mackenzie, A. Nurr1, an orphan nuclear receptor with essential functions in developing dopamine cells. *Cell Tissue Res.* **2004**, *318*, 45–52. [CrossRef] [PubMed]
102. Boyer, L.A.; Lee, T.I.; Cole, M.F.; Johnstone, S.E.; Levine, S.S.; Zucker, J.P.; Guenther, M.G.; Kumar, R.M.; Murray, H.L.; Jenner, R.G.; et al. Core transcriptional regulatory circuitry in human embryonic stem cells. *Cell* **2005**, *122*, 947–956. [CrossRef] [PubMed]
103. Hack, M.A.; Sugimori, M.; Lundberg, C.; Nakafuku, M.; Götz, M. Regionalization and fate specification in neurospheres: The role of Olig2 and Pax6. *Mol. Cell. Neurosci.* **2004**, *25*, 664–678. [CrossRef]
104. Heins, N.; Malatesta, P.; Cecconi, F.; Nakafuku, M.; Tucker, K.L.; Hack, M.A.; Chapouton, P.; Barde, Y.A.; Gotz, M. Glial cells generate neurons: The role of the transcription factor Pax6. *Nat. Neurosci.* **2002**, *5*, 308–315. [CrossRef]
105. Panman, L.; Andersson, E.; Alekseenko, Z.; Hedlund, E.; Kee, N.; Mong, J.; Uhde, C.W.; Deng, Q.; Sandberg, R.; Stanton, L.W.; et al. Transcription factor-induced lineage selection of stem-cell-derived neural progenitor cells. *Cell Stem Cell* **2011**, *8*, 663–675. [CrossRef]
106. Jin, K.; Zou, M.; Xiao, D.; Xiang, M. Reprogramming Fibroblasts to Neural Stem Cells by Overexpression of the Transcription Factor Ptf1a. *Methods Mol. Biol.* **2020**, *2117*, 245–263. [CrossRef]
107. Mu, L.F.; Berti, L.; Masserdotti, G.; Covic, M.; Michaelidis, T.M.; Doberauer, K.; Merz, K.; Rehfeld, F.; Haslinger, A.; Wegner, M.; et al. SoxC Transcription Factors Are Required for Neuronal Differentiation in Adult Hippocampal Neurogenesis. *J. Neurosci.* **2012**, *32*, 3067–3080. [CrossRef] [PubMed]
108. Wang, L.; Wang, L.; Huang, W.; Su, H.; Xue, Y.; Su, Z.; Liao, B.; Wang, H.; Bao, X.; Qin, D.; et al. Generation of integration-free neural progenitor cells from cells in human urine. *Nat. Methods* **2013**, *10*, 84–89. [CrossRef]
109. Cheng, L.; Arata, A.; Mizuguchi, R.; Qian, Y.; Karunaratne, A.; Gray, P.A.; Arata, S.; Shirasawa, S.; Bouchard, M.; Luo, P.; et al. Tlx3 and Tlx1 are post-mitotic selector genes determining glutamatergic over GABAergic cell fates. *Nat. Neurosci.* **2004**, *7*, 510–517. [CrossRef]
110. Yan, L.; Li, Y.; Shi, Z.; Lu, X.; Ma, J.; Hu, B.; Jiao, J.; Wang, H. The zinc finger E-box-binding homeobox 1 (Zeb1) promotes the conversion of mouse fibroblasts into functional neurons. *J. Biol. Chem.* **2017**, *292*, 12959–12970. [CrossRef]
111. Damberg, M. Transcription factor AP-2 and monoaminergic functions in the central nervous system. *J. Neural Transm.* **2005**, *112*, 1281–1296. [CrossRef]
112. Bertrand, N.; Castro, D.S.; Guillemot, F. Proneural genes and the specification of neural cell types. *Nat. Rev. Neurosci.* **2002**, *3*, 517–530. [CrossRef]
113. Liu, F.; Zhang, Y.; Chen, F.; Yuan, J.; Li, S.; Han, S.; Lu, D.; Geng, J.; Rao, Z.; Sun, L.; et al. Neurog2 directly converts astrocytes into functional neurons in midbrain and spinal cord. *Cell Death Dis.* **2021**, *12*, 225. [CrossRef]
114. Mattugini, N.; Bocchi, R.; Scheuss, V.; Russo, G.L.; Torper, O.; Lao, C.L.; Gotz, M. Inducing Different Neuronal Subtypes from Astrocytes in the Injured Mouse Cerebral Cortex. *Neuron* **2019**, *103*, 1086–1095. [CrossRef]
115. Kronenberg, G.; Gertz, K.; Cheung, G.; Buffo, A.; Kettenmann, H.; Götz, M.; Endres, M. Modulation of fate determinants Olig2 and Pax6 in resident glia evokes spiking neuroblasts in a model of mild brain ischemia. *Stroke* **2010**, *41*, 2944–2949. [CrossRef] [PubMed]
116. Kim, J.; Efe, J.A.; Zhu, S.; Talantova, M.; Yuan, X.; Wang, S.; Lipton, S.A.; Zhang, K.; Ding, S. Direct reprogramming of mouse fibroblasts to neural progenitors. *Proc. Natl. Acad. Sci. USA* **2011**, *108*, 7838–7843. [CrossRef] [PubMed]
117. Tian, C.; Wang, Y.; Sun, L.; Ma, K.; Zheng, J.C. Reprogrammed mouse astrocytes retain a "memory" of tissue origin and possess more tendencies for neuronal differentiation than reprogrammed mouse embryonic fibroblasts. *Protein Cell* **2011**, *2*, 128–140. [CrossRef]
118. Blum, R.; Heinrich, C.; Sánchez, R.; Lepier, A.; Gundelfinger, E.D.; Berninger, B.; Götz, M. Neuronal network formation from reprogrammed early postnatal rat cortical glial cells. *Cereb. Cortex* **2011**, *21*, 413–424. [CrossRef]
119. Krol, J.; Loedige, I.; Filipowicz, W. The widespread regulation of microRNA biogenesis, function and decay. *Nat. Rev. Genet.* **2010**, *11*, 597–610. [CrossRef] [PubMed]

120. Meza-Sosa, K.F.; Valle-Garcia, D.; Pedraza-Alva, G.; Perez-Martinez, L. Role of microRNAs in central nervous system development and pathology. *J. Neurosci. Res.* **2012**, *90*, 1–12. [CrossRef]

121. Zhao, Y.; Srivastava, D. A developmental view of microRNA function. *Trends Biochem. Sci.* **2007**, *32*, 189–197. [CrossRef] [PubMed]

122. Li, X.; Jin, P. Roles of small regulatory RNAs in determining neuronal identity. *Nat. Rev. Neurosci.* **2010**, *11*, 329–338. [CrossRef] [PubMed]

123. Visvanathan, J.; Lee, S.; Lee, B.; Lee, J.W.; Lee, S.K. The microRNA miR-124 antagonizes the anti-neural REST/SCP1 pathway during embryonic CNS development. *Genes Dev.* **2007**, *21*, 744–749. [CrossRef] [PubMed]

124. Xue, Y.; Ouyang, K.; Huang, J.; Zhou, Y.; Ouyang, H.; Li, H.; Wang, G.; Wu, Q.; Wei, C.; Bi, Y.; et al. Direct conversion of fibroblasts to neurons by reprogramming PTB-regulated microRNA circuits. *Cell* **2013**, *152*, 82–96. [CrossRef] [PubMed]

125. Wohl, S.G.; Reh, T.A. miR-124-9-9* potentiates Ascl1-induced reprogramming of cultured Müller glia. *Glia* **2016**, *64*, 743–762. [CrossRef] [PubMed]

126. Hou, P.; Li, Y.; Zhang, X.; Liu, C.; Guan, J.; Li, H.; Zhao, T.; Ye, J.; Yang, W.; Liu, K.; et al. Pluripotent stem cells induced from mouse somatic cells by small-molecule compounds. *Science* **2013**, *341*, 651–654. [CrossRef]

127. Quist, E.; Trovato, F.; Avaliani, N.; Zetterdahl, O.G.; Gonzalez-Ramos, A.; Hansen, M.G.; Kokaia, M.; Canals, I.; Ahlenius, H. Transcription factor-based direct conversion of human fibroblasts to functional astrocytes. *Stem Cell Rep.* **2022**, *17*, 1620–1635. [CrossRef]

128. Lee, C.; Robinson, M.; Willerth, S.M. Direct Reprogramming of Glioblastoma Cells into Neurons Using Small Molecules. *ACS Chem. Neurosci.* **2018**, *9*, 3175–3185. [CrossRef]

129. Yang, Z.; Xu, X.; Gu, C.; Nielsen, A.V.; Chen, G.; Guo, F.; Tang, C.; Zhao, Y. Chemical Pretreatment Activated a Plastic State Amenable to Direct Lineage Reprogramming. *Front. Cell Dev. Biol.* **2022**, *10*, 865038. [CrossRef]

130. Ruetz, T.; Pfisterer, U.; Di Stefano, B.; Ashmore, J.; Beniazza, M.; Tian, T.V.; Kaemena, D.F.; Tosti, L.; Tan, W.; Manning, J.R.; et al. Constitutively Active SMAD2/3 Are Broad-Scope Potentiators of Transcription-Factor-Mediated Cellular Reprogramming. *Cell Stem Cell* **2017**, *21*, 791–805.e799. [CrossRef]

131. Liu, C.; Hu, X.; Li, Y.; Lu, W.; Li, W.; Cao, N.; Zhu, S.; Cheng, J.; Ding, S.; Zhang, M. Conversion of mouse fibroblasts into oligodendrocyte progenitor-like cells through a chemical approach. *J. Mol. Cell Biol.* **2019**, *11*, 489–495. [CrossRef]

132. Nakano, R.; Kitanaka, T.; Namba, S.; Kitanaka, N.; Sato, M.; Shibukawa, Y.; Masuhiro, Y.; Kano, K.; Matsumoto, T.; Sugiya, H. All-trans retinoic acid induces reprogramming of canine dedifferentiated cells into neuron-like cells. *PLoS ONE* **2020**, *15*, e0229892. [CrossRef] [PubMed]

133. Schweinfurth, N.; Hohmann, S.; Deuschle, M.; Lederbogen, F.; Schloss, P. Valproic acid and all trans retinoic acid differentially induce megakaryopoiesis and platelet-like particle formation from the megakaryoblastic cell line MEG-01. *Platelets* **2010**, *21*, 648–657. [CrossRef] [PubMed]

134. Zhao, Y.; Zhao, T.; Guan, J.; Zhang, X.; Fu, Y.; Ye, J.; Zhu, J.; Meng, G.; Ge, J.; Yang, S.; et al. A XEN-like State Bridges Somatic Cells to Pluripotency during Chemical Reprogramming. *Cell* **2015**, *163*, 1678–1691. [CrossRef]

135. Kogo, Y.; Seto, C.; Totani, Y.; Mochizuki, M.; Nakahara, T.; Oka, K.; Yoshioka, T.; Ito, E. Rapid differentiation of human dental pulp stem cells to neuron-like cells by high K(+) stimulation. *Biophys. Physicobiol.* **2020**, *17*, 132–139. [CrossRef] [PubMed]

136. He, Z.Q.; Li, Y.H.; Feng, G.H.; Yuan, X.W.; Lu, Z.B.; Dai, M.; Hu, Y.P.; Zhang, Y.; Zhou, Q.; Li, W. Pharmacological Perturbation of Mechanical Contractility Enables Robust Transdifferentiation of Human Fibroblasts into Neurons. *Adv. Sci.* **2022**, *9*, e2104682. [CrossRef]

137. Lee, T.H.; Liu, P.S.; Wang, S.J.; Tsai, M.M.; Shanmugam, V.; Hsieh, H.L. Bradykinin, as a Reprogramming Factor, Induces Transdifferentiation of Brain Astrocytes into Neuron-like Cells. *Biomedicines* **2021**, *9*, 923. [CrossRef]

138. Li, X.; Liu, D.; Ma, Y.; Du, X.; Jing, J.; Wang, L.; Xie, B.; Sun, D.; Sun, S.; Jin, X.; et al. Direct Reprogramming of Fibroblasts via a Chemically Induced XEN-like State. *Cell Stem Cell* **2017**, *21*, 264–273.e267. [CrossRef]

139. Solomon, E.; Davis-Anderson, K.; Hovde, B.; Micheva-Viteva, S.; Harris, J.F.; Twary, S.; Iyer, R. Global transcriptome profile of the developmental principles of in vitro iPSC-to-motor neuron differentiation. *BMC Mol. Cell Biol.* **2021**, *22*, 13. [CrossRef]

140. Yin, J.C.; Zhang, L.; Ma, N.X.; Wang, Y.; Lee, G.; Hou, X.Y.; Lei, Z.F.; Zhang, F.Y.; Dong, F.P.; Wu, G.Y.; et al. Chemical Conversion of Human Fetal Astrocytes into Neurons through Modulation of Multiple Signaling Pathways. *Stem Cell Rep.* **2019**, *12*, 488–501. [CrossRef]

141. Tian, E.; Sun, G.; Sun, G.; Chao, J.; Ye, P.; Warden, C.; Riggs, A.D.; Shi, Y. Small-Molecule-Based Lineage Reprogramming Creates Functional Astrocytes. *Cell Rep.* **2016**, *16*, 781–792. [CrossRef]

142. Yang, C.; Chen, Z.; Yu, H.; Liu, X. Inhibition of Disruptor of Telomeric Silencing 1-Like Alleviated Renal Ischemia and Reperfusion Injury-Induced Fibrosis by Blocking PI3K/AKT-Mediated Oxidative Stress. *Drug. Des. Devel. Ther.* **2019**, *13*, 4375–4387. [CrossRef] [PubMed]

143. Pan, Z.; Oh, J.; Huang, L.; Zeng, Z.; Duan, P.; Li, Z.; Yun, Y.; Kim, J.; Ha, Y.; Cao, K. The combination of forskolin and VPA increases gene expression efficiency to the hypoxia/neuron-specific system. *Ann. Transl. Med.* **2020**, *8*, 933. [CrossRef]

144. Liu, D.; Rychkov, G.; Al-Hawwas, M.; Manaph, N.P.A.; Zhou, F.; Bobrovskaya, L.; Liao, H.; Zhou, X.F. Conversion of human urine-derived cells into neuron-like cells by small molecules. *Mol. Biol. Rep.* **2020**, *47*, 2713–2722. [CrossRef]

145. Chang, J.H.; Tsai, P.H.; Wang, K.Y.; Wei, Y.T.; Chiou, S.H.; Mou, C.Y. Generation of Functional Dopaminergic Neurons from Reprogramming Fibroblasts by Nonviral-based Mesoporous Silica Nanoparticles. *Sci. Rep.* **2018**, *8*, 11. [CrossRef] [PubMed]

146. Qin, H.; Zhao, A.D.; Sun, M.L.; Ma, K.; Fu, X.B. Direct conversion of human fibroblasts into dopaminergic neuron-like cells using small molecules and protein factors. *Mil. Med. Res.* **2020**, *7*, 52. [CrossRef]
147. Ma, N.X.; Yin, J.C.; Chen, G. Transcriptome Analysis of Small Molecule-Mediated Astrocyte-to-Neuron Reprogramming. *Front. Cell Dev. Biol.* **2019**, *7*, 82. [CrossRef] [PubMed]
148. Sun, R.; Gong, T.; Liu, H.; Shen, J.; Wu, B.; Jiang, Q.; Wang, Q.; Zhang, Y.; Duan, L.; Hu, J.; et al. Identification of microRNAs related with neural germ layer lineage-specific progenitors during reprogramming. *J. Mol. Histol.* **2022**, *53*, 623–634. [CrossRef] [PubMed]
149. Chen, Z.; Huang, Y.; Yu, C.; Liu, Q.; Qiu, C.; Wan, G. Cochlear Sox2(+) Glial Cells Are Potent Progenitors for Spiral Ganglion Neuron Reprogramming Induced by Small Molecules. *Front. Cell Dev. Biol.* **2021**, *9*, 728352. [CrossRef]
150. Drouin-Ouellet, J.; Lau, S.; Brattås, P.L.; Rylander Ottosson, D.; Pircs, K.; Grassi, D.A.; Collins, L.M.; Vuono, R.; Andersson Sjöland, A.; Westergren-Thorsson, G.; et al. REST suppression mediates neural conversion of adult human fibroblasts via microRNA-dependent and -independent pathways. *EMBO Mol. Med.* **2017**, *9*, 1117–1131. [CrossRef]
151. Seibler, P.; Graziotto, J.; Jeong, H.; Simunovic, F.; Klein, C.; Krainc, D. Mitochondrial Parkin Recruitment Is Impaired in Neurons Derived from Mutant PINK1 Induced Pluripotent Stem Cells. *J. Neurosci.* **2011**, *31*, 5970–5976. [CrossRef] [PubMed]
152. Zhu, S.; Li, W.; Zhou, H.; Wei, W.; Ambasudhan, R.; Lin, T.; Kim, J.; Zhang, K.; Ding, S. Reprogramming of human primary somatic cells by OCT4 and chemical compounds. *Cell Stem Cell* **2010**, *7*, 651–655. [CrossRef]
153. Zhao, A.D.; Qin, H.; Sun, M.L.; Ma, K.; Fu, X.B. Efficient and rapid conversion of human astrocytes and ALS mouse model spinal cord astrocytes into motor neuron-like cells by defined small molecules. *Mil. Med. Res.* **2020**, *7*, 42. [CrossRef]
154. Wan, X.Y.; Xu, L.Y.; Li, B.; Sun, Q.H.; Ji, Q.L.; Huang, D.D.; Zhao, L.; Xiao, Y.T. Chemical conversion of human lung fibroblasts into neuronal cells. *Int. J. Mol. Med.* **2018**, *41*, 1463–1468. [CrossRef] [PubMed]
155. Zheng, Y.; Huang, Z.; Xu, J.; Hou, K.; Yu, Y.; Lv, S.; Chen, L.; Li, Y.; Quan, C.; Chi, G. MiR-124 and Small Molecules Synergistically Regulate the Generation of Neuronal Cells from Rat Cortical Reactive Astrocytes. *Mol. Neurobiol.* **2021**, *58*, 2447–2464. [CrossRef] [PubMed]
156. Zhang, M.; Lin, Y.H.; Sun, Y.J.; Zhu, S.; Zheng, J.; Liu, K.; Cao, N.; Li, K.; Huang, Y.; Ding, S. Pharmacological Reprogramming of Fibroblasts into Neural Stem Cells by Signaling-Directed Transcriptional Activation. *Cell Stem Cell* **2016**, *18*, 653–667. [CrossRef]
157. Kirchenwitz, M.; Stahnke, S.; Grunau, K.; Melcher, L.; van Ham, M.; Rottner, K.; Steffen, A.; Stradal, T.E.B. The autophagy inducer SMER28 attenuates microtubule dynamics mediating neuroprotection. *Sci. Rep.* **2022**, *12*, 17805. [CrossRef]
158. Fernandes, G.S.; Singh, R.D.; Kim, K.K. Generation of a Pure Culture of Neuron-like Cells with a Glutamatergic Phenotype from Mouse Astrocytes. *Biomedicines* **2022**, *10*, 928. [CrossRef]
159. Wang, J.; Lu, J.; Bond, M.C.; Chen, M.; Ren, X.R.; Lyerly, H.K.; Barak, L.S.; Chen, W. Identification of select glucocorticoids as Smoothened agonists: Potential utility for regenerative medicine. *Proc. Natl. Acad. Sci. USA* **2010**, *107*, 9323–9328. [CrossRef]
160. Ye, W.; Shimamura, K.; Rubenstein, J.L.; Hynes, M.A.; Rosenthal, A. FGF and Shh signals control dopaminergic and serotonergic cell fate in the anterior neural plate. *Cell* **1998**, *93*, 755–766. [CrossRef]
161. Rhee, Y.H.; Ko, J.Y.; Chang, M.Y.; Yi, S.H.; Kim, D.; Kim, C.H.; Shim, J.W.; Jo, A.Y.; Kim, B.W.; Lee, H.; et al. Protein-based human iPS cells efficiently generate functional dopamine neurons and can treat a rat model of Parkinson disease. *J. Clin. Investig.* **2011**, *121*, 2326–2335. [CrossRef]
162. Sotthibundhu, A.; Nopparat, C.; Natphopsuk, S.; Phuthong, S.; Noisa, P.; Govitrapong, P. Combination of Melatonin and Small Molecules Improved Reprogramming Neural Cell Fates via Autophagy Activation. *Neurochem. Res.* **2022**, *47*, 2580–2590. [CrossRef]
163. Roussa, E.; Oehlke, O.; Rahhal, B.; Heermann, S.; Heidrich, S.; Wiehle, M.; Krieglstein, K. Transforming growth factor beta cooperates with persephin for dopaminergic phenotype induction. *Stem. Cells.* **2008**, *26*, 1683–1694. [CrossRef]
164. Zhang, L.; Yin, J.-C.; Yeh, H.; Ma, N.-X.; Lee, G.; Chen, X.A.; Wang, Y.; Lin, L.; Chen, L.; Jin, P.; et al. Small Molecules Efficiently Reprogram Human Astroglial Cells into Functional Neurons. *Cell Stem Cell* **2015**, *17*, 735–747. [CrossRef]
165. Alari, V.; Scalmani, P.; Ajmone, P.F.; Perego, S.; Avignone, S.; Catusi, I.; Lonati, P.A.; Borghi, M.O.; Finelli, P.; Terragni, B.; et al. Histone Deacetylase Inhibitors Ameliorate Morphological Defects and Hypoexcitability of iPSC-Neurons from Rubinstein-Taybi Patients. *Int. J. Mol. Sci.* **2021**, *22*, 5777. [CrossRef]
166. Kanai, H.; Sawa, A.; Chen, R.W.; Leeds, P.; Chuang, D.M. Valproic acid inhibits histone deacetylase activity and suppresses excitotoxicity-induced GAPDH nuclear accumulation and apoptotic death in neurons. *Pharmacogenom. J.* **2004**, *4*, 336–344. [CrossRef] [PubMed]
167. Klemm, S.L.; Shipony, Z.; Greenleaf, W.J. Chromatin accessibility and the regulatory epigenome. *Nat. Rev. Genet.* **2019**, *20*, 207–220. [CrossRef] [PubMed]
168. He, S.; Chen, J.; Zhang, Y.; Zhang, M.; Yang, X.; Li, Y.; Sun, H.; Lin, L.; Fan, K.; Liang, L.; et al. Sequential EMT-MET induces neuronal conversion through Sox2. *Cell Discov.* **2017**, *3*, 17017. [CrossRef] [PubMed]
169. Liu, Z.; Cai, Y.; Wang, Y.; Nie, Y.; Zhang, C.; Xu, Y.; Zhang, X.; Lu, Y.; Wang, Z.; Poo, M.; et al. Cloning of Macaque Monkeys by Somatic Cell Nuclear Transfer. *Cell* **2018**, *172*, 881–887.e887. [CrossRef] [PubMed]
170. Willadsen, S.M. Nuclear transplantation in sheep embryos. *Nature* **1986**, *320*, 63–65. [CrossRef]
171. Luo, C.; Lee, Q.Y.; Wapinski, O.; Castanon, R.; Nery, J.R.; Mall, M.; Kareta, M.S.; Cullen, S.M.; Goodell, M.A.; Chang, H.Y.; et al. Global DNA methylation remodeling during direct reprogramming of fibroblasts to neurons. *eLife* **2019**, *8*, e40197. [CrossRef]

172. Yang, D.W.; Moon, J.S.; Ko, H.M.; Shin, Y.K.; Fukumoto, S.; Kim, S.H.; Kim, M.S. Direct reprogramming of fibroblasts into diverse lineage cells by DNA demethylation followed by differentiating cultures. *Korean J. Physiol. Pharmacol.* **2020**, *24*, 463–472. [CrossRef]

173. Chung, Y.G.; Matoba, S.; Liu, Y.; Eum, J.H.; Lu, F.; Jiang, W.; Lee, J.E.; Sepilian, V.; Cha, K.Y.; Lee, D.R.; et al. Histone Demethylase Expression Enhances Human Somatic Cell Nuclear Transfer Efficiency and Promotes Derivation of Pluripotent Stem Cells. *Cell Stem Cell* **2015**, *17*, 758–766. [CrossRef] [PubMed]

174. Patel, T.; Hobert, O. Coordinated control of terminal differentiation and restriction of cellular plasticity. *eLife* **2017**, *6*, e24100. [CrossRef] [PubMed]

175. Treutlein, B.; Lee, Q.Y.; Camp, J.G.; Mall, M.; Koh, W.; Shariati, S.A.; Sim, S.; Neff, N.F.; Skotheim, J.M.; Wernig, M.; et al. Dissecting direct reprogramming from fibroblast to neuron using single-cell RNA-seq. *Nature* **2016**, *534*, 391–395. [CrossRef]

176. Ma, N.X.; Puls, B.; Chen, G. Transcriptomic analyses of NeuroD1-mediated astrocyte-to-neuron conversion. *Dev. Neurobiol.* **2022**, *82*, 375–391. [CrossRef]

177. Wang, L.L.; Su, Z.; Tai, W.; Zou, Y.; Xu, X.M.; Zhang, C.L. The p53 Pathway Controls SOX2-Mediated Reprogramming in the Adult Mouse Spinal Cord. *Cell Rep.* **2016**, *17*, 891–903. [CrossRef]

178. McKay, N.D.; Robinson, B.; Brodie, R.; Rooke-Allen, N. Glucose transport and metabolism in cultured human skin fibroblasts. *Biochim. Biophys. Acta* **1983**, *762*, 198–204. [CrossRef] [PubMed]

179. Tsacopoulos, M.; Magistretti, P.J. Metabolic coupling between glia and neurons. *J. Neurosci.* **1996**, *16*, 877–885. [CrossRef]

180. Liu, H.; He, Z.; April, S.L.; Trefny, M.P.; Rougier, J.S.; Salemi, S.; Olariu, R.; Widmer, H.R.; Simon, H.U. Biochemical reprogramming of human dermal stem cells to neurons by increasing mitochondrial membrane potential. *Cell Death Differ.* **2019**, *26*, 1048–1061. [CrossRef]

181. Riechers, S.P.; Mojsilovic-Petrovic, J.; Belton, T.B.; Chakrabarty, R.P.; Garjani, M.; Medvedeva, V.; Dalton, C.; Wong, Y.C.; Chandel, N.S.; Dienel, G.; et al. Neurons undergo pathogenic metabolic reprogramming in models of familial ALS. *Mol. Metab.* **2022**, *60*, 101468. [CrossRef]

182. Moutaoufik, M.T.; Malty, R.; Amin, S.; Zhang, Q.; Phanse, S.; Gagarinova, A.; Zilocchi, M.; Hoell, L.; Minic, Z.; Gagarinova, M.; et al. Rewiring of the Human Mitochondrial Interactome during Neuronal Reprogramming Reveals Regulators of the Respirasome and Neurogenesis. *iScience* **2019**, *19*, 1114–1132. [CrossRef]

183. Hämäläinen, R.H.; Manninen, T.; Koivumäki, H.; Kislin, M.; Otonkoski, T.; Suomalainen, A. Tissue- and cell-type-specific manifestations of heteroplasmic mtDNA 3243A>G mutation in human induced pluripotent stem cell-derived disease model. *Proc. Natl. Acad. Sci. USA* **2013**, *110*, E3622–E3630. [CrossRef]

184. Liu, Y.; Lopez-Santiago, L.F.; Yuan, Y.; Jones, J.M.; Zhang, H.; O'Malley, H.A.; Patino, G.A.; O'Brien, J.E.; Rusconi, R.; Gupta, A.; et al. Dravet syndrome patient-derived neurons suggest a novel epilepsy mechanism. *Ann. Neurol.* **2013**, *74*, 128–139. [CrossRef] [PubMed]

185. Chanda, S.; Marro, S.; Wernig, M.; Sudhof, T.C. Neurons generated by direct conversion of fibroblasts reproduce synaptic phenotype caused by autism-associated neuroligin-3 mutation. *Proc. Natl. Acad. Sci. USA* **2013**, *110*, 16622–16627. [CrossRef] [PubMed]

186. Marchetto, M.C.; Belinson, H.; Tian, Y.; Freitas, B.C.; Fu, C.; Vadodaria, K.; Beltrao-Braga, P.; Trujillo, C.A.; Mendes, A.P.D.; Padmanabhan, K.; et al. Altered proliferation and networks in neural cells derived from idiopathic autistic individuals. *Mol. Psychiatry* **2017**, *22*, 820–835. [CrossRef] [PubMed]

187. Tang, Y.; Liu, M.L.; Zang, T.; Zhang, C.L. Direct Reprogramming Rather than iPSC-Based Reprogramming Maintains Aging Hallmarks in Human Motor Neurons. *Front. Mol. Neurosci.* **2017**, *10*, 359. [CrossRef] [PubMed]

188. Kumamaru, H.; Kadoya, K.; Adler, A.F.; Takashima, Y.; Graham, L.; Coppola, G.; Tuszynski, M.H. Generation and post-injury integration of human spinal cord neural stem cells. *Nat. Methods* **2018**, *15*, 723–731. [CrossRef]

189. Kenvin, S.; Torregrosa-Muñumer, R.; Reidelbach, M.; Pennonen, J.; Turkia, J.J.; Rannila, E.; Kvist, J.; Sainio, M.T.; Huber, N.; Herukka, S.K.; et al. Threshold of heteroplasmic truncating MT-ATP6 mutation in reprogramming, Notch hyperactivation and motor neuron metabolism. *Hum. Mol. Genet.* **2022**, *31*, 958–974. [CrossRef]

190. Doers, M.E.; Musser, M.T.; Nichol, R.; Berndt, E.R.; Baker, M.; Gomez, T.M.; Zhang, S.C.; Abbeduto, L.; Bhattacharyya, A. iPSC-derived forebrain neurons from FXS individuals show defects in initial neurite outgrowth. *Stem Cells Dev.* **2014**, *23*, 1777–1787. [CrossRef]

191. Begentas, O.C.; Koc, D.; Kiris, E. Establishment of Human Induced Pluripotent Stem Cells from Multiple Sclerosis Patients. *Methods Mol. Biol.* **2022**, *2549*, 43–67. [CrossRef]

192. Almeida, S.; Gascon, E.; Tran, H.; Chou, H.J.; Gendron, T.F.; DeGroot, S.; Tapper, A.R.; Sellier, C.; Charlet-Berguerand, N.; Karydas, A.; et al. Modeling key pathological features of frontotemporal dementia with C9ORF72 repeat expansion in iPSC-derived human neurons. *Acta Neuropathol.* **2013**, *126*, 385–399. [CrossRef] [PubMed]

193. Camnasio, S.; Delli Carri, A.; Lombardo, A.; Grad, I.; Mariotti, C.; Castucci, A.; Rozell, B.; Lo Riso, P.; Castiglioni, V.; Zuccato, C.; et al. The first reported generation of several induced pluripotent stem cell lines from homozygous and heterozygous Huntington's disease patients demonstrates mutation related enhanced lysosomal activity. *Neurobiol. Dis.* **2012**, *46*, 41–51. [CrossRef] [PubMed]

194. Brennand, K.J.; Simone, A.; Jou, J.; Gelboin-Burkhart, C.; Tran, N.; Sangar, S.; Li, Y.; Mu, Y.; Chen, G.; Yu, D.; et al. Modelling schizophrenia using human induced pluripotent stem cells. *Nature* **2011**, *473*, 221–225. [CrossRef] [PubMed]

195. Tan, Z.; Qin, S.; Yuan, Y.; Hu, X.; Huang, X.; Liu, H.; Pu, Y.; He, C.; Su, Z. NOTCH1 signaling regulates the latent neurogenic program in adult reactive astrocytes after spinal cord injury. *Theranostics* **2022**, *12*, 4548–4563. [CrossRef]
196. Kim, K.Y.; Hysolli, E.; Park, I.H. Neuronal maturation defect in induced pluripotent stem cells from patients with Rett syndrome. *Proc. Natl. Acad. Sci. USA* **2011**, *108*, 14169–14174. [CrossRef]
197. Petit, I.; Kesner, N.S.; Karry, R.; Robicsek, O.; Aberdam, E.; Müller, F.J.; Aberdam, D.; Ben-Shachar, D. Induced pluripotent stem cells from hair follicles as a cellular model for neurodevelopmental disorders. *Stem Cell Res.* **2012**, *8*, 134–140. [CrossRef]
198. Gatto, N.; Dos Santos Souza, C.; Shaw, A.C.; Bell, S.M.; Myszczynska, M.A.; Powers, S.; Meyer, K.; Castelli, L.M.; Karyka, E.; Mortiboys, H.; et al. Directly converted astrocytes retain the ageing features of the donor fibroblasts and elucidate the astrocytic contribution to human CNS health and disease. *Aging Cell* **2021**, *20*, e13281. [CrossRef]
199. Qin, H.; Zhao, A.; Ma, K.; Fu, X. Chemical conversion of human and mouse fibroblasts into motor neurons. *Sci. China Life Sci.* **2018**, *61*, 1151–1167. [CrossRef]
200. Zhao, T.; Zhang, Z.N.; Rong, Z.; Xu, Y. Immunogenicity of induced pluripotent stem cells. *Nature* **2011**, *474*, 212–215. [CrossRef]
201. Hu, K. Vectorology and factor delivery in induced pluripotent stem cell reprogramming. *Stem Cells Dev.* **2014**, *23*, 1301–1315. [CrossRef]
202. Yu, J.; Hu, K.; Smuga-Otto, K.; Tian, S.; Stewart, R.; Slukvin, I.I.; Thomson, J.A. Human induced pluripotent stem cells free of vector and transgene sequences. *Science* **2009**, *324*, 797–801. [CrossRef]
203. Rouhani, F.J.; Zou, X.; Danecek, P.; Badja, C.; Amarante, T.D.; Koh, G.; Wu, Q.; Memari, Y.; Durbin, R.; Martincorena, I.; et al. Substantial somatic genomic variation and selection for BCOR mutations in human induced pluripotent stem cells. *Nat. Genet.* **2022**, *54*, 1406–1416. [CrossRef]
204. Tanaka, Y.; Hysolli, E.; Su, J.; Xiang, Y.; Kim, K.Y.; Zhong, M.; Li, Y.; Heydari, K.; Euskirchen, G.; Snyder, M.P.; et al. Transcriptome Signature and Regulation in Human Somatic Cell Reprogramming. *Stem Cell Rep.* **2015**, *4*, 1125–1139. [CrossRef]
205. Dhanasekaran, R.; Deutzmann, A.; Mahauad-Fernandez, W.D.; Hansen, A.S.; Gouw, A.M.; Felsher, D.W. The MYC oncogene—The grand orchestrator of cancer growth and immune evasion. *Nat. Rev. Clin. Oncol.* **2022**, *19*, 23–36. [CrossRef] [PubMed]
206. Nakagawa, M.; Koyanagi, M.; Tanabe, K.; Takahashi, K.; Ichisaka, T.; Aoi, T.; Okita, K.; Mochiduki, Y.; Takizawa, N.; Yamanaka, S. Generation of induced pluripotent stem cells without Myc from mouse and human fibroblasts. *Nat. Biotechnol.* **2008**, *26*, 101–106. [CrossRef] [PubMed]
207. Assinck, P.; Duncan, G.J.; Hilton, B.J.; Plemel, J.R.; Tetzlaff, W. Cell transplantation therapy for spinal cord injury. *Nat. Neurosci.* **2017**, *20*, 637–647. [CrossRef]
208. Rubio, A.; Luoni, M.; Giannelli, S.G.; Radice, I.; Iannielli, A.; Cancellieri, C.; Di Berardino, C.; Regalia, G.; Lazzari, G.; Menegon, A.; et al. Rapid and efficient CRISPR/Cas9 gene inactivation in human neurons during human pluripotent stem cell differentiation and direct reprogramming. *Sci. Rep.* **2016**, *6*, 37540. [CrossRef] [PubMed]
209. Ge, L.J.; Yang, F.H.; Li, W.; Wang, T.; Lin, Y.; Feng, J.; Chen, N.H.; Jiang, M.; Wang, J.H.; Hu, X.T.; et al. In vivo Neuroregeneration to Treat Ischemic Stroke Through NeuroD1 AAV-Based Gene Therapy in Adult Non-human Primates. *Front. Cell Dev. Biol.* **2020**, *8*, 590008. [CrossRef]

Article

Evaluation of the Autologous Genetically Enriched Leucoconcentrate on the Lumbar Spinal Cord Morpho-Functional Recovery in a Mini Pig with Thoracic Spine Contusion Injury

Ravil Garifulin [1], Maria Davleeva [1], Andrei Izmailov [1], Filip Fadeev [1], Vage Markosyan [1], Roman Shevchenko [1], Irina Minyazeva [1], Tagir Minekayev [1], Igor Lavrov [2] and Rustem Islamov [1,*]

[1] Department of Histology, Cytology and Embryology, Kazan State Medical University, 420012 Kazan, Russia
[2] Department of Neurology, Mayo Clinic, Rochester, MN 55905, USA
* Correspondence: rustem.islamov@gmail.com

Abstract: Background: Pathological changes associated with spinal cord injury (SCI) can be observed distant, rostral, or caudal to the epicenter of injury. These remote areas represent important therapeutic targets for post-traumatic spinal cord repair. The present study aimed to investigate the following in relation to SCI: distant changes in the spinal cord, peripheral nerve, and muscles. Methods: The changes in the spinal cord, the tibial nerve, and the hind limb muscles were evaluated in control SCI animals and after intravenous infusion of autologous leucoconcentrate enriched with genes encoding neuroprotective factors (VEGF, GDNF, and NCAM), which previously demonstrated a positive effect on post-traumatic restoration. Results: Two months after thoracic contusion in the treated mini pigs, a positive remodeling of the macro- and microglial cells, expression of PSD95 and Chat in the lumbar spinal cord, and preservation of the number and morphological characteristics of the myelinated fibers in the tibial nerve were observed and were aligned with hind limb motor recovery and reduced soleus muscle atrophy. Conclusion: Here, we show the positive effect of autologous genetically enriched leucoconcentrate-producing recombinant neuroprotective factors on targets distant to the primary lesion site in mini pigs with SCI. These findings open new perspectives for the therapy of SCI.

Keywords: spinal cord injuries; swine; autologous blood transfusion; leukocytes; genetic vectors; genetic therapy; vascular endothelial growth factor A; glial cell line-derived neurotrophic factor; neural cell adhesion molecules; spinal cord regeneration

Citation: Garifulin, R.; Davleeva, M.; Izmailov, A.; Fadeev, F.; Markosyan, V.; Shevchenko, R.; Minyazeva, I.; Minekayev, T.; Lavrov, I.; Islamov, R. Evaluation of the Autologous Genetically Enriched Leucoconcentrate on the Lumbar Spinal Cord Morpho-Functional Recovery in a Mini Pig with Thoracic Spine Contusion Injury. *Biomedicines* **2023**, *11*, 1331. https://doi.org/10.3390/biomedicines11051331

Academic Editor: Masaru Tanaka

Received: 22 March 2023
Revised: 27 April 2023
Accepted: 27 April 2023
Published: 30 April 2023

1. Introduction

The functional restoration and effect of rehabilitation after thoracic SCI depends on the morpho-functional state of the lumbar spinal cord, distant from the epicenter of the injury. Disruptions of the descending and ascending pathways interrupt the neural connections between the brain and spinal cord neurons innervating target organs, leading to motor and sensory dysfunctions [1,2]. Lumbar α-motoneurons deprived of efferent signaling from the pyramidal and extrapyramidal systems lose control of the skeletal muscles of the hind limbs, which depend on the impulse activity of motor neurons and neurogenic molecules released from the nerve terminal at the neuromuscular junction [3]. In addition to post-traumatic changes, SCI causes neuroinflammation-related damage to segments of the spinal cord remote from the epicenter of injury. Numerous alterations in the lumbar region were found after thoracic SCI, with changes in the number of neurons [4–6], synapses [6,7], as well as in the response of microglia [8–11] and astrocytes [12,13]. Thus, pathological changes in SCI affect large areas around the epicenter of the lesion [8,14,15]; therefore, after thoracic SCI, the distant areas, i.e., lumbosacral segments of the spinal cord, are important therapeutic targets.

Currently, the most promising therapeutic regenerative strategies for SCI rely on cell and gene therapies, small biomolecules, biodegradable materials, and neuromodulation [16]. Numerous novel regenerative approaches are under intensive preclinical research. At the same time, only a few therapeutic strategies were advanced to clinical trials and focused on the following areas: (i) neurotransplantation of mesenchymal stem cells (MSC) derived from the umbilical cord, adipose tissue, bone marrow, olfactory ensheathing cells (OEC), and Schwann cells [17–19], (ii) administration of small biomolecules neutralizing inhibitors of neuroregeneration, such as the Rho-associated protein kinase (ROCK) inhibitor and anti-NOGO (neurite outgrowth inhibitor) monoclonal antibody [20,21], and (iii) implantation of neuro-spinal scaffolds [22,23]. Unfortunately, these clinical studies showed minimal or no progress and are still far from practical implementation [21,22]. The other innovative cell-based approach in preclinical studies of SCI treatment is based on the application of exosomes derived from MSC and neural stem cells [24]. Preclinical studies demonstrated promising results in using engineered exosomes carrying microRNA derived from gene-modified MSC [25]. However, future clinical trials will require compliance with good manufacturing practice to ensure a standardized protocol of exosome biogenesis, purification and quality control [26].

Delivery of therapeutic genes expressing growth, neurotrophic, anti-apoptotic, and anti-inflammatory factors to the injured spinal cord using viral vectors (in vivo gene therapy) or cell carriers (ex vivo gene therapy) has demonstrated positive effects on post-traumatic spinal cord regeneration in various animal models [27,28]. However, to date, no gene therapy in vivo or ex vivo has been tested in clinical trials. A long list of potential genes applicable for stimulation of post-traumatic neuroregeneration, and the variety of viral vectors used to target cell transduction and delivered genetic constructs create enormous variations, which slow down the progress in the field of gene therapies [29,30]. Moreover, testing these therapies on large animals should be considered before translation to humans, as the differences in regeneration capacity and anatomy between small and large animals impact the therapeutic outcome [31,32]. A recently developed approach with autologous genetically enriched leucoconcentrate simultaneously producing recombinant vascular endothelial growth factor (VEGF), glial cell line-derived neurotrophic factor (GDNF), and neural cell adhesion molecule (NCAM) demonstrated functional and morphological improvement in SCI mini pigs [33] similarly to previously reported results using other cell-mediated therapies [34]. The positive effect of the genetically enriched leucoconcentrate on the molecular and cellular changes was found in the epicenter and adjacent supra- and sublesion parts of the spinal cord after contusion injury at Th8-Th9 [33]. Intravenous infusion of the autologous leucoconcentrate producing recombinant VEGF, GDNF, and NCAM in the acute phase of spinal cord injury demonstrated increased survivability of the spinal cord cells, recovery of synaptic protein expression, prevention of astrogliosis in the grey matter and higher growth rates of regenerating axons accompanied by a higher number of oligodendroglial cells in the lateral corticospinal tract region in the white matter. Following these reports, we expanded our study on thoracic SCI mini pigs treated with autologous leucoconcentrate enriched with genes encoding VEGF, GDNF, and NCAM [33] to evaluate the recovery of the lumbar spinal cord.

In this study, molecular and cellular changes in the lumbar spinal cord and peripheral nerves were evaluated in relation to restoration of motor performance and electrophysiological and histological examination of the hind limb skeletal muscles in mini pigs with thoracic SCI after intravenous infusion of the autologous genetically enriched leucoconcentrate simultaneously producing recombinant VEGF, GDNF, and NCAM (Figure 1).

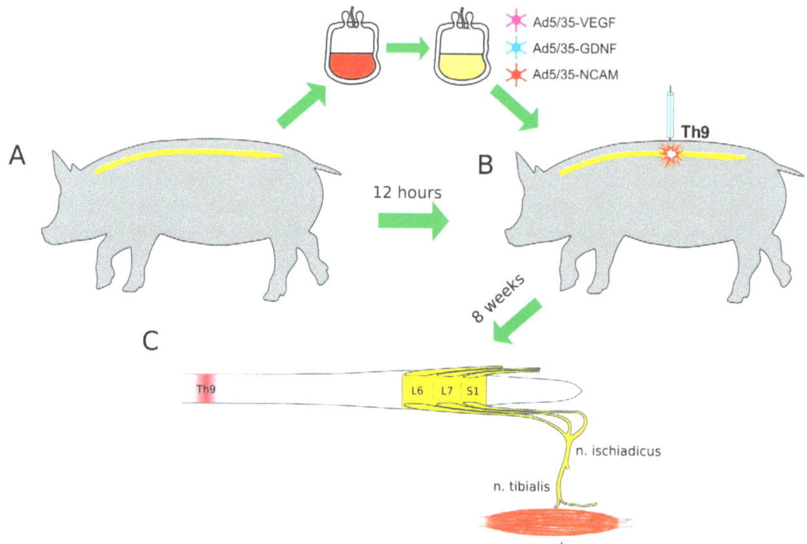

Figure 1. Study design. (**A**)—Collection of 50 mL of venous blood and preparation of the autologous genetically enriched leucoconcentrate. (**B**)—Spinal cord contusion injury at the Th9 segment followed by intravenous autoinfusion of the genetically enriched leucoconcentrate 4 h after surgery. At 2, 4, and 8 weeks after surgery, behavioral tests and electrophysiological studies were performed. (**C**)—The lumbar spinal cords (L6-S1), tibial nerves and soleus muscles were harvested at 8 weeks after injury. Ad5/35—adenovirus serotype 5 with fibers derived from adenovirus serotype 35 fiber gene; VEGF—vascular endothelial growth factor; GDNF—glial cell line-derived neurotrophic factor; NCAM—neural cell adhesion molecule.

2. Materials and Methods

This study aligns with our long-term efforts to stimulate post-traumatic regeneration after spinal cord injury (SCI) [35]. To test our hypothesis that autologous leucoconcentrate enriched with *vegf165, gdnf, ncam1* may have positive effects on therapeutic targets distant to the primary lesion site after SCI, assuming its paracrine mechanism of action after intravenous infusion [36], in support of this concept, after low thoracic SCI, we evaluated the efficacy of the autologous genetically enriched leucoconcentrate on the lumbar spine, peripheral nerves, and hind limb skeletal muscles' morpho-functional characteristics in relation with our previous studies [33,37].

Female adult miniature Vietnamese pot-bellied pigs (25–30 kg; $n = 11$) were used in the study. Two weeks before investigation, mini pigs were kept one per housing area in a 12 h light/dark regimen at a temperature of 24–25 °C with controlled air conditioning and properly organized access to food and water. Animal treatments were conducted according to the Animal Care and Use Committee of Kazan State Medical University (approval No. 5, dated 26 May 2020). Preparation of the autologous genetically enriched leucoconcentrate (GEL) and treatment of the animals are presented briefly, in view of the detailed descriptions of those in our recent publication, in which the same animals were employed [33].

2.1. Preparation of the Autologous Genetically Enriched Leucoconcentrate

Recombinant replication-defective viral vectors carrying *vegf165, gdnf,* and *ncam1* were constructed based on the human adenovirus serotype 5 with fibers derived from adenovirus serotype 35 fiber gene (Ad5/35) and nucleotide sequences encoding VEGF165 (Gene Bank NM_001171626.1), GDNF (Gene Bank NM_000514.4), and NCAM1 (Gene Bank NM_001076682.2) [37]. In the study, we used Ad5/35 to provide effective transduction of leucocytes in the blood bag via the 35 fiber, which has a high affinity for the cluster of differentiation 46 (CD46) that is expressed on all leucocytes [36]. Transduction of the leucoconcentrate directly in the blood bag is a safe and economic approach that prevents usage of antibiotics and biological materials of animal origin.

From each experimental mini pig, 50 mL of peripheral blood was collected in a plastic blood bag from v. subclavia in a 14–16 h period before SCI modelling. The procedure for leucoconcentrate preparation was performed according to our original protocol [36] and subsequently included sedimentation of erythrocytes with 6% hydroxyethyl starch, centrifugation at $34\times g$ for 10 min at 10 °C (DP-2065 R PLUS, Centrifugal Presvac RV; Presvac, Buenos Aires, Argentina) and washing with saline. The obtained leucoconcentrate was immediately transduced with a mixture of chimeric adenoviral vectors carrying cDNA encoding VEGF165, GDNF, and NCAM1. Transduction was performed according to the nucleated cell count in the leucoconcentrate, the titers and the equal ratio of each viral vector correspondingly: 1/3 Ad5/35-VEGF165 (2.0×10^9 PFU/mL), 1/3 Ad5/35-GDNF (7.0×10^{10} PFU/mL), and 1/3 Ad5/35-NCAM1 (5.0×10^{10} PFU/mL) for 12 h in the blood bag with a multiplicity of infection (MOI) equal to 10 [37]. After washing with saline and centrifugation, the solution (30 mL) in the blood bag was considered as the leucoconcentrate enriched with *vegf165, gdnf* and *ncam1* (GEL-VGN). The efficacy of the leucocyte transduction was confirmed as described earlier [33,37]. In vitro comparative molecular analysis of naïve and gene-modified leucocytes 72 h after culturing confirmed effective expression of *vegf165, gdnf* and *ncam1* by RT-PCR (synthesis of transgenes mRNA), immunofluorescence method (synthesis of the recombinant VEGF, GDNF, and NCAM), and ELISE (secretion of recombinant VEGF, GDNG, and NCAM by leucocytes in the culture medium).

2.2. Animals Treatment

A moderate contusion injury of the spinal cord was performed a day after blood collection for GEL preparation. SCI was carried out under deep anesthesia induced by intramuscular administration of Zoletil 100 (Virbac Sante Animale, Carros, France) at a dose of 10 mg/kg and maintained using an inhalation apparatus (Minor Vet Optima, Zoomed, Moscow, Russia) with isoflurane (Laboratorios Karizoo, S.A., Barcelona, Spain) as a 2.0–2.5% mixture with oxygen. After laminectomy at the Th8-Th9 vertebral level, the dura matter was exposed, and a 50 g metal rod with a diameter of 9 mm was dropped on the spinal cord from a 50 cm height as described in our previous study [34]. Under anesthesia, the mini pigs were kept on the operating table with a plate maintaining their body temperature at 38 °C, and 4 h after SCI, the mini pigs in the therapeutic group ($n = 4$), were infused via the auricular vein with 30 mL of the genetically enriched autologous leucoconcentrate. Animals from the control group ($n = 4$) received an infusion of the naïve autologous leucoconcentrate (Figure 1). Intact (healthy) mini pigs ($n = 3$) were employed to collect basic electrophysiological and histological data for comparative analysis of the morpho-functional changes in experimental animals (Figure 1). The rationale for using only autologous leucoconcentrate in control experiments was to exclude the influence on neuroregeneration by naïve leukocytes used as cell carriers for the delivery of the therapeutic genes. Based on our previous comparative analysis, which showed no impact of the direct or leucocyte-mediated delivery of the adenoviral vector carrying reporter gfp on the spinal cord recovery [38], in this study we did not employ these control groups.

2.3. Behavioral Assessment

Post-traumatic recovery of motor functions in mini pigs with SCI was evaluated using the Porcine Thoracic Injury Behaviour Scale (PTIBS) [39]. Behavioral assessment was performed when the experimental animal was placed in the center of the arena (2.5 m × 2.5 m) and evaluated according to the 10-point PTIBS score. In general, scores of 1 to 3 indicated "hind limb dragging", the scores of 4 to 6 corresponded to various degrees of "stepping", and the scores of 7–10 were related to "walking" ability. The PTIBS score was estimated by two observers in a blinded manner with respect to the experimental groups a week before SCI modeling and 2, 4, and 8 weeks after surgery.

2.4. Electrophysiological Study

Before the electrophysiological assessment, mini pigs were anesthetized and connected to an inhalation apparatus, as described above. A Digitimer DS7A (Digitimer Ltd., Welwyn Garden, UK), amplifier with filters ranging from 5 Hz to 2 kHz (Biosignal amplifier, g.tec medical engineering GmbH, Schieldberg, Austria) and LabChart data collection and analysis systems (AD Instruments Inc., Colorado Springs, CO, USA) were used to evaluate M-response in the soleus muscle of both hind limbs evoked by electrical stimulation of the sciatic nerve. Stimulating stainless steel electrodes (0.6 mm in diameter and 50 mm length) were inserted into the area of the sciatic nerve projection at 2 cm below the large trochanter of the femur. M-response was recorded using the similar needle electrodes injected into the soleus muscle before and 2, 4, and 8 weeks after SCI. Stimulation was performed with a single rectangular pulse with a frequency of 0.6 Hz, duration of 0.2 ms, and a current range of 4–72 mA. The parameters of M-responses were obtained by stimulating the sciatic nerve with various current intensities. The absolute values of the duration and amplitude of the M-responses were averaged in each animal by current intensity and then averaged per group. Data are presented as a percentage for comparison with intact animals, the absolute values of which were taken as 100%.

2.5. Sample Collection

Animals from the therapeutic and control groups were sacrificed 60 days after SCI. The spinal cord was removed from the vertebral column, and the lumbar part was fixed in 4% paraformaldehyde (Sigma, St. Louis, MO, USA) in phosphate-buffered saline (pH 7.4) and processed for immunofluorescence staining. The 5 mm long fragments from both tibial nerves were taken, fixed in 2.5% solution of glutaraldehyde for 4 h, incubated in 1% solution of osmium tetroxide for 24 h, and embedded in EMbed 812 (Electron Microscopy Sciences, Hatfield, PA, USA) for preparation of the semi-thin sections. Skeletal muscles (*m. soleus*) from both hind limbs were weighed, fixed in 4% paraformaldehyde and processed for immunofluorescence phenotyping and morphometric analysis of the skeletal muscle fibers.

2.6. Immunofluorescence Study of Lumbar Spinal Cord

Frozen free-floating cross-sections of 20 μm thickness were prepared from L6-S1 level with a cryostat (Microm HM 560, Thermo Scientific, Waltham, MA, USA). Antibodies (Ab) against potassium-chloride cotransporter protein (KCC2), choline acetyltransferase (Chat), synaptophysin and postsynaptic density protein of 95 kDa (PSD95) were employed to analyze functional activity of the motor neurons. Astrocytes and oligodendroglial and microglial cells were identified with Abs to glial fibrillary acidic protein (GFAP), oligodendrocyte transcription factor (Olig2), and ionized calcium binding adaptor molecule 1 (Iba1), correspondingly. The appropriate secondary Abs were used to visualize the target molecules (Table 1). Nuclei were counterstained with DAPI (10 μg/mL in PBS, Sigma, Burlington, MA, USA), and sections were embedded in glycerol (GalenoPharm, Saint Petersburg, Russia). The slides were investigated with a luminescence microscope Axioscope A1 (Carl Zeiss, Oberkochen, Germany) and a confocal microscope Leica TCS SP5 MP (Leica Microsystems, Wetzlar, Germany) using identical settings. Obtained digital images were analyzed using ImageJ (NIH) software. Expression of the target molecules was evaluated in the ventral and dorsal horns with an area of 0.05 mm^2. The count of Olig2-positive cells was performed in regard to specific nuclear immunostaining and DAPI nuclear counterstaining. Expression of neural (KCC2, Chat, Synaptophysin, PSD95) and glial markers (GFAP, Iba1) was estimated as the relative immunopositive area and presented as percentage considering the studied area [37].

Table 1. Antibodies used in immunofluorescence staining.

Antibody against:	Host	Dilution
Choline Acetyltransferase (ChAT)	Rabbit	1:100
Glial fibrillary acidic protein (GFAP)	Mouse	1:200
Ionized calcium binding adaptor molecule 1 (Iba1)	Rabbit	1:150
The K+–Cl− cotransporter isoform 2 (KCC2)	Rabbit	1:100
Oligodendrocyte transcription factor 2 (Olig2)	Rabbit	1:100
Postsynaptic density protein 95 kDa (PSD95)	Rabbit	1:200
Slow Skeletal Myosin Heavy chain	Rabbit	1:100
Synaptophysin	Rabbit	1:100
Mouse IgG conjugated with Alexa 488	Donkey	1:200
Rabbit IgG conjugated with Alexa 488	Donkey	1:200

2.7. Morphometric Analysis of the Sciatic Nerve Myelinated Fibers

Semi-thin cross-sections of the tibial nerves from both hind limbs were cut using an ultramicrotome (LKB-3; LKB, Sollentuna, Sweden) and stained with methylene blue dye. Digitized images of the tibial nerves were collected for morphometric analysis. Total count of myelinated fibers, thickness of the myelin sheet and axon diameter were measured in an area of 0.05 mm^2 in each sample using ImageJ (NIH) software.

2.8. Hind Limb Skeletal Muscle Study

Slow soleus muscles were investigated from both hind limbs of the experimental animals. Phenotyping of the skeletal muscle fibers was performed using immunofluorescence staining of the 10 μm frozen cross-sections with Ab to slow myosin heavy chains. The number of slow skeletal muscles was expressed as percentage in regard to the total count of 200 muscle fibers in each studied muscle. Morphometric analysis of the skeletal muscle fibers area was performed in digital images captured with a 20-fold magnification of the microscope. The area of 200 skeletal muscle fibers in each soleus muscle was measured using the ImageJ (NIH) software.

2.9. Statistics

Statistical data analysis and visualizations were performed using R version 4.1.2 (R Foundation for Statistical Computing, Vienna, Austria). Sample distributions of quantitative values were visualized using box plots, and descriptive statistics are presented as: (median [1st quartile; 3rd quartile]). The results of M-response were presented as mean ± SEM. The Kruskal–Wallis test was used to compare experimental groups. Dunn's test was used as the post hoc method. Differences were considered statistically significant, where $p < 0.05$.

3. Results

3.1. Hind Limb Skeletal Muscle Recovery

Behavioral tests and electrophysiological and histological studies were performed to evaluate the recovery of the hind limb skeletal muscles innervated by lumbar motor neurons.

3.1.1. Motor Activity

Two weeks after SCI, control and treated mini pigs demonstrated a drastic decline in motor activity. PTIBS score was equal to 1 point, animals were not able to stand up, and the sacrum and knees of the hind limbs were on the floor (Figure 2A). Four weeks after SCI, the PTIBS score increased to 2.0 (2.0; 2.75) points in the therapeutic group; the animals tried to pull their hind limbs to the belly and stand on them. The control mini pigs were trying to move their hind limbs by this time; the PTIBS score was 2.0 (1.75; 2.0). At 8 weeks, a further increase in motor activity was observed in the therapeutic group, with the PTIBS score corresponding to 3.0 (2.25; 3.0). Treated animals demonstrated active flexion in hind limbs' hips, knees, and ankles, and mini pigs could stand on four legs and tried to make steps. Control animals at 8 weeks post-SCI could drag their hind limbs while moving with their front limbs; the PTIBS score (2.0 (1.75; 2.0)) was lower than in the therapeutic group (p = 0.009). The recovery of locomotor activity was consistent with the results of our pilot study in SCI pigs treated with genetically modified human umbilical cord blood cells overexpressing VEGF, GDNF, and NCAM [40].

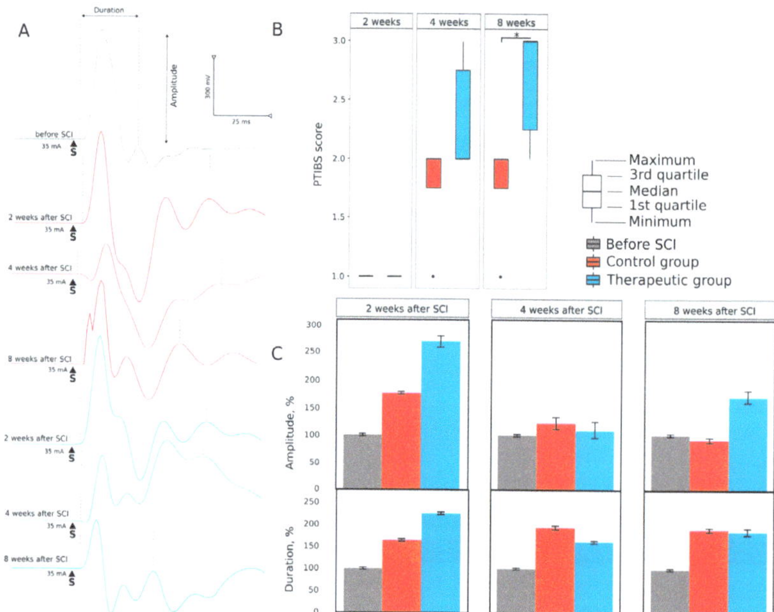

Figure 2. Hind limb skeletal muscles recovery in mini pigs after low thoracic spinal cord injury (SCI). (**A**)—Motor activity of mini pigs in control and therapeutic groups at 2, 4, and 8 weeks after SCI assessed with the Porcine Thoracic Injury Behavioural Scale (PTIBS). (**B**)—Evaluation of amplitude and duration of motor-evoked potential (M-response) in the soleus muscle evoked by electrical stimulation of the sciatic nerve in control and treated mini pigs at 2, 4, and 8 weeks after SCI, *—$p < 0.05$. (**C**)—Analysis of the amplitude and duration of motor-evoked potential.

3.1.2. Electrophysiology

The soleus muscle response to electrical stimulation of the sciatic nerve at 2, 4, and 8 weeks after SCI revealed the changes in the pattern of M-response in experimental animals (Figure 2B). In control mini pigs, M-response had a polyphasic shape and was associated with an increased duration at 2 (165 $\pm$ 7.1%), 4 (194 $\pm$ 11%) and 8 (190 $\pm$ 8.5%) weeks after SCI (Figure 2C). In the treated animals, the M-response also had a multiphase pattern with increased duration at 2 (225 $\pm$ 6%), 4 (161 $\pm$ 5.9%), and 8 (185 $\pm$ 15.7%) weeks after SCI. The amplitude of M-response was increased 2 weeks after SCI both in control (175 $\pm$ 4.6%) and treated (274 $\pm$ 19.4%) animals. At the following 4 and 8 weeks, the amplitude in the control group decreased to 121 $\pm$ 13.6% and 91 $\pm$ 6.3%, and in the therapeutic group to 108 $\pm$ 22.6% and 169 $\pm$ 15%, correspondingly. The threshold and latency in all intact and experimental animals were not significantly different.

3.1.3. Histology

At 8 weeks after SCI, the raw weight (g) of soleus muscles in the control group was decreased (3.52 (3.47–5.52)) when compared to intact (9.90 (9.32–11.32), p = 0.0031) and therapeutic (9.76 (9.38–12.90), p = 0.0013) groups, which were no different from each other (Figure 3B). At 8 weeks, after SCI the content of slow skeletal fibers in the intact slow soleus muscles was 92% (Figure 3A,D) and decreased in the control group (45%, p = 0.0011736)

8 weeks after SCI (Figure 3A′,D). In treated animals, the percentage of slow muscle fibers in *m. soleus* was 63% (Figure 3A″) and did not differ from that of intact animals (Figure 3D). Morphometric analysis of muscle fibers area (μm^2) in the soleus muscles revealed that the average area of muscle fibers in the control mini pigs (1058.142 ± 158.093 μm^2) was lower than in the intact animals (1366.473 ± 139.559 μm^2) ($p = 0.1304793$). In the therapeutic group, the average area of muscle fibers in the soleus (1293.308 ± 189.384 μm^2) did not differ from that of the intact group (Figure 3C). Thus, the improvement of motor activity, inhibition of skeletal muscle atrophy, and preservation of soleus muscle phenotype) suggest that autologous leucoconcentrate enriched with *vegf*, *gdnf*, and *ncam* has a positive effect on the recovery of skeletal muscles of hind limbs in mini pigs with SCI.

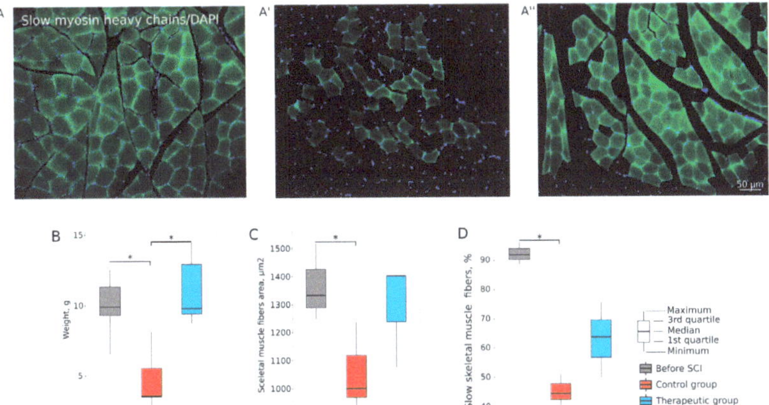

Figure 3. Assessment of soleus muscles in mini pigs 8 weeks after low thoracic spinal cord injury (SCI). (**A,A′,A″**) Cross-sections of the muscle stained with antibodies against slow myosin heavy chains (green glow) from intact, control, and therapeutic groups, correspondingly. Nuclei were counterstained with DAPI (blue glow). (**B**) Soleus muscle weighing analysis. (**C**) Morphometric analysis of skeletal muscle fiber area. (**D**) Relative content of slow skeletal muscle fibers, *—$p < 0.05$.

3.2. Lumbar Spinal Cord and Tibial Nerve Plasticity

Immunofluorescence study of the lumbar spinal cord and morphometric analysis of myelinated nerve fibers of both tibial nerves were conducted in relation to the hind limbs motor function recovery in the mini pigs at 8 weeks after SCI.

3.2.1. Motor Neurons

Expression of Chat, PSD95, Synaptophysin, and KCC2 was evaluated in the lumbar motor neurons. Analysis of Chat immunoreactivity in the ventral horns of the lumbar spinal cord revealed that the relative immunopositive area for Chat was higher in the treated animals (3.569 (3.520; 4.416)%, $p = 0.0173$) in comparison with control mini pigs (1.766 (0.789; 2.632)%) and was similar to that of the intact (3.382 (2.343; 3.752)%) group (Figure 4A). The decreased immunoreactivity for PSD95 (Figure 4B) as well was found in the control animals (38.308 (34.474; 42.513)%) when compared to intact mini pigs (44.132 (42.856; 48.104)%) ($p = 0.0235$). In the therapeutic group, expression of PSD95 (42.463 (39.947; 47.107)%) was not different from that in the intact group. Analysis of the relative immunopositive area for synaptophysin (Figure 4C) and KCC2 (Figure 4D) in the ventral horns of the lumbar spinal cord did not reveal essential changes in the therapeutic and control groups when compared to the intact group.

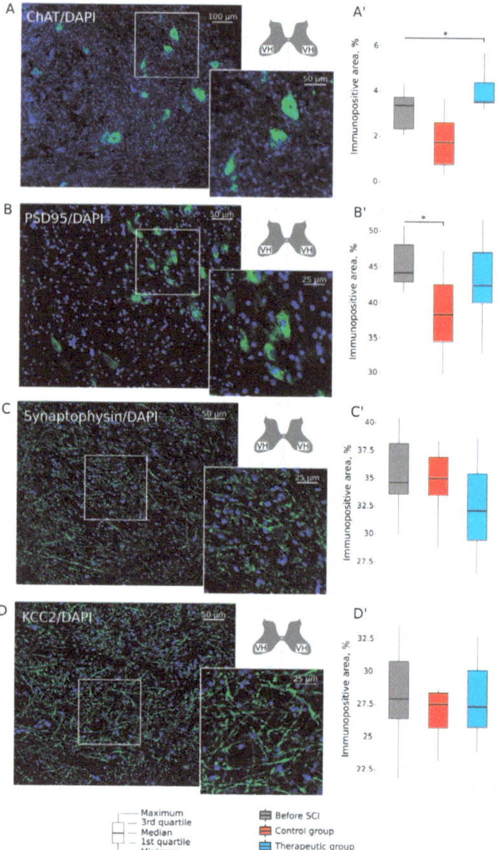

Figure 4. Immunofluorescence staining of the ventral horns of the lumbar spinal cord in mini pigs 8 weeks after low thoracic spinal cord injury (SCI). (**A**) Immunopositive reaction with an antibody against a choline acetyltransferase (Chat) (green glow) in an intact animal. (**A′**) Morphometric analysis demonstrates relative Chat-positive area in the experimental groups. (**B**) Immunopositive reaction with an antibody against a postsynaptic density protein of 95 kDa (PSD95) (green glow) in an intact animal. (**B′**) Morphometric analysis demonstrates relative PSD95-positive area in the experimental groups. (**C**) Immunopositive reaction with an antibody against a synaptic vesicle protein synaptophysin (green glow) in an intact animal. (**C′**) Morphometric analysis demonstrates relative synaptophysin-positive area in the experimental groups. (**D**) Immunopositive reaction with an antibody against a potassium-chloride cotransporter protein (KCC2) (green glow) in an intact animal. (**D′**) Morphometric analysis demonstrates relative KCC2-positive area in the experimental groups. Nuclei were counterstained with DAPI (blue glow). Inserts demonstrate the zooming part of images. The squares inserted in the schematic transverse spinal cord indicate the areas used for immunofluorescence analysis, VH—ventral horn. Data are visualized using box plots, *—$p < 0.05$.

3.2.2. Neuroglial Cells

Remodeling of astrocytes, oligodendroglial, and microglial cells was investigated in the ventral and dorsal horns in the lumbar spinal cord of the experimental animals. Analysis of GFAP expression found an increase in relative GFAP-positive area in dorsal horns in control animals (19.871 (15.677; 23.700)%, $p = 0.002$), in comparison with intact animals (10.095 (9.165; 12.101)). In the therapeutic group, GFAP-positive area was not different in the ventral and dorsal horns relative to the intact group (Figure 5A). Evaluation of Iba1 expression demonstrated increased Iba1-positive area in control and treated mini pigs both in ventral (25.409 (20.503; 26.254)%, $p = 0.0001$; 14.657 (12.976; 15.744)%, $p = 0.003$) and dorsal (22.190(14.292; 23.533)%, $p = 0.001$; 15.386 (13.332; 17.112)%, $p = 0.0006$) horns when compared to intact animals (9.802 (7.347; 10.611)% and 8.118 (7.580; 8.794)%) (Figure 5B). The number of Olig2-positive nuclei in intact mini pigs was 12.0 (11.0; 13.0) in the ventral and 16.5 (16.0; 20.75) in the dorsal horns. In the ventral horns, the number of oligodendroglial cells was decreased in the control (5.0 (4.0; 6.0), $p = 0.0001$) and treated (7.5 (6.0; 8.0), $p = 0.0055$) mini pigs. In the dorsal horns, the decreased number of the Olig2-positive nuclei was shown in the control group 3.0 (2.0; 4.25) when compared both with intact ($p = 0.0001$) and therapeutic group 13.0 (11.0; 14.75) ($p = 0.0006$) (Figure 5C).

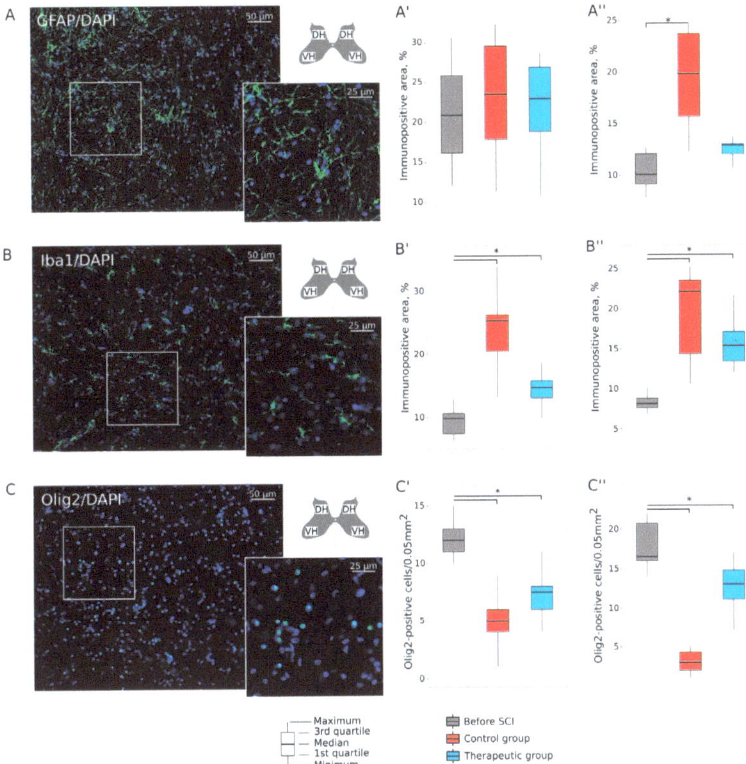

Figure 5. Immunofluorescence staining of the ventral and dorsal horns of the lumbar spinal cord in mini pigs 8 weeks after low thoracic spinal cord injury (SCI). (**A**) Immunopositive reaction with an antibody against a glial fibrillary acidic protein (GFAP) (green glow) in an intact animal. (**A′,A″**) Morphometric analysis demonstrates relative GFAP-positive area in the ventral and dorsal horns, correspondingly, in the experimental groups. (**B**) Immunopositive reaction with an antibody against an ionized calcium binding adaptor molecule 1 (Iba1) (green glow) in an intact animal. (**B′,B″**) Morphometric analysis demonstrates relative Iba1-positive area in the ventral and dorsal horns, correspondingly, in the experimental groups. (**C**) Immunopositive reaction with an antibody against an oligodendrocyte transcription factor (Olig2) (green glow) in an intact animal. (**C′,C″**) Morphometric analysis demonstrates the number of Olig2-positive nuclei in the ventral and dorsal horns, correspondingly, in the experimental groups. Nuclei were counterstained with DAPI (blue glow). Inserts demonstrate the zooming part of images. The squares inserted in the schematic transverse spinal cord indicate the areas used for immunofluorescence analysis, VH—ventral horn and DH—dorsal horn. Data are visualized using box plots, *—$p < 0.05$.

3.2.3. Tibial Nerve Myelinated Fibers

Myelinated fibers in both tibial nerves (Figure 6A) were studied 8 weeks after SCI with respect to axons of the lumbar spinal cord motor neurons innervating soleus muscles. The count of myelinated fibers found a decrease in the nerve fibers in the control group (277.180 (177.037; 379.340)) when compared to the intact (940.625 (856.108; 987.652), $p = 0.0013749$) and therapeutic (485.850 (433.360; 656.345), $p = 0.0551393$) groups (Figure 6B). No difference was observed between intact and treated animals. The thickness of myelin sheath (1.935 (1.595; 2.298)) and axon diameter (8.69 (7.850; 10.050)) in the control mini pigs was increased when compared to the intact mini pigs (1.240 (1.120; 1.370), $p = 0.0209560$ and 4.78 (4.690; 5.040), $p = 0.0033348$), correspondingly (Figure 6C,D). In the therapeutic group, the thickness of the myelin sheath was 1.170 (1.055; 1.372), and the axon diameter was 5.94 (5.365; 6.195), which was not differ from the intact group (Figure 6C,D). Thus, the preserved number of myelinated fibers and their morphology (axon diameter and thickness of myelin sheath) in the treated animals suggest that autologous leucoconcentrate producing recombinant VEGF, GDNF, and NCAM may affect lumbar motor neurons, dorsal root ganglion neurons, and their neurites in the tibial nerve.

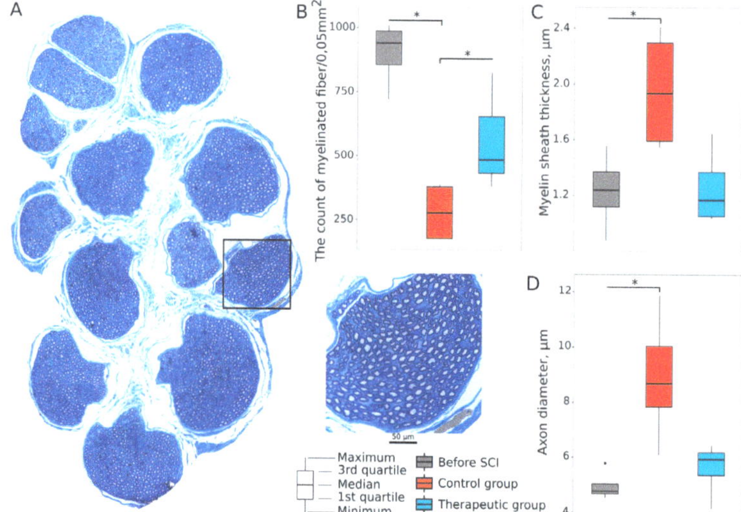

Figure 6. Preservation of the tibial nerve myelinated fibers in mini pigs 8 weeks after low thoracic spinal cord injury (SCI). (**A**) Semi-thin cross-section of the tibial nerve stained with methylene blue dye in an intact animal. (**B**) Number of the myelinated fibers. (**C**) Thickness of myelin sheath (µm) in the myelinated fibers. (**D**) Axon diameter (µm) of the myelinated fibers. Data are visualized using box plots, *—$p < 0.05$.

Thus, the study of lumbar spinal cord, peripheral nerves, and hind limb skeletal muscles in mini pigs with low thoracic SCI revealed the essential changes in the structures distant to the primary lesion site, which may be considered as important therapeutic targets in the SCI treatment. The complex analysis of the obtained data demonstrated correlations between the positive changes in the lumbar spinal cord (positive remodeling of the neuroglia and increased expression of synaptic proteins), tibial nerves (preservation of the number and morphological characteristics of the myelinated nerve fibers) and hind limbs skeletal muscles (prevention of soleus muscles atrophy and sparing its phenotype) in the condition of the treatment with GEL.

4. Discussion

Spinal cord injury results in the loss of motor, sensory, and autonomic functions of various organs and has a lifelong devastating impact on independence and activities of daily living. Pathological processes following SCI include mass death of spinal cord cells, nerve fibers tears, hemorrhages, and ischaemic and inflammatory lesions, which subsequently form cavities and cysts. Neuroinflammation (activation of astroglia and microglia and infiltration of myeloid cells) is a key factor of the secondary injury involving the entire spinal cord in the pathologic process [1,41]. The post-traumatic spinal cord remodeling takes place not only at the epicenter of the injury but also in large areas around the epicenter [10,42]. Spatiotemporal neurodegeneration in rostral and caudal regions of the spinal cord distant from the primary injury suggests that these areas should be considered as important therapeutic targets, particularly when analyzing the effectiveness of SCI therapies [14,43]. However, the effect of the therapies for SCI has generally been considered mainly regarding the therapeutic targets at the epicenter of the lesion and in the adjacent segments [37].

Earlier, we studied the effect of autologous genetically enriched leucoconcentrate producing recombinant VEGF, GDNF, and NCAM on molecular and cellular changes in the epicenter of the injury [33]. Cavitation in the grey and white matter, astrogliosis, decreased number of oligodendroglial cells, and affected expression of the synaptic proteins were revealed at 60 days after SCI at the Th8-Th9 level in SCI mini pigs. Intravenous infusion of the autologous genetically enriched leucoconcentrate had a positive effect on the sparing of the grey matter, recovery of synaptic protein expression, prevention of astrogliosis, and growth of regenerating axons accompanied by a higher number of oligodendrocytes in the rostral and caudal segments near the epicenter of the injury.

The positive effect of GEL-VGN auto-infusion four hours after SCI modeling was based on the neuroprotective action of VEGF [44], GDNF [45,46], and NCAM [37,47–49] on affected nervous tissue. The neuroprotective action of VEGF promotes the survival of neuronal cells mediated via the activation of high-affinity receptor tyrosine kinases by the activation of mitogen-activated protein kinase (MAPK) [50,51] and phosphatidylinositol 3-kinase (PI3K) signaling cascades [52,53]. One of the key functions of GDNF is the maintenance of neuron–neuron and neuron–target tissue interactions and prevention of cell death [54]. GDNF triggers phosphorylation of the RET (rearranged during transfection) receptor tyrosine kinase responsible for activation of the MAPK and/or PI3K pathway [55,56]. NCAM (CD56) is expressed on the surface of neurons and neuroglia cells. NCAM-mediated intercellular interactions in neuroontogenesis and post-traumatic regeneration ensure neuronal survival, directed neurite growth, and synaptogenesis [49]. NCAM recruits non-receptor Fyn tyrosine kinase, leading to MAPK and transcription factor CREB (cAMP response element binding protein) activation [47]. NCAM also serves as a signaling co-receptor for GDNF [57].

Based on in vitro studies, we confirmed effective expression of *vegf165*, *gdnf*, and *ncam1* in genetically modified leucocytes 72 h after transduction with adenoviral vectors [33,37]. Delivery of *vegf165*, *gdnf*, and *ncam1* with autologous leucocytes proposes effective production of the recombinant molecules in about 2–3 weeks, which is limited by the expression of adenovirus vectors [38,58]. After intravenous infusion of GEL-VGN, recombinant VEGF, GDNF, and NCAM produced by gene-modified leucocytes in the bloodstream may reach the injury site via the impaired blood–brain barrier after SCI [59]. The role of NCAM in the GEL-VGN is also proposed to stimulate migration of the gene-modified leucocytes into the damaged spinal cord and provide the local production of the neurotrophic factors [60]. Active expression of the transgenes in the acute phase of SCI may reduce the negative consequences of the neurotrauma and stimulate post-traumatic spinal cord regeneration. However, the role of each therapeutic molecule in this combination has to be further explored.

In this work, we expanded our previous study and investigated dysfunction of the lumbar spinal cord in mini pigs with low thoracic SCI. We specifically evaluated the effect of the autologous genetically enriched leucoconcentrate on the morphological and functional restoration of the lumbar spinal cord in regard to the condition of the myelinated fibers of the peripheral nerves (*n. tibialis*) and the skeletal muscles of the hind limbs (*m. soleus*) innervated by tibial nerve branches.

The rationality of the study is based on our hypothesis that intravenous infusion of the autologous genetically enriched leucoconcentrate may affect not only nervous tissue at the epicenter of the lesion by paracrine action of the recombinant VEGF, GDNF, and NCAM produced by genetically modified leucocytes migrated into the damaged region, but also the areas distant from the injury site via the endocrine mechanism of action of the recombinant therapeutic molecules secreted by the genetically modified leucocytes into the blood stream.

4.1. Lumbar Spinal Cord

The negative consequences reached the lumbar spinal cord about 60 days after low thoracic SCI in control animals in the control group with no therapy. Notably, the pathological distribution of neuroglial cells found in the lumbar spinal cord was similar to that near the epicenter of the injury [33]. Immunofluorescence analysis of the GFAP expression revealed an increase in the immunopositive area occupied by astrocytes in the dorsal horns of the lumbar spinal cord. The Iba1-immunopositive area corresponding to the localization of microglial cells was increased in both the ventral and dorsal horns. The number of oligodendroglial Olig2-positive cells significantly decreased in the ventral and dorsal horns of the lumbar spinal cord. In treated mini pigs, infusion of the autologous genetically enriched leucoconcentrate producing recombinant VEGF, GDNF, and NCAM positively affected the remodeling of microglial cells in the ventral horns and astrocytes and oligodendroglial cells in the dorsal horns of the lumbar spinal cord. Immunofluorescence analysis of synaptic protein expression in the ventral horns near the neurotrauma revealed a reduced immunopositive area for synaptophysin and PSD95. At the same time, in the lumbar part, we found restoration of synaptophysin expression and still-reduced PSD95 expression. The autologous genetically enriched leucoconcentrate increased the immunopositive area for PSD95 and Chat in the lumbar spinal cord. Thus, in the lumbar spinal cord of SCI mini pigs at the low thoracic level, molecular and cellular changes demonstrated the efficacy of the autologous genetically enriched leucoconcentrate producing recombinant VEGF, GDNF, and NCAM to overcome these negative consequences (recovery of synaptic protein expression by motor neurons and positive neuroglial cell remodeling).

4.2. Peripheral Nerves

After SCI, peripheral nerves carrying motor, sensory, and autonomic axons have severe morphological disorders [61,62]. In our study, morphometric analysis of the tibial fascicle of the sciatic nerve in control mini pigs revealed a decrease in the number of myelinated fibers and an increase in myelin thickness and nerve fiber axon diameter. Such changes may be associated with degeneration of the sensory and sympathetic fibers due to direct damage of the dorsal and lateral columns of the spinal cord at the low thoracic level. The possible preservation of the motor axons in control animals was evident by sparing motor neurons in the lumbar region 60 days after SCI, which had an equal expression of KCC2 compared to intact mini pigs. In the therapeutic group, intravenous infusion of the autologous genetically enriched leucoconcentrate producing recombinant VEGF, GDNF, and NCAM positively affected the morphological sparing of tibial nerves myelinated fibers.

4.3. Hind Limb Skeletal Muscle

Disrupted connections between central neurons of the pyramidal and extrapyramidal systems and motor neurons of the spinal cord directly reflect the condition of the hind limbs skeletal muscles innervated by motor neurons. After low thoracic SCI in control mini pigs without therapy, the PTIBS score was equal to 1 point, and by 60 days, it had reached 2.0 points. Behavioral results were consistent with electrophysiological (polyphasic M-response associated with increased duration and amplitude) and morphological (decreased muscle weight, decreased average skeletal muscle fiber area, and a number of slow skeletal muscle fibers) characteristics of the soleus muscle. In mini pigs treated with autologous genetically enriched leucoconcentrate, better recovery of voluntary locomotion scored 3.0 by the PTIBS test was accompanied by prevention of soleus muscle atrophy and slow to fast transformation of muscle fibers, which suggests the effectiveness of the neurotrophic control of skeletal muscle fibers via axons of the lumbar motor neurons. The observed recovery of the hind limbs motor function was strongly supported by the beneficial effect of autologous genetically enriched leucoconcentrate on: (1) the damaged spinal cord at the thoracic level (sparing of the grey matter, inhibition of astrogliosis, growth of axons accompanied by increased numbers of oligodendroglial cells), as was shown earlier [33]; (2) the lumbar spinal cord (recovery of PSD95 and Chat expression in motor neurons, positive neuroglial cell remodeling); and (3) the tibial nerves morphological preservation. It is evident that regeneration of axons through the epicenter of injury, restoration of motor neurons of the lumbar region and preservation of their axons providing neurotrophic control of the skeletal muscles generally results in improved locomotion of the hind limbs, which is in line with inhibition of skeletal muscle atrophy (preservation of raw muscle weight and muscle fiber area) and sparing the muscle phenotype (slow and fast muscle fiber content).

The recombinant VEGF, GDNF, and NCAM secreted into the bloodstream may have a positive effect on the neuromuscular junctions as well. Thus, it was shown that genetically modified human mesenchymal stem cells, simultaneously overexpressing GDNF and VEGF, when injected intramuscularly into rats with a model of amyotrophic lateral sclerosis (ALS), maintain the neuromuscular synapse structure and prolong animal life [63]. In another study, ALS mice injected intramuscularly with a combination of two adenoviral vectors carrying cDNA of VEGF and angiogenin demonstrated a delay in the manifestation of the disease, higher motor activity, and increased lifespan [64]. Intravenous injection of human umbilical cord blood mononuclear cells simultaneously co-transduced with Ad5 carrying cDNA of *vegf165*, *gdnf*, and *ncam1* resulted in a prominent increase in life span and improved performance in behavioral tests in ALS mice [60]. In the model of hypogravity on Earth, administration of a composition of three adenoviral vectors carrying *vegf165*, *gdnf*, and *ncam1* into mice hind limbs skeletal muscles before the hind limbs unloading had a positive effect on myelinated fibers in the anterior spinocerebellar tract and sciatic nerve [65]. In our study, the recombinant VEGF, GDNF, and NCAM via the endocrine mechanism may have affected post-traumatic spinal cord recovery at lumbar spine and neuromuscular junction levels as well.

Employment of autologous leucocytes in gene therapy for the temporary production of biologically active molecules by demand represents one of the breakthrough directions in gene therapy for SCI [37]. Recombinant therapeutic molecules produced by genetically modified leucocytes at the site of injury may have a local effect (paracrine) on neuron survival, regrowth of the axons, and recovery of neural circuitry, with a simultaneously positive systemic (endocrine) effect on the preservation of lumbar motor neurons, their axons, and neuromuscular junctions. Standing with the efficacy of the SCI treatment obtained in this study, we provided evidence that auto-infusion of genetically enriched leucoconcentrate producing recombinant neurotrophic factors and neuronal cell adhesion molecules is a novel, potentially successful approach for multitarget treatment of SCI. Meanwhile, the exploratory results obtained in the study indicate the importance of performing further experiments with a higher number and different gender of large animals with anatomical, physiological, and biochemical characteristics close to humans in preclinical investigations. The complex nature of the GEL-VGN raises important questions about dose, mode of delivery, pharmacokinetics of the therapeutic molecules, etc. Further research with human blood is needed to develop a protocol for the preparation of genetically enriched leucoconcentrate before its translation to clinical trials.

5. Conclusions

This research aimed to test the hypothesis that intravenous infusion of an autologous leucoconcentrate enriched with therapeutic genes is a potentially successful approach for multitarget treatment of thoracic SCI in clinical practice. Our results demonstrate that in mini pigs with low thoracic SCI, autologous genetically enriched leucoconcentrate producing recombinant neurotrophic factors (VEGF and GDNF) and neural cell adhesion molecules (NCAM) affect not only the epicenter of injury but also remote structures, including lumbar spinal cord, peripheral nerves, and hind limbs skeletal muscles. We believe that the simple, safe, and economical method of the genetically enriched leucoconcentrate preparation from the patient's peripheral blood and chimeric adenoviral vectors (Ad5/35F) carrying cDNA of the therapeutic genes might represent a novel avenue for future personalized precision gene therapy research for other CNS disorders, including neurodegenerative diseases and stroke.

Author Contributions: Conceptualization, R.I. and I.L.; methodology, R.G. and M.D.; validation, F.F. and R.S.; formal analysis, R.G.; investigation, A.I., F.F., R.S. and I.M.; resources, V.M.; data curation, T.M.; writing—original draft preparation, R.I.; writing—review and editing, R.I. and I.L.; visualization, A.I.; supervision, V.M.; project administration, R.I. All authors have read and agreed to the published version of the manuscript.

Funding: This research was funded by a grant from the Russian Science Foundation, no. 16-15-00010.

Institutional Review Board Statement: Animal treatments were conducted according to the Animal Care and Use Committee of Kazan State Medical University (approval No. 5 dated 26 May 2020).

Informed Consent Statement: Not applicable.

Data Availability Statement: The data presented in this study are available on request from the corresponding author.

Acknowledgments: This research was supported by the International Scientific Council Grant for Young Scientists of Kazan State Medical University (Ravil Garifulin). Authors are grateful to the students of Kazan State Medical University (Alukaev A.R., Ibragimov D.R., Golovko D.I., Kalistratova Y.A., Khalitova A.T.) for their assistance in the study.

Conflicts of Interest: The authors declare no conflict of interest.

References

1. O'Shea, T.M.; Burda, J.E.; Sofroniew, M.V. Cell Biology of Spinal Cord Injury and Repair. *J. Clin. Investig.* **2017**, *127*, 3259–3270. [CrossRef]
2. Wang, D.; Fawcett, J. The Perineuronal Net and the Control of CNS Plasticity. *Cell Tissue Res.* **2012**, *349*, 147–160. [CrossRef]
3. Chai, R.J.; Vukovic, J.; Dunlop, S.; Grounds, M.D.; Shavlakadze, T. Striking Denervation of Neuromuscular Junctions without Lumbar Motoneuron Loss in Geriatric Mouse Muscle. *PLoS ONE* **2011**, *6*, e28090. [CrossRef] [PubMed]
4. Eidelberg, E.; Nguyen, L.H.; Polich, R.; Walden, J.G. Transsynaptic Degeneration of Motoneurones Caudal to Spinal Cord Lesions. *Brain Res. Bull.* **1989**, *22*, 39–45. [CrossRef]
5. McBride, R.L.; Feringa, E.R. Ventral Horn Motoneurons 10, 20 and 52 Weeks after T-9 Spinal Cord Transection. *Brain Res. Bull.* **1992**, *28*, 57–60. [CrossRef] [PubMed]
6. Yokota, K.; Kubota, K.; Kobayakawa, K.; Saito, T.; Hara, M.; Kijima, K.; Maeda, T.; Katoh, H.; Ohkawa, Y.; Nakashima, Y.; et al. Pathological Changes of Distal Motor Neurons after Complete Spinal Cord Injury. *Mol. Brain* **2019**, *12*, 4. [CrossRef]
7. Wang, F.; Yuan, Y.; Li, J. Reconstruction of Lower Extremity Function of Complete Spinal Cord Injury Rats by First Neuron Connection. *Chin. J. Reparative Reconstr. Surg.* **2015**, *29*, 1528–1533.
8. Detloff, M.R.; Fisher, L.C.; McGaughy, V.; Longbrake, E.E.; Popovich, P.G.; Basso, D.M. Remote Activation of Microglia and Pro-Inflammatory Cytokines Predict the Onset and Severity of below-Level Neuropathic Pain after Spinal Cord Injury in Rats. *Exp. Neurol.* **2008**, *212*, 337–347. [CrossRef]
9. McKay, S.M.; Brooks, D.J.; Hu, P.; McLachlan, E.M. Distinct Types of Microglial Activation in White and Grey Matter of Rat Lumbosacral Cord after Mid-Thoracic Spinal Transection. *J. Neuropathol. Exp. Neurol.* **2007**, *66*, 698–710. [CrossRef] [PubMed]
10. Nakajima, H.; Honjoh, K.; Watanabe, S.; Kubota, A.; Matsumine, A. Distribution and Polarization of Microglia and Macrophages at Injured Sites and the Lumbar Enlargement after Spinal Cord Injury. *Neurosci. Lett.* **2020**, *737*, 135152. [CrossRef] [PubMed]

11. Honjoh, K.; Nakajima, H.; Hirai, T.; Watanabe, S.; Matsumine, A. Relationship of Inflammatory Cytokines From M1-Type Microglia/Macrophages at the Injured Site and Lumbar Enlargement with Neuropathic Pain after Spinal Cord Injury in the CCL21 Knockout (Plt) Mouse. *Front. Cell. Neurosci.* **2019**, *13*, 525. [CrossRef]
12. Gwak, Y.S.; Kang, J.; Unabia, G.C.; Hulsebosch, C.E. Spatial and Temporal Activation of Spinal Glial Cells: Role of Gliopathy in Central Neuropathic Pain Following Spinal Cord Injury in Rats. *Exp. Neurol.* **2012**, *234*, 362–372. [CrossRef] [PubMed]
13. Pallottie, A.; Ratnayake, A.; Ni, L.; Acioglu, C.; Li, L.; Mirabelli, E.; Heary, R.F.; Elkabes, S. A Toll-like Receptor 9 Antagonist Restores below-Level Glial Glutamate Transporter Expression in the Dorsal Horn Following Spinal Cord Injury. *Sci. Rep.* **2018**, *8*, 8723. [CrossRef]
14. Min, K.-J.; Jeong, H.-K.; Kim, B.; Hwang, D.H.; Shin, H.Y.; Nguyen, A.T.; Kim, J.-H.; Jou, I.; Kim, B.G.; Joe, E.-H. Spatial and Temporal Correlation in Progressive Degeneration of Neurons and Astrocytes in Contusion-Induced Spinal Cord Injury. *J. Neuroinflammation* **2012**, *9*, 100. [CrossRef] [PubMed]
15. Bisicchia, E.; Sasso, V.; Catanzaro, G.; Leuti, A.; Besharat, Z.M.; Chiacchiarini, M.; Molinari, M.; Ferretti, E.; Viscomi, M.T.; Chiurchiù, V. Resolvin D1 Halts Remote Neuroinflammation and Improves Functional Recovery after Focal Brain Damage via ALX/FPR2 Receptor-Regulated MicroRNAs. *Mol. Neurobiol.* **2018**, *55*, 6894–6905. [CrossRef]
16. Courtine, G.; Sofroniew, M.V. Spinal Cord Repair: Advances in Biology and Technology. *Nat. Med.* **2019**, *25*, 898–908. [CrossRef]
17. Cofano, F.; Boido, M.; Monticelli, M.; Zenga, F.; Ducati, A.; Vercelli, A.; Garbossa, D. Mesenchymal Stem Cells for Spinal Cord Injury: Current Options, Limitations, and Future of Cell Therapy. *Int. J. Mol. Sci.* **2019**, *20*, 2698. [CrossRef] [PubMed]
18. Anna, Z.; Katarzyna, J.-W.; Joanna, C.; Barczewska, M.; Joanna, W.; Wojciech, M. Therapeutic Potential of Olfactory Ensheathing Cells and Mesenchymal Stem Cells in Spinal Cord Injuries. *Stem Cells Int.* **2017**, *2017*, 3978595. [CrossRef]
19. Kanno, H.; Pearse, D.D.; Ozawa, H.; Itoi, E.; Bunge, M.B. Schwann Cell Transplantation for Spinal Cord Injury Repair: Its Significant Therapeutic Potential and Prospectus. *Rev. Neurosci.* **2015**, *26*, 121–128. [CrossRef]
20. Zörner, B.; Schwab, M.E. Anti-Nogo on the Go: From Animal Models to a Clinical Trial. *Ann. N. Y. Acad. Sci.* **2010**, *1198*, E22–E34. [CrossRef]
21. Fehlings, M.G.; Kim, K.D.; Aarabi, B.; Rizzo, M.; Bond, L.M.; McKerracher, L.; Vaccaro, A.R.; Okonkwo, D.O. Rho Inhibitor VX-210 in Acute Traumatic Subaxial Cervical Spinal Cord Injury: Design of the SPinal Cord Injury Rho INhibition InvestiGation (SPRING). *Clin. Trial. J. Neurotrauma* **2018**, *35*, 1049–1056. [CrossRef]
22. Xiao, Z.; Tang, F.; Zhao, Y.; Han, G.; Yin, N.; Li, X.; Chen, B.; Han, S.; Jiang, X.; Yun, C.; et al. Significant Improvement of Acute Complete Spinal Cord Injury Patients Diagnosed by a Combined Criteria Implanted with NeuroRegen Scaffolds and Mesenchymal Stem Cells. *Cell Transplant.* **2018**, *27*, 907–915. [CrossRef]
23. Siddiqui, A.M.; Islam, R.; Cuellar, C.A.; Silvernail, J.L.; Knudsen, B.; Curley, D.E.; Strickland, T.; Manske, E.; Suwan, P.T.; Latypov, T.; et al. Newly Regenerated Axons via Scaffolds Promote Sub-Lesional Reorganization and Motor Recovery with Epidural Electrical Stimulation. *NPJ Regen. Med.* **2021**, *6*, 66. [CrossRef]
24. Yi, H.; Wang, Y. A Meta-Analysis of Exosome in the Treatment of Spinal Cord Injury. *Open Med.* **2021**, *16*, 1043–1060. [CrossRef] [PubMed]
25. Zhou, X.; Chu, X.; Yuan, H.; Qiu, J.; Zhao, C.; Xin, D.; Li, T.; Ma, W.; Wang, H.; Wang, Z.; et al. Mesenchymal Stem Cell Derived EVs Mediate Neuroprotection after Spinal Cord Injury in Rats via the MicroRNA-21-5p/FasL Gene Axis. *Biomed. Pharmacother.* **2019**, *115*, 108818. [CrossRef]
26. Chen, Y.-S.; Lin, E.-Y.; Chiou, T.-W.; Harn, H.-J. Exosomes in Clinical Trial and Their Production in Compliance with Good Manufacturing Practice. *Tzu-Chi Med. J.* **2020**, *32*, 113–120. [CrossRef]
27. Fehlings, M.; Zavvarian, M.M.; Toossi, A.; Khazaei, M.; Hong, J. Novel Innovations in Cell and Gene Therapies for Spinal Cord Injury. *F1000Research* **2020**, *9*. [CrossRef]
28. Lavrov, I.; Islamov, R. Implementing Principles of Neuroontogenesis and Neuroplasticity for Spinal Cord Injury Therapy. *Front. Biosci.* **2022**, *27*, 163. [CrossRef] [PubMed]
29. Cunningham, C.J.; Viskontas, M.; Janowicz, K.; Sani, Y.; Håkansson, M.E.; Heidari, A.; Huang, W.; Bo, X. The Potential of Gene Therapies for Spinal Cord Injury Repair: A Systematic Review and Meta-Analysis of Pre-Clinical Studies. *Neural Regen. Res.* **2023**, *18*, 299–305. [CrossRef]
30. Hanna, E.; Rémuzat, C.; Auquier, P.; Toumi, M. Gene Therapies Development: Slow Progress and Promising Prospect. *J. Mark. Access Health Policy* **2017**, *5*, 1265293. [CrossRef] [PubMed]
31. Nardone, R.; Florea, C.; Höller, Y.; Brigo, F.; Versace, V.; Lochner, P.; Golaszewski, S.; Trinka, E. Rodent, Large Animal and Non-Human Primate Models of Spinal Cord Injury. *Zoology* **2017**, *123*, 101–114. [CrossRef]
32. Sharif-Alhoseini, M.; Khormali, M.; Rezaei, M.; Safdarian, M.; Hajighadery, A.; Khalatbari, M.M.; Safdarian, M.; Meknatkhah, S.; Rezvan, M.; Chalangari, M.; et al. Animal Models of Spinal Cord Injury: A Systematic Review. *Spinal Cord* **2017**, *55*, 714–721. [CrossRef]
33. Davleeva, M.A.; Garifulin, R.R.; Bashirov, F.V.; Izmailov, A.A.; Nurullin, L.F.; Salafutdinov, I.I.; Gatina, D.Z.; Shcherbinin, D.N.; Lysenko, A.A.; Tutykhina, I.L.; et al. Molecular and Cellular Changes in the Post-Traumatic Spinal Cord Remodeling after Autoinfusion of a Genetically-Enriched Leucoconcentrate in a Mini-Pig Model. *Neural Regen. Res.* **2023**, *18*, 1505–1511. [CrossRef]
34. Islamov, R.; Bashirov, F.; Fadeev, F.; Shevchenko, R.; Izmailov, A.; Markosyan, V.; Sokolov, M.; Kuznetsov, M.; Davleeva, M.; Garifulin, R.; et al. Epidural Stimulation Combined with Triple Gene Therapy for Spinal Cord Injury Treatment. *Int. J. Mol. Sci.* **2020**, *21*, 8896. [CrossRef]

35. Islamov, R.R.; Izmailov, A.A.; Sokolov, M.E.; Fadeev, F.O.; Bashirov, F.V.; Eremeev, A.A.; Shmarov, M.M.; Naroditskiy, B.S.; Chelyshev, Y.A.A.; Lavrov, I.A.; et al. Evaluation of Direct and Cell-Mediated Triple-Gene Therapy in Spinal Cord Injury in Rats. *Brain Res. Bull.* **2017**, *132*, 44–52. [CrossRef] [PubMed]
36. Islamov, R.R.; Bashirov, F.V.; Sokolov, M.E.; Izmailov, A.A.; Fadeev, F.O.; Markosyan, V.A.; Davleeva, M.A.; Zubkova, O.V.; Smarov, M.M.; Logunov, D.Y.; et al. Gene-Modified Leucoconcentrate for Personalized Ex Vivo Gene Therapy in a Mini Pig Model of Moderate Spinal Cord Injury. *Neural Regen. Res.* **2021**, *16*, 357–361. [CrossRef] [PubMed]
37. Islamov, R.; Bashirov, F.; Izmailov, A.; Fadeev, F.; Markosyan, V.; Sokolov, M.; Shmarov, M.; Logunov, D.; Naroditsky, B.; Lavrov, I. New Therapy for Spinal Cord Injury: Autologous Genetically-Enriched Leucoconcentrate Integrated with Epidural Electrical Stimulation. *Cells* **2022**, *11*, 144. [CrossRef]
38. Izmailov, A.A.; Povysheva, T.V.; Bashirov, F.V.; Sokolov, M.E.; Fadeev, F.O.; Garifulin, R.R.; Naroditsky, B.S.; Logunov, D.Y.; Salafutdinov, I.I.; Chelyshev, Y.A.; et al. Spinal Cord Molecular and Cellular Changes Induced by Adenoviral Vector- and Cell-Mediated Triple Gene Therapy after Severe Contusion. *Front. Pharmacol.* **2017**, *8*, 813. [CrossRef]
39. Lee, J.H.T.; Jones, C.F.; Okon, E.B.; Anderson, L.; Tigchelaar, S.; Kooner, P.; Godbey, T.; Chua, B.; Gray, G.; Hildebrandt, R.; et al. A Novel Porcine Model of Traumatic Thoracic Spinal Cord Injury. *J. Neurotrauma* **2013**, *30*, 142–159. [CrossRef] [PubMed]
40. Islamov, R.R.; Sokolov, M.E.; Bashirov, F.V.; Fadeev, F.O.; Shmarov, M.M.; Naroditskiy, B.S.; Povysheva, T.V.; Shaymardanova, G.F.; Yakupov, R.A.; Chelyshev, Y.A.; et al. A Pilot Study of Cell-Mediated Gene Therapy for Spinal Cord Injury in Mini Pigs. *Neurosci. Lett.* **2017**, *644*, 67–75. [CrossRef] [PubMed]
41. Alizadeh, A.; Dyck, S.M.; Karimi-Abdolrezaee, S. Traumatic Spinal Cord Injury: An Overview of Pathophysiology, Models and Acute Injury Mechanisms. *Front. Neurol.* **2019**, *10*, 282. [CrossRef]
42. Chelyshev, Y. More Attention on Segments Remote from the Primary Spinal Cord Lesion Site. *Front. Biosci.* **2022**, *27*, 235. [CrossRef]
43. Kabdesh, I.M.; Mukhamedshina, Y.O.; Arkhipova, S.S.; Sabirov, D.K.; Kuznecov, M.S.; Vyshtakalyuk, A.B.; Rizvanov, A.A.; James, V.; Chelyshev, Y.A. Cellular and Molecular Gradients in the Ventral Horns with Increasing Distance from the Injury Site after Spinal Cord Contusion. *Front. Cell. Neurosci.* **2022**, *16*, 817752. [CrossRef]
44. Calvo, P.M.; Hernández, R.G.; de la Cruz, R.R.; Pastor, A.M. Role of Vascular Endothelial Growth Factor as a Critical Neurotrophic Factor for the Survival and Physiology of Motoneurons. *Neural Regen. Res.* **2023**, *18*, 1691–1696. [CrossRef] [PubMed]
45. Walker, M.J.; Xu, X.M. History of Glial Cell Line-Derived Neurotrophic Factor (GDNF) and Its Use for Spinal Cord Injury Repair. *Brain Sci.* **2018**, *8*, 109. [CrossRef]
46. Nicoletti, V.G.; Pajer, K.; Calcagno, D.; Pajenda, G.; Nógrádi, A. The Role of Metals in the Neuroregenerative Action of BDNF, GDNF, NGF and Other Neurotrophic Factors. *Biomolecules* **2022**, *12*, 1015. [CrossRef] [PubMed]
47. Schmid, R.S.; Maness, P.F. L1 and NCAM Adhesion Molecules as Signaling Coreceptors in Neuronal Migration and Process Outgrowth. *Curr. Opin. Neurobiol.* **2008**, *18*, 245–250. [CrossRef] [PubMed]
48. Saini, V.; Loers, G.; Kaur, G.; Schachner, M.; Jakovcevski, I. Impact of Neural Cell Adhesion Molecule Deletion on Regeneration after Mouse Spinal Cord Injury. *Eur. J. Neurosci.* **2016**, *44*, 1734–1746. [CrossRef] [PubMed]
49. Sytnyk, V.; Leshchyns'ka, I.; Schachner, M. Neural Cell Adhesion Molecules of the Immunoglobulin Superfamily Regulate Synapse Formation, Maintenance, and Function. *Trends Neurosci.* **2017**, *40*, 295–308. [CrossRef]
50. De Almodovar, C.R.; Lambrechts, D.; Mazzone, M.; Carmeliet, P. Role and Therapeutic Potential of VEGF in the Nervous System. *Physiol. Rev.* **2009**, *89*, 607–648. [CrossRef]
51. Ogunshola, O.O.; Antic, A.; Donoghue, M.J.; Fan, S.-Y.; Kim, H.; Stewart, W.B.; Madri, J.A.; Ment, L.R. Paracrine and Autocrine Functions of Neuronal Vascular Endothelial Growth Factor (VEGF) in the Central Nervous System. *J. Biol. Chem.* **2002**, *277*, 11410–11415. [CrossRef] [PubMed]
52. Zachary, I. Neuroprotective Role of Vascular Endothelial Growth Factor: Signalling Mechanisms, Biological Function, and Therapeutic Potential. *Neurosignals* **2005**, *14*, 207–221. [CrossRef] [PubMed]
53. Li, B.; Xu, W.; Luo, C.; Gozal, D.; Liu, R. VEGF-Induced Activation of the PI3-K/Akt Pathway Reduces Mutant SOD1-Mediated Motor Neuron Cell Death. *Mol. Brain Res.* **2003**, *111*, 155–164. [CrossRef] [PubMed]
54. Ibáñez, C.F.; Andressoo, J.O. Biology of GDNF and Its Receptors—Relevance for Disorders of the Central Nervous System. *Neurobiol. Dis.* **2017**, *97*, 80–89. [CrossRef]
55. Nicole, O.; Ali, C.; Docagne, F.; Plawinski, L.; MacKenzie, E.T.; Vivien, D.; Buisson, A. Neuroprotection Mediated by Glial Cell Line-Derived Neurotrophic Factor: Involvement of a Reduction of NMDA-Induced Calcium Influx by the Mitogen-Activated Protein Kinase Pathway. *J. Neurosci.* **2001**, *21*, 3024–3033. [CrossRef]
56. Cintrón-Colón, A.F.; Almeida-Alves, G.; Boynton, A.M.; Spitsbergen, J.M. GDNF Synthesis, Signaling, and Retrograde Transport in Motor Neurons. *Cell Tissue Res.* **2020**, *382*, 47–56. [CrossRef]
57. Paratcha, G.; Ledda, F.; Ibáñez, C.F. The Neural Cell Adhesion Molecule NCAM Is an Alternative Signaling Receptor for GDNF Family Ligands. *Cell* **2003**, *113*, 867–879. [CrossRef]
58. Crystal, R.G. Adenovirus: The First Effective In Vivo Gene Delivery Vector. *Hum. Gene Ther.* **2014**, *25*, 3–11. [CrossRef]
59. Nasser, M.; Bejjani, F.; Raad, M.; Abou-El-Hassan, H.; Mantash, S.; Nokkari, A.; Ramadan, N.; Kassem, N.; Mondello, S.; Hamade, E.; et al. Traumatic Brain Injury and Blood-Brain Barrier Cross-Talk. *CNS Neurol. Disord. Drug Targets* **2016**, *15*, 1030–1044. [CrossRef]

60. Islamov, R.R.; Rizvanov, A.A.; Fedotova, V.Y.; Izmailov, A.A.; Safiullov, Z.Z.; Garanina, E.E.; Salafutdinov, I.I.; Sokolov, M.E.; Mukhamedyarov, M.A.; Palotás, A. Tandem Delivery of Multiple Therapeutic Genes Using Umbilical Cord Blood Cells Improves Symptomatic Outcomes in ALS. *Mol. Neurobiol.* **2017**, *54*, 4756–4763. [CrossRef]
61. Galea, M.P.; van Zyl, N.; Messina, A. Peripheral Nerve Dysfunction after Spinal Cord Injury. *OBM Neurobiol.* **2020**, *4*, 17. [CrossRef]
62. Messina, A.; van Zyl, N.; Weymouth, M.; Flood, S.; Nunn, A.; Cooper, C.; Hahn, J.; Galea, M.P. Morphology of Donor and Recipient Nerves Utilised in Nerve Transfers to Restore Upper Limb Function in Cervical Spinal Cord Injury. *Brain Sci.* **2016**, *6*, 42. [CrossRef] [PubMed]
63. Krakora, D.; Mulcrone, P.; Meyer, M.; Lewis, C.; Bernau, K.; Gowing, G.; Zimprich, C.; Aebischer, P.; Svendsen, C.N.; Suzuki, M. Synergistic Effects of GDNF and VEGF on Lifespan and Disease Progression in a Familial ALS Rat Model. *Mol. Ther.* **2013**, *21*, 1602–1610. [CrossRef] [PubMed]
64. Ismailov, S.M.; Barykova, I.A.; Shmarov, M.M.; Tarantul, V.Z.; Barskov, I.V.; Kucherianu, V.G.; Brylev, L.V.; Logunov, D.I.; Tutykhina, I.L.; Bocharov, E.V.; et al. Experimental Approach to the Gene Therapy of Motor Neuron Disease with the Use of Genes Hypoxia-Inducible Factors. *Genetika* **2014**, *50*, 591–601. [CrossRef]
65. Lisyukov, A.N.; Kuznetsov, M.S.; Saitov, V.R.; Salnikova, M.M.; Bikmullina, I.A.; Koshpaeva, E.S.; Tyapkina, O.V.; Valiullin, V.V.; Islamov, R.R. Morphological Changes in Myelinated Fibers of the Spinal Cord and the Sciatic Nerve in Mice after Modeling of the Hypogravity and the Approach of Their Correction by Preventive Gene Therapy. *Genes Cells* **2021**, *16*, 75–80. [CrossRef]

Article

ERK1/2 Signalling Pathway Regulates Tubulin-Binding Cofactor B Expression and Affects Astrocyte Process Formation after Acute Foetal Alcohol Exposure

Yin Zheng [1],*,†, Jiechao Huo [1],*,†, Mei Yang [1], Gaoli Zhang [2], Shanshan Wan [3], Xiaoqiao Chen [1], Bingqiu Zhang [1], and Hui Liu [1]

[1] Institute of Neuroscience, Chongqing Medical University, Chongqing 400016, China; yangmei503@cqmu.edu.cn (M.Y.); 2020110047@stu.cqmu.edu.cn (X.C.); 2021110072@stu.cqmu.edu.cn (B.Z.); huiliu@cqmu.edu.cn (H.L.)

[2] Institute for Viral Hepatitis, Second Affiliated Hospital of Chongqing Medical University, Chongqing 400063, China; 305974@hospital.cqmu.edu.cn

[3] Department of Blood Transfusion, Sichuan Cancer Hospital & Institute, Chengdu 610044, China; 2020110060@stu.cqmu.edu.cn

* Correspondence: 2019110041@stu.cqmu.edu.cn (Y.Z.); 2018110043@stu.cqmu.edu.cn (J.H.);

† These authors contributed equally to this work.

Citation: Zheng, Y.; Huo, J.; Yang, M.; Zhang, G.; Wan, S.; Chen, X.; Zhang, B.; Liu, H. ERK1/2 Signalling Pathway Regulates Tubulin-Binding Cofactor B Expression and Affects Astrocyte Process Formation after Acute Foetal Alcohol Exposure. *Brain Sci.* **2022**, *12*, 813. https://doi.org/10.3390/brainsci12070813

Academic Editors: Masaru Tanaka, Lydia Giménez-Llort, Simone Battaglia, Chong Chen, Piril Hepsomali and Mark Burke

Received: 23 April 2022
Accepted: 20 June 2022
Published: 22 June 2022

Publisher's Note: MDPI stays neutral with regard to jurisdictional claims in published maps and institutional affiliations.

Abstract: Foetal alcohol spectrum disorders (FASDs) are a spectrum of neurological disorders whose neurological symptoms, besides the neuronal damage caused by alcohol, may also be associated with neuroglial damage. Tubulin-binding cofactor B (TBCB) may be involved in the pathogenesis of FASD. To understand the mechanism and provide new insights into the pathogenesis of FASD, acute foetal alcohol exposure model on astrocytes was established and the interference experiments were carried out. First, after alcohol exposure, the nascent astrocyte processes were reduced or lost, accompanied by the absence of TBCB expression and the disruption of microtubules (MTs) in processes. Subsequently, TBCB was silenced with siRNA. It was severely reduced or lost in nascent astrocyte processes, with a dramatic reduction in astrocyte processes, indicating that TBCB plays a vital role in astrocyte process formation. Finally, the regulating mechanism was studied and it was found that the extracellular signal-regulated protease 1/2 (ERK1/2) signalling pathway was one of the main pathways regulating TBCB expression in astrocytes after alcohol injury. In summary, after acute foetal alcohol exposure, the decreased TBCB in nascent astrocyte processes, regulated by the ERK1/2 signalling pathway, was the main factor leading to the disorder of astrocyte process formation, which could contribute to the neurological symptoms of FASD.

Keywords: tubulin-binding cofactor B; astrocyte processes; microtubules; extracellular signal-regulated protease 1/2 signalling pathway; foetal alcohol spectrum disorders; acute foetal alcohol exposure

1. Introduction

Foetal alcohol spectrum disorder (FASD) is an umbrella term that encompasses many congenital abnormalities resulting from prenatal alcohol exposure, including foetal cognitive impairment, behavioural abnormalities, and physical impairment [1]. Foetal alcohol syndrome (FAS) is one of the most severe types of FASD [2]. Many aspects of FASD disorder have been successfully summarised in vitro and in vivo, including morphological and behavioural defects, but the pathogenesis of the disease is still unclear [3,4]. The unawareness of pregnancy or addictive alcoholism in women during pregnancy has led to FASD in recent years.

FASD patients' brains usually are reduced in size, have prominent abnormalities in shape, with reduced brain growth [5], significantly fewer cells [6], and greater cortical thickness in the frontal lobes [7]. In rodent neonates, the altered morphology of pyramidal neuron dendrites in the medial prefrontal cortex was also found [8]. The structural alterations, especially the medial prefrontal cortex after alcohol exposure, can lead to numerous cognitive and behavioural outcomes [5,6,9]; the medial prefrontal cortex is critical for behaviour through the prefrontal–thalamo–hippocampal circuit [10] and for cognition, including fear, through prefrontal cortex–amygdala–hippocampus circuitry [11]. In recent years, heart rate variability has been used as a biological marker of cognitive changes [12] and the neurovisceral integration model of fear was used to clarify the fear conditioning: the prefrontal cortex, cognitive structures, influence the activity of amygdala and hippocampus, eliciting a neurovisceral fear response through sympathetic and parasympathetic projections that mediate heart-related dynamics [13]. Although the above research showed us some data about FASD, the mechanism of FASD is still not clear.

Almost all current research on the pathogenesis of FASD has focused on neurons [4]. Alcohol can trigger a robust neurodegenerative response and neuronal apoptosis [14,15] and impact synaptic formation [1,16]. In addition to neurons, astrocytes are also required for synaptic formation [17,18]. Cultured neurons formed fewer weak synapses in the absence of astrocytes than neurons cocultured with astrocytes [16,19]. Indeed, during central nervous system development, astrocytes are more sensitive than neurons to alcohol toxicity [20,21]. Moderate levels of alcohol can delay their growth and maturation [20,22,23]. As the most abundant cells in the nervous system, astrocyte morphology and function alterations must be closely related to the neurological dysfunction of FASD. However, no related studies have been reported thus far. Moreover, almost all of the research focused on chronic alcohol injury [24–27] causing neurological symptoms of FASD that appeared after birth or even in adolescence. In fact, damage to the foetal nervous system by alcohol occurs during maternal alcohol consumption [28]. However, there have been few reports about acute alcohol injury on nerve cells.

Tubulin-binding cofactor B (TBCB) is a cofactor that helps α-tubulin fold during the de novo synthesis of microtubules [29–32], which participates in and regulates microtubule assembly [33–35], growth, and dynamic stability. Recently, Feltes et al. found an upregulation of TBCB RNA in FASD mice, suggesting a possible relationship between alcohol and TBCB in FASD [29,36]. Interestingly, alcohol was found to cause impaired process formation in astrocytes [20,22,23]. TBCB was shown to affect the growth of axons, the essential processes of neurons, suggesting that TBCB might be involved in the formation of cell processes [37], possibly by acting on the positive end of microtubules, which contributes to the formation of cell processes [30,37]. In summary, there may be some links between TBCB, astrocyte processes, and alcohol in FASD patients and this is worth studying.

Mitogen-activated protein kinase (MAPK) is an evolutionarily conserved major signalling pathway that transmits extracellular stimuli into cells and includes three signalling pathways: extracellular signal-regulated kinase 1/2 (ERK1/2), c-Jun amino-terminal kinase (JNK), and P38 [37]. Previous studies have shown that ethanol can impact the expression of some proteins in neurons [38,39], microglia [40], and other cells [41]. Alcohol, as an exogenous stimulant, is associated with the MAPK pathway [37–40] and the ERK1/2 signalling pathway is sensitive to alcohol [42]. Therefore, it is worth investigating whether the MAPK signalling pathway regulates TBCB expression in astrocytes after alcohol exposure.

In summary, we hypothesised that some neurological symptoms in FASD patients may be associated with morphological and functional alterations of astrocytes, which may be related to alterations of TBCB expression induced by alcohol exposure, at least partly, and may be regulated by the MAPK signalling pathway. To confirm this speculation, we performed three experiments. First, we studied the relationship between a reduction in TBCB expression and astrocyte morphological alterations after acute foetal alcohol exposure. Then, we silenced the TBCB gene to confirm that the astrocyte morphological alterations were related to the reduction in TBCB expression. Finally, the possible signalling pathway that regulates TBCB expression after acute foetal alcohol exposure was revealed.

Our studies will provide new experimental data to fill in the gaps in the pathological mechanisms during the acute injury stage of FASD.

2. Materials and Methods

2.1. Culture of Primary Astrocytes

Astrocytes were prepared from the cerebral cortex of C57BL/6J mice on postnatal Day 0 [43]. Briefly, newborn mice were sterilised with 75% ethanol and decapitated in sterile condition. Brain tissue was taken out and the meninges were carefully removed. The cerebral cortex was cut into pieces and digested with 0.25% trypsin at 37 °C for 10 min and then digestion was stopped by a mixture of Dulbecco's modified Eagle's medium with Ham's F-12 medium (DMEM/F12; HyClone) supplemented with 10% foetal bovine serum (FBS, Gemini). After centrifugation, the supernatant was discarded and the pellet was resuspended in a mixture of DMEM/F12 supplemented with 10% FBS. Then, the cell suspension was filtered with a 200-mesh filter (Saimike), placed in flasks, and incubated at 37 °C with 5% CO_2 and 95% air. Cultures were passaged every 4–5 days at least 3 times (G3) to achieve highly pure astrocyte culture [44,45]. The culture medium was replaced every two days. Finally, the cells were transferred to 6-well plates for Western blotting (WB) or reverse transcription real-time PCR (RT–PCR) and in 12-well plates covered with glass slides for immunofluorescence staining (IF). All experiments were performed on astrocytes at passage 3 after being grown for 1 day.

2.2. Establishment of the Astrocyte Alcohol Interference Model

Mouse astrocytes were randomly divided into a normal control group (Con) and an acute foetal alcohol exposure group (E). Cells in the control group were cultured conventionally, while cells in the alcohol group were cultured in a conventional medium containing 100 mM alcohol [28,40,46–51]. At 1, 6, 12, and 24 h after alcohol exposure, cells in each group were collected for IF ($n = 6$), WB ($n = 6$), and RT–PCR ($n = 6$) detection.

2.3. Small Interfering RNA (siRNA) Transfection

To clarify the effect of the TBCB decrease on the formation of astrocyte processes, a TBCB silencing experiment was carried out. The astrocytes were randomly divided into a blank control group (Con, conventional culture), negative control (NC, transfected with negative control sequence), and a TBCB silencing group (siTBCB, transfected with a TBCB siRNA sequence). TBCB siRNA and scrambled sequences were synthesised by Chongqing Maobai Technology Co. (Chongqing, China). Astrocytes were transfected with TBCB siRNA or negative oligonucleotides in 6-well or 24-well plates for 6 h using the Lipofectamine™ 3000 transfection kit (Invitrogen, USA). Each well of a 6-well plate contained 0.8×10^6 cells, 5 μL siRNA, 3.75 μL Lipofectamine 3000, and 250 μL Opti-MEM (Gibco, Grand Island, NY, USA). Each well of 24-well plates contained 0.6×10^5 cells per well, 1.25 μL siRNA, 0.75 μL Lipofectamine 3000, and 50 μL Opti-MEM [52]. The sequence of TBCB was as follows: forward 5′-GCAUCCAUGUCAUUGACCATT-3′ and reverse 5′-UGGUCAAUGACAUGGAUGCTT-3′. The negative control sequence was forward 5′-UUCUCCGAACGUGUCACGUTT-3′, reverse 5′-ACGUGACGUUCGGAGAATT-3′. Six hours after transfection, the media was replaced with fresh DMEM/F12 supplemented with 10% FBS. At 48 h following transfection, the cells were collected and analysed by RT–PCR ($n = 6$). Then, 72 h after transfection, the cells were collected and analysed by WB ($n = 6$) and IF ($n = 6$). All operations were carried out in strict accordance with the manufacturer's instructions.

2.4. ERK1/2 Signalling Pathway Interference Assay

To confirm whether the ERK1/2 signalling pathway regulated TBCB expression after alcohol exposure, we interfered with it and observed the effects. Astrocytes were randomly divided into three groups: the solvent control group (Con) was cultured in medium containing 2 μL DMSO (Saimike, Chongqing, China) [52]; the ERK1/2 agonist group (TPA) was cultured in medium containing 200 μM TAP (ERK1/2 agonist, CST [52]) dissolved in 2 μL DMSO; and the ERK1/2 inhibitor group (U0126) was cultured in medium containing 10 mM U0126 (ERK1/2 inhibitor, Selleck, Shanghai, China [52]) dissolved in 2 μL DMSO. After 1 h, the medium containing the treatments was removed and complete fresh medium was added. After 12 h, the three groups of cells were collected separately and analysed by WB ($n = 6$).

To test whether the alcohol-induced TBCB alterations were related to the ERK1/2 pathway, we divided the astrocytes into three groups and pretreated them as described above. After 1 h, the medium containing the treatments was removed, complete fresh medium was added to the solvent control group (Con) and complete fresh medium containing 100 mM alcohol was added to the agonist and inhibitor groups. After 12 h, all of the cells were collected and detected by WB ($n = 6$).

2.5. Western Blotting

The harvested astrocytes were lysed on ice in RIPA buffer containing 1% PMSF (Beyotime, Guangzhou, China) and the total protein concentration was measured with a BCA protein assay kit (Beyotime, Shanghai, China). After dilution in the sample loading buffer, 20 μg of protein was added to each lane. The proteins were then separated on a 10% SDS–PAGE gel and transferred to a 0.2 μm polyvinylidene difluoride (PVDF) membrane (Millipore, Bedford, MA, USA). The membranes were blocked with blocking buffer (Beyotime, Shanghai, China) at room temperature for 30 min. Then, they were probed with the properly diluted primary antibodies followed by HRP-labelled anti-mouse or anti-rabbit IgG secondary antibody (ZB-2305 or ZB2301, ZSGBBIO, Beijing, China). They were visualised by Western Bright ECL (Advansta, San Jose, CA, USA) and imaged using a Western blotting detection system (Bio-Rad, Hercules, CA, USA) or X-ray film. Each sample was repeated 3 times and each blot was imaged 3 times. Then, the densities of the bands in each image were quantified 3 times by Quantity-One software. The value of the target protein was normalised to the value of the housekeeping protein from the same sample within the same blot. Then, all of the corresponding values from the different groups were statistically analysed by GraphPad Prism 6.0 software (GraphPad Software, San Diego, CA, USA) [53].

The locations of all proteins detected by the antibodies used in WB are shown in the full-length blots in Supplementary S1. The primary antibodies used in WB were as follows: anti-TBCB (1:500, A13248, ABclonal, Wuhan, China), anti-p38 (1:1000, #8690, Cell Signalling Technology, Danvers, MA, USA), anti-pp38 (1:800, #4511, Cell Signalling Technology, Danvers, MA, USA), anti-JNK (1:1000, #9252, Cell Signalling Technology, Danvers, MA, USA), anti-p-JNK (1:1000, #4668, Cell Signalling Technology, Danvers, MA, USA), anti-β-actin (1:5000, 20536-1-AP, Proteintech, Wuhan, China), anti-α-tubulin (1:5000, GTX628802, GeneTex, Irvine, CA, USA), anti-β-T (1:5000, TA503129, OriGene, Rockville, MD, USA), anti-ERK1/2 (1:1000, #4695, Cell Signalling Technology, Danvers, MA, USA), anti-pERK1/2 (1:1000, #4370, Cell Signalling Technology, Danvers, MA, USA), and anti-GAPDH (1:5000, 60004-1-lg, Proteintech, Wuhan, China).

2.6. Reverse Transcription Real-Time PCR

Total RNA was extracted from harvested astrocytes ($n = 6$) by using RANiso plus (#9108, TaKaRa, Beijing, China) [54] and the concentration of RNA was measured by spectrophotometer. A total of 1 μg of RNA was reverse transcribed to generate cDNA using the PrimeScript™ II 1st Strand cDNA Synthesis Kit (TaKaRa, Beijing, China). Messenger expression of TBCB as a housekeeping gene was assessed by real-time PCR. PCR amplification was performed using a T100 thermal cycler (BIO-RAD) and Premix Taq™ (TaKaRa, Beijing, China). The PCR mixture consisted of 1 μL of each primer, 25 μL of Premix Taq, and 1 μL of cDNA in a final volume of 50 μL. The PCR conditions were denaturation at 94 °C for 3 min, followed by 34 cycles of denaturation at 94 °C for 30 s, annealing at 55 °C for 30 s, and extension at 72 °C for 30 s. All qPCRs were run on a CFX96 real-time system

(Bio-Rad). The $2^{-\Delta\Delta Ct}$ method was used to calculate the RNA or miRNA level fold change compared to the control samples [52]. The primer sequences (5'- > 3') were as follows: TBCB, forward ATGGAGCAGACGACAAGTTCT, reverse CCGTCATCCACAGGATAG-GAG, product size (77 bp); β-actin (control), forward CAGCCTTCCTTCTTGGGTA, reverse TTTACGGATGTCAACGTCACAC, product size (87 bp). All operations were carried out in strict accordance with the manufacturer's instructions.

2.7. Immunofluorescence Staining

The astrocytes were fixed in $-20°$ precooled acetone and methanol (1:1) for 5 min and then blocked with 5% bovine serum albumin (BSA) at room temperature for 30 min. The astrocytes were probed with the indicated primary antibodies (anti-TBCB, 1:50, A13248, ABclonal, China; anti-TBCB, 1:250, sc-377139, Santa Cruz, USA; anti-α- tubulin, 1:5000, GTX628802, GeneTex, Irvine, California, USA) properly diluted at $4°$ overnight. Then, the cells were incubated with secondary antibodies (FITC goat anti-rabbit IgG, 1:200, E031220-01, EARTH, China; Cy3 goat anti-mouse IgG, 1:200, Abbkine, Wuhan, China) and stained with DAPI (C1005, Beyotime, Shanghai, China). Subsequently, the cells were mounted in Fluorescence Mounting Medium (ab104135, Abcam, Cambridge, UK) and sealed with nail polish. Images were obtained by confocal laser scanning microscopy (Leica DMI8, Germany) and the intensity of the fluorescence was analysed by ImageJ (1.53 c) software ($n = 6$) [53,55]. The astrocyte processes in the high magnification images were also counted by ImageJ (1.53 c) software (approximately 50 cells were counted in each group). IF was mainly used to observe the changes in cell morphology and protein distribution in this study.

2.8. Statistical Analysis

Statistical analyses were performed, and the corresponding graphs were drawn using GraphPad Prism 6.0 software (GraphPad Software, San Diego, CA, USA). All experimental data were expressed as means ± standard deviation (SD). Differences between the treatment group and control group were compared using analysis of a two-sample unpaired t-test. All reported p values were two-sided and a value of $p < 0.05$ was considered statistically significant. All the data presented in the manuscript is presented in Table 1.

Table 1. Statistical tests used in the study.

Corresponding Figures	Prerequisites	Main Test
Figure 1 F1	Single factor, unpaired	Unpaired t test, 1 h group, t(10) = 1.922, $p = 0.1029$ 6 h group, t(10) = 2.523, $p = 0.0451$ 12 h group, t(10) = 7.421, $p = 0.0003$ 24 h group, t(10) = 5.733, $p = 0.0012$
Figure 1 F2	Single factor, unpaired	Unpaired t test, 1 h group, t(10) = 5.200, $p = 0.0004$ 6 h group, t(10) = 8.538, $p < 0.0001$ 12 h group, t(10) = 8.906, $p < 0.0001$ 24 h group, t(10) = 7.085, $p < 0.0001$
Figure 1 F3	Single factor, unpaired	Unpaired t test, 1 h group, t(10) = 0.3504, $p = 0.7333$ 6 h group, t(10) = 2.326, $p = 0.0424$ 12 h group, t(10) = 4.234, $p = 0.0017$ 24 h group, t(10) = 2.651, $p = 0.0243$
Figure 1 G2	Single factor, unpaired	Unpaired t test, 1 h group, t(10) = 0.4869, $p = 0.6518$ 6 h group, t(10) = 10.23, $p = 0.0005$ 12 h group, t(10) = 17.09, $p < 0.0001$ 24 h group, t(10) = 5.931, $p = 0.0040$

Table 1. *Cont.*

Corresponding Figures	Prerequisites	Main Test
Figure 1 G3	Single factor, unpaired	Unpaired *t* test, 1 h group, t(10) = 0.4869, *p* = 0.6518 6 h group, t(10) = 2.177, *p* = 0.0950 12 h group, t(10) = 12.23, *p* = 0.0003 24 h group, t(10) = 4.363, *p* = 0.0120
Figure 1 H1	Single factor, unpaired	Unpaired *t* test, 1 h group, t(10) = 1.117, *p* = 0.3265 6 h group, t(10) = 11.57, *p* = 0.0003 12 h group, t(10) = 4.453, *p* = 0.0112 24 h group, t(10) = 1.665, *p* = 0.1713
Figure 1 H2	Single factor, unpaired	Unpaired *t* test, 1 h group, t(10) = 1.896, *p* = 0.1309 6 h group, t(10) = 28.15, *p* < 0.0001 12 h group, t(10) = 13.14, *p* = 0.0002 24 h group, t(10) = 10.48, *p* = 0.0005
Figure 2 I1	Single factor, unpaired	Unpaired *t* test, t(10) = 6.148, *p* = 0.0008
Figure 2 I2	Single factor, unpaired	Unpaired *t* test, t(10) = 8.803, *p* < 0.0001
Figure 2 I3	Single factor, unpaired	Unpaired *t* test, t(10) = 4.585, *p* = 0.0010
Figure 2 J	Single factor, unpaired	Unpaired *t* test, TBCB: t(10) = 6.911, *p* < 0.0001 α-tubulin: t(10) = 9.021, *p* = 0.0001
Figure 2 K	Single factor, unpaired	Unpaired *t* test, TBCB: t(10) = 8.661, *p* = 0.0010 α-tubulin: t(10) = 12.73, *p* = 0.0002
Figure 3 A2	Single factor, unpaired	Unpaired *t* test, t(10) = 9.365, *p* = 0.0007
Figure 3 A3	Single factor, unpaired	Unpaired *t* test, t(10) = 18.61, *p* < 0.0001
Figure 3 A4	Single factor, unpaired	Unpaired *t* test, t(10) = 1.169, *p* = 0.3072
Figure 3 A5	Single factor, unpaired	Unpaired *t* test, t(10) = 2.412, *p* = 0.2318
Figure 3 B2	Single factor, unpaired	Unpaired *t* test, Con vs. TPA: t(10) = 6.544, *p* = 0.0028 Con vs. U0126: t(10) = 0.0059, *p* = 0.0059
Figure 3 B3	Single factor, unpaired	Unpaired *t* test, Con vs. TPA: t(10) = 7.650, *p* = 0.0016 Con vs. U0126: t(10) = 10.30, *p* = 0.0005
Figure 3 C2	Single factor, unpaired	Unpaired *t* test, Con vs. TPA: t(10) = 3.278, *p* = 0.0083 Con vs. U0126: t(10) = 23.99, *p* < 0.0001
Figure 3 C3	Single factor, unpaired	Unpaired *t* test, Con vs. TPA: t(10) = 7.623, *p* = 0.0016 Con vs. U0126: t(10) = 18.46, *p* < 0.0001

3. Results

3.1. Alcohol Inhibited the Formation of Astrocyte Processes and TBCB Expression in Nascent Processes

We established an acute foetal alcohol exposure astrocyte model to detect the morphological changes in astrocytes and to reveal the possible link between the change in astrocytes and TBCB. The basic morphology, number, status, and structure of the astrocytes had no noticeable difference except for an increased cell volume and the elongating processes at different time points in the control group (Figures S2 and 1F), so only the IF images at 12 h are shown in Figure 1 as the control group (Figure 1A).

In the control group, the astrocytes grew well, with flat and plump bodies and plentiful processes (Figure 1A1–A3). Most of the MTs in the cells were arranged in a linear, filamentous, and radial pattern from the MT organising centre to the edge of the cell cortex, with a uniform and dense distribution and they were especially abundant in the nascent processes (Figure 1A1,A4, arrows). TBCB was distributed with a diffuse punctiform pattern in the cytoplasm or along the MTs. It was highly abundant in the nascent processes (Figure 1A2, arrows), suggesting that TBCB might have an important role in the formation of the nascent processes of astrocytes. It was also rich around the MT organising centre (Figure 1A2, arrowheads).

Compared with the control group, in the alcohol exposure group, the astrocyte cell body gradually shrank and collapsed and the processes were significantly reduced (Figure 1B–E,F1). The sharp tips of newly generated processes became blunt or began to retract (Figure 1B, arrows) at 1 h after treatment with alcohol. The processes very clearly retracted (Figure 1C, arrows and Figure 1F1) at 6 h and few new processes were formed, with the cells collapsing significantly after 12 h (Figure 1D, arrows and Figure 1F1). The density of MT was gradually reduced in processes (Figure 1F3). The MTs were disordered or intertwined into bundles (Figure 1D,E, arrowheads) and most of the positive ends became curly (Figure 1D, arrows), especially in the nascent processes, suggesting a growth disorder at the positive end of the MTs. Along with the retraction and disappearance of the astrocyte processes, TBCB, which was originally abundant in the processes, was also significantly weakened or it disappeared (Figure 1F2), suggesting a close relationship between TBCB and astrocyte processes (Figure 1C2,C3, arrows). WB (Figure 1G) and PCR (Figure 1H) showed that TBCB and α-tubulin expression decreased at 6 h, reaching a trough at 12 h and still lower than the control group at 24 h, suggesting that alcohol could cause a decrease in TBCB and α-tubulin expression.

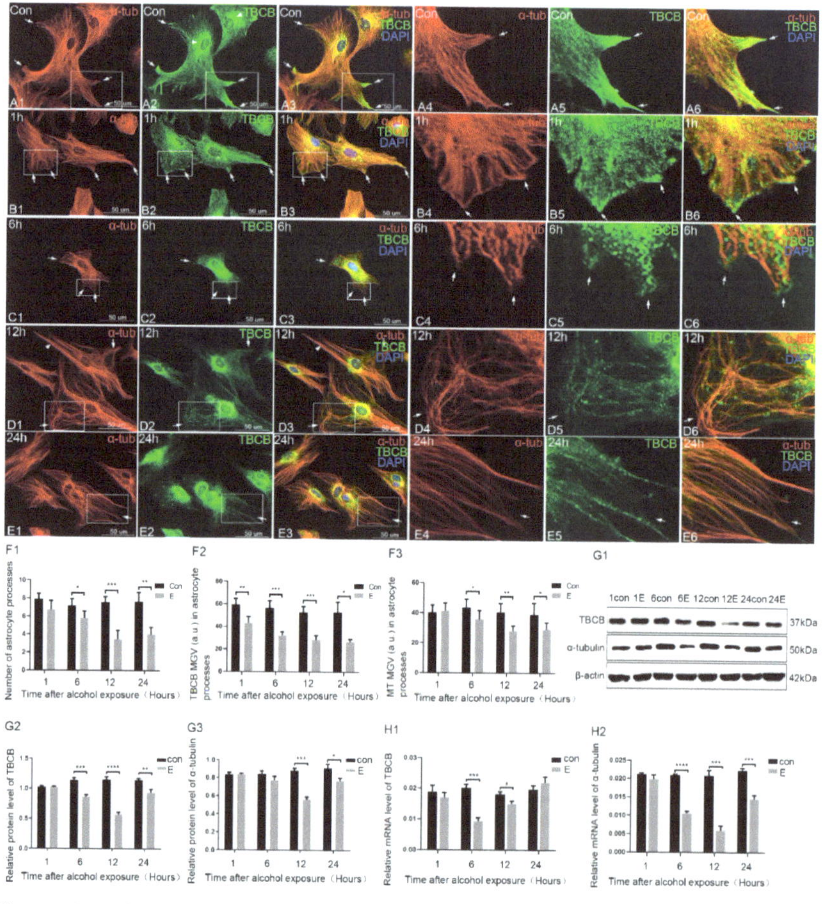

Figure 1. Disorder of astrocyte formation and decreased expression of TBCB after 1, 6, 12, and 24 h

of acute foetal alcohol exposure. (**A–E**) Immunofluorescence showing changes in the expression of α-tubulin (red signal) and TBCB (green signal) in astrocytes after 1, 6, 12, and 24 h of acute foetal alcohol exposure. Panels A-E 4-6 are magnified images from the square area of Panels A-E 1-3. (**F**) The number of astrocyte processes and mean grey value (MGV) of TBCB and MT in astrocyte processes (MGV = integrated density/area) after 1, 6, 12, and 24 h of acute foetal alcohol exposure. (**G**) Western blot and (**H**) RT–PCR analysis of the protein and mRNA levels of TBCB and α-tubulin after 1, 6, 12, and 24 h of acute foetal alcohol exposure ($n = 6$, $p < 0.05$). *: $p < 0.05$, **: $p < 0.01$, ***: $p < 0.001$, ****: $p < 0.0001$.

The above results made it clear that alcohol could impede the formation of astrocyte processes and TBCB expression in these processes. It also suggested that there should be some links between the impaired formation of nascent astrocyte processes and the loss of TBCB expression. To verify this speculation, we used siRNA to silence TBCB expression in astrocytes.

3.2. Silencing TBCB Led to Inhibition of Astrocyte Process Formation

To clarify the relationship between the reduction in TBCB and the inhibition of astrocyte process formation, we performed TBCB silencing experiments with siRNA. In this experiment, the transfection efficiency of TBCB siRNA was higher than 80% and the positive astrocytes showed green fluorescence (Figure 2A,B1–B3). In the silencing group, the mRNA and protein expression of TBCB was significantly downregulated, as shown by PCR (Figure 2K) and WB (Figure 2J), which proved that the silencing effect of TBCB was good.

After silencing, TBCB expression was obviously reduced in astrocytes (Figure 2D,F1,F3), most notably in the nascent processes (Figure 2H1,H3,I2), where originally TBCB expression was extremely abundant (Figure 2C,E,G1,G3, arrows), along with the loss of most nascent cell processes (Figure 2D,F1,F3,I1). These results indicated that the absence of TBCB in the nascent processes of astrocytes led to disordered astrocyte process formation. In addition, it was accompanied by a severe reduction in intracellular MT density (Figure 2F2), most obviously in the nascent processes where MTs were reduced (Figure 2H2,I3), disordered, and intertwined with curly plus ends (Figure 2H2, arrows), suggesting that the severe disorder or dysfunction at the positive end of MTs and TBCB in astrocyte processes was related to the formation and growth of the plus ends of MTs.

This experiment confirmed that TBCB was involved in the formation of astrocyte processes by regulating the plus end of MT. Thus, combined with the first part's results, it was revealed that the disorder of astrocyte process formation after alcohol exposure was associated with a reduction in TBCB expression in the processes caused by alcohol through regulating the MT plus end. These results provide an essential mechanism for understanding the pathogenesis of FASD and the signalling pathway involved in this process is worth studying.

3.3. ERK1/2 Signalling Pathway Regulated TBCB Expression after Acute Foetal Alcohol Exposure in Astrocytes

Compared with the control group, TBCB expression was reduced at 12 h after alcohol exposure (Figure 3A1,A2, $p < 0.001$). In the MAPK signalling pathway as shown by WB, p-ERK1/2 expression was significantly reduced (Figure 3A1,A3, $p < 0.0001$), while the changes in p-JNK and p-P38 proteins were not noticeable (Figure 3A1,A4,A5, $p > 0.05$). This result suggested that the ERK1/2 signalling pathway may regulate TBCB expression after alcohol exposure.

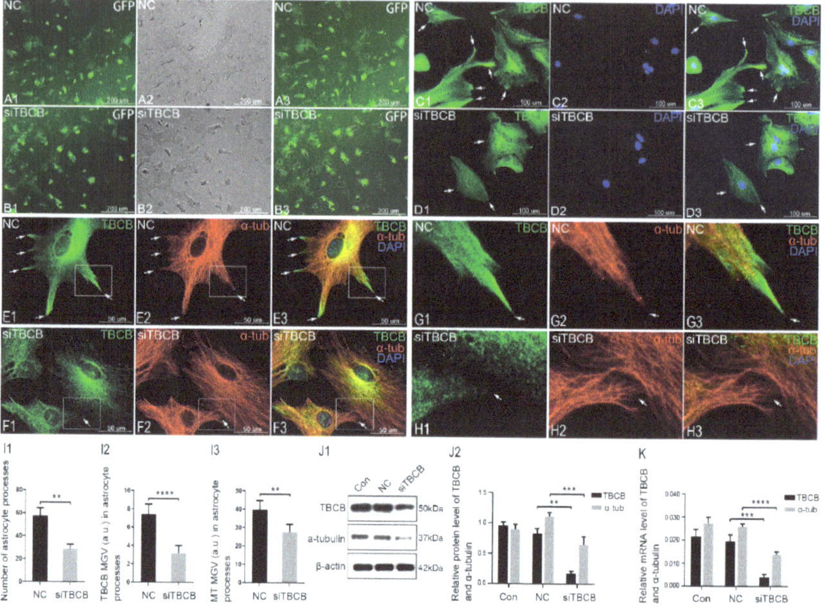

Figure 2. Inhibition of astrocyte process formation after TBCB was silenced. (**A,B**) Immunofluorescence shows that the positive transfection efficiency with TBCB siRNA with green fluorescence was 80%. (**C,D**) In lower magnification images, immunofluorescence shows a change in TBCB expression at nascent astrocyte processes (arrows) after TBCB silencing. (**E–H**) Immunofluorescence shows changes in the expression of TBCB (green signal) and α-tubulin (red signal) in astrocytes in both the control group (**E,G**) and the TBCB-silenced group (**F,H**). Panels G and H are magnified images from the square area of Panels E and F. (**I**) The number of astrocyte processes and mean grey value (MGV) of TBCB and MT in astrocyte processes (MGV = integrated density/area) after 1, 6, 12, and 24 h of acute foetal alcohol exposure. (**J**) Western blot and (**K**) RT–PCR analyses of the protein and mRNA levels of TBCB and α-tubulin after TBCB was silenced ($n = 6$, $p < 0.05$). NC: negative control group; siTBCB: TBCB siRNA group. **: $p < 0.01$, ***: $p < 0.001$, ****: $p < 0.0001$.

To verify that the ERK1/2 signalling pathway is one of the main signalling pathways regulating TBCB expression in astrocytes, the cells were pretreated with the ERK1/2-specific inhibitor U0126 and agonist TPA for 1 h. Compared with the control group, in the agonist pretreatment group, p-ERK1/2 expression was significantly upregulated (Figure 3B1,B2, $p < 0.01$), accompanied by a significant enhancement of TBCB (Figure 3B1,B3, $p < 0.001$). In the inhibitor pretreatment group, p-ERK1/2 expression was significantly decreased (Figure 3B1,B2, $p < 0.01$), accompanied by a significant decrease in TBCB (Figure 3B1,B3, $p < 0.001$). These results confirmed that the ERK1/2 signalling pathway regulates the expression of TBCB in astrocytes.

To further confirm that the ERK1/2 pathway was related to the regulation of TBCB expression in astrocytes after alcohol exposure, astrocytes were pretreated with U0126 or TPA for 1 h and then exposed to alcohol for 12 h, the time point at which the reduction in TBCB was most obvious. Similarly, TBCB expression was significantly lower in the inhibition pretreatment group and higher in the agonist pretreatment group (Figure 3C1–C3, $p < 0.05$) compared with the single alcohol exposure group. These data further demonstrated that the ERK1/2 signalling pathway is one of the main pathways regulating TBCB expression in astrocytes after alcohol injury.

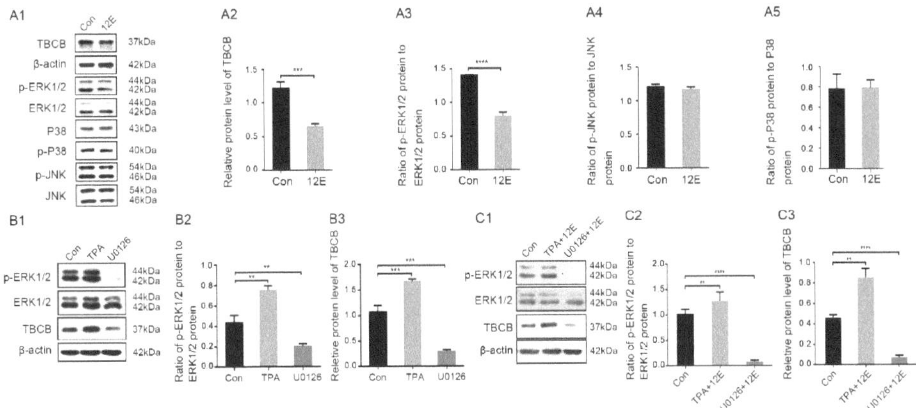

Figure 3. Changes in protein levels in the TBCB and ERK1/2 signalling pathways after alcohol exposure and the protein levels of TBCB and α-tubulin after interfering with the ERK1/2 signalling pathway. (**A**) Western blot analysis of phosphorylation levels and relative quantitative analysis of TBCB and MAPK (ERK1/2, p38, JNK) signalling pathways in astrocytes after 12 h of alcohol exposure ($n = 6$, $p < 0.05$). (**B**) Western blot analysis of the p-ERK1/2 to ERK1/2 ratio and protein levels of TBCB and α-tubulin in astrocytes after pretreatment with the ERK1/2-specific inhibitor U0126 or agonist TPA ($n = 6$, $p < 0.05$). (**C**) Western blot analysis of the p-ERK1/2/ERK1/2 ratio and protein levels of TBCB and α-tubulin in astrocytes after pretreatment with U0126 or TPA before alcohol exposure ($n = 6$, $p < 0.05$). Con: control group; 12E: ethanol or alcohol exposure group; TAP: ERK1/2 agonist group; U0126: ERK1/2 inhibitor group; TPA + 12E: ERK1/2 agonist and ethanol exposure group; U0126 + 12E: ERK1/2 inhibitor and ethanol exposure group. **: $p < 0.01$; ***: $p < 0.001$, ****: $p < 0.0001$.

4. Discussion

In order to clarify the role of TBCB in the morphological changes of astrocytes after acute foetal alcohol exposure, we carried out the above experiments. In this study, we confirmed that decreased TBCB was one of the critical factors for the formation and growth disorder of astrocyte processes after chronic alcohol exposure; TBCB, which could be regulated by the ERK signalling pathway, regulated the growth of MT plus-ends through binding with EB1 and EB3; thus, regulating the formation and growth of astrocyte processes. In addition, our study also showed the following interesting findings.

4.1. Possible Functions of TBCB

Previous studies have demonstrated that TBCB is involved in the de novo synthesis of MT [29–32]. MTs are polar structures of the cytoskeleton formed by the head-to-tail association of αβ heterodimers. The heterodimers bind to protofilaments associated laterally, forming a hollow cylindrical wall, the microtubule [30,56,57]. MTs have different polymerisation rates at their two ends. The slower-growing minus end binds to the perinuclear centrosome-based MT-organising centre and the faster-growing plus end radiates to the cell edges [57,58]. In general, the plus end undergoes cycles of rapid growth and disassembly, known as dynamic instability, which allows microtubules to reorganise rapidly and differentiate spatially and temporally in accordance with the cell context, generate pushing and pulling forces, searching the cell's three-dimensional space and leading to cell asymmetry [59].

TBCB was reported to play an essential role in the dynamics of MT in early oocytes [60] and microglia [61]. It was also found to increase at the end of the axon, the essential neural protrusions, characterised by an assembled MT plus end [37]. These data suggest that TBCB might be involved in the MT plus end's dynamic changes and be related to the cell processes. In our experiments, TBCB was highly expressed in the nascent processes of astrocytes and was coexpressed with the plus end of MT. Moreover, silencing TBCB reduced the expression of TBCB in the processes of astrocytes, accompanied by disordered formation and growth of astrocyte processes, revealing an essential role of TBCB in the formation and growth of astrocyte processes by regulating the plus end of MT. These results may provide new ideas for the study of neurological diseases.

In our experiment, in addition to its diffuse high expression in the nascent process, TBCB could distribute along the MTs and was abundant in the perinuclear area and MT organising centre. This result was similar to Vadlamudi RK's results. They observed colocalisation between TBCB and tubulin during interphase and mitosis, in which TBCB was associated with staining of the mitotic spindle [62]. These results suggested that TBCB might be involved in MT growth at the minus-end and cell division.

The diffuse expression of TBCB decreased sharply after TBCB was silenced (Figure 2B). However, TBCB arranged along preexisting MTs was still visible, with these MTs maintaining their original structure (Figure 2D,F1), suggesting that TBCB might be associated with the stability of the preexisting MT. Although some functions of TBCB were found, it remains a mystery that needs more study.

4.2. Regulating Mechanism of TBCB

We found that ERK1/2 and the PAK1 pathway were inhibited in alcohol-exposed astrocytes. These results were similar to previous studies on neurons of the hippocampus [38] and microglia [40], in which alcohol also caused a decrease in p-ERK1/2 expression. To confirm that the ERK1/2 signalling pathway regulates TBCB expression in normal astrocytes and in astrocytes after alcohol exposure, we detected TBCB expression after treatment with an inhibitor or agonist of ERK1/2. As expected, in normal and alcohol-exposed astrocytes, TBCB expression was regulated by the ERK1/2 signalling pathway. Therefore, after alcohol exposure, inhibiting the ERK1/2 pathway caused by alcohol led to the downregulation of TBCB expression, resulting in disordered MT plus end growth and finally disrupting the astrocyte process formation. This study provides a new view and experimental basis for revealing the neurological symptoms caused by astrocyte injury in FASD.

4.3. Astrocytes and FASD

The upregulated TBCB mRNA expression found by Feltes et al. when analysing transcriptomic data from mice suffering from FASD [36], seemed to conflict with our experimental results, which showed downregulation of TBCB protein expression after alcohol exposure in astrocytes. The upregulated mRNA data came from the brain of newborn or one-month old mice, which suffered alcohol injury when they were foetuses by intraperitoneal injection of alcohol into the mother mice. However, the nervous system injury of the foetal mice caused by alcohol occurs immediately after the mother mice are subjected to intraperitoneal injection. Studies have shown that alcohol can immediately cause the loss of proteins [63], severe DNA damage [64], significant ultrastructure injury [65], and so on from 1 to 48 h after alcohol exposure. These results coincided with ours and proved that the damage to cell structures and the loss of protein, such as TBCB in astrocytes, occurred immediately after alcohol exposure. Then, the cells that suffered from alcohol toxicity would upregulate various mRNAs to compensate for the lost proteins to help the organism recover over time after alcohol injury, unless the cell damage was too severe to recover from. Therefore, the upregulated TBCB mRNA in newborn or one-month old mice [29,30] compensated for the loss of TBCB caused by alcohol toxicity during the foetal stage. These results also highlighted the important role of TBCB in the development of the nervous system.

Astrocyte processes are involved in most astrocyte functions. They contact nodes of Ranvier in brain white matter to regulate nerve conduction [66], contact with blood vessels to control blood flow, and form gap junctions with neighbouring astrocytes [67]. Particularly, in the brain the processes envelop and cover about 60% of synapses, called perisynaptic processes and form the astroglial synaptic cradle, which contributes to synaptogenesis, synaptic maturation, synaptic maintenance, and synaptic extinction [18,68].

The first postnatal week in rodents is known as the 'brain growth spurt' period, also called the synaptogenesis period, with the marked formation of synapses and neuronal networks [1,15]. Our experiments were performed during this period and we found that a decrease in TBCB expression in astrocyte processes was one of the leading causes of the disorder of astrocyte process formation induced by alcohol toxicity. The absence or disappearance of perisynaptic processes, due to the formation disordered of astrocyte process, would certainly lead to the impairments of neuronal synapse, such as formation, substrate transmission, energy metabolism [18], glutamate metabolism [19,69], and so on. These finally could cause neurological symptoms. Combined with our experimental results, these neurological symptoms could be caused by disordered astrocyte process formation because of TBCB loss after alcohol exposure, providing a new idea to study the pathogenic mechanism of FASD.

Although we believe that astrocyte injury can cause the neurological impairment in FASD, we still cannot address which symptom is caused by neuronal injury and which by astrocytic injury. As well, in this study, we confirmed that TBCB could regulate the formation of astrocyte processes; however, the mechanism is still unclear. Future research should explore how TBCB regulates the astrocyte process formation, maybe through regulating the MT plus-end by interacting with some other factors.

5. Conclusions

Our results show that acute foetal alcohol exposure of astrocytes leads to a decrease in TBCB expression, which first affects the positive end of MT and then leads to disorders in the formation, growth, and development of astrocyte processes. The ERK1/2 signalling pathway positively regulates TBCB expression; thus, affecting the growth of astrocytes. Our findings provide new insights into the mechanism of alcohol on nerve cells and provides an experimental and theoretical basis for exploring the pathogenesis of FASD.

Supplementary Materials: The following are available online at https://www.mdpi.com/article/10.3390/brainsci12070813/s1.

Author Contributions: Conceptualisation, H.L.; methodology, H.L. and Y.Z.; software, G.Z.; formal analysis, J.H.; investigation, Y.Z. and J.H.; data curation, M.Y.; writing—original draft preparation, Y.Z. and J.H.; writing—review and editing, Y.Z. and J.H. and H.L.; visualisation, S.W., X.C. and B.Z.; supervision, Y.Z. and H.L.; project administration, H.L.; funding acquisition, M.Y. and H.L. All authors have read and agreed to the published version of the manuscript.

Funding: This work was supported by the National Natural Science Foundation of China (81971230, 81500978, 81671312, 81000566), the Natural Science Foundation Project of Chong Qing (cstc2016jcyjA0229, cstc2017jcyjAX0414, cstc2015jcyja10018, cstc2011jjA10093), and the Foundation of Chongqing Municipal Education Commission (KJ1600213).

Institutional Review Board Statement: All animals were obtained from the Animal Center of Chongqing Medical University and the procedures were approved by the Ethics Committee of Animal Care of Chongqing Medical University. All animal experiments in this study conformed to the standards of the National Institutes of Health Guide for the Care and Use of Laboratory Animals (NIH Publication No. 85–23, revised 1996).

Informed Consent Statement: Informed consent was obtained from all subjects involved in the study.

Data Availability Statement: The data presented in this study are available on request from the corresponding author.

Acknowledgments: In this section, you can acknowledge any support given which is not covered by the author contribution or funding sections. This may include administrative and technical support or donations in kind (e.g., materials used for experiments).

Conflicts of Interest: The authors declare no conflict of interest.

References

1. Trindade, P.; Hampton, B.; Manhaes, A.C.; Medina, A.E. Developmental alcohol exposure leads to a persistent change on astrocyte secretome. *J. Neurochem.* **2016**, *137*, 730–743. [CrossRef]
2. Wilhoit, L.F.; Scott, D.A.; Simecka, B.A. Fetal Alcohol Spectrum Disorders: Characteristics, Complications, and Treatment. *Community Ment. Health J.* **2017**, *53*, 711–718. [CrossRef] [PubMed]
3. Lowery, R.L.; Cealie, M.Y.; Lamantia, C.E.; Mendes, M.S.; Drew, P.D.; Majewska, A.K. Microglia and astrocytes show limited, acute alterations in morphology and protein expression following a single developmental alcohol exposure. *J. Neurosci. Res.* **2021**, *99*, 2008–2025. [CrossRef] [PubMed]
4. Fischer, M.; Chander, P.; Kang, H.; Mellios, N.; Weick, J.P. Transcriptomic changes due to early, chronic intermittent alcohol exposure during forebrain development implicate WNT signaling, cell-type specification, and cortical regionalization as primary determinants of fetal alcohol syndrome. *Alcohol. Clin. Exp. Res.* **2021**, *45*, 979–995. [CrossRef] [PubMed]
5. Norman, A.L.; Crocker, N.; Mattson, S.N.; Riley, E.P. Neuroimaging and fetal alcohol spectrum disorders. *Dev. Disabil. Res. Rev.* **2009**, *15*, 209–217. [CrossRef] [PubMed]
6. Burke, M.W.; Palmour, R.M.; Ervin, F.R.; Ptito, M. Neuronal reduction in frontal cortex of primates after prenatal alcohol exposure. *Neuroreport* **2009**, *20*, 13–17. [CrossRef] [PubMed]
7. Fernandez-Jaen, A.; Fernandez-Mayoralas, D.M.; Quinones Tapia, D.; Calleja-Perez, B.; Garcia-Segura, J.M.; Arribas, S.L.; Munoz Jareno, N. Cortical thickness in fetal alcohol syndrome and attention deficit disorder. *Pediatr. Neurol.* **2011**, *45*, 387–391. [CrossRef] [PubMed]
8. Hamilton, G.F.; Criss, K.J.; Klintsova, A.Y. Voluntary exercise partially reverses neonatal alcohol-induced deficits in mPFC layer II/III dendritic morphology of male adolescent rats. *Synapse* **2015**, *69*, 405–415. [CrossRef]
9. Green, J.H. Fetal Alcohol Spectrum Disorders: Understanding the effects of prenatal alcohol exposure and supporting students. *J. Sch. Health* **2007**, *77*, 103–108. [CrossRef]
10. Gursky, Z.H.; Savage, L.M.; Klintsova, A.Y. Executive functioning-specific behavioral impairments in a rat model of human third trimester binge drinking implicate prefrontal-thalamo-hippocampal circuitry in Fetal Alcohol Spectrum Disorders. *Behav. Brain Res.* **2021**, *405*, 113208. [CrossRef]
11. Maren, S.; Quirk, G.J. Neuronal signalling of fear memory. *Nat. Rev. Neurosci.* **2004**, *5*, 844–852. [CrossRef] [PubMed]
12. Richter, J.; Pietzner, A.; Koenig, J.; Thayer, J.F.; Pane-Farre, C.A.; Gerlach, A.L.; Gloster, A.T.; Wittchen, H.U.; Lang, T.; Alpers, G.W.; et al. Vagal control of the heart decreases during increasing imminence of interoceptive threat in patients with panic disorder and agoraphobia. *Sci. Rep.* **2021**, *11*, 7960. [CrossRef] [PubMed]
13. Battaglia, S.; Thayer, J.F. Functional interplay between central and autonomic nervous systems in human fear conditioning. *Trends Neurosci.* **2022**, *in press.* [CrossRef]
14. Olney, J.W.; Tenkova, T.; Dikranian, K.; Qin, Y.Q.; Labruyere, J.; Ikonomidou, C. Ethanol-induced apoptotic neurodegeneration in the developing C57BL/6 mouse brain. *Brain Res. Dev. Brain Res.* **2002**, *133*, 115–126. [CrossRef]
15. Olney, J.W.; Wozniak, D.F.; Jevtovic-Todorovic, V.; Ikonomidou, C. Glutamate signaling and the fetal alcohol syndrome. *Ment. Retard. Dev. Disabil. Res. Rev.* **2001**, *7*, 267–275. [CrossRef] [PubMed]
16. Olney, J.W.; Ishimaru, M.J.; Bittigau, P.; Ikonomidou, C. Ethanol-induced apoptotic neurodegeneration in the developing brain. *Apoptosis* **2000**, *5*, 515–521. [CrossRef]
17. Santello, M.; Toni, N.; Volterra, A. Astrocyte function from information processing to cognition and cognitive impairment. *Nat. Neurosci.* **2019**, *22*, 154–166. [CrossRef] [PubMed]
18. Allen, N.J.; Barres, B.A. Neuroscience: Glia—More than just brain glue. *Nature* **2009**, *457*, 675–677. [CrossRef] [PubMed]
19. Bacci, A.; Verderio, C.; Pravettoni, E.; Matteoli, M. The role of glial cells in synaptic function. *Philos. Trans. R. Soc. Lond. B Biol. Sci.* **1999**, *354*, 403–409. [CrossRef]
20. Guerri, C.; Renau-Piqueras, J. Alcohol, astroglia, and brain development. *Mol. Neurobiol.* **1997**, *15*, 65–81. [CrossRef]
21. Lokhorst, D.K.; Druse, M.J. Effects of ethanol on cultured fetal astroglia. *Alcohol. Clin. Exp. Res.* **1993**, *17*, 810–815. [CrossRef] [PubMed]
22. Renau-Piqueras, J.; Zaragoza, R.; De Paz, P.; Baguena-Cervellera, R.; Megias, L.; Guerri, C. Effects of prolonged ethanol exposure on the glial fibrillary acidic protein-containing intermediate filaments of astrocytes in primary culture: A quantitative immunofluorescence and immunogold electron microscopic study. *J. Histochem. Cytochem.* **1989**, *37*, 229–240. [CrossRef] [PubMed]
23. Davies, D.L.; Cox, W.E. Delayed growth and maturation of astrocytic cultures following exposure to ethanol: Electron microscopic observations. *Brain Res.* **1991**, *547*, 53–61. [CrossRef]
24. Tomas, M.; Marin, P.; Megias, L.; Egea, G.; Renau-Piqueras, J. Ethanol perturbs the secretory pathway in astrocytes. *Neurobiol. Dis.* **2005**, *20*, 773–784. [CrossRef]
25. Eckardt, M.J.; File, S.E.; Gessa, G.L.; Grant, K.A.; Guerri, C.; Hoffman, P.L.; Kalant, H.; Koob, G.F.; Li, T.K.; Tabakoff, B. Effects of moderate alcohol consumption on the central nervous system. *Alcohol. Clin. Exp. Res.* **1998**, *22*, 998–1040. [CrossRef]
26. Tomas, M.; Lazaro-Dieguez, F.; Duran, J.M.; Marin, P.; Renau-Piqueras, J.; Egea, G. Protective effects of lysophosphatidic acid (LPA) on chronic ethanol-induced injuries to the cytoskeleton and on glucose uptake in rat astrocytes. *J. Neurochem.* **2003**, *87*, 220–229. [CrossRef]
27. Guasch, R.M.; Tomas, M.; Minambres, R.; Valles, S.; Renau-Piqueras, J.; Guerri, C. RhoA and lysophosphatidic acid are involved in the actin cytoskeleton reorganization of astrocytes exposed to ethanol. *J. Neurosci. Res.* **2003**, *72*, 487–502. [CrossRef]

28. Minambres, R.; Guasch, R.M.; Perez-Arago, A.; Guerri, C. The RhoA/ROCK-I/MLC pathway is involved in the ethanol-induced apoptosis by anoikis in astrocytes. *J. Cell Sci.* **2006**, *119*, 271–282. [CrossRef]
29. Tian, G.; Cowan, N.J. Tubulin-specific chaperones: Components of a molecular machine that assembles the alpha/beta heterodimer. *Methods Cell Biol.* **2013**, *115*, 155–171. [CrossRef]
30. Carranza, G.; Castano, R.; Fanarraga, M.L.; Villegas, J.C.; Goncalves, J.; Soares, H.; Avila, J.; Marenchino, M.; Campos-Olivas, R.; Montoya, G.; et al. Autoinhibition of TBCB regulates EB1-mediated microtubule dynamics. *Cell. Mol. Life Sci.* **2013**, *70*, 357–371. [CrossRef]
31. Nithianantham, S.; Le, S.; Seto, E.; Jia, W.; Leary, J.; Corbett, K.D.; Moore, J.K.; Al-Bassam, J. Tubulin cofactors and Arl2 are cage-like chaperones that regulate the soluble alphabeta-tubulin pool for microtubule dynamics. *eLife* **2015**, *4*, e08811. [CrossRef]
32. Al-Bassam, J. Revisiting the tubulin cofactors and Arl2 in the regulation of soluble alphabeta-tubulin pools and their effect on microtubule dynamics. *Mol. Biol. Cell* **2017**, *28*, 359–363. [CrossRef] [PubMed]
33. Kortazar, D.; Fanarraga, M.L.; Carranza, G.; Bellido, J.; Villegas, J.C.; Avila, J.; Zabala, J.C. Role of cofactors B (TBCB) and E (TBCE) in tubulin heterodimer dissociation. *Exp. Cell Res.* **2007**, *313*, 425–436. [CrossRef] [PubMed]
34. Kortazar, D.; Carranza, G.; Bellido, J.; Villegas, J.C.; Fanarraga, M.L.; Zabala, J.C. Native tubulin-folding cofactor E purified from baculovirus-infected Sf9 cells dissociates tubulin dimers. *Protein Expr. Purif.* **2006**, *49*, 196–202. [CrossRef]
35. Martin, L.; Fanarraga, M.L.; Aloria, K.; Zabala, J.C. Tubulin folding cofactor D is a microtubule destabilizing protein. *FEBS Lett.* **2000**, *470*, 93–95. [CrossRef]
36. Feltes, B.C.; de Faria Poloni, J.; Nunes, I.J.; Bonatto, D. Fetal alcohol syndrome, chemo-biology and OMICS: Ethanol effects on vitamin metabolism during neurodevelopment as measured by systems biology analysis. *OMICS J. Integr. Biol.* **2014**, *18*, 344–363. [CrossRef] [PubMed]
37. Lopez-Fanarraga, M.; Carranza, G.; Bellido, J.; Kortazar, D.; Villegas, J.C.; Zabala, J.C. Tubulin cofactor B plays a role in the neuronal growth cone. *J. Neurochem.* **2007**, *100*, 1680–1687. [CrossRef] [PubMed]
38. Qiao, X.; Sun, M.; Chen, Y.; Jin, W.; Zhao, H.; Zhang, W.; Lai, J.; Yan, H. Ethanol-Induced Neuronal and Cognitive/Emotional Impairments are Accompanied by Down-Regulated NT3-TrkC-ERK in Hippocampus. *Alcohol Alcohol.* **2021**, *56*, 220–229. [CrossRef] [PubMed]
39. Zamora-Martinez, E.R.; Edwards, S. Neuronal extracellular signal-regulated kinase (ERK) activity as marker and mediator of alcohol and opioid dependence. *Front. Integr. Neurosci.* **2014**, *8*, 24. [CrossRef]
40. Gofman, L.; Fernandes, N.C.; Potula, R. Relative Role of Akt, ERK and CREB in Alcohol-Induced Microglia P2X4R Receptor Expression. *Alcohol Alcohol.* **2016**, *51*, 647–654. [CrossRef]
41. Aroor, A.R.; Shukla, S.D. MAP kinase signaling in diverse effects of ethanol. *Life Sci.* **2004**, *74*, 2339–2364. [CrossRef] [PubMed]
42. Peng, J.; Wagle, M.; Mueller, T.; Mathur, P.; Lockwood, B.L.; Bretaud, S.; Guo, S. Ethanol-modulated camouflage response screen in zebrafish uncovers a novel role for cAMP and extracellular signal-regulated kinase signaling in behavioral sensitivity to ethanol. *J. Neurosci.* **2009**, *29*, 8408–8418. [CrossRef]
43. McCarthy, K.D.; de Vellis, J. Preparation of separate astroglial and oligodendroglial cell cultures from rat cerebral tissue. *J. Cell Biol.* **1980**, *85*, 890–902. [CrossRef] [PubMed]
44. Tarassishin, L.; Suh, H.S.; Lee, S.C. LPS and IL-1 differentially activate mouse and human astrocytes: Role of CD14. *Glia* **2014**, *62*, 999–1013. [CrossRef] [PubMed]
45. John, G.R. Investigation of astrocyte–oligodendrocyte interactions in human cultures. *Methods Mol. Biol.* **2012**, *814*, 401–414. [CrossRef] [PubMed]
46. Liu, L.; Sun, T.; Xin, F.; Cui, W.; Guo, J.; Hu, J. Nerve Growth Factor Protects Against Alcohol-Induced Neurotoxicity in PC12 Cells via PI3K/Akt/mTOR Pathway. *Alcohol Alcohol.* **2017**, *52*, 12–18. [CrossRef]
47. Smith, A.M.; Zeve, D.R.; Dohrman, D.P.; Chen, W.J. The interactive effect of alcohol and nicotine on NGF-treated pheochromocytoma cells. *Alcohol* **2006**, *39*, 65–72. [CrossRef]
48. Loureiro, S.O.; Heimfarth, L.; Reis, K.; Wild, L.; Andrade, C.; Guma, F.T.; Goncalves, C.A.; Pessoa-Pureur, R. Acute ethanol exposure disrupts actin cytoskeleton and generates reactive oxygen species in c6 cells. *Toxicol. In Vitro* **2011**, *25*, 28–36. [CrossRef]
49. Nagy, J.; Kolok, S.; Dezso, P.; Boros, A.; Szombathelyi, Z. Differential alterations in the expression of NMDA receptor subunits following chronic ethanol treatment in primary cultures of rat cortical and hippocampal neurones. *Neurochem. Int.* **2003**, *42*, 35–43. [CrossRef]
50. Ledig, M.; Kopp, P.; Mandel, P. Effect of ethanol on adenosine triphosphatase and enolase activities in rat brain and in cultured nerve cells. *Neurochem. Res.* **1985**, *10*, 1311–1324. [CrossRef]
51. Noraberg, J.; Zimmer, J. Ethanol induces MAP2 changes in organotypic hippocampal slice cultures. *Neuroreport* **1998**, *9*, 3177–3182. [CrossRef] [PubMed]
52. Zhang, G.; Ma, P.; Wan, S.; Xu, J.; Yang, M.; Qiu, G.; Zhuo, F.; Xu, S.; Huo, J.; Ju, Y.; et al. Dystroglycan is involved in the activation of ERK pathway inducing the change of AQP4 expression in scratch-injured astrocytes. *Brain Res.* **2019**, *1721*, 146347. [CrossRef]
53. He, X.; Gao, F.; Hou, J.; Li, T.; Tan, J.; Wang, C.; Liu, X.; Wang, M.; Liu, H.; Chen, Y.; et al. Metformin inhibits MAPK signaling and rescues pancreatic aquaporin 7 expression to induce insulin secretion in type 2 diabetes mellitus. *J. Biol. Chem.* **2021**, *297*, 101002. [CrossRef] [PubMed]
54. Liu, Y.; Xu, M.; Zhang, H.; Li, X.; Su, Z.; Zhang, C. SEC2-induced superantigen and antitumor activity is regulated through calcineurin. *Appl. Microbiol. Biotechnol.* **2013**, *97*, 9695–9703. [CrossRef] [PubMed]

55. Jensen, E.C. Quantitative analysis of histological staining and fluorescence using ImageJ. *Anat. Rec.* **2013**, *296*, 378–381. [CrossRef]
56. Nogales, E.; Wang, H.W. Structural mechanisms underlying nucleotide-dependent self-assembly of tubulin and its relatives. *Curr. Opin. Struct. Biol.* **2006**, *16*, 221–229. [CrossRef]
57. Desai, A.; Mitchison, T.J. Microtubule polymerization dynamics. *Annu. Rev. Cell Dev. Biol.* **1997**, *13*, 83–117. [CrossRef]
58. Vinogradova, T.; Miller, P.M.; Kaverina, I. Microtubule network asymmetry in motile cells: Role of Golgi-derived array. *Cell Cycle* **2009**, *8*, 2168–2174. [CrossRef]
59. Conde, C.; Caceres, A. Microtubule assembly, organization and dynamics in axons and dendrites. *Nat. Rev. Neurosci.* **2009**, *10*, 319–332. [CrossRef]
60. Baffet, A.D.; Benoit, B.; Januschke, J.; Audo, J.; Gourhand, V.; Roth, S.; Guichet, A. Drosophila tubulin-binding cofactor B is required for microtubule network formation and for cell polarity. *Mol. Biol. Cell* **2012**, *23*, 3591–3601. [CrossRef]
61. Fanarraga, M.L.; Villegas, J.C.; Carranza, G.; Castano, R.; Zabala, J.C. Tubulin cofactor B regulates microtubule densities during microglia transition to the reactive states. *Exp. Cell Res.* **2009**, *315*, 535–541. [CrossRef] [PubMed]
62. Vadlamudi, R.K.; Barnes, C.J.; Rayala, S.; Li, F.; Balasenthil, S.; Marcus, S.; Goodson, H.V.; Sahin, A.A.; Kumar, R. p21-activated kinase 1 regulates microtubule dynamics by phosphorylating tubulin cofactor B. *Mol. Cell. Biol.* **2005**, *25*, 3726–3736. [CrossRef]
63. Xu, T.; Zhang, S.Y.; Xu, X.M.; Zhao, S.; Zhu, K.H.; Zhang, W.W.; Zhao, L.L. Alcohol inhibits the proliferation of Neuro2a cells via promoting the asymmetric cell division through down-regulation of the expression of centrosome protein-J. *Toxicol. Lett.* **2018**, *294*, 177–183. [CrossRef] [PubMed]
64. Lacaille, H.; Duterte-Boucher, D.; Liot, D.; Vaudry, H.; Naassila, M.; Vaudry, D. Comparison of the deleterious effects of binge drinking-like alcohol exposure in adolescent and adult mice. *J. Neurochem.* **2015**, *132*, 629–641. [CrossRef]
65. Zimatkin, S.M.; Fedina, E.M.; Kuznetsova, V.B. Brain histaminergic neurons in rats subjected to the acute effect of alcohol. *Morfologiia* **2012**, *142*, 17–22. [PubMed]
66. Montgomery, D.L. Astrocytes: Form, functions, and roles in disease. *Vet. Pathol.* **1994**, *31*, 145–167. [CrossRef]
67. Sofroniew, M.V.; Vinters, H.V. Astrocytes: Biology and pathology. *Acta Neuropathol.* **2010**, *119*, 7–35. [CrossRef]
68. Verkhratsky, A.; Augusto-Oliveira, M.; Pivoriunas, A.; Popov, A.; Brazhe, A.; Semyanov, A. Astroglial asthenia and loss of function, rather than reactivity, contribute to the ageing of the brain. *Pflugers Arch.* **2021**, *473*, 753–774. [CrossRef]
69. Rothstein, J.D.; Dykes-Hoberg, M.; Pardo, C.A.; Bristol, L.A.; Jin, L.; Kuncl, R.W.; Kanai, Y.; Hediger, M.A.; Wang, Y.; Schielke, J.P.; et al. Knockout of glutamate transporters reveals a major role for astroglial transport in excitotoxicity and clearance of glutamate. *Neuron* **1996**, *16*, 675–686. [CrossRef]

Article

Buffy Coat Score as a Biomarker of Treatment Response in Neuronal Ceroid Lipofuscinosis Type 2

Siyamini Sivananthan [1], Laura Lee [1], Glenn Anderson [1,2], Barbara Csanyi [1], Ruth Williams [3] and Paul Gissen [1,2,*]

[1] Department of Inherited Metabolic Diseases, Great Ormond Street Hospital, London WC1N 1EH, UK
[2] Institute for Health Research Great Ormond Street Hospital Biomedical Research Centre, University College London, London WC1N 1EH, UK
[3] Department of Children's Neurosciences, Evelina London Children's Hospital, London SE1 7EH, UK
* Correspondence: p.gissen@ucl.ac.uk

Abstract: The introduction of intracerebroventricular (ICV) enzyme replacement therapy (ERT) for treatment of neuronal ceroid lipofuscinosis type 2 (CLN2) disease has produced dramatic improvements in disease management. However, assessments of therapeutic effect for ICV ERT are limited to clinical observational measures, namely the CLN2 Clinical Rating Scale, a subjective measure of motor and language performance. There is a need for an objective biomarker to enable assessments of disease progression and response to treatment. To address this, we investigated whether the proportion of cells with abnormal storage inclusions on electron microscopic examination of peripheral blood buffy coats could act as a biomarker of disease activity in CLN2 disease. We conducted a prospective longitudinal analysis of six patients receiving ICV ERT. We demonstrated a substantial and continuing reduction in the proportion of abnormal cells over the course of treatment, whereas symptomatic scores revealed little or no change over time. Here, we proposed the use of the proportion of cells with abnormal storage as a biomarker of response to therapy in CLN2. In the future, as more tissue-specific biomarkers are developed, the buffy coats may form part of a panel of biomarkers in order to give a more holistic view of a complex disease.

Keywords: neuronal ceroid lipofuscinosis type 2 (CLN2) disease; lysosomal storage disorder; intracerebroventricular; enzyme replacement therapy; disease progression; neurodegeneration; biomarker; blood buffy coat; electron microscopy; curvilinear inclusions

Citation: Sivananthan, S.; Lee, L.; Anderson, G.; Csanyi, B.; Williams, R.; Gissen, P. Buffy Coat Score as a Biomarker of Treatment Response in Neuronal Ceroid Lipofuscinosis Type 2. *Brain Sci.* **2023**, *13*, 209. https://doi.org/10.3390/brainsci13020209

Academic Editors: Masaru Tanaka, Lydia Giménez-Llort, Simone Battaglia, Chong Chen and Piril Hepsomali

Received: 2 November 2022
Revised: 19 January 2023
Accepted: 22 January 2023
Published: 27 January 2023

1. Introduction

Neuronal ceroid lipofuscinosis type 2 (CLN2) is a rare, rapidly progressive neurodegenerative lysosomal storage disorder. It is 1 of 13 different subtypes of neuronal ceroid lipofuscinoses (NCL) [1–3]. The reported incidence of CLN2 is 0.1–7/100,000, with variation depending on geographical region [4,5]. It has an autosomal recessive pattern of inheritance and is caused by deficient activity of the enzyme tripeptidyl peptidase 1 (TPP1) [6,7]. Loss of TPP1 enzyme activity leads to accumulation of ceroid lipofuscin [8,9], an autofluorescent lysosomal storage material [10]. This results in catastrophic neuronal degeneration throughout the CNS and retina [11–13].

The most common CLN2 genetic variants largely lead to a late-infantile phenotype which display a high genotype–phenotype correlation [1,14]. There are also some rarer variants that have been linked to delayed onset or prolonged disease course [15] and are described as "atypical" or "non-classical" phenotypes [16,17]. In late-infantile CLN2 disease, affected children typically have delayed language acquisition but otherwise are functionally normal until two to four years of age, and subsequently manifest various different types of seizures [18,19]. This is followed by a rapid decline in motor, language, cognitive, and visual function over a period of four to six years, with death by early adolescence [13,20]. Seizure photosensitivity detected by EEG using low-frequency photic stimulation is regarded as an early hallmark of the disease but is not often identified [21]. Many of these children are classified as having "childhood dementia" owing to the loss of previously acquired developmental skills and intellectual disability [22–24]. Whilst CLN2 disease predominantly affects the central nervous system, there is some evidence that the disease has important peripheral effects [25,26]. Cardiac co-morbidities, including cardiomyopathy and conduction defects, are well described in the juvenile onset form of NCL (CLN3 disease) [26] but have also been reported in atypical CLN2 [27].

Until recently, treatment of CLN2 was largely supportive [8,28], with pharmacological management of seizures and movement disorder alongside multimodal therapies including speech and language therapy, physiotherapy, and occupational therapy [5,29]. Psychosocial, family, and palliative care support is also of vital importance as the disease progresses [30]. Recently, in a trial of intracerebroventricular (ICV) cerliponase alfa (a recombinant proenzyme form of human TPP1), children with CLN2 disease showed a significant improvement in clinical course, with a less severe decline in motor and language function than in historical controls [31]. This treatment was given as a fortnightly infusion of 300 mg cerliponase alfa directly into an ICV device and was well tolerated [31]. Several subsequent independent case series of children receiving cerliponase alfa have supported these findings, concluding that the ICV treatment is safe and stabilises disease progression [32,33]. Importantly, this is the first disease modifying treatment that has been shown to be effective in any of the NCLs [31,32].

However, assessments of therapeutic effect have been largely limited to clinical observational measures, namely the CLN2 rating scale [31,34]. This rating scale is a crude measure of motor and language performance [35] and does not give any detailed information on the progress of the patients. These clinical examinations are often performed by the patients' long-term physician and are therefore open to bias, limiting assessment of treatment efficacy on an individual level in clinical practice. Furthermore, determining the presence or absence of any therapeutic effect is typically delayed if symptoms alone are relied upon, because changes in symptoms occur relatively slowly (it is often unclear whether the absence of change is due to a slow rate of progression or because the treatment has been effective). A biomarker that reflects the underlying disease process more directly would therefore enable earlier and more objective assessment of therapeutic effect (or failure). This could act as an adjunct to clinical ratings to support timely decisions regarding modification of treatment.

From a disease characterisation perspective, one of the main consequences of aberrant accumulation of metabolic by-products is the development of abnormal cytoplasmic vacuolation of lymphocytes [36,37]. This feature can be detected by microscopic and ultrastructural examination of peripheral blood buffy coat [38,39]. A study examining 2500 blood films looking for vacuolated lymphocytes from patients with clinical indications that included developmental delay/regression, ataxia, seizures, and cardiomyopathy demonstrated that the most common pathology was NCL (typically CLN3) [36]. However, patients with CLN2 disease do not have vacuolated lymphocytes but instead have characteristic abnormal curvilinear storage inclusions [36,39]. These abnormal inclusions are present in lymphocytes and can only be visualised using electron microscopy since conventional light microscopy lacks the resolution needed to characterise such fine detail [36,40]. Visualisation of curvilinear storage inclusions in peripheral blood samples has circumvented the historical need for more invasive tests such as skin and rectal biopsies and has made it easier to detect cases earlier [41]. Electron microscopy is readily available in centres providing NCL diagnosis. With appropriate setup and regular use, peripheral buffy coat electron microscopy can be applied to routinely obtained peripheral blood samples.

Classic CLN2 disease demonstrates pure membrane-bound curvilinear profiles, which are a hallmark of the disease [36,42,43]. This specific pattern is not found in any other NCL disorder [36,40,41]. Those with atypical CLN2 disease often have a later onset of symptoms [16,17] and tend to have a mixed pattern of curvilinear storage material and fingerprint stacks seen on electron microscopy of their lymphocytes [40,44,45]. Whilst the presence of these storage inclusions is a useful diagnostic marker for CLN2 disease [36,39], to date, the inclusions have not been evaluated in detail following ERT treatment. We hypothesised that the percentage of curvilinear storage material in peripheral blood buffy coats would reduce after treatment and continue to do so over the course of treatment. Here, we investigate the change in appearance of the peripheral blood storage material over time whilst on treatment. Our main objective is to identify whether the amount of curvilinear storage material reduces over the course of treatment and could therefore act as a biomarker of therapeutic response in patients with CLN2 disease.

2. Methods

A retrospective cohort study of CLN2 patients receiving cerliponase alfa on compassionate grounds at our centre was undertaken. All included patients were given cerliponase alfa as part of an expanded access scheme and underwent yearly collection of peripheral blood samples (which were used for buffy coat testing in addition to routine monitoring) as part of standard care. These samples were taken prior to the ICV infusion on an annual basis. The collection of samples for this study has appropriate regulatory approvals (13/LO/0168; IRAS ID 95005; London-Bloomsbury Research Ethics Committee) and all participants provided informed written consent to participate.

2.1. Study Patients

We included all patients referred to a single tertiary centre who (i) had a diagnosis of CLN2 disease based on genetic testing and (ii) underwent ICV ERT treatment through an expanded access scheme. Exclusion criteria were (i) any contraindication to neurosurgery and (ii) known hypersensitivity to any component of the study drug.

To enable administration of cerliponase alpha, a ventricular reservoir was surgically implanted in all patients, with the reservoir placed under the scalp and the catheter placed in the cerebral lateral ventricle, with placement confirmed using brain magnetic resonance imaging (MRI). Cerliponase alfa was administered as an infusion of 300 mg every 14 days [31,46,47]. Antihistamine and paracetamol were administered orally approximately 30 min before each infusion.

2.2. Electron Micrographs

We investigated the percentage of cells containing curvilinear storage inclusions in the peripheral blood buffy coat layer (a concentrated layer of leucocytes and platelets found at the erythrocyte–plasma junction, following centrifugation) [39,48–51] and in skin cells [52]. A sample of EDTA blood was collected from each patient alongside routine disease surveillance bloods. In two patients, a skin biopsy was performed two years after initiation of treatment. Blood buffy coat and skin biopsy samples were fixed in 2.5% glutaraldehyde solution in 0.1 M cacodylate buffer at pH 7.2, followed by secondary fixation in 1% osmium tetroxide [51]. Samples were dehydrated in graded ethanol transferred to an intermediate reagent, propylene oxide, and then infiltrated and embedded in Agar 100 resin. Polymerisation was undertaken at 60 °C for 48 h. Semithin sections were stained with toluidine blue to identify appropriate areas. Moreover, 90 nm ultrathin sections were cut using a Diatome diamond knife on a Leica ultramicrotome. Sections were transferred to copper grids and contrasted with uranyl acetate and lead citrate. The sections were examined using a JEOL 1400 transmission electron microscope [36,39]. One hundred cells were evaluated per sample and any lymphocyte identified was used for analysis [40].

We recorded the percentage of abnormal cells on electron micrographs of the buffy coat prior to starting treatment and at yearly intervals after commencing treatment. To determine the effect of therapy in different cell types other than those observed in peripheral blood, two patients also underwent skin biopsies at two years post-treatment, and we recorded the percentage of abnormal skin cells on electron micrographs. These results were reported by the lead clinical histopathologist at our centre.

2.3. Clinical Assessment

Treatment outcomes were recorded using the motor and language domains of the CLN2 Clinical Rating Scale [31].

For all patients, age at diagnosis (defined as date of confirmatory enzyme assay), genotype, treatment start date, and adverse events whilst on treatment were recorded. We documented basic haematology and biochemistry (renal and liver function), ECG, EEG, and MRI brain results. We recorded the time until first unreversed two-point decline in the score on the CLN2 Clinical Rating Scale measuring motor and language skills or until the attainment of a combined motor–language score of 0. The performance was assessed over a period of 3 years during which each patient received a fortnightly 300 mg dose of cerliponase alfa.

The clinical rating scale was performed by the lead medical practitioner, a consultant specialist in inherited metabolic disorders. Data regarding adverse events and concomitant medications were reported at every visit. The attribution of adverse events was determined as part of standard care. The baseline measurement was the last observation preceding the first administration of cerliponse alfa. The reports from any MRI head scans that the patients had undergone (requested as part of standard care) were summarised descriptively.

2.4. Statistics

Buffy coat data were analysed using repeated measures one-way analysis of various (ANOVA) with Dunnett's multiple comparison test [53], which treated the pre-treatment data as a baseline and the mean compared the mean of each subsequent data point against this baseline. Data from the final year were excluded from the ANOVA analysis as data were only available for two of the patients. Nonparametric data (e.g., CLN2 scores) were compared using the Friedman test [54], again comparing each post-treatment column with the pre-treatment column as a baseline.

3. Results

3.1. Demographics

Patient demographics are summarised in Table 1. A total of 6 patients received ERT as part of the compassionate use programme (2M:4F), with a median age at diagnosis of 4 years 2 months (2 years 2 months–13 years 3 months) and age at treatment initiation of 4 years 7 months (3 years 11 months–14 years 11 months). One patient (patient C) had an attenuated phenotype and was thus at the upper end of the range at diagnosis and treatment initiation (13 years 3 months and 14 years 11 months, respectively). The mean (SD) duration of treatment with cerliponase alfa was 2 years 5 months (3 months). Three patients were homozygous for a common mutation, one patient had one common and one uncommon allele, and two patients had two uncommon alleles. All the patients had a combined score of 1 to 5 on the motor and language domains of the CLN2 Clinical Rating Scale (the motor and language domains each have a 0–3 point subscale, with 0 representing no function and 3 representing normal function). The sum of the motor and language scores therefore range from 0 to 6 [18,55]. Two patients had a language score of 0 at the start of treatment, while the remaining four patients had a language score of at least 1. All patients had a motor score of 1 or above at the start of treatment.

Table 1. Patient demographics. Six patients (A–F) are each shown in a different row.

Patient	Current Age	Age at Diagnosis	Mutation	Age at Treatment Start	CLN2 Scores at Start of Treatment (Motor, Language)
A	10 years 2 months	4 years 7 months	c.89 + 5G > A, c 509. 1G > C	4 years 10 months	1, 0
B	10 years 1 months	4 years 4 months	homozygous c.509-1G > C	4 years 7 months	1, 2
C	20 years 3 months	13 years 3 months	c.89 + 5G > C c. 1340G > A	14 years 11 months	2, 2
D	9 years 4 months	4 years 2 months	homozygous c.1052 G > T	4 years 5 months	2, 0
E	11 years 4 months	4 years	homozygous c.1052 G > T	5 years 10 months	1, 1
F	9 years 2 months	2 years 2 months	homozygous c.1052 G > T	3 years 11 months	3, 2

3.2. Buffy Coats

Examples of electron micrograph images of peripheral blood buffy coats before and after treatment, in a single patient, are shown in Figure 1. These demonstrate clusters of membrane-bound, regularly arranged curvilinear inclusions, typical of CLN2 disease, taken prior to treatment initiation (pre-ERT). Following ERT, there are poorly defined curvilinear-like storage inclusions with prominent lipid droplets and electron dense inclusions.

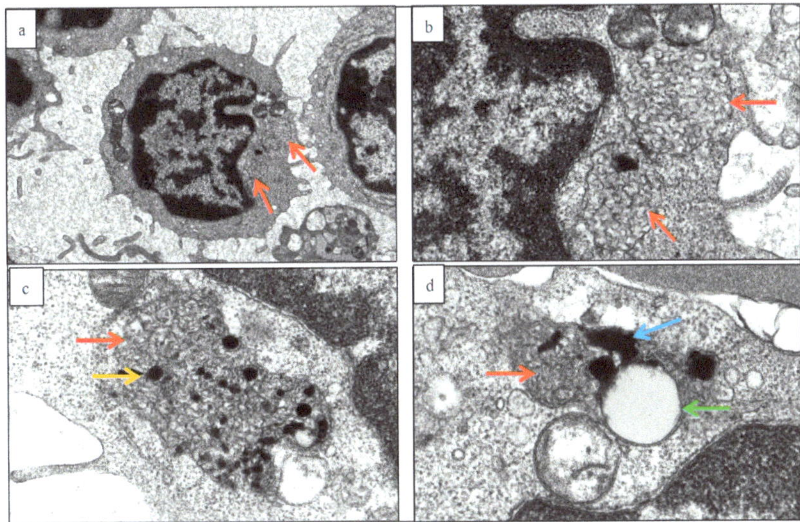

Figure 1. Electron micrographs of peripheral blood buffy coats pre- and post-treatment in patient E: (**a,b**) show low- and high-power images (mag. ×3000 and mag. ×10,000), respectively. These demonstrate clusters of membrane-bound, regularly arranged curvilinear inclusions (red arrow), typical of CLN2 disease, taken prior to treatment initiation (pre-ERT). (**c**) shows a high-power image (mag. ×12,000) one year post-ERT. There are irregular curvilinear-like inclusions (red arrow) with electron dense material representing fragments. There is also a fingerprint stack (yellow arrow) which is seen occasionally in CLN2 lymphocytes. (**d**) shows a high-power image (mag. ×15,000) two years post-ERT. There are poorly defined curvilinear-like storage inclusions (red arrow) with prominent lipid droplets (green arrow) and electron dense debris (blue arrow) (mag. ×15,000).

The relationship between time since treatment initiation and the percentage of abnormal cells in the peripheral blood buffy coats is shown in Figure 2. Overall, a substantial reduction in the percentage of cells with abnormal storage was observed. The mean (95% CI) differences in percentage of abnormal cells in the buffy coats compared to the baseline measurements were as follows: one year: -3.5 (-10.8 to $+3.8$, $p = 0.36$), two years: -6.5 (-3.5 to -9.5, $p = 0.002$), and three years: -7.5 (-3.3 to -11.8, $p = 0.005$).

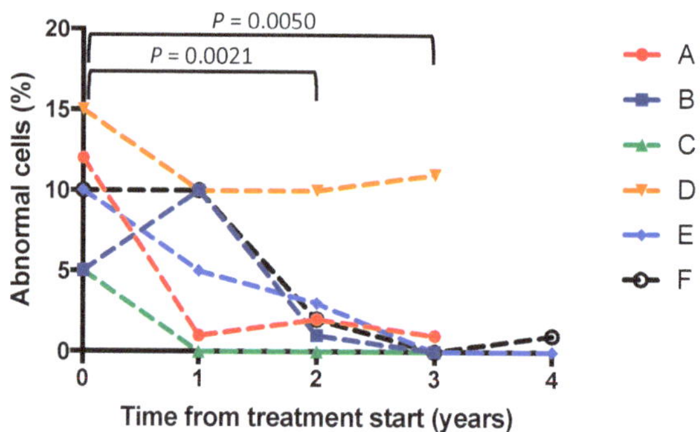

Figure 2. Change in percentage of cells showing abnormal storage inclusions after treatment for each of the six patients (A–F, labelled in separate colours as per the legend).

3.3. Clinical Outcomes

The relationship between time since treatment initiation and the CLN2 scores is shown in Figure 3. The median (range) CLN2 score at the baseline was 2.5 (1–5). Two of the six patients underwent reductions in CLN2 scores with treatment (one declining from 2 to 0 and the other from 5 to 4); the other four patients remained stable. There was no significant difference in CLN2 scores over the time points in the study ($p = 0.25$). This was in line with the findings of the clinical trial, which demonstrated stabilisation of CLN2 scores in patients treated with cerliponase alfa compared to natural historical controls [31].

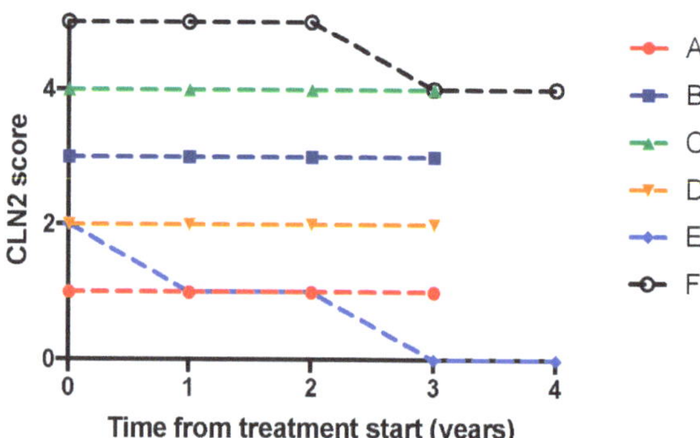

Figure 3. Change in CLN2 motor and language scores after treatment. Each of the six patients are represented individually (A–F, labelled in separate colours as per the legend). The dotted lines link the individual scores for each patient.

Full blood count and renal and liver function remained stable throughout the study period. ECG and EEG results were grossly unchanged. Brain MRI showed reduced volume in the first year, with more stable appearances in the second year post-treatment.

3.4. Skin Biopsies

For two patients (E and F), skin biopsies were performed (two years after starting treatment) in order to assess response to treatment in various cell populations other than those present in peripheral blood. These samples were taken two years after treatment had commenced. The effect of treatment was different in different skin cells. Patient E (who underwent a 50% reduction in peripheral buffy coat storage material in the first year after treatment) had profuse deposits of curvilinear inclusions in secretory epithelial cells from eccrine sweat glands, with a similar (indistinct and non-uniform) morphological profile to the patient's post-treatment buffy coat inclusions (Figures 4 and 5). In patient F (who showed no change in percentage of buffy coat inclusions in the first year after treatment), curvilinear storage inclusions were present in the sweat glands but were more distinct in appearance.

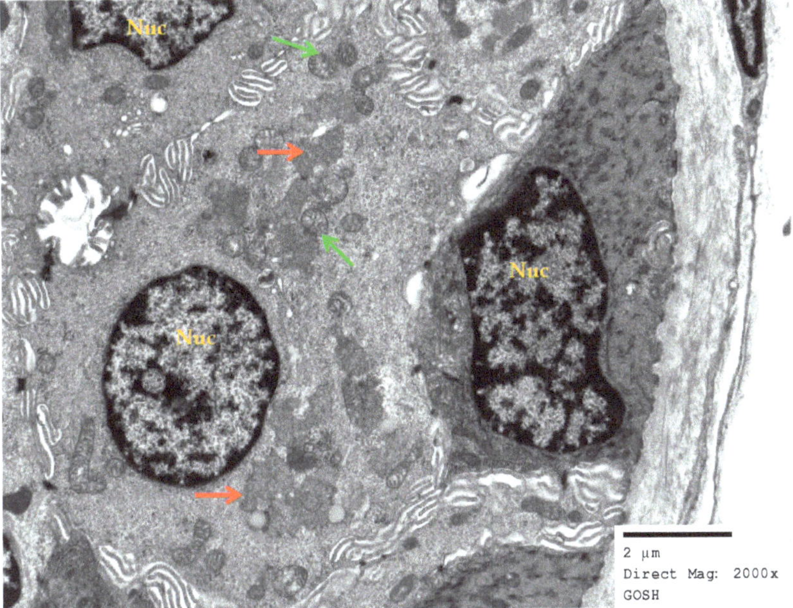

Figure 4. Low-power electron micrograph, mag. ×2000, of secretory epithelial cells from an eccrine sweat gland with abnormal storage inclusions present in clusters of expanded lysosomes (red arrow). Numerous mitochondria with normal morphology present (green arrow) and prominent nuclei (Nuc) also with normal appearances.

There was no indication of storage in fibroblasts or endothelial cells in either of the two patients.

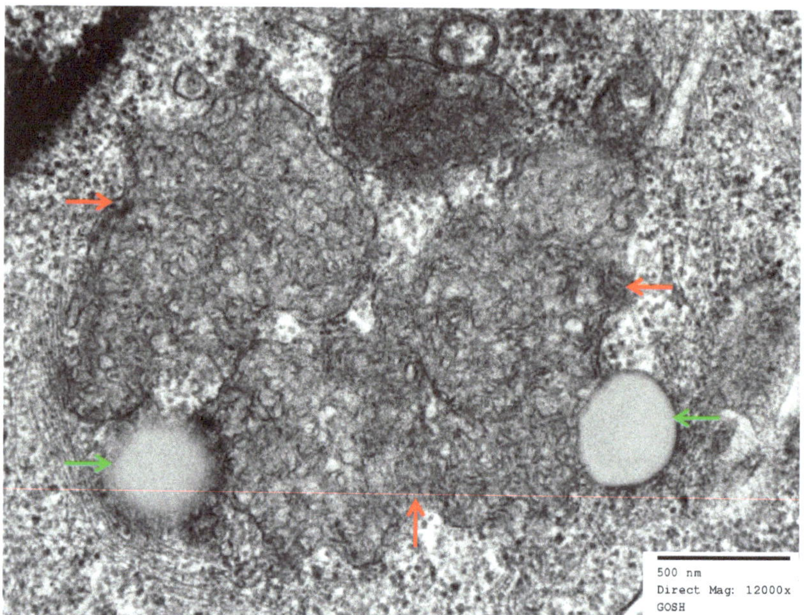

Figure 5. High-power electron micrograph, mag. ×12,000, of membrane-bound storage inclusions in secretory epithelial cells from eccrine sweat gland with curvilinear-like bodies (red arrow) and lipid droplets (green arrow).

4. Discussion

In this study, we found a significant reduction in abnormal storage material seen on electron micrographs of peripheral blood buffy coats following regular ERT over 3 years, supporting our hypothesis of the biological validity of this as a marker of cellular response to treatment.

Our data suggest that the percentage of abnormal storage inclusions in CLN2 disease could be of value as an objective, visual marker of therapeutic response and a biomarker of disease activity. By comparison, the CLN2 clinical score is much better suited to the assessment of functional loss rather than any gain or normalisation of function.

We found that patients receiving ICV cerliponase alfa for CLN2 disease underwent little to no deterioration in symptoms over a three-year period, providing further evidence for the efficacy of this treatment. Our data are consistent with those recently reported [31,33]. Additionally, these outcomes represent a significant improvement when compared to historical untreated patient controls. An observational cohort study of CLN2 patients showed a rapid annual decline of 1.81 score units, with the combined motor–language scores declining from normal (score of 6) to no function (score of 0) in approximately 30 months [20]. The treatment was well tolerated within this group with no terminations, providing further evidence for the safety of this therapy.

There has been concern that ICV-delivered therapy would not have any effect on peripheral tissues. The demonstration of a reduction in the percentage of abnormal cells in the peripheral blood is significant because it suggests that therapy delivered directly to the CNS does also have peripheral organ effects in addition to previously documented neurological improvement. The exact mechanism is unclear, but it is possible that passive transfer of the enzyme from the CSF into the systemic circulation underlies this effect.

Ultrastructural examination of peripheral blood buffy coats is a reliable, minimally invasive test used for the diagnosis of CLN2 disease in childhood and has been shown to be particularly useful with harder-to-diagnose NCL variants [39]. The use of buffy coat analysis to determine response to treatment has not previously been reported.

The reduction in storage material present in peripheral blood white cells, despite the fact that these cells replenish after approximately 20 days, supports the notion that there is an excess of storage material in bone marrow progenitor cells (and/or in other reservoirs elsewhere) which takes several years to be cleared fully.

There are reports of cardiac manifestations of NCL disease, affecting conduction pathways or the myocardium itself and leading to development of a cardiomyopathy [22,24]. Pathological investigation of these structures in NCL disorders which have a high incidence of cardiac disease, such as CLN3, has demonstrated an accumulation of ceroid lipopigment [53,54,56,57]. Although the patients in our cohort did not, to our knowledge, have overt cardiac manifestations, the reduction of inclusions in both peripheral blood and skin biopsies raises the possibility that ICV therapy can have peripherally beneficial effects which are detectable by buffy coat analysis, offering a potential means to monitor non-neurological manifestations of the disease in the future.

Although only available in two patients, the skin biopsies provide a useful mechanism to correlate the peripheral blood buffy coat results with those in other cell types. The similarity in morphology of the inclusions observed in the buffy coats and in skin biopsies suggests that the buffy coats can provide a reliable insight into abnormalities in a range of cell types. However, our results also suggest that the effect of treatment may have been heterogenous across tissues, with some tissues responding better to ERT. For example, the lack of abnormal storage in endothelial cells suggests that these tissues may have responded better than sweat glands. This may reflect differences in enzyme availability between tissues. This has further implications regarding the non-neuronal effect of ICV therapy, namely the effect on cardiac pathology that is associated with CLN2 disease.

One major limitation of this study is its small sample size. However, this is a rare disease, and it is therefore difficult to obtain large cohorts for such studies. No contemporary control group is available for comparison as the ERT was delivered as part of a compassionate use program. Similarly, we were not able to obtain quantitative measurements of brain volume as this was a retrospective study and the necessary sequences were not consistently available. To address this and to increase the cohort size, we plan to collect this data routinely as part of standard care in the future.

The future of biomarker discovery arguably lies in creating a panel of biomarkers, which work synergistically to have better combined accuracy and capture more aspects of the phenotype [58]. It is increasingly clear that a single biomarker is not enough to capture and monitor both neurological and peripheral disease processes when dealing with complex neuro-metabolic diseases. A wide spectrum of technology may help to achieve this goal and could enable evaluation of disease at various scales. For example, whole organ pathology may be monitored using magnetic resonance imaging (MRI) with the use of quantitative brain MRI [59] and cardiac MRI [60]. Cellular pathology can be investigated and monitored using electron microscopy [39,40,51], whilst abnormalities in small molecules and specific proteins can be evaluated using mass spectrometry [58,61].

Approximately five new patients are diagnosed with CLN2 in the UK every year and become part of the ever-increasing cohort for the prospective follow-up study, which will include assessment of the peripheral blood buffy coat levels. This biomarker will be correlated with the continuous assessments of the CNS and peripheral organ manifestations of the disease. Clearly, buffy coat levels provide limited information on the individual organ damage and can only be used as a proxy measure. Recent publications have suggested that neurofilament light (NFL) chain can be a possible biomarker of axonal disease response in CLN2 [62,63]. All CLN2 patients with the classical form of CLN2 disease had high plasma and cerebrospinal fluid (CSF) NFL levels, which normalised after 2 years of treatment with ERT. The search for better biomarkers of central and peripheral CLN2 disease response to ERT is still in progress and we anticipate the use of a biomarker panel that will be representative of different cell- and tissue-specific responses.

5. Conclusions

We show that the percentage of cells demonstrating curvilinear inclusions in peripheral blood buffy coats undergoes a statistically significant reduction after ERT over 3 years in patients with CLN2 disease, supporting further investigation of this measurement as a biomarker of therapeutic response. Furthermore, the demonstration of a reduction in abnormal peripheral blood cells is significant because it suggests that ERT delivered directly to the CNS has peripheral tissue effects. It is important to evaluate these peripheral effects further both in order to develop non-invasive biomarkers for assessing treatment efficacy and to understand the potential positive effect of ERT on extra-CNS organ function in the long-term. Our take-home point is that the amount of abnormal storage material in lymphocytes can act as an objective marker of response to treatment in children receiving ICV ERT for CLN2 disease.

Author Contributions: Conceptualization, P.G.; methodology, P.G. and G.A.; formal analysis, P.G. and S.S.; data curation, G.A., S.S., B.C., L.L. and P.G.; writing—original draft preparation, S.S.; writing—review and editing, P.G., G.A. and R.W.; supervision, P.G. All authors have read and agreed to the published version of the manuscript.

Funding: P.G. is an NIHR senior investigator (Reference NIHR202370) and L.L. is a BDFA clinical nurse specialist. This work was supported by funding from the United Kingdom Medical Research Council grants MR/N019075/1 and MR/R026084/1 to P.G.

Institutional Review Board Statement: The study was conducted in accordance with the Declaration of Helsinki and has ethical approval (13/LO/0168; IRAS ID 95005; London-Bloomsbury Research Ethics Committee) and Health Research Authority (HRA) approval.

Informed Consent Statement: Informed consent was obtained from all subjects involved in the study.

Data Availability Statement: Not applicable.

Conflicts of Interest: P.G. receives grant funding from BioMarin but did not receive any funding associated with this study.

References

1. Mole, S.E.; Cotman, S.L. Genetics of the Neuronal Ceroid Lipofuscinoses (Batten Disease). *Biochim. Biophys. Acta (BBA) Mol. Basis Dis.* **2015**, *1852*, 2237–2241. [CrossRef] [PubMed]
2. Warrier, V.; Vieira, M.; Mole, S.E. Genetic Basis and Phenotypic Correlations of the Neuronal Ceroid Lipofusinoses. *Biochim. Biophys. Acta (BBA) Mol. Basis Dis.* **2013**, *1832*, 1827–1830. [CrossRef] [PubMed]
3. Kohlschütter, A.; Schulz, A. Towards Understanding the Neuronal Ceroid Lipofuscinoses. *Brain Dev.* **2009**, *31*, 499–502. [CrossRef]
4. Moore, S.J.; Buckley, D.J.; Macmillan, A.; Marshall, H.D.; Steele, L.; Ray, P.N.; Nawaz, Z.; Baskin, B.; Frecker, M.; Carr, S.M.; et al. The Clinical and Genetic Epidemiology of Neuronal Ceroid Lipofuscinosis in Newfoundland. *Clin. Genet.* **2008**, *74*, 213–222. [CrossRef]
5. Williams, R.E.; Adams, H.R.; Blohm, M.; Cohen-Pfeffer, J.L.; de los Reyes, E.; Denecke, J.; Drago, K.; Fairhurst, C.; Frazier, M.; Guelbert, N.; et al. Management Strategies for CLN2 Disease. *Pediatr. Neurol.* **2017**, *69*, 102–112. [CrossRef] [PubMed]
6. Sleat, D.E.; Donnelly, R.J.; Lackland, H.; Liu, C.G.; Sohar, I.; Pullarkat, R.K.; Lobel, P. Association of Mutations in a Lysosomal Protein with Classical Late- Infantile Neuronal Ceroid Lipofuscinosis. *Science* **1997**, *277*, 1802–1805. [CrossRef] [PubMed]
7. Simpson, N.A.; Wheeler, E.D.; Pearce, D.A. Screening, Diagnosis and Epidemiology of Batten Disease. *Expert Opin. Orphan Drugs* **2014**, *2*, 903–910. [CrossRef]
8. Haltia, M. The Neuronal Ceroid-Lipofuscinoses. *J. Neuropathol. Exp. Neurol.* **2003**, *62*, 1–13. [CrossRef]
9. Mole, S.E.; Williams, R.E.; Goebel, H.H. Correlations between Genotype, Ultrastructural Morphology and Clinical Phenotype in the Neuronal Ceroid Lipofuscinoses. *Neurogenetics* **2005**, *6*, 107–126. [CrossRef]
10. Radke, J.; Stenzel, W.; Goebel, H.H. Human NCL Neuropathology. *Biochim. Biophys. Acta (BBA) Mol. Basis Dis.* **2015**, *1852*, 2262–2266. [CrossRef]
11. Mole, S.E.; Haltia, M. The Neuronal Ceroid-Lipofuscinoses (Batten Disease). In *Rosenberg's Molecular and Genetic Basis of Neurological and Psychiatric Disease*, 5th ed.; Academic Press: Cambridge, MA, USA, 2015; pp. 793–808. [CrossRef]
12. Chang, M.; Cooper, J.D.; Davidson, B.L.; van Diggelen, O.P.; Elleder, M.; Goebel, H.H.; Golabek, A.A.; Kida, E.; Kohlschütter, A.; Lobel, P.; et al. CLN2. In *The Neuronal Ceroid Lipofuscinoses (Batten Disease)*; Oxford University Press: Oxford, UK, 2011; pp. 80–109. [CrossRef]

13. Schulz, A.; Kohlschütter, A.; Mink, J.; Simonati, A.; Williams, R. NCL Diseases—Clinical Perspectives. *Biochim. Biophys. Acta Mol. Basis Dis.* **2013**, *1832*, 1801–1806. [CrossRef] [PubMed]
14. Nickel, M.; Schulz, A. Natural History Studies in NCL and Their Expanding Role in Drug Development: Experiences from CLN2 Disease and Relevance for Clinical Trials. *Front. Neurol.* **2022**, *13*, 785841. [CrossRef]
15. Gardner, E.; Bailey, M.; Schulz, A.; Aristorena, M.; Miller, N.; Mole, S.E. Mutation Update: Review of *TPP1* Gene Variants Associated with Neuronal Ceroid Lipofuscinosis CLN2 Disease. *Hum. Mutat.* **2019**, *40*, 1924–1938. [CrossRef] [PubMed]
16. Lourenço, C.M.; Pessoa, A.; Mendes, C.C.; Rivera-Nieto, C.; Vergara, D.; Troncoso, M.; Gardner, E.; Mallorens, F.; Tavera, L.; Lizcano, L.A.; et al. Revealing the Clinical Phenotype of Atypical Neuronal Ceroid Lipofuscinosis Type 2 Disease: Insights from the Largest Cohort in the World. *J. Paediatr. Child Health* **2020**, *57*, 519–525. [CrossRef] [PubMed]
17. Elleder, M.; Dvořáková, L.; Stolnaja, L.; Vlášková, H.; Hůlková, H.; Druga, R.; Poupětová, H.; Košťálová, E.; Mikuláštík, J. Atypical CLN2 with Later Onset and Prolonged Course: A Neuropathologic Study Showing Different Sensitivity of Neuronal Subpopulations to TPP1 Deficiency. *Acta Neuropathol.* **2008**, *116*, 119–124. [CrossRef] [PubMed]
18. Worgall, S.; Kekatpure, M.V.; Heier, L.; Ballon, D.; Dyke, J.P.; Shungu, D.; Mao, X.; Kosofsky, B.; Kaplitt, M.G.; Souweidane, M.M.; et al. Neurological Deterioration in Late Infantile Neuronal Ceroid Lipofuscinosis. *Neurology* **2007**, *69*, 521–535. [CrossRef]
19. Nickel, M.; Jacoby, D.; Lezius, S.; Down, M.; Simonati, A.; Genter, F.; Wittes, J.; Kohlschütter, A.; Schulz, A. Natural History of CLN2 Disease: Quantitative Assessment of Disease Characteristics and Rate of Progression. *Neuropediatrics* **2016**, *47* (Suppl. S1), FV04-03. [CrossRef]
20. Nickel, M.; Simonati, A.; Jacoby, D.; Lezius, S.; Kilian, D.; Van de Graaf, B.; Pagovich, O.E.; Kosofsky, B.; Yohay, K.; Downs, M.; et al. Disease Characteristics and Progression in Patients with Late-Infantile Neuronal Ceroid Lipofuscinosis Type 2 (CLN2) Disease: An Observational Cohort Study. *Lancet Child Adolesc. Health* **2018**, *2*, 582–590. [CrossRef]
21. Specchio, N.; Bellusci, M.; Pietrafusa, N.; Trivisano, M.; de Palma, L.; Vigevano, F. Photosensitivity Is an Early Marker of Neuronal Ceroid Lipofuscinosis Type 2 Disease. *Epilepsia* **2017**, *58*, 1380–1388. [CrossRef] [PubMed]
22. Shapiro, E.G.; Klein, K.A. Dementia in Childhood: Issues in Neuropsychological Assessment with Application to the Natural History and Treatment of Degenerative Storage Diseases. *Adv. Child Neuropsychol.* **1994**, *1*, 119–171. [CrossRef]
23. Schulz, A.; Kohlschütter, A. NCL Disorders: Frequent Causes of Childhood Dementia. *Iran. J. Child Neurol.* **2013**, *7*, 1–8. [CrossRef] [PubMed]
24. Mink, J.W. Natural History Data for Childhood Neurodegenerative Disease. *The Lancet Child and Adolescent Health.* **2018**, *2*, 547–548. [CrossRef]
25. Mole, S.E.; Schulz, A.; Badoe, E.; Berkovic, S.F.; de Los Reyes, E.C.; Dulz, S.; Gissen, P.; Guelbert, N.; Lourenco, C.M.; Mason, H.L.; et al. Guidelines on the Diagnosis, Clinical Assessments, Treatment and Management for CLN2 Disease Patients. *Orphanet J. Rare Dis.* **2021**, *16*, 185. [CrossRef] [PubMed]
26. Ostergaard, J.R.; Rasmussen, T.B.; Molgaard, H. Cardiac Involvement in Juvenile Neuronal Ceroid Lipofuscinosis (Batten Disease). *Neurology* **2011**, *76*, 1245–1251. [CrossRef] [PubMed]
27. Fukumura, S.; Saito, Y.; Saito, T.; Komaki, H.; Nakagawa, E.; Sugai, K.; Sasaki, M.; Oka, A.; Takamisawa, I. Progressive Conduction Defects and Cardiac Death in Late Infantile Neuronal Ceroid Lipofuscinosis. *Dev. Med. Child Neurol.* **2011**, *54*, 663–666. [CrossRef]
28. Specchio, N.; Pietrafusa, N.; Trivisano, M. Changing Times for CLN2 Disease: The Era of Enzyme Replacement Therapy. *Ther. Clin. Risk Manag.* **2020**, *16*, 213–222. [CrossRef] [PubMed]
29. Mazurkiewicz-Bełdzińska, M.; del Toro, M.; Haliloğlu, G.; Huidekoper, H.H.; Kravljanac, R.; Mühlhausen, C.; Andersen, B.N.; Prpić, I.; Striano, P.; Auvin, S. Managing CLN2 Disease: A Treatable Neurodegenerative Condition among Other Treatable Early Childhood Epilepsies. *Expert Rev. Neurother.* **2021**, *21*, 1275–1282. [CrossRef] [PubMed]
30. Labbé, E.E.; Lopez, I.; Murphy, L.; O'Brien, C. Optimism and Psychosocial Functioning in Caring for Children with Battens and Other Neurological Diseases. *Psychol. Rep.* **2002**, *90*, 1129. [CrossRef]
31. Schulz, A.; Ajayi, T.; Specchio, N.; de Los Reyes, E.; Gissen, P.; Ballon, D.; Dyke, J.P.; Cahan, H.; Slasor, P.; Jacoby, D.; et al. Study of Intraventricular Cerliponase Alfa for CLN2 Disease. *N. Engl. J. Med.* **2018**, *378*, 1898–1907. [CrossRef]
32. Wibbeler, E.; Wang, R.; de los Reyes, E.; Specchio, N.; Gissen, P.; Guelbert, N.; Nickel, M.; Schwering, C.; Lehwald, L.; Trivisano, M.; et al. Cerliponase Alfa for the Treatment of Atypical Phenotypes of CLN2 Disease: A Retrospective Case Series. *J. Child Neurol.* **2020**, *129*, S161–S162. [CrossRef]
33. Espitia Segura, O.M.; Hernández, Z.; Mancilla, N.I.; Naranjo, R.A.; Tavera, L. Real World Effectiveness of Cerliponase Alfa in Classical and Atypical Patients. A Case Series. *Mol. Genet. Metab. Rep.* **2021**, *27*, 100718. [CrossRef]
34. Wyrwich, K.W.; Schulz, A.; Nickel, M.; Slasor, P.; Ajayi, T.; Jacoby, D.R.; Kohlschütter, A. An Adapted Clinical Measurement Tool for the Key Symptoms of CLN2 Disease. *J. Inborn Errors Metab. Screen.* **2018**, *6*, 232640981878838. [CrossRef]
35. Bellettato, C.M.; Scarpa, M. Pathophysiology of Neuropathic Lysosomal Storage Disorders. *J. Inherit. Metab. Dis.* **2010**, *33*, 347–362. [CrossRef] [PubMed]
36. Anderson, G.; Smith, V.V.; Malone, M.; Sebire, N.J. Blood Film Examination for Vacuolated Lymphocytes in the Diagnosis of Metabolic Disorders; Retrospective Experience of More than 2500 Cases from a Single Centre. *J. Clin. Pathol.* **2005**, *58*, 1305–1310. [CrossRef]
37. Bajel, A.; Bazargan, A.; Doery, J.; Roy, S. Lymphocyte Vacuolation: Clue to Inherited Metabolic Disease. *Pathology* **2010**, *42*, 699–700. [CrossRef] [PubMed]

38. Baumann, R.J.; Markesbery, W.R. Juvenile Amaurotic Idiocy (Neuronal Ceroid Lipofuscinosis) and Lymphocyte Fingerprint Profiles. *Ann. Neurol.* **1978**, *4*, 531–536. [CrossRef] [PubMed]
39. Anderson, G.W.; Smith, V.V.; Brooke, I.; Malone, M.; Sebire, N.J. Diagnosis of Neuronal Ceroid Lipofuscinosis (Batten Disease) by Electron Microscopy in Peripheral Blood Specimens. *Ultrastruct. Pathol.* **2006**, *30*, 373–378. [CrossRef]
40. Anderson, G.W.; Goebel, H.H.; Simonati, A. Human Pathology in NCL. *Biochim. Biophys. Acta Mol. Basis Dis.* **2013**, *1832*, 1807–1826. [CrossRef] [PubMed]
41. Markesbery, W.R. Late-Infantile Neuronal Ceroid-Lipofuscinosis. *Arch. Neurol.* **1976**, *33*, 630. [CrossRef]
42. Vogler, C.; Rosenberg, H.S.; Williams, J.C.; Butler, I. Electron Microscopy in the Diagnosis of Lysosomal Storage Diseases. *Am. J. Med. Genet. Suppl.* **1987**, *3*, 243–255. [CrossRef] [PubMed]
43. Zeman, W.; Dyken, P. Neuronal Ceroid-Lipofuscinosis (Batten's Disease): Relationship to Amaurotic Family Idiocy? *Pediatrics* **1969**, *44*, 570–583. [CrossRef] [PubMed]
44. Fietz, M.; AlSayed, M.; Burke, D.; Cohen-Pfeffer, J.; Cooper, J.D.; Dvořáková, L.; Giugliani, R.; Izzo, E.; Jahnová, H.; Lukacs, Z.; et al. Diagnosis of Neuronal Ceroid Lipofuscinosis Type 2 (CLN2 Disease): Expert Recommendations for Early Detection and Laboratory Diagnosis. *Mol. Genet. Metab.* **2016**, *119*, 160–167. [CrossRef] [PubMed]
45. Kousi, M.; Lehesjoki, A.E.; Mole, S.E. Update of the Mutation Spectrum and Clinical Correlations of over 360 Mutations in Eight Genes That Underlie the Neuronal Ceroid Lipofuscinoses. *Hum. Mutat.* **2012**, *33*, 42–63. [CrossRef]
46. Katz, M.L.; Coates, J.R.; Sibigtroth, C.M.; Taylor, J.D.; Carpentier, M.; Young, W.M.; Wininger, F.A.; Kennedy, D.; Vuillemenot, B.R.; O'Neill, C.A. Enzyme Replacement Therapy Attenuates Disease Progression in a Canine Model of Late-infantile Neuronal Ceroid Lipofuscinosis (CLN2 Disease). *J. Neurosci. Res* **2014**, *92*, 1591–1598. [CrossRef]
47. Schwering, C.; Kammler, G.; Wibbeler, E.; Christner, M.; Knobloch, J.K.-M.; Nickel, M.; Denecke, J.; Baehr, M.; Schulz, A. Development of the "Hamburg Best Practice Guidelines for ICV−Enzyme Replacement Therapy (ERT) in CLN2 Disease" Based on 6 Years Treatment Experience in 48 Patients. *J. Child Neurol.* **2021**, *36*, 635–641. [CrossRef]
48. Fagerland, J.A. Preparation of Buffy Coat Leukocytes for Transmission Electron Microscopy. *Microsc. Today* **1999**, *7*, 29. [CrossRef]
49. Choudhary, O.P.; Sarkar, R.; Priyanka; Chethan, G.E.; Doley, P.J.; Chandra Kalita, P.; Kalita, A. Preparation of Blood Samples for Electron Microscopy: The Standard Protocol. *Ann. Med. Surg.* **2021**, *70*, 102895. [CrossRef] [PubMed]
50. Koster, A.J.; Klumperman, J. Electron Microscopy in Cell Biology: Integrating Structure and Function. *Nat. Rev. Mol. Cell Biol.* **2003**, *4*, SS6–SS9. Available online: https://pubmed.ncbi.nlm.nih.gov/14587520/ (accessed on 5 January 2023).
51. Kuper, W.F.E.; Oostendorp, M.; van den Broek, B.T.A.; van Veghel, K.; Nonkes, L.J.P.; Nieuwenhuis, E.E.S.; Fuchs, S.A.; Veenendaal, T.; Klumperman, J.; Huisman, A.; et al. Quantifying Lymphocyte Vacuolization Serves as a Measure of CLN3 Disease Severity. *JIMD Rep.* **2020**, *54*, 87–97. [CrossRef]
52. Dolman, C.L.; MacLeod, P.M.; Chang, E. Skin Punch Biopsies and Lymphocytes in the Diagnosis of Lipidoses. *Can. J. Neurol. Sci.* **1975**, *2*, 67–73. [CrossRef]
53. Dunnett, C.W. A Multiple Comparison Procedure for Comparing Several Treatments with a Control. *J. Am. Stat. Assoc.* **1955**, *50*, 1096–1121. [CrossRef]
54. Friedman, M. The Use of Ranks to Avoid the Assumption of Normality Implicit in the Analysis of Variance. *J. Am. Stat. Assoc.* **1937**, *32*, 675–701. [CrossRef]
55. Wisniewski, K.E. The Diagnostic Value of Ultrastructural Studies of Skin-Punch Biopsies and Buffy Coat for the Early Diagnosis of Some Neurodegenerative Diseases. *Ann. N. Y. Acad. Sci.* **1986**, *477*, 285–311. [CrossRef] [PubMed]
56. Martin, J.J.; de Groote, C. Involvement of the Skin in Late Infantile and Juvenile Amaurotic Idiocies (Neuronal Ceroid-Lipofuscinoses). *Pathol. Eur.* **1974**, *9*, 263–272. Available online: https://europepmc.org/article/MED/4457780 (accessed on 6 January 2023).
57. Steinfeld, R.; Heim, P.; Von Gregory, H.; Meyer, K.; Ullrich, K.; Goebel, H.H.; Kohlschütter, A. Late Infantile Neuronal Ceroid Lipofuscinosis: Quantitative Description of the Clinical Course in Patients with CLN2 Mutations. *Am. J. Med. Genet.* **2002**, *112*, 347–354. [CrossRef]
58. Papandreou, A.; Doykov, I.; Spiewak, J.; Komarov, N.; Habermann, S.; Kurian, M.A.; Mills, P.B.; Mills, K.; Gissen, P.; Heywood, W.E. Niemann−Pick Type C Disease as Proof-of-concept for Intelligent Biomarker Panel Selection in Neurometabolic Disorders. *Dev. Med. Child Neurol.* **2022**, *64*, 1539–1546. [CrossRef]
59. Baker, E.H.; Levin, S.W.; Zhang, Z.; Mukherjee, A.B. MRI Brain Volume Measurements in Infantile Neuronal Ceroid Lipofuscinosis. *Am. J. Neuroradiol.* **2017**, *38*, 376–382. [CrossRef]
60. Todiere, G.; della Vecchia, S.; Morales, M.A.; Barison, A.; Ricca, I.; Tessa, A.; Colombi, E.; Santorelli, F.M. Cardiac Magnetic Resonance Findings in Neuronal Ceroid Lipofuscinosis: A Case Report. *Front. Neurol.* **2022**, *13*, 942667. [CrossRef]
61. Brudvig, J.J.; Swier, V.J.; Johnson, T.B.; Cain, J.C.; Pratt, M.; Rechtzigel, M.; Leppert, H.; Dang Do, A.N.; Porter, F.D.; Weimer, J.M. Glycerophosphoinositol Is Elevated in Blood Samples from *CLN3* $^{\Delta ex7-8}$ Pigs, *Cln3* $^{\Delta ex7-8}$ Mice, and CLN3-Affected Individuals. *Biomark. Insights* **2022**, *17*, 117727192211077. [CrossRef]

62. Iwan, K.; Patel, N.; Heslegrave, A.; Borisova, M.; Lee, L.; Bower, R.; Mole, S.E.; Mills, P.B.; Zetterberg, H.; Mills, K.; et al. Cerebrospinal Fluid Neurofilament Light Chain Levels in CLN2 Disease Patients Treated with Enzyme Replacement Therapy Normalise after Two Years on Treatment. *F1000Res* **2022**, *10*, 614. [CrossRef]
63. Ru, Y.; Corado, C.; Soon, R.K.; Melton, A.C.; Harris, A.; Yu, G.K.; Pryer, N.; Sinclair, J.R.; Katz, M.L.; Ajayi, T.; et al. Neurofilament Light Is a Treatment-responsive Biomarker in CLN2 Disease. *Ann. Clin. Transl. Neurol.* **2019**, *6*, 2437–2447. [CrossRef] [PubMed]

pharmaceuticals

Review

Efficacy, Tolerability, and Safety of Toludesvenlafaxine for the Treatment of Major Depressive Disorder—A Narrative Review

Octavian Vasiliu

Department of Psychiatry, Dr. Carol Davila University Emergency Central Military Hospital, 010825 Bucharest, Romania; octavvasiliu@yahoo.com

Abstract: The estimated rate of treatment-resistant major depressive disorder (TRD) remains higher than 30%, even after the discovery of multiple classes of antidepressants in the last 7 decades. Toludesvenlafaxine (ansofaxine, LY03005, or LPM570065) is a first-in-class triple monoaminergic reuptake inhibitor (TRI) that has reached clinical use. The objective of this narrative review was to summarize clinical and preclinical evidence about the efficacy, tolerability, and safety of toludesvenlafaxine. Based on the results of 17 reports retrieved in the literature, the safety and tolerability profiles of toludesvenlafaxine were good in all clinical trials, and the pharmacokinetic parameters were well described in the phase 1 trials. The efficacy of toludesvenlafaxine was demonstrated in one phase 2 and one phase 3 trial, both on primary and secondary outcomes. In conclusion, this review highlights the favorable clinical results of toludesvenlafaxine in only two short-term trials that enrolled patients with major depressive disorder (MDD) (efficacy and tolerability were good for up to eight weeks), indicating the need for more good quality, larger-sample, and longer-term trials. Exploring new antidepressants, such as TRI, can be considered a priority for clinical research due to the high rates of TRD, but also due to the significant percentages of relapse in patients with MDD.

Keywords: toludesvenlafaxine; ansofaxine; LY03005; LPM570065; triple monoamine reuptake inhibitors; major depressive disorders; efficacy; tolerability; treatment-resistant depression; first-in-class antidepressant

Citation: Vasiliu, O. Efficacy, Tolerability, and Safety of Toludesvenlafaxine for the Treatment of Major Depressive Disorder—A Narrative Review. *Pharmaceuticals* **2023**, *16*, 411. https://doi.org/10.3390/ph16030411

Academic Editors: Masaru Tanaka, Lydia Giménez-Llort, Simone Battaglia, Chong Chen and Piril Hepsomali

Received: 23 January 2023
Revised: 5 March 2023
Accepted: 6 March 2023
Published: 8 March 2023

1. Introduction

In the context of the constant debate in the literature about the unmet needs of patients with major depressive disorder (MDD), which usually revolves around the high incidence of treatment-resistant depression (TRD) and the negative functional consequences of this resistance [1], the urgency of finding new antidepressants becomes obvious. Another important aspect in the treatment of MDD and especially TRD is represented by the "antidepressant polypharmacy" (i.e., concomitant administration of ≥2 antidepressants), a strategy considered to improve the outcomes by activating multiple neurotransmitter pathways at the same time [2]. Combining multiple antidepressants or adding different agents to the ongoing antidepressant in order to improve its efficacy or to speed its onset is not without risks, and adverse events might appear more often in these patients, especially in vulnerable populations (i.e., patients with late-life depression, multiple somatic or psychiatric comorbidities, etc.) [3–7]. The risk of negative pharmacokinetic and pharmacodynamic interactions also needs to be considered when multiple medications are concomitantly administered in patients with MDD, potentially triggering more frequent treatment tolerability and tolerance problems than in the case of monotherapy [5,8–10].

The monoamine hypothesis was launched in the 1950s, and it stipulates that the deficit of serotonin, dopamine, and norepinephrine in the central nervous system (CNS) is the pathophysiological basis of depression [11–13]. Based on this hypothesis, antidepressants can raise the levels of monoamine neurotransmitters in the brain and, as a consequence, effectively decrease the severity of depressive symptoms [11]. *Serotonin* has been presumed to have a causal role in the pathogenesis of depression based on multiple pieces of evidence:

tryptophan depletion, which is an acute dietary manipulation, leads to lower CNS serotonin activity and may produce brief, but clinically significant, depressive symptoms in recovered depressed patients; peripheral inflammation has been involved in the onset of depression in vulnerable individuals by lowering plasma tryptophan level; serotoninergic antidepressants have a long history of clinical use, and they have traditionally been associated with serotonin pathway activation; and serotonin deficiency has been related to anxiety, obsessions, and compulsions [11–19]. *Norepinephrine* is considered a key element in the pathogenesis of MDD because locus coeruleus is the major anatomical target for monoaminergic antidepressants; a rapid decrease in the catecholamine level in the CNS has been associated with relapses of MDD; the mechanisms of the inhibition of monoamine oxidase, antagonism of presynaptic norepinephrine receptors, and inhibition of norepinephrine reuptake from the synaptic cleft increase the catecholamine neurotransmission and improve the depressive symptoms; and norepinephrine deficits have been related to the onset of low energy, inattention, executive dysfunction, and decreased alertness [20–23]. *Dopamine* neurotransmission dysfunctions are considered responsible for core symptoms of depression, e.g., anhedonia and cognitive deficits; translational and clinical studies demonstrate deficits of the dopaminergic system in depression; and antidepressants enhancing dopaminergic transmission improved symptoms of energy, pleasure, and interest in MDD patients [24–27].

This pathogenetic model of depression has evolved since its initial formulation, including now adaptive changes in receptors and the participation of second messengers/signal transduction pathways (cAMP, phosphokinase A, cAMP-response element-binding protein, neurotrophin-mediated pathway, etc.), elements which are explored in connection with the original hypothesis [28,29]. Although heavily criticized from diverse perspectives, e.g., not explaining the latency of response to antidepressants, not answering the bothersome question of why there is still a high rate of nonresponse even after multiple trials of antidepressants, etc. [29,30], the monoamine hypothesis is still productive, and new agents are researched starting from its core concept [31]. Moreover, it is important to observe that new pathogenetic models for depression have been formulated in the last decades (e.g., based on glutamatergic neurotransmission and neuroplasticity), and new antidepressants, with a faster onset of their therapeutic action, have already been marketed, i.e., brexanolone and esketamine [29,32,33]. As observed by several authors, "in the era of neural networks and systems-level neuroscience, 'single' neurotransmitter theories of depression look increasingly implausible" [34].

Translational research is expected to play an important role in finding new pathophysiological models for MDD and, therefore, new therapeutic agents. In this context, preclinical research is essential for connecting theoretical knowledge with clinical data by exploring various paradigms, such as genetic models, acute/chronic stress models, brain neurotransmitters/specific brain injury models, and models induced by pharmacological interventions [35–38]. Although none of these models completely reproduce human MDD (i.e., face validity, construct validity, and predictive validity are still far from being perfect), they are considered relevant because of their capacity to mimic certain features of this pathological condition and thus allowing for the systematic investigation of etiology, pathogenesis, and treatment of depression [36,39]. For example, the antidepressant effects of kynurenic acid, an L-tryptophan metabolite with neuromodulatory activities, have been explored in preclinical studies [40]. In a modified forced swimming test in mice, kynurenic acid successfully reversed immobility, climbing, and swimming times [40]. Moreover, this study showed that the antidepressant effects of kynurenic acid are modeled by this pathway's interactions with GABA-ergic (through GABA-A receptors), dopaminergic (i.e., D2, D3, and D4 receptors), and serotonergic (i.e., 5-HT2 receptors) neurotransmission [40]. The stress vulnerability model has been proven useful in separating depressive-like and anxious-like phenotypes during benzodiazepine administration in mice; this model has further supported the utility of brain-derived neurotrophic factor (BDNF) in developing neural and behavioral plasticity and the effects of antidepressants [39,41]. The chronic unpredictable mild stress model is another robust tool for the pharmacological research of MDD, and various antidepressants have been explored using this paradigm; for example, antidepressant-induced reduction in hippocampal cytogenesis has been correlated with symptoms amelioration in a study investigating the effects of escitalopram in rats with this chronic stress model [39,42].

Triple reuptake inhibitors (TRI) are a class of antidepressants that target the serotonin, norepinephrine, and dopamine neurotransmission simultaneously (Figure 1), thus representing a step forward in treating MDD, at least at the conceptual level [11,43]. It is expected that by administering these agents, the need to augment antidepressants that enhance the activity of only one or two monoaminergic systems (e.g., the selective serotonin reuptake inhibitors, or serotonin and norepinephrine reuptake inhibitors) would be less frequently met. In turn, this would present multiple possible advantages, such as the elimination of potentially negative pharmacokinetic interactions; decreasing the risks for adverse events or toxic reactions due to polypharmacy; reducing the latency of therapeutic effect by eliminating the need for a monotherapy trial followed by the addition of/switching to another agent; and mitigating the adverse effects due to serotonin activation by adding dopamine-activatory features, etc. (Figure 2).

Many agents within the TRI class have been described in the last two decades, some of them reaching the clinical stage of research while others being discontinued in the preclinical phase. An exhaustive presentation of these agents is beyond the aim of this review, and it has been carried out very well elsewhere [44,45]. It is also to be noted that TRIs have been explored in clinical trials for different indications, starting from MDD to attention-deficit hyperactivity disorder (ADHD), and from pain disorder to cocaine use disorder [44,45].

To illustrate the complexity of this pharmacological category, several agents investigated for their antidepressant effect will be presented further, emphasizing their pharmacodynamic profile and results from clinical/preclinical studies. For example, DOV 216,303 is a TRI with IC_{50} values of 14 nM, 20 nM, and 78 nM for serotonin, norepinephrine, and dopamine uptake, which was explored in healthy volunteers and was found safe and well tolerated; in a phase 2 study, this agent was compared with citalopram in patients with MDD and both antidepressants decreased the Hamilton Depression Rating Scale (HAMD) scores compared with the baseline; no placebo arm existed in this phase 2 trial [46].

DOV 21,947 (amitifadine, EB-1010) is the (+) enantiomer of DOV-216,303 and it is a serotonin-preferring TRI with a potency to block the reuptake of serotonin, norepinephrine, and dopamine of 12 nM, 23 nM, and 96 nM, respectively; in a phase 2 study, the change in Montgomery–Asberg Depression Rating Scale (MADRS) scores was significant vs. placebo at week 6, and the tolerability was good; and in a phase 2b/3a study, the efficacy was not confirmed in patients with MDD at 6 weeks [47,48].

Ro 8-4650 (Diclofensine) is a TRI with IC_{50} for serotonin, norepinephrine, and dopamine of 4.8 µM, 2.5 µM, and 4.5 µM, respectively; it was found to be well tolerated in studies with healthy volunteers and MDD patients, with an efficacy comparable to that of maprotiline [49,50].

SKF83959 is a competitive inhibitor of serotonin transporter and a non-competitive blocker of norepinephrine and dopamine transporters, with K_i of 1.43 ± 0.45 µmol/L, 0.6 ± 0.07 µmol/L, and 9.01 ± 0.8 µmol/L, respectively. In preclinical studies (i.e., the chronic social defeat stress model of depression), this molecule has proven antidepressant effects and also mitigated the decrease in hippocampal brain-derived neurotrophic factor (BDNF) signaling pathway, dendritic spine density, and neurogenesis induced by stress [51,52].

BMS-820836 is another TRI with IC_{50} values of 0.2 nM, 26.7 nM, and 6.19 nM for serotonin, norepinephrine, and dopamine transporters, respectively. In two randomized, phase 2b studies, this investigational product was well tolerated in patients with TRD during six weeks, but failed to show superiority versus continuation of an existing antidepressant (i.e., duloxetine or citalopram) [53,54].

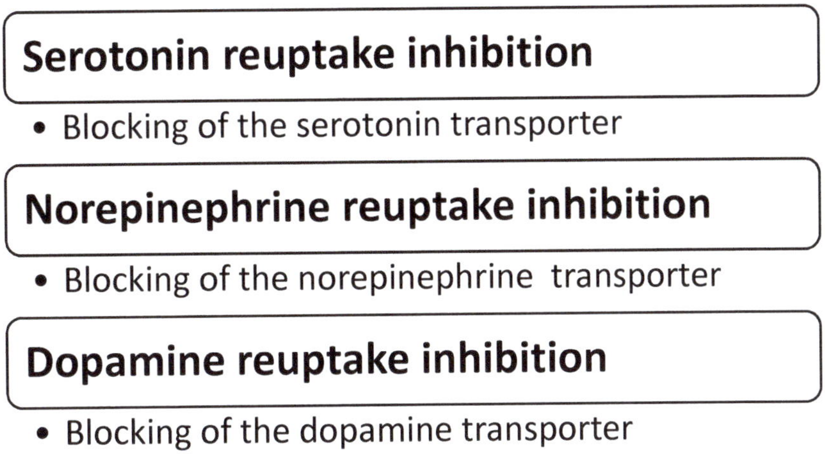

Figure 1. Mechanisms of action for TRI as antidepressants.

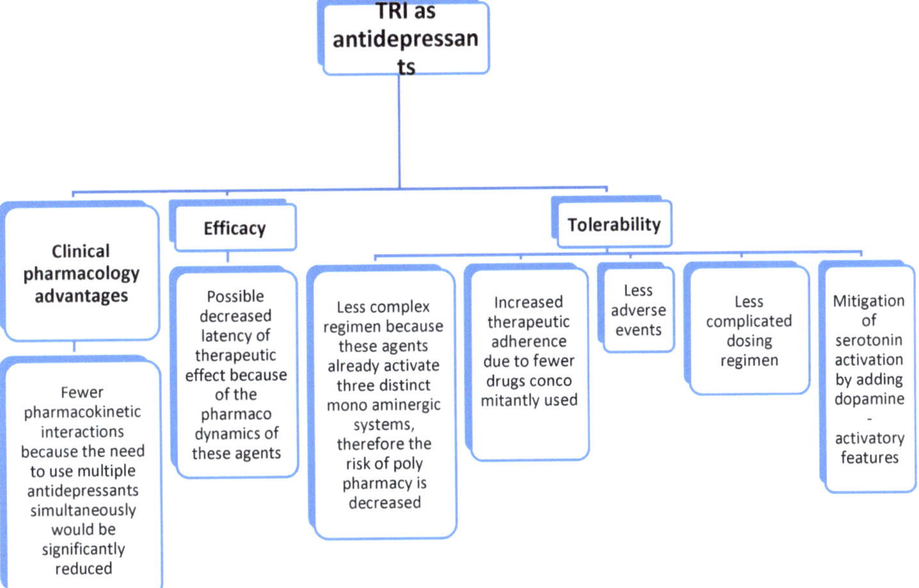

Figure 2. Potential benefits of TRI as antidepressants use in clinical practice.

Regarding the currently marketed antidepressants, which target all three monoamine transporters in different degrees, it is considered useful to compare at least some of them with TRIs from the perspective of their affinity for these transporters. For example, the inhibition of serotonin, norepinephrine, and dopamine transporters in the case of amitriptyline provides IC_{50} values of 67 nM, 63 nM, and 7500 nM, respectively; for duloxetine, the values are 13 nM, 42 nM, and 439 nM; for venlafaxine the values are 145 nM, 2483 nM, and 7647 nM; and for milnacipran the values are 151, 200, and >100,000, respectively [55]. In the case of desvenlafaxine, the IC_{50} was 47.3 ± 19.4 and 531.3 ± 113 nM for serotonin and norepinephrine transporters, respectively [56]. These distinct affinities for monoamine transporters are important for the clinical effects of these agents and explain their use not only in the treatment of MDD or TRD but also in fibromyalgia, diabetic peripheral neuropathy, migraine, anxiety disorders, etc. [57]. Moreover, these pharmacodynamic profiles have an impact on the tolerability of the explored drugs; for example, serotonergic activity enhancement may cause sexual dysfunctions, nausea, and sleep dysfunctions, while the activation of noradrenaline pathways can cause urinary hesitancy and insomnia, but are less likely to induce sexual dysfunction [7,58–60]. The activation of dopamine neurotransmission has been explored in relation to an addictive potential (which was found not significant for antidepressants administered in usual doses) [59].

Toludesvenlafaxine (LY03005, LPM570065, ansofaxine, and anshufaxine) is a TRI that is able to block the reuptake of serotonin, dopamine, and norepinephrine in the CNS [61,62]. The formula of this drug is 4-methyl benzoate desvenlafaxine hydrochloride, and it could be converted rapidly into desvenlafaxine in vivo, but both substances may coexist in the brain due to their liposolubility [62]. The product was authorized for marketing in China as toludesvenlafaxine hydrochloride by the National Medicinal Products Administration (NMPA) in November 2022, and it is the only available triple monoamine reuptake inhibitor (first-in-class product) [63]. As stated in a press release, "the application for marketing authorization is based on clinical data generated from six clinical studies in China" [64]. Moreover, a New Drug Application (NDA) file was reviewed and accepted by the Food and Drug Administration (FDA) in 2020 for LY03005, and a phase 1 study with this pharmacological agent has been completed in Japan [64,65]. Ansofaxine has been explored in clinical trials in the United States, Japan, the European Union, and China [66].

The importance of TRI investigation is derived from theoretical and pragmatic reasons. As previously mentioned, the monoamine hypothesis is under extensive scrutiny for its clinical limitations, and new theoretical frameworks for MDD have been formulated, based on other neurotransmitters' dysfunction, gut microbiome alterations, inflammatory factors, immune or hypothalamic–pituitary–adrenal axis dysregulation, and neurogenesis or cholesterol biosynthesis pathway impairments [29,67–71]. Still, the monoaminergic system is considered a central element in the pathogenesis of MDD, although multiple interactions with other CNS pathways are also important for the understanding of this disorder [72,73]. At a practical level, despite quite a large armamentarium of pharmacological agents for MDD, there is an important percentage of patients who did not reach remission or have a high relapse rate [74,75]. Therefore, the validation through clinical research of new treatments for MDD is paramount for the quality of life, clinical evolution, and prognosis of these patients.

The main objective of this narrative review was to search for data regarding the pharmacological and clinical properties of a new TRI, toludesvenlafaxine. Efficacy, tolerability, and safety, as well as the pharmacological profile of this TRI, were considered important elements to integrate into the case management of patients with MDD or TRD.

A narrative review dedicated to finding evidence for the efficacy and tolerability of toludesvenlafaxine has been conducted by searching five electronic databases (PubMed, Cochrane, Clarivate/Web of Science, Google Scholar, and EMBASE) using the paradigm "toludesvenlafaxine" OR "ansofaxine" OR "LY03005" OR "LPM570065" AND "efficacy" OR "tolerability" OR "safety" OR "pharmacology" OR "clinical studies" OR "preclinical studies". Moreover, references within the main reviewed papers were searched for supplementary information when it was considered consistent with the objective of this review.

Due to the novelty of the researched pharmacological agent, grey literature was included according to the Luxembourg definition [76]. More specifically, three sources of information were targeted: (a) main repositories of clinical trials run by the United States National Library of Medicine and the National Institutes of Health—www.clinicaltrials.gov (accessed on 26 December 2022), World Health Organization (International Clinical Trials Registry Platform)—www.who.int/clinical-trials-registry-platform (accessed on 26 December 2022), the European Union (EU Clinical Trial Register)—www.clinicaltrialsregister.eu (accessed on 26 December 202), and the Chinese Clinical Trial Registry (ChiCTR)—http://www.chictr.org.cn/index.aspx (accessed on 26 December 2022); and (b) industry and commercial press releases containing references to ansofaxine/toludesvenlafaxine; (c) annual reports, news articles, presentations, and other non-peer-reviewed materials. For the grey literature, all sources including references to "ansofaxine", "LY03005", "LPM570065", or "toludesvenlafaxine" were explored.

No inferior time limit was established for the retrieved papers included in the review, while the superior limit was December 2022. Moreover, no limitations regarding the language of the reports, the study environment (inpatient, outpatient, or mixed milieu for clinical trials), the population characteristics, or the type of research (preclinical or clinical studies) were established. As exclusion criteria, (1) reports that could not be attributed with certainty to the individual author(s), research institutes, commercial entities, and governmental or non-governmental institution(s), and (2) sources not specifying clinical, preclinical, or pharmacological data about toludesvenlafaxine were excluded.

2. Preclinical and Clinical Data on the Efficacy, Tolerability, Safety, and Pharmacological Profile of Toludesvenlafaxine

Based on the analysis of the retrieved data, 5 preclinical studies and 12 clinical studies were identified in the literature. However, while all the preclinical studies had published results (Table 1), the clinical trials presented a more nuanced status—only two clinical trials had published results, five trials had results disclosed by the manufacturer, but not published in peer-reviewed journals, while another five trials had undisclosed results (Table 2).

2.1. Preclinical Studies

A preclinical study explored the acute (single dose administration) and long-term effects of LPM570065 in Sprague–Dawley rats, concluding the following: (a) the maximum tolerated dose in the acute-administration study was 500 mg/kg, and the lethal dose was 1000 mg/kg; (b) the 13-week study led to no significant adverse events in doses higher than 300 mg/kg for rats, with no mutagenic or clastogenic effects; (c) the maximum tolerated dose in clinical conditions was deduced to be 300 mg/day; and (d) the sexual functioning in patients who might receive this agent should be monitored because changes in the prolactin and testosterone levels were detected in this preclinical study [77].

Another preclinical study explored the effects of LPM570065 on fertility and early embryonic development in Sprague–Dawley rats, concluding that (a) no observable adverse effect level was established at 100 mg (female rats) and 300 mg/kg (male rats); (b) the same parameter determined this time for fertility and early embryonic development was established at 300 mg/kg (female rats) and 100 mg/kg (male rats) [78].

In another preclinical study, acute administration of LPM570065 or desvenlafaxine was initiated in rats using an oral or intravenous solution [79]. High-performance liquid chromatography analysis showed that ansofaxine can rapidly penetrate the striatum and is converted into desvenlafaxine while presenting larger total exposure vs. desvenlafaxine [79]. Long-term administration of ansofaxine (up to 14 days) via the oral route increased all three monoamine levels more than desvenlafaxine, especially dopamine levels (detected by microdialysis) [79]. During the forced swim test, acute and chronic administration of ansofaxine decreased the immobility time more than desvenlafaxine, suggesting a higher efficacy and/or a more rapid onset of antidepressant effect than desvenlafaxine [79].

In a "two-hit" stress mouse model (early life maternal separation and social defeat stress), three behavioral models were applied—sucrose preference test, tail suspension test, and forced swimming test—in adult mice receiving LPM570065 [80]. According to this animal study, ansofaxine significantly reversed depressive-like behaviors in all three paradigms used and increased the density of dendritic spines in the CA1 hippocampal neurons [80]. LPM570065 also reduced the hypermethylation of the oxytocin receptor gene (*Oxtr*) in the hippocampus of mice experiencing the "two hits" stress [80]. This last observation suggests that LPM570065 may reduce depression vulnerability via epigenetic mechanisms involving the *Oxtr* expression [80].

Toludesvenlafaxine has a high binding affinity for serotonin, norepinephrine, and dopamine transporters and significantly inhibited the reuptake of all three monoamines-

$IC_{50} = 31.4 \pm 0.4$ nM for serotonin, $IC_{50} = 586.7 \pm 83.6$ nM for norepinephrine, and $IC_{50} = 733.2 \pm 10.3$ nM for dopamine in vitro (serotonin:norepinephrine:dopamine = 23.3:1.2:1) [46]. The highest inhibition for serotonin transporters was reported in in vitro assays [46]. The antidepressant effects were observed in rodent models at 8–16 mg/kg [46]. The absorption was good after oral administration, and it was converted to O-desvenlafaxine due to the action of esterases in vivo, both reaching the hypothalamus in high concentration [81]. The plasma exposure was proportional to the dose after oral administration [81]. While desvenlafaxine does not increase the striatal level of dopamine, toludesvenlafaxine has this effect, which indicates supplementary benefits vs. the older drug [81]. The preclinical data do not support the existence of CNS excessive activation (manifested as irritation or hyperreactivity)/depression (reflected in sedation or muscle relaxation) and no significant abuse potential [77,81].

Table 1. Summary of the preclinical studies focused on toludesvenlafaxine.

Design	Results	Observations	Reference
Single and 13-week repeated-dose oral toxicity assessment and mutagenicity assays. Acute dose: 500 mg/kg, 1000 mg/kg, and 2000 mg/kg LPM570065 in SD rats. A 13-week toxicity study: 30 mg/kg, 100 mg/kg, or 300 mg/kg LPM570065 for 13 consecutive weeks + 4-week recovery period. N = 80 rats (40 males and 40 females).	In a single-dose acute study: 2 out of 20 rats died in the 1000 mg/kg group vs. seven out of 20 in the 2000 mg/kg group vs. none in the 500 mg/kg group. In the 13-week toxicity study: transitory salivation and minor body weight decrease was reported in the 300 mg/kg group in males. Serum PRL levels ↓ by 43% and 78% in male rats in 100 mg/kg and 300 mg/kg groups, respectively. Serum TST ↑ by 37% in the 30 mg/kg and 100 mg/kg males.	MTD = 500 mg/kg and lethal dose = 1000 mg/kg in the acute administration. In the long-term administration, no observed AE level was ≥300 mg/kg for rats; no mutagenic or clastogenic effects. MTD = 3000 mg/patient/day in clinical conditions. The effects of LPM570065 on sexual function are to be monitored.	Li C, Jiang W, Gao Y, et al. [77]
Acute phase: 30 mg/kg, 100 mg/kg, and 300 mg/kg LPM570065 vs. control. Female rats received 2 weeks of the investigational product + mating up to the 7th gestation day. Male rats received 4 weeks of investigational product + mating with treated female rats. Following this stage, all males were treated up to the ninth week and a new mating period was initiated with non-treated female rats. Mortality, toxicity symptoms, body weight, amount of food consumed, sexual cycle, mating behavior, pregnancy, sperm production, gross necropsy, and weight of organs. N = 264 rats were distributed in 4 groups (44 females and 22 males in each group).	Excessive salivation post-treatment in all females and males on 100 mg/kg and 300 mg/kg LPM570065 groups. BW gain ↓ in gravid rats with 300 mg/kg investigational product during gestation days 0–6. Decreased fertility rates were associated with a 300 mg/kg dose of investigational product in male rates. Sperm concentration and count were higher in all three groups treated with LPM570065 vs. controls. Duration of mating ↓ significantly to 37.5% after 9 weeks of treatment with 300 mg/kg.	The no observable AE level was established at 100 mg/kg (female rats) and 300 mg/kg (male rats). The no observable AE level for fertility and early embryonic development was established at 300 mg/kg (female rats) and 100 mg/kg (male rats).	Guo W, Gao Y, Jiang W, et al. [78]
Exploring extracellular 5HT, NE, and DA levels in the rat striatum after acute and chronic administration of LPM570065 vs. DSVLFX. The methods used were HPLC and microdialysis. N = 72 rats divided into 9 equal groups.	HPLC results showed that LPM570065 rapidly penetrates the striatum and converts into DSVLFX while presenting larger total exposure vs. DSVLFX. Long-term administration of LPM570065 (up to 14 days) via the oral route increased all three monoamine levels more than DSVLFX, and especially dopamine levels (detected by microdialysis). During the forced swim test, acute and chronic administration of LPM570065 ↓ the immobility time more than DSVLFX.	LPM570065 may possess an ↑ efficacy and/or a more rapid onset of antidepressant effect than DSVLFX. LPM570065 counterbalances the negative effects of DSVLFX on 5HT neurotransmission related to the 5HT1A autoreceptors.	Zhang R, Li X, Shi Y, et al. [79]

Table 1. *Cont.*

Design	Results	Observations	Reference
Adult male and female C57BL/6J mice, 5 groups, each group had 24 animals: control vs. single-stress vs. double-stress vs. LPM570065 vs. fluoxetine groups. Sucrose preference test, forced swimming test, and tail suspension test.	LPM570065 reduced susceptibility to depression-like behaviors in adult mice + maternal separation. LPM570065 protected against the reduced number of dendritic spines in the hippocampal CA1 of mice subjected to stress. LPM570065 regulated the expression of DNMTs in the mouse hippocampus.	LPM570065 may reduce depression vulnerability via epigenetic mechanisms involving the *Oxtr* expression.	Meng P, Li C, Duan S, et al. [80]
Male and female Wistar and Sprague–Dawley rats (total of 12/sex/group and 5/sex/group, respectively); affinity for monoamine transporters was determined by radioligand membrane binding assay; chronic unpredictable mild stress procedure; rat olfactory bulbectomized model, open field test, sucrose consumption test, serum corticosterone, and testosterone levels. Toludesvenlafaxine 10 μM.	The highest inhibition for serotonin transporters was reported in in vitro assays. The absorption was good after oral administration, and it was converted to O-desvenlafaxine due to the action of esterases in vivo, both reaching the hypothalamus in high concentration.	While desvenlafaxine does not increase the striatal level of dopamine, toludesvenlafaxine has this effect, which indicates supplementary benefits vs. the older drug.	Zhu H, Wang W, Sha C, et al. [81]

5HT = serotonin; AE = adverse effect; BW = body weight; DA = dopamine; DNMT = DNA methyltransferases; DSVLFX = desvenlafaxine; HPLC = high-performance liquid chromatography; MTD = maximum tolerated dose; NE = norepinephrine; *Oxtr* = oxytocin receptor; PRL = prolactine; SD = Sprague–Dawley; and TST = testosterone.

2.2. Clinical Trials

Out of the 12 references found for clinical trials, from phase 1 to 3, only a phase 2 and a phase 3 study had published results [62,82]. A total of 5 phase 1 trials conducted in different sites in the U.S. and China had results presented in manufacturer press releases [83–86]. Another 5 phase 1 trials had undisclosed results [64,87–90].

A total of 3 phase 1 studies were conducted in China: a single dose ascending trial (N = 72 healthy volunteers) explored the pharmacokinetics and tolerability of 20–200 mg LY03005, another study (N = 12 subjects) investigated the effects of food on the pharmacokinetics of 120 mg investigational product, and yet another (N = 48 participants) received multiple ascending doses of LY03005 [85]. No significant effect of the food on the bioavailability was observed, and the concentrations of the main active metabolite of LY03005 were dose-proportional for the 20–200 mg range [85]. The multiple-dose ascending dose study concluded that the steady state of the main active metabolite could be reached on the third day after daily dosing, and the concentrations of this metabolite were dose-proportional at the steady state for the dose of LY03005 ranging from 40–160 mg/day [85]. A good safety and tolerability profile of LY03005 was supported by all these trials (two main trials and a substudy) [85]. Besides these studies, another 2 phase 1 trials took place in the U.S. and enrolled a total of 120 healthy volunteers [48,49,51]. In the randomized, double-blind, single-ascending dose study, 72 subjects received 20 mg, 40 mg, 80 mg, 120 mg, 160 mg, or 200 mg LY03005 or placebo, and the results supported a good safety profile and linear dose proportionality on the plasma exposure after a unique dose of investigational product [84,86]. A substudy explored the effect of food on the pharmacokinetics of LY03005 in 10 subjects, but no impact on the bioavailability was detected [84,86]. In the multiple-ascending-dose, 48 healthy volunteers received daily 1 of the 4 regimens—40 mg, 80 mg, 120 mg, or 160—or placebo for 8 consecutive days [83,86]. The overall safety profile was good and linear dose proportionality on the plasma exposure was confirmed after repeated dose administration; the steady state of plasma exposure was reached after the third or fourth oral intake of the investigational product [83,86].

Orally administered ansofaxine extended-release (ER) was explored in a phase 2, multicenter, randomized, double-blind, placebo-controlled, dose-finding trial, which enrolled 260 patients with MDD (18–65 years old) [62]. Fixed doses of ansofaxine, i.e., 40 mg/day,

80 mg/day, 120 mg/day, or 160 mg/day or placebo, were administered for 6 weeks, and the primary outcome measure was the change in total HAMD-17 items from baseline to week 6 [62]. At the endpoint, significant changes were reported in all the groups that received the active drug vs. placebo, and the tolerability of all doses of ansofaxine was good [62]. The incidence of treatment-related adverse events was 52% (vs. 38.8% in the placebo group), reported by 141 patients and totalizing 303 cases [62].

In a phase 3 trial conducted by the manufacturer in China, which enrolled 558 adults with MDD (according to the DSM-5 criteria), ansofaxine doses of 80 mg and 160 mg were compared with placebo, and efficacy and tolerability parameters were monitored for 8 weeks [82]. This double-blind, randomized trial showed significant improvement in primary and secondary outcomes in patients who received the active drug vs. placebo at the end point [82]. Montgomery–Asberg Depression Scale (MADRS) scores decreased significantly at week 8 vs. baseline values and vs. placebo in the 2 groups which received 80 mg and 160 mg of active drug, respectively [82]. HAMD-17 items total score and "anxiety/somatization", "cognitive impairment", and "blocking" factors, Clinical Global Impression (CGI), Hamilton Anxiety Scale (HAM-A), Sheehan Disability Scale (SDS), and MADRS "anhedonia factor" scores were also significantly improved vs. placebo at the end of week 8 [82]. Most of the adverse events were mild and moderate, and no serious adverse event was reported [82]. Nausea, vomiting, headache, and drowsiness were the most frequently reported (over 5%) adverse events in the active treatment groups [82].

Table 2. Registered clinical trials exploring the efficacy and/or tolerability of toludesvenlafaxine (ansofaxine).

Methodology	Primary Outcome(s) and Measures	Secondary Outcome(s) and Measures	Sponsor of the Clinical Trial	The Country Where the Clinical Trial Took Place	Status of the Trial	Results and Observations	Registration of Clinical Trial and/or Reference(s)
LY03005 (40 mg, 80 mg, 120 mg, and 160 mg) vs. placebo, DBRCT, phase 2, dose-finding study, N = 260 MDD patients (18–65 years old), 2 weeks wash out + 6 weeks treatment	HAMD-17 scores at week 8	MADRS and CGI-I at week 8	Luye Pharma Group Ltd. (China)	China	Completed	HAMD-17 scores were significantly changed by the intervention vs. placebo at week 6 in all active treatment groups vs. placebo ($p < 0.05$). All doses were generally well tolerated, but the % of AEs was superior to the placebo group in each active treatment group.	NCT03785652 [62,91]
LY03005 (80 mg or 160 mg) vs. placebo, DBRCT, phase 3, N = 558 MDD patients (18–65 years old), 1-week screening + 8-week double-blind treatment	MADRS scores at 8 weeks	HAMD-17 at week 8	Luye Pharma Group Ltd. (China)	China	Completed	HAMD-17 total score and "anxiety/somatization", "cognitive impairment", and "blocking" factors, CGI, HAM-A, SDS, and MADRS "anhedonia factor" scores were significantly improved vs. placebo at week 8. Most of the adverse events were mild and moderate, and no SAE was reported. Nausea, vomiting, headache, and drowsiness were the most frequently reported (over 5%) AEs in the active treatment groups.	NCT04853407 [82,92]
LY03005 (20 mg, 40 mg, 80 mg, 120 mg, 160 mg, 200 mg, and 120 mg + fed) vs. DSVLFX (50 mg) vs. placebo, phase 1, RDBCT, N = 72 healthy participants in the SAD study + 12 subjects in food effect study (18–45 years old)	Number of participants with AEs during 11 days	PK parameters-C_{max} up to 4 days	Luye Pharma Group Ltd. (China)	United States	Completed	Unpublished results. No obvious effect of food on the bioavailability of LY03005. Good safety profile and linear dose proportionality on the plasma exposure after single oral dose administration.	NCT02055300 [84,86]

Table 2. *Cont.*

Methodology	Primary Outcome(s) and Measures	Secondary Outcome(s) and Measures	Sponsor of the Clinical Trial	The Country Where the Clinical Trial Took Place	Status of the Trial	Results and Observations	Registration of Clinical Trial and/or Reference(s)
LY03005 (40 mg, 80 mg, 120 mg, or 160 mg) vs. placebo, phase 1, RDBCT, MAD, $N = 48$ healthy subjects (18–45 years old), 8 consecutive days	Number of participants with AEs during 3 to 4 months	The PK of MAD	Luye Pharma Group Ltd. (China)	United States	Completed	Unpublished results. Good safety profile of LY03005 treatment and linear dose proportionality on the plasma exposure after multiple oral administrations. The steady state of plasma exposure was reached after 3rd or 4th oral intake of the investigational product.	NCT02271412 [83,86]
Phase 1 trials, healthy volunteers ($N = 132$), single or multiple doses of oral LY03005 administration; $N_1 = 72$ subjects in SAD study, $N_2 = 12$ subjects in the food-effect study, and $N_3 = 48$ subjects in the MAD study.	Safety and PK profiles for LY03005		Luye Pharma Group Ltd. (China)	China	Completed	Unpublished results. SAD study: concentrations of the main active metabolite were dose proportional for the dose range of 20–200 mg LY03005. Food-effect study: food did not affect the bioavailability in healthy subjects. MAD study: the steady state of the main active metabolite could be achieved on the 3rd day following multiple dosing; concentrations of the main active metabolite were dose proportional at the steady state for the dose range 40–160 mg/day LY03005. All studies: good tolerability and safety.	CTR20130364, CTR20140333, and CTR20140418 [85]
LY03005 (80 mg) vs. DSVLFX (50 mg), phase 1, pilot study, open-label study, single dose, $N = 20$ healthy subjects (18–50 years old)	Bioavailability of oral tablets under fasting conditions: AUC-PK samples were drawn at t0 (i.e., 30 min prior to dosing), 1 h, 2 h, 3 h, 4 h, 6 h, 8 h, 10 h, 12 h, 23 h, 32 h, 48 h, and 72 h after dosing		Luye Pharma Group Ltd. (China)	United States	Completed	Undisclosed	NCT02988024 [87]

Table 2. *Cont.*

Methodology	Primary Outcome(s) and Measures	Secondary Outcome(s) and Measures	Sponsor of the Clinical Trial	The Country Where the Clinical Trial Took Place	Status of the Trial	Results and Observations	Registration of Clinical Trial and/or Reference(s)
LY03005 (80 mg) fasted vs. fed crossover open-label randomized trial, single dose, phase 1, N = 34 participants (18–50 years old)	AUC and Cmax:PK parameters (predose and after dose) of parent and active metabolite		Luye Pharma Group Ltd. (China)	United States	Completed	Undisclosed	NCT03822065 [88]
LY03005 (80 mg) vs. DSVLFX (50 mg) 2-sequence, 2-period crossover open-label randomized trial, phase 1, single dose, N = 56 healthy participants (18–50 years)	AUC 15 days, parent drug and its active metabolite		Luye Pharma Group Ltd. (China)	United States	Completed	Undisclosed	NCT03733574 [89]
LY03005 (80 mg) vs. DSVLFX (50 mg), phase 1, randomized, open-label, cross-over, 2-period study, single dose, N = 20 healthy participants (18–50 years old)	AUC assessment up to 72 h after dosing in both trial periods. C_{max} assessment up to 72 h after dosing in both trial periods.	AEs assessment up to 35 days	Luye Pharma Group Ltd. (China)	United States	Completed	Undisclosed	NCT03357796 [90]
Phase 1 trial, healthy volunteers	Safety and tolerability of LY03005		Luye Pharma Group Ltd. (China)	Japan	Completed	Undisclosed	[64]

AEs = adverse events; AUC = area under the curve; CGI-I = Clinical Global Impression Scale—Improvement; DSVLFX = desvenlafaxine; HAMA = Hamilton Anxiety Scale; HAMD-17 = Hamilton Depression Rating Scale—17-item version; MAD = multiple ascending dose; MADRS = Montgomery–Asberg Depression Rating Scale; MDD = major depressive disorder; PK = pharmacokinetics; RDBCT = randomized double-blind controlled trial; SAE = severe adverse events; SAD = single ascending dose; and SDS = Sheehan Disability Scale.

3. Discussion

The findings of this review indicate there are practical and theoretical reasons for further exploration of the clinical properties of toludesvenlafaxine in larger and longer trials, focused both on efficacy and tolerability. The available data seem promising; therefore, this further exploration is expected to increase our knowledge of the TRI utility in MDD patients and would be expected to improve their chances of recovery. However, this process is not without challenges since the monoaminergic hypothesis is undergoing extensive criticism, and new pathogenetic mechanisms have already produced clinically available antidepressants, as mentioned in the introductory chapter. Still, the field of antidepressant research is a very intensively explored one, and there are multiple reasons to expect this interest will not decrease any time soon. The social and economic burden of treatment-resistant MDD makes the subject of finding new antidepressants very urgent [93–96].

Based on the reviewed data, toludesvenlafaxine possesses a high affinity for all three monoaminergic transporters, and preclinical data showed increased levels of serotonin, dopamine, and norepinephrine in rat striatum after acute or chronic administration of this drug [62,79]. A pharmacodynamic comparison between toludesvenlafaxine and other antidepressants that enhance serotonergic and norepinephrinergic $\pm$ dopaminergic neurotransmission is presented in Table 3.

Table 3. Pharmacodynamic characteristics of TRIs and other antidepressants [54–56,81].

Pharmacological Agent	SERT	NET	DAT	Observations
Toludesvenlafaxine	$IC_{50} = 31.4 \pm 0.4$ nM	$IC_{50} = 586.7 \pm 83.6$ nM	$IC_{50} = 733.2 \pm 10.3$ nM	Prodrug of desvenlafaxine, TRI
Desvenlafaxine	$IC_{50} = 47.3 \pm 19.4$ nM	$IC_{50} = 531.3 \pm 113$ nM	-	Major metabolite of venlafaxine, SNRI
Venlafaxine	$IC_{50} = 145$ nM	$IC_{50} = 2483$ nM	$IC_{50} = 7647$ nM	SNRI
Duloxetine	$IC_{50} = 13$ nM	$IC_{50} = 42$ nM	$IC_{50} = 439$ nM	SNRI
Milnacipran	$IC_{50} = 151$ nM	$IC_{50} = 200$ nM	$IC_{50} > 100,000$	SNRI
DOV 216,303	$IC_{50} = 14$ nM	$IC_{50} = 20$ nM	$IC_{50} = 78$ nM	TRI
DOV 21,947	$IC_{50} = 12$ nM	$IC_{50} = 23$ nM	$IC_{50} = 96$ nM	(+)-DOV-216,303, TRI
Ro 8-4650	$IC_{50} = 4.8$ μM	$IC_{50} = 2.5$ μM	$IC_{50} = 4.5$ μM	TRI
SKF83959	$K_i = 1.43 \pm 0.45$ μmol/L	$K_i = 0.6 \pm 0.07$ μmol/L	$K_i = 9.01 \pm 0.8$ μmol/L	TRI
BMS-820836	$IC_{50} = 0.2$	$IC_{50} = 26.7$ nM	$IC_{50} = 6.19$ nM	TRI
Amitriptyline	$IC_{50} = 67$ nM	$IC_{50} = 63$ nM	$IC_{50} = 7500$ nM	Tricyclic antidepressant

DAT = dopamine transporter; NET = norepinephrine transporter; SERT = serotonin transporter; SNRI = serotonin and norepinephrine reuptake inhibitor; and TRI = triple reuptake inhibitor.

Changes in prolactin and testosterone levels were reported in preclinical studies [77]. No significant adverse effects were observed, including mutagenic and clastogenic events up to 13 weeks and above 300 mg/day in rats [77]. Fertility and early embryonic development were not affected by the investigational product at doses up to 300 mg/kg for female rats and 100 mg/kg for male rats [78]. Acute injection of toludesvenlafaxine led to larger exposure vs. desvenlafaxine, and long-term administration of this agent (up to two weeks) increased all three monoamines more than desvenlafaxine, but especially dopamine [79]. A higher efficacy and/or more rapid onset of antidepressant effect than desvenlafaxine was also deduced based on the preclinical models of depression [79]. Toludesvenlafaxine significantly reversed depressive-like behaviors in all three paradigms used and increased the density of dendritic spines in the CA1 hippocampal neurons and reduced the hypermethylation of the oxytocin receptor gene (*Oxtr*) in the hippocampus of mice experiencing the "two hits" stress; therefore, it may reduce depression vulnerability via epigenetic mechanisms involving the *Oxtr* expression [80]. The plasma exposure was proportional to the dose after oral administration [81]. While desvenlafaxine does not increase the striatal level of dopamine, toludesvenlafaxine has this effect, which indicates supplementary benefits vs. the older drug [81]. The preclinical data do not support the existence of CNS excessive activation (manifested as irritation or hyperreactivity)/depression (reflected in sedation or muscle relaxation) and no significant abuse potential [77,81].

All phase 1 studies conducted in China and the U.S. supported a good tolerability and safety profile without a significant effect of the food on the bioavailability, and the concentrations of the main active metabolite of LY03005 were dose proportional for the 20–200 mg range [83–86]. The steady state of the main active metabolite could be reached on the third day after daily dosing, and the concentrations of this metabolite were dose-proportional at the steady state for the dose of LY03005, ranging from 40–160 mg/day [85].

Orally administered toludesvenlafaxine extended-release (ER) was efficient in a phase 2 study and significantly decreased the HAMD-17 scores after 6 weeks in patients with MDD, while its tolerability was good [62]. In a phase 3 trial, toludesvenlafaxine reduced MADRS scores significantly at week 8 vs. baseline values and vs. placebo after 8 weeks in the 2 groups which received 80 mg and 160 mg of active drug, respectively [82]. HAMD-17 items total score and "anxiety/somatization", "cognitive impairment", and "blocking" factors, CGI, HAM-A, SDS, and MADRS "anhedonia factor" scores were also significantly improved vs. placebo at the end of week 8 [82]. Most of the adverse events were mild and moderate, and no serious adverse event was reported [82].

Medication-specific aspects that have been correlated with treatment adherence were adverse effects, delayed onset of action, and subtherapeutic doses, but also complicated dosing schedules or titration strategies [97,98]. It is expected that TRI will associate a lower burden associated with pharmacological treatment with a lower risk of pharmacokinetic interactions, better tolerability, and higher treatment adherence because fewer drugs will be administered for the enhancement of all three monoamine transmitters. Moreover, a higher impact on dopamine neurotransmission observed in the case of toludesvenlafaxine could be beneficial in patients with MDD and substance use disorders, hyposexual desire disorder, or serotonin-induced sexual dysfunctions [44,45,64].

The main advantage of this review is the novelty of the topic because no other synthesis of preclinical and clinical data on toludesvenlafaxine has been identified in the literature during the initial search. All retrieved data on efficacy, tolerability, safety, and clinical pharmacology were analyzed, regardless of the research stage.

As a limitation of this review, it should be mentioned that it was only narrative, which could be considered a minus. However, because a broad search paradigm was used and extensive sources (white and grey) were included, it is not expected that important reports on toludesvenlafaxine were missed. An important number of trials did not have published results, and yet another study had undisclosed results. However, the major limitation of this review is represented by the fact that only two short-term trials that enrolled MDD patients were found [62,63], indicating that further research is needed in order to confirm the clinical utility of toludesvenlafaxine.

Future perspectives are related to the need to find the most appropriate target population for this antidepressant, following the principles of personalized medicine. For this purpose, pharmacogenetic studies are needed and comparative trials with other antidepressants in MDD patients are expected to provide the proper information for delineating the pharmacodynamic unicity of toludesvenlafaxine. Moreover, it is worth exploration of additional mechanisms of action for toludesvenlafaxine through translational research, in order to verify if enhancing the three monoaminergic pathways is related to other factors, such as immune or inflammatory markers, neurotrophic factors concentration, etc. Moreover, exploration of toludesvenlafaxine's potential use in other psychiatric disorders, besides MDD, which share monoaminergic dysfunctions, e.g., anxiety disorders, fibromyalgia, obsessive-compulsive spectrum disorders, etc. [99,100], is considered to be highly important.

4. Conclusions and Future Perspectives

Toludesvenlafaxine is a first-in-class TRI that became available recently on the market in China. It is important to note the theoretical importance of toludesvenlafaxine due to its mechanism of action, in a moment when the monoaminergic hypothesis of depression is considered obsolete by part of the scientific community [29,34,101–103]. There are many hopes related to each new class of antidepressants that has been launched in the last decade, i.e., glutamatergic modulators, such as esketamine for treatment-resistant depression as an add-on to ongoing agents, or brexanolone for post-partum depression [104–107]. Although many other TRI agents have been explored, until now the results have not been encouraging, highlighting the importance of toludesvenlafaxine. Available data about the efficacy and tolerability of this antidepressant are encouraging, but it is still too early to fully describe its clinical properties because of the lack of long-term studies. It is expected that toludesvenlafaxine will improve the functional outcome of patients diagnosed with MDD and treatment-resistant MDD if the efficacy and safety of this new antidepressant will be confirmed by further translational and clinical research. This expectation is related to the fact that the simultaneous activation of all three monoaminergic systems (i.e., serotonin, dopamine, and norepinephrine) will have a favorable impact on mood (e.g., anhedonia, irritability, and depressive affect), cognitive (e.g., reduced attention, impaired learning, and memory deficits) and somatic (e.g., fatigue, loss of energy, and insomnia) dimensions of the depression.

Funding: This research received no external funding.

Institutional Review Board Statement: Not applicable.

Informed Consent Statement: Not applicable.

Data Availability Statement: Not applicable.

Acknowledgments: As the sole author of this review, I assume entire responsibility for the selection, analysis, and presentation of the data herein.

Conflicts of Interest: The author declares no conflict of interest.

References

1. Liu, X.; Mukai, Y.; Furtek, C.I.; Bortnichak, E.A.; Liaw, K.-L.; Zhong, W. Epidemiology of Treatment-Resistant Depression in the United States. *J. Clin. Psychiatry* **2021**, *83*, 38389. [CrossRef]
2. Si, T.; Wang, P. When is antidepressant polypharmacy appropriate in the treatment of depression? *Shanghai Arch. Psychiatry* **2014**, *26*, 357–359. [CrossRef] [PubMed]
3. Vasiliu, O.; Vasile, D. Risk factors and quality of life in late-life depressive disorders. *Rom. J. Mil. Med.* **2016**, *119*, 24–28. [CrossRef]
4. Wiersema, C.; Voshaar, R.C.O.; Brink, R.H.S.V.D.; Wouters, H.; Verhaak, P.; Comijs, H.C.; Jeuring, H.W. Determinants and consequences of polypharmacy in patients with a depressive disorder in later life. *Acta Psychiatr. Scand.* **2022**, *146*, 85–97. [CrossRef]
5. Vasiliu, O. Effects of the selective serotonin reuptake inhibitors over coagulation in patients with depressive dis-orders—A systematic review and retrospective analysis. *RJMM* **2019**, *122*, 7–11. [CrossRef]
6. Rhee, T.G.; Rosenheck, R.A. Psychotropic polypharmacy reconsidered: Between-class polypharmacy in the context of multimor-bidity in the treatment of depressive disorders. *J. Affect. Disord.* **2019**, *252*, 450–457. [CrossRef]
7. Vasile, D.; Vasiliu, O.; Vasile, M.L.; Terpan, M.; Ojog, D.G. P.2.c.002 Agomelatine versus selective serotoninergic reuptake inhibitors in major depressive disorder and comorbid diabetes mellitus. *Eur. Neuropsychopharmacol.* **2011**, *21*, S383–S384. [CrossRef]
8. Paulzen, M.; Haen, E.; Hiemke, C.; Fay, B.; Unholzer, S.; Gründer, G.; Schoretsanitis, G. Antidepressant polypharmacy and the potential of pharmacokinetic interactions: Doxepin but not mirtazapine causes clinically relevant changes in venlafaxine metabolism. *J. Affect. Disord.* **2018**, *227*, 506–511. [CrossRef]
9. Wolff, J.; Hefner, G.; Normann, C.; Kaier, K.; Binder, H.; Hiemke, C.; Toto, S.; Domschke, K.; Marschollek, M.; Klimke, A. Polyphar-macy and the risk of drug–drug interactions and potentially inappropriate medications in hospital psychiatry. *Pharmacoepidemiol. Drug Saf.* **2021**, *30*, 1258–1268. [CrossRef]
10. Fond, G. A comparative analysis of effectiveness, tolerance and cost of second generation antidepressants in France. *La Tunis. Med.* **2015**, *93*, 123–128. [PubMed]
11. Delgado, P.L. Depression: The case for a monoamine deficiency. *J. Clin. Psychiatry* **2000**, *61* (Suppl. S6), 7–11. [PubMed]

12. Heninger, G.R.; Delgado, P.L.; Charney, D.S. The Revised Monoamine Theory of Depression: A Modulatory Role for Monoamines, Based on New Findings From Monoamine Depletion Experiments in Humans. *Pharmacopsychiatry* **1996**, *29*, 2–11. [CrossRef]

13. Charney, D.S. Monoamine dysfunction and the pathophysiology and treatment of depression. *J. Clin. Psychiatry* **1998**, *59*, 11–14.

14. Smith, K.; Fairburn, C.; Cowen, P. Relapse of depression after rapid depletion of tryptophan. *Lancet* **1997**, *349*, 915–919. [CrossRef] [PubMed]

15. Moncrieff, J.; Cooper, R.E.; Stockmann, T.; Amendola, S.; Hengartner, M.P.; Horowitz, M.A. The serotonin theory of depression: A systematic umbrella review of the evidence. *Mol. Psychiatry* **2022**, 1–14. [CrossRef] [PubMed]

16. Wichers, M.C.; Koek, G.H.; Robaeys, G.; Verkerk, R.; Scharpé, S.; Maes, M. IDO and interferon-α-induced depressive symptoms: A shift in hypothesis from tryptophan depletion to neurotoxicity. *Mol. Psychiatry* **2004**, *10*, 538–544. [CrossRef]

17. Köhler, S.; Cierpinsky, K.; Kronenberg, G.; Adli, M. The serotonergic system in the neurobiology of depression: Relevance for novel antidepressants. *J. Psychopharmacol.* **2016**, *30*, 13–22. [CrossRef]

18. Nutt, D. Relationship of neurotransmitters to the symptoms of major depressive disorder. *J. Clin. Psychiatry* **2008**, *69* (Suppl. E1), 4–7.

19. Nutt, D.J. The role of dopamine and norepinephrine in depression and antidepressant treatment. *J. Clin. Psychiatry* **2006**, *67* (Suppl. S6), 3–8. [PubMed]

20. Chandley, M.J.; Ordway, G.A. Noradrenergic dysfunction in depression and suicide. In *The Neurobiological Basis of Suicide*; Dwivedi, Y., Ed.; CRC Press: Boca Raton, FL, USA; Taylor & Francis: Boca Raton, FL, USA, 2012. Available online: https://www.ncbi.nlm.nih.gov/books/NBK107205/#top (accessed on 23 February 2023).

21. Moret, C.; Briley, M. The importance of norepinephrine in depression. *Neuropsychiatr. Dis. Treat.* **2011**, *7* (Suppl. S1), 9–13. [CrossRef]

22. Dailly, E.; Chenu, F.; Renard, C.E.; Bourin, M. Dopamine, depression and antidepressants. *Fundam. Clin. Pharmacol.* **2004**, *18*, 601–607. [CrossRef] [PubMed]

23. Dunlop, B.; Nemeroff, C.B. The Role of Dopamine in the Pathophysiology of Depression. *Arch. Gen. Psychiatry* **2007**, *64*, 327–337. [CrossRef]

24. Belujon, P.; Grace, A.A. Dopamine System Dysregulation in Major Depressive Disorders. *Int. J. Neuropsychopharmacol.* **2017**, *20*, 1036–1046. [CrossRef] [PubMed]

25. Nutt, D.J.; Demyttenaere, K.; Janka, Z.; Aarre, T.; Bourin, M.; Canonico, P.L.; Carrasco, J.L.; Stahl, S. The other face of depression, reduced positive affect: The role of catecholamines in causation and cure. *J. Psychopharmacol.* **2006**, *21*, 461–471. [CrossRef]

26. Budisteanu, M.; Andrei, E.; Linca, F.; Hulea, D.S.; Velicu, A.C.; Mihailescu, I.; Riga, S.; Arghir, A.; Papuc, S.M.; Sirbu, C.A.; et al. Predictive factors in early onset schizophrenia. *Exp. Ther. Med.* **2020**, *20*, 210. [CrossRef] [PubMed]

27. Peitl, V.; Štefanović, M.; Karlović, D. Depressive symptoms in schizophrenia and dopamine and serotonin gene polymorphisms. *Prog. Neuro-Psychopharmacol. Biol. Psychiatry* **2017**, *77*, 209–215. [CrossRef] [PubMed]

28. Hirschfeld, R.M. History and evolution of the monoamine hypothesis of depression. *J. Clin. Psychiatry* **2000**, *61* (Suppl. S6), 4–6.

29. Boku, S.; Nakagawa, S.; Toda, H.; Hishimoto, A. Neural basis of major depressive disorder: Beyond monoamine hypothesis. *Psychiatry Clin. Neurosci.* **2017**, *72*, 3–12. [CrossRef]

30. Pigott, H.E. The STAR*D Trial: It is Time to Reexamine the Clinical Beliefs That Guide the Treatment of Major Depression. *Can. J. Psychiatry* **2015**, *60*, 9–13. [CrossRef]

31. Vasiliu, O. Investigational Drugs for the Treatment of Depression (Part 1): Monoaminergic, Orexinergic, GABA-Ergic, and Anti-Inflammatory Agents. *Front. Pharmacol.* **2022**, *13*, 884143. [CrossRef]

32. Khoodoruth, M.A.S.; Estudillo-Guerra, M.A.; Pacheco-Barrios, K.; Nyundo, A.; Chapa-Koloffon, G.; Ouanes, S. Glutamatergic System in Depression and Its Role in Neuromodulatory Techniques Optimization. *Front. Psychiatry* **2022**, *13*, 886918. [CrossRef]

33. Vasiliu, O. Investigational Drugs for the Treatment of Depression (Part 2): Glutamatergic, Cholinergic, Sestrin Modulators, and Other Agents. *Front. Pharmacol.* **2022**, *13*, 884155. [CrossRef]

34. Cowen, P.J.; Browning, M. What has serotonin to do with depression? *World Psychiatry* **2015**, *14*, 158–160. [CrossRef] [PubMed]

35. Tanaka, M.; Szabó, A.; Vécsei, L. Integrating Armchair, Bench, and Bedside Research for Behavioral Neurology and Neuropsychiatry: Editorial. *Biomedicines* **2022**, *10*, 2999. [CrossRef] [PubMed]

36. Becker, M.; Pinhasov, A.; Ornoy, A. Animal Models of Depression: What Can They Teach Us about the Human Disease? *Diagnostics* **2021**, *11*, 123. [CrossRef] [PubMed]

37. Slaney, C.; Hinchcliffe, J.K.; Robinson, E.S.J. Translational Shifts in Preclinical Models of Depression: Implications for Biomarkers for Improved Treatments. *Curr. Top. Behav. Neurosci.* **2018**, *40*, 169–193. [CrossRef]

38. von Mücke-Heim, I.-A.; Urbina-Treviño, L.; Bordes, J.; Ries, C.; Schmidt, M.V.; Deussing, J.M. Introducing a depression-like syndrome for translational neuropsychiatry: A plea for taxonomical validity and improved comparability between humans and mice. *Mol. Psychiatry* **2022**, *28*, 329–340. [CrossRef]

39. Petković, A.; Chaudhury, D. Encore: Behavioural animal models of stress, depression and mood disorders. *Front. Behav. Neurosci.* **2022**, *16*, 931964. [CrossRef]

40. Tanaka, M.; Bohár, Z.; Martos, D.; Telegdy, G.; Vécsei, L. Antidepressant-like effects of kynurenic acid in a modified forced swim test. *Pharmacol. Rep.* **2020**, *72*, 449–455. [CrossRef]

41. Berton, O.; McClung, C.A.; Dileone, R.J.; Krishnan, V.; Renthal, W.; Russo, S.J.; Graham, D.; Tsankova, N.M.; Bolanos, C.A.; Rios, M.; et al. Essential Role of BDNF in the Mesolimbic Dopamine Pathway in Social Defeat Stress. *Science* **2006**, *311*, 864–868. [CrossRef]

42. Jayatissa, M.N.; Bisgaard, C.; Tingström, A.; Papp, M.; Wiborg, O. Hippocampal Cytogenesis Correlates to Escitalopram-Mediated Recovery in a Chronic Mild Stress Rat Model of Depression. *Neuropsychopharmacology* **2006**, *31*, 2395–2404. [CrossRef]

43. Voineskos, D.; Daskalakis, Z.J.; Blumberger, D.M. Management of Treatment-Resistant Depression: Challenges and Strategies. *Neuropsychiatr. Dis. Treat.* **2020**, *16*, 221–234. [CrossRef] [PubMed]

44. Sharma, H.; Santra, S.; Dutta, A. Triple reuptake inhibitors as potential next-generation antidepressants: A new hope? *Futur. Med. Chem.* **2015**, *7*, 2385–2406. [CrossRef]

45. Marks, D.M.; Pae, C.-U.; Patkar, A.A. Triple Reuptake Inhibitors: The Next Generation of Antidepressants. *Curr. Neuropharmacol.* **2008**, *6*, 338–343. [CrossRef]

46. Skolnick, P.; Krieter, P.; Tizzano, J.; Basile, A.; Popik, P.; Czobor, P.; Lippa, A. Preclinical and Clinical Pharmacology of DOV 216,303, a "Triple" Reuptake Inhibitor. *CNS Drug Rev.* **2006**, *12*, 123–134. [CrossRef]

47. Tran, P.; Skolnick, P.; Czobor, P.; Huang, N.; Bradshaw, M.; McKinney, A.; Fava, M. Efficacy and tolerability of the novel triple reuptake inhibitor amitifadine in the treatment of patients with major depressive disorder: A randomized, double-blind, placebo-controlled trial. *J. Psychiatr. Res.* **2012**, *46*, 64–71. [CrossRef]

48. Euthymics Bioscience Inc. Euthymics Reports Top-Line Results from TRIADE Trial of Amitifadine for Major Depressive Disorder. Available online: https://web.archive.org/web/20170924095823/http://euthymics.com/wp-content/uploads/2013/05/FINAL_Euthymics_TRIADE_Results_052913.pdf (accessed on 27 December 2022).

49. Luethi, D.; Hoener, M.; Liechti, M.E. Effects of the new psychoactive substances diclofensine, diphenidine, and methoxphenidine on monoaminergic systems. *Eur. J. Pharmacol.* **2018**, *819*, 242–247. [CrossRef] [PubMed]

50. Cherpillod, C.; Omer, L.M.O. A Controlled Trial with Diclofensine, a New Psychoactive Drug, in the Treatment of Depression. *J. Int. Med. Res.* **1981**, *9*, 324–329. [CrossRef] [PubMed]

51. Fang, X.; Guo, L.; Jia, J.; Jin, G.-Z.; Zhao, B.; Zheng, Y.-Y.; Li, J.-Q.; Zhang, A.; Zhen, X.-C. SKF83959 is a novel triple reuptake inhibitor that elicits anti-depressant activity. *Acta Pharmacol. Sin.* **2013**, *34*, 1149–1155. [CrossRef]

52. Jiang, B.; Wang, F.; Yang, S.; Fang, P.; Deng, Z.-F.; Xiao, J.-L.; Hu, Z.-L.; Chen, J.-G. SKF83959 Produces Antidepressant Effects in a Chronic Social Defeat Stress Model of Depression through BDNF-TrkB Pathway. *Int. J. Neuropsychopharmacol.* **2015**, *18*, pyu096. [CrossRef]

53. Risinger, R.; Bhagwagar, Z.; Luo, F.; Cahir, M.; Miler, L.; Mendonza, A.E.; Meyer, J.H.; Zheng, M.; Hayes, W. Evaluation of safety and tolerability, pharmacokinetics, and pharmacodynamics of BMS-820836 in healthy subjects: A placebo-controlled, ascending single-dose study. *Psychopharmacology* **2013**, *231*, 2299–2310. [CrossRef]

54. Bhagwagar, Z.; Torbeyns, A.; Hennicken, D.; Zheng, M.; Dunlop, B.W.; Mathew, S.J.; Khan, A.; Weisler, R.; Nelson, C.; Shelton, R.; et al. Assessment of the Efficacy and Safety of BMS-820836 in Patients with Treatment-Resistant Major Depression. *J. Clin. Psychopharmacol.* **2015**, *35*, 454–459. [CrossRef] [PubMed]

55. Lawson, K. A Brief Review of the Pharmacology of Amitriptyline and Clinical Outcomes in Treating Fibromyalgia. *Biomedicines* **2017**, *5*, 24. [CrossRef]

56. Deecher, D.C.; Beyer, C.E.; Johnston, G.; Bray, J.; Shah, S.; Abou-Gharbia, M.; Andree, T.H. Desvenlafaxine Succinate: A New Serotonin and Norepinephrine Reuptake Inhibitor. *Experiment* **2006**, *318*, 657–665. [CrossRef] [PubMed]

57. Moraczewski, J.; Aedma, K.K. Tricyclic antidepressants. StatPearls (Internet). Available online: https://www.ncbi.nlm.nih.gov/books/NBK557791/ (accessed on 27 December 2022).

58. Ferguson, J.M. SSRI Antidepressant Medications: Adverse effects and tolerability. *Prim. Care Companion J. Clin. Psychiatry* **2001**, *3*, 22–27. [CrossRef]

59. Haddad, P. Do antidepressants have any potential to cause addiction? *J. Psychopharmacol.* **1999**, *13*, 300–307. [CrossRef]

60. Whiskey, E.; Taylor, D. A review of the adverse effects and safety of noradrenergic antidepressants. *J. Psychopharmacol.* **2013**, *27*, 732–739. [CrossRef] [PubMed]

61. Peitl, V.; Vlahović, D. Ansofaxine Hydrochloride. *Arch. Psychiatry Res.* **2020**, *57*, 87–90. [CrossRef]

62. Mi, W.; Yang, F.; Li, H.; Xu, X.; Li, L.; Tan, Q.; Wang, G.; Zhang, K.; Tian, F.; Luo, J.; et al. Efficacy, Safety, and Tolerability of Ansofaxine (LY03005) Extended-Release Tablet for Major Depressive Disorder: A Randomized, Double-Blind, Placebo-Controlled, Dose-Finding, Phase 2 Clinical Trial. *Int. J. Neuropsychopharmacol.* **2021**, *25*, 252–260. [CrossRef]

63. Luye Pharma. Luye Pharma's Class 1 Innovative Antidepressant Ruoxinlin®Approved for Launch in China. Available online: https://www.luye.cn/lvye_en/view.php?id=2108 (accessed on 26 December 2022).

64. Luye Pharma. Marketing Authorization Application Accepted by CDE for Luye Pharma's Antidepressant Anshu-Faxine Hydrochloride Extended-Release Tablets. Available online: https://www.luye.cn/lvye_en/view.php?id=1954 (accessed on 26 December 2022).

65. Luye Pharma. NDA Filing for Luye Pharma's Antidepressant Drug LY03005 Accepted by the U.S. FDA. Available online: https://www.luye.cn/lvye_en/view.php?id=1809 (accessed on 26 December 2022).

66. Fasipe, O.J. The emergence of new antidepressants for clinical use: Agomelatine paradox versus other novel agents. *IBRO Rep.* **2019**, *6*, 95–110. [CrossRef]

67. Park, S.-C. Neurogenesis and antidepressant action. *Cell Tissue Res.* **2019**, *377*, 95–106. [CrossRef] [PubMed]

68. Roohi, E.; Jaafari, N.; Hashemian, F. On inflammatory hypothesis of depression: What is the role of IL-6 in the middle of the chaos? *J. Neuroinflammation* **2021**, *18*, 45. [CrossRef] [PubMed]

69. Miyata, S.; Ishino, Y.; Shimizu, S.; Tohyama, M. Involvement of inflammatory responses in the brain to the onset of major depressive disorder due to stress exposure. *Front. Aging Neurosci.* **2022**, *14*, 934346. [CrossRef]

70. Elias, E.; Zhang, A.Y.; Manners, M.T. Novel Pharmacological Approaches to the Treatment of Depression. *Life* **2022**, *12*, 196. [CrossRef]

71. Richardson, B.; MacPherson, A.; Bambico, F. Neuroinflammation and neuroprogression in depression: Effects of alternative drug treatments. *Brain Behav. Immun.-Health* **2022**, *26*, 100554. [CrossRef]

72. Brigitta, B. Pathophysiology of depression and mechanisms of treatment. *Dialog. Clin. Neurosci.* **2002**, *4*, 7–20. [CrossRef]

73. Nemeroff, C.B. The State of Our Understanding of the Pathophysiology and Optimal Treatment of Depression: Glass Half Full or Half Empty? *Am. J. Psychiatry* **2020**, *177*, 671–685. [CrossRef] [PubMed]

74. Touya, M.; Lawrence, D.F.; Kangethe, A.; Chrones, L.; Evangelatos, T.; Polson, M. Incremental burden of relapse in patients with major depressive disorder: A real-world, retrospective cohort study using claims data. *BMC Psychiatry* **2022**, *22*, 152. [CrossRef]

75. Kurimoto, N.; Inagaki, T.; Aoki, T.; Kadotani, H.; Kurimoto, F.; Kuriyama, K.; Yamada, N.; Ozeki, Y. Factors causing a relapse of major depressive disorders following successful electroconvulsive therapy: A retrospective cohort study. *World J. Psychiatry* **2021**, *11*, 841–853. [CrossRef]

76. Schöpfel, J.; Farace, D.J. Grey literature. In *Encyclopedia of Library and Information Sciences*, 3rd ed.; Bates, M.J., Maack, M.N., Eds.; CRC Press: Boca Raton, FL, USA, 2010; pp. 2029–2039.

77. Li, C.; Jiang, W.; Gao, Y.; Lin, F.; Zhu, H.; Wang, H.; Ye, L.; Qi, J.G.; Tian, J. Acute, subchronic oral toxicity, and genotoxicity evaluations of LPM570065, a new potent triple reuptake inhibitor. *Regul. Toxicol. Pharmacol.* **2018**, *98*, 129–139. [CrossRef]

78. Guo, W.; Gao, Y.; Jiang, W.; Li, C.; Lin, F.; Zhu, H.; Wang, H.; Ye, L.; Qi, J.G.; Cen, X.; et al. Toxicity effects of a novel potent triple reuptake inhibitor, LPM570065, on the fertility and early embryonic development in Sprague-Dawley rats. *Regul. Toxicol. Pharmacol.* **2018**, *100*, 45–51. [CrossRef]

79. Zhang, R.; Li, X.; Shi, Y.; Shao, Y.; Sun, K.; Wang, A.; Sun, F.; Liu, W.; Wang, D.; Jin, J.; et al. The Effects of LPM570065, a Novel Triple Reuptake Inhibitor, on Extracellular Serotonin, Dopamine and Norepinephrine Levels in Rats. *PLoS ONE* **2014**, *9*, e91775. [CrossRef]

80. Meng, P.; Li, C.; Duan, S.; Ji, S.; Xu, Y.; Mao, Y.; Wang, H.; Tian, J. Epigenetic Mechanism of 5-HT/NE/DA Triple Reuptake Inhibitor on Adult Depression Susceptibility in Early Stress Mice. *Front. Pharmacol.* **2022**, *13*, 848251. [CrossRef]

81. Zhu, H.; Wang, W.; Sha, C.; Guo, W.; Li, C.; Zhao, F.; Wang, H.; Jiang, W.; Tian, J. Pharmacological Characterization of Toludesvenlafaxine as a Triple Reuptake Inhibitor. *Front. Pharmacol.* **2021**, *12*, 741794. [CrossRef] [PubMed]

82. Luye Pharma. Luye Pharma's Class 1 New Drug Anshufaxine Hydrochloride Extended-Release Tablets Meets Pre-Defined Endpoints in Phase III Trial. March 2021. Available online: https://www.luye.cn/lvye_en/view.php?id=1922 (accessed on 26 December 2022).

83. National Library of Medicine (U.S.). Multiple Ascending Dose Study in Healthy Subjects to Evaluate the Safety, Tolerability, and Pharmacokinetics of LY03005. Available online: https://clinicaltrials.gov/ct2/show/NCT02271412 (accessed on 26 December 2022).

84. National Library of Medicine (U.S.). Safety, Tolerability and Pharmacokinetics Study of LY03005 (LY03005SAD). Available online: https://clinicaltrials.gov/ct2/show/NCT02055300 (accessed on 26 December 2022).

85. FiercePharma. Completion of Phase 1 Clinical Studies of Ansofaxine Hydrochloride Extended. 30 March 2015. Available online: https://www.fiercepharma.com/pharma-asia/completion-of-phase-1-clinical-studies-of-ansofaxine-hydrochloride-extended-release (accessed on 27 December 2022).

86. Chua, S. Phase I Study Program for Ansofaxine (LY03005) Is Completed in USA. Available online: https://www.linkedin.com/pulse/phase-i-study-program-ansofaxine-ly03005-completed-usa-sam-chua (accessed on 27 December 2022).

87. National Library of Medicine (U.S.). Pilot BA Study of New LY03005 vs. Pristiq. Available online: https://clinicaltrials.gov/ct2/show/NCT02988024 (accessed on 26 December 2022).

88. National Library of Medicine (U.S.). A Relative Bioavailability Food Effect Study of LY03005. Available online: https://clinicaltrials.gov/ct2/show/NCT03822065 (accessed on 26 December 2022).

89. National Library of Medicine (U.S.). A Study of LY03005 vs. Pristiq. Available online: https://clinicaltrials.gov/ct2/show/NCT03733574 (accessed on 26 December 2022).

90. National Library of Medicine (U.S.). Relative Bioavailability (RBA) Study of LY03005 vs. Pristiq. Available online: https://clinicaltrials.gov/ct2/show/NCT03357796 (accessed on 26 December 2022).

91. National Library of Medicine (U.S.). Dose-Finding Clinical Trial to Evaluate the Efficacy and Safety of LY03005 Extended-Release Tablets in the Treatment of Major Depressive Disorder (MDD). Available online: https://clinicaltrials.gov/ct2/show/NCT03785652 (accessed on 26 December 2022).

92. National Library of Medicine (U.S.). A Study to Evaluate the Efficacy and Safety of Annsofaxine Hydrochloride Extended-Release Tablets in the Treatment of Major Depressive Disorder (MDD). Available online: https://clinicaltrials.gov/ct2/show/NCT04853407 (accessed on 26 December 2022).

93. Greenberg, P.E.; Fournier, A.-A.; Sisitsky, T.; Simes, M.; Berman, R.; Koenigsberg, S.H.; Kessler, R.C. The Economic Burden of Adults with Major Depressive Disorder in the United States (2010 and 2018). *Pharmacoeconomics* **2021**, *39*, 653–665. [CrossRef] [PubMed]

94. Tanner, J.-A.; Hensel, J.; Davies, P.E.; Brown, L.; DeChairo, B.M.; Mulsant, B.H. Economic Burden of Depression and Associated Resource Use in Manitoba, Canada. *Can. J. Psychiatry* **2019**, *65*, 338–346. [CrossRef]

95. Touloumis, C. The burden and the challenge of treatment-resistant depression. *Psychiatriki* **2021**, *32* (Suppl. SI), 11–14. [CrossRef] [PubMed]

96. Jaffe, D.H.; Rive, B.; Denee, T.R. The humanistic and economic burden of treatment-resistant depression in Europe: A cross-sectional study. *BMC Psychiatry* **2019**, *247*, 19. [CrossRef] [PubMed]

97. Masand, P.S. Tolerability and adherence issues in antidepressant therapy. *Clin. Ther.* **2003**, *25*, 2289–2304. [CrossRef]

98. Ho, S.C.; Jacob, S.A.; Tangiisuran, B. Barriers and facilitators of adherence to antidepressants among outpatients with major depressive disorder: A qualitative study. *PLoS ONE* **2017**, *12*, e0179290. [CrossRef]

99. Wood, J.; LaPalombara, Z.; Ahmari, S.E. Monoamine abnormalities in the SAPAP3 knockout model of obsessive-compulsive disorder-related behaviour. *Philos. Trans. R. Soc. B Biol. Sci.* **2018**, *373*, 20170023. [CrossRef] [PubMed]

100. Calandre, E.P.; Rico-Villademoros, F.; Slim, M. An update on pharmacotherapy for the treatment of fibromyalgia. *Expert Opin. Pharmacother.* **2015**, *16*, 1347–1368. [CrossRef] [PubMed]

101. Goldberg, J.S.; Bell, C.E.; Pollard, D.A.; Bell, J.C.E. Revisiting the Monoamine Hypothesis of Depression: A New Perspective. *Perspect. Med. Chem.* **2014**, *6*, 1–8. [CrossRef] [PubMed]

102. Liu, B.; Liu, J.; Wang, M.; Zhang, Y.; Li, L. From Serotonin to Neuroplasticity: Evolvement of Theories for Major Depressive Disorder. *Front. Cell. Neurosci.* **2017**, *11*, 305. [CrossRef]

103. Massart, R.; Mongeau, R.; Lanfumey, L. Beyond the monoaminergic hypothesis: Neuroplasticity and epigenetic changes in a transgenic mouse model of depression. *Philos. Trans. R. Soc. B Biol. Sci.* **2012**, *367*, 2485–2494. [CrossRef]

104. Epperson, C.N.; Rubinow, D.R.; Meltzer-Brody, S.; Deligiannidis, K.M.; Riesenberg, R.; Krystal, A.D.; Bankole, K.; Huang, M.-Y.; Li, H.; Brown, C.; et al. Effect of brexanolone on depressive symptoms, anxiety, and insomnia in women with postpartum depression: Pooled analyses from 3 double-blind, randomized, placebo-controlled clinical trials in the HUMMINGBIRD clinical program. *J. Affect. Disord.* **2023**, *320*, 353–359. [CrossRef]

105. Cooper, M.C.; Kilvert, H.S.; Hodgkins, P.; Roskell, N.S.; Eldar-Lissai, A. Using Matching-Adjusted Indirect Comparisons and Network Meta-analyses to Compare Efficacy of Brexanolone Injection with Selective Serotonin Reuptake Inhibitors for Treating Postpartum Depression. *CNS Drugs* **2019**, *33*, 1039–1052. [CrossRef]

106. Bahji, A.; Vazquez, G.H.; Zarate, C.A. Comparative efficacy of racemic ketamine and esketamine for depression: A systematic review and meta-analysis. *J. Affect. Disord.* **2020**, *278*, 542–555. [CrossRef]

107. Xiong, J.; Lipsitz, O.; Chen-Li, D.; Rosenblat, J.D.; Rodrigues, N.B.; Carvalho, I.; Lui, L.M.; Gill, H.; Narsi, F.; Mansur, R.B.; et al. The acute antisuicidal effects of single-dose intravenous ketamine and intranasal esketamine in individuals with major depression and bipolar disorders: A systematic review and meta-analysis. *J. Psychiatr. Res.* **2020**, *134*, 57–68. [CrossRef]

Review

Inhibition of Microglial GSK3β Activity Is Common to Different Kinds of Antidepressants: A Proposal for an In Vitro Screen to Detect Novel Antidepressant Principles

Hans O. Kalkman

Child and Adolescent Psychiatry Research Centre, Department of Child and Adolescent Psychiatry and Psychotherapy, Psychiatric University Hospital, University of Zurich, CH-8032 Zurich, Switzerland; hans.kalkman@bluewin.ch

Abstract: Depression is a major public health concern. Unfortunately, the present antidepressants often are insufficiently effective, whilst the discovery of more effective antidepressants has been extremely sluggish. The objective of this review was to combine the literature on depression with the pharmacology of antidepressant compounds, in order to formulate a conceivable pathophysiological process, allowing proposals how to accelerate the discovery process. Risk factors for depression initiate an infection-like inflammation in the brain that involves activation microglial Toll-like receptors and glycogen synthase kinase-3β (GSK3β). GSK3β activity alters the balance between two competing transcription factors, the pro-inflammatory/pro-oxidative transcription factor NFκB and the neuroprotective, anti-inflammatory and anti-oxidative transcription factor NRF2. The antidepressant activity of tricyclic antidepressants is assumed to involve activation of G_S-coupled microglial receptors, raising intracellular cAMP levels and activation of protein kinase A (PKA). PKA and similar kinases inhibit the enzyme activity of GSK3β. Experimental antidepressant principles, including cannabinoid receptor-2 activation, opioid μ receptor agonists, 5HT2 agonists, valproate, ketamine and electrical stimulation of the Vagus nerve, all activate microglial pathways that result in GSK3β-inhibition. An in vitro screen for NRF2-activation in microglial cells with TLR-activated GSK3β activity, might therefore lead to the detection of totally novel antidepressant principles with, hopefully, an improved therapeutic efficacy.

Keywords: depression risk factor; microglia; toll-like receptor; GSK3β; NRF2; G_S-coupled receptor; ketamine; cannabinoid CBR2; psilocybin; 5-HT2B

Citation: Kalkman, H.O. Inhibition of Microglial GSK3β Activity Is Common to Different Kinds of Antidepressants: A Proposal for an In Vitro Screen to Detect Novel Antidepressant Principles. *Biomedicines* **2023**, *11*, 806. https://doi.org/10.3390/biomedicines11030806

Academic Editors: Masaru Tanaka, Lydia Giménez-Llort, Simone Battaglia, Chong Chen, Piril Hepsomali and Andrea Mattarei

Received: 3 February 2023
Revised: 17 February 2023
Accepted: 4 March 2023
Published: 7 March 2023

1. Introduction

Major depressive disorder is a psychiatric syndrome involving persistent low mood, anhedonia, fatigue, loss of energy, sleep disturbances, and deficits in the cognitive domain, including impaired ability to concentrate, poor attention, and memory [1]. The view on major depression has changed during the last three decades from being a mental disorder caused by a lack of monoamine neurotransmitters [2] to a concept where the mental disorder is secondary to an immune activation in the brain [3,4]. The earliest antidepressants were serendipitously discovered, and, as a common denominator, they all raise the extracellular levels of dopamine, noradrenaline, and serotonin (5-hydroxytryptamine) in the brain [5–7]. The indirect monoaminergic activity [8] provides a moderately effective, central anti-inflammatory effect and initiates a slowly-developing improvement in the physical and mental symptoms of depression [9]. Major depression ranks high on the list of diseases posing the strongest burden on society [10], and unfortunately, the global situation has further deteriorated recently due to infections with SARS-Cov2 [11]. The discovery of antidepressants with a novel mode of action has stagnated for many decades, and recent progress is very modest. There is a pressing need to identify novel antidepressant principles and develop alternatives to the currently available insufficiently effective medications. Of course, there has been a been continuous progress in the understanding of the pathophysiology of depression and the mechanism of action of existing antidepressants. In the current review the novel insights in the pathophysiology of major depression and the pharmacology of antidepressants are integrated. Based on this, it is possible to propose a relatively simple in vitro screening method to identify new pharmacological principles with anti-inflammatory and antidepressant potential.

2. Activation of Microglia Cells by Risk Factors for Depression

Risk factors for depression (see below) activate the immune system in the brain [4,9,12]. These risk factors activate Toll-like receptors (TLRs) expressed by the brain's resident immune cells (microglia) and trigger an intracellular signaling pathway involving Nuclear Factor-κB (NFκB) [13] that promotes the transcription of genes that encode pro-oxidative and pro-inflammatory mediators. The serine/threonine kinase glycogen-synthase kinase 3 (GSK3) acts as a master switch between, on the one hand, a pro-inflammatory/pro-oxidative and, on the other hand, a proliferative, protective, and restorative biological program [14,15]. This process, which is discussed in more detail below, is depicted in Figure 1.

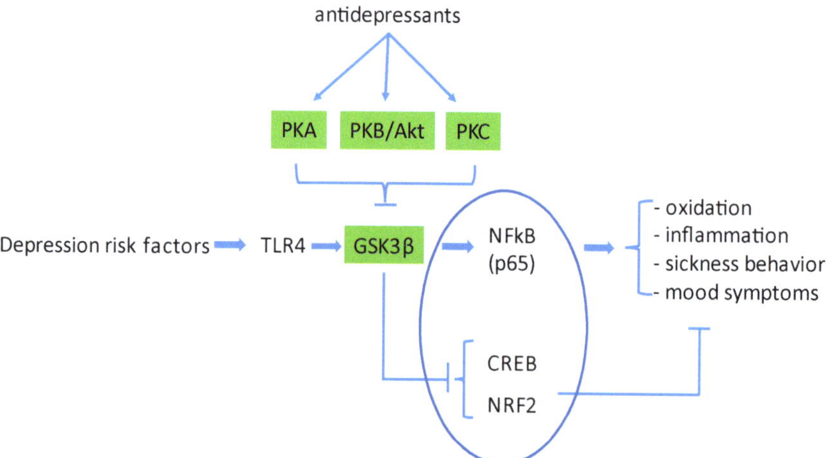

Figure 1. Schematic representation of the process by which risk factors for depression increase inflammatory and pro-oxidative processes in the brain and lead to physical and psychological symptoms of depressive disorder. Depression risk factors activate Toll-like receptor-4 (TLR4) on microglia cells, which causes activation of the kinase GSK3β and an increase in gene transcription by NFκB (p65/RelA). Antidepressant compounds, via activation of different members of the AGC protein-kinase family, including protein kinase A (PKA), PKB/Akt, and PKC, inhibit the enzyme activity of GSK3β. This, on the one hand, inhibits NFκB-signaling and, on the other hand, increases gene transcription by CREB and NRF2, two transcription factors that promote the transcription of neuroprotective, anti-inflammatory, and anti-oxidant proteins. This way, antidepressants neutralize the negative effects evoked by TLR4 activation. CREB: cyclic-AMP responsive element binding protein; NRF2: nuclear factor erythroid 2-related factor 2. The transcription factors are enclosed in an ellipse. For further details, the reader is referred to the text.

The evidence that inflammation plays a subtle role in the pathophysiology of major depressive disorder (MDD) comes from three observations: (1) one-third of those with major depression show elevated peripheral inflammatory biomarkers, even in the absence of an additional medical illness; (2) inflammatory illnesses are associated with greater rates of MDD; and (3) patients treated with pro-inflammatory cytokines are at greater risk for developing major depressive illness (reviewed by [16,17]).

Infection [18] and inflammatory disorders such as IBD (inflammatory bowel disease) [19], metabolic disorders [19,20], cardiovascular disorders [21], neurological disorders [22–26], autoimmune diseases [18], and even alcohol abuse [27,28] are among the likely causes of immune dysfunction that contribute to the pathogenesis of depression [29]. In these diseases, affected cell populations generate and release so-called "danger associated molecular-pattern molecules" (DAMPs), which activate the innate immune system [30]. For instance, DAMPs such as HMGB1, modified lipids, heat shock proteins, S100, or hyaluronan-oligosaccharides activate the same signaling cascade as LPS, a cell wall fragment from Gram-negative bacteria [29–31]. Pathogens are recognized by specific immune receptors called Toll-like receptors (TLRs). Like LPS [30], DAMPs activate TLR2 and, in particular, TLR4 [32,33]. Notably, alcohol triggers a TLR4 response too [34].

Inflammation is thought to affect brain signaling causing mood symptoms, cognition dysfunction, and the production of a constellation of symptoms, termed 'sickness behavior'. Sickness behavior is a set of symptoms (fatigue, immobility, fever, and loss of appetite) induced by an infection and mediated by pro-inflammatory cytokines. It is conceptualized as an adaptive response that enhances recovery by conserving energy to combat acute inflammation [35,36]. Mood symptoms in depression are anxiety, apathy, general discontent, guilt, hopelessness, loss of interest, anhedonia, and sadness [35].

Chronic stress greatly contributes to the development of major depressive disorder [14,37–40]. The pathophysiological sequelae to chronic stress have been evaluated in great detail in an array of rodent models, including chronic mild stress, social defeat stress, restraint stress, prenatal stress, and others (for review, see [41,42]). In each of these models, microglial cells were activated in a manner that resembled a bacterial infection [41,43–48], in particular in mood-relevant areas such as the hippocampus, frontal cortex, nucleus accumbens, and amygdala [31,38,49–51]. Chronic stress in laboratory animals thus increased the expression of pro-inflammatory cytokines and chemokines (IL1β, IL6, TNFα, TLR4, CCL2, and CX3CR1) by microglia and elevated the levels of corticosterone and IL6 in the blood circulation [52,53]. In further studies in chronic stress models, it was noted that stress increased inflammation markers (CD14, CD86, and TLR4) on microglia and macrophages [53,54]. Moreover, the morphology of microglia changed from ramified to hypertrophic/activated, with short and thick processes [42,53,55]. Similar to stress, chronic exposure to glucocorticoids also provoked microglial activation in rodents [39]. The process by which stress results in activation of microglia may involve the release of cellular distress signals such as HMGB1 (an agonist at TLR2 and TLR4 [32,56]) and ATP by abnormally active neurons [41,56]. Since attempts to understand the pathophysiology of depression have traditionally focused on neuronal dysfunction, the functionality of other types of brain cells has received (too) little attention [42]. The firm awareness that microglial cells play a crucial role in the development and maintenance of major depression has developed only quite recently [41,42,57–59].

Transcription in cells is limited by two important histone acetyltransferases, CBP and p300. In microglia cells, inflammatory signaling is to a large extent determined by the competition for CBP/p300-occupancy between the pro-inflammatory transcription factor NFκB (in particular the p65/RelA subunit) and the anti-inflammatory transcription factors, NRF2 and CREB [60–64]. As mentioned above, the balance between these pro- and anti-inflammatory pathways, which is essential for major depressive disorder [65,66], is influenced by the kinase activity of GSK3. Active GSK3 enhances the affinity of p65 for CBP and p300 [61] and, on the other hand, promotes the nuclear export of NRF2 [67], inhibits the nuclear import of CREB [60], and promotes the catabolism of CREB and NRF2 [68]. Although GSK3 comes in two distinct isoforms, modulation of CREB and NRF2 seems to be mediated by the GSK3β isoform only [60]. NRF2 increases the transcription of genes involved in oxidant-defense [63] and factors that stimulate mitochondrial biosynthesis [66,69]. Downstream of CREB are anti-inflammatory mediators such as brain-derived neurotrophic factor (BDNF), IL1-receptor antagonist (IL1-RA), IL10, and DUSP1 [62,67]. Notably, DUSP1 is an inhibitor of the pro-inflammatory kinase p38 [70–72]. Activation of NRF2 provides transcriptional repression of TNFα, IL1β, IL6, IL8, and CCL2 in microglia, monocytes, and macrophages [66,69]. For instance, activation of NRF2 reduced TLR4-induced IL1β, IL6, iNOS, and COX2 expression in primary mouse peritoneal macrophages [66,69]. Stimulation of Toll-like receptors promotes the enzyme activity of GSK3β [14] (for review, see [73]), whereas active GSK3 inhibits the production of the anti-inflammatory cytokine IL10, of IL1-RA, BDNF, and DUSP1 [72,74]. Activators of TLR-signaling are mutant presenilin-1 [75], α-synuclein fibrils [76], ethanol [77], and pathogens [73]. Importantly, repeated stress results in activation of GSK3 [73]. A group of kinases with similar substrate preferences, the so-called 'AGC' kinases are able to phosphorylate a serine residue in the N-terminal part of GSK3 (Ser-9 in GSK3β), which generates a pseudo-substrate for GSK3 and thereby inhibits the phosphorylation of other protein substrates (so inhibits GSK3 activity). AGC-kinases include the protein kinases PKA, PKB/Akt, PKC, PKG, S6K, and others [15,78,79]. Data from patients indicate that the cAMP-PKA pathway is hypoactive in MDD [80–82], so GSK3 is tendentially hyperactive [83]. This prediction is consistent with observations in MDD patients that both the nuclear translocation of NRF2 and the NRF2-mediated gene transcriptions are diminished [84], whereas biomarkers reflecting oxidative damage are elevated [85,86].

3. Classical Antidepressants Activate PKA to Inhibit GSK3β

Interestingly, as in MDD patients, also in the chronic mild stress depression model there is a reduction in cortical cAMP-PKA signaling and consequently a decrease in CREB activation and BDNF transcription [39]. Chronic but not acute treatment with 'classical' antidepressants attenuates the decrease in BDNF [8]. Classical antidepressants are typically inhibitors of monoamine reuptake, inhibitors of monoamine metabolism, or antagonists at presynaptic monoamine receptors, so their overall effect is an increase in extracellular levels of monoamine neurotransmitters. It is conceivable that this will lead to stimulation of monoamine receptors on microglia. Some of these receptors (dopamine D1/D5 [76,87–89], noradrenaline β2 [90,91], and serotonin 5HT7 [92]) are coupled to G_S and generate high cAMP levels. Elevated cAMP levels activate PKA, which in turn inhibits GSK3β. As outlined above and depicted in Figure 1, the balance between pro-inflammatory TLR-NFκB signaling and NRF2/CREB signaling would be shifted away from inflammation (seen as a reduction in pro-inflammatory cytokines) and towards transcription growth factors such as insulin-like growth factor-1, BDNF, and anti-inflammatory cytokines such as IL10 and IL1-RA [93,94]. In accordance with this, it has been proposed that classical antidepressants limit the sequelae of TLR-activation by activation of cAMP-PKA signaling [8,95,96] and reduce the production of pro-inflammatory cytokines and oxygen radicals by microglia cells [52,96–100]. Importantly, chronic antidepressants also ameliorate the behavioral effects of stress [42,52]. So, these antidepressants limit an ongoing inflammation in the brain, and this is presumably the reason why they reduce depression symptoms. However, since the classical antidepressants elevate monoamine levels in the brain, any inflammatory process in the periphery will remain unopposed [67]. It is conceivable that a continuing peripheral inflammation will hinder full recovery and might represent a precipitating factor for relapses [16].

G_S-coupled monoamine receptors that are expressed by microglial cells are D1 and D5 [88], β1, and β2, and 5HT7 (notably two other G_S-coupled serotonin receptors, 5HT4 and 5HT6, are not expressed by microglia [101,102]). Direct agonists for these microglial receptors would provide anti-inflammatory and antidepressant activity, possibly with a more rapid onset of action. In this context, it is worth noting that already in 1996, Shimizu and colleagues suggested that activation of 5HT7 receptors might underlie the therapeutic response to classical antidepressants [95]. However, there is no reason for a restriction to monoamine receptors only. Extensive reviews have been published that report on the types of G-protein-coupled receptors (GPCRs) expressed by microglia [101,103,104]. Those GPCRs that couple to G_S might offer alternative targets for novel anti-inflammatory and antidepressant drugs. In fact, there is not even a reason to stick to only the cAMP-PKA pathway. Essential is the activity of GSK3β, and this enzyme can be inhibited by, for instance, PKB or PKC. The next paragraphs deal with compound classes with certain evidence for clinical efficacy against depression, for which activation of PKB or PKC could represent the underlying mechanism for therapeutic activity.

4. Activation of PKB/Akt to Inhibit GSK3β

Electrical stimulation of the vagus nerve is an exploratory method for the treatment of depression and inflammation [105–108]. The mode of action involves activation of nicotinic α7 receptors [109–111], which results in increases in Ser9-phosphorylation of GSK3β [112–114]. This pharmacological response involves a pathway where stimulation of the α7 receptor activates the kinase JAK2, which leads to subsequent activation of PI3K, PKB, and phosphorylation/inhibition of GSK3 [115,116]. Consistent with the inhibition of GSK3, NRF2 signaling was promoted, as seen by increases in gene transcription of antioxidant and anti-apoptotic genes and by a reduction in the levels of pro-inflammatory cytokines [112,115,116]. These effects occur in microglia cells [117,118]. Importantly, treatment with an experimental nicotinic α7 agonist mitigated both the biochemical changes and the depressive behavior of mice subjected to the chronic mild stress procedure [119]. It may be noted that responses to nicotinic-α7 agonists resemble the effects of the anti-inflammatory cytokine IL10. Both activate the JAK2/STAT3 and the PI3K-Akt pathways to improve NRF2-mediated HO-1 (heme oxygenase-1) transcription, and both inhibit NFκB and suppress the hypertrophic/activated microglia polarization [116].

BDNF is produced by neurons [120–122], astrocytes [123,124], endothelial cells [93], as well as by reparative microglia cells [125,126]. The receptor for BDNF, TrkB (tropomyosin kinase receptor B), is expressed by neurons [120,122], astrocytes, and microglia [127–130]. TrkB signaling leads to activation of the PI3K-Akt pathway [122,131,132] and, as a consequence thereof, to inhibition of GSK3β [133] and activation of CREB and NRF2 [134]. Since long, BDNF is suspected to play an important role in the antidepressant effect of both classical and novel antidepressant medications [135–139]. HDAC inhibition by valproate, butyrate, trichostatin, or fingolimod increases the expression of BDNF [121,140,141]. HDAC inhibitors may act as antidepressants [142,143], which is consistent with their proposed effect on the BDNF-TrkB-Akt-GSK3 signal transduction cascade.

5. Gq-PLC-PKC-Induced GSK3β-Inhibition
5.1. Fatty Acids and Endocannabinoids

Fatty acids and their oxidized metabolites can form endogenous cannabinoids when they are esterified to ethanolamine, glycerin, or amino acids such as glycine or serine. These mediators are produced by astrocytes and microglia [144] and activate several GPCRs [145]. Microglia and macrophages particularly express CBR2 and the 'orphan' cannabinoid receptor, GPR18 [144,146–148]. Activation of both receptors inhibits the production of pro-inflammatory cytokines and increases the levels of IL10 and IL1-RA [147–149]. GPR18 seems to couple to G$_S$ [150,151], whereas the CB2 receptor causes PLC activation [147], although G$_S$-cAMP-PKA signaling has been reported too [152]. The CB2 receptor is inducible and becomes expressed in microglia cells following inflammation or injury (reviewed by [153]). Overexpression of CBR2 in mice resulted in resistance to chronic mild stress-induced depression [39,154]. The ethanolamines (EA's) formed from the omega-3 polyunsaturated fatty acids DHA and EPA (DHA-EA and EPEA, respectively) display anti-inflammatory activity in microglia, both in vitro and in vivo [155]. DHA-EA (known also as "synaptamide") activates cAMP-PKA signaling in microglia and is a potent suppressor of LPS-induced neuroinflammation in mice [156–158]. Synaptamide is an agonist at the Gs-coupled orphan receptor GPR110/ADGRF1 [159]. Likewise, the epoxides of EPA and DHA (EEQ and EDP, respectively) can be esterified to ethanolamine, generating the endogenous cannabinoids EEQ-EA and EDP-EA. These products are produced by BV-2 microglial cells and are known to inhibit IL6 and NO production while raising the levels of IL10 [160]. These responses are in part mediated through CBR2 [160], however, any agonistic activity at GPR18 or GPR110 remains to be tested. Such experiments are indicated, as there is an evident discrepancy between the potent anti-inflammatory effects of 19,20-EDP-EA and its weak CB2 affinity in the data reported by McDougle et al. [160]. To summarize, the data collected in this paragraph indicate that agonists for microglial endocannabinoid receptors could represent novel approaches for the treatment of depression.

5.2. Opioid μ-Receptor Agonists

Other examples of receptors that are Gq-coupled and activate PLC-PKC are the opioid μ-receptor and the class of 5HT2-receptors (see Table 1).

Table 1. Gq-coupled receptors that mediate anti-inflammatory and antidepressant activity.

Receptor	Citations for Expression by Microglia	Citations for Gq-PLC-PKC Coupling	Citations for Inflammatory Cytokines Decrease	Citation for Antidepressant Activity
Cannabinoid CBR2	[146,153]	[147,161]	[148,149,161]	[144,162]
Opioid μ-receptor	[103,163]	[164]	[165]	[166–168]
Serotonin 5HT2B	[101,102]	[169–171]	[101,102]	[172–174]

Literature citations for the expression of the cannabinoid receptor CBR2, the opioid μ receptor, and the serotonin receptor 5HT2B by microglia cells. Agonists at these receptors activate phospholipase-C (PLC) and protein kinase-C (PKC), which results in inhibition of brain inflammation and improvement of depression symptoms.

The endogenous opioid system is known to play a central role in the pathophysiology of affective disorders [175–177]. Opioid μ-receptors, in contrast to δ- or κ-opioid receptors, are involved in immune modulation in the brain [178]. Furthermore, depression-scores in healthy subjects were inversely correlated with the binding capacity of an μ-receptor PET-ligand in cortical and subcortical brain regions [179]. Microglia cells express μ-receptors [103,163], and the prototypic opioid receptor agonist morphine decreases the production of IL1β, IL2, TNFα, and IFNγ, and increases TGFβ1 and IL10 [163]. Similarly, tianeptine, a selective μ-receptor receptor agonist [180], inhibited LPS-induced microglia M1-polarization and reduced the LPS-induced expression of IL1β, IL18, IL6, TNFα, and CCL2, as well as the production of NO and ROS [165]. In animal models, tianeptine reversed stress-induced dystrophy of hippocampal dendrites and produced antidepressant-like effects [176]. These effects were blocked by a selective μ-receptor antagonist and were absent in mice with a genetic μ-receptor deficiency [176]. In contrast to morphine, which has off-target pro-inflammatory activity via TLR4 in BV-2 microglia cells [181], tianeptine did not induce tolerance to the analgesic and anti-inflammatory effects [176]. Tianeptine reduces the depression symptoms in patients with mild to moderate-to-severe major depression [167], but given its inherent abuse liability, the risk-benefit ratio of tianeptine remains uncertain [182]. The mixed μ-receptor agonist/κ-antagonist buprenorphine is antidepressant too, and similar to tianeptine, it remains a therapeutic option for patients with treatment-resistant major depression [166,168]. The μ-receptor primarily couples to G_i/G_0 but additionally to G_q-mediated PLC-activation [164]. Activation of PLC leads to PKC activity and thereby, to phosphorylation/inhibition of GSK3.

5.3. Hallucinogenic 5HT2 Agonists

Anecdotal reports of the antidepressant activity of psilocybin and LSD [173] have led to the exploration of the antidepressant effects of psilocybin in randomized clinical trials [172,174]. Although the psychedelic effect precludes an effective study blinding, the antidepressant effect was encouraging. The hallucinogenic activity is ascribed to stimulation of central 5HT2A receptors [183,184], and in the neuron-centric view on depression, 5HT2A receptor stimulation is considered the likely mechanism for the improvement of mood [173,185]. Notably, serotonergic hallucinogens are potent activators of the 5HT receptor of the rat stomach fundus [186], a model for 5HT2B receptor activation of PKC [170].

Importantly, the 5HT2B receptor is strongly expressed in microglia [101], and its stimulation suppresses the release of pro-inflammatory cytokines [187]. It is conceivable that the antidepressant activity of psilocybin is related to the activation of 5HT2B receptors on microglia. This would imply that it should be possible to dissociate the hallucinogenic activity from the antidepressant activity by increasing the 5HT2B/5HT2A selectivity.

6. Discussion

Stress and infection activate the NFκB pathway in phagocytic cells and lead to the transcription of pro-inflammatory cytokines and oxidizing enzymes such as COX2 and iNOS [44,188–191]. The opposite process is initiated by the transcription factors CREB and NRF2, which compete with NFκB for access to the gene transcription machinery and activate the production of anti-oxidant proteins and anti-inflammatory cytokines [192–194]. Both CREB and NRF2 are stabilized when the pro-inflammatory kinase GSK3β is inhibited. GSK3β becomes blocked when kinases such as PKA, PKB, and PKC phosphorylate a distinct serine residue (Ser-9) in the N-terminal part of GSK3β. Any process that activates PKA, PKB, or PKC therefore has the potential to limit the consequences of stress and infection. In the foregoing sections, several pharmacological principles with antidepressant potential were discussed that, in each case, inhibited GSK3β, either by activating PKA, PKB, or PKC in microglial cells. A screen in microglial cells for compounds and pharmacological principles that inhibit GSK3β or activate CREB and NRF2 transcription has the potential to discover novel antidepressants. The pharmacological principles that were discussed in the foregoing sections were mostly investigated in LPS-stimulated human or rodent microglia cell lines [92,128,195], rat and mouse microglia primary cultures, or primary microglia acutely isolated after in vivo exposure to LPS or stress [94,118]. A good alternative to consider would be to screen induced pluripotent stem cell-derived microglial precursor cells [196]. However, a simple screen in mouse BV-2 microglial cells would already detect α-lipoic acid [197], 9-*cis*-retinoic acid [198], stable cAMP analogues [94,199], PDE4 inhibitors [200,201], agonists of the fatty acid receptor FFAR1/GPR40 [202], the 'specialized pro-resolving mediators' lipoxin A4, resolvin D1 and resolvin E1 [59,203–206], and inhibitors of the enzyme soluble epoxide-hydrolase [207]. Presumably, even ketamine might be detected as a hit.

6.1. Ketamine

The mechanism by which ketamine exerts its antidepressant effects is still a matter of debate [208,209]. The two mechanisms that are discussed are glutamate NMDA-receptor blockade and opioid μ-receptor activation. The issue with NMDA-blockade is that other NMDA blockers do not induce a rapid antidepressant effect (for instance, memantine [210]), whereas the ketamine-metabolite 2R,6R-HNK provoked rapid antidepressant-like activity in preclinical models that was independent of NMDA-blockade [211]. Moreover, ketamine was still 'antidepressant' in transgenic mice that lack the essential NR1-subunit of the NMDA channel, whereas glycine-site agonists that *activate* the functioning of the NMDA-channel are also antidepressants (reviewed in [209]). Ketamine has a significant affinity for opioid receptors and is antinociceptive [210]. In a pilot study in humans with major depression, the antidepressant but not the dissociative effects of ketamine were blocked by naltrexone [212]. Notably, the antinociceptive effects of 2R,6R-HNK were *not* affected by the opioid μ-receptor antagonist naltrexone [213]. Independent of the precise mechanism of action, ketamine induced a robust inhibitory phosphorylation of GSK3 in preclinical models, whereas the antidepressant-like effect of ketamine was lost in transgenic mice that carried a non-inhibitable GSK3-variant [214,215]. Consistent with this activity, ketamine increased the expression of BDNF [216,217], increased the expression of the NRF2-target gene heme oxygenase-1 [218], suppressed chronic stress-induced depressive behaviors and pro-inflammatory cytokine levels [219–221], as well as LPS-induced depressive behaviors [222]. In a human monocyte cell line, ketamine inhibited NFκB and reduced pro-inflammatory cytokine levels [223,224]. Ketamine administration to humans before or during surgery inhibited the increase in post-operative IL6 and TNFα levels [225,226]. Similarly, in two small studies in patients suffering from MDD, blood IL6 levels decreased after ketamine treatment in the responders but not in the non-responders [227,228]. These results suggest that suppression of inflammation is relevant for the antidepressant effects of ketamine. The remaining question is: by what mechanism does ketamine provoke inhibition of GSK3β. An interesting proposal was made by Wray and co-authors [124]. G_S is singly palmitoylated on its N-terminus, which targets the protein to lipid rafts. Ketamine and 2R,6R-HNK are lipophilic and remove G_s from the lipid raft. This has a positive effect on the interaction between G_S and adenylate cyclase and thus on the production of cAMP, the activation of PKA, and the CREB-target gene BDNF [82,124]. The authors consider it likely that other lipid-soluble anesthetics act similarly to ketamine and 2R,6R-HNK. This proposal is consistent with the observation that intravenous and inhaled anesthetics (ketamine, propofol, opioids, isoflurane, sevoflurane, desflurane, etc.) to varying extents upregulate heme oxygenase-1 expression [218]. The ketamine-induced displacement of G_S from the lipid rafts occurs much more rapidly than with, for instance, tricyclic antidepressants [82] and could be an explanation for ketamine's faster onset of antidepressant activity. It is conceivable that microglial cells are the targets for these effects [41,229,230].

6.2. GSK3β Blockade by Antipsychotics

Although dopamine D2 receptor antagonists (antipsychotics) are known to elevate the levels of Ser-9 phosphorylated GSK3β in the brain [231], such compounds are not genuinely known to act as antidepressants. A conceivable explanation for this apparent discrepancy is provided hereafter.

Dopamine and serotonin not only influence microglia cells, they also modulate the function of neurons. However, the downstream signaling pathways in neurons are distinctly different from those in microglia. Activation of dopamine D2 receptors expressed on neurons (by, for instance, amphetamine) leads to activation of GSK3. The signaling cascade involves the formation of a molecular complex consisting of β-arrestin2 (βArr2), PKB, and the phosphatase PP2A [232]. Formation of this complex leads to PP2A-mediated deactivation of PKB and thus to the disinhibition of GSK3 [232]. Conversely, D2-receptor antagonists such as haloperidol, raclopride, and atypical antipsychotics such as clozapine, risperidone, olanzapine, quetiapine, and ziprasidone (their common profile is dual antagonism of 5HT2A receptors and D2-receptors [233,234]) all induce inhibition of GSK3, owing to the inhibitory phosphorylation of GSK3 by PKB [235,236]. Similar to dopamine, activation of neuronal 5HT2A receptors involves the formation of a molecular complex, here consisting of βArr2, PKB, and Src [237] and ultimately results in activation of GSK3 [238]. In agreement, 5HT2A receptor antagonists provoke an inhibition of GSK3 [239]. Therefore, a GSK3 inhibitor will block amphetamine-induced mania (GSK3 inhibition in neurons) and will generate antidepressant responses (via GSK3 inhibition in microglia). The solution to the apparent discrepancy mentioned above is quite simple. Whereas a GKS3 inhibitor will presumably limit both the manic and depressive symptoms, a D2-receptor antagonist will affect the signaling in neurons [231], but will not produce a profound antidepressant effect (see Figure 2). Under the assumption that this explanation is proven to be correct, it would emphasize the point that only GSK3 blockade in microglia results in antidepressant activity.

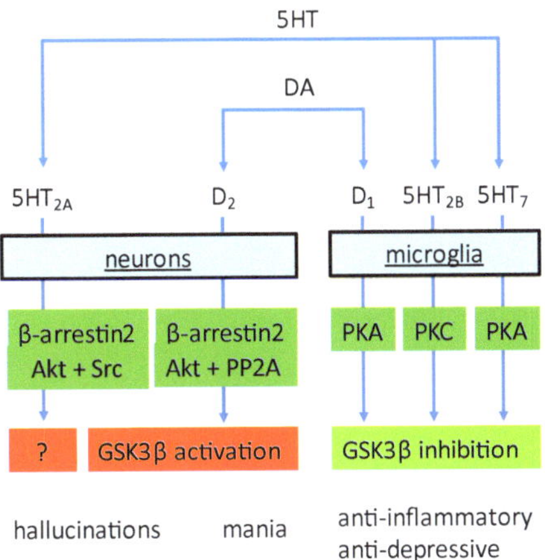

Figure 2. Dopamine (DA) and serotonin (5HT) activate distinct receptor subtypes and signaling pathways in microglia and neurons. In microglial cells, the functional outcome is an inhibition of GSK3β, whereas activation of the D2-receptor in neurons results in activation of GSK3β. Whether activation of the 5HT2A receptor in neurons leads to GSK3β activation has not been investigated.

6.3. Conclusions

Depression is a major public health concern. The risk-benefit ratio of the currently available antidepressants is quite unsatisfactory, since either their therapeutic effectiveness is insufficient or they cause unacceptable side effects. Consequently, there is a pressing need to discover better treatments. Literature data indicate that risk factors for depression initiate an infection-like inflammation in the brain that involves activation of microglial Toll-like receptors and glycogen synthase kinase-3β (GSK3β). GSK3β activity alters the balance between two major competing transcription factors, the pro-inflammatory/pro-oxidative transcription factor NFκB and the neuroprotective, anti-inflammatory, and anti-oxidative transcription factor NRF2. The negative consequences of depression risk factors could thus be counteracted by procedures that lead to GSK3β inhibition. As an example, the antidepressant activity of tricyclic antidepressants is assumed to involve activation of G$_S$-coupled microglial receptors, raising intracellular cAMP levels, and activation of protein kinase A (PKA). PKA inhibits the enzyme activity of GSK3β. Importantly, PKA is just one of a series of AGC-kinases that is able to inhibit GSK3β. Literature information is provided to show that each pharmacological principle for which evidence for antidepressant activity in humans exists results in the inhibition of GSK3β in microglial cells, albeit via different signaling cascades and involving different AGC-kinases. Ketamine and lipid-soluble anesthetics act somewhat distinct in that they do lead to GSK3β inhibition; however, not via a pharmacological effect but in response to a physical property (lipid solubility) that alters the lipid raft composition. Furthermore, it is argued that in order to achieve antidepressant activity, GSK3β inhibition has to occur in microglia rather than in neurons. Based on these considerations, an in vitro screen testing for GSK3β inhibition/NRF2 activation in microglial cells with TLR4-stimulated GSK3β activity might ultimately lead to the detection of novel antidepressant principles. This would, hopefully, end the stagnation in the discovery and development of safe and effective treatments for major depressive disorders.

Funding: This research received no external funding.

Institutional Review Board Statement: Not applicable.

Informed Consent Statement: Not applicable.

Data Availability Statement: Not applicable.

Conflicts of Interest: The author declares no conflict of interest.

Abbreviations

5HT (5-hydroxytryptamine); ATP (adenosine triphosphate); BDNF (brain-derived neurotrophic factor); cAMP (cyclic adenosine monophosphate); CCL2 (CC-chemokine ligand-2); CNS (central nervous system); CBP (CREB-binding protein); COX2 (cyclooxygenase-2); CREB (cyclic-AMP responsive element binding protein); CX3CR1 (CX3C motif chemokine receptor-1); DHA (docosahexaenoic acid); DUSP1 (dual specificity phosphatase-1); EA (ethanolamine); EPA (Eicosapentaenoic acid); FFAR1 (free fatty acid receptor-1); GPCR (G-protein coupled receptor); GSK3β (glycogen synthase kinase-3β); HDAC (histone deacetylase); HMGB1 (high mobility group box-1); HO-1 (heme oxygenase-1); IFNγ (interferon-γ); IL (interleukin); IL1-RA (interleukin-1 receptor antagonist); iNOS (inducible nitric oxide synthase); JAK2 jJanus kinase-2); LPS (lipopolysaccharide/TLR4 agonist); MDD (major depressive disorder); NFκB (nuclear Factor-κB); NMDA (N-methyl-D-aspartate); NO (nitric oxide); NRF2 (nuclear factor erythroid 2-related factor-2); PDE4 (phosphodiesterase-4); PET (positron emission tomography); PKA (protein kinase-A); PKB (protein kinase-B); PKC (protein kinase-C); PLC (phospholipase-C); PP2A (protein phosphatase-2A); ROS (reactive oxygen species); TGFβ1 (transforming growth factor-β1); TLR (toll-like receptor); TNFα (tumor necrosis factor-α); TrkB (tropomyosin kinase receptor-B).

References

1. Malhi, G.S.; Mann, J.J. Depression. *Lancet* **2018**, *392*, 2299–2312. [CrossRef]
2. Schildkraut, J.J. The Catecholamine Hypothesis of Affective Disorders: A Review of Supporting Evidence. *Am. J. Psychiatry* **1965**, *122*, 509–522. [CrossRef]
3. Capuron, L.; Miller, A.H. Immune system to brain signaling: Neuropsychopharmacological implications. *Pharmacol. Ther.* **2011**, *130*, 226–238. [CrossRef]
4. Berk, M.; Williams, L.J.; Jacka, F.N.; O'Neil, A.; Pasco, J.A.; Moylan, S.; Allen, N.B.; Stuart, A.L.; Hayley, A.; Byrne, M.L.; et al. So depression is an inflammatory disease, but where does the inflammation come from? *BMC Med.* **2013**, *11*, 200. [CrossRef]
5. Ögren, S.O.; Fuxe, K.; Agnati, L. The Importance of Brain Serotonergic Receptor Mechanisms for the Action of Antidepressant Drugs. *Pharmacopsychiatry* **1985**, *18*, 209–213. [CrossRef]
6. Palazidou, E.; Beer, M.S.; Checkley, S.; Stahl, S.M. Pharmacologic exploitation of neurotransmitter receptors for the design of novel antidepressant drugs. *Drug Des. Deliv.* **1988**, *2*, 247–256.
7. Broekkamp, C.L.E.; Leysen, D.; Peeters, B.W.; Pinder, R.M. Prospects for Improved Antidepressants. *J. Med. Chem.* **1995**, *38*, 4615–4633. [CrossRef]
8. Kenis, G.; Maes, M. Effects of antidepressants on the production of cytokines. *Int. J. Neuropsychopharmacol.* **2002**, *5*, 401–412. [CrossRef]
9. Haroon, E.; Raison, C.L.; Miller, A.H. Psychoneuroimmunology Meets Neuropsychopharmacology: Translational Implications of the Impact of Inflammation on Behavior. *Neuropsychopharmacology* **2011**, *37*, 137–162. [CrossRef]
10. GBD 2019 Mental Disorders Collaborators. Global, regional, and national burden of 12 mental disorders in 204 countries and territories, 1990–2019: A systematic analysis for the Global Burden of Disease Study 2019. *Lancet Psychiatry* **2022**, *9*, 137–150. [CrossRef]
11. COVID-19 Mental Disorders Collaborators. Global prevalence and burden of depressive and anxiety disorders in 204 countries and territories in 2020 due to the COVID-19 pandemic. *Lancet* **2021**, *398*, 1700–1712. [CrossRef]
12. Lee, S.; Jeong, J.; Kwak, Y.; Park, S.K. Depression research: Where are we now? *Mol. Brain* **2010**, *3*, 8. [CrossRef]
13. Koo, J.W.; Russo, S.J.; Ferguson, D.; Nestler, E.J.; Duman, R.S. Nuclear factor-κB is a critical mediator of stress-impaired neurogenesis and depressive behavior. *Proc. Natl. Acad. Sci. USA* **2010**, *107*, 2669–2674. [CrossRef]
14. Cheng, Y.; Pardo, M.; Armini, R.D.S.; Martinez, A.; Mouhsine, H.; Zagury, J.-F.; Jope, R.S.; Beurel, E. Stress-induced neuroinflammation is mediated by GSK3-dependent TLR4 signaling that promotes susceptibility to depression-like behavior. *Brain Behav. Immun.* **2016**, *53*, 207–222. [CrossRef]
15. Duda, P.; Hajka, D.; Wójcicka, O.; Rakus, D.; Gizak, A. GSK3β: A Master Player in Depressive Disorder Pathogenesis and Treatment Responsiveness. *Cells* **2020**, *9*, 727. [CrossRef]
16. Krishnadas, R.; Cavanagh, J. Depression: An inflammatory illness? *J. Neurol. Neurosurg. Psychiatry* **2012**, *83*, 495–502. [CrossRef]
17. Strawbridge, R.; Arnone, D.; Danese, A.; Papadopoulos, A.; Vives, A.H.; Cleare, A.J. Inflammation and clinical response to treatment in depression: A meta-analysis. *Eur. Neuropsychopharmacol.* **2015**, *25*, 1532–1543. [CrossRef]
18. Benros, M.E.; Waltoft, B.L.; Nordentoft, M.; Østergaard, S.D.; Eaton, W.W.; Krogh, J.; Mortensen, P.B. Autoimmune Diseases and Severe Infections as Risk Factors for Mood Disorders: A nationwide study. *JAMA Psychiatry* **2013**, *70*, 812–820. [CrossRef]
19. Bhandari, S.; Larson, M.E.; Kumar, N.; Stein, D. Association of Inflammatory Bowel Disease (IBD) with Depressive Symptoms in the United States Population and Independent Predictors of Depressive Symptoms in an IBD Population: A NHANES Study. *Gut Liver* **2017**, *11*, 512–519. [CrossRef]

20. Beurel, E.; Toups, M.; Nemeroff, C.B. The Bidirectional Relationship of Depression and Inflammation: Double Trouble. *Neuron* **2020**, *107*, 234–256. [CrossRef]
21. Kalkman, H.O. The Association Between Vascular Inflammation and Depressive Disorder. Causality, Biomarkers and Targeted Treatment. *Pharmaceuticals* **2020**, *13*, 92. [CrossRef] [PubMed]
22. Vezzani, A. Epilepsy and Inflammation in the Brain: Overview and Pathophysiology. *Epilepsy Curr.* **2014**, *14*, 3–7. [CrossRef]
23. Chung, Y.-C.; Ko, H.-W.; Bok, E.-G.; Park, E.-S.; Huh, S.-H.; Nam, J.-H.; Jin, B.-K. The role of neuroinflammation on the pathogenesis of Parkinson's disease. *BMB Rep.* **2010**, *43*, 225–232. [CrossRef]
24. Ouchi, Y.; Yagi, S.; Yokokura, M.; Sakamoto, M. Neuroinflammation in the living brain of Parkinson's disease. *Park. Relat. Disord.* **2009**, *15* (Suppl. S3), S200–S204. [CrossRef]
25. Tan, Z.S.; Seshadri, S. Inflammation in the Alzheimer's disease cascade: Culprit or innocent bystander? *Alzheimer's Res. Ther.* **2010**, *2*, 6. [CrossRef]
26. Sandberg, M.; Patil, J.; D'Angelo, B.; Weber, S.G.; Mallard, C. NRF2-regulation in brain health and disease: Implication of cerebral inflammation. *Neuropharmacology* **2014**, *79*, 298–306. [CrossRef]
27. Sullivan, L.E.; Fiellin, D.A.; O'Connor, P.G. The prevalence and impact of alcohol problems in major depression: A systematic review. *Am. J. Med.* **2005**, *118*, 330–341. [CrossRef]
28. Fergusson, D.M.; Boden, J.M.; Horwood, L.J. Tests of Causal Links Between Alcohol Abuse or Dependence and Major Depression. *Arch. Gen. Psychiatry* **2009**, *66*, 260–266. [CrossRef]
29. Beurel, E.; Jope, R.S. Inflammation and lithium: Clues to mechanisms contributing to suicide-linked traits. *Transl. Psychiatry* **2014**, *4*, e488. [CrossRef]
30. Mogensen, T.H. Pathogen Recognition and Inflammatory Signaling in Innate Immune Defenses. *Clin. Microbiol. Rev.* **2009**, *22*, 240–273. [CrossRef]
31. Kaufmann, F.N.; Costa, A.P.; Ghisleni, G.; Diaz, A.P.; Rodrigues, A.L.S.; Peluffo, H.; Kaster, M.P. NLRP3 inflammasome-driven pathways in depression: Clinical and preclinical findings. *Brain Behav. Immun.* **2017**, *64*, 367–383. [CrossRef]
32. Park, J.S.; Svetkauskaite, D.; He, Q.; Kim, J.Y.; Strassheim, D.; Ishizaka, A.; Abraham, E. Involvement of toll-like receptors 2 and 4 in cellular activation by high mobility group box 1 protein. *J. Biol. Chem.* **2004**, *279*, 7370–7377. [CrossRef]
33. Kim, S.Y.; Son, M.; Lee, S.E.; Park, I.H.; Kwak, M.S.; Han, M.; Lee, H.S.; Kim, E.S.; Kim, J.-Y.; Lee, J.E.; et al. High-Mobility Group Box 1-Induced Complement Activation Causes Sterile Inflammation. *Front. Immunol.* **2018**, *9*, 705. [CrossRef]
34. Alfonso-Loeches, S.; Pascual-Lucas, M.; Blanco, A.M.; Sanchez-Vera, I.; Guerri, C. Pivotal Role of TLR4 Receptors in Alcohol-Induced Neuroinflammation and Brain Damage. *J. Neurosci.* **2010**, *30*, 8285–8295. [CrossRef]
35. Dantzer, R.; O'Connor, J.C.; Freund, G.G.; Johnson, R.W.; Kelley, K.W. From inflammation to sickness and depression: When the immune system subjugates the brain. *Nat. Rev. Neurosci.* **2008**, *9*, 46–56. [CrossRef]
36. Maes, M.; Berk, M.; Goehler, L.; Song, C.; Anderson, G.; Gałecki, P.; Leonard, B. Depression and sickness behavior are Janus-faced responses to shared inflammatory pathways. *BMC Med.* **2012**, *10*, 66. [CrossRef]
37. Kendler, K.S.; Hettema, J.M.; Butera, F.; Gardner, C.O.; Prescott, C.A. Life Event Dimensions of Loss, Humiliation, Entrapment, and Danger in the Prediction of Onsets of Major Depression and Generalized Anxiety. *Arch. Gen. Psychiatry* **2003**, *60*, 789–796. [CrossRef]
38. Calcia, M.A.; Bonsall, D.R.; Bloomfield, P.S.; Selvaraj, S.; Barichello, T.; Howes, O.D. Stress and neuroinflammation: A systematic review of the effects of stress on microglia and the implications for mental illness. *Psychopharmacology* **2016**, *233*, 1637–1650. [CrossRef]
39. Willner, P. The chronic mild stress (CMS) model of depression: History, evaluation and usage. *Neurobiol. Stress* **2017**, *6*, 78–93. [CrossRef]
40. Rohleder, N. Stimulation of Systemic Low-Grade Inflammation by Psychosocial Stress. *Psychosom. Med.* **2014**, *76*, 181–189. [CrossRef]
41. Yirmiya, R.; Rimmerman, N.; Reshef, R. Depression as a Microglial Disease. *Trends Neurosci.* **2015**, *38*, 637–658. [CrossRef] [PubMed]
42. Kreisel, T.; Frank, M.G.; Licht, T.; Reshef, R.; Ben-Menachem-Zidon, O.; Baratta, M.V.; Maier, S.F.; Yirmiya, R. Dynamic microglial alterations underlie stress-induced depressive-like behavior and suppressed neurogenesis. *Mol. Psychiatry* **2014**, *19*, 699–709. [CrossRef] [PubMed]
43. Beumer, W.; Gibney, S.M.; Drexhage, R.C.; Pont-Lezica, L.; Doorduin, J.; Klein, H.C.; Steiner, J.; Connor, T.J.; Harkin, A.; Versnel, M.A.; et al. The immune theory of psychiatric diseases: A key role for activated microglia and circulating monocytes. *J. Leukoc. Biol.* **2012**, *92*, 959–975. [CrossRef] [PubMed]
44. Hinwood, M.; Morandini, J.; Day, T.A.; Walker, F.R. Evidence that Microglia Mediate the Neurobiological Effects of Chronic Psychological Stress on the Medial Prefrontal Cortex. *Cereb. Cortex* **2012**, *22*, 1442–1454. [CrossRef]
45. Delpech, J.-C.; Madore, C.; Nadjar, A.; Joffre, C.; Wohleb, E.S.; Layé, S. Microglia in neuronal plasticity: Influence of stress. *Neuropharmacology* **2015**, *96*, 19–28. [CrossRef]
46. Lehmann, M.L.; Cooper, H.A.; Maric, D.; Herkenham, M. Social defeat induces depressive-like states and microglial activation without involvement of peripheral macrophages. *J. Neuroinflamm.* **2016**, *13*, 224. [CrossRef]
47. Lehmann, M.L.; Weigel, T.K.; Poffenberger, C.N.; Herkenham, M. The Behavioral Sequelae of Social Defeat Require Microglia and Are Driven by Oxidative Stress in Mice. *J. Neurosci.* **2019**, *39*, 5594–5605. [CrossRef]

48. Tang, J.; Yu, W.; Chen, S.; Gao, Z.; Xiao, B. Microglia Polarization and Endoplasmic Reticulum Stress in Chronic Social Defeat Stress Induced Depression Mouse. *Neurochem. Res.* **2018**, *43*, 985–994. [CrossRef]
49. McKim, D.B.; Niraula, A.; Tarr, A.J.; Wohleb, E.S.; Sheridan, J.F.; Godbout, J.P. Neuroinflammatory Dynamics Underlie Memory Impairments after Repeated Social Defeat. *J. Neurosci.* **2016**, *36*, 2590–2604. [CrossRef] [PubMed]
50. Ohgidani, M.; Kato, T.A.; Sagata, N.; Hayakawa, K.; Shimokawa, N.; Sato-Kasai, M.; Kanba, S. TNF-α from hippocampal microglia induces working memory deficits by acute stress in mice. *Brain Behav. Immun.* **2016**, *55*, 17–24. [CrossRef]
51. Wang, Y.-L.; Han, Q.-Q.; Gong, W.-Q.; Pan, D.-H.; Wang, L.-Z.; Hu, W.; Yang, M.; Li, B.; Yu, J.; Liu, Q. Microglial activation mediates chronic mild stress-induced depressive- and anxiety-like behavior in adult rats. *J. Neuroinflamm.* **2018**, *15*, 21. [CrossRef]
52. Ramirez, K.; Sheridan, J.F. Antidepressant imipramine diminishes stress-induced inflammation in the periphery and central nervous system and related anxiety- and depressive- like behaviors. *Brain Behav. Immun.* **2016**, *57*, 293–303. [CrossRef] [PubMed]
53. Zhang, C.; Zhang, Y.-P.; Li, Y.-Y.; Liu, B.-P.; Wang, H.-Y.; Li, K.-W.; Zhao, S.; Song, C. Minocycline ameliorates depressive behaviors and neuro-immune dysfunction induced by chronic unpredictable mild stress in the rat. *Behav. Brain Res.* **2019**, *356*, 348–357. [CrossRef]
54. Ślusarczyk, J.; Trojan, E.; Głombik, K.; Budziszewska, B.; Kubera, M.; Lasoń, W.; Popiołek-Barczyk, K.; Mika, J.; Wędzony, K.; Basta-Kaim, A. Prenatal stress is a vulnerability factor for altered morphology and biological activity of microglia cells. *Front. Cell. Neurosci.* **2015**, *9*, 82. [CrossRef]
55. Wohleb, E.S.; Hanke, M.L.; Corona, A.W.; Powell, N.D.; Stiner, L.M.; Bailey, M.T.; Nelson, R.J.; Godbout, J.P.; Sheridan, J.F. β-Adrenergic Receptor Antagonism Prevents Anxiety-Like Behavior and Microglial Reactivity Induced by Repeated Social Defeat. *J. Neurosci.* **2011**, *31*, 6277–6288. [CrossRef] [PubMed]
56. Weber, M.D.; Frank, M.G.; Tracey, K.J.; Watkins, L.R.; Maier, S.F. Stress Induces the Danger-Associated Molecular Pattern HMGB-1 in the Hippocampus of Male Sprague Dawley Rats: A Priming Stimulus of Microglia and the NLRP3 Inflammasome. *J. Neurosci.* **2015**, *35*, 316–324. [CrossRef] [PubMed]
57. Deng, S.-L.; Chen, J.-G.; Wang, F. Microglia: A Central Player in Depression. *Curr. Med. Sci.* **2020**, *40*, 391–400. [CrossRef]
58. Wang, H.; He, Y.; Sun, Z.; Ren, S.; Liu, M.; Wang, G.; Yang, J. Microglia in depression: An overview of microglia in the pathogenesis and treatment of depression. *J. Neuroinflamm.* **2022**, *19*, 132. [CrossRef]
59. Rahimian, R.; Belliveau, C.; Chen, R.; Mechawar, N. Microglial Inflammatory-Metabolic Pathways and Their Potential Therapeutic Implication in Major Depressive Disorder. *Front. Psychiatry* **2022**, *13*, 871997. [CrossRef]
60. Martin, M.; Rehani, K.; Jope, R.S.; Michalek, S.M. Toll-like receptor–mediated cytokine production is differentially regulated by glycogen synthase kinase 3. *Nat. Immunol.* **2005**, *6*, 777–784. [CrossRef]
61. Viatour, P.; Merville, M.-P.; Bours, V.; Chariot, A. Phosphorylation of NF-κB and IκB proteins: Implications in cancer and inflammation. *Trends Biochem. Sci.* **2005**, *30*, 43–52. [CrossRef] [PubMed]
62. Wen, A.Y.; Sakamoto, K.M.; Miller, L.S. The Role of the Transcription Factor CREB in Immune Function. *J. Immunol.* **2010**, *185*, 6413–6419. [CrossRef]
63. Gill, A.J.; Kolson, D.L. Dimethyl Fumarate Modulation of Immune and Antioxidant Responses: Application to HIV Therapy. *Crit. Rev. Immunol.* **2013**, *33*, 307–359. [CrossRef] [PubMed]
64. Kim, S.-W.; Lee, H.-K.; Shin, J.-H.; Lee, J.-K. Up-down Regulation of HO-1 and iNOS Gene Expressions by Ethyl Pyruvate via Recruiting p300 to Nrf2 and Depriving It from p65. *Free. Radic. Biol. Med.* **2013**, *65*, 468–476. [CrossRef] [PubMed]
65. Morris, G.; Walker, A.J.; Walder, K.; Berk, M.; Marx, W.; Carvalho, A.F.; Maes, M.; Puri, B.K. Increasing Nrf2 Activity as a Treatment Approach in Neuropsychiatry. *Mol. Neurobiol.* **2021**, *58*, 2158–2182. [CrossRef]
66. Zuo, C.; Cao, H.; Song, Y.; Gu, Z.; Huang, Y.; Yang, Y.; Miao, J.; Zhu, L.; Chen, J.; Jiang, Y.; et al. Nrf2: An all-rounder in depression. *Redox Biol.* **2022**, *58*, 102522. [CrossRef]
67. Maes, M.; Fišar, Z.; Medina, M.; Scapagnini, G.; Nowak, G.; Berk, M. New drug targets in depression: Inflammatory, cell-mediated immune, oxidative and nitrosative stress, mitochondrial, antioxidant, and neuroprogressive pathways. And new drug candidates—Nrf2 activators and GSK-3 inhibitors. *Inflammopharmacology* **2012**, *20*, 127–150. [CrossRef]
68. Rada, P.; Rojo, A.I.; Chowdhry, S.; McMahon, M.; Hayes, J.D.; Cuadrado, A. SCF/β-TrCP Promotes Glycogen Synthase Kinase 3-Dependent Degradation of the Nrf2 Transcription Factor in a Keap1-Independent Manner. *Mol. Cell. Biol.* **2011**, *31*, 1121–1133. [CrossRef]
69. Dinkova-Kostova, A.T.; Kostov, R.V.; Kazantsev, A.G. The role of Nrf2 signaling in counteracting neurodegenerative diseases. *FEBS J.* **2018**, *285*, 3576–3590. [CrossRef]
70. Kondoh, K.; Nishida, E. Regulation of MAP kinases by MAP kinase phosphatases. *Biochim. Biophys. Acta (BBA) Mol. Cell Res.* **2007**, *1773*, 1227–1237. [CrossRef]
71. Kasuya, Y.; Umezawa, H.; Hatano, M. Stress-Activated Protein Kinases in Spinal Cord Injury: Focus on Roles of p38. *Int. J. Mol. Sci.* **2018**, *19*, 867. [CrossRef]
72. Hoppstädter, J.; Ammit, A. Role of Dual-Specificity Phosphatase 1 in Glucocorticoid-Driven Anti-inflammatory Responses. *Front. Immunol.* **2019**, *10*, 1446. [CrossRef] [PubMed]
73. Jope, R.S.; Cheng, Y.; Lowell, J.A.; Worthen, R.J.; Sitbon, Y.H.; Beurel, E. Stressed and Inflamed, Can GSK3 Be Blamed? *Trends Biochem. Sci.* **2017**, *42*, 180–192. [CrossRef] [PubMed]
74. Beurel, E.; Michalek, S.M.; Jope, R.S. Innate and adaptive immune responses regulated by glycogen synthase kinase-3 (GSK3). *Trends Immunol.* **2010**, *31*, 24–31. [CrossRef]

75. Engel, T.; Hernández, F.; Avila, J.; Lucas, J.J. Full Reversal of Alzheimer's Disease-Like Phenotype in a Mouse Model with Conditional Overexpression of Glycogen Synthase Kinase-3. *J. Neurosci.* **2006**, *26*, 5083–5090. [CrossRef]

76. Guo, H.; Callaway, J.B.; Ting, J.P.-Y. Inflammasomes: Mechanism of action, role in disease, and therapeutics. *Nat. Med.* **2015**, *21*, 677–687. [CrossRef]

77. Qin, L.; He, J.; Hanes, R.N.; Pluzarev, O.; Hong, J.-S.; Crews, F.T. Increased systemic and brain cytokine production and neuroinflammation by endotoxin following ethanol treatment. *J. Neuroinflamm.* **2008**, *5*, 10. [CrossRef]

78. Forlenza, O.V.; Spink, J.M.; Dayanandan, R.; Anderton, B.H.; Olesen, O.F.; Lovestone, S. Muscarinic agonists reduce tau phosphorylation in non-neuronal cells via GSK-3β inhibition and in neurons. *J. Neural Transm.* **2000**, *107*, 1201–1212. [CrossRef]

79. Pearce, L.R.; Komander, D.; Alessi, D.R. The nuts and bolts of AGC protein kinases. *Nat. Rev. Mol. Cell Biol.* **2010**, *11*, 9–22. [CrossRef] [PubMed]

80. Dwivedi, Y.; Pandey, G.N. Adenylyl cyclase-cyclicAMP signaling in mood disorders: Role of the crucial phosphorylating enzyme protein kinase A. *Neuropsychiatr. Dis. Treat.* **2008**, *4*, 161–176. [CrossRef] [PubMed]

81. Fujita, M.; Richards, E.M.; Niciu, M.; Ionescu, D.F.; Zoghbi, S.S.; Hong, J.; Telu, S.; Hines, C.S.; Pike, V.W.; Zarate, C.A.; et al. cAMP signaling in brain is decreased in unmedicated depressed patients and increased by treatment with a selective serotonin reuptake inhibitor. *Mol. Psychiatry* **2017**, *22*, 754–759. [CrossRef] [PubMed]

82. Schappi, J.M.; Rasenick, M.M. Gαs, adenylyl cyclase, and their relationship to the diagnosis and treatment of depression. *Front. Pharmacol.* **2022**, *13*, 1012778. [CrossRef]

83. Karege, F.; Perroud, N.; Burkhardt, S.; Fernandez, R.; Ballmann, E.; La Harpe, R.; Malafosse, A. Protein levels of β-catenin and activation state of glycogen synthase kinase-3β in major depression. A study with postmortem prefrontal cortex. *J. Affect. Disord.* **2012**, *136*, 185–188. [CrossRef]

84. Bansal, Y.; Singh, R.; Parhar, I.; Kuhad, A.; Soga, T. Quinolinic Acid and Nuclear Factor Erythroid 2-Related Factor 2 in Depression: Role in Neuroprogression. *Front. Pharmacol.* **2019**, *10*, 452. [CrossRef] [PubMed]

85. Forlenza, M.J.; Miller, G.E. Increased Serum Levels of 8-Hydroxy-2′-Deoxyguanosine in Clinical Depression. *Psychosom. Med.* **2006**, *68*, 1–7. [CrossRef] [PubMed]

86. Savage, K.; Gogarty, L.; Lea, A.; Deleuil, S.; Nolidin, K.; Croft, K.; Stough, C. The Relationship between F_2-Isoprostanes Plasma Levels and Depression Symptoms in Healthy Older Adults. *Antioxidants* **2022**, *11*, 822. [CrossRef]

87. Yan, Y.; Jiang, W.; Liu, L.; Wang, X.; Ding, C.; Tian, Z.; Zhou, R. Dopamine Controls Systemic Inflammation through Inhibition of NLRP3 Inflammasome. *Cell* **2015**, *160*, 62–73. [CrossRef] [PubMed]

88. Wang, B.; Chen, T.; Li, G.; Jia, Y.; Wang, J.; Xue, L.; Chen, Y. Dopamine Alters Lipopolysaccharide-Induced Nitric Oxide Production in Microglial Cells via Activation of D1-Like Receptors. *Neurochem. Res.* **2019**, *44*, 947–958. [CrossRef]

89. Perreault, M.L.; Jones-Tabah, J.; O'Dowd, B.F.; George, S.R. A physiological role for the dopamine D5 receptor as a regulator of BDNF and Akt signalling in rodent prefrontal cortex. *Int. J. Neuropsychopharmacol.* **2013**, *16*, 477–483. [CrossRef]

90. Fujita, H.; Tanaka, J.; Maeda, N.; Sakanaka, M. Adrenergic agonists suppress the proliferation of microglia through β2-adrenergic receptor. *Neurosci. Lett.* **1998**, *242*, 37–40. [CrossRef]

91. Russo, C.D.; Boullerne, A.I.; Gavrilyuk, V.; Feinstein, D.L. Inhibition of microglial inflammatory responses by norepinephrine: Effects on nitric oxide and interleukin-1β production. *J. Neuroinflamm.* **2004**, *1*, 9. [CrossRef] [PubMed]

92. Mahé, C.; Loetscher, E.; Dev, K.K.; Bobirnac, I.; Otten, U.; Schoeffter, P. Serotonin 5-HT7 receptors coupled to induction of interleukin-6 in human microglial MC-3 cells. *Neuropharmacology* **2005**, *49*, 40–47. [CrossRef]

93. Nakahashi, T.; Fujimura, H.; Altar, C.; Li, J.; Kambayashi, J.-I.; Tandon, N.N.; Sun, B. Vascular endothelial cells synthesize and secrete brain-derived neurotrophic factor. *FEBS Lett.* **2000**, *470*, 113–117. [CrossRef] [PubMed]

94. Ghosh, M.; Xu, Y.; Pearse, D.D. Cyclic AMP is a key regulator of M1 to M2a phenotypic conversion of microglia in the presence of Th2 cytokines. *J. Neuroinflamm.* **2016**, *13*, 9. [CrossRef] [PubMed]

95. Shimizu, M.; Nishida, A.; Zensho, H.; Yamawaki, S. Chronic antidepressant exposure enhances 5-hydroxytryptamine7 receptor-mediated cyclic adenosine monophosphate accumulation in rat frontocortical astrocytes. *J. Pharmacol. Exp. Ther.* **1996**, *279*, 1551–1558.

96. Tynan, R.J.; Weidenhofer, J.; Hinwood, M.; Cairns, M.J.; Day, T.A.; Walker, F.R. A comparative examination of the anti-inflammatory effects of SSRI and SNRI antidepressants on LPS stimulated microglia. *Brain Behav. Immun.* **2012**, *26*, 469–479. [CrossRef]

97. Hashioka, S.; Klegeris, A.; Monji, A.; Kato, T.; Sawada, M.; McGeer, P.L.; Kanba, S. Antidepressants inhibit interferon-γ-induced microglial production of IL-6 and nitric oxide. *Exp. Neurol.* **2007**, *206*, 33–42. [CrossRef]

98. Bielecka, A.M.; Paul-Samojedny, M.; Obuchowicz, E. Moclobemide exerts anti-inflammatory effect in lipopolysaccharide-activated primary mixed glial cell culture. *Naunyn-Schmiedeberg's Arch. Pharmacol.* **2010**, *382*, 409–417. [CrossRef]

99. Wang, L.; Wang, R.; Liu, L.; Qiao, D.; Baldwin, D.S.; Hou, R. Effects of SSRIs on peripheral inflammatory markers in patients with major depressive disorder: A systematic review and meta-analysis. *Brain Behav. Immun.* **2019**, *79*, 24–38. [CrossRef]

100. Dionisie, V.; Filip, G.A.; Manea, M.C.; Manea, M.; Riga, S. The anti-inflammatory role of SSRI and SNRI in the treatment of depression: A review of human and rodent research studies. *Inflammopharmacology* **2021**, *29*, 75–90. [CrossRef]

101. Krabbe, G.; Matyash, V.; Pannasch, U.; Mamer, L.; Boddeke, H.W.; Kettenmann, H. Activation of serotonin receptors promotes microglial injury-induced motility but attenuates phagocytic activity. *Brain Behav. Immun.* **2012**, *26*, 419–428. [CrossRef] [PubMed]

102. de las Casas-Engel, M.; Domínguez-Soto, A.; Sierra-Filardi, E.; Bragado, R.; Nieto, C.; Puig-Kroger, A.; Samaniego, R.; Loza, M.; Corcuera, M.T.; Gómez-Aguado, F.; et al. Serotonin Skews Human Macrophage Polarization through HTR2B and HTR7. *J. Immunol.* **2013**, *190*, 2301–2310. [CrossRef]

103. Kettenmann, H.; Hanisch, U.-K.; Noda, M.; Verkhratsky, A. Physiology of Microglia. *Physiol. Rev.* **2011**, *91*, 461–553. [CrossRef] [PubMed]

104. Harry, G.J. Microglia during development and aging. *Pharmacol. Ther.* **2013**, *139*, 313–326. [CrossRef] [PubMed]

105. Rush, A.J.; Sackeim, H.A.; Marangell, L.B.; George, M.S.; Brannan, S.K.; Davis, S.M.; Lavori, P.; Howland, R.; Kling, M.A.; Rittberg, B.; et al. Effects of 12 Months of Vagus Nerve Stimulation in Treatment-Resistant Depression: A Naturalistic Study. *Biol. Psychiatry* **2005**, *58*, 355–363. [CrossRef]

106. Bajbouj, M.; Merkl, A.; Schlaepfer, T.; Frick, C.; Zobel, A.; Maier, W.; O'Keane, V.; Corcoran, C.; Adolfsson, R.; Trimble, M.; et al. Two-Year Outcome of Vagus Nerve Stimulation in Treatment-Resistant Depression. *J. Clin. Psychopharmacol.* **2010**, *30*, 273–281. [CrossRef]

107. Howland, R.H. Vagus Nerve Stimulation. *Curr. Behav. Neurosci. Rep.* **2014**, *1*, 64–73. [CrossRef]

108. Bonaz, B.; Sinniger, V.; Hoffmann, D.; Clarençon, D.; Mathieu, N.; Dantzer, C.; Vercueil, L.; Picq, C.; Trocmé, C.; Faure, P.; et al. Chronic vagus nerve stimulation in Crohn's disease: A 6-month follow-up pilot study. *Neurogastroenterol. Motil.* **2016**, *28*, 948–953. [CrossRef]

109. Borovikova, L.V.; Ivanova, S.; Zhang, M.; Yang, H.; Botchkina, G.I.; Watkins, L.R.; Wang, H.; Abumrad, N.; Eaton, J.W.; Tracey, K.J. Vagus nerve stimulation attenuates the systemic inflammatory response to endotoxin. *Nature* **2000**, *405*, 458–462. [CrossRef] [PubMed]

110. De Jonge, W.J.; Van Der Zanden, E.P.; The, F.O.; Bijlsma, M.F.; Van Westerloo, D.J.; Bennink, R.J.; Berthoud, H.-R.; Uematsu, S.; Akira, S.; van den Wijngaard, R.M.; et al. Stimulation of the vagus nerve attenuates macrophage activation by activating the Jak2-STAT3 signaling pathway. *Nat. Immunol.* **2005**, *6*, 844–851. [CrossRef]

111. Bencherif, M.; Lippiello, P.M.; Lucas, R.; Marrero, M.B. Alpha7 nicotinic receptors as novel therapeutic targets for inflammation-based diseases. *Cell. Mol. Life Sci.* **2011**, *68*, 931–949. [CrossRef] [PubMed]

112. Rehani, K.; Scott, D.A.; Renaud, D.; Hamza, H.; Williams, L.R.; Wang, H.; Martin, M. Cotinine-induced convergence of the cholinergic and PI3 kinase-dependent anti-inflammatory pathways in innate immune cells. *Biochim. Biophys. Acta (BBA) Mol. Cell Res.* **2008**, *1783*, 375–382. [CrossRef]

113. Bitner, R.S.; Nikkel, A.L.; Markosyan, S.; Otte, S.; Puttfarcken, P.; Gopalakrishnan, M. Selective α7 nicotinic acetylcholine receptor activation regulates glycogen synthase kinase3β and decreases tau phosphorylation in vivo. *Brain Res.* **2009**, *1265*, 65–74. [CrossRef]

114. Echeverria, V.; Zeitlin, R.; Burgess, S.; Patel, S.; Barman, A.; Thakur, G.; Mamcarz, M.; Wang, L.; Sattelle, D.B.; Kirschner, D.A.; et al. Cotinine Reduces Amyloid-β Aggregation and Improves Memory in Alzheimer's Disease Mice. *J. Alzheimer's Dis.* **2011**, *24*, 817–835. [CrossRef] [PubMed]

115. Bertrand, D.; Lee, C.-H.L.; Flood, D.; Marger, F.; Donnelly-Roberts, D. Therapeutic Potential of α7 Nicotinic Acetylcholine Receptors. *Pharmacol. Rev.* **2015**, *67*, 1025–1073. [CrossRef] [PubMed]

116. Egea, J.; Buendia, I.; Parada, E.; Navarro, E.; León, R.; Lopez, M.G. Anti-inflammatory role of microglial alpha7 nAChRs and its role in neuroprotection. *Biochem. Pharmacol.* **2015**, *97*, 463–472. [CrossRef]

117. Parada, E.; Egea, J.; Buendia, I.; Negredo, P.; Cunha, A.C.; Cardoso, S.; Soares, M.P.; López, M.G. The Microglial α7-Acetylcholine Nicotinic Receptor Is a Key Element in Promoting Neuroprotection by Inducing Heme Oxygenase-1 via Nuclear Factor Erythroid-2-Related Factor 2. *Antioxid. Redox Signal.* **2013**, *19*, 1135–1148. [CrossRef]

118. Navarro, E.; Gonzalez-Lafuente, L.; Pérez-Liébana, I.; Buendia, I.; López-Bernardo, E.; Sánchez-Ramos, C.; Prieto, I.; Cuadrado, A.; Satrustegui, J.; Cadenas, S.; et al. Heme-Oxygenase I and PCG-1α Regulate Mitochondrial Biogenesis *via* Microglial Activation of Alpha7 Nicotinic Acetylcholine Receptors Using PNU282987. *Antioxidants Redox Signal.* **2017**, *27*, 93–105. [CrossRef]

119. Zhao, D.; Xu, X.; Pan, L.; Zhu, W.; Fu, X.; Guo, L.; Lu, Q.; Wang, J. Pharmacologic activation of cholinergic alpha7 nicotinic receptors mitigates depressive-like behavior in a mouse model of chronic stress. *J. Neuroinflamm.* **2017**, *14*, 234. [CrossRef] [PubMed]

120. Kerschensteiner, M.; Gallmeier, E.; Behrens, L.; Leal, V.V.; Misgeld, T.; Klinkert, W.E.F.; Kolbeck, R.; Hoppe, E.; Oropeza-Wekerle, R.-L.; Bartke, I.; et al. Activated Human T Cells, B Cells, and Monocytes Produce Brain-derived Neurotrophic Factor In Vitro and in Inflammatory Brain Lesions: A Neuroprotective Role of Inflammation? *J. Exp. Med.* **1999**, *189*, 865–870. [CrossRef]

121. Yasuda, S.; Liang, M.-H.; Marinova, Z.I.; Yahyavi, A.; Chuang, D.-M. The mood stabilizers lithium and valproate selectively activate the promoter IV of brain-derived neurotrophic factor in neurons. *Mol. Psychiatry* **2009**, *14*, 51–59. [CrossRef] [PubMed]

122. Ghosal, S.; Bang, E.; Yue, W.; Hare, B.; Lepack, A.E.; Girgenti, M.J.; Duman, R.S. Activity-Dependent Brain-Derived Neurotrophic Factor Release Is Required for the Rapid Antidepressant Actions of Scopolamine. *Biol. Psychiatry* **2018**, *83*, 29–37. [CrossRef]

123. Wiese, S.; Karus, M.; Faissner, A. Astrocytes as a Source for Extracellular Matrix Molecules and Cytokines. *Front. Pharmacol.* **2012**, *3*, 120. [CrossRef] [PubMed]

124. Wray, N.H.; Schappi, J.M.; Singh, H.; Senese, N.B.; Rasenick, M.M. NMDAR-independent, cAMP-dependent antidepressant actions of ketamine. *Mol. Psychiatry* **2019**, *24*, 1833–1843. [CrossRef] [PubMed]

125. Chen, P.-S.; Peng, G.-S.; Li, G.; Yang, S.; Wu, X.; Wang, C.-C.; Wilson, B.A.; Lu, R.-B.; Gean, P.-W.; Chuang, D.-M.; et al. Valproate protects dopaminergic neurons in midbrain neuron/glia cultures by stimulating the release of neurotrophic factors from astrocytes. *Mol. Psychiatry* **2006**, *11*, 1116–1125. [CrossRef]

126. Cherry, J.D.; Olschowka, J.A.; O'Banion, M.K. Neuroinflammation and M2 microglia: The good, the bad, and the inflamed. *J. Neuroinflamm.* **2014**, *11*, 98. [CrossRef]

127. Nakajima, K.; Kikuchi, Y.; Ikoma, E.; Honda, S.; Ishikawa, M.; Liu, Y.; Kohsaka, S. Neurotrophins regulate the function of cultured microglia. *Glia* **1998**, *24*, 272–289. [CrossRef]

128. Zhang, J.; Geula, C.; Lu, C.; Koziel, H.; Hatcher, L.M.; Roisen, F.J. Neurotrophins regulate proliferation and survival of two microglial cell lines in vitro. *Exp. Neurol.* **2003**, *183*, 469–481. [CrossRef]

129. Parkhurst, C.N.; Yang, G.; Ninan, I.; Savas, J.N.; Yates, J.R., 3rd; Lafaille, J.J.; Hempstead, B.L.; Littman, D.R.; Gan, W.-B. Microglia promote learning-dependent synapse formation through brain-derived neurotrophic factor. *Cell* **2013**, *155*, 1596–1609. [CrossRef]

130. Gomes, C.; Ferreira, R.; George, J.; Sanches, R.; Rodrigues, D.I.; Gonçalves, N.; A Cunha, R. Activation of microglial cells triggers a release of brain-derived neurotrophic factor (BDNF) inducing their proliferation in an adenosine A2A receptor-dependent manner: A2A receptor blockade prevents BDNF release and proliferation of microglia. *J. Neuroinflamm.* **2013**, *10*, 16. [CrossRef]

131. Jiang, Y.; Wei, N.; Zhu, J.; Lu, T.; Chen, Z.; Xu, G.; Liu, X. Effects of Brain-Derived Neurotrophic Factor on Local Inflammation in Experimental Stroke of Rat. *Mediat. Inflamm.* **2010**, *2010*, 372423. [CrossRef]

132. Porcher, C.; Medina, I.; Gaiarsa, J.-L. Mechanism of BDNF Modulation in GABAergic Synaptic Transmission in Healthy and Disease Brains. *Front. Cell. Neurosci.* **2018**, *12*, 273. [CrossRef] [PubMed]

133. Li, X.; Frye, M.A.; Shelton, R.C. Review of Pharmacological Treatment in Mood Disorders and Future Directions for Drug Development. *Neuropsychopharmacology* **2012**, *37*, 77–101. [CrossRef] [PubMed]

134. Tao, X.; Finkbeiner, S.; Arnold, D.B.; Shaywitz, A.J.; Greenberg, M.E. Ca^{2+} Influx Regulates BDNF Transcription by a CREB Family Transcription Factor-Dependent Mechanism. *Neuron* **1998**, *20*, 709–726. [CrossRef]

135. Krishnan, V.; Nestler, E.J. The molecular neurobiology of depression. *Nature* **2008**, *455*, 894–902. [CrossRef] [PubMed]

136. Lee, B.-H.; Kim, Y.-K. BDNF mRNA expression of peripheral blood mononuclear cells was decreased in depressive patients who had or had not recently attempted suicide. *J. Affect. Disord.* **2010**, *125*, 369–373. [CrossRef]

137. McEwen, B.S.; Chattarji, S.; Diamond, D.M.; Jay, T.M.; Reagan, L.P.; Svenningsson, P.; Fuchs, E. The neurobiological properties of tianeptine (Stablon): From monoamine hypothesis to glutamatergic modulation. *Mol. Psychiatry* **2010**, *15*, 237–249. [CrossRef]

138. Schmidt, H.D.; Shelton, R.C.; Duman, R.S. Functional Biomarkers of Depression: Diagnosis, Treatment, and Pathophysiology. *Neuropsychopharmacology* **2011**, *36*, 2375–2394. [CrossRef]

139. Björkholm, C.; Monteggia, L.M. BDNF—A Key Transducer of Antidepressant Effects. *Neuropharmacology* **2016**, *102*, 72–79. [CrossRef]

140. Hait, N.C.; Wise, L.; Allegood, J.C.; O'Brien, M.; Avni, D.; Reeves, T.M.; Knapp, P.; Lu, J.; Luo, C.; Miles, M.F.; et al. Active, phosphorylated fingolimod inhibits histone deacetylases and facilitates fear extinction memory. *Nat. Neurosci.* **2014**, *17*, 971–980. [CrossRef]

141. di Nuzzo, L.; Orlando, R.; Tognoli, C.; Di Pietro, P.; Bertini, G.; Miele, J.; Bucci, D.; Motolese, M.; Scaccianoce, S.; Caruso, A.; et al. Antidepressant activity of fingolimod in mice. *Pharmacol. Res. Perspect.* **2015**, *3*, e00135. [CrossRef] [PubMed]

142. Covington, H.E., 3rd; Vialou, V.F.; LaPlant, Q.; Ohnishi, Y.N.; Nestler, E.J. Hippocampal-dependent antidepressant-like activity of histone deacetylase inhibition. *Neurosci. Lett.* **2011**, *493*, 122–126. [CrossRef] [PubMed]

143. Lima, I.V.D.A.; Almeida-Santos, A.F.; Ferreira-Vieira, T.H.; Aguiar, D.C.; Ribeiro, F.M.; Campos, A.C.; de Oliveira, A.C.P. Antidepressant-like effect of valproic acid—Possible involvement of PI3K/Akt/mTOR pathway. *Behav. Brain Res.* **2017**, *329*, 166–171. [CrossRef] [PubMed]

144. Boorman, E.; Zajkowska, Z.; Ahmed, R.; Pariante, C.M.; Zunszain, P.A. Crosstalk between endocannabinoid and immune systems: A potential dysregulation in depression? *Psychopharmacology* **2016**, *233*, 1591–1604. [CrossRef]

145. Biringer, R.G. Endocannabinoid signaling pathways: Beyond CB1R and CB2R. *J. Cell Commun. Signal.* **2021**, *15*, 335–360. [CrossRef] [PubMed]

146. Núñez, E.; Benito, C.; Pazos, M.R.; Barbachano, A.; Fajardo, O.; González, S.; Tolón, R.M.; Romero, J. Cannabinoid CB_2 receptors are expressed by perivascular microglial cells in the human brain: An immunohistochemical study. *Synapse* **2004**, *53*, 208–213. [CrossRef] [PubMed]

147. Turcotte, C.; Blanchet, M.-R.; LaViolette, M.; Flamand, N. The CB2 receptor and its role as a regulator of inflammation. *Cell. Mol. Life Sci.* **2016**, *73*, 4449–4470. [CrossRef]

148. Benyó, Z.; Ruisanchez, É.; Leszl-Ishiguro, M.; Sándor, P.; Pacher, P. Endocannabinoids in cerebrovascular regulation. *Am. J. Physiol. Heart Circ. Physiol.* **2016**, *310*, H785–H801. [CrossRef] [PubMed]

149. Roche, M.; Finn, D.P. Brain CB2 Receptors: Implications for Neuropsychiatric Disorders. *Pharmaceuticals* **2010**, *3*, 2517–2553. [CrossRef]

150. Chiang, N.; de la Rosa, X.; Libreros, S.; Serhan, C.N. Novel Resolvin D2 Receptor Axis in Infectious Inflammation. *J. Immunol.* **2017**, *198*, 842–851. [CrossRef]

151. Chiang, N.; Dalli, J.; Colas, R.A.; Serhan, C.N. Identification of resolvin D2 receptor mediating resolution of infections and organ protection. *J. Exp. Med.* **2015**, *212*, 1203–1217. [CrossRef] [PubMed]

152. Tao, Y.; Li, L.; Jiang, B.; Feng, Z.; Yang, L.; Tang, J.; Chen, Q.; Zhang, J.; Tan, Q.; Feng, H.; et al. Cannabinoid receptor-2 stimulation suppresses neuroinflammation by regulating microglial M1/M2 polarization through the cAMP/PKA pathway in an experimental GMH rat model. *Brain Behav. Immun.* **2016**, *58*, 118–129. [CrossRef] [PubMed]

153. Bie, B.; Wu, J.; Foss, J.F.; Naguib, M. An overview of the cannabinoid type 2 receptor system and its therapeutic potential. *Curr. Opin. Anaesthesiol.* **2018**, *31*, 407–414. [CrossRef]

154. Gutiérrez, G.; Pérez-Ortiz, J.; Gutiérrez-Adán, A.; Manzanares, J. Depression-resistant endophenotype in mice overexpressing cannabinoid CB $_2$ receptors. *Br. J. Pharmacol.* **2010**, *160*, 1773–1784. [CrossRef]

155. Meijerink, J.; Balvers, M.; Witkamp, R. *N*-acyl amines of docosahexaenoic acid and other *n*-3 polyunsatured fatty acids—From fishy endocannabinoids to potential leads. *Br. J. Pharmacol.* **2013**, *169*, 772–783. [CrossRef] [PubMed]

156. Park, T.; Chen, H.; Kevala, K.; Lee, J.-W.; Kim, H.-Y. N-Docosahexaenoylethanolamine ameliorates LPS-induced neuroinflammation via cAMP/PKA-dependent signaling. *J. Neuroinflamm.* **2016**, *13*, 284. [CrossRef]

157. Park, T.; Chen, H.; Kim, H.-Y. GPR110 (ADGRF1) mediates anti-inflammatory effects of N-docosahexaenoylethanolamine. *J. Neuroinflamm.* **2019**, *16*, 225. [CrossRef]

158. Tyrtyshnaia, A.; Konovalova, S.; Bondar, A.; Ermolenko, E.; Sultanov, R.; Manzhulo, I. Anti-Inflammatory Activity of *N*-Docosahexaenoylethanolamine and *N*-Eicosapentaenoylethanolamine in a Mouse Model of Lipopolysaccharide-Induced Neuroinflammation. *Int. J. Mol. Sci.* **2021**, *22*, 10728. [CrossRef]

159. Lee, J.-W.; Huang, B.X.; Kwon, H.; Rashid, A.; Kharebava, G.; Desai, A.; Patnaik, S.; Marugan, J.; Kim, H.-Y. Orphan GPR110 (ADGRF1) targeted by N-docosahexaenoylethanolamine in development of neurons and cognitive function. *Nat. Commun.* **2016**, *7*, 13123. [CrossRef]

160. McDougle, D.R.; Watson, J.E.; Abdeen, A.A.; Adili, R.; Caputo, M.P.; Krapf, J.E.; Johnson, R.W.; Kilian, K.A.; Holinstat, M.; Das, A. Anti-inflammatory ω-3 endocannabinoid epoxides. *Proc. Natl. Acad. Sci. USA* **2017**, *114*, E6034–E6043. [CrossRef] [PubMed]

161. Ma, L.; Jia, J.; Liu, X.; Bai, F.; Wang, Q.; Xiong, L. Activation of murine microglial N9 cells is attenuated through cannabinoid receptor CB2 signaling. *Biochem. Biophys. Res. Commun.* **2015**, *458*, 92–97. [CrossRef]

162. Yang, B.; Lin, L.; Bazinet, R.P.; Chien, Y.-C.; Chang, J.P.-C.; Satyanarayanan, S.K.; Su, H.; Su, K.-P. Clinical Efficacy and Biological Regulations of ω–3 PUFA-Derived Endocannabinoids in Major Depressive Disorder. *Psychother. Psychosom.* **2019**, *88*, 215–224. [CrossRef] [PubMed]

163. Ninković, J.; Roy, S. Role of the mu-opioid receptor in opioid modulation of immune function. *Amino Acids* **2013**, *45*, 9–24. [CrossRef]

164. Lee, J.W.; Joshi, S.; Chan, J.S.; Wong, Y.H. Differential coupling of mu-, delta-, and kappa-opioid receptors to G alpha16-mediated stimulation of phospholipase C. *J. Neurochem.* **1998**, *70*, 2203–2211. [CrossRef]

165. Ślusarczyk, J.; Trojan, E.; Głombik, K.; Piotrowska, A.; Budziszewska, B.; Kubera, M.; Popiołek-Barczyk, K.; Lasoń, W.; Mika, J.; Basta-Kaim, A. Targeting the NLRP3 Inflammasome-Related Pathways via Tianeptine Treatment-Suppressed Microglia Polarization to the M1 Phenotype in Lipopolysaccharide-Stimulated Cultures. *Int. J. Mol. Sci.* **2018**, *19*, 1965. [CrossRef] [PubMed]

166. Emrich, H.M.; Vogt, P.; Herz, A. Possible Antidepressive Effects of Opioids: Action of Buprenorphine. *Ann. N. Y. Acad. Sci.* **1982**, *398*, 108–112. [CrossRef] [PubMed]

167. Kasper, S.; McEwen, B.S. Neurobiological and Clinical Effects of the Antidepressant Tianeptine. *CNS Drugs* **2008**, *22*, 15–26. [CrossRef] [PubMed]

168. Peciña, M.; Karp, J.F.; Mathew, S.; Todtenkopf, M.S.; Ehrich, E.W.; Zubieta, J.-K. Endogenous opioid system dysregulation in depression: Implications for new therapeutic approaches. *Mol. Psychiatry* **2019**, *24*, 576–587. [CrossRef]

169. Loric, S.; Maroteaux, L.; Kellermann, O.; Launay, J.M. Functional serotonin-2B receptors are expressed by a teratocarcinoma-derived cell line during serotoninergic differentiation. *Mol. Pharmacol.* **1995**, *47*, 458–466.

170. Cox, D.A.; Cohen, M.L. 5-HT2B receptor signaling in the rat stomach fundus: Dependence on calcium influx, calcium release and protein kinase C. *Behav. Brain Res.* **1996**, *73*, 289–292. [CrossRef] [PubMed]

171. Naito, K.; Tanaka, C.; Mitsuhashi, M.; Moteki, H.; Kimura, M.; Natsume, H.; Ogihara, M. Signal Transduction Mechanism for Serotonin 5-HT2B Receptor-Mediated DNA Synthesis and Proliferation in Primary Cultures of Adult Rat Hepatocytes. *Biol. Pharm. Bull.* **2016**, *39*, 121–129. [CrossRef]

172. Carhart-Harris, R.; Giribaldi, B.; Watts, R.; Baker-Jones, M.; Murphy-Beiner, A.; Murphy, R.; Martell, J.; Blemings, A.; Erritzoe, D.; Nutt, D.J. Trial of Psilocybin versus Escitalopram for Depression. *N. Engl. J. Med.* **2021**, *384*, 1402–1411. [CrossRef] [PubMed]

173. Carhart-Harris, R.L.; Bolstridge, M.; Rucker, J.; Day, C.M.J.; Erritzoe, D.; Kaelen, M.; Bloomfield, M.; Rickard, J.A.; Forbes, B.; Feilding, A.; et al. Psilocybin with psychological support for treatment-resistant depression: An open-label feasibility study. *Lancet Psychiatry* **2016**, *3*, 619–627. [CrossRef] [PubMed]

174. Gukasyan, N.; Davis, A.K.; Barrett, F.S.; Cosimano, M.P.; Sepeda, N.D.; Johnson, M.W.; Griffiths, R.R. Efficacy and safety of psilocybin-assisted treatment for major depressive disorder: Prospective 12-month follow-up. *J. Psychopharmacol.* **2022**, *36*, 151–158. [CrossRef] [PubMed]

175. Yovell, Y.; Bar, G.; Mashiah, M.; Baruch, Y.; Briskman, I.; Asherov, J.; Lotan, A.; Rigbi, A.; Panksepp, J. Ultra-Low-Dose Buprenorphine as a Time-Limited Treatment for Severe Suicidal Ideation: A Randomized Controlled Trial. *Am. J. Psychiatry* **2016**, *173*, 491–498. [CrossRef]

176. Samuels, B.A.; Nautiyal, K.M.; Kruegel, A.C.; Levinstein, M.R.; Magalong, V.M.; Gassaway, M.M.; Grinnell, S.G.; Han, J.; Ansonoff, M.A.; E Pintar, J.; et al. The Behavioral Effects of the Antidepressant Tianeptine Require the Mu-Opioid Receptor. *Neuropsychopharmacology* **2017**, *42*, 2052–2063. [CrossRef]

177. Serafini, G.; Adavastro, G.; Canepa, G.; De Berardis, D.; Valchera, A.; Pompili, M.; Nasrallah, H.; Amore, M. The Efficacy of Buprenorphine in Major Depression, Treatment-Resistant Depression and Suicidal Behavior: A Systematic Review. *Int. J. Mol. Sci.* **2018**, *19*, 2410. [CrossRef]

178. Nelson, C.J.; Schneider, G.M.; Lysle, D.T. Involvement of Central μ- but Not δ- or κ-Opioid Receptors in Immunomodulation. *Brain Behav. Immun.* **2000**, *14*, 170–184. [CrossRef]

179. Nummenmaa, L.; Karjalainen, T.; Isojärvi, J.; Kantonen, T.; Tuisku, J.; Kaasinen, V.; Joutsa, J.; Nuutila, P.; Kalliokoski, K.; Hirvonen, J.; et al. Lowered endogenous mu-opioid receptor availability in subclinical depression and anxiety. *Neuropsychopharmacology* **2020**, *45*, 1953–1959. [CrossRef]

180. Gassaway, M.M.; Rives, M.-L.; Kruegel, A.C.; Javitch, J.A.; Sames, D. The atypical antidepressant and neurorestorative agent tianeptine is a μ-opioid receptor agonist. *Transl. Psychiatry* **2014**, *4*, e411. [CrossRef] [PubMed]

181. Wang, X.; Loram, L.C.; Ramos, K.; de Jesus, A.J.; Thomas, J.; Cheng, K.; Reddy, A.; Somogyi, A.A.; Hutchinson, M.R.; Watkins, L.R.; et al. Morphine activates neuroinflammation in a manner parallel to endotoxin. *Proc. Natl. Acad. Sci. USA* **2012**, *109*, 6325–6330. [CrossRef] [PubMed]

182. Lutz, P.-E.; Kieffer, B.L. Opioid receptors: Distinct roles in mood disorders. *Trends Neurosci.* **2013**, *36*, 195–206. [CrossRef] [PubMed]

183. Halberstadt, A.L.; Geyer, M.A. Multiple receptors contribute to the behavioral effects of indoleamine hallucinogens. *Neuropharmacology* **2011**, *61*, 364–381. [CrossRef]

184. Ling, S.; Ceban, F.; Lui, L.M.W.; Lee, Y.; Teopiz, K.M.; Rodrigues, N.B.; Lipsitz, O.; Gill, H.; Subramaniapillai, M.; Mansur, R.B.; et al. Molecular Mechanisms of Psilocybin and Implications for the Treatment of Depression. *CNS Drugs* **2022**, *36*, 17–30. [CrossRef] [PubMed]

185. Desouza, L.A.; Benekareddy, M.; Fanibunda, S.E.; Mohammad, F.; Janakiraman, B.; Ghai, U.; Gur, T.; Blendy, J.A.; Vaidya, V.A. The Hallucinogenic Serotonin2A Receptor Agonist, 2,5-Dimethoxy-4-Iodoamphetamine, Promotes cAMP Response Element Binding Protein-Dependent Gene Expression of Specific Plasticity-Associated Genes in the Rodent Neocortex. *Front. Mol. Neurosci.* **2021**, *14*, 790213. [CrossRef] [PubMed]

186. Glennon, R.A.; Rosecrans, J.A. Speculations on the mechanism of action of hallucinogenic indolealkylamines. *Neurosci. Biobehav. Rev.* **1981**, *5*, 197–207. [CrossRef] [PubMed]

187. Szabo, A.; Gogolak, P.; Koncz, G.; Foldvari, Z.; Pazmandi, K.; Miltner, N.; Poliska, S.; Bacsi, A.; Djurovic, S.; Rajnavolgyi, E. Immunomodulatory capacity of the serotonin receptor 5-HT2B in a subset of human dendritic cells. *Sci. Rep.* **2018**, *8*, 1765. [CrossRef]

188. Bierhaus, A.; Wolf, J.; Andrassy, M.; Rohleder, N.; Humpert, P.M.; Petrov, D.; Ferstl, R.; von Eynatten, M.; Wendt, T.; Rudofsky, G.; et al. A mechanism converting psychosocial stress into mononuclear cell activation. *Proc. Natl. Acad. Sci. USA* **2003**, *100*, 1920–1925. [CrossRef]

189. Bailey, M.T.; Engler, H.; Powell, N.D.; Padgett, D.A.; Sheridan, J.F. Repeated social defeat increases the bactericidal activity of splenic macrophages through a Toll-like receptor-dependent pathway. *Am. J. Physiol. Regul. Integr. Comp. Physiol.* **2007**, *293*, R1180–R1190. [CrossRef]

190. Diz-Chaves, Y.; Pernía, O.; Carrero, P.; Garcia-Segura, L.M. Prenatal stress causes alterations in the morphology of microglia and the inflammatory response of the hippocampus of adult female mice. *J. Neuroinflamm.* **2012**, *9*, 71. [CrossRef]

191. Lindqvist, D.; Dhabhar, F.S.; James, S.J.; Hough, C.M.; Jain, F.A.; Bersani, F.S.; Reus, V.I.; Verhoeven, J.E.; Epel, E.S.; Mahan, L.; et al. Oxidative stress, inflammation and treatment response in major depression. *Psychoneuroendocrinology* **2017**, *76*, 197–205. [CrossRef] [PubMed]

192. Vilhardt, F.; Haslund-Vinding, J.; Jaquet, V.; McBean, G. Microglia antioxidant systems and redox signalling. *Br. J. Pharmacol.* **2017**, *174*, 1719–1732. [CrossRef]

193. Bellezza, I.; Giambanco, I.; Minelli, A.; Donato, R. Nrf2-Keap1 signaling in oxidative and reductive stress. *Biochim. Biophys. Acta (BBA) Mol. Cell Res.* **2018**, *1865*, 721–733. [CrossRef] [PubMed]

194. Martín-Hernández, D.; Caso, J.R.; Meana, J.J.; Callado, L.F.; Madrigal, J.L.M.; García-Bueno, B.; Leza, J.C. Intracellular inflammatory and antioxidant pathways in postmortem frontal cortex of subjects with major depression: Effect of antidepressants. *J. Neuroinflamm.* **2018**, *15*, 251. [CrossRef] [PubMed]

195. Nadjar, A.; Leyrolle, Q.; Joffre, C.; Laye, S. Bioactive lipids as new class of microglial modulators: When nutrition meets neuroimunology. *Prog. Neuro-Psychopharmacol. Biol. Psychiatry* **2017**, *79*, 19–26. [CrossRef]

196. Hsiao, C.-C.; Sankowski, R.; Prinz, M.; Smolders, J.; Huitinga, I.; Hamann, J. GPCRomics of Homeostatic and Disease-Associated Human Microglia. *Front. Immunol.* **2021**, *12*, 674189. [CrossRef]

197. Wang, Q.; Lv, C.; Sun, Y.; Han, X.; Wang, S.; Mao, Z.; Xin, Y.; Zhang, B. The Role of Alpha-Lipoic Acid in the Pathomechanism of Acute Ischemic Stroke. *Cell. Physiol. Biochem.* **2018**, *48*, 42–53. [CrossRef]

198. Xu, J.; Drew, P.D. 9-Cis-retinoic acid suppresses inflammatory responses of microglia and astrocytes. *J. Neuroimmunol.* **2006**, *171*, 135–144. [CrossRef]

199. Suh, H.-S.; Zhao, M.-L.; Derico, L.; Choi, N.; Lee, S.C. Insulin-like growth factor 1 and 2 (IGF1, IGF2) expression in human microglia: Differential regulation by inflammatory mediators. *J. Neuroinflamm.* **2013**, *10*, 37. [CrossRef]

200. Zhang, B.; Yang, L.; Konishi, Y.; Maeda, N.; Sakanaka, M.; Tanaka, J. Suppressive effects of phosphodiesterase type IV inhibitors on rat cultured microglial cells: Comparison with other types of cAMP-elevating agents. *Neuropharmacology* **2002**, *42*, 262–269. [CrossRef]

201. Dinarello, C.A. Anti-inflammatory Agents: Present and Future. *Cell* **2010**, *140*, 935–950. [CrossRef] [PubMed]

202. Nishimura, Y.; Moriyama, M.; Kawabe, K.; Satoh, H.; Takano, K.; Azuma, Y.-T.; Nakamura, Y. Lauric Acid Alleviates Neuroinflammatory Responses by Activated Microglia: Involvement of the GPR40-Dependent Pathway. *Neurochem. Res.* **2018**, *43*, 1723–1735. [CrossRef]

203. Wang, Y.-P.; Wu, Y.; Li, L.-Y.; Zheng, J.; Liu, R.-G.; Zhou, J.-P.; Yuan, S.-Y.; Shang, Y.; Yao, S.-L. Aspirin-triggered lipoxin A4attenuates LPS-induced pro-inflammatory responses by inhibiting activation of NF-κB and MAPKs in BV-2 microglial cells. *J. Neuroinflamm.* **2011**, *8*, 95. [CrossRef] [PubMed]

204. Rey, C.; Nadjar, A.; Buaud, B.; Vaysse, C.; Aubert, A.; Pallet, V.; Layé, S.; Joffre, C. Resolvin D1 and E1 promote resolution of inflammation in microglial cells in vitro. *Brain Behav. Immun.* **2016**, *55*, 249–259. [CrossRef] [PubMed]

205. Gu, Z.; Lamont, G.J.; Lamont, R.J.; Uriarte, S.M.; Wang, H.; Scott, D.A. Resolvin D1, resolvin D2 and maresin 1 activate the GSK3β anti-inflammatory axis in TLR4-engaged human monocytes. *J. Endotoxin Res.* **2016**, *22*, 186–195. [CrossRef]

206. Tylek, K.; Trojan, E.; Leśkiewicz, M.; Regulska, M.; Bryniarska, N.; Curzytek, K.; Lacivita, E.; Leopoldo, M.; Basta-Kaim, A. Time-Dependent Protective and Pro-Resolving Effects of FPR2 Agonists on Lipopolysaccharide-Exposed Microglia Cells Involve Inhibition of NF-κB and MAPKs Pathways. *Cells* **2021**, *10*, 2373. [CrossRef] [PubMed]

207. Hung, T.-H.; Shyue, S.-K.; Wu, C.-H.; Chen, C.-C.; Lin, C.-C.; Chang, C.-F.; Chen, S.-F. Deletion or inhibition of soluble epoxide hydrolase protects against brain damage and reduces microglia-mediated neuroinflammation in traumatic brain injury. *Oncotarget* **2017**, *8*, 103236–103260. [CrossRef]

208. Abdallah, C.G.; Sanacora, G.; Duman, R.S.; Krystal, J.H. The neurobiology of depression, ketamine and rapid-acting antidepressants: Is it glutamate inhibition or activation? *Pharmacol. Ther.* **2018**, *190*, 148–158. [CrossRef]

209. Zanos, P.; Gould, T.D. Mechanisms of ketamine action as an antidepressant. *Mol. Psychiatry* **2018**, *23*, 801–811. [CrossRef]

210. Sanacora, G.; Schatzberg, A.F. Ketamine: Promising Path or False Prophecy in the Development of Novel Therapeutics for Mood Disorders? *Neuropsychopharmacology* **2015**, *40*, 259–267. [CrossRef]

211. Lumsden, E.W.; Troppoli, T.A.; Myers, S.J.; Zanos, P.; Aracava, Y.; Kehr, J.; Lovett, J.; Kim, S.; Wang, F.-H.; Schmidt, S.; et al. Antidepressant-relevant concentrations of the ketamine metabolite (2 *R*,6 *R*)-hydroxynorketamine do not block NMDA receptor function. *Proc. Natl. Acad. Sci. USA* **2019**, *116*, 5160–5169. [CrossRef] [PubMed]

212. Williams, N.R.; Heifets, B.; Blasey, C.; Sudheimer, K.; Pannu, J.; Pankow, H.; Hawkins, J.; Birnbaum, J.; Lyons, D.M.; Rodriguez, C.I.; et al. Attenuation of Antidepressant Effects of Ketamine by Opioid Receptor Antagonism. *Am. J. Psychiatry* **2018**, *175*, 1205–1215. [CrossRef]

213. Yost, J.G.; Wulf, H.A.; Browne, C.A.; Lucki, I. Antinociceptive and Analgesic Effects of (2R,6R)-Hydroxynorketamine. *J. Pharmacol. Exp. Ther.* **2022**, *382*, 256–265. [CrossRef] [PubMed]

214. Beurel, E.; Song, L.; Jope, R.S. Inhibition of glycogen synthase kinase-3 is necessary for the rapid antidepressant effect of ketamine in mice. *Mol. Psychiatry* **2011**, *16*, 1068–1070. [CrossRef] [PubMed]

215. Zhou, W.; Dong, L.; Wang, N.; Shi, J.-Y.; Yang, J.-J.; Zuo, Z.-Y.; Zhou, Z.-Q. Akt Mediates GSK-3β Phosphorylation in the Rat Prefrontal Cortex during the Process of Ketamine Exerting Rapid Antidepressant Actions. *Neuroimmunomodulation* **2014**, *21*, 183–188. [CrossRef]

216. Choi, M.; Lee, S.H.; Park, M.H.; Kim, Y.-S.; Son, H. Ketamine induces brain-derived neurotrophic factor expression via phosphorylation of histone deacetylase 5 in rats. *Biochem. Biophys. Res. Commun.* **2017**, *489*, 420–425. [CrossRef]

217. Hashimoto, K. Essential Role of Keap1-Nrf2 Signaling in Mood Disorders: Overview and Future Perspective. *Front. Pharmacol.* **2018**, *9*, 1182. [CrossRef]

218. Hoetzel, A.; Schmidt, R. Regulatory role of anesthetics on heme oxygenase-1. *Curr. Drug Targets* **2010**, *11*, 1495–1503. [CrossRef]

219. Wang, N.; Yu, H.-Y.; Shen, X.-F.; Gao, Z.-Q.; Yang, C.; Yang, J.-J.; Zhang, G.-F. The rapid antidepressant effect of ketamine in rats is associated with down-regulation of pro-inflammatory cytokines in the hippocampus. *Upsala J. Med. Sci.* **2015**, *120*, 241–248. [CrossRef]

220. Tan, S.; Wang, Y.; Chen, K.; Long, Z.; Zou, J. Ketamine Alleviates Depressive-Like Behaviors via Down-Regulating Inflammatory Cytokines Induced by Chronic Restraint Stress in Mice. *Biol. Pharm. Bull.* **2017**, *40*, 1260–1267. [CrossRef]

221. Qu, Y.; Shan, J.; Wang, S.; Chang, L.; Pu, Y.; Wang, X.; Tan, Y.; Yamamoto, M.; Hashimoto, K. Rapid-acting and long-lasting antidepressant-like action of (R)-ketamine in Nrf2 knock-out mice: A role of TrkB signaling. *Eur. Arch. Psychiatry Clin. Neurosci.* **2021**, *271*, 439–446. [CrossRef] [PubMed]

222. Walker, A.K.; Budac, D.P.; Bisulco, S.; Lee, A.W.; Smith, R.A.; Beenders, B.; Kelley, K.W.; Dantzer, R. NMDA Receptor Blockade by Ketamine Abrogates Lipopolysaccharide-Induced Depressive-Like Behavior in C57BL/6J Mice. *Neuropsychopharmacology* **2013**, *38*, 1609–1616. [CrossRef] [PubMed]

223. Wu, G.-J.; Chen, T.-L.; Ueng, Y.-F.; Chen, R.-M. Ketamine inhibits tumor necrosis factor-α and interleukin-6 gene expressions in lipopolysaccharide-stimulated macrophages through suppression of toll-like receptor 4-mediated c-Jun N-terminal kinase phosphorylation and activator protein-1 activation. *Toxicol. Appl. Pharmacol.* **2008**, *228*, 105–113. [CrossRef] [PubMed]

224. Welters, I.D.; Hafer, G.; Menzebach, A.; Mühling, J.; Neuhäuser, C.; Browning, P.; Goumon, Y. Ketamine Inhibits Transcription Factors Activator Protein 1 and Nuclear Factor-κB, Interleukin-8 Production, as well as CD11b and CD16 Expression: Studies in Human Leukocytes and Leukocytic Cell Lines. *Anesth. Analg.* **2010**, *110*, 934–941. [CrossRef]
225. Beilin, B.; Rusabrov, Y.; Shapira, Y.; Roytblat, L.; Greemberg, L.; Yardeni, I.Z.; Bessler, H. Low-dose ketamine affects immune responses in humans during the early postoperative period. *Br. J. Anaesth.* **2007**, *99*, 522–527. [CrossRef]
226. Zunszain, P.A.; Horowitz, M.A.; Cattaneo, A.; Lupi, M.M.; Pariante, C.M. Ketamine: Synaptogenesis, immunomodulation and glycogen synthase kinase-3 as underlying mechanisms of its antidepressant properties. *Mol. Psychiatry* **2013**, *18*, 1236–1241. [CrossRef]
227. Yang, J.-J.; Wang, N.; Yang, C.; Shi, J.-Y.; Yu, H.-Y.; Hashimoto, K. Serum Interleukin-6 Is a Predictive Biomarker for Ketamine's Antidepressant Effect in Treatment-Resistant Patients With Major Depression. *Biol. Psychiatry* **2015**, *77*, e19–e20. [CrossRef]
228. Zhan, Y.; Zhou, Y.; Zheng, W.; Liu, W.; Wang, C.; Lan, X.; Deng, X.; Xu, Y.; Zhang, B.; Ning, Y. Alterations of multiple peripheral inflammatory cytokine levels after repeated ketamine infusions in major depressive disorder. *Transl. Psychiatry* **2020**, *10*, 246. [CrossRef]
229. Yuskaitis, C.J.; Jope, R.S. Glycogen synthase kinase-3 regulates microglial migration, inflammation, and inflammation-induced neurotoxicity. *Cell. Signal.* **2009**, *21*, 264–273. [CrossRef]
230. VanderZwaag, J.; Halvorson, T.; Dolhan, K.; Šimončičová, E.; Ben-Azu, B.; Tremblay, M. The Missing Piece? A Case for Microglia's Prominent Role in the Therapeutic Action of Anesthetics, Ketamine, and Psychedelics. *Neurochem. Res.* **2022**, 1–38. [CrossRef]
231. Urs, N.M.; Snyder, J.C.; Jacobsen, J.P.R.; Peterson, S.M.; Caron, M.G. Deletion of GSK3β in D2R-expressing neurons reveals distinct roles for β-arrestin signaling in antipsychotic and lithium action. *Proc. Natl. Acad. Sci. USA* **2012**, *109*, 20732–20737. [CrossRef]
232. Beaulieu, J.-M. A role for Akt and glycogen synthase kinase-3 as integrators of dopamine and serotonin neurotransmission in mental health. *J. Psychiatry Neurosci.* **2012**, *37*, 7–16. [CrossRef]
233. Meltzer, H.Y.; Matsubara, S.; Lee, J.C. The ratios of serotonin2 and dopamine2 affinities differentiate atypical and typical antipsychotic drugs. *Psychopharmacol. Bull.* **1989**, *25*, 390–392. [PubMed]
234. Kapur, S.; Zipursky, R.B.; Remington, G. Clinical and Theoretical Implications of 5-HT$_2$ and D$_2$ Receptor Occupancy of Clozapine, Risperidone, and Olanzapine in Schizophrenia. *Am. J. Psychiatry* **1999**, *156*, 286–293. [CrossRef] [PubMed]
235. Alimohamad, H.; Rajakumar, N.; Seah, Y.-H.; Rushlow, W. Antipsychotics alter the protein expression levels of β-catenin and GSK-3 in the rat medial prefrontal cortex and striatum. *Biol. Psychiatry* **2005**, *57*, 533–542. [CrossRef]
236. Li, X.; Rosborough, K.M.; Friedman, A.B.; Zhu, W.; Roth, K. Regulation of mouse brain glycogen synthase kinase-3 by atypical antipsychotics. *Int. J. Neuropsychopharmacol.* **2007**, *10*, 7–19. [CrossRef] [PubMed]
237. Schmid, C.L.; Bohn, L.M. Serotonin, But Not N-Methyltryptamines, Activates the Serotonin 2A Receptor Via a β-Arrestin2/Src/Akt Signaling Complex In Vivo. *J. Neurosci.* **2010**, *30*, 13513–13524. [CrossRef] [PubMed]
238. Li, X.; Zhu, W.; Roh, M.-S.; Friedman, A.B.; Rosborough, K.; Jope, R.S. In Vivo Regulation of Glycogen Synthase Kinase-3β (GSK3β) by Serotonergic Activity in Mouse Brain. *Neuropsychopharmacology* **2004**, *29*, 1426–1431. [CrossRef]
239. Polter, A.M.; Li, X. Glycogen Synthase Kinase-3 is an Intermediate Modulator of Serotonin Neurotransmission. *Front. Mol. Neurosci.* **2011**, *4*, 31. [CrossRef]

Review

The Role of Alpha Oscillations among the Main Neuropsychiatric Disorders in the Adult and Developing Human Brain: Evidence from the Last 10 Years of Research

Giuseppe Ippolito [1], Riccardo Bertaccini [1], Luca Tarasi [1], Francesco Di Gregorio [2], Jelena Trajkovic [1], Simone Battaglia [1,3] and Vincenzo Romei [1,*]

1 Centro Studi e Ricerche in Neuroscienze Cognitive, Dipartimento di Psicologia, Alma Mater Studiorum—Università di Bologna, 47521 Cesena, Italy
2 UO Medicina Riabilitativa e Neuroriabilitazione, Azienda Unità Sanitaria Locale, 40133 Bologna, Italy
3 Dipartimento di Psicologia, Università di Torino, 10124 Torino, Italy
* Correspondence: vincenzo.romei@unibo.it

Abstract: Alpha oscillations (7–13 Hz) are the dominant rhythm in both the resting and active brain. Accordingly, translational research has provided evidence for the involvement of aberrant alpha activity in the onset of symptomatological features underlying syndromes such as autism, schizophrenia, major depression, and Attention Deficit and Hyperactivity Disorder (ADHD). However, findings on the matter are difficult to reconcile due to the variety of paradigms, analyses, and clinical phenotypes at play, not to mention recent technical and methodological advances in this domain. Herein, we seek to address this issue by reviewing the literature gathered on this topic over the last ten years. For each neuropsychiatric disorder, a dedicated section will be provided, containing a concise account of the current models proposing characteristic alterations of alpha rhythms as a core mechanism to trigger the associated symptomatology, as well as a summary of the most relevant studies and scientific contributions issued throughout the last decade. We conclude with some advice and recommendations that might improve future inquiries within this field.

Keywords: Schizophrenia Spectrum Disorder (SSD); Major Depressive Disorder (MDD); Attention Deficit Hyperactivity Disorder (ADHD); Autistic Spectrum Disorder (ASD); neuropsychiatric disorders; EEG; alpha oscillations; alpha frequency; alpha amplitude; connectivity

Citation: Ippolito, G.; Bertaccini, R.; Tarasi, L.; Di Gregorio, F.; Trajkovic, J.; Battaglia, S.; Romei, V. The Role of Alpha Oscillations among the Main Neuropsychiatric Disorders in the Adult and Developing Human Brain: Evidence from the Last 10 Years of Research. *Biomedicines* **2022**, *10*, 3189. https://doi.org/10.3390/biomedicines10123189

Academic Editor: Willibald Wonisch

Received: 31 October 2022
Accepted: 5 December 2022
Published: 8 December 2022

Publisher's Note: MDPI stays neutral with regard to jurisdictional claims in published maps and institutional affiliations.

1. Introduction

Neuropsychiatric disorders are currently one of the most important sanitary emergences to watch for in high- and middle-income countries. Recent data collected on the European population [1] reports that approximately 165 million people are affected each year by mental disorders, and it is estimated that more than 50% of the general population in middle- and high-income countries will suffer from at least one mental disorder at some point in their lives. Some pieces of evidence [2] suggest that these results may be an underestimation, which would further increase these numbers. This would have enormous consequences from an economic perspective, and in terms of years of life lived with disability. This burden involves children as well, especially those from economically developed countries, even though an increase in other countries is soon expected [3]. Nonetheless, despite its relevance, it is still difficult to make a proper diagnosis because of the vast range of manifestations and overlap among the symptoms that may exist even inside the same cultural group, making these symptoms somewhat elusive to behavioral inspection [4–6]. In fact, despite the availability of numerous disorder assessment scales, the misdiagnosis rates are still high, with some reports indicating that over one third of patients are misdiagnosed [7–10]. This has obvious consequences on the patient's well-being. Therefore, it is crucial to find a more reliable way of accurately making differential diagnoses, thus enabling prompt and proper intervention.

To this aim, various efforts have been put towards identifying neural markers capable of discriminating between neuropsychiatric disorders, treatment response, and the outcome prediction [11,12]. The electroencephalogram (EEG) has been very helpful, since it allows the brain electrical activity to be recorded in a totally non-invasive manner, with a relative fast montage and a high temporal precision [13,14]. For these reasons, it has been widely used to assess brain function in both the healthy and pathological populations. Indeed, the EEG is actually used in several clinical settings with diagnostic and prognostic purposes [15,16]. In the research field, it is commonly used to link a modulation in its signal during a task execution (ERP, event related potential) to a specific cognitive process [17,18]. Further, it is possible to investigate the cognitive functions through the analysis of the brain's oscillatory activity.

The Use of Alpha Rhythm for the Study of the Cognitive Functioning

Since Hans Berger's [19] early studies, it has become evident that oscillatory patterns can be extracted from the brain's electrophysiological signal, resulting from the nearly simultaneous firing of large ensembles of neurons. This rhythmic activity has been differentiated into five main functional categories, according to its frequency [20,21]: delta (δ; 0.5–3 Hz), theta (θ; 4–7 Hz), alpha (α; 8–13 Hz), beta (β; 14–30 Hz), and gamma (γ; 30–50 Hz). Their power is progressively reduced by increasing frequency, with a $1/f$ ratio [22].

These rhythmic oscillations can be detected by analyzing the signal obtained from the spontaneous brain activity, or in response to a sensory stimulation, internally or externally driven [20]. Thus, these frequencies have been linked to sensory and cognitive aspects, such as perception, attention, memory, or even consciousness [23–25]. Consequently, their disruptions have been associated with alterations such as the ones occurring in brain lesions or neuropsychiatric disorders [20,26]. It follows that the brain's oscillatory activity can possibly represent an electrophysiological marker of these conditions, in which, although there is a biological connotate, it is still difficult to make a diagnosis after a behavioral evaluation. In particular, alpha oscillations have been linked to several cognitive processes such as memory, attention, distractor suppression, or even language [25,27].

Alpha oscillations are the most prominent rhythm in the human electroencephalographic signal [28]. A significant modulation of the alpha amplitude after external or internal events is called event-related synchronization (or desynchronization) (ERS/ERD), and is ascribable to an increase (or a decrease) in the rhythmic activity of a large number of neurons. Alpha synchronization seems to have a role in maintaining an active and adaptive inhibitory mechanism during perceptual suppression of upcoming information [29]. This inhibitory mechanism is enacted through a reduction in the cortical excitability that lessens the processing capacity of a particular area which is irrelevant to the ongoing processing [30,31]. Therefore, a synchronization within the alpha band may be an electrophysiological correlate of an information suppression mechanism [32]. Moreover, alpha amplitude has an important role in predictive processes, since the ability to predict the identity [33] or even the probability of occurrence [34] of the incoming stimulus modulates alpha power, which plays a role in preparing the brain for forthcoming stimulus perception [35,36]. Indeed, in the perceptual decision-making domain, alpha power appears to drive choice bias, as alpha ERD correlates with the adoption of a liberal criterion, increased decision confidence, and visual awareness [37]. On the other hand, it has been causally demonstrated [38] that the frequency of the maximum power within the alpha band (i.e., individual alpha frequency, IAF) is a critical parameter in defining sensory accuracy. Indeed, reducing (or, otherwise, increasing) IAF would result in a worsening (or an improvement) of the individual's accuracy level [39–41]. Altogether, these results highlight the functional role of alpha activity in perceptual and cognitive processes. Moreover, the oscillatory activity can be used as a connectivity index between two or more brain regions, comparing the synchronization in phase or in amplitude over the areas of interest [42]. This kind of information is particularly useful when considering the oscillatory activity within the alpha band, given the empirical evidence supporting its involvement in diverse cognitive functions.

It follows that the alpha-based indices can be used to assess cognitive functioning in a broad range of pathological conditions [43–45]. In particular, neuropsychiatric diseases seem to be associated with anatomical and functional changes in the brain architecture, including connectivity alterations [46,47]. For instance, a reduction in the alpha power over the occipital regions is reported in children with Attention Deficit Hyperactivity Disorder (ADHD) [48,49]. This alteration seems to interfere with the processing of relevant and irrelevant stimuli, contributing to the attentional deficits in ADHD patients. Therefore, the behavioral symptoms of these disorders seem to be accompanied by alterations in the alpha-band oscillatory activity. For this reason, alpha-based parameters have been useful for the diagnosis of diverse neuropsychiatric disorders, as highlighted by the number of publications linking anomalous alpha activity to these conditions. In particular, among the conditions included in the DSM-5, the literature mostly focused on disorders such as schizophrenia, depression, ADHD, and autism. This also emphasizes the relevance of alpha band parameters as potential neuromarkers when exploring the disorders' expression. Thus, a better understanding of the role of alpha waves for the manifestation of the symptoms would help to better comprehend the disorder and its associated biotypes. This could allow for a proper diagnosis and help to build more effective treatments and interventions.

However, even if the EEG-derived indices are commonly used in the clinical practice, there is still a lack of agreement about the diagnostic accuracy of these measures. In fact, even if there are plenty of EEG indices available in the literature, the results are often conflicting, if not opposing [50]. This is mostly caused by the heterogeneity of the studies and the different pathological biotypes [51], instead of the fast technological and methodological advances which this field of research has witnessed in the last few years. Indeed, while sometimes these new indices confirm the previous results, other times, they do not. Thus, the aim of the present work is to examine the existing literature to find reliable indices that would support the operator during the diagnosis and intervention procedures. To better identify how these disorders differ at the neural level, we took into account both local and connectivity electrophysiological indices within the alpha band, in order to depict the issue from a broader perspective (see Table 1).

Table 1. Table showing the main electrophysiological indices used for the review, along with a brief description.

Table of EEG Indices		
Index	**Full Name**	**Description**
ERD/ERS	event-related synchronization/desynchronization	Amplitude up/down-modulation in response to a specific event, due to a synchronized activity of a large number of neurons
FAA (PAA)	frontal (or posterior) alpha asymmetry	Relatively higher alpha power recorded from the left (as compared to the right) hemisphere over frontal (or posterior) regions
IAF	individual alpha frequency	The frequency bin displaying the highest power value within the alpha band (8–13 Hz)
ITC	intertrial coherence	The degree of oscillatory phase-synchronization across different trials
PAC	phase amplitude coupling	Coupling between the phase of slower oscillations with the amplitude of faster oscillations (i.e., reflecting integrative mechanisms of neural activity within the brain)
PSD	power spectral density	Measure of signal's power content versus frequency

Hence, given the abundance of work referring specifically to alpha oscillations, we aimed to summarize the literature regarding the main neuropsychiatric disorders in both adult and developing age groups which have emerged over the last 10 years. This work's objective is to give the reader an overall perspective on how the alpha-based indices can be useful when investigating neuropsychiatric conditions, to help isolate the core features that better depict them. This may help with finding more reliable diagnostic and prognostic indices for each condition. For this reason, we excluded from the current work articles including patients with multiple diagnoses, or investigating comorbidities. Furthermore, while inspecting the publications, we also excluded those investigating drug effects or animal studies, in order to focus specifically on those studies which could help to define a more concise view of these conditions.

2. Schizophrenia Spectrum Disorder (SSD)

Although over a hundred years have passed since Kraepelin's pioneering works on dementia praecox, the nosographical profile of Schizophrenia Spectrum Disorder (SSD) still displays some core clinical features that place this disorder among the most debilitating psychiatric syndromes. These hallmark symptoms include psychotic-like manifestations, such as delusions and hallucinations (positive symptoms), amotivation, anhedonia, and social withdrawal (negative symptoms), along with severe deficits in the cognitive (i.e., attention, working memory [WM], executive functions) and perceptual domain [52].

Inquiries aiming to elucidate the pathophysiological mechanics behind the emergence of SSD have recently focused on providing an oscillatory account of the aforementioned impairments, implying that such deficiencies might be engendered by a systematic failure to temporally integrate local neural activities into large-scale networks [53,54]. These considerations substantiate a conceptual framework based on the notion of SSD as a disconnection syndrome, where an aberrant decrease (or increase) in cross-regional synchronization might be responsible for the behavioral abnormalities witnessed in SSD patients [55]. Accordingly, it has been proposed that this rhythmic dysconnectivity may elicit primary deficits at a cognitive and perceptual level, whereas the subsequent pathological attempts at their resolution should lead to the onset of positive symptomatology [53].

On a molecular level, such oscillatory alterations are construed as a by-product of a malfunctioning of neurotransmitters' dynamics. For instance, changes in inhibitory mechanisms mediated by gamma-aminobutyric acid (GABA)ergic interneurons (especially parvalbumin-positive cells) and reduced efficiency of N-methyl-D-aspartate receptors (NMDAR) have been tied to the emergence of SSD symptoms, which has also been associated with dopaminergic disturbances in mesolimbic and mesocortical loci [56]. While the complex interplay between these different dysregulations has been suggested to mainly affect activity in the gamma band, mounting evidence garnered in the last few years points toward their involvement in the generation of alpha rhythms as well [57–59].

2.1. Resting State Data

A seemingly common finding in SSD patients is a slightly reduced parieto-occipital spontaneous alpha power (resulting in an increased cortical excitability) as compared to healthy individuals [60,61]. This kind of evidence has been replicated in several studies exploiting resting-state data, which showed a tendency toward a reduction in local alpha activity as measured via power spectral density (PSD, i.e., the signal's power content versus frequency) not only posteriorly, but also in frontal and central regions [61–64]. Moreover, a direct link has been found between this decrease in alpha power at the parietal and left frontotemporal sites and the gravity of the positive SSD symptomatology [65,66]. Accordingly, treatments capitalizing on alpha-tuned transcranial alternating current stimulation (tACS), which were administered to patients experiencing auditory hallucinations, were proven to restore such imbalances by boosting resting alpha power, consequently reducing the severity of hallucinatory symptoms [67]. However, recent evidence appears to depict a more nuanced portrait of the matter. While some studies described no reduction in terms of posterior alpha activity in SSD patients [68–70], one reported higher parieto-central alpha power in medicated patients as compared to healthy controls [71]. Such an increase was also outlined by data collected in three additional papers, focusing on, rather than task-positive regions (i.e., neural areas more active during attention-demanding tasks), task-negative areas, which are known to increase their level of excitability at rest [72]. A globally heightened spectral power was reported in SSD individuals [73,74] when compared to controls, whereas in a similar work, the same abnormal increase was found with regard to an alpha component located over parietal and temporal areas largely overlapping with the posterior portions of the Default Mode Network (DMN) [75]. Interestingly, spectral inquiries carried out on DMN sites via EEG and magnetoencephalography (MEG) showed comparable results, namely increased alpha power in the medial prefrontal and posterior cingulate cortices (mPFC and PCC) of SSD patients [76,77].

Another oscillatory parameter that can be extrapolated from resting-state power analyses is IAF, namely the exact frequency within the alpha band at the maximum amplitude value. The existing literature suggests that a faster (rather than slower) IAF entails better perceptual acuity and efficiency [38,78], with individuals displaying pronounced schizotypal traits characterized by slower IAFs [79]. Findings gathered over the last decade unilaterally uphold previous knowledge, namely slower resting-state IAFs in SSD patients [68–70,73,77], with the degree of deceleration positively correlating with visuo-attentional performance and scorings at cognitive scales [80]. Furthermore, faster IAFs in SSD individuals undergoing multisession cognitive training was found to predict protocol outcome (i.e., responders vs. non-responders). Along this line of reasoning, it has also been found that occipital IAFs of patients affected by negative symptoms cycle more slowly than in healthy participants, while the opposite held for individuals with SSD displaying positive symptoms [81]. Accordingly, IAFs in both groups of patients correlated with the severity of symptomatology, as measured via behavioral scales [62].

At-rest connectivity metrics have also been largely exploited to gain some insights into the network architecture of SSD patients. Heightened synchronization within the alpha band has been reported in first-episode schizophrenic patients, with cortical hubs sited along frontocentral, occipital, and right temporo-parietal regions displaying the highest levels of interconnections, which was negatively correlated with scorings at cognitive scales [82]. In a similar study, alpha coherence parameters were also found to be enhanced both in an inter- and intra-hemispheric manner in SSD patients [83]. Notably, synchronization measures gathered from DMN nodes are in line with such results (i.e., similar or increased inter-areal alpha coupling in SSD) [75,76,84]. On the other hand, several studies provided a different perspective on the way alpha oscillations behave on a network level in SSD. Evidence for globally reduced alpha synchrony over frontal areas has been outlined by means of different indices, such as phase-locking value (PLV) and phase-lag index (PLI), in patients compared to healthy controls, with a concurrent increase in the information flow from occipital to anterior sites (and a decrease in the opposite direction) [85]. A study adopting non-negative matrix factorization, as well as both energy and entropy measures of connectivity, in individuals with SSD unveiled a general decrease in alpha-band coherence within four spread clusters centered on the bilateral cingulate, left temporal-parietal, precuneal-PCC, and right prefrontal cortices [86]. These patterns of reduced rhythmic interaction were reported to be associated with patients' psychiatric symptoms. Another work uncovered attenuation of inter-hemispheric alpha connections in SSD (but not healthy individuals) at multiple sites (frontal, parietal, and temporal) [68]. A decrease in alpha-tuned inter-hemispheric anterior connectivity and frontoposterior cross-talking was also highlighted in two further studies adopting connectivity analyses [87,88].

To summarize, a significant slowdown of IAFs appears to be a core pathological feature characterizing SSD. As for oscillatory power, spontaneous alpha activity appears to be decreased and increased in SSD patients over, respectively, task-positive and task-negative regions (Figure 1). On a network level, dysregulations in both directions (higher vs. lower connectivity) in various neural clusters have been found, perhaps due to the different methodological approaches adopted (scalp- or source-based computations), or sample variability (first-episode vs. chronic or medicated vs. unmedicated patients).

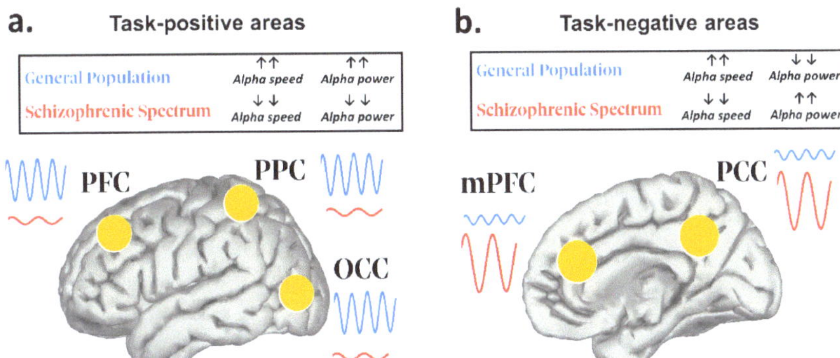

Figure 1. Graphical representation summarizing the main findings on resting-state alpha power in SSD (relative to healthy individuals). The left panel (**a**) depicts alpha oscillatory patterns relative to task-positive areas, while the right panel (**b**) depicts those relative to task-negative regions (largely overlapping with DMN's nodes). Upward and downward arrows indicate respectively an increase or a decrease in alpha power or frequency speed. As for the former areas, alpha power is decreased, while the latter show an overall increase of such oscillatory index. PFC (prefrontal cortex); PPC (posterior parietal cortex); OCC (occipital cortex); mPFC (medial prefrontal cortex); PCC (posterior cingulate cortex).

2.2. Perceptual Impairments

Alpha oscillatory dynamics in SSD have been scrutinized in patients asked to perform perceptual tasks. Indeed, a plethora of deficits involving sensory processing has been reported in SSD, the severity of which has been hypothesized to trigger the onset of positive symptomatology [53,89].

Auditory hallucinations correspond to the most salient psychopathological manifestation in SSD. As such, defective mechanisms in the rhythmic signaling underlying auditory perception have been widely investigated to pinpoint some of the putative features driving this kind of hallucinations. Auditory steady-state response (i.e., the electrical response recorded from the auditory cortex to the entrainment induced by repetitive acoustic stimuli) paradigms have been frequently employed to probe the electrophysiological malfunctioning in SSD [90]. These studies reported aberrations in the low frequency bands (theta and alpha) in SSD (as compared to controls) during the task, such as lower evoked power, inter-trial coherence (ITC), and increased theta–alpha phase–amplitude coupling (PAC) [90–92]. Similar tasks capitalizing on the presentation of multiple pairs of auditory stimuli yielded interesting results about whether and how alpha oscillations mediate perceptual impairment in auditory-related sensory gating. Specifically, evoked alpha power over posterior and midline sites has been shown to undergo a reduced suppression in response to the administered acoustic pairs in SSD as compared to controls [71,93,94], with the degree of (deficient) suppression being associated with GABAergic levels over frontocentral areas [95]. However, evidence for a demeaned alpha suppression over midline posterior areas, which tended toward a relative increase (i.e., more alpha suppression) over frontocentral sites, has also been reported in SSD [96]. Moreover, audio-verbal training was found to be effective in boosting such impaired alpha suppression in response to the second (but not the first) acoustic stimulus in each pair, leading to an oscillatory improvement that correlated with better scoring on verbal learning scales [97]. Impaired sensory gating was also shown to be accompanied by reduced frontocentral alpha ITC between the first and second auditory stimulus, which was inversely correlated with the negative symptoms assessed using a behavioral scale [98]. Consistent with this finding, a lower alpha-tuned inter-hemispheric coherence between temporal and parietal electrodes was outlined during the completion of a passive auditory task in patients with SSD experiencing auditory hallucinations (as compared to controls and SSD patients not affected by such symptoms).

Likewise, aberrant rhythmic patterns within the alpha band have also been found through paradigms presenting a series of standard acoustic stimuli intermingled with deviant tones. Two studies showed that SSD patients exhibit lower ITC and evoked power than healthy participants over central regions in response to standard tones, with the magnitude of the power increase among patients correlating with verbal learning and working memory capacities [99,100]. However, higher evoked alpha power, after both standard and deviant stimuli, was also reported [101].

Overall, alpha rhythmic activity appears to be less reactive and susceptible to task-relevant suppression in patients engaged in perceptual tasks, which suggest both an over-increased power at rest and diminished control during the performance.

2.3. Cognitive Deficits

Impaired cognition is an additional feature enriching the already complex clinical phenotype of SSD. Subtle derangements at this level often tend to occur many years before the onset of psychotic symptoms, fueling the idea that delusions and hallucinations might represent a pathological attempt to make sense of erratic and vague information provided by deficient cognitive mechanisms [53]. While it remains to be clarified whether these deficits result from a disruption of lower-level perceptual mechanisms, various inquiries were set to further illuminate the oscillatory contributions to such phenomena.

The auditory oddball task has been among the most widely adopted paradigms to disentangle the neurophysiological correlates underpinning the way in which patients with SSD handle novel stimuli and cope with irrelevant information. For instance, SSD patients and individuals at a high risk of developing psychosis were found to display a reduced posterior alpha ERD as compared to healthy controls in response to target stimuli [102,103]. On the other hand, a similar study capitalizing on MEG recordings reported a significant decrease in alpha ERS over occipital and posterior midline regions [104]. Transcranial magnetic stimulation (TMS) treatments administered along the left frontoparietal axis to restore these disbalances showed that, in most SSD patients, an increase in task-related alpha power (recorded after, as compared to before, the TMS protocol) occurred in response to both rare and frequent stimuli. This degree of spectral increase also displayed a slight association with improvements in positive and negative symptomatology, as assessed via behavioral scales [105].

Alpha ERD/ERS dysregulations in SSD have been explored even via working memory (WM) paradigms. Indeed, individuals with SSD exhibit a reduced contralateral alpha suppression (ERD-like response) in trials with higher cognitive loads. This aberrant reduction correlated with worse WM performance both between (lower SSD relative to controls) and within (lower SSD with less alpha suppression relative to those displaying greater reduction) groups. In the latter case, a significant relationship with diminished contralateral ERD and the psychiatric symptoms was also uncovered [106]. Furthermore, a comparable reduction in contralateral alpha ERD, coupled with impaired behavioral performance, was reported in SSD in two similar studies [107,108], while weaker inter-areal connectivity within the alpha range has been outlined between occipital regions and several temporo-parietal and frontal clusters (the magnitude of which was associated with the severity of positive symptoms) [109].

Maladaptive alpha dynamics also appear to arise in SSD individuals instantiating inhibition-related mechanisms (or a release from inhibition). Patients with SSD showed higher alpha ERD over frontal and temporal sites during the inhibitory task [110]. Conversely, reduced alpha suppression over sensorimotor cortices was also found in SSD during incongruent trials of a Stroop task (in association with slower reaction times) [111].

Lastly, mounting evidence suggests that a disruption in alpha dynamics might also be involved in social cognition impairments (i.e., poor empathy judgment and mental-izing skills, along with pathological social withdrawal) [112]. Specifically, SSD patients engaged in a facial emotion recognition task displayed lower alpha power and higher connectivity within the alpha band over frontocentral sites, together with lower accuracy in recognizing both happy and fearful emotional expressions [113]. When performing an ecological face-to-face interaction with a confederate, the SSD group displayed an alpha connectivity increase undetected in healthy controls during the more affiliative task condition (*closeness* condition). Such an increase was found to be positively correlated with negative symptoms [114]. Aberrant modulation of alpha power in SSD was also noted during the Ultimatum Game, a social decision-making task involving a fair split of a sum of money between other humans or a computer [115]. Patients displayed a more robust upregulation of alpha power over midfrontal spots during the anticipation phase when playing with a computer vs. a human agent. This power difference was proven to be negatively correlated with positive symptoms [116]. Moreover, in a self-referential memory task, SSD displayed higher scores when presented with self-related items (as compared to neutral or other-related items), which was paralleled by a demeaned alpha ERS over the midline and right frontocentral regions during the encoding phase, and a massive reduction (relative to healthy subjects) in long-range alpha synchronization across multiple cortical electrodes [117].

Altogether, these findings (Table 2) resemble those concerning perceptual processing, namely a pervasive dysregulation in the way alpha oscillations instantiate phasic fluctuations in event-related cortical excitability (ERD vs. ERS). Regarding connectivity metrics, the results are more interspersed, even though they seem to point toward a reduction in long-range alpha coherence across different cognitive tasks, suggesting a bioelectrical disruption in the way higher-level neuronal firing modulates the activity of lower areas.

Table 2. Table representing the main electrophysiological findings in SSD, with the methods and the studies who contributed to them.

Schizophrenia Spectrum Disorder (SSD)		
Studies	**Analytic Method**	**Main Findings**
[68–70,73,77,80]	Resting state IAF	Slower IAF over posterior regions
[60–66]	Resting state PSD	Posterior and frontal Alpha power reduction
[75–77]	Resting state PSD	Alpha power increase in the DMN
[68,75,76,82–88,97,109,114,117]	Functional connectivity	Aberrant long-range functional connectivity
[71,93,94,96,99–105]	Auditory evoked response	Aberrant Alpha ERD/ERS over posterior and frontocentral areas
[106–108,110,111]	Evoked response during WM and attentive tasks	Aberrant Alpha ERD modulation

3. Major Depressive Disorder (MDD)

Major depression has been described as a psychiatric syndrome whose core features entail persistently low mood and anhedonia, coupled with sleep and psychomotor disturbances, fatigue, or loss of energy. Alterations in the affective domain might also involve feelings of worthlessness or guilt and suicidal thoughts, suggesting a pervasive tendency toward negative self-referential thinking and ruminations. In addition, subclinical cognitive deficits (i.e., diminished ability to concentrate, impaired attentional and memory functioning) have often been reported in patients suffering from this disorder [52].

Oscillatory insights into the pathophysiological dynamics underlying MDD suggest that individuals diagnosed with this disorder exhibit rhythmic aberrations, encompassing lower frequency bands, which is likely due to alterations occurring at cortical and subcortical loops that coalesce into a neuropathological phenotype known as thalamocortical dysrhythmia [64,118,119]. Consistent with this notion, MDD patients display higher alpha power levels over the left (relative to the right) frontal lobe, an oscillatory pattern called frontal alpha asymmetry (FAA; Figure 2) [120]. Given the inverse relationship between alpha rhythms and cortical excitability, this electrophysiological abnormality has been framed within the approach–withdrawal model [121], where an increased activation of the right vs. left frontal cortices has been thought to occur in individuals more prone to behavioral withdrawal vs. approach (i.e., proneness toward a negative vs. positive affective style). As such, FAA has been deemed as one of the most reliable biomarkers (although for a different account, see [122–124]) indexing MDD-related affective asymmetries responsible for symptoms such as hopelessness or helplessness and anhedonia [118,120,125].

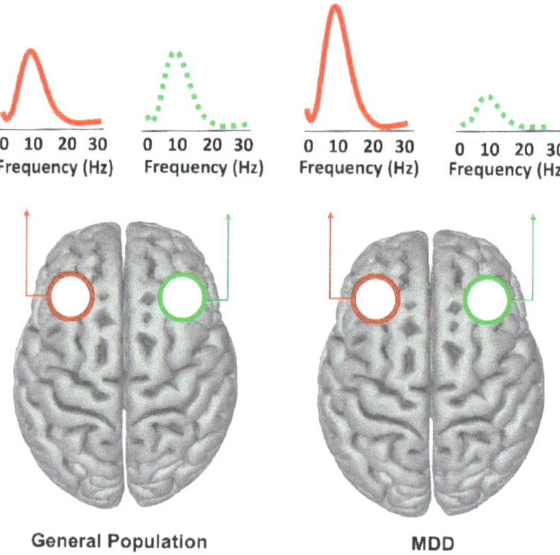

Figure 2. Schematic depiction of FFA patterns witnessed in depressive patients. Relative to healthy individuals (left panel), MDD patients (right panel) display higher relative alpha power over the left (vs. right) frontal cortex.

Aside from FAA, while it has been recently proposed that MDD might be also characterized by interhemispheric asymmetries over posterior sites (i.e., posterior, or parietal, alpha asymmetry, PAA) [118], interareal communication seems to be altered as well. For instance, MDD patients displayed higher alpha-band coherence within the DMN and between anterior midline areas and the frontoparietal network [126,127]. These aberrant patterns of interregional cross-talking are assumed to underlie many functional dysregulations, putatively resulting in ruminative thoughts, increased self-focus, and a reduced ability to concentrate and properly deploy attentional resources [127].

Thus, a general disruption in alpha rhythms, occurring on several different levels, seems to crucially contribute to the pathogenic mechanics behind MDD syndrome and some of its associated affective and cognitive symptoms.

3.1. Resting State Data

Much of the resting-state literature gathered over the last decade has been devoted to the investigation of FAA, with the aim of affirming its role as a neuromarker able to provide an early diagnosis in individuals at risk, but also to anticipate the clinical outcomes. A stronger activation in the alpha band of the right frontal cortex in MDD was replicated in several studies [128–135], some of them showing a robust relationship between the degree of FAA and scores on depression behavioral scales [136,137]. FAA has also been proven to predict treatment effectiveness. For instance, FAA seems to display a negative correlation with sensitivity to specific psychotherapeutic protocols [138], while bifrontal alpha-tuned tACS was capable of reducing both alpha power over the left prefrontal cortex (i.e., asymmetry decrease) and the depressive symptoms [139,140].

Lateralized patterns of rhythmic activation have been reported over posterior sites as well. Although not as straightforward as FAA, PAA appears to be a recurrent oscillatory feature of MDD [118]. Specifically, alpha power was found to be enhanced over the left parietal cortex of female patients suffering from MDD [131], whereas a sample of depressed adolescents tested in a different study displayed an opposite lateralization trend, which was associated with rumination and anhedonia symptoms [141]. Further, MDD patients less sensitive to deep-brain stimulation treatments seem to show an increased parietal alpha activity over the left hemisphere, with the magnitude of PAA negatively correlating with behavioral depression scores [142]. In addition, anodal transcranial direct-current stimulation (tDCS) over the dorsomedial PFC was shown to induce a greater reduction in anxiety scores in patients displaying lower and higher baseline alpha power, as recorded from, respectively, the left parietal lobe and the precuneus [143].

Apart from oscillatory asymmetries, canonical spectral aberrations within the alpha band have also been described in patients exhibiting suicidal ideations [144]. Higher parieto-central alpha power was also found to characterize elders suffering from late-life depression [145], and to better discriminate MDD individuals from healthy participants [146]. Moreover, a positive clinical outcome in depressed patients has been associated with a decrease in frontal alpha power triggered by 10-Hz repetitive TMS protocols and electroconvulsive therapy [147,148]. However, some studies also reported lower posterior oscillatory power within the alpha band in MDD [149,150], the degree of which was linked with depression severity and attentional impairments [151,152]. Conversely, symptomatologic improvements in MDD have been proven to correlate with frontal and midline alpha power increases stemming from alpha-tuned repetitive TMS protocols [153,154].

Resting-state paradigms have also been exploited to address whether and how MDD might alter rhythmic connectivity dynamics. To begin with, proneness to brooding rumination appeared to go along with reduced alpha–gamma PAC over posterior brain areas [155], and global reduction in functional connectivity within the alpha band has been proven to correlate with depression severity [156]. In a network-based study adopting both nodal and global measurements, lower alpha connectivity was uncovered in MDD. Additional analyses also revealed decreased nodal clustering in several cortical hubs spread over frontal regions, as well as the temporal lobe and the visual cortex [157]. Multi-layer analyses run on MEG data to explore interareal oscillatory cross-talking showed similar results, namely a decrease in alpha connectivity in depressed participants [158]. Conversely, alpha-tuned repetitive TMS protocols were shown to ignite an increase in connectivity in MDD, displaying symptom improvements [159]. Still, these results are challenged by plenty of data suggesting an opposite (positive) relationship between levels of alpha connectivity and depression severity. Increased interhemispheric (central) and intrahemispheric (right frontocentral) alpha coherence was found to discriminate depressed individuals from those diagnosed with bipolar disorder [160]. Similarly, MDD displays higher alpha coupling, linking frontopolar loci to the temporal and parieto-occipital regions [161]. Moreover, depressed patients also displayed increased cross-regional alpha connectivity between the ventromedial PFC and both the left mPFC and left dlPFC, as well as between the subgenual anterior cingulate cortex (ACC) and the left dlPFC [162,163]. Furthermore, positive outcomes yielded by antidepressant-based treatments have been associated with weaker alpha oscillatory connectivity bridging the insular cortex to the rostral ACC [164], whereas the magnitude of the TMS-driven decrease in spectral correlation metrics relative to the left hemisphere was shown to be correlated with clinical ameliorations [165].

Taken together, resting-state data concur in assigning FAA (and to a lesser degree PAA) a pathophysiological role in the emergence of MDD. As for more generic impairments in oscillatory power and connectivity, the aforementioned patterns appear more difficult to interpret. These controversial findings might be construed as the result of heterogeneous clinical samples in terms of age or gender, not to mention the different frequency bins (within the canonical alpha band) adopted throughout the analyses. Lastly, a scenario where MDD might exhibit diverging rhythmic phenotypes should not be ruled out; this jeopardizes a unified interpretation of the matter.

3.2. Cognition, MDD and Alpha Rhythms

Oscillatory fingerprints underpinning impaired alpha activity in MDD are not restricted to the resting brain. Electrophysiological investigations during cognitive tasks are consistent with the notion that cortical circuits responsible for attentional deployment, memory storage, and executive functions display demeaned activation in MDD [127].

For instance, during the auditory oddball task, MDD patients who exhibited lower IAFs and alpha power at rest were shown to upregulate such parameters during tasks in response to deviant stimuli [150], along with increased event-related alpha phase synchronization between frontocentral and parieto-occipital electrodes in response to target stimuli during a visual oddball paradigm [166].

With regard to the WM domain, MDD patients displayed poorer behavioral performance on a Sternberg task, which was associated with decreased posterior alpha power during the retention period [167,168]. Moreover, lower increases and decreases in alpha phase synchronization within, respectively, the frontoparietal route and occipito-central cortical clusters have been reported in MDD during the n-back task [169].

Altered patterns of alpha activity appear to also underlie impairments in inhibitory mechanisms, as measured using Go/No-Go tasks. For instance, both reaction times and accuracy rates were reported to be significantly lower in depressed patients displaying suicidal behavior. These data were linked to abnormally high alpha power levels recorded from the ventromedial PFC and ACC during, respectively, Go and No-Go trials [170]. Further, mindfulness-based therapeutic protocols have been reported to enhance task-related alpha ERD and left frontoparietal coherence during the same task, with the degree of oscillatory realignment shown to correlate with clinical ameliorations [171].

Defective emotional processing is an additional symptomatological feature characterizing major depression, with disrupted alpha rhythm dynamics contributing to this pathological phenomenon. In a sample composed of MDD patients with and without dysphoria, higher bilateral frontal alpha power was reported in response to both pleasant and unpleasant emotional stimuli only in MDD patients with dysphoric symptoms [172]. During an emotion self-regulation study employing happy emotion induction training, a power decrease in the upper alpha band over the frontal and temporal sites was reported during the first session of the protocol. Intriguingly, changes in task-related FAA (asymmetry decrease) resulting from such self-regulation training were proven to correlate with scoring at the behavioral scales [173]. Similarly, bifrontal alpha tACS applied to MDD patients was reported to attenuate left alpha power, and thus FAA, during the passive viewing of emotionally positive images [139]. In addition, a weaker posterior alpha ERD was found in MDD patients engaged by WM items superimposed on negative emotional pictures, whereas the opposite pattern was witnessed when positive emotional pictures were presented in the background [174].

Biases in emotional processing have also been investigated through face recognition paradigms. For instance, FAA appeared to unfold in MDD in response to both happy and sad faces [175]. Moreover, alpha–gamma PAC unveiled an attenuated interplay between right-lateralized cortical (i.e., orbitofrontal cortex) and subcortical (i.e., thalamus and amygdala) neural nodes in non-responders to antidepressant treatments, which was likely to reflect abnormal sensitivity to negative emotional stimuli [170].

Lastly, impaired alpha activity has been proven to play a pivotal role in higher cognitive abilities. Patients with MDD were found to display FAA patterns when performing a reinforcement learning task, the occurrence of which has been deemed to underlie losses in approach-related motivation [176]. Reduced alpha power over the left hemisphere was also reported during a multi-stage decision-making task in MDD patients [134], while a neurofeedback training was able to boost alpha power as measured during the completion of a counting task over the ACC and parieto-occipital sites [177].

In brief, task-relative alpha dysregulations tend to be often witnessed in MDD. Aside from FAA, generic alterations in oscillatory power and inter-regional communication have been found to underpin different cognitive deficits. However, the direction of these alterations is still unclear (Table 3), and seemingly depends on each individual's patho-physiological biotype, as well as the specific cognitive process.

Table 3. Table representing the main electrophysiological findings in the MDD, with the methods and the studies who contributed to them.

Major Depressive Disorder (MDD)		
Studies	**Analytic Method**	**Main Findings**
[118,128–137]	Resting state PSD	Frontal Alpha asymmetry
[118,131,141,142]	Resting state PSD	Posterior Alpha asymmetry
[144–146,149–152]	Resting state PSD	Aberrant posterior Alpha power
[126,127,156–158,160–164,169]	Functional connectivity	Aberrant short and long-range functional connectivity
[134,167,168,170]	Evoked response during WM and attentive tasks	Aberrant Alpha power over posterior (reduced) and midfrontal (increased) areas

4. Autism Spectrum Disorder (ASD)

Autism spectrum disorder (ASD) is a complex neurological and developmental disorder characterized by impairments in social cognition and interaction other than sensory and perceptual abnormalities [178,179]. The behavioral and cognitive symptoms associated with ASD are extremely heterogeneous, and the diagnosis is based on different patterns of behavior, including: (a) persistent deficits in social communication and social interaction, such as socio-emotional reciprocity, non-verbal communication, or difficulties adjusting behavior according to various social contexts; (b) restricted interests, repetitive patterns of behavior, such as stereotyped motor movements, inflexible adherence to routines, or unusually intense interests [179].

4.1. The Role of Alpha Oscillations in ASD's Symtomatology

Studies on the neural underpinnings of ASD show functional and anatomical alterations within the perceptual neural network [180–182]. ASD individuals showed alterations in long-range structural and functional connectivity (rather than in a local areas) [180,183–185]. Specifically, ASD individuals seem to display a reduction in interhemispheric long-range synchronization within the alpha band. Indeed, it has been demonstrated that ASD individuals have a reduction in the alpha phase coherence between temporal regions [181]. Other authors [186] showed that frontal alpha asymmetry in 6-month-old children correlates with ASD diagnosis at 24 months. Similarly, connectivity measures within the alpha band have also been linked to sensory symptoms assessed using behavioral scales [180]. These results point toward a decrease in long-range cross-talking, although additional evidence also reports an increase in short-range connectivity among the ASD population (Figure 3) [180,181,187]. In general, these connectivity patterns seem to significantly differ from the ones found in neurotypical individuals, reflecting an atypical brain network development in ASD. Specifically, while neurotypicals show a strengthening of long-range connections and a weakening of short-range ones with aging, the opposite tendency can be seen in ASD [188,189]. Furthermore, several studies based on connectivity measures [182,190,191] highlight the role of the directional interactions among brain areas in ASD. In more detail, ASD individuals show a prevalence of ascending connections from posterior to anterior areas, pointing to a tendency to convey more bottom-up information. Altogether, and consistent with previous literature [192,193], these pieces of evidence highlight electrophysiological abnormalities in ASD (i.e., reduced interhemispheric connectivity over temporal and frontal regions). Since these aberrant modulations appear at early stages of development, they tend to be accompanied by anatomical disturbances, including atypical axon numbers, synaptogenesis, and pruning [187,194]. Altogether, these alterations seem to significantly contribute to the perceptual deficits and clinical manifestations of ASD.

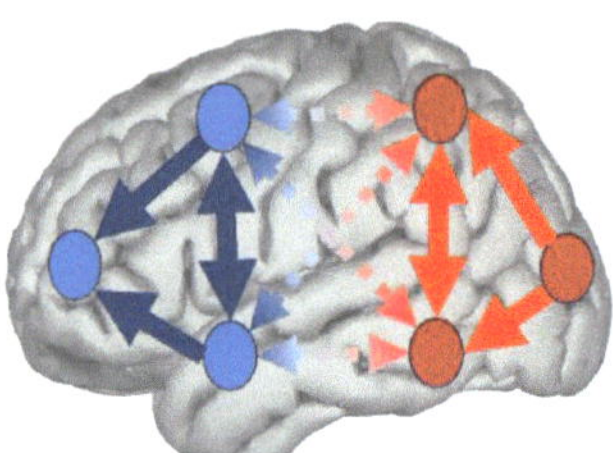
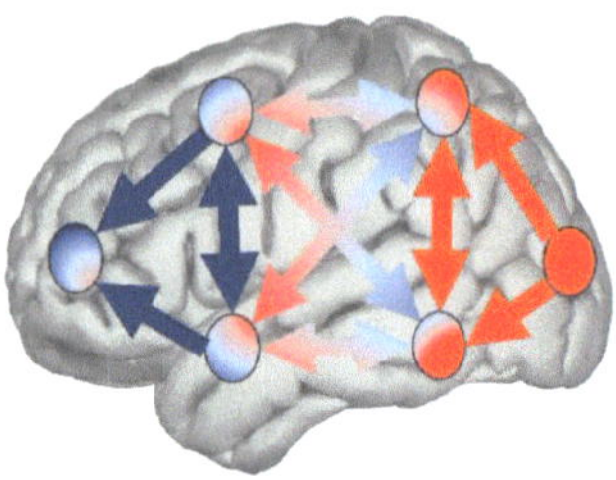

Figure 3. Imbalance in alpha connectivity in ASD relative to typically developing individuals. Persons with ASD show a local increase in connectivity in both posterior and anterior areas, while long-range connectivity between these two regions is reduced (dashed lines). Conversely, neurotypical individuals show the opposite pattern, resulting in a minor local integration (shaded circles).

A large number of studies also associated brain oscillatory activity with ASD symptoms. Specifically, Machado and colleagues [195] investigated the role of PSD in sensory elaboration in ASD, using a visual and audio-visual passive task. They reported a reduction in alpha power and an increase in slow-delta, high-beta, and gamma bands. This seems to be mainly due to the role that the alpha phase plays for the integration of cortical information, supporting executive function and the response to sensorial stimuli via the modulation of neuronal excitability [196]. Coherently, Han et al. [180] reported that brain connectivity within the alpha band correlates with symptom severity in ASD, specifically with the "sensory stimuli and relating behavior" subscale of the Autism Behavior Checklist. This alteration in brain activity may be responsible for a disruption in the excitatory/inhibitory balance and, consequently, may affect the neural response, leading to biased behavioral responses to sensory stimuli. Such imbalance could partly explain the unusual interest in simple objects characterizing people with ASD [180,196]. In line with this evidence, studies addressing how ASD individuals allocate attentional resources have revealed abnormal processing of the upcoming stimuli. Specifically, during an intersensory attention task, in which the suppression of a distractor is relevant to reaching good performance levels, ASD individuals do not show the anticipatory power increase in the alpha band over parieto-occipital areas that is seen in neurotypical individuals before the distracting stimuli [197]. The lack of this preparatory activity has been associated with impaired performance due to the presence of task-irrelevant sensory information. Furthermore, people with ASD did not exhibit any posterior alpha desynchronization (with the resulting reduction in power) during the appearance of a relevant target, which was associated with impaired behavioral performance and increased ASD symptomatology [198]. Thus, these aberrant alpha power modulation patterns seem to be linked to impairments in perceptual suppression of irrelevant sensory information, contributing to difficulties in focusing on relevant stimuli [180,196].

Another EEG measure linked to ASD symptoms is the mu–alpha suppression, i.e., a reduction in the alpha power over sensorimotor areas during action execution or observation of actions and facial expressions [199]. Consistent with the mirror neurons hypothesis, mu suppression reflects the internal simulations of others' actions, allowing one to better understand their intentions [200]. Several studies [201,202] have shown less consistent mu–alpha suppression in ASD, which is possibly related to social deficits.

In conclusion, due to its heterogeneous manifestations, ASD can be challenging to diagnose using behavioral scales. The findings described here indicate that some of its symptoms can be linked to specific neural alterations, thus legitimatizing the search for neural markers of ASD. Thus, finding an electrophysiological correlate of ASD would considerably support and ease the diagnostical evaluation.

4.2. EEG Indices for an Early ASD Diagnosis

Despite the significant number of electrophysiological measures available in the literature, there is still a lack of concordance regarding which index would best represent an anatomical marker of ASD to assist an early diagnosis [178,203]. In fact, several MEG and EEG studies have reported anomalies in one or more frequency bands in ASD, linking these indices to symptom severity. Yet, the results are often conflicting [50,204–211]. This seems to be attributable to the great discrepancies between experimental design and procedures, as well as sample discrepancies (e.g., some studies were conducted on children, while others on adults, some on persons with a low- or high-functioning profile, under or without medications, etc.) [204,212–214].

An EEG study from Matlis and colleagues [203], conducted on a large sample of children with ASD, showed a robust reduction in the peak alpha-ratio (i.e., reduced posterior to anterior power ratio) in persons with ASD. The authors also used this index as an electrophysiological marker for ASD, reaching high accuracy levels. These results suggest that ASD individuals may have higher power values over the anterior areas, which correlates with behavioral inhibition and sociability [215,216]. A recent study [188] used the EEG data obtained from the spontaneous brain activity in 3-month-old infants to predict a later ASD diagnosis. This study highlighted that the best predictors of a later ASD diagnosis at 18 months are a lower frontal and a higher fronto-temporal connectivity. These results are consistent with the literature indicating the presence of hypoconnectivity within the frontal regions, which is in line with the executive and social difficulties among the ASD population [217,218]. Similarly, Orekhova et al. [219] demonstrated that high-risk 14-month-old infants later diagnosed with ASD showed higher alpha-based connectivity over the fronto-central areas. These results are in line with the aforementioned studies [187,220,221] suggesting that ASD individuals are associated with a shift from early white matter maturation during infancy to hypoconnectivity with aging.

Altogether, these data suggest the presence of an excitation–inhibition unbalance in the neural excitability in persons with ASD [178,222,223]. This feature could be responsible for a modified signal-to-noise ratio, resulting in an altered sensorial experience. In fact, due to the abnormal levels of cortical excitability, these individuals could be characterized by hypo- or hyper-responsiveness to sensorial stimuli, thus affecting several cognitive and perceptive domains. Therefore, such electrophysiological alterations may underlie perceptual alterations, contributing to explanations for some of the distinctive symptoms of ASD. The gamma band has been demonstrated to have a key role in these dynamics, given its involvement in the binding of perceptual information into one coherent whole via the integration of responses from near areas [224]. However, the results involving the measurement of the gamma frequency are often conflicting, since its large frequency spread makes it difficult to track the phase between brain areas [178]. To avoid these difficulties, it can be useful to use cross-coupling indices, in which the phase of a lower frequency oscillation in one area has been shown to modulate the amplitude of a higher frequency in another area [225]. Due to its involvement in top-down processing and the perceptual experience [226,227], alpha power has been demonstrated to be a promising index for the study of ASD-related difficulties. In addition, this index is highly sensitive to alterations in long-range connectivity [228,229]. Thus, integrating both frequencies using the alpha–gamma PAC could provide an accurate depiction of the local and global connectivity [196]. Several studies report altered alpha–gamma PAC among the ASD population [185,230,231], achieving a diagnostic accuracy of 90% [185]. However, while Berman and colleagues [231] found an increased alpha–gamma PAC within the ASD group, other studies [185,230] reported the opposite result. This discrepancy is likely due to differences in the experimental designs (resting state vs. visual task EEG recording) [185,230,231]. In general, as suggested by Kessler et al. [178], alterations in the alpha–gamma PAC may reflect an imbalance between excitatory/inhibitory activity in the perceptual brain network, resulting in a hypo- or hyper-sensitivity to various classes of stimuli and, thus, different clinical manifestations among the ASD population. This explanation could clarify the conflicting results (Table 4) about the alpha–gamma PAC, and then help to understand the nature of the perceptive alterations in individuals with ASD.

Table 4. Table representing the main electrophysiological findings in the ASD, with the methods and the studies who contributed to them.

Autistic Spectrum Disorder (ASD)		
Studies	**Analytic Method**	**Main Findings**
[195,197,198]	Task-induced PSD	Impaired modulation of Alpha power
[186,203,215]	Topographical distribution of Alpha power	Aberrant Alpha power in frontal regions
[185,230,231]	PAC	Aberrant Alpha–Gamma PAC
[180,181,184,185,187,189]	Functional connectivity	Aberrant short- (enhanced) and long-range (reduced) functional connectivity
[201,202]	Evoked response during motor tasks	Less suppressed Mu–Alpha ERD/ERS over sensorimotor areas

5. Attention Deficit Hyperactivity Disorder (ADHD)

ADHD is a neurodevelopmental disorder defined by two main pathological clusters [179]: (A) marked difficulties in the deployment and maintenance of the attentional focus (likely due to ineffective suppression of distracting stimuli) during most daily-life activities; (B) hyperactivity and poor impulsiveness control (i.e., logorrheic behavior, interrupting or intruding on others, blurting out answers before questions are over, squirming in seat). These two different core symptoms might occur separately or jointly, and have been associated with impairments in sensorimotor mechanisms, reward processing, affective self-regulation, and executive functions [232,233]. While the onset of such symptoms occurs before the age of 12, they tend to persist in adulthood, often in comorbidity with anxiety disorders, depression, or substance abuse [234,235].

5.1. The Role of Alpha Oscillations in ADHD's Symtomatology

EEG analyses have played a pivotal role in the exploration of ADHD oscillatory biomarkers [43,46,47]. Although some findings might appear to be controversial due to the heterogeneity of ADHD phenotypes, existing literature suggests that individuals diagnosed with ADHD exhibit several alterations in oscillatory mechanisms crucial for cognition (see for instance [236]). Mid-frontal theta [237] and motoric beta alterations are often reported [238,239] together with impairments involving posterior alpha rhythms. In particular, it has been observed that the anticipation of visual distractors is linked to an alpha activity decrease over the visual cortex in typically developing individuals, whereas the anticipation of relevant stimuli increases it [30,49]. Moreover, an increase in the alpha power has been reported over relevant regions during high-demanding cognitive tasks, which is thought to suppress external inputs in order to support the relevant ones [49].

Consequently, ADHD's symptoms may depend on abnormal oscillatory neural activity. Such alterations may also help to explain the WM deficit in ADHD. For instance, Lenartowicz and colleagues [240] recorded EEG activity in a large sample of ADHD children during a spatial WM task. EEG parameters associated with encoding, vigilance, and maintenance functions were analyzed. During the encoding, reduced occipital alpha power was reported, while the maintenance phase was accompanied by a greater power increase in the alpha band, interpreted as a compensatory response to weak alpha activity during the previous encoding stage. Such a failure in the encoding process was associated with poorer reading comprehension and executive functioning, as well with more severe ADHD symptomatology. Similarly, Hasler and colleagues [241] found reduced alpha and theta anticipatory activity in adults with ADHD when engaged in bottom-up and top-down attentive tasks. Such patterns were thought to reflect dysfunctional neural dynamics underlying the suppression of distractors and the prioritization of relevant information (Figure 4). The reduced ability to inhibit task-irrelevant stimuli may prompt or result from the enhanced tendency to mind-wander in ADHD. Some studies [48,242] linked alpha and theta reduction to spontaneous mind-wandering, namely the attentional shift from the task at hand to inner and unrelated thoughts [243]. The reduced ability to inhibit task-irrelevant stimuli may prompt or result from the enhanced tendency to mind-wander, and may trigger some of the core cognitive symptoms characterizing ADHD. In line with these considerations, Bozhilova and colleagues [242] asked adults with ADHD to perform a Go/No-Go task, while mind-wandering reports and EEG data (relative to both response execution and inhibition) were collected. The authors reported a higher error rate in the ADHD group and increased reaction time variability, along with reduced event-related alpha and beta suppression during No-Go trials. A hierarchical regression model applied to these measurements unveiled that ADHD diagnosis and proneness to mind-wandering might share a common oscillatory deficit, consistent with the notion that mind-wandering may be a supplemental pathological facet of this disorder. These pieces of evidence have been further supported by a later study [48], in which the authors reported how the reduction in the alpha and theta modulation in ADHD patients during WM and attentional paradigms is linked to mind-wandering episodes. Altogether, these findings suggest that alpha band alterations responsible for the impaired inhibition of task-irrelevant information might also underlie the increase in mind-wandering episodes.

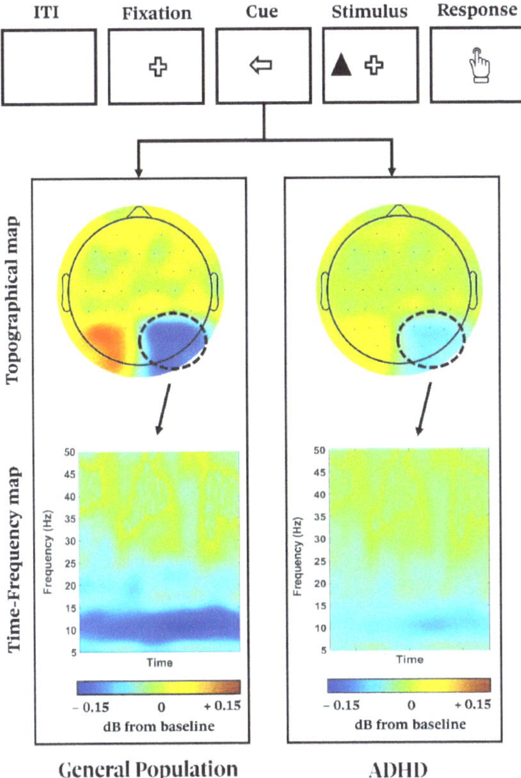

Figure 4. Simulated data showing a demeaned modulation in the inhibitory alpha power in ADHD (relative to controls) during the anticipation of a target stimulus in a visual task. Noteworthily, the lack of modulation comprises both contralateral ERD and ipsilateral ERS.

Accordingly, a review by Lenartowicz and colleagues [49] highlighted how such attenuation in alpha suppression during visuo-attentional tasks is primarily linked to the ADHD inattentive profile, along with atypical lateralization patterns. Neurotypical individuals engaged in visuo-attentional paradigms show inter-hemispheric modulations in posterior alpha power. In particular, ERD is commonly reported in the hemisphere contralateral to the attended stimulus, while ERS is reported ipsilaterally. This inter-hemispheric imbalance is altered in ADHD patients, who do not display atypical modulations over posterior regions [244]. Similar results have been replicated by several different studies [245–250]. Specifically, Guo et al. [247] adopted a visuospatial attentional task in which the stimulus onset could be primed by a cue consisting of a gaze pointing toward either the left or right hemifield. The authors reported that, in the control group, an alpha lateralization with ERD was present in the hemisphere contralateral to the hemifield containing the to-be-attended upcoming stimuli, while children diagnosed with ADHD did not exhibit such a lateralization. This aberrant modulation was more pronounced in the left hemisphere, and was proven to correlate with both behavioral performance and severity of the inattentive symptoms. Interestingly, this lateralization pattern has been observed even over the sensorimotor areas when employing a motor task [248]. In this case, a reduction in the mu–alpha power over sensorimotor regions was reported to occur within the hemisphere contralateral to the hand performing the task in the neurotypical group, but not in ADHD. Moreover, in the ADHD group, a correlation was found between the aberrant lateralization in the oscillatory pattern and both the behavioral performance and difficulties to control disruptive motor activity and attentional processes in daily life.

Altogether, these results emphasize once more how the pathophysiological mechanisms triggering the symptoms of such a disorder might ensue from a deficient modulation of posterior alpha power, which may hinder the proper suppression of distracting information and, as a consequence, the allocation of attentional resources toward relevant stimuli. An interesting line of research, consequently, attempted to artificially modulate alpha activity with the aim of inducing shift in performance and in the symptoms' severity in individuals with ADHD.

5.2. Normalizing Alpha Power Using the Neurofeedback Technique

As mentioned above [49,244,247], imbalanced alpha oscillations are an important biomarker of ADHD symptomatology. Accordingly, many authors used neurofeedback to normalize alpha power imbalances in ADHD patients. Neurofeedback-based protocols consist of training sessions where the participants learn to self-modulate their brain oscillations through real-time feedback, with the aim of concurrently reshaping specific behavioral routines [251,252]. The effectiveness of such approach in ADHD patients has been demonstrated by Escolano and colleagues [253], who were able to enhance fronto-midline upper alpha power in ADHD children undergoing 18 training sessions. The authors reported improvements in neuropsychological tests assessing WM, concentration, and impulsivity. Indeed, boosting alpha power appeared to generate a rebound effect that entailed a robust task-related alpha power decrease. A similar approach has been employed in another study [254] in which the authors induced a rebound effect through a neurofeedback protocol aimed at desynchronizing the alpha power during a Go/No-Go task. This resulted in a subsequent power normalization, and in improvements in terms of motor inhibition. Furthermore, results highlighting the relevance of alpha desynchronization in attentive visual paradigms [49] have been adopted to reduce the power in this frequency range via neurofeedback protocols in ADHD individuals, prompting significant improvements to sustained attention [255].

In conclusion, the literature revealed how some electrophysiological features such as the posterior alpha suppression seem to be linked to the occurrence of ADHD symptomatology [49,51,240,244]. These findings are further strengthened by neurofeedback studies, which highlighted how normalizing these aberrant oscillatory patterns could benefit ADHD patients. Altogether, these pieces of evidence (Table 5) underline the benefits of looking into oscillatory activity in the alpha band, as well as how this kind of information can be used to induce an amelioration.

Table 5. Table representing the main electrophysiological findings in ADHD, with the methods and the studies who contributed to them.

Attention Deficit Hyperactivity Disorder (ADHD)		
Studies	**Analytic Method**	**Main Findings**
[49,240,241,244,246,247,250]	Topographical distribution of Alpha power during attentional tasks	Aberrant Alpha ERD/ERS lateralization
[240,245]	Topographical distribution of Alpha power during motor tasks	Reduced Mu–Alpha ERD/ERS lateralization
[48,242]	Evoked response during spontaneous mind wandering and tasks	Reduced Alpha ERD/ERS

6. General Conclusions

The present work aimed to review the last 10 years of research on the role of alpha oscillations among the main neuropsychiatric disorders, such as SSD, MDD, ASD, and ADHD, to summarize the high volume of publications on these topics.

The main findings suggest that individuals with SSD may display, in task-positive regions, a reduction in both the alpha power and frequency compared to healthy individuals, while in task-negative areas, the oscillatory power appears to be increased [76,80]. Further, FAA has proven to be a valuable proxy of MDD emergence, since these patients seem to display an increase in the alpha power over the left frontal hemisphere, and a decrease in the right one compared to controls [118,131]. In individuals with ASD, ground evidence suggests the presence of an imbalance in the alpha band's neural connectivity, with a local increase in short-range connectivity in both the posterior and anterior networks, while the long-range connectivity between these two regions is reduced [180,181,194]. Lastly, a lower modulation of contralateral alpha ERD and ipsilateral alpha ERS has been found in ADHD during the anticipation of target stimuli in a visual task, when compared to typically developing individuals [48,49].

Nonetheless, robust electrophysiological biomarkers of neuropsychiatric disorders are still difficult to identify. This seems to be due to the enormous differences in the clinical conditions and their behavioral manifestations. Furthermore, in the studies we considered, several methodological discrepancies emerged. In particular, some studies use high-density EEG while others do not, resulting in a range between 16 and 256 electrodes, and, similarly, important discrepancies can be found regarding the number of individuals included in the studies. Other relevant discrepancies concern the participants' age (children vs. adults), the disorder onset (first or not), whether the person is under treatment (behavioral/pharmacological), or the choice to record the EEG signal during a task or in a resting state condition. Moreover, a consistent heterogeneity can be seen in the biotypes included in the research (for instance, considering the inattentive, the hyperactive, or both subtypes in the ADHD sample). Altogether, these factors might lead to discrepancies in the scientific literature, despite searching for the same condition. Thus, even the line of research focusing specifically on the role of alpha bands found conflicting results with small differences in the experimental methodology.

The current work also highlights how the availability of several EEG indices to address the same issue allows for a deeper understanding of certain specific aspects of the phenomena. Indeed, most of these measures are often based on the investigation of different alpha parameters, not allowing for a direct comparison between studies. This is often regarded as controversial evidence, making it difficult to draw parallels even with regard to studies sharing methodological similarities. Therefore, herein, we emphasize the importance for future research to adopt a standardized methodological procedure, allowing for a better comparison of these fragmented pieces of evidence. Such an approach might also help provide an overall conclusion regarding the magnitude of the effects found herein. This may strengthen our highlights, allowing researchers and clinicians to use the aforementioned finding to plan more effective evidence-based treatments tailored to each specific clinical phenotype. In this paper, we attempted to circumscribe these conditions and their electrophysiological peculiarities, for example by excluding from the current work studies investigating comorbidities. Forthcoming studies should fill this gap, since tracing links between these pathological conditions would provide a more exhaustive picture, possibly illustrating how these behavioral and neurophysiological features overlap. This would also help us to understand which alpha-based indices better explain the commonalities between these disorders.

In conclusion, the last 10 years of research have brought several proofs on the role of alpha oscillations in neuropsychiatric disorders. In particular, alterations in the alpha power and frequency have been reported in patients with SSD [76,80], whereas FAA has been proven to be able to discriminate MDD from healthy controls [118,131]. Furthermore, an imbalance in the alpha band brain connectivity between long- and short-range regions has been observed in ASD [180,181,194]. Finally, ADHD seems to display reduced alpha power during stimuli processing relative to controls [48,49].

Hence, the alpha band indices may represent reliable and practical measures to support the clinician during the diagnosis formulation, the choice and the evaluation of a treatment, or the assessment of the symptoms.

Author Contributions: Conceptualization, G.I., R.B., and V.R.; investigation, G.I. and R.B.; data curation, G.I. and R.B.; writing—original draft preparation, G.I. and R.B.; writing—review and editing, F.D.G., L.T., and J.T.; visualization, L.T. and F.D.G.; supervision, S.B. and V.R.; project administration, G.I. and V.R. All authors have read and agreed to the published version of the manuscript.

Funding: This research received no external funding.

Abbreviations

ACC	Anterior cingulate cortex
ADHD	attention deficit hyperactivity disorder
ASD	autism spectrum disorders
dlPFC	dorsolateral prefrontal cortex
DMN	default mode network
DSM-5	diagnostic and statistical manual of mental disorders
EEG	electroencephalogram
ERD	event-related desynchronization
ERP	event related potential
ERS	event-related synchronization
FAA	frontal alpha asymmetry
IAF	individual alpha frequency
ITC	intertrial coherence
MDD	major depressive disorders
MEG	magnetoencephalography
mPFC	medial prefrontal cortex
OCC	occipital cortex
PAA	posterior/parietal alpha asymmetry
PAC	phase amplitude coupling
PCC	posterior cingulate cortex
PFC	prefrontal cortex
PPC	posterior parietal cortex
PSD	power spectral density
SSD	schizophrenia spectrum disorder

tACS	transcranial alternating current stimulation
tDCS	transcranial direct-current stimulation
TMS	transcranial magnetic stimulation
WM	working memory

References

1. Trautmann, S.; Rehm, J.; Wittchen, H.-U. The Economic Costs of Mental Disorders. *EMBO Rep.* **2016**, *17*, 1245–1249. [CrossRef] [PubMed]
2. Vigo, D.; Thornicroft, G.; Atun, R. Estimating the True Global Burden of Mental Illness. *Lancet Psychiatry* **2016**, *3*, 171–178. [CrossRef] [PubMed]
3. Baranne, M.L.; Falissard, B. Global Burden of Mental Disorders among Children Aged 5–14 Years. *Child Adolesc. Psychiatry Ment. Health Health* **2018**, *12*, 19. [CrossRef] [PubMed]
4. Pies, R. How "Objective" Are Psychiatric Diagnoses? *Psychiatry* **2007**, *4*, 18–22. [PubMed]
5. Sadock, B.J.; Sadock, V.A.; Ruiz, P.; Kaplan, H.I. *Kaplan & Sadock's Comprehensive Textbook of Psychiatry*, 9th ed.; Wolters Kluwer Health/Lippincott Williams & Wilkins: Philadelphia, PA, USA, 2009; ISBN 978-0-7817-6899-3.
6. Kohrt, B.A.; Rasmussen, A.; Kaiser, B.N.; Haroz, E.E.; Maharjan, S.M.; Mutamba, B.B.; de Jong, J.T.; Hinton, D.E. Cultural Concepts of Distress and Psychiatric Disorders: Literature Review and Research Recommendations for Global Mental Health Epidemiology. *Int. J. Epidemiol.* **2014**, *43*, 365–406. [CrossRef]
7. Ayano, G.; Demelash, S.; Yohannes, Z.; Haile, K.; Tulu, M.; Assefa, D.; Tesfaye, A.; Haile, K.; Solomon, M.; Chaka, A.; et al. Misdiagnosis, Detection Rate, and Associated Factors of Severe Psychiatric Disorders in Specialized Psychiatry Centers in Ethiopia. *Ann. Gen. Psychiatry* **2021**, *20*, 10. [CrossRef]
8. Stahnke, B. A Systematic Review of Misdiagnosis in Those with Obsessive-Compulsive Disorder. *J. Affect. Disord. Rep.* **2021**, *6*, 100231. [CrossRef]
9. Vermani, M.; Marcus, M.; Katzman, M.A. Rates of Detection of Mood and Anxiety Disorders in Primary Care: A Descriptive, Cross-Sectional Study. *Prim Care Companion CNS Disord* **2011**, *13*, PCC.10m01013. [CrossRef]
10. Phillips, M.L.; Kupfer, D.J. Bipolar Disorder Diagnosis: Challenges and Future Directions. *Lancet* **2013**, *381*, 1663–1671. [CrossRef]
11. Rivera, M.J.; Teruel, M.A.; Maté, A.; Trujillo, J. Diagnosis and Prognosis of Mental Disorders by Means of EEG and Deep Learning: A Systematic Mapping Study. *Artif. Intell. Rev.* **2022**, *55*, 1209–1251. [CrossRef]
12. Yahata, N.; Kasai, K.; Kawato, M. Computational Neuroscience Approach to Biomarkers and Treatments for Mental Disorders. *Psychiatry Clin. Neurosci.* **2017**, *71*, 215–237. [CrossRef] [PubMed]
13. Rosenberg, S.D.; Périn, B.; Michel, V.; Debs, R.; Navarro, V.; Convers, P. EEG in Adults in the Laboratory or at the Patient's Bedside. *Neurophysiol. Clin. Clin. Neurophysiol.* **2015**, *45*, 19–37. [CrossRef] [PubMed]
14. Bathelt, J.; O'Reilly, H.; de Haan, M. Cortical Source Analysis of High-Density EEG Recordings in Children. *J. Vis. Exp.* **2014**, *88*, e51705. [CrossRef] [PubMed]
15. Di Gregorio, F.; La Porta, F.; Petrone, V.; Battaglia, S.; Orlandi, S.; Ippolito, G.; Romei, V.; Piperno, R.; Lullini, G. Accuracy of EEG Biomarkers in the Detection of Clinical Outcome in Disorders of Consciousness after Severe Acquired Brain Injury: Preliminary Results of a Pilot Study Using a Machine Learning Approach. *Biomedicines* **2022**, *10*, 1897. [CrossRef]
16. Jaworska, N.; de la Salle, S.; Ibrahim, M.-H.; Blier, P.; Knott, V. Leveraging Machine Learning Approaches for Predicting Antidepressant Treatment Response Using Electroencephalography (EEG) and Clinical Data. *Front. Psychiatry* **2019**, *9*, 768. [CrossRef]
17. Norton, E.S.; MacNeill, L.A.; Harriott, E.M.; Allen, N.; Krogh-Jespersen, S.; Smyser, C.D.; Rogers, C.E.; Smyser, T.A.; Luby, J.; Wakschlag, L. EEG/ERP as a Pragmatic Method to Expand the Reach of Infant-Toddler Neuroimaging in HBCD: Promises and Challenges. *Dev. Cogn. Neurosci.* **2021**, *51*, 100988. [CrossRef] [PubMed]
18. Horvath, A.; Szucs, A.; Csukly, G.; Sakovics, A.; Stefanics, G.; Kamondi, A. EEG and ERP biomarkers of Alzheimer's disease: A critical review. *Front. Biosci.* **2018**, *23*, 183–220. [CrossRef]
19. Berger, H. Über das Elektrenkephalogramm des Menschen. *Arch. Psychiatr.* **1929**, *87*, 527–570. [CrossRef]
20. Başar, E. Brain Oscillations in Neuropsychiatric Disease. *Dialogues Clin. Neurosci.* **2013**, *15*, 291–300. [CrossRef]
21. Mahjoory, K.; Schoffelen, J.-M.; Keitel, A.; Gross, J. The Frequency Gradient of Human Resting-State Brain Oscillations Follows Cortical Hierarchies. *eLife* **2020**, *9*, e53715. [CrossRef]
22. Bak, P.; Tang, C.; Wiesenfeld, K. Self-Organized Criticality: An Explanation of the 1/f Noise. *Phys. Rev. Lett.* **1987**, *59*, 381–384. [CrossRef] [PubMed]
23. Karakaş, S.; Barry, R.J. A Brief Historical Perspective on the Advent of Brain Oscillations in the Biological and Psychological Disciplines. *Neurosci. Biobehav. Rev.* **2017**, *75*, 335–347. [CrossRef]
24. Gupta, D.S.; Chen, L. Brain Oscillations in Perception, Timing and Action. *Curr. Opin. Behav. Sci.* **2016**, *8*, 161–166. [CrossRef]
25. Ward, L.M. Synchronous Neural Oscillations and Cognitive Processes. *Trends Cogn. Sci.* **2003**, *7*, 553–559. [CrossRef]
26. Başar, E.; Güntekin, B. Chapter 19—Review of Delta, Theta, Alpha, Beta, and Gamma Response Oscillations in Neuropsychiatric Disorders. In *Supplements to Clinical Neurophysiology. Application of Brain Oscillations in Neuropsychiatric Diseases*; Başar, E., Başar-Eroğlu, C., Özerdem, A., Rossini, P.M., Yener, G.G., Eds.; Elsevier: Amsterdam, The Netherlands, 2013; Volume 62, pp. 303–341.
27. Klimesch, W. EEG Alpha and Cognitive Processes. In *Time and the Brain*; CRC Press: Boca Raton, FL, USA, 2000; ISBN 978-0-429-17831-3.
28. Klimesch, W. EEG Alpha and Theta Oscillations Reflect Cognitive and Memory Performance: A Review and Analysis. *Brain Res. Rev.* **1999**, *29*, 169–195. [CrossRef]

29. Jensen, O.; Mazaheri, A. Shaping Functional Architecture by Oscillatory Alpha Activity: Gating by Inhibition. *Front. Hum. Neurosci.* **2010**, *4*, 186. [CrossRef] [PubMed]

30. Klimesch, W. Alpha-Band Oscillations, Attention, and Controlled Access to Stored Information. *Trends Cogn. Sci.* **2012**, *16*, 606–617. [CrossRef] [PubMed]

31. Sigala, R.; Haufe, S.; Roy, D.; Dinse, H.; Ritter, P. The Role of Alpha-Rhythm States in Perceptual Learning: Insights from Experiments and Computational Models. *Front. Comput. Neurosci.* **2014**, *8*, 36. [CrossRef] [PubMed]

32. Pfurtscheller, G.; Stancák, A.; Neuper, C. Event-Related Synchronization (ERS) in the Alpha Band—An Electrophysiological Correlate of Cortical Idling: A Review. *Int. J. Psychophysiol.* **1996**, *24*, 39–46. [CrossRef]

33. Mayer, A.; Schwiedrzik, C.M.; Wibral, M.; Singer, W.; Melloni, L. Expecting to See a Letter: Alpha Oscillations as Carriers of Top-Down Sensory Predictions. *Cereb. Cortex* **2016**, *26*, 3146–3160. [CrossRef]

34. Tarasi, L.; di Pellegrino, G.; Romei, V. Are You an Empiricist or a Believer? Neural Signatures of Predictive Strategies in Humans. *Prog. Neurobiol.* **2022**, *219*, 102367. [CrossRef] [PubMed]

35. Romei, V.; Gross, J.; Thut, G. On the Role of Prestimulus Alpha Rhythms over Occipito-Parietal Areas in Visual Input Regulation: Correlation or Causation? *J. Neurosci.* **2010**, *30*, 8692–8697. [CrossRef] [PubMed]

36. Samaha, J.; Iemi, L.; Postle, B.R. Prestimulus Alpha-Band Power Biases Visual Discrimination Confidence, but Not Accuracy. *Conscious. Cogn.* **2017**, *54*, 47–55. [CrossRef]

37. Iemi, L.; Chaumon, M.; Crouzet, S.M.; Busch, N.A. Spontaneous Neural Oscillations Bias Perception by Modulating Baseline Excitability. *J. Neurosci.* **2017**, *37*, 807–819. [CrossRef] [PubMed]

38. Di Gregorio, F.; Trajkovic, J.; Roperti, C.; Marcantoni, E.; Di Luzio, P.; Avenanti, A.; Thut, G.; Romei, V. Tuning Alpha Rhythms to Shape Conscious Visual Perception. *Curr. Biol.* **2022**, *32*, 988–998.e6. [CrossRef]

39. Cecere, R.; Rees, G.; Romei, V. Individual Differences in Alpha Frequency Drive Crossmodal Illusory Perception. *Curr. Biol.* **2015**, *25*, 231–235. [CrossRef]

40. Minami, S.; Amano, K. Illusory Jitter Perceived at the Frequency of Alpha Oscillations. *Curr. Biol.* **2017**, *27*, 2344–2351.e4. [CrossRef] [PubMed]

41. Cooke, J.; Poch, C.; Gillmeister, H.; Costantini, M.; Romei, V. Oscillatory Properties of Functional Connections Between Sensory Areas Mediate Cross-Modal Illusory Perception. *J. Neurosci.* **2019**, *39*, 5711–5718. [CrossRef]

42. Friston, K.J. Functional and Effective Connectivity: A Review. *Brain Connect.* **2011**, *1*, 13–36. [CrossRef]

43. Lejko, N.; Larabi, D.I.; Herrmann, C.S.; Aleman, A.; Ćurčić-Blake, B. Alpha Power and Functional Connectivity in Cognitive Decline: A Systematic Review and Meta-Analysis. *J. Alzheimer's Dis.* **2020**, *78*, 1047–1088. [CrossRef]

44. Douw, L.; de Groot, M.; van Dellen, E.; Heimans, J.J.; Ronner, H.E.; Stam, C.J.; Reijneveld, J.C. 'Functional Connectivity' Is a Sensitive Predictor of Epilepsy Diagnosis after the First Seizure. *PLoS ONE* **2010**, *5*, e10839. [CrossRef] [PubMed]

45. Morand-Beaulieu, S.; Wu, J.; Mayes, L.C.; Grantz, H.; Leckman, J.F.; Crowley, M.J.; Sukhodolsky, D.G. Increased Alpha-Band Connectivity During Tic Suppression in Children with Tourette Syndrome Revealed by Source Electroencephalography Analyses. *Biol. Psychiatry: Cogn. Neurosci. Neuroimaging* **2021**, in press. [CrossRef] [PubMed]

46. Hinkley, L.B.N.; Vinogradov, S.; Guggisberg, A.G.; Fisher, M.; Findlay, A.M.; Nagarajan, S.S. Clinical Symptoms and Alpha Band Resting-State Functional Connectivity Imaging in Patients with Schizophrenia: Implications for Novel Approaches to Treatment. *Biol. Psychiatry* **2011**, *70*, 1134–1142. [CrossRef]

47. Fingelkurts, A.A.; Fingelkurts, A.A.; Rytsälä, H.; Suominen, K.; Isometsä, E.; Kähkönen, S. Impaired Functional Connectivity at EEG Alpha and Theta Frequency Bands in Major Depression. *Hum. Brain Mapp.* **2007**, *28*, 247–261. [CrossRef] [PubMed]

48. Bozhilova, N.; Kuntsi, J.; Rubia, K.; Asherson, P.; Michelini, G. Event-Related Brain Dynamics during Mind Wandering in Attention-Deficit/Hyperactivity Disorder: An Experience-Sampling Approach. *NeuroImage Clin.* **2022**, *35*, 103068. [CrossRef] [PubMed]

49. Lenartowicz, A.; Mazaheri, A.; Jensen, O.; Loo, S.K. Aberrant Modulation of Brain Oscillatory Activity and Attentional Impairment in Attention-Deficit/Hyperactivity Disorder. *Biol. Psychiatry Cogn. Neurosci. Neuroimaging* **2018**, *3*, 19–29. [CrossRef] [PubMed]

50. Lefebvre, A.; Delorme, R.; Delanoë, C.; Amsellem, F.; Beggiato, A.; Germanaud, D.; Bourgeron, T.; Toro, R.; Dumas, G. Alpha Waves as a Neuromarker of Autism Spectrum Disorder: The Challenge of Reproducibility and Heterogeneity. *Front. Neurosci.* **2018**, *12*, 662. [CrossRef] [PubMed]

51. Loo, S.K.; McGough, J.J.; McCracken, J.T.; Smalley, S.L. Parsing Heterogeneity in Attention-deficit Hyperactivity Disorder Using EEG-based Subgroups. *J. Child Psychol. Psychiatry* **2018**, *59*, 223–231. [CrossRef] [PubMed]

52. Edition, F. Diagnostic and Statistical Manual of Mental Disorders. *Am. Psychiatric Assoc.* **2013**, *21*, 591–643.

53. Uhlhaas, P.J.; Singer, W. Oscillations and Neuronal Dynamics in Schizophrenia: The Search for Basic Symptoms and Translational Opportunities. *Biol. Psychiatry* **2015**, *77*, 1001–1009. [CrossRef] [PubMed]

54. Uhlhaas, P.J. Dysconnectivity, Large-Scale Networks and Neuronal Dynamics in Schizophrenia. *Curr. Opin. Neurobiol.* **2013**, *23*, 283–290. [CrossRef]

55. Stephan, K.E.; Friston, K.J.; Frith, C.D. Dysconnection in Schizophrenia: From Abnormal Synaptic Plasticity to Failures of Self-Monitoring. *Schizophr. Bull.* **2009**, *35*, 509–527. [CrossRef] [PubMed]

56. Pittman-Polletta, B.R.; Kocsis, B.; Vijayan, S.; Whittington, M.A.; Kopell, N.J. Brain Rhythms Connect Impaired Inhibition to Altered Cognition in Schizophrenia. *Biol. Psychiatry* **2015**, *77*, 1020–1030. [CrossRef] [PubMed]

57. Lozano-Soldevilla, D.; ter Huurne, N.; Cools, R.; Jensen, O. GABAergic Modulation of Visual Gamma and Alpha Oscillations and Its Consequences for Working Memory Performance. *Curr. Biol.* **2014**, *24*, 2878–2887. [CrossRef] [PubMed]

58. Puig, M.V.; Antzoulatos, E.G.; Miller, E.K. Prefrontal Dopamine in Associative Learning and Memory. *Neuroscience* **2014**, *282*, 217–229. [CrossRef] [PubMed]
59. Lemercier, C.E.; Holman, C.; Gerevich, Z. Aberrant Alpha and Gamma Oscillations Ex Vivo after Single Application of the NMDA Receptor Antagonist MK-801. *Schizophr. Res.* **2017**, *188*, 118–124. [CrossRef]
60. Başar, E.; Schmiedt-Fehr, C.; Mathes, B.; Femir, B.; Emek-Savaş, D.D.; Tülay, E.; Tan, D.; Düzgün, A.; Güntekin, B.; Özerdem, A.; et al. What Does the Broken Brain Say to the Neuroscientist? Oscillations and Connectivity in Schizophrenia, Alzheimer's Disease, and Bipolar Disorder. *Int. J. Psychophysiol.* **2016**, *103*, 135–148. [CrossRef]
61. Garakh, Z.; Zaytseva, Y.; Kapranova, A.; Fiala, O.; Horacek, J.; Shmukler, A.; Ya Gurovich, I.; Strelets, V.B. EEG Correlates of a Mental Arithmetic Task in Patients with First Episode Schizophrenia and Schizoaffective Disorder. *Clin. Neurophysiol.* **2015**, *126*, 2090–2098. [CrossRef] [PubMed]
62. Goldstein, M.R.; Peterson, M.J.; Sanguinetti, J.L.; Tononi, G.; Ferrarelli, F. Topographic Deficits in Alpha-Range Resting EEG Activity and Steady State Visual Evoked Responses in Schizophrenia. *Schizophr. Res.* **2015**, *168*, 145–152. [CrossRef]
63. Kim, J.W.; Lee, Y.S.; Han, D.H.; Min, K.J.; Lee, J.; Lee, K. Diagnostic Utility of Quantitative EEG in Un-Medicated Schizophrenia. *Neurosci. Lett.* **2015**, *589*, 126–131. [CrossRef]
64. Canali, P.; Sarasso, S.; Rosanova, M.; Casarotto, S.; Sferrazza-Papa, G.; Gosseries, O.; Fecchio, M.; Massimini, M.; Mariotti, M.; Cavallaro, R.; et al. Shared Reduction of Oscillatory Natural Frequencies in Bipolar Disorder, Major Depressive Disorder and Schizophrenia. *J. Affect. Disord.* **2015**, *184*, 111–115. [CrossRef]
65. Candelaria-Cook, F.T.; Schendel, M.E.; Ojeda, C.J.; Bustillo, J.R.; Stephen, J.M. Reduced Parietal Alpha Power and Psychotic Symptoms: Test-Retest Reliability of Resting-State Magnetoencephalography in Schizophrenia and Healthy Controls. *Schizophr. Res.* **2020**, *215*, 229–240. [CrossRef] [PubMed]
66. Mitra, S.; Nizamie, S.H.; Goyal, N.; Tikka, S.K. Electroencephalogram Alpha-to-Theta Ratio over Left Fronto-Temporal Region Correlates with Negative Symptoms in Schizophrenia. *Asian J. Psychiatry* **2017**, *26*, 70–76. [CrossRef] [PubMed]
67. Ahn, S.; Mellin, J.M.; Alagapan, S.; Alexander, M.L.; Gilmore, J.H.; Jarskog, L.F.; Fröhlich, F. Targeting Reduced Neural Oscillations in Patients with Schizophrenia by Transcranial Alternating Current Stimulation. *NeuroImage* **2019**, *186*, 126–136. [CrossRef] [PubMed]
68. Yeum, T.-S.; Kang, U.G. Reduction in Alpha Peak Frequency and Coherence on Quantitative Electroencephalography in Patients with Schizophrenia. *J. Korean Med. Sci.* **2018**, *33*, e179. [CrossRef] [PubMed]
69. Murphy, M.; Öngür, D. Decreased Peak Alpha Frequency and Impaired Visual Evoked Potentials in First Episode Psychosis. *NeuroImage Clin.* **2019**, *22*, 101693. [CrossRef]
70. Freche, D.; Naim-Feil, J.; Hess, S.; Peled, A.; Grinshpoon, A.; Moses, E.; Levit-Binnun, N. Phase-Amplitude Markers of Synchrony and Noise: A Resting-State and TMS-EEG Study of Schizophrenia. *Cereb. Cortex Commun.* **2020**, *1*, tgaa013. [CrossRef] [PubMed]
71. Hong, L.E.; Summerfelt, A.; Mitchell, B.D.; O'Donnell, P.; Thaker, G.K. A Shared Low-Frequency Oscillatory Rhythm Abnormality in Resting and Sensory Gating in Schizophrenia. *Clin. Neurophysiol.* **2012**, *123*, 285–292. [CrossRef]
72. Fox, M.D.; Snyder, A.Z.; Vincent, J.L.; Corbetta, M.; Van Essen, D.C.; Raichle, M.E. The Human Brain Is Intrinsically Organized into Dynamic, Anticorrelated Functional Networks. *Proc. Natl. Acad. Sci. USA* **2005**, *102*, 9673–9678. [CrossRef] [PubMed]
73. Nakhnikian, A.; Oribe, N.; Hirano, S.; Hirano, Y.; Levin, M.; Spencer, K. Lower Peak Alpha Frequency Accounts for Elevated Theta-Alpha Power in Resting State EEG in Schizophrenia. *Biol. Psychiatry* **2020**, *87*, S411. [CrossRef]
74. Narayanan, B.; O'Neil, K.; Berwise, C.; Stevens, M.C.; Calhoun, V.D.; Clementz, B.A.; Tamminga, C.A.; Sweeney, J.A.; Keshavan, M.S.; Pearlson, G.D. Resting State Electroencephalogram Oscillatory Abnormalities in Schizophrenia and Psychotic Bipolar Patients and Their Relatives from the Bipolar and Schizophrenia Network on Intermediate Phenotypes Study. *Biol. Psychiatry* **2014**, *76*, 456–465. [CrossRef] [PubMed]
75. Nakhnikian, A.; Oribe, N.; Hirano, S.; Hirano, Y.; Levin, M.; Spencer, K. Increased Theta/Alpha Source Activity and Default Mode Network Connectivity in Schizophrenia During Eyes-Closed Rest. *Biol. Psychiatry* **2021**, *89*, S150–S151. [CrossRef]
76. Kim, J.S.; Shin, K.S.; Jung, W.H.; Kim, S.N.; Kwon, J.S.; Chung, C.K. Power Spectral Aspects of the Default Mode Network in Schizophrenia: An MEG Study. *BMC Neurosci.* **2014**, *15*, 104. [CrossRef] [PubMed]
77. Koshiyama, D.; Miyakoshi, M.; Tanaka-Koshiyama, K.; Joshi, Y.B.; Sprock, J.; Braff, D.L.; Light, G.A. Abnormal Phase Discontinuity of Alpha- and Theta-Frequency Oscillations in Schizophrenia. *Schizophr. Res.* **2021**, *231*, 73–81. [CrossRef]
78. Bertaccini, R.; Ellena, G.; Macedo-Pascual, J.; Carusi, F.; Trajkovic, J.; Poch, C.; Romei, V. Parietal Alpha Oscillatory Peak Frequency Mediates the Effect of Practice on Visuospatial Working Memory Performance. *Vision* **2022**, *6*, 30. [CrossRef]
79. Trajkovic, J.; Di Gregorio, F.; Ferri, F.; Marzi, C.; Diciotti, S.; Romei, V. Resting State Alpha Oscillatory Activity Is a Valid and Reliable Marker of Schizotypy. *Sci. Rep.* **2021**, *11*, 10379. [CrossRef]
80. Ramsay, I.S.; Lynn, P.A.; Schermitzler, B.; Sponheim, S.R. Individual Alpha Peak Frequency Is Slower in Schizophrenia and Related to Deficits in Visual Perception and Cognition. *Sci. Rep.* **2021**, *11*, 17852. [CrossRef]
81. Garakh, Z.V.; Novototsky-Vlasov, V.Y.; Zaitseva, Y.S.; Rebreikina, A.B.; Strelets, V.B. Frequency of the Alpha Activity Spectral Peak and Psychopathological Symptoms in Schizophrenia. *Neurosci. Behav. Physiol.* **2012**, *42*, 1068–1073. [CrossRef]
82. Liu, T.; Zhang, J.; Dong, X.; Li, Z.; Shi, X.; Tong, Y.; Yang, R.; Wu, J.; Wang, C.; Yan, T. Occipital Alpha Connectivity During Resting-State Electroencephalography in Patients with Ultra-High Risk for Psychosis and Schizophrenia. *Front. Psychiatry* **2019**, *10*, 553. [CrossRef] [PubMed]
83. Kam, J.W.Y.; Bolbecker, A.R.; O'Donnell, B.F.; Hetrick, W.P.; Brenner, C.A. Resting State EEG Power and Coherence Abnormalities in Bipolar Disorder and Schizophrenia. *J. Psychiatr. Res.* **2013**, *47*, 1893–1901. [CrossRef]

84. Baenninger, A.; Palzes, V.A.; Roach, B.J.; Mathalon, D.H.; Ford, J.M.; Koenig, T. Abnormal Coupling Between Default Mode Network and Delta and Beta Band Brain Electric Activity in Psychotic Patients. *Brain Connect.* **2017**, *7*, 34–44. [CrossRef] [PubMed]

85. Olejarczyk, E.; Jernajczyk, W. Graph-Based Analysis of Brain Connectivity in Schizophrenia. *PLoS ONE* **2017**, *12*, e0188629. [CrossRef]

86. Phalen, H.; Coffman, B.A.; Ghuman, A.; Sejdić, E.; Salisbury, D.F. Non-Negative Matrix Factorization Reveals Resting-State Cortical Alpha Network Abnormalities in the First-Episode Schizophrenia Spectrum. *Biol. Psychiatry: Cogn. Neurosci. Neuroimaging* **2020**, *5*, 961–970. [CrossRef]

87. Di Lorenzo, G.; Daverio, A.; Ferrentino, F.; Santarnecchi, E.; Ciabattini, F.; Monaco, L.; Lisi, G.; Barone, Y.; Di Lorenzo, C.; Niolu, C.; et al. Altered Resting-State EEG Source Functional Connectivity in Schizophrenia: The Effect of Illness Duration. *Front. Hum. Neurosci.* **2015**, *9*, 234. [CrossRef]

88. Lehmann, D.; Faber, P.L.; Pascual-Marqui, R.D.; Milz, P.; Herrmann, W.M.; Koukkou, M.; Saito, N.; Winterer, G.; Kochi, K. Functionally Aberrant Electrophysiological Cortical Connectivities in First Episode Medication-Naive Schizophrenics from Three Psychiatry Centers. *Front. Hum. Neurosci.* **2014**, *8*, 635. [CrossRef] [PubMed]

89. Postmes, L.; Sno, H.N.; Goedhart, S.; van der Stel, J.; Heering, H.D.; de Haan, L. Schizophrenia as a Self-Disorder Due to Perceptual Incoherence. *Schizophr. Res.* **2014**, *152*, 41–50. [CrossRef] [PubMed]

90. Parker, D.A.; Hamm, J.P.; McDowell, J.E.; Keedy, S.K.; Gershon, E.S.; Ivleva, E.I.; Pearlson, G.D.; Keshavan, M.S.; Tamminga, C.A.; Sweeney, J.A.; et al. Auditory Steady-State EEG Response across the Schizo-Bipolar Spectrum. *Schizophr. Res.* **2019**, *209*, 218–226. [CrossRef]

91. Hirano, S.; Nakhnikian, A.; Hirano, Y.; Oribe, N.; Kanba, S.; Onitsuka, T.; Levin, M.; Spencer, K.M. Phase-Amplitude Coupling of the Electroencephalogram in the Auditory Cortex in Schizophrenia. *Biol. Psychiatry: Cogn. Neurosci. Neuroimaging* **2018**, *3*, 69–76. [CrossRef]

92. Edgar, J.C.; Chen, Y.-H.; Lanza, M.; Howell, B.; Chow, V.Y.; Heiken, K.; Liu, S.; Wootton, C.; Hunter, M.A.; Huang, M.; et al. Cortical Thickness as a Contributor to Abnormal Oscillations in Schizophrenia? *NeuroImage Clin.* **2014**, *4*, 122–129. [CrossRef]

93. Carolus, A.M.; Schubring, D.; Popov, T.G.; Popova, P.; Miller, G.A.; Rockstroh, B.S. Functional Cognitive and Cortical Abnormalities in Chronic and First-Admission Schizophrenia. *Schizophr. Res.* **2014**, *157*, 40–47. [CrossRef]

94. Hamm, J.P.; Ethridge, L.E.; Shapiro, J.R.; Stevens, M.C.; Boutros, N.N.; Summerfelt, A.T.; Keshavan, M.S.; Sweeney, J.A.; Pearlson, G.; Tamminga, C.A.; et al. Spatiotemporal and Frequency Domain Analysis of Auditory Paired Stimuli Processing in Schizophrenia and Bipolar Disorder with Psychosis. *Psychophysiology* **2012**, *49*, 522–530. [CrossRef] [PubMed]

95. Rowland, L.M.; Edden, R.A.E.; Kontson, K.; Zhu, H.; Barker, P.B.; Hong, L.E. GABA Predicts Inhibition of Frequency-Specific Oscillations in Schizophrenia. *J. Nucl. Phys.* **2013**, *25*, 83–87. [CrossRef]

96. Başar-Eroğlu, C.; Schmiedt-Fehr, C.; Mathes, B. Auditory-Evoked Alpha Oscillations Imply Reduced Anterior and Increased Posterior Amplitudes in Schizophrenia. In *Supplements to Clinical Neurophysiology*; Elsevier: Amsterdam, The Netherlands, 2013; Volume 62, pp. 121–129. ISBN 978-0-7020-5307-8.

97. Popov, T.G.; Carolus, A.; Schubring, D.; Popova, P.; Miller, G.A.; Rockstroh, B.S. Targeted Training Modifies Oscillatory Brain Activity in Schizophrenia Patients. *NeuroImage Clin.* **2015**, *7*, 807–814. [CrossRef] [PubMed]

98. Keil, J.; Roa Romero, Y.; Balz, J.; Henjes, M.; Senkowski, D. Positive and Negative Symptoms in Schizophrenia Relate to Distinct Oscillatory Signatures of Sensory Gating. *Front. Hum. Neurosci.* **2016**, *10*, 104. [CrossRef] [PubMed]

99. Lee, M.; Sehatpour, P.; Dias, E.C.; Silipo, G.S.; Kantrowitz, J.T.; Martinez, A.M.; Javitt, D.C. A Tale of Two Sites: Differential Impairment of Frequency and Duration Mismatch Negativity across a Primarily Inpatient versus a Primarily Outpatient Site in Schizophrenia. *Schizophr. Res.* **2018**, *191*, 10–17. [CrossRef] [PubMed]

100. Lee, M.; Sehatpour, P.; Hoptman, M.J.; Lakatos, P.; Dias, E.C.; Kantrowitz, J.T.; Martinez, A.M.; Javitt, D.C. Neural Mechanisms of Mismatch Negativity Dysfunction in Schizophrenia. *Mol. Psychiatry* **2017**, *22*, 1585–1593. [CrossRef] [PubMed]

101. Hong, L.E.; Moran, L.V.; Du, X.; O'Donnell, P.; Summerfelt, A. Mismatch Negativity and Low Frequency Oscillations in Schizophrenia Families. *Clin. Neurophysiol.* **2012**, *123*, 1980–1988. [CrossRef] [PubMed]

102. Kayser, J.; Tenke, C.E.; Kroppmann, C.J.; Alschuler, D.M.; Fekri, S.; Ben-David, S.; Corcoran, C.M.; Bruder, G.E. Auditory Event-Related Potentials and Alpha Oscillations in the Psychosis Prodrome: Neuronal Generator Patterns during a Novelty Oddball Task. *Int. J. Psychophysiol.* **2014**, *91*, 104–120. [CrossRef] [PubMed]

103. Núñez, P.; Poza, J.; Bachiller, A.; Gomez-Pilar, J.; Lubeiro, A.; Molina, V.; Hornero, R. Exploring Non-Stationarity Patterns in Schizophrenia: Neural Reorganization Abnormalities in the Alpha Band. *J. Neural. Eng.* **2017**, *14*, 046001. [CrossRef]

104. Fujimoto, T.; Okumura, E.; Takeuchi, K.; Kodabashi, A.; Tanaka, H.; Otsubo, T.; Nakamura, K.; Sekine, M.; Kamiya, S.; Higashi, Y.; et al. Changes in Event-Related Desynchronization and Synchronization during the Auditory Oddball Task in Schizophrenia Patients. *Open Neuroimag. J.* **2012**, *6*, 26–36. [CrossRef]

105. Aubonnet, R.; Banea, O.C.; Sirica, R.; Wassermann, E.M.; Yassine, S.; Jacob, D.; Magnúsdóttir, B.B.; Haraldsson, M.; Stefansson, S.B.; Jónasson, V.D.; et al. P300 Analysis Using High-Density EEG to Decipher Neural Response to RTMS in Patients With Schizophrenia and Auditory Verbal Hallucinations. *Front. Neurosci.* **2020**, *14*, 575538. [CrossRef] [PubMed]

106. Coffman, B.A.; Haas, G.; Olson, C.; Cho, R.; Ghuman, A.S.; Salisbury, D.F. Reduced Dorsal Visual Oscillatory Activity During Working Memory Maintenance in the First-Episode Schizophrenia Spectrum. *Front. Psychiatry* **2020**, *11*, 743. [CrossRef] [PubMed]

107. Erickson, M.A.; Albrecht, M.A.; Robinson, B.; Luck, S.J.; Gold, J.M. Impaired Suppression of Delay-Period Alpha and Beta Is Associated with Impaired Working Memory in Schizophrenia. *Biol. Psychiatry Cogn. Neurosci. Neuroimaging* **2017**, *2*, 272–279. [CrossRef] [PubMed]

108. Kustermann, T.; Rockstroh, B.; Kienle, J.; Miller, G.A.; Popov, T. Deficient Attention Modulation of Lateralized Alpha Power in Schizophrenia. *Psychophysiology* **2016**, *53*, 776–785. [CrossRef]

109. Sklar, A.L.; Coffman, B.A.; Longenecker, J.M.; Curtis, M.; Salisbury, D.F. Load-Dependent Functional Connectivity Deficits during Visual Working Memory in First-Episode Psychosis. *J. Psychiatr. Res.* **2022**, *153*, 174–181. [CrossRef]

110. Cooper, P.S.; Hughes, M.E. Impaired Theta and Alpha Oscillations Underlying Stopsignal Response Inhibition Deficits in Schizophrenia. *Schizophr. Res.* **2018**, *193*, 474–476. [CrossRef]

111. Popov, T.; Kustermann, T.; Popova, P.; Miller, G.A.; Rockstroh, B. Oscillatory Brain Dynamics Supporting Impaired Stroop Task Performance in Schizophrenia-Spectrum Disorder. *Schizophr. Res.* **2019**, *204*, 146–154. [CrossRef] [PubMed]

112. Green, M.F.; Horan, W.P.; Lee, J. Social Cognition in Schizophrenia. *Nat. Rev. Neurosci.* **2015**, *16*, 620–631. [CrossRef]

113. Popov, T.G.; Rockstroh, B.S.; Popova, P.; Carolus, A.M.; Miller, G.A. Dynamics of Alpha Oscillations Elucidate Facial Affect Recognition in Schizophrenia. *Cogn. Affect. Behav. Neurosci.* **2014**, *14*, 364–377. [CrossRef]

114. Li, L.Y.; Schiffman, J.; Hu, D.K.; Lopour, B.A.; Martin, E.A. An Effortful Approach to Social Affiliation in Schizophrenia: Preliminary Evidence of Increased Theta and Alpha Connectivity during a Live Social Interaction. *Brain. Sci.* **2021**, *11*, 1346. [CrossRef]

115. Gabay, A.S.; Radua, J.; Kempton, M.J.; Mehta, M.A. The Ultimatum Game and the Brain: A Meta-Analysis of Neuroimaging Studies. *Neurosci. Biobehav. Rev.* **2014**, *47*, 549–558. [CrossRef]

116. Billeke, P.; Armijo, A.; Castillo, D.; López, T.; Zamorano, F.; Cosmelli, D.; Aboitiz, F. Paradoxical Expectation: Oscillatory Brain Activity Reveals Social Interaction Impairment in Schizophrenia. *Biol. Psychiatry* **2015**, *78*, 421–431. [CrossRef] [PubMed]

117. Jia, S.; Liu, M.; Huang, P.; Zhao, Y.; Tan, S.; Go, R.; Yan, T.; Wu, J. Abnormal Alpha Rhythm During Self-Referential Processing in Schizophrenia Patients. *Front. Psychiatry* **2019**, *10*, 691. [CrossRef] [PubMed]

118. Fingelkurts, A.A.; Fingelkurts, A.A. Altered Structure of Dynamic Electroencephalogram Oscillatory Pattern in Major Depression. *Biol. Psychiatry* **2015**, *77*, 1050–1060. [CrossRef] [PubMed]

119. Leuchter, A.; Cook, I.; Jin, Y.; Phillips, B. The Relationship between Brain Oscillatory Activity and Therapeutic Effectiveness of Transcranial Magnetic Stimulation in the Treatment of Major Depressive Disorder. *Front. Hum. Neurosci.* **2013**, *7*, 37. [CrossRef]

120. Eidelman-Rothman, M.; Levy, J.; Feldman, R. Alpha Oscillations and Their Impairment in Affective and Post-Traumatic Stress Disorders. *Neurosci. Biobehav. Rev.* **2016**, *68*, 794–815. [CrossRef]

121. Davidson, R.J.; Saron, C.D.; Senulis, J.A.; Ekman, P.; Friesen, W.V. Approach-Withdrawal and Cerebral Asymmetry: Emotional Expression and Brain Physiology. *J. Pers. Soc. Psychol.* **1990**, *58*, 330–341. [CrossRef] [PubMed]

122. Horato, N.; Quagliato, L.A.; Nardi, A.E. The Relationship between Emotional Regulation and Hemispheric Lateralization in Depression: A Systematic Review and a Meta-Analysis. *Transl. Psychiatry* **2022**, *12*, 162. [CrossRef] [PubMed]

123. Kołodziej, A.; Magnuski, M.; Ruban, A.; Brzezicka, A. No Relationship between Frontal Alpha Asymmetry and Depressive Disorders in a Multiverse Analysis of Five Studies. *eLife* **2021**, *10*, e60595. [CrossRef] [PubMed]

124. Van der Vinne, N.; Vollebregt, M.A.; van Putten, M.J.A.M.; Arns, M. Frontal Alpha Asymmetry as a Diagnostic Marker in Depression: Fact or Fiction? A Meta-Analysis. *NeuroImage Clin.* **2017**, *16*, 79–87. [CrossRef] [PubMed]

125. De Aguiar Neto, F.S.; Rosa, J.L.G. Depression Biomarkers Using Non-Invasive EEG: A Review. *Neurosci. Biobehav. Rev.* **2019**, *105*, 83–93. [CrossRef] [PubMed]

126. Alamian, G.; Hincapié, A.-S.; Combrisson, E.; Thiery, T.; Martel, V.; Althukov, D.; Jerbi, K. Alterations of Intrinsic Brain Connectivity Patterns in Depression and Bipolar Disorders: A Critical Assessment of Magnetoencephalography-Based Evidence. *Front. Psychiatry* **2017**, *8*, 41. [CrossRef] [PubMed]

127. Northoff, G. Spatiotemporal Psychopathology I_ No Rest for the Brain's Resting State Activity in Depression? Spatiotemporal Psychopathology of Depressive Symptoms. *J. Affect. Disord.* **2016**, *13*, 854–866. [CrossRef]

128. Barros, C.; Pereira, A.R.; Sampaio, A.; Buján, A.; Pinal, D. Frontal Alpha Asymmetry and Negative Mood: A Cross-Sectional Study in Older and Younger Adults. *Symmetry* **2022**, *14*, 1579. [CrossRef]

129. Cantisani, A.; Koenig, T.; Stegmayer, K.; Federspiel, A.; Horn, H.; Müller, T.J.; Wiest, R.; Strik, W.; Walther, S. EEG Marker of Inhibitory Brain Activity Correlates with Resting-State Cerebral Blood Flow in the Reward System in Major Depression. *Eur. Arch. Psychiatry Clin. Neurosci.* **2016**, *266*, 755–764. [CrossRef]

130. Jesulola, E.; Sharpley, C.F.; Agnew, L.L. The Effects of Gender and Depression Severity on the Association between Alpha Asymmetry and Depression across Four Brain Regions. *Behav. Brain Res.* **2017**, *321*, 232–239. [CrossRef] [PubMed]

131. Jaworska, N.; Blier, P.; Fusee, W.; Knott, V. Alpha Power, Alpha Asymmetry and Anterior Cingulate Cortex Activity in Depressed Males and Females. *J. Psychiatr. Res.* **2012**, *46*, 1483–1491. [CrossRef] [PubMed]

132. Quinn, C.R.; Rennie, C.J.; Harris, A.W.F.; Kemp, A.H. The Impact of Melancholia versus Non-Melancholia on Resting-State, EEG Alpha Asymmetry: Electrophysiological Evidence for Depression Heterogeneity. *Psychiatry Res.* **2014**, *215*, 614–617. [CrossRef] [PubMed]

133. Koo, P.C.; Berger, C.; Kronenberg, G.; Bartz, J.; Wybitul, P.; Reis, O.; Hoeppner, J. Combined Cognitive, Psychomotor and Electrophysiological Biomarkers in Major Depressive Disorder. *Eur. Arch. Psychiatry Clin. Neurosci.* **2019**, *269*, 823–832. [CrossRef]

134. Kustubayeva, A.; Kamzanova, A.; Kudaibergenova, S.; Pivkina, V.; Matthews, G. Major Depression and Brain Asymmetry in a Decision-Making Task with Negative and Positive Feedback. *Symmetry* **2020**, *12*, 2118. [CrossRef]

135. Auerbach, R.P.; Stewart, J.G.; Stanton, C.H.; Mueller, E.M.; Pizzagalli, D.A. Emotion-Processing Biases and Resting Eeg Activity in Depressed Adolescents. *Depress. Anxiety* **2015**, *32*, 693–701. [CrossRef] [PubMed]

136. Park, Y.; Jung, W.; Kim, S.; Jeon, H.; Lee, S.-H. Frontal Alpha Asymmetry Correlates with Suicidal Behavior in Major Depressive Disorder. *Clin. Psychopharmacol. Neurosci.* **2019**, *17*, 377–387. [CrossRef]

137. Smith, E.E.; Cavanagh, J.F.; Allen, J.J.B. Intracranial Source Activity (ELORETA) Related to Scalp-Level Asymmetry Scores and Depression Status. *Psychophysiology* **2018**, *55*, e13019. [CrossRef]
138. Gollan, J.K.; Hoxha, D.; Chihade, D.; Pflieger, M.E.; Rosebrock, L.; Cacioppo, J. Frontal Alpha EEG Asymmetry before and after Behavioral Activation Treatment for Depression. *Biol. Psychol.* **2014**, *99*, 198–208. [CrossRef] [PubMed]
139. Riddle, J.; Alexander, M.L.; Schiller, C.E.; Rubinow, D.R.; Frohlich, F. Reduction in Left Frontal Alpha Oscillations by Transcranial Alternating Current Stimulation in Major Depressive Disorder Is Context Dependent in a Randomized Clinical Trial. *Biol. Psychiatry Cogn. Neurosci. Neuroimaging* **2022**, *7*, 302–311. [CrossRef] [PubMed]
140. Alexander, M.L.; Alagapan, S.; Lugo, C.E.; Mellin, J.M.; Lustenberger, C.; Rubinow, D.R.; Fröhlich, F. Double-Blind, Randomized Pilot Clinical Trial Targeting Alpha Oscillations with Transcranial Alternating Current Stimulation (TACS) for the Treatment of Major Depressive Disorder (MDD). *Transl. Psychiatry* **2019**, *9*, 1–12. [CrossRef] [PubMed]
141. Umemoto, A.; Panier, L.Y.X.; Cole, S.L.; Kayser, J.; Pizzagalli, D.A.; Auerbach, R.P. Resting Posterior Alpha Power and Adolescent Major Depressive Disorder. *J. Psychiatry Res.* **2021**, *141*, 233–240. [CrossRef] [PubMed]
142. Quraan, M.A.; Protzner, A.B.; Daskalakis, Z.J.; Giacobbe, P.; Tang, C.W.; Kennedy, S.H.; Lozano, A.M.; McAndrews, M.P. EEG Power Asymmetry and Functional Connectivity as a Marker of Treatment Effectiveness in DBS Surgery for Depression. *Neuropsychopharmacology* **2014**, *39*, 1270–1281. [CrossRef] [PubMed]
143. Nishida, K.; Koshikawa, Y.; Morishima, Y.; Yoshimura, M.; Katsura, K.; Ueda, S.; Ikeda, S.; Ishii, R.; Pascual-Marqui, R.; Kinoshita, T. Pre-Stimulus Brain Activity Is Associated with State-Anxiety Changes During Single-Session Transcranial Direct Current Stimulation. *Front. Hum. Neurosci.* **2019**, *13*, 266. [CrossRef] [PubMed]
144. Benschop, L.; Baeken, C.; Vanderhasselt, M.-A.; de Steen, F.V.; Heeringen, K.V.; Arns, M. Electroencephalogram Resting State Frequency Power Characteristics of Suicidal Behavior in Female Patients with Major Depressive Disorder. *J. Clin. Psychiatry* **2019**, *80*, 5459. [CrossRef]
145. Wu, Z.; Zhong, X.; Lin, G.; Peng, Q.; Zhang, M.; Zhou, H.; Wang, Q.; Chen, B.; Ning, Y. Resting-State Electroencephalography of Neural Oscillation and Functional Connectivity Patterns in Late-Life Depression. *J. Affect. Disord.* **2022**, *316*, 169–176. [CrossRef] [PubMed]
146. Lee, P.F.; Kan, D.P.X.; Croarkin, P.; Phang, C.K.; Doruk, D. Neurophysiological Correlates of Depressive Symptoms in Young Adults: A Quantitative EEG Study. *J. Clin. Neurosci.* **2018**, *47*, 315–322. [CrossRef] [PubMed]
147. Woźniak-Kwaśniewska, A.; Szekely, D.; Harquel, S.; Bougerol, T.; David, O. Resting Electroencephalographic Correlates of the Clinical Response to Repetitive Transcranial Magnetic Stimulation: A Preliminary Comparison between Unipolar and Bipolar Depression. *J. Affect. Disord.* **2015**, *183*, 15–21. [CrossRef]
148. Hill, A.T.; Hadas, I.; Zomorrodi, R.; Voineskos, D.; Fitzgerald, P.B.; Blumberger, D.M.; Daskalakis, Z.J. Characterizing Cortical Oscillatory Responses in Major Depressive Disorder Before and After Convulsive Therapy: A TMS-EEG Study. *J. Affect. Disord.* **2021**, *287*, 78–88. [CrossRef] [PubMed]
149. Kan, D.P.X.; Lee, P.F. Decrease Alpha Waves in Depression: An Electroencephalogram (EEG) Study. In Proceedings of the 2015 International Conference on BioSignal Analysis, Kuala Lumpur, Malaysia, 26-28 May 2015; Processing and Systems (ICBAPS). pp. 156–161.
150. Wolff, A.; de la Salle, S.; Sorgini, A.; Lynn, E.; Blier, P.; Knott, V.; Northoff, G. Atypical Temporal Dynamics of Resting State Shapes Stimulus-Evoked Activity in Depression-An EEG Study on Rest-Stimulus Interaction. *Front. Psychiatry* **2019**, *10*, 719. [CrossRef] [PubMed]
151. Jiang, H.; Popov, T.; Jylänki, P.; Bi, K.; Yao, Z.; Lu, Q.; Jensen, O.; van Gerven, M.A.J. Predictability of Depression Severity Based on Posterior Alpha Oscillations. *Clin. Neurophysiol.* **2016**, *127*, 2108–2114. [CrossRef] [PubMed]
152. Keller, A.S.; Ball, T.M.; Williams, L.M. Deep Phenotyping of Attention Impairments and the 'Inattention Biotype' in Major Depressive Disorder. *Psychol. Med.* **2020**, *50*, 2203–2212. [CrossRef]
153. Cook, I.A.; Wilson, A.C.; Corlier, J.; Leuchter, A.F. Brain Activity and Clinical Outcomes in Adults with Depression Treated with Synchronized Transcranial Magnetic Stimulation: An Exploratory Study. *Neuromodulation: Technol. Neural Interface* **2019**, *22*, 894–897. [CrossRef]
154. Noda, Y.; Nakamura, M.; Saeki, T.; Inoue, M.; Iwanari, H.; Kasai, K. Potentiation of Quantitative Electroencephalograms Following Prefrontal Repetitive Transcranial Magnetic Stimulation in Patients with Major Depression. *Neurosci. Res.* **2013**, *77*, 70–77. [CrossRef] [PubMed]
155. Wang, J.; Liu, Q.; Tian, F.; Zhou, S.; Parra, M.A.; Wang, H.; Yu, X. Disrupted Spatiotemporal Complexity of Resting-State Electroencephalogram Dynamics Is Associated with Adaptive and Maladaptive Rumination in Major Depressive Disorder. *Front. Neurosci.* **2022**, *16*, 829755. [CrossRef]
156. Mohammadi, Y.; Moradi, M.H. Prediction of Depression Severity Scores Based on Functional Connectivity and Complexity of the EEG Signal. *Clin. EEG Neurosci.* **2021**, *52*, 52–60. [CrossRef]
157. Shim, M.; Im, C.-H.; Kim, Y.-W.; Lee, S.-H. Altered Cortical Functional Network in Major Depressive Disorder: A Resting-State Electroencephalogram Study. *NeuroImage: Clin.* **2018**, *19*, 1000–1007. [CrossRef] [PubMed]
158. Nugent, A.C.; Ballard, E.D.; Gilbert, J.R.; Tewarie, P.K.; Brookes, M.J.; Zarate, C.A. Multilayer MEG Functional Connectivity as a Potential Marker for Suicidal Thoughts in Major Depressive Disorder. *NeuroImage Clin.* **2020**, *28*, 102378. [CrossRef]
159. Corlier, J.; Wilson, A.; Hunter, A.M.; Vince-Cruz, N.; Krantz, D.; Levitt, J.; Minzenberg, M.J.; Ginder, N.; Cook, I.A.; Leuchter, A.F. Changes in Functional Connectivity Predict Outcome of Repetitive Transcranial Magnetic Stimulation Treatment of Major Depressive Disorder. *Cereb. Cortex* **2019**, *29*, 4958–4967. [CrossRef] [PubMed]
160. Tas, C.; Cebi, M.; Tan, O.; Hızlı-Sayar, G.; Tarhan, N.; Brown, E.C. EEG Power, Cordance and Coherence Differences between Unipolar and Bipolar Depression. *J. Affect. Disord.* **2015**, *172*, 184–190. [CrossRef]

161. Leuchter, A.F.; Cook, I.A.; Hunter, A.M.; Cai, C.; Horvath, S. Resting-State Quantitative Electroencephalography Reveals Increased Neurophysiologic Connectivity in Depression. *PLoS ONE* **2012**, *7*, e32508. [CrossRef] [PubMed]
162. Olbrich, S.; Tränkner, A.; Chittka, T.; Hegerl, U.; Schönknecht, P. Functional Connectivity in Major Depression: Increased Phase Synchronization between Frontal Cortical EEG-Source Estimates. *Psychiatry Res. Neuroimaging* **2014**, *222*, 91–99. [CrossRef] [PubMed]
163. Wang, Q.; Tian, S.; Tang, H.; Liu, X.; Yan, R.; Hua, L.; Shi, J.; Chen, Y.; Zhu, R.; Lu, Q.; et al. Identification of Major Depressive Disorder and Prediction of Treatment Response Using Functional Connectivity between the Prefrontal Cortices and Subgenual Anterior Cingulate: A Real-World Study. *J. Affect. Disord.* **2019**, *252*, 365–372. [CrossRef] [PubMed]
164. Minami, S.; Kato, M.; Ikeda, S.; Yoshimura, M.; Ueda, S.; Koshikawa, Y.; Takekita, Y.; Kinoshita, T.; Nishida, K. Association between the Rostral Anterior Cingulate Cortex and Anterior Insula in the Salience Network on Response to Antidepressants in Major Depressive Disorder as Revealed by Isolated Effective Coherence. *Neuropsychobiology* **2022**, 1–9. [CrossRef]
165. Leuchter, A.F.; Wilson, A.C.; Vince-Cruz, N.; Corlier, J. Novel Method for Identification of Individualized Resonant Frequencies for Treatment of Major Depressive Disorder (MDD) Using Repetitive Transcranial Magnetic Stimulation (RTMS): A Proof-of-Concept Study. *Brain Stimul.* **2021**, *14*, 1373–1383. [CrossRef] [PubMed]
166. Li, Y.; Kang, C.; Qu, X.; Zhou, Y.; Wang, W.; Hu, Y. Depression-Related Brain Connectivity Analyzed by EEG Event-Related Phase Synchrony Measure. *Front. Hum. Neurosci.* **2016**, *10*, 477. [CrossRef] [PubMed]
167. Bailey, N.W. Impaired Upper Alpha Synchronisation during Working Memory Retention in Depression and Depression Following Traumatic Brain Injury. *Biol. Psychol.* **2014**, *10*, 115–124. [CrossRef]
168. Murphy, O.W.; Hoy, K.E.; Wong, D.; Bailey, N.W.; Fitzgerald, P.B.; Segrave, R.A. Individuals with Depression Display Abnormal Modulation of Neural Oscillatory Activity during Working Memory Encoding and Maintenance. *Biol. Psychol.* **2019**, *148*, 107766. [CrossRef]
169. Li, Y.; Kang, C.; Wei, Z.; Qu, X.; Liu, T.; Zhou, Y.; Hu, Y. Beta Oscillations in Major Depression—Signalling a New Cortical Circuit for Central Executive Function. *Sci. Rep.* **2017**, *7*, 18021. [CrossRef] [PubMed]
170. Dai, Z.; Zhou, H.; Zhang, W.; Tang, H.; Wang, T.; Chen, Z.; Yao, Z.; Lu, Q. Alpha-Beta Decoupling Relevant to Inhibition Deficits Leads to Suicide Attempt in Major Depressive Disorder. *J. Affect. Disord.* **2022**, *314*, 168–175. [CrossRef]
171. Schoenberg, P.L.A.; Speckens, A.E.M. Multi-Dimensional Modulations of α and γ Cortical Dynamics Following Mindfulness-Based Cognitive Therapy in Major Depressive Disorder. *Cogn. Neurodyn.* **2015**, *9*, 13–29. [CrossRef]
172. Messerotti Benvenuti, S.; Buodo, G.; Mennella, R.; Dal Bò, E.; Palomba, D. Appetitive and Aversive Motivation in Depression: The Temporal Dynamics of Task-Elicited Asymmetries in Alpha Oscillations. *Sci. Rep.* **2019**, *9*, 17129. [CrossRef]
173. Zotev, V.; Yuan, H.; Misaki, M.; Phillips, R.; Young, K.D.; Feldner, M.T.; Bodurka, J. Correlation between Amygdala BOLD Activity and Frontal EEG Asymmetry during Real-Time FMRI Neurofeedback Training in Patients with Depression. *NeuroImage: Clin.* **2016**, *11*, 224–238. [CrossRef] [PubMed]
174. Segrave, R.A.; Thomson, R.H.; Cooper, N.R.; Croft, R.J.; Sheppard, D.M.; Fitzgerald, P.B. Emotive Interference during Cognitive Processing in Major Depression: An Investigation of Lower Alpha 1 Activity. *J. Affect. Disord.* **2012**, *141*, 185–193. [CrossRef] [PubMed]
175. Koller-Schlaud, K.; Ströhle, A.; Bärwolf, E.; Behr, J.; Rentzsch, J. EEG Frontal Asymmetry and Theta Power in Unipolar and Bipolar Depression. *J. Affect. Disord.* **2020**, *276*, 501–510. [CrossRef] [PubMed]
176. Gheza, D.; Bakic, J.; Baeken, C.; De Raedt, R.; Pourtois, G. Abnormal Approach-Related Motivation but Spared Reinforcement Learning in MDD: Evidence from Fronto-Midline Theta Oscillations and Frontal Alpha Asymmetry. *Cogn. Affect. Behav. Neurosci.* **2019**, *19*, 759–777. [CrossRef] [PubMed]
177. Escolano, C.; Navarro-Gil, M.; Garcia-Campayo, J.; Congedo, M.; De Ridder, D.; Minguez, J. A Controlled Study on the Cognitive Effect of Alpha Neurofeedback Training in Patients with Major Depressive Disorder. *Front. Behav. Neurosci.* **2014**, *8*, 296. [CrossRef] [PubMed]
178. Kessler, K.; Seymour, R.A.; Rippon, G. Brain Oscillations and Connectivity in Autism Spectrum Disorders (ASD): New Approaches to Methodology, Measurement and Modelling. *Neurosci. Biobehav. Rev.* **2016**, *71*, 601–620. [CrossRef] [PubMed]
179. Association, A.P. *Diagnostic and Statistical Manual of Mental Disorders: Dsm-5*, 5th ed.; Amer Psychiatric Pub Inc.: Washington, DC, USA, 2013; ISBN 978-0-89042-555-8.
180. Han, J.; Zeng, K.; Kang, J.; Tong, Z.; Cai, E.; Chen, H.; Ding, M.; Gu, Y.; Ouyang, G.; Li, X. Development of Brain Network in Children with Autism from Early Childhood to Late Childhood. *Neuroscience* **2017**, *367*, 134–146. [CrossRef] [PubMed]
181. Dickinson, A.; DiStefano, C.; Lin, Y.-Y.; Scheffler, A.W.; Senturk, D.; Jeste, S.S. Interhemispheric Alpha-Band Hypoconnectivity in Children with Autism Spectrum Disorder. *Behav. Brain Res.* **2018**, *348*, 227–234. [CrossRef] [PubMed]
182. Tarasi, L.; Trajkovic, J.; Diciotti, S.; di Pellegrino, G.; Ferri, F.; Ursino, M.; Romei, V. Predictive Waves in the Autism-Schizophrenia Continuum: A Novel Biobehavioral Model. *Neurosci. Biobehav. Rev.* **2022**, *132*, 1–22. [CrossRef]
183. Delorme, R.; Ey, E.; Toro, R.; Leboyer, M.; Gillberg, C.; Bourgeron, T. Progress toward Treatments for Synaptic Defects in Autism. *Nat. Med.* **2013**, *19*, 685–694. [CrossRef]
184. Wolff, J.J.; Swanson, M.R.; Elison, J.T.; Gerig, G.; Pruett, J.R.; Styner, M.A.; Vachet, C.; Botteron, K.N.; Dager, S.R.; Estes, A.M.; et al. Neural Circuitry at Age 6 Months Associated with Later Repetitive Behavior and Sensory Responsiveness in Autism. *Mol. Autism* **2017**, *8*, 8. [CrossRef]
185. Khan, S.; Gramfort, A.; Shetty, N.R.; Kitzbichler, M.G.; Ganesan, S.; Moran, J.M.; Lee, S.M.; Gabrieli, J.D.E.; Tager-Flusberg, H.B.; Joseph, R.M.; et al. Local and Long-Range Functional Connectivity Is Reduced in Concert in Autism Spectrum Disorders. *Proc. Natl. Acad. Sci. USA* **2013**, *110*, 3107–3112. [CrossRef]

186. Riva, V.; Marino, C.; Piazza, C.; Riboldi, E.M.; Mornati, G.; Molteni, M.; Cantiani, C. Paternal—But Not Maternal—Autistic Traits Predict Frontal EEG Alpha Asymmetry in Infants with Later Symptoms of Autism. *Brain Sci.* **2019**, *9*, 342. [CrossRef] [PubMed]
187. Zhou, T.; Kang, J.; Cong, F.; Li, D.X. Early Childhood Developmental Functional Connectivity of Autistic Brains with Non-Negative Matrix Factorization. *Neuroimage Clin.* **2020**, *26*, 102251. [CrossRef]
188. Dickinson, A.; Daniel, M.; Marin, A.; Gaonkar, B.; Dapretto, M.; McDonald, N.M.; Jeste, S. Multivariate Neural Connectivity Patterns in Early Infancy Predict Later Autism Symptoms. *Biol. Psychiatry Cogn. Neurosci. Neuroimaging* **2021**, *6*, 59–69. [CrossRef] [PubMed]
189. Ghanbari, Y.; Bloy, L.; Christopher Edgar, J.; Blaskey, L.; Verma, R.; Roberts, T.P.L. Joint Analysis of Band-Specific Functional Connectivity and Signal Complexity in Autism. *J. Autism Dev. Disord.* **2015**, *45*, 444–460. [CrossRef] [PubMed]
190. Tarasi, L.; Magosso, E.; Ricci, G.; Ursino, M.; Romei, V. The Directionality of Fronto-Posterior Brain Connectivity Is Associated with the Degree of Individual Autistic Traits. *Brain Sci.* **2021**, *11*, 1443. [CrossRef] [PubMed]
191. Ursino, M.; Serra, M.; Tarasi, L.; Ricci, G.; Magosso, E.; Romei, V. Bottom-up vs. Top-down Connectivity Imbalance in Individuals with High-Autistic Traits: An Electroencephalographic Study. *Front. Syst. Neurosci.* **2022**, *16*, 932128. [CrossRef] [PubMed]
192. Courchesne, E.; Pierce, K.; Schumann, C.M.; Redcay, E.; Buckwalter, J.A.; Kennedy, D.P.; Morgan, J. Mapping Early Brain Development in Autism. *Neuron* **2007**, *56*, 399–413. [CrossRef] [PubMed]
193. Wass, S. Distortions and Disconnections: Disrupted Brain Connectivity in Autism. *Brain Cogn.* **2011**, *75*, 18–28.
194. Courchesne, E.; Pierce, K. Why the Frontal Cortex in Autism Might Be Talking Only to Itself: Local over-Connectivity but Long-Distance Disconnection. *Curr. Opin. Neurobiol.* **2005**, *15*, 225–230. [CrossRef]
195. Machado, C.; Estévez, M.; Leisman, G.; Melillo, R.; Rodríguez, R.; DeFina, P.; Hernández, A.; Pérez-Nellar, J.; Naranjo, R.; Chinchilla, M.; et al. QEEG Spectral and Coherence Assessment of Autistic Children in Three Different Experimental Conditions. *J. Autism Dev. Disord.* **2015**, *45*, 406–424. [CrossRef] [PubMed]
196. Palva, S.; Palva, J.M. Functional Roles of Alpha-Band Phase Synchronization in Local and Large-Scale Cortical Networks. *Front. Psychol.* **2011**, *2*, 204. [CrossRef] [PubMed]
197. Murphy, J.W.; Foxe, J.J.; Peters, J.B.; Molholm, S. Susceptibility to Distraction in Autism Spectrum Disorder: Probing the Integrity of Oscillatory Alpha-Band Suppression Mechanisms. *Autism Res.* **2014**, *7*, 442–458. [CrossRef] [PubMed]
198. Keehn, B.; Westerfield, M.; Müller, R.-A.; Townsend, J. Autism, Attention, and Alpha Oscillations: An Electrophysiological Study of Attentional Capture. *Biol. Psychiatry: Cogn. Neurosci. Neuroimaging* **2017**, *2*, 528–536. [CrossRef]
199. Niedermeyer, E.; Da Silva, L. *Electroencephalography: Basic Principles, Clinical Applications, and Related Fields*, 5th ed.; Lippincott Williams & Wilkins: Philadelphia, PA, USA, 2005.
200. Wood, A.; Rychlowska, M.; Korb, S.; Niedenthal, P. Fashioning the Face: Sensorimotor Simulation Contributes to Facial Expression Recognition. *Trends Cogn. Sci.* **2016**, *20*, 227–240. [CrossRef] [PubMed]
201. Liu, A.; Harris, A.M.; Atkinson, A.P.; Reed, C.L. Dissociable Processing of Emotional and Neutral Body Movements Revealed by μ-Alpha and Beta Rhythms. *Soc. Cogn. Affect. Neurosci.* **2018**, *13*, 1269–1279. [CrossRef]
202. Ewen, J.B.; Lakshmanan, B.M.; Pillai, A.S.; McAuliffe, D.; Nettles, C.; Hallett, M.; Crone, N.E.; Mostofsky, S.H. Decreased Modulation of EEG Oscillations in High-Functioning Autism during a Motor Control Task. *Front. Hum. Neurosci.* **2016**, *10*, 198. [CrossRef] [PubMed]
203. Matlis, S.; Boric, K.; Chu, C.J.; Kramer, M.A. Robust Disruptions in Electroencephalogram Cortical Oscillations and Large-Scale Functional Networks in Autism. *BMC Neurol.* **2015**, *15*, 97. [CrossRef]
204. Cornew, L.; Roberts, T.P.L.; Blaskey, L.; Edgar, J.C. Resting-State Oscillatory Activity in Autism Spectrum Disorders. *J. Autism Dev. Disord.* **2012**, *42*, 1884–1894. [CrossRef]
205. Takesaki, N.; Kikuchi, M.; Yoshimura, Y.; Hiraishi, H.; Hasegawa, C.; Kaneda, R.; Nakatani, H.; Takahashi, T.; Mottron, L.; Minabe, Y. The Contribution of Increased Gamma Band Connectivity to Visual Non-Verbal Reasoning in Autistic Children: A MEG Study. *PLoS ONE* **2016**, *11*, e0163133. [CrossRef] [PubMed]
206. Ronconi, L.; Vitale, A.; Federici, A.; Pini, E.; Molteni, M.; Casartelli, L. Altered Neural Oscillations and Connectivity in the Beta Band Underlie Detail-Oriented Visual Processing in Autism. *NeuroImage Clin.* **2020**, *28*, 102484. [CrossRef]
207. Levin, A.R.; Naples, A.J.; Scheffler, A.W.; Webb, S.J.; Shic, F.; Sugar, C.A.; Murias, M.; Bernier, R.A.; Chawarska, K.; Dawson, G.; et al. Day-to-Day Test-Retest Reliability of EEG Profiles in Children with Autism Spectrum Disorder and Typical Development. *Front. Integr. Neurosci.* **2020**, *14*, 21. [CrossRef]
208. Fauzan, N.; Amran, N.H. Brain Waves and Connectivity of Autism Spectrum Disorders. *Procedia Soc. Behav. Sci.* **2015**, *171*, 882–890. [CrossRef]
209. Gregory, M.D.; Mandelbaum, D.E. Evidence of a Faster Posterior Dominant EEG Rhythm in Children with Autism. *Res. Autism Spectr. Disord.* **2012**, *6*, 1000–1003. [CrossRef]
210. Kareem, A.S.; Kadhim, Z.M. Use of Quantitative Electroencephalography as a Marker of Severity of Patients with Autism Spectrum Disorder. *Int. J. Health Sci.* **2022**, *6(S2)*, 4418–4428. [CrossRef]
211. Carter Leno, V.; Pickles, A.; van Noordt, S.; Huberty, S.; Desjardins, J.; Webb, S.J.; Elsabbagh, M. 12-Month Peak Alpha Frequency Is a Correlate but Not a Longitudinal Predictor of Non-Verbal Cognitive Abilities in Infants at Low and High Risk for Autism Spectrum Disorder. *Dev. Cogn. Neurosci.* **2021**, *48*, 100938. [CrossRef] [PubMed]
212. Edgar, J.C.; Dipiero, M.; McBride, E.; Green, H.L.; Berman, J.; Ku, M.; Liu, S.; Blaskey, L.; Kuschner, E.; Airey, M.; et al. Abnormal Maturation of the Resting-state Peak Alpha Frequency in Children with Autism Spectrum Disorder. *Hum. Brain Mapp.* **2019**, *40*, 3288–3298. [CrossRef] [PubMed]

213. Mohammad-Rezazadeh, I.; Frohlich, J.; Loo, S.K.; Jeste, S.S. Brain Connectivity in Autism Spectrum Disorder. *Curr. Opin. Neurol.* **2016**, *29*, 137–147. [CrossRef]
214. Wang, J.; Barstein, J.; Ethridge, L.E.; Mosconi, M.W.; Takarae, Y.; Sweeney, J.A. Resting State EEG Abnormalities in Autism Spectrum Disorders. *J. Neurodev. Disord.* **2013**, *5*, 24. [CrossRef] [PubMed]
215. Knyazev, G.G.; Bocharov, A.V.; Pylkova, L.V. Extraversion and Fronto-Posterior EEG Spectral Power Gradient: An Independent Component Analysis. *Biol. Psychol.* **2012**, *89*, 515–524. [CrossRef]
216. Knyazev, G.G. Antero-Posterior EEG Spectral Power Gradient as a Correlate of Extraversion and Behavioral Inhibition. *Open Neuroimag. J.* **2010**, *4*, 114–120. [CrossRef]
217. Just, M.A.; Keller, T.A.; Malave, V.L.; Kana, R.K.; Varma, S. Autism as a Neural Systems Disorder: A Theory of Frontal-Posterior Underconnectivity. *Neurosci. Biobehav. Rev.* **2012**, *36*, 1292–1313. [CrossRef]
218. Lombardo, M.V.; Chakrabarti, B.; Bullmore, E.T.; MRC AIMS Consortium; Baron-Cohen, S. Specialization of Right Temporo-Parietal Junction for Mentalizing and Its Relation to Social Impairments in Autism. *Neuroimage* **2011**, *56*, 1832–1838. [CrossRef] [PubMed]
219. Orekhova, E.V.; Elsabbagh, M.; Jones, E.J.; Dawson, G.; Charman, T.; Johnson, M.H. EEG Hyper-Connectivity in High-Risk Infants Is Associated with Later Autism. *J. Neurodev. Disord.* **2014**, *6*, 11. [CrossRef] [PubMed]
220. Fair, D.A.; Cohen, A.L.; Power, J.D.; Dosenbach, N.U.F.; Church, J.A.; Miezin, F.M.; Schlaggar, B.L.; Petersen, S.E. Functional Brain Networks Develop from a "Local to Distributed" Organization. *PLOS Comput. Biol.* **2009**, *5*, e1000381. [CrossRef] [PubMed]
221. Supekar, K.; Musen, M.; Menon, V. Development of Large-Scale Functional Brain Networks in Children. *PLOS Biol.* **2009**, *7*, e1000157. [CrossRef] [PubMed]
222. Rubenstein, J.L.R.; Merzenich, M.M. Model of Autism: Increased Ratio of Excitation/Inhibition in Key Neural Systems. *Genes Brain Behav.* **2003**, *2*, 255–267. [CrossRef] [PubMed]
223. Simon, D.M.; Wallace, M.T. Dysfunction of Sensory Oscillations in Autism Spectrum Disorder. *Neurosci. Biobehav. Rev.* **2016**, *68*, 848–861. [CrossRef] [PubMed]
224. Singer, W.; Gray, C.M. Visual Feature Integration and the Temporal Correlation Hypothesis. *Annu. Rev. Neurosci.* **1995**, *18*, 555–586. [CrossRef] [PubMed]
225. Canolty, R.T.; Knight, R.T. The Functional Role of Cross-Frequency Coupling. *Trends Cogn. Sci.* **2010**, *14*, 506–515. [CrossRef] [PubMed]
226. Arnal, L.H.; Giraud, A.-L. Cortical Oscillations and Sensory Predictions. *Trends Cogn. Sci.* **2012**, *16*, 390–398. [CrossRef]
227. Van Driel, J.; Knapen, T.; van Es, D.M.; Cohen, M.X. Interregional Alpha-Band Synchrony Supports Temporal Cross-Modal Integration. *NeuroImage* **2014**, *101*, 404–415. [CrossRef] [PubMed]
228. Jensen, O.; Bonnefond, M.; Marshall, T.R.; Tiesinga, P. Oscillatory Mechanisms of Feedforward and Feedback Visual Processing. *Trends Neurosci.* **2015**, *38*, 192–194. [CrossRef] [PubMed]
229. Engel, A.K.; Fries, P.; Singer, W. Dynamic Predictions: Oscillations and Synchrony in Top-down Processing. *Nat. Rev. Neurosci.* **2001**, *2*, 704–716. [CrossRef]
230. Seymour, R.A.; Rippon, G.; Gooding-Williams, G.; Schoffelen, J.M.; Kessler, K. Dysregulated Oscillatory Connectivity in the Visual System in Autism Spectrum Disorder. *Brain* **2019**, *142*, 3294–3305. [CrossRef] [PubMed]
231. Berman, J.I.; Liu, S.; Bloy, L.; Blaskey, L.; Roberts, T.P.L.; Edgar, J.C. Alpha-to-Gamma Phase-Amplitude Coupling Methods and Application to Autism Spectrum Disorder. *Brain Connect.* **2015**, *5*, 80–90. [CrossRef]
232. Cortese, S.; Kelly, C.; Chabernaud, C.; Proal, E.; Di Martino, A.; Milham, M.P.; Castellanos, F.X. Toward Systems Neuroscience of ADHD: A Meta-Analysis of 55 FMRI Studies. *Am. J. Psychiatry* **2012**, *169*, 1038–1055. [CrossRef] [PubMed]
233. Willcutt, E.G.; Doyle, A.E.; Nigg, J.T.; Faraone, S.V.; Pennington, B.F. Validity of the Executive Function Theory of Attention-Deficit/Hyperactivity Disorder: A Meta-Analytic Review. *Biol. Psychiatry* **2005**, *57*, 1336–1346. [CrossRef]
234. McCarthy, S.; Wilton, L.; Murray, M.L.; Hodgkins, P.; Asherson, P.; Wong, I.C.K. The Epidemiology of Pharmacologically Treated Attention Deficit Hyperactivity Disorder (ADHD) in Children, Adolescents and Adults in UK Primary Care. *BMC Pediatr.* **2012**, *12*, 78. [CrossRef] [PubMed]
235. Franke, B.; Michelini, G.; Asherson, P.; Banaschewski, T.; Bilbow, A.; Buitelaar, J.K.; Cormand, B.; Faraone, S.V.; Ginsberg, Y.; Haavik, J.; et al. Live Fast, Die Young? A Review on the Developmental Trajectories of ADHD across the Lifespan. *Eur. Neuropsychopharmacol.* **2018**, *28*, 1059–1088. [CrossRef]
236. Michelini, G.; Salmastyan, G.; Vera, J.D.; Lenartowicz, A. Event-Related Brain Oscillations in Attention-Deficit/Hyperactivity Disorder (ADHD): A Systematic Review and Meta-Analysis. *Int. J. Psychophysiol.* **2022**, *174*, 29–42. [CrossRef]
237. Bickel, S.; Dias, E.C.; Epstein, M.L.; Javitt, D.C. Expectancy-Related Modulations of Neural Oscillations in Continuous Performance Tasks. *NeuroImage* **2012**, *62*, 1867–1876. [CrossRef]
238. Spitzer, B.; Haegens, S. Beyond the Status Quo: A Role for Beta Oscillations in Endogenous Content (Re)Activation. *eNeuro* **2017**, *4*. [CrossRef] [PubMed]
239. Neuper, C.; Pfurtscheller, G. Event-Related Dynamics of Cortical Rhythms: Frequency-Specific Features and Functional Correlates. *Int. J. Psychophysiol.* **2001**, *43*, 41–58. [CrossRef] [PubMed]
240. Lenartowicz, A.; Truong, H.; Salgari, G.C.; Bilder, R.M.; McGough, J.; McCracken, J.T.; Loo, S.K. Alpha Modulation during Working Memory Encoding Predicts Neurocognitive Impairment in ADHD. *J. Child Psychol. Psychiatry* **2019**, *60*, 13042. [CrossRef]
241. Hasler, R.; Perroud, N.; Meziane, H.B.; Herrmann, F.; Prada, P.; Giannakopoulos, P.; Deiber, M.-P. Attention-Related EEG Markers in Adult ADHD. *Neuropsychologia* **2016**, *87*, 120–133. [CrossRef]

242. Bozhilova, N.; Cooper, R.; Kuntsi, J.; Asherson, P.; Michelini, G. Electrophysiological Correlates of Spontaneous Mind Wandering in Attention-Deficit/Hyperactivity Disorder. *Behav. Brain Res.* **2020**, *391*, 112632. [CrossRef]
243. Mooneyham, B.W.; Schooler, J.W. The Costs and Benefits of Mind-Wandering: A Review. *Can. J. Exp. Psychol. Rev. Can. Psychol. Exp.* **2013**, *67*, 11–18. [CrossRef] [PubMed]
244. Vollebregt, M.A.; Zumer, J.M.; Ter Huurne, N.; Buitelaar, J.K.; Jensen, O. Posterior Alpha Oscillations Reflect Attentional Problems in Boys with Attention Deficit Hyperactivity Disorder. *Clin. Neurophysiol.* **2016**, *127*, 2182–2191. [CrossRef]
245. Yordanova, J.; Kolev, V.; Rothenberger, A. Event-Related Oscillations Reflect Functional Asymmetry in Children with Attention Deficit/Hyperactivity Disorder. *Suppl. Clin. Neurophysiol.* **2013**, *62*, 289–301. [CrossRef]
246. Guo, J.; Luo, X.; Li, B.; Chang, Q.; Sun, L.; Song, Y. Abnormal Modulation of Theta Oscillations in Children with Attention-Deficit/Hyperactivity Disorder. *NeuroImage Clin.* **2020**, *27*, 102314. [CrossRef]
247. Guo, J.; Luo, X.; Wang, E.; Li, B.; Chang, Q.; Sun, L.; Song, Y. Abnormal Alpha Modulation in Response to Human Eye Gaze Predicts Inattention Severity in Children with ADHD. *Dev. Cogn. Neurosci.* **2019**, *38*, 100671. [CrossRef]
248. Ter Huurne, N.; Lozano-Soldevilla, D.; Onnink, M.; Kan, C.; Buitelaar, J.; Jensen, O. Diminished Modulation of Preparatory Sensorimotor Mu Rhythm Predicts Attention-Deficit/Hyperactivity Disorder Severity. *Psychol. Med.* **2017**, *47*, 1947–1956. [CrossRef] [PubMed]
249. Longarzo, M.; Cavaliere, C.; Alfano, V.; Mele, G.; Salvatore, M.; Aiello, M. Electroencephalographic and Neuroimaging Asymmetry Correlation in Patients with Attention-Deficit Hyperactivity Disorder. *Neural Plast.* **2020**, *2020*, 1–9. [CrossRef]
250. Mazaheri, A.; Fassbender, C.; Coffey-Corina, S.; Hartanto, T.A.; Schweitzer, J.B.; Mangun, G.R. Differential Oscillatory Electroencephalogram Between Attention-Deficit/Hyperactivity Disorder Subtypes and Typically Developing Adolescents. *Biol. Psychiatry* **2014**, *76*, 422–429. [CrossRef] [PubMed]
251. Loo, S.K.; Makeig, S. Clinical Utility of EEG in Attention-Deficit/Hyperactivity Disorder: A Research Update. *Neurotherapeutics* **2012**, *9*, 569–587. [CrossRef] [PubMed]
252. Arns, M.; Heinrich, H.; Strehl, U. Evaluation of Neurofeedback in ADHD: The Long and Winding Road. *Biol. Psychol.* **2014**, *95*, 108–115. [CrossRef]
253. Escolano, C.; Navarro-Gil, M.; Garcia-Campayo, J.; Congedo, M.; Minguez, J. The Effects of Individual Upper Alpha Neurofeedback in ADHD: An Open-Label Pilot Study. *Appl. Psychophysiol. Biofeedback* **2014**, *39*, 193–202. [CrossRef]
254. Deiber, M.-P.; Hasler, R.; Colin, J.; Dayer, A.; Aubry, J.-M.; Baggio, S.; Perroud, N.; Ros, T. Linking Alpha Oscillations, Attention and Inhibitory Control in Adult ADHD with EEG Neurofeedback. *NeuroImage Clin.* **2020**, *25*, 102145. [CrossRef]
255. Mishra, J.; Lowenstein, M.; Campusano, R.; Hu, Y.; Diaz-Delgado, J.; Ayyoub, J.; Jain, R.; Gazzaley, A. Closed-Loop Neurofeedback of α Synchrony during Goal-Directed Attention. *J. Neurosci.* **2021**, *41*, 5699–5710. [CrossRef]

Article

The Cost of Imagined Actions in a Reward-Valuation Task

Manuela Sellitto [1,2], Damiano Terenzi [3,4], Francesca Starita [1], Giuseppe di Pellegrino [1,*] and Simone Battaglia [1,2,*]

1 Centre for Studies and Research in Cognitive Neuroscience, Department of Psychology, University of Bologna, 40126 Bologna, Italy; manuela.sellitto@psypec.it (M.S.); francesca.starita2@unibo.it (F.S.)
2 School of Psychology, Bangor University, Bangor LL572AS, UK
3 Department of Decision Neuroscience and Nutrition, German Institute of Human Nutrition (DIfE), 14558 Potsdam-Rehbrücke, Germany; damiano.terenzi@dife.de
4 Charité—Universitätsmedizin Berlin, Corporate Member of Freie Universität Berlin, Humboldt-Universität zu Berlin, Berlin Institute of Health, Neuroscience Research Center, 10117 Berlin, Germany
* Correspondence: g.dipellegrino@unibo.it (G.d.P.); simone.battaglia@unibo.it (S.B.)

Abstract: Growing evidence suggests that humans and other animals assign value to a stimulus based not only on its inherent rewarding properties, but also on the costs of the action required to obtain it, such as the cost of time. Here, we examined whether such cost also occurs for mentally simulated actions. Healthy volunteers indicated their subjective value for snack foods while the time to imagine performing the action to obtain the different stimuli was manipulated. In each trial, the picture of one food item and a home position connected through a path were displayed on a computer screen. The path could be either large or thin. Participants first rated the stimulus, and then imagined moving the mouse cursor along the path from the starting position to the food location. They reported the onset and offset of the imagined movements with a button press. Two main results emerged. First, imagery times were significantly longer for the thin than the large path. Second, participants liked significantly less the snack foods associated with the thin path (i.e., with longer imagery time), possibly because the passage of time strictly associated with action imagery discounts the value of the reward. Importantly, such effects were absent in a control group of participants who performed an identical valuation task, except that no action imagery was required. Our findings hint at the idea that imagined actions, like real actions, carry a cost that affects deeply how people assign value to the stimuli in their environment.

Keywords: delay discounting; effort discounting; Fitts' law; motor imagery; mental simulation; reward value; visual imagery

Citation: Sellitto, M.; Terenzi, D.; Starita, F.; di Pellegrino, G.; Battaglia, S. The Cost of Imagined Actions in a Reward-Valuation Task. *Brain Sci.* **2022**, *12*, 582. https://doi.org/10.3390/brainsci12050582

Academic Editor: Michela Balconi

Received: 15 April 2022
Accepted: 27 April 2022
Published: 29 April 2022

Publisher's Note: MDPI stays neutral with regard to jurisdictional claims in published maps and institutional affiliations.

1. Introduction

What is the relation between the value that the brain assigns to a rewarding stimulus and the cost of the action required to obtain it? The value an agent gives to a reward is well known to be determined by its immediate sensory characteristics (e.g., the amount of a food product or its taste), as well as by individual preferences and experiences. However, among the majority of animals, the decision to forage often implies the computation of benefits against predictable costs, including the effort required to obtain the reward [1–4]. Recent behavioural and neuroeconomic studies have suggested that the value assigned to a reward stands in inverse relation to the amount of effort required to obtain it [5–8]. In other words, people give more value to those rewards that can be achieved with less effort [9–11]. Such effort can be physical but also cognitive (e.g., by manipulating visuospatial task demands) [12]. Therefore, it is plausible to expect that value computation can be influenced by the degree of cognitive effort as well [13,14].

Similar to effort, another cost that is often incurred during decision-making is the delay that one has to endure before receiving a reward. This phenomenon is known as temporal discounting (TD) and reflects the decrease in the subjective value of a reward [15–19].

Recently, TD has been deemed to affect the motor commands responsible for the movements of the human body necessary to reach a stimulus [20–22]. When a stimulus is associated with a future reward, lower dopaminergic neurons firing is observed, as well as slower saccades for the visual reaching of the stimulus are recorded [23,24]. In addition, when a stimulus is associated with an immediate reward, higher dopaminergic neuron firing usually precedes faster movements [25,26]. Therefore, the kinematics of movements seem to reflect the processes with which the brain temporally discounts rewards, suggesting a link between motor and valuation processes [20,27,28]. Movements and time share a well-known relation as well: according to Fitts' law [29], the time needed by an individual to move between a starting point and a target point within a given space increases linearly with the difficulty to execute the movement. In addition, the time for mentally simulating movements (motor imagery, MI) [30–35] is highly correlated with the time to actually make such movements. Thus, mental movements closely mimic real movements in their temporal organization, involve the same planning programs [36] and rely on similar neural processes [37–39]. Furthermore, Fitts' law accounts equally well for imagined and executed movements [40–43].

A substantial body of neurophysiological and neuroimaging studies suggests that there are brain areas and neural populations responsible for both the decisional process and the selection of an action [44–47]. These areas represent a common substrate for the different kinds of decision-making processes. In particular, the orbitofrontal cortex (OFC) has a critical role in representing the subjective value of a reward, thus employing such value to condition the choice among different options [5,48–50]. In addition to the OFC, value is represented in other regions, including those related to motor preparation and execution [51–56]. In line with these studies, decisions based on value would affect the competition between representations of actions and the movement itself before the decision-making processes are completed [57,58]. Moreover, action representation has also been observed to influence the value of an item during the decision-making processes. Such influence could be interpreted as a sort of value signal from the motor and pre-motor regions [44,55,59–61]. Based on this evidence, it can be argued that the mechanisms underlying the computation of reward value are extremely sensitive to the cost of an action needed to obtain such reward. Importantly, the preference for an action is inversely proportional to its effort, even in terms of time. This can, in turn, influence the preference of a reward such that when the action is more difficult and requires more time to be executed, the reward may be perceived as less valuable.

In the present study, we investigated, for the first time, whether imagined movements, like real ones [23,62], would carry a cost that affects how people assign value to food items. These processes are currently unexplored, and they can be useful in better understanding diseases such as apathy, which is characterized by an alteration of abilities to anticipate the effort or difficulty. To this end, via a novel experimental procedure, participants performed an experimental task in which the pictures of food and drink item were associated with two different path widths (large or thin) that were really and imaginarily navigated via a mouse cursor. After navigation, participants rated how much they liked and wanted those items on a Likert scale. We expected participants to assign a higher value score (i.e., wanting and liking) to food items that required a shorter time (lower cost) to be reached (large path), relative to those items requiring a longer time (higher cost, thin path). Thus, thin path was expected to take more time, incurring a higher cost, while large path was expected to require less time and hence lower cost.

2. Materials and Methods

2.1. Participants

A power analysis based on previously published studies [63–66] indicated that a sample size of ~20 participants was necessary to achieve a statistical power of >95% (2-tailed = 0.05). Thus, a total of 40 young adult right-handed volunteers (22 female) with a mean age of 23.6 years (sd = 2.76) and mean education of 16.87 years (sd = 2.16) were recruited from the student population of the University of Bologna for a single-session experiment (see Table 1 for further demographic information). Participants were randomly assigned either to the experimental group or to the control group. Body Mass Index (BMI) was calculated for each participant and then overall for each group [67]. Moreover, hunger (on a 5-point Likert scale ranging from 0 "not at all hungry" to 5 "extremely hungry") and fasting level (hours without eating) were collected. Subjects remained naïve as to the purpose of the study until debriefing at the end of the experimental session. The two groups did not show any significant difference in terms of BMI index, hunger and fasting level (all ps > 0.2). Participants had no history of psychiatric or neurological diseases, had normal or corrected-to-normal vision and all gave written informed consent before the beginning of the experiment. The experimental protocol was conducted in accordance with the Declaration of Helsinki (2008) and approved by the Institutional Review Board (or Ethics Committee) of the University of Bologna (protocol number 14, 8 February 2013).

Table 1. Participants' demographic data.

Group	n	Age	Education	Hunger	Fasting	BMI
Experimental	20	23.35 (2.16)	16.8 (1.85)	2.25 (1.45)	2.08 (1.34)	21.49 (2.49)
Control	20	23.85 (3.36)	16.95 (2.46)	2.75 (1.16)	2.18 (1.32)	21.52 (3.16)

n is the number of participants. Age and Education are expressed in years. Hunger is expressed on a five-point scale. Fasting is expressed in hours. Body Mass Index (BMI) is expressed in kg/m^2. Mean is represented out of the brackets, while the standard deviation is within them.

2.2. Apparatus and Stimuli

The experiment was implemented in E-prime 2.0 (Psychology Software Tool, Sharpsburg, PA, USA) and run on a Windows-based PC (Lenovo ThinkCentre Desktop Computer, Beijing, China). Participants sat in front of a 15-inch colour LCD monitor (1024 × 768 pixels) with an unconstrained viewing distance of approximately 65 cm.

Two rectangles were displayed centred on the screen in two different positions. One rectangle (55 × 75 mm) was positioned in the upper part, and the other (15 × 75 mm) in the lower part of the screen. They could be connected either through a large (155 × 65 mm) or a thin (155 × 20 mm) path (Figure 1). These paths induced participants to perform a potential movement over the path. In particular, the thin path forced participants to perform more careful movements compared to the large one [68]. Pictures of food and drink items (50 × 50 mm) appeared one at the time inside the top rectangle, while the bottom one was used as the starting point. The size of the paths was randomized among trials (see Figure 1). Pictures of Food and drink items were coloured pictures of 28 different snack food and drink items consisting of candy bars (e.g., Bounty or KitKat), potato crisps (e.g., Fonzies), crackers (e.g., Ritz), fruit juices (e.g., Pago) and carbonated drinks (e.g., Sprite). Such items were available at local convenience stores, each with a mean price of EUR 1.36 (sd = 0.54). To reduce the influence of personal food preference, the presentation of the stimuli was counterbalanced across participants. All stimuli were balanced for luminance, complexity and colour saturation using photo editing software (Adobe Photoshop CS6).

Figure 1. Examples of visual stimuli and paths. Shown brands for demonstrative purpose only.

2.3. Tasks

2.3.1. Experimental Valuation Task

In this task, and for each trial, participants first had to answer to two questions. Second, they had to imagine themselves dragging the mouse along the path shown on the screen (thin or large) from the starting point rectangle to the upper rectangle. Specifically, they were required to imagine keeping the mouse cursor inside the path until the food item was reached.

Trials started with a black fixation cross, centrally displayed on a white background (1000 ms), followed by the presentation of a food item, together with one of the two paths. Participants could observe the screen for as long as they needed. When they felt ready, they were instructed to press the spacebar so that one question and its rating scale appeared at one bottom corner of the screen. The order and screen position of questions and their respective scales were counterbalanced across participants. The two questions (in Italian) were "How pleasant would it be to experience eating this item now?" (liking ratings), and "How much do you want to eat it now?" (wanting ratings) [69]. Participants answered through a seven-point Likert scale (ranging from 1 "not at all" to 7 "extremely") [70] by pressing the corresponding number on the keyboard. After the last rating was submitted, the question and the scale faded out. Then, mouse cursor appeared, centred within the bottom starting rectangle. At this point, participants had to imagine, in a first-person perspective, dragging the mouse along the path from the starting point to the food item picture in the upper rectangle at the end of the path. They had to press the left mouse button with the index finger of the right hand to signal the start of their imagined movement, and the button had to be pressed again to flag the imagined reaching of the food item. Trials were separated by a 1000 ms inter-trial interval (ITI) with a blank screen (see Figure 2). The experimental valuation task consisted of 56 randomized trials, 28 for each path width. Before starting the task, participants had to perform three trials for training. They had to press the left mouse button with the index finger of the right hand, drag the mouse inside the path and reach the food item. Finally, the mouse button had to be pressed again to complete the trial. It is important to note that during the training, participants had to physically move the mouse along the path to ensure that participants understood the kinetic characteristics of the actions [68]. Indeed, the training aimed at familiarising participants with the reaching movement that they had to imagine completing across the paths, where the large path was assumed to be easier than the thin one.

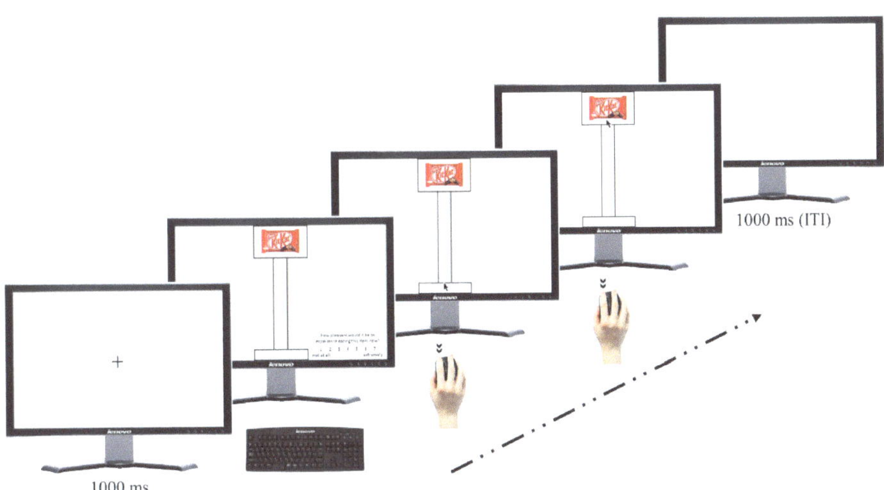

Figure 2. Schematic representation of the experimental trial sequences. Participants had to imagine, in a first-person perspective, dragging the mouse along the path, from the starting point to the food item picture in the upper rectangle at the end of the path. They had to press the left mouse button with the index finger of the right hand to signal the start of their imagined movement, and the button had to be pressed again to flag the imagined reaching of the food item. Trials were separated by a 1000 ms inter-trial interval (ITI) with a blank screen.

2.3.2. Control Valuation Task

The control valuation task was designed to ensure that only the imagery time required to perform the reaching movement affected the value assigned to the item, as performed during the experimental task. Before participants began the control valuation task, a short training (e.g., 3 trials) was performed, during which they were familiarised with the food and drink items, as well as with the two different paths, to ensure that they could understand the characteristics of the task (e.g., the different paths). After the initial set of training trials, participants started the task. The sequence of trial events was equivalent to the experimental task. There was a 1000 ms fixation period, followed by the presentation of a food item and one of the two paths. They could observe the path on the screen for as long as they wanted and, as in the experimental task, they were provided with liking and wanting rating scales. Subsequently, participants had to press the left mouse button twice, as done in the experimental task, although here it had no relation with path navigation. Finally, a 1000 ms ITI blank screen appeared. Notice that, differently from the experimental task, participants were not requested to make or imagine any action.

2.4. Experimental Procedures

The study was performed at the Centre for Studies and Research in Cognitive Neuroscience (CsrCN), Cesena, Italy. Participants sat comfortably in a silent room, and their demographic data, including height and weight with self-report questionnaires, were collected in order to calculate their BMI. They were also asked to rate their hunger at the moment of the experiment and estimate their fasting level (hours). In addition, participants filled out two questionnaires assessing imagery abilities: The Vividness of Visual Imagery Questionnaire (VVIQ; [71]) and the Vividness of Movement Imagery Questionnaire (VMIQ [72]). The main purpose of these questionnaires was, on the one hand, to assess participants' ability to imagine movements. On the other hand, they raised the vividness of the motor imagination, which was necessary for the experimental task. The VVIQ consists of 16 items, divided into four groups of four items each, in which participants are invited to vividly imagine specific scenes and situations, e.g., a sunrise evolving into a rainstorm.

The VMIQ consists of 24 items asking participants to describe a simple movement, such as walking, or a complex movement, such as rope jumping. The VVIQ and VMIQ consist of items triggering the vivid imagination of movement, rated on a five-point scale. The order of questionnaire submission was counterbalanced across participants. After filling out the questionnaires, participants performed the task. At the end of the session, they were debriefed on the purpose of the study.

2.5. Statistical Analyses

The aim of this study was to investigate how the effort, associated with imagining moving along each path, influenced the value assigned to the food and drink items. The significance of the experimental factors was tested using 2 × 2 ANOVAs—one for the liking rating, one for the wanting rating and one for the reaction times as dependent variables—using group (experimental, control) as the between-participants independent variable and the path width (large, thin) as the within-participants independent variable. The alpha-level of all analyses was set at $p < 0.05$ using a univariate approach. Furthermore, scores related to the questionnaires were analysed. All statistical analyses were performed with SPSS Statistics (IBM Corp, released March 2013, IBM SPSS Statistics for Windows, version 22.0, Armonk, New York, USA) and STATISTICA (Dell Software, released September 2015, StatSoft STATISTICA for Windows, version 13.0, Round Rock, TX, USA).

3. Results

3.1. Liking

The ANOVA performed on the liking rating revealed a significant main effect of the path width ($F(1, 38) = 22.461$, $p < 0.001$, part.$\eta^2 = 0.4$), a significant group X path width interaction ($F(1, 38) = 5.958$, $p = 0.019$, part.$\eta^2 = 0.1$; Figure 3) and no significant group effect ($p = 0.778$). Newman–Keuls post-hoc analysis revealed a significant difference ($p < 0.001$) between the large path (mean = 4.74; sd = 0.89) and the thin path (mean = 4.15; sd = 0.88) in the experimental group, but not in the control group ($p = 0.112$; large path: mean = 4.62, sd = 0.95; thin path: mean = 4.43, sd = 1.02). These results indicate that the path widths significantly influenced the liking ratings only in the experimental group, with greater liking when items appeared at the end of the large path than the thin path.

3.2. Wanting

The ANOVA performed on the wanting rating revealed a significant main effect of the path width ($F(1, 38) = 8.219$ $p = 0.006$ part.$\eta^2 = 0.17$), but no significant group X path width interaction ($F(1, 38) = 0.259$ $p = 0.613$) and no group effect ($p = 0.281$; Figure 4). These results indicate that the path widths influenced the wanting rating such that for the large path (mean = 3.44; sd = 1.02), participants showed higher wanting score compared to the thin path (mean = 3.22; sd = 1.04), regardless of the group.

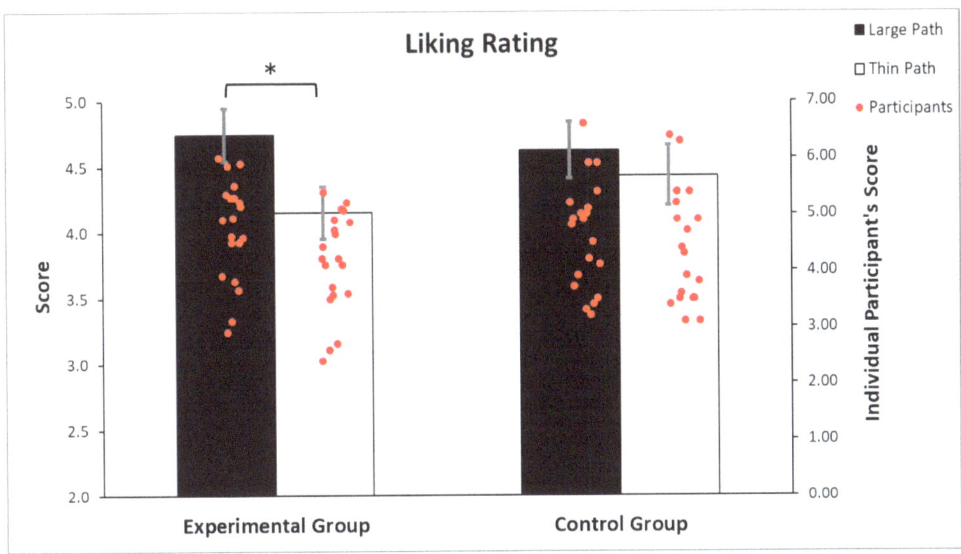

Figure 3. Analysis of the liking scores illustrates that the liking ratings were influenced by the path widths only in the experimental group (*p* < 0.001). Vertical lines indicate the standard error of mean (SEM), * *p* < 0.05.

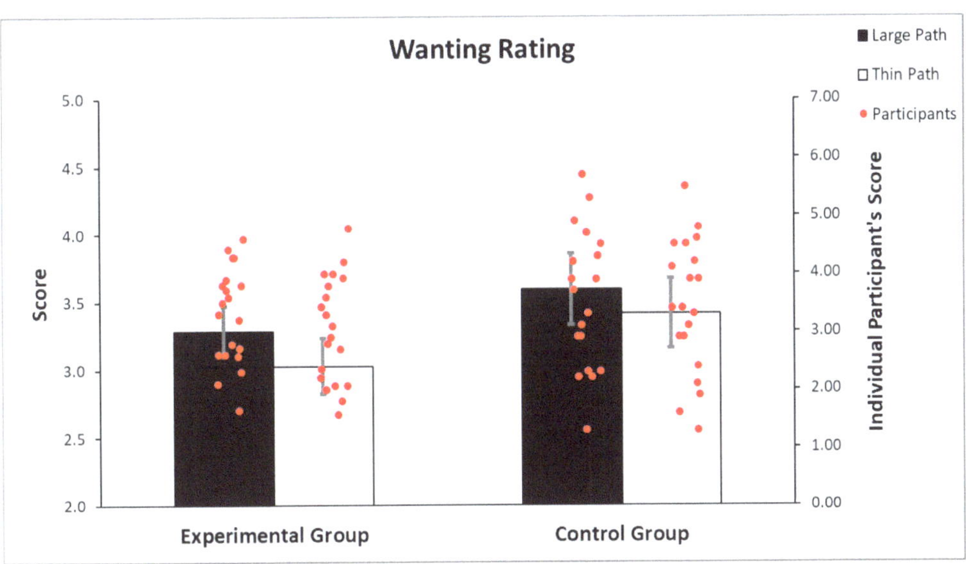

Figure 4. Analysis of the wanting scores illustrates that the wanting ratings were not differently influenced by the path widths in both groups. Vertical lines indicate the standard error of mean (SEM).

3.3. Response Time

The ANOVA performed on the participants' response times revealed a significant main effect of the path (F(1, 38) = 37.459, *p* < 0.001, part.η^2 = 0.49), a significant main effect of the group (F(1, 38) = 34.360, *p* < 0.001, part.η^2 = 0.47) and a significant group X path width interaction (F(1, 38) = 34.283, *p* < 0.001, part.η^2 = 0.48; Figure 5). Newman–

Keuls post-hoc analysis revealed a significant difference ($p < 0.001$) between the large path (mean = 3152.91 ms; sd = 2437.17 ms) and thin path (mean = 3994.39 ms; sd = 2665.35 ms) on the response times in the experimental group, but not in the control group ($p = 0.852$; large path trials: mean = 236.25 ms, sd = 140.93 ms; thin path trials: mean = 254.88 ms, sd = 166.54 ms). These findings indicate that the path width had a significant influence on response times only in the experimental group, with greater response times when imagining navigating the thin path than the large path.

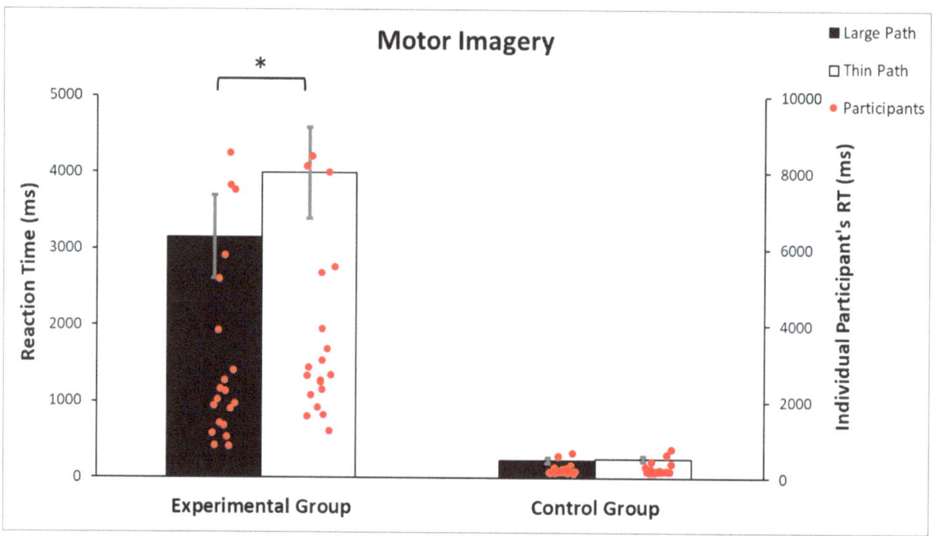

Figure 5. Analysis of the motor imagery time illustrates that the path width had an influence on the response time only in the experimental group. Vertical lines indicate the standard error of mean (SEM), * $p < 0.001$.

3.4. Questionnaires and Correlations

The analysis of the questionnaires revealed that the experimental group's VVIQ mean score was 58.4 (sd = 7.93), and control group's VVIQ mean score was 63.9 (sd = 8.41) out of a maximum of 80. The experimental group's VMIQ mean score was 84.7 (sd = 15.75), and control group's VMIQ mean score was 89.3 (sd = 14.31) out of a maximum of 120 (Table 2). Two-tailed Pearson correlations were performed between the participants' scores on the questionnaires and response times. The VVIQ scores negatively correlated with the response times ($r = -0.363$ ($p = 0.021$)), while no correlation resulted for the VMIQ scores ($r = -0.17$ ($p = 0.283$)). The significant negative correlation suggests that lower response times were associated with higher scores of vividness of visual imagery but not of motor imagery.

Table 2. Participants' questionnaire scoring.

	VVIQ		VMIQ	
Group	Experimental	Control	Experimental	Control
Mean	58.4	63.9	84.7	89.3
SD	7.93	8.41	15.75	14.31

The VVIQ score out of a maximum of 80, and the VMIQ score out of a maximum of 120.

4. Discussion

Although different studies have investigated multiple factors involved in value computation [47,52,73–76], including risk [77,78], spatial distance [79–81], effort [9–12,82] and temporal discounting [15–18], little is known about how the cost of an action impacts on valuation processes. This study describes a new experimental protocol for studying and quantifying how such cost can affect valuation processes and, thus, behaviour. In particular, this research explores how the subjective value assigned to an item can be affected by manipulating the difficulty of an imaged action to reach it.

A previous study showed that the cost of real actions could affect how people assign value to the stimuli in their environment [23]; however, to the best of our knowledge, no study to date has tested the cost of imagined actions. To this end, first, participants familiarized with physically moving the mouse cursor inside each of two different paths. Second, they were asked to imagine dragging the mouse inside these paths in order to reach a food item. According to Fitts' law [29], given the different path widths, participants' imagined movements implied a variable difficulty in reaching the target; such difficulty was revealed by the discrepant timings needed to navigate each path. In addition, participants answered two questions via a seven-point Likert scale, which was designed to evaluate the liking and wanting of each food item. Participants took shorter time to imagine navigating the large path than the thin path. In addition, they liked more foods and drinks at the end of the large path than the thin path, but only if they were performing the motor imagery task. These results suggest that the cost (in terms of time) of the imagined action influences value ratings, with less value attributed to items that involve higher action cost. The rating differences were obtained by increasing or decreasing the path widths and, thus, manipulating the cost of the imagined action required to reach the item. The higher liking to the lower-difficulty and short-time items at the end of the large path reveals that an action's cost is computed even before its execution [83–85]. Furthermore, these data show how the value of an item is influenced by the time required to reach it, even if the reaching is only imagined. In other words, presenting a food item with a large or thin path induced participants to anticipate the cost of the action, consequently affecting their evaluation. This is known as effort discounting [6].

The results reveal that an increased difficulty in mentally performing an action implied lower scores on the liking ratings for each item; the same expected result was not found across the wanting ratings. We explained that this discrepancy was caused by how the question may have been interpreted by participants. In line with this hypothesis, a recent systematic review of the human literature revealed a contradictory operationalization of the wanting and liking constructs [86]. Pool and colleagues [86] argued that expected pleasantness represents a major conceptual confound. In particular, expected pleasantness is an evaluation of how good or how bad a particular reward is going to be. This prediction involves the active reconstruction of past episodic memories of liking experiences with the current reward and the use of these episodic memories to anticipate or predict a future experience [87–89]. Whereas expected pleasantness drives cognitive desires, the interaction between the individual's physiological state and the perception of the relevant reward-associated cue determines incentive salience or wanting [90]. Therefore, cognitive desires driven by memories of past liking experiences are not completely independent from liking, whereas wanting is underlain by a mechanism that is independent from the liking component [91–93]. Furthermore, the cognitive domain, including memory, is in constant complex interaction with other behavioural domains such as the negative valence domain, including depression and anxiety, which is under the control of endogenous neuropeptides, neurohormones and metabolites [94–99]. In addition, the functions of the domains are subject to the close influence of the adjacent biosystems [100–104].

In other words, it may be possible that our experimental manipulation had an effect on the expected pleasantness of eating or drinking the items, which is more related to self-reported measures of liking rather than wanting. In this regard, it has been shown in the human literature that self-reported rating scales seem to be the most reliable index of the hedonic experience [105,106]. Therefore, our results suggest that the cost of an imagined action directly influences the food item's value computation, even though the action is not related to the hedonic features of such item. Specifically, the difficulty to reach an item is also included in the computation of value, in addition to the main sensory features that define that item. Crucially, in this study, the participants' rating was affected because they were anticipating the specific difficulty of each action while performing the experimental trials. Our results support the hypothesis that imagining actions with variable difficulties automatically influences the item value.

Concerning response time, the results showed that the experimental group took more time to imagine making a movement along the thin path rather than the large path. These data revealed that simply imagining such movement influenced the participants' response time, as well as their ratings. As previously described, we also found some correlations. The significant negative correlations between the VVIQ score and response time revealed that an increased participants' vividness of imagination resulted in a reduced motor imagery time and, therefore, in reaching the food item faster.

These findings indicate that the value of the item is influenced by the cost of the action required to obtain it. In line with previous studies, items that can be obtained with minor difficulty have a higher value because they can be reached more easily [9,10,92,107], and this occurs even when performing the reaching of the object is not necessary [108]. According to these data, the value of the reward is inversely proportional to the amount of effort required to reach it. In general, this evidence confirms the main principle that governs effort discounting: a higher effort leads to a devaluation of the reward.

Our results could be interpreted according to an evolutionary perspective, which makes plausible the interaction between the computation processes underlying the value of an item and those codifying the difficulty to obtain it. Therefore, our evidence could be included along the theories of embodied cognition [109–111]. These theories emphasize the role of the body in cognition, arguing that the body's condition influences cognitive states as they, at the same time, influence the body. According to these theories, participants' knowledge of the world would be represented by neural structures which store sensory, motor and affective information. In agreement with this idea, the simple reaching of an object, or anticipating the difficulty to obtain it, activates such neural structures which are also involved in one's own perceptual experience. In line with our results, the anticipation of an action's difficulty could be considered as a simulation of the effort required for executing the action. The process governing the value computation could thus be influenced by the physical or cognitive effort required to reach it. Our data are further confirmed by, and in line with, several neurophysiological and neuroimaging data. These data show a possible representation of the subjective value not only in areas directly related to value computation and decision-making processes, such as ventromedial prefrontal cortex (vmPFC) [112–119], but also in other cortices related to motor preparation and action [51,52,120,121]. According to this evidence, the economic choice would be embodied in the motor processes of action selection [27,122].

5. Conclusions

In conclusion, our data show that the difficulty of an imagined action can influence valuation judgements. Moreover, our results support the hypothesis of the close relationship between motor and valuation processes. These links remain largely unexplored in current investigations, and they can be useful to better understand neurological or psychiatric conditions, such as schizophrenia, depression or frontal-apathy, which are characterized by an alteration of the ability to anticipate the effort necessary to obtain rewarding items in the environment [123–128]. Finally, future studies should also investigate the neural basis of this phenomenon; consequently, it would be very informative to submit a similar experimental paradigm to patients with focal brain lesions, for example, lesions affecting the medial orbitofrontal cortex (mOFC) or the anterior cingulate cortex.

Author Contributions: Conceptualization, M.S. and G.d.P.; methodology, M.S., G.d.P. and S.B.; software, M.S.; formal analysis, M.S., G.d.P. and S.B.; investigation, M.S., D.T. and S.B.; data collection, D.T.; writing—original draft preparation, M.S., G.d.P. and S.B.; writing—review and editing, M.S., D.T., F.S., G.d.P. and S.B.; visualization, S.B.; supervision G.d.P. and S.B.; project administration, G.d.P.

and S.B.; funding acquisition, G.d.P., F.S. and S.B. All authors have read and agreed to the published version of the manuscript.

Funding: This work was supported by grants from the RFO of the University of Bologna to G.d.P. Bial Foundation Grant for Scientific Research 2020/2021 [Grant Number 47/20] to F.S. and private funding by S.B.

Institutional Review Board Statement: The study was conducted in accordance with the Declaration of Helsinki, and approved by the Institutional Review Board (or Ethics Committee) of the University of Bologna (protocol code #14 on 8 February 2013).

Informed Consent Statement: Informed consent was obtained from all subjects involved in the study, see Section 2.1 for more details.

Data Availability Statement: The datasets used and analysed during the current study are available from the corresponding authors on reasonable request.

Acknowledgments: The authors thank Chiara Di Fazio and Claudio Nazzi for their help in searching the relevant literature, and Vincenza Iacono and Nicholas Menghi for recruiting participants and helping with data acquisition.

Conflicts of Interest: The authors declare no conflict of interest.

References

1. Walton, M.E.; Kennerley, S.W.; Bannerman, D.M.; Phillips, P.E.M.; Rushworth, M.F.S. Weighing up the Benefits of Work: Behavioral and Neural Analyses of Effort-Related Decision Making. *Neural Netw.* **2006**, *19*, 1302–1314. [CrossRef]
2. Kolling, N.; Behrens, T.E.; Mars, R.B.; Rushworth, M.F. Neural Mechanisms of Foraging. *Science* **2012**, *336*, 95–98. [CrossRef] [PubMed]
3. Shenhav, A.; Straccia, M.A.; Cohen, J.D.; Botvinick, M.M. Anterior Cingulate Engagement in a Foraging Context Reflects Choice Difficulty, Not Foraging Value. *Nat. Neurosci.* **2014**, *17*, 1249–1254. [CrossRef]
4. Seinstra, M.S.; Sellitto, M.; Kalenscher, T. Rate Maximization and Hyperbolic Discounting in Human Experiential Intertemporal Decision Making. *Behav. Ecol.* **2018**, *29*, 193–203. [CrossRef]
5. Rangel, A.; Camerer, C.; Montague, P.R. A Framework for Studying the Neurobiology of Value-Based Decision Making. *Nat. Rev. Neurosci.* **2008**, *9*, 545–556. [CrossRef] [PubMed]
6. Botvinick, M.M.; Huffstetler, S.; McGuire, J.T. Effort Discounting in Human Nucleus Accumbens. *Cogn. Affect. Behav. Neurosci.* **2009**, *9*, 16–27. [CrossRef]
7. Massar, S.A.A.; Lim, J.; Sasmita, K.; Chee, M.W.L. Rewards Boost Sustained Attention through Higher Effort: A Value-Based Decision Making Approach. *Biol. Psychol.* **2016**, *120*, 21–27. [CrossRef]
8. Frömer, R.; Lin, H.; Wolf, C.K.D.; Inzlicht, M.; Shenhav, A. Expectations of Reward and Efficacy Guide Cognitive Control Allocation. *Nat. Commun.* **2021**, *12*, 1030. [CrossRef]
9. Phillips, P.E.M.; Walton, M.E.; Jhou, T.C. Calculating Utility: Preclinical Evidence for Cost-Benefit Analysis by Mesolimbic Dopamine. *Psychopharmacol* **2007**, *191*, 483–495. [CrossRef]
10. Rudebeck, P.H.; Walton, M.E.; Smyth, A.N.; Bannerman, D.M.; Rushworth, M.F.S. Separate Neural Pathways Process Different Decision Costs. *Nat. Neurosci.* **2006**, *9*, 1161–1168. [CrossRef] [PubMed]
11. Prévost, C.; Pessiglione, M.; Metereau, E.; Clery-Melin, M.-L.; Dreher, J.-C. Separate Valuation Subsystems for Delay and Effort Decision Costs. *J. Neurosci.* **2010**, *30*, 14080–14090. [CrossRef] [PubMed]
12. Garofalo, S.; Finotti, G.; Starita, F.; Bouchard, A.E.; Fecteau, S. Effort-Based Decision Making. *The Sage Handbook of Cognitive and Systems Neuroscience II*, 2022; *in press.*
13. Apps, M.A.J.; Grima, L.L.; Manohar, S.; Husain, M. The Role of Cognitive Effort in Subjective Reward Devaluation and Risky Decision-Making. *Sci. Rep.* **2015**, *5*, 16880. [CrossRef]
14. Vassena, E.; Deraeve, J.; Alexander, W.H. Predicting Motivation: Computational Models of PFC Can Explain Neural Coding of Motivation and Effort-Based Decision-Making in Health and Disease. *J. Cogn. Neurosci.* **2017**, *29*, 1633–1645. [CrossRef]
15. Ainslie, G. Specious Reward: A Behavioral Theory of Impulsiveness and Impulse Control. *Psychol. Bull.* **1975**, *82*, 463–496. [CrossRef]
16. Myerson, J.; Green, L. Discounting of Delayed Rewards: Models of Individual Choice. *J. Exp. Anal. Behav.* **1995**, *64*, 263–276. [CrossRef] [PubMed]
17. Frederick, S.; Loewenstein, G.; O'Donoghue, T. Time Discounting and Preference: A Critical Review. *J. Econ. Lit.* **2002**, *40*, 351–401. [CrossRef]
18. Sellitto, M.; Ciaramelli, E.; Pellegrino, G. The Neurobiology of Intertemporal Choice: Insight from Imaging and Lesion Studies. *Rev. Neurosci.* **2011**, *22*, 565–574. [CrossRef]
19. Guleken, Z.; Sutcubasi, B.; Metin, B. The Cognitive Dynamics of Small-Sooner over Large-Later Preferences during Temporal Discounting Task through Event-Related Oscillations (EROs). *Neuropsychologia* **2021**, *162*, 108046. [CrossRef]

20. Shadmehr, R.; Orban de Xivry, J.J.; Xu-Wilson, M.; Shih, T.-Y. Temporal Discounting of Reward and the Cost of Time in Motor Control. *J. Neurosci.* **2010**, *30*, 10507–10516. [CrossRef]

21. Berret, B.; Jean, F. Why Don't We Move Slower? The Value of Time in the Neural Control of Action. *J. Neurosci.* **2016**, *36*, 1056–1070. [CrossRef]

22. Shadmehr, R.; Huang, H.J.; Ahmed, A.A. A Representation of Effort in Decision-Making and Motor Control. *Curr. Biol.* **2016**, *26*, 1929–1934. [CrossRef] [PubMed]

23. Shadmehr, R. Control of Movements and Temporal Discounting of Reward. *Curr. Opin. Neurobiol.* **2010**, *20*, 726–730. [CrossRef]

24. Choi, J.E.S.; Vaswani, P.A.; Shadmehr, R. Vigor of Movements and the Cost of Time in Decision Making. *J. Neurosci.* **2014**, *34*, 1212–1223. [CrossRef] [PubMed]

25. Schultz, W. Dopamine Signals for Reward Value and Risk: Basic and Recent Data. *Behav. Brain Funct.* **2010**, *6*, 24. [CrossRef] [PubMed]

26. Grogan, J.P.; Sandhu, T.R.; Hu, M.T.; Manohar, S.G. Dopamine Promotes Instrumental Motivation, but Reduces Reward-Related Vigour. *Elife* **2020**, *9*, e58321. [CrossRef]

27. Sedaghat-Nejad, E.; Herzfeld, D.J.; Shadmehr, R. Reward Prediction Error Modulates Saccade Vigor. *J. Neurosci. Off. J. Soc. Neurosci.* **2019**, *39*, 5010–5017. [CrossRef]

28. Shadmehr, R.; Reppert, T.R.; Summerside, E.M.; Yoon, T.; Ahmed, A.A. Movement Vigor as a Reflection of Subjective Economic Utility. *Trends Neurosci.* **2019**, *42*, 323–336. [CrossRef]

29. Fitts, P.M. The Information Capacity of the Human Motor System in Controlling the Amplitude of Movement. *J. Exp. Psychol.* **1954**, *47*, 381–391. [CrossRef]

30. Jeannerod, M. The Representing Brain: Neural Correlates of Motor Intention and Imagery. *Behav. Brain Sci.* **1994**, *17*, 187–202. [CrossRef]

31. Paszkiel, S.; Dobrakowski, P. Brain—Computer Technology-Based Training System in the Field of Motor Imagery. *IET Sci. Meas. Technol.* **2021**, *14*, 1014–1018. [CrossRef]

32. Hwang, H.-J.; Kwon, K.; Im, C.-H. Neurofeedback-Based Motor Imagery Training for Brain-Computer Interface (BCI). *J. Neurosci. Methods* **2009**, *179*, 150–156. [CrossRef] [PubMed]

33. Paszkiel, S.; Rojek, R.; Lei, N.; Castro, M. Review of Solutions for the Application of Example of Machine Learning Methods for Motor Imagery in Correlation with Brain-Computer Interfaces. *Przegląd Elektrotechniczny* **2021**, *1*, 113–118. [CrossRef]

34. Mane, R.; Robinson, N.; Vinod, A.P.; Lee, S.-W.; Guan, C. A Multi-View CNN with Novel Variance Layer for Motor Imagery Brain Computer Interface. In Proceedings of the 2020 42nd Annual International Conference of the IEEE Engineering in Medicine & Biology Society (EMBC), Montreal, QC, Canada, 20–24 July 2020; pp. 2950–2953. [CrossRef]

35. Pawus, D.; Paszkiel, S. The Application of Integration of EEG Signals for Authorial Classification Algorithms in Implementation for a Mobile Robot Control Using Movement Imagery—Pilot Study. *Appl. Sci.* **2022**, *12*, 2161. [CrossRef]

36. Decety, J.; Jeannerod, M.; Prablanc, C. The Timing of Mentally Represented Actions. *Behav. Brain Res.* **1989**, *34*, 35–42. [CrossRef]

37. Porro, C.A.; Francescato, M.P.; Cettolo, V.; Diamond, M.E.; Baraldi, P.; Zuiani, C.; Bazzocchi, M.; Di Prampero, P.E. Primary Motor and Sensory Cortex Activation during Motor Performance and Motor Imagery: A Functional Magnetic Resonance Imaging Study. *J. Neurosci.* **1996**, *16*, 7688–7698. [CrossRef]

38. Decety, J. The Neurophysiological Basis of Motor Imagery. *Behav. Brain Res.* **1996**, *77*, 45–52. [CrossRef]

39. Batula, A.M.; Mark, J.A.; Kim, Y.E.; Ayaz, H. Comparison of Brain Activation during Motor Imagery and Motor Movement Using FNIRS. *Comput. Intell. Neurosci.* **2017**, *2017*, 1–12. [CrossRef]

40. Decety, J.; Jeannerod, M. Mentally Simulated Movements in Virtual Reality: Does Fitt's Law Hold in Motor Imagery? *Behav. Brain Res.* **1995**, *72*, 127–134. [CrossRef]

41. Sirigu, A.; Duhamel, J.; Cohen, L.; Pillon, B.; Dubois, B.; Agid, Y. The Mental Representation of Hand Movements after Parietal Cortex Damage. *Science* **1996**, *273*, 1564–1568. [CrossRef]

42. Dahm, S.F.; Rieger, M. Cognitive Constraints on Motor Imagery. *Psychol. Res.* **2016**, *80*, 235–247. [CrossRef]

43. Bertucco, M.; Cesari, P.; Latash, M.L. Fitts' Law in Early Postural Adjustments. *Neuroscience* **2013**, *231*, 61–69. [CrossRef] [PubMed]

44. Cisek, P.; Kalaska, J.F. Neural Mechanisms for Interacting with a World Full of Action Choices. *Annu. Rev. Neurosci.* **2010**, *33*, 269–298. [CrossRef]

45. Huda, R.; Goard, M.J.; Pho, G.N.; Sur, M. Neural Mechanisms of Sensorimotor Transformation and Action Selection. *Eur. J. Neurosci.* **2019**, *49*, 1055–1060. [CrossRef]

46. Borgomaneri, S.; Serio, G.; Battaglia, S. Please, Don't Do It! Fifteen Years of Progress of Non-Invasive Brain Stimulation in Action Inhibition. *Cortex* **2020**, *132*, 404–422. [CrossRef] [PubMed]

47. Battaglia, S.; Serio, G.; Scarpazza, C.; D'Ausilio, A.; Borgomaneri, S. Frozen in (e)Motion: How Reactive Motor Inhibition Is Influenced by the Emotional Content of Stimuli in Healthy and Psychiatric Populations. *Behav. Res. Ther.* **2021**, *146*, 103963. [CrossRef] [PubMed]

48. O'Doherty, J.; Dayan, P.; Schultz, J.; Deichmann, R.; Friston, K.; Dolan, R.J. Dissociable Roles of Ventral and Dorsal Striatum in Instrumental Conditioning. *Science* **2004**, *304*, 452–454. [CrossRef]

49. Rich, E.L.; Wallis, J.D. Decoding Subjective Decisions from Orbitofrontal Cortex. *Nat. Neurosci.* **2016**, *19*, 973–980. [CrossRef]

50. Ballesta, S.; Shi, W.; Conen, K.E.; Padoa-Schioppa, C. Values Encoded in Orbitofrontal Cortex Are Causally Related to Economic Choices. *Nature* **2020**, *588*, 450–453. [CrossRef]

51. Gluth, S.; Rieskamp, J.; Büchel, C. Classic EEG Motor Potentials Track the Emergence of Value-Based Decisions. *Neuroimage* **2013**, *79*, 394–403. [CrossRef]
52. Gold, J.I.; Shadlen, M.N. The Neural Basis of Decision Making. *Annu. Rev. Neurosci.* **2007**, *30*, 535–574. [CrossRef]
53. Stuphorn, V.; Taylor, T.L.; Schall, J.D. Performance Monitoring by the Supplementary Eye Field. *Nature* **2000**, *408*, 857–860. [CrossRef] [PubMed]
54. Ikeda, T.; Hikosaka, O. Reward-Dependent Gain and Bias of Visual Responses in Primate Superior Colliculus. *Neuron* **2003**, *39*, 693–700. [CrossRef]
55. Roesch, M.R.; Olson, C.R. Impact of Expected Reward on Neuronal Activity in Prefrontal Cortex, Frontal and Supplementary Eye Fields and Premotor Cortex. *J. Neurophysiol.* **2003**, *90*, 1766–1789. [CrossRef] [PubMed]
56. Galaro, J.K.; Celnik, P.; Chib, V.S. Motor Cortex Excitability Reflects the Subjective Value of Reward and Mediates Its Effects on Incentive-Motivated Performance. *J. Neurosci.* **2019**, *39*, 1236–1248. [CrossRef]
57. Donner, T.H.; Siegel, M.; Fries, P.; Engel, A.K. Buildup of Choice-Predictive Activity in Human Motor Cortex during Perceptual Decision Making. *Curr. Biol.* **2009**, *19*, 1581–1585. [CrossRef]
58. Aitken, F.; Turner, G.; Kok, P. Prior Expectations of Motion Direction Modulate Early Sensory Processing. *J. Neurosci.* **2020**, *40*, 6389–6397. [CrossRef]
59. Pastor-Bernier, A.; Cisek, P. Neural Correlates of Biased Competition in Premotor Cortex. *J. Neurosci.* **2011**, *31*, 7083–7088. [CrossRef]
60. Gupta, N.; Aron, A.R. Urges for Food and Money Spill over into Motor System Excitability before Action Is Taken. *Eur. J. Neurosci.* **2011**, *33*, 183–188. [CrossRef]
61. Borgomaneri, S.; Vitale, F.; Battaglia, S.; Avenanti, A. Early Right Motor Cortex Response to Happy and Fearful Facial Expressions: A TMS Motor-Evoked Potential Study. *Brain Sci.* **2021**, *11*, 1203. [CrossRef]
62. Domínguez-Zamora, F.J.; Marigold, D.S. Motor Cost Affects the Decision of When to Shift Gaze for Guiding Movement. *J. Neurophysiol.* **2019**, *122*, 378–388. [CrossRef]
63. Vourvopoulos, A.; Bermudez, S.; Liarokapis, F. EEG Correlates of Video Game Experience and User Profile in Motor-Imagery-Based Brain—Computer Interaction. *Vis. Comput.* **2017**, *33*, 533–546. [CrossRef]
64. Klement, J.; Kubera, B.; Eggeling, J.; Rädel, C.; Wagner, C.; Park, S.Q.; Peters, A. Effects of Blood Glucose on Delay Discounting, Food Intake and Counterregulation in Lean and Obese Men. *Psychoneuroendocrinology* **2018**, *89*, 177–184. [CrossRef] [PubMed]
65. Massar, S.A.A.; Pu, Z.; Chen, C.; Chee, M.W.L. Losses Motivate Cognitive Effort More Than Gains in Effort-Based Decision Making and Performance. *Front. Hum. Neurosci.* **2020**, *14*, 287. [CrossRef]
66. Bowyer, C.; Brush, C.J.; Threadgill, H.; Harmon-Jones, E.; Treadway, M.; Patrick, C.J.; Hajcak, G. The Effort-Doors Task: Examining the Temporal Dynamics of Effort-Based Reward Processing Using ERPs. *Neuroimage* **2021**, *228*, 117656. [CrossRef]
67. Smalley, K.; Knerr, N.; Kendrick, V.; Colliver, A.; Owen, O. Reassessment of Body Mass Indices. *Am. J. Clin. Nutr.* **1990**, *52*, 405–408. [CrossRef]
68. Bakker, M.; De Lange, F.P.; Stevens, J.A.; Toni, I.; Bloem, B.R. Motor Imagery of Gait: A Quantitative Approach. *Exp. Brain Res.* **2007**, *179*, 497–504. [CrossRef] [PubMed]
69. Finlayson, G.; King, N.; Blundell, J. The Role of Implicit Wanting in Relation to Explicit Liking and Wanting for Food: Implications for Appetite Control. *Appetite* **2008**, *50*, 120–127. [CrossRef]
70. Born, J.M.; Lemmens, S.G.T.; Martens, M.J.I.; Goebel, R. Differences between Liking and Wanting Signals in the Human Brain and Relations with Cognitive Dietary Restraint and Body Mass Index. *Am. J. Clin. Nutr.* **2011**, *94*, 393–403. [CrossRef]
71. Marks, D.F. Visual Imagery Differences in the Recall of Pictures. *Br. J. Psychol.* **1973**, *64*, 17–24. [CrossRef]
72. Isaac, A.; Marks, D.F.; Russell, D.G. An Instrument for Assessing Imagery of Movement: The Vividness of Movement Imagery Questionnaire (VMIQ). *J. Ment. Imag.* **1986**, *10*, 23–30.
73. Glimcher, P.W. Decisions, Decisions, Decisions: Choosing a Biological Science of Choice. *Neuron* **2002**, *36*, 323–332. [CrossRef]
74. Kennerley, S.W.; Dahmubed, A.F.; Lara, A.H.; Wallis, J.D. Neurons in the Frontal Lobe Encode the Value of Multiple Decision Variables. *J. Cogn. Neurosci.* **2009**, *21*, 1162–1178. [CrossRef] [PubMed]
75. Garofalo, S.; Battaglia, S.; Di Pellegrino, G. Individual Differences in Working Memory Capacity and Cue-Guided Behavior in Humans. *Sci. Rep.* **2019**, *9*, 7327. [CrossRef]
76. Garofalo, S.; Battaglia, S.; Starita, F.; di Pellegrino, G. Modulation of Cue-Guided Choices by Transcranial Direct Current Stimulation. *Cortex* **2021**, *137*, 124–137. [CrossRef] [PubMed]
77. Kahneman, D.; Tversky, A.; Kahneman, B.Y.D.; Tversky, A. Prospect Theory: An Analysis of Decision under Risk Linked. *Econometrica* **1979**, *47*, 263–292. [CrossRef]
78. Rigoli, F.; Martinelli, C.; Shergill, S.S. The Role of Expecting Feedback during Decision-Making under Risk. *Neuroimage* **2019**, *202*, 116079. [CrossRef] [PubMed]
79. Mühlhoff, N.; Stevens, J.R.; Reader, S.M. Spatial Discounting of Food and Social Rewards in Guppies (Poecilia Reticulata). *Front. Psychol.* **2011**, *2*, 68. [CrossRef]
80. Stevens, J.R.; Rosati, A.G.; Ross, K.R.; Hauser, M.D. Will Travel for Food: Spatial Discounting in Two New World Monkeys. *Curr. Biol.* **2005**, *15*, 1855–1860. [CrossRef]
81. Dezfouli, A.; Balleine, B.W. Learning the Structure of the World: The Adaptive Nature of State-Space and Action Representations in Multi-Stage Decision-Making. *PLoS Comput. Biol.* **2019**, *15*, e1007334. [CrossRef]

82. Salamone, J.D.; Correa, M.; Ferrigno, S.; Yang, J.-H.; Rotolo, R.A.; Presby, R.E. The Psychopharmacology of Effort-Related Decision Making: Dopamine, Adenosine, and Insights into the Neurochemistry of Motivation. *Pharmacol. Rev.* **2018**, *70*, 747–762. [CrossRef]

83. Cos, I.; Bélanger, N.; Cisek, P. The Influence of Predicted Arm Biomechanics on Decision Making. *J. Neurophysiol.* **2011**, *105*, 3022–3033. [CrossRef] [PubMed]

84. Cos, I.; Duque, J.; Cisek, P. Rapid Prediction of Biomechanical Costs during Action Decisions. *J. Neurophysiol.* **2014**, *112*, 1256–1266. [CrossRef] [PubMed]

85. Pierrieau, E.; Lepage, J.-F.; Bernier, P.-M. Action Costs Rapidly and Automatically Interfere with Reward-Based Decision-Making in a Reaching Task. *Eneuro* **2021**, *8*, ENEURO.0247-21.2021. [CrossRef] [PubMed]

86. Pool, E.; Sennwald, V.; Delplanque, S.; Brosch, T.; Sander, D. Measuring Wanting and Liking from Animals to Humans: A Systematic Review. *Neurosci. Biobehav. Rev.* **2016**, *63*, 124–142. [CrossRef] [PubMed]

87. Dickinson, A.; Balleine, B. Motivational Control of Goal-Directed Action. *Anim. Learn. Behav.* **1994**, *22*, 1–18. [CrossRef]

88. Kahneman, D. Choices, Values, and Frames. *Am. Psychol.* **1984**, *39*, 341–350. [CrossRef]

89. Di Gregorio, F.; Ernst, B.; Steinhauser, M. Differential effects of instructed and objective feedback reliability on feedback-related brain activity. *Psychophysiology* **2019**, *56*, e13399. [CrossRef]

90. Berridge, K.C.; O'Doherty, J.P. From Experienced Utility to Decision Utility. *Neuroeconomics* **2013**, 335–354. [CrossRef]

91. Berridge, K.C.; Robinson, T.E. Parsing Reward. *Trends Neurosci.* **2003**, *26*, 507–513. [CrossRef]

92. Klein-Flügge, M.C.; Kennerley, S.W.; Friston, K.; Bestmann, S. Neural Signatures of Value Comparison in Human Cingulate Cortex during Decisions Requiring an Effort-Reward Trade-Off. *J. Neurosci.* **2016**, *36*, 10002–10015. [CrossRef]

93. Finlayson, G.; King, N.; Blundell, J.E. Is It Possible to Dissociate 'Liking' and 'Wanting' for Foods in Humans? A Novel Experimental Procedure. *Physiol. Behav.* **2007**, *90*, 36–42. [CrossRef] [PubMed]

94. Telegdy, G.; Adamik, A.; Tanaka, M.; Schally, A.V. Effects of the LHRH Antagonist Cetrorelix on Affective and Cognitive Functions in Rats. *Regul. Pept.* **2010**, *159*, 142–147. [CrossRef] [PubMed]

95. Telegdy, G.; Tanaka, M.; Schally, A.V. Effects of the LHRH Antagonist Cetrorelix on the Brain Function in Mice. *Neuropeptides* **2009**, *43*, 229–234. [CrossRef] [PubMed]

96. Martos, D.; Tuka, B.; Tanaka, M.; Vécsei, L.; Telegdy, G. Memory Enhancement with Kynurenic Acid and Its Mechanisms in Neurotransmission. *Biomedicines* **2022**, *10*, 849. [CrossRef] [PubMed]

97. Tanaka, M.; Csabafi, K.; Telegdy, G. Neurotransmissions of Antidepressant-like Effects of Kisspeptin-13. *Regul. Pept.* **2013**, *180*, 1–4. [CrossRef]

98. Tanaka, M.; Vécsei, L. Monitoring the Kynurenine System: Concentrations, Ratios or What Else? *Adv. Clin. Exp. Med.* **2021**, *30*, 775–778. [CrossRef]

99. Telegdy, G.; Tanaka, M.; Schally, A.V. Effects of the Growth Hormone-Releasing Hormone (GH-RH) Antagonist on Brain Functions in Mice. *Behav. Brain Res.* **2011**, *224*, 155–158. [CrossRef]

100. Tanaka, M.; Vécsei, L. Editorial of Special Issue "Crosstalk between Depression, Anxiety, and Dementia: Comorbidity in Behavioral Neurology and Neuropsychiatry". *Biomedicines* **2021**, *9*, 517. [CrossRef]

101. Spekker, E.; Tanaka, M.; Szabó, Á.; Vécsei, L. Neurogenic Inflammation: The Participant in Migraine and Recent Advancements in Translational Research. *Biomedicines* **2021**, *10*, 76. [CrossRef]

102. Tanaka, M.; Török, N.; Vécsei, L. Are 5-HT(1) Receptor Agonists Effective Anti-Migraine Drugs? *Expert Opin. Pharmacother.* **2021**, *22*, 1221–1225. [CrossRef]

103. Tanaka, M.; Bohár, Z.; Vécsei, L. Are Kynurenines Accomplices or Principal Villains in Dementia? Maintenance of Kynurenine Metabolism. *Molecules* **2020**, *25*, 564. [CrossRef] [PubMed]

104. Tanaka, M.; Tóth, F.; Polyák, H.; Szabó, Á.; Mándi, Y.; Vécsei, L. Immune Influencers in Action: Metabolites and Enzymes of the Tryptophan-Kynurenine Metabolic Pathway. *Biomedicines* **2021**, *9*, 734. [CrossRef] [PubMed]

105. Bartoshuk, L. The Measurement of Pleasure and Pain. *Perspect. Psychol. Sci.* **2014**, *9*, 91–93. [CrossRef] [PubMed]

106. Pichon, A.M.; Coppin, G.; Cayeux, I.; Porcherot, C.; Sander, D.; Delplanque, S. Sensitivity of Physiological Emotional Measures to Odors Depends on the Product and the Pleasantness Ranges Used. *Front. Psychol.* **2015**, *6*, 1821. [CrossRef]

107. Kivetz, R. The Effects of Effort and Intrinsic Motivation on Risky Choice. *Mark. Sci.* **2003**, *22*, 477–502. [CrossRef]

108. Bushong, B.; King, L.M.; Camerer, C.F.; Rangel, A. Pavlovian Processes in Consumer Choice: The Physical Presence of a Good Increases Willingness-to-Pay. *Am. Econ. Rev.* **2010**, *100*, 1556–1571. [CrossRef]

109. Barsalou, L.W. Perceptual Symbol Systems. *Behav. Brain Sci.* **1999**, *22*, 577–660. [CrossRef]

110. Barsalou, L.W. Simulation, Situated Conceptualization, and Prediction. *Philos. Trans. R. Soc. B Biol. Sci.* **2009**, *364*, 1281–1289. [CrossRef]

111. Niedenthal, P.M.; Barsalou, L.W.; Winkielman, P.; Krauth-Gruber, S.; Ric, F. Embodiment in Attitudes, Social Perception, and Emotion. *Personal. Soc. Psychol. Rev.* **2005**, *9*, 184–211. [CrossRef]

112. Padoa-Schioppa, C. Neurobiology of Economic Choice: A Good-Based Model. *Annu. Rev. Neurosci.* **2011**, *34*, 333–359. [CrossRef]

113. Plassmann, H.; O'Doherty, J.; Rangel, A. Orbitofrontal Cortex Encodes Willingness to Pay in Everyday Economic Transactions. *J. Neurosci.* **2007**, *27*, 9984–9988. [CrossRef] [PubMed]

114. Wallis, J.D. Orbitofrontal Cortex and Its Contribution to Decision-Making. *Annu. Rev. Neurosci.* **2007**, *30*, 31–56. [CrossRef] [PubMed]

115. Battaglia, S.; Garofalo, S.; di Pellegrino, G.; Starita, F. Revaluing the Role of VmPFC in the Acquisition of Pavlovian Threat Conditioning in Humans. *J. Neurosci.* **2020**, *40*, 8491–8500. [CrossRef]
116. Battaglia, S.; Harrison, B.J.; Fullana, M.A. Does the Human Ventromedial Prefrontal Cortex Support Fear Learning, Fear Extinction or Both? A Commentary on Subregional Contributions. *Mol. Psychiatry* **2021**, 1–3. [CrossRef]
117. Battaglia, S. Neurobiological Advances of Learned Fear in Humans. *Adv. Clin. Exp. Med.* **2022**, *31*, 217–221. [CrossRef] [PubMed]
118. Maier, M.E.; Di Gregorio, F.; Muricchio, T.; Di Pellegrino, G. Impaired rapid error monitoring but intact error signaling following rostral anterior cingulate cortex lesions in humans. *Front. Hum. Neurosci.* **2015**, *9*, 339. [CrossRef] [PubMed]
119. Battaglia, S.; Thayer, J.F. Functional interplay between central and autonomic nervous systems in human fear conditioning. *Trends Neurosci.* 2022; *in press.*
120. Platt, M.L.; Glimcher, P.W. Neural Correlates of Decision Variables in Parietal Cortex. *Nature* **1999**, *400*, 233–238. [CrossRef]
121. Bari, B.A.; Cohen, J.Y. Dynamic Decision Making and Value Computations in Medial Frontal Cortex. In *What Does Medial Frontal Cortex Signal During Behavior? Insights from Behavioral Neurophysiology*; Brockett, A.T., Amarante, L.M., Laubach, M., Roesch, M.R., Eds.; Academic Press: Cambridge, MA, USA, 2021; Volume 158, pp. 83–113. [CrossRef]
122. Glimcher, P.W.; Dorris, M.C.; Bayer, H.M. Physiological Utility Theory and the Neuroeconomics of Choice. *Games Econ. Behav.* **2005**, *52*, 213–256. [CrossRef]
123. Denk, F.; Walton, M.E.; Jennings, K.A.; Sharp, T.; Rushworth, M.F.S.; Bannerman, D.M. Differential Involvement of Serotonin and Dopamine Systems in Cost-Benefit Decisions about Delay or Effort. *Psychopharmacology* **2005**, *179*, 587–596. [CrossRef]
124. Floresco, S.B.; Ghods-Sharifi, S. Amygdala-Prefrontal Cortical Circuitry Regulates Effort-Based Decision Making. *Cereb. Cortex* **2007**, *17*, 251–260. [CrossRef]
125. Salamone, J.D.; Correa, M.; Mingote, S.; Weber, S.M. Nucleus Accumbens Dopamine and the Regulation of Effort in Food-Seeking Behavior: Implications for Studies of Natural Motivation, Psychiatry, and Drug Abuse. *J. Pharmacol. Exp. Ther.* **2003**, *305*, 1–8. [CrossRef] [PubMed]
126. Walton, M.E.; Devlin, J.T.; Rushworth, M.F.S. Interactions between Decision Making and Performance Monitoring within Prefrontal Cortex. *Nat. Neurosci.* **2004**, *7*, 1259–1265. [CrossRef] [PubMed]
127. Trajkovic, J.; Di Gregorio, F.; Ferri, F.; Marzi, C.; Diciotti, S.; Romei, V. Resting State Alpha Oscillatory Activity Is a Valid and Reliable Marker of Schizotypy. *Sci. Rep.* **2021**, *11*, 10379. [CrossRef]
128. Marciniak, M.A.; Shanahan, L.; Binder, H.; Kalisch, R.; Kleim, B. Positive Prospective Mental Imagery Characteristics in Young Adults and Their Associations with Depressive Symptoms. *PsyArXiv*, 2022; *pre-print.* [CrossRef]

Article

NGFR Gene and Single Nucleotide Polymorphisms, rs2072446 and rs11466162, Playing Roles in Psychiatric Disorders

Longyou Zhao [1,†], Binyin Hou [2,3,†], Lei Ji [2,3], Decheng Ren [2,3], Fan Yuan [2,3], Liangjie Liu [2,3], Yan Bi [2,3], Fengping Yang [2,3], Shunying Yu [3], Zhenghui Yi [3], Chuanxin Liu [4], Bo Bai [4], Tao Yu [2,3], Changqun Cai [5], Lin He [2,3], Guang He [2,3], Yi Shi [2,3], Xingwang Li [2,3] and Shaochang Wu [1,*]

[1] Lishui No.2 People's Hospital, 69 Beihuan Road, Liandu District, Lishui 323000, China
[2] Bio-X Institutes, Key Laboratory for the Genetics of Developmental and Neuropsychiatric Disorders, Shanghai Jiao Tong University, 1954 Huashan Road, Shanghai 200030, China
[3] Shanghai Key Laboratory of Psychotic Disorders, and Brain Science and Technology Research Center, Shanghai Jiao Tong University, 1954 Huashan Road, Shanghai 200030, China
[4] School of Mental Health, Jining Medical University, 16 Hehua Road, Taibaihu New District, Jining 272067, China
[5] Wuhu Fourth People's Hospital, 1 Xuxiashan Road, Wuhu 241002, China
[*] Correspondence: lseywsc@163.com; Tel./Fax: +86-(0)-578-2299912
[†] These authors contributed equally to this work.

Abstract: Psychiatric disorders are a class of complex disorders characterized by brain dysfunction with varying degrees of impairment in cognition, emotion, consciousness and behavior, which has become a serious public health issue. The *NGFR* gene encodes the p75 neurotrophin receptor, which regulates neuronal growth, survival and plasticity, and was reported to be associated with depression, schizophrenia and antidepressant efficacy in human patient and animal studies. In this study, we investigated its association with schizophrenia and major depression and its role in the behavioral phenotype of adult mice. Four *NGFR* SNPs were detected based on a study among 1010 schizophrenia patients, 610 patients with major depressive disorders (MDD) and 1034 normal controls, respectively. We then knocked down the expression of *NGFR* protein in the hippocampal dentate gyrus of the mouse brain by injection of shRNA lentivirus to further investigate its behavioral effect in mice. We found significant associations of s2072446 and rs11466162 for schizophrenia. *Ngfr* knockdown mice showed social and behavioral abnormalities, suggesting that it is linked to the etiology of neuropsychiatric disorders. We found significant associations between *NGFR* and schizophrenia and that *Ngfr* may contribute to the social behavior of adult mice in the functional study, which provided meaningful clues to the pathogenesis of psychiatric disorders.

Keywords: psychiatric disorders; schizophrenia; major depressive disorders (MDD); association analyses; *NGFR* gene; p75 neurotrophin receptor; shRNA lentivirus; knock down; social behavior; functional study

Citation: Zhao, L.; Hou, B.; Ji, L.; Ren, D.; Yuan, F.; Liu, L.; Bi, Y.; Yang, F.; Yu, S.; Yi, Z.; et al. *NGFR* Gene and Single Nucleotide Polymorphisms, rs2072446 and rs11466162, Playing Roles in Psychiatric Disorders. *Brain Sci.* **2022**, *12*, 1372. https://doi.org/10.3390/brainsci12101372

Academic Editors: Julia A. Chester, Masaru Tanaka, Lydia Giménez-Llort, Simone Battaglia, Chong Chen and Piril Hepsomali

Received: 25 August 2022
Accepted: 3 October 2022
Published: 9 October 2022

Publisher's Note: MDPI stays neutral with regard to jurisdictional claims in published maps and institutional affiliations.

1. Introduction

Psychiatric disorders are a class of complex disorders characterized by brain dysfunction with varying degrees of impairment in cognition, emotion, consciousness and behavior, including schizophrenia and major depressive disorders (MDD). Psychiatric disorders are becoming a serious public health issue due to their first rank in disease burden among non-fatal diseases worldwide. Previous genetic studies, including twin, adoption and family studies [1–3], showed that the heritability of psychiatric disorders was around 80%, and the gene-environment interactions played an important role in their pathogenesis.

Evidence from previous research implicated that Corpus Callosum (CC) white matter tracts deficits, dissociative symptoms and abnormal fatty acid (FAs) metabolism relates to schizophrenia [4–6]. Recent studies found that mitochondrial functions compromise influenced by the loss of mitochondrial stress resilience, activation of the tryptophan (Trp)–kynurenine (KYN) metabolic system and vitamin D deficiency may contribute to the development of psychiatric disorders [7,8]. Some other researchers found that lipophilic statins (including simvastatin), bioactive kynurenines and their analogs can be neuroprotective agents [9,10]. It was reported that synaptic plasticity defects were possibly due to neurodevelopmental and neurodegenerative abnormalities and could contribute to cognitive impairment underlying schizophrenia and some other psychiatric disorders [11]. Synaptic plasticity regulation genes, especially the ones that encode neurotrophic factor and their receptors, are thus believed to be involved in neuronal development, synapse generation, and response to stress/anxiety stimuli [12,13] and may have important roles in the molecular mechanism of cognition impairment in schizophrenia, MDD and some other psychiatric disorders, including nerve growth factor receptor (*NGFR*).

The *NGFR* gene encodes the p75 neurotrophin receptor (p75NTR), which participates in the controlling of many signaling pathways by interacting with all kinds of TRK receptors, sorting proteins and NOGO receptors. Previous research showed that *NGFR* is involved in neurogenesis, regulation of sprouting, synaptogenesis and pruning, which contributes to altered neural functions, and is thought to be the basis of psychiatric disorders [14–16]. Studies showed that the serum *NGFR* levels in patients with depression [17,18], schizophrenia [19] and bipolar disorder [20] were significantly different from those in healthy controls. *NGFR* gene polymorphisms were reported to be associated with depression, schizophrenia and antidepressant efficacy [21,22]. Increasing evidence based on animal studies found that genetic variants in *NGFR* could alter the brain's susceptibility to psychiatric disorders. For instance, a study showed that in the brain of rats with unpredictable chronic mild stress (UCMS), the apoptosis signal of proBDNF/*NGFR*/sorting protein was activated, and the mRNA and protein expression levels of *NGFR* were increased [23]). *Ngfr*-knockout mice exhibited or alleviated behavioral deficits such as anxiety, spatial memory impairment and depression [24–26]. It was reported that the *NGFR* SNPs rs11466155 and rs734194 were strongly associated with schizophrenia based on the Armenia Caucasus populations [19].

In view of a series of research above indicating that the *NGFR* gene may be closely related to the pathogenesis of psychiatric disorders such as depression and schizophrenia, we performed an association study of the Chinese Han population to investigate the role of the four *NGFR* SNPs (rs575791, rs1035050, rs2072446, rs11466162, Table 1) in schizophrenia and MDD. Furthermore, we knocked down the expression of *Ngfr* in the hippocampal dentate gyrus of adult mice brain by injection of shRNA lentivirus to examine the behavioral alterations it may cause in mice.

Table 1. *NGFR* SNPs analyzed in this study.

	SNP ID [a]	Chromosome [b]	Location	Polymorphisms [c]
	rs1035050	17:49486650	promoter	T/C
	rs575791	17:49497393	intron	A/G
NGFR	rs2072446	17:49510457	missense	C/T
	rs11466162	17:49513533	3 Prime UTR Variant	G/A

Note. [a] According to the dbSNP database, [b] the SNP Chromosome positions are based on the NCBI human genome build GRCh38; [c] The allele under the slash is the minor allele.

2. Materials and Methods

2.1. Participants and Animals

In this research, 1010 schizophrenic patients (396 females and 614 males, age: 43.04 ± 12.91, onset age: 25.93 ± 9.62) and 1034 healthy controls (446 females and 588 males, age: 34.06 ± 10.00) were recruited from Chinese Han population. All patients were diagnosed on the basis of DSM-IV criteria by two independent, experienced psychiatrists in two separate clinical interviews. All controls were in good health, and none of them showed symptoms of any psychiatric disorders.

In order to analyze the candidate site of major depression, we recruited 610 unrelated major depressive disorder patients (324 females and 286 males, age: 36.31 ± 12.10) and used the same 1034 individuals as healthy controls. Each of the patients was strictly diagnosed by two experienced psychiatrists adopting DSM-IV criteria independently. The degree of depression was assessed by a 17-item Hamilton Depression Rating Scale (HAMD). Patients pregnant with other psychiatric disorders and substance abuse were excluded.

All the participants were of unrelated Chinese Han origin. All of them signed the informed consent, and the study was appraised and confirmed by the Ethics Committee of the Bio-X center, Shanghai Jiao Tong University (M16033, 2016). Six- to eight-week-old male mice (C57BL/6) were obtained from Shanghai Lingchang Biotech Co., Ltd. All experiments were performed using standard protocols and approved by the Animal Care and Ethics Committee of Shanghai Jiao Tong University, China (202006003, 2020).

2.2. DNA Extraction and Genotyping

In the present study, we collected peripheral venous blood from each participant and applied AxyPrep Blood Genomic DNA Miniprep Kit to DNA extractions. *NGFR* SNPs were selected from the HapMap database (release #24) of the Chinese Han population (minor allele frequency, MAF ≥ 0.05). The detailed information on SNPs is summarized in Table 1. Four SNPs (rs575791, rs1035050, rs2072446, rs11466162) were genotyped by MassARRAY® Analyzer platform (Agena, San Diego, CA, USA). All the probes and primers were designed by My-sequenom online software Assay Design Suite v2.0 (Agena, San Diego, CA, USA). Each tube in the polymerase chain reaction contained 10 ng genomic DNA dissolved in 5 μL buffer.

2.3. Statistical Analysis

For all analyses, we set statistical significance at a *p*-value < 0.05. In this study, we analyzed the allelic and genotypic distributions, Hardy–Weinberg equilibrium and pairwise linkage disequilibrium (LD) by SHEsis (http://analysis.bio-x.cn/myAnalysis.php) (accessed on 24 February 2022). Comparisons of allele and genotype frequencies between cases and controls were performed through the Chi-square test. Linkage disequilibrium of the four pairs of SNPs within *NGFR* was measured by standardized D′. The association between the candidate SNPs with the MDD/schizophrenia risk in five genetic models (codominant, dominant, recessive, over-dominant and log-additive models, respectively) was assessed by "genetic" packages in R software (version 4.1.3., accessed on 10 March 2022, https://www.r-project.org/) with the odds ratios (ORs) and their 95% confidence intervals (CIs). Pairwise linkage disequilibrium (LD) and haplotype constitution were tested using Haploview 4.2 (accessed on 15 March 2022).

2.4. Behavioral Tests on Ngfr-Knock-Down Mice

2.4.1. Construction of *Ngfr*-Interference Plasmid

We designed and selected three shRNA with the highest scores on the Invitrogen website (https://rnaidesigner.thermofisher.com/) (accessed on 25 June 2022) for mice *Ngfr*-interference plasmid construction and subsequent interference efficiency verification. The plasmid skeleton was obtained from Li Weidong's research group at Shanghai Jiao Tong University, and the restriction enzyme cutting sites used were BamHI and XbaI. The control plasmid and the inserted control sequence were proved not to interfere with any known genes after whole-genome alignment in mice. The DNA sequence was synthesized by Shanghai Jieli Bio-Technology Co., Ltd. ShRNA and control sequences are shown in Table 2 (5′-3′ direction).

Table 2. *NGFR* shRNA and control sequences.

Ngfr-sh1-F.	GATCCCGGGCCTTGTGGCCTATATTCTCAAGAGAAATATAGGCCACAAGGCCCTTTTTT
Ngfr-sh1-R	CTAGAAAAAAGGGCCTTGTGGCCTATATTTCTCTTGAGAATATAGGCCACAAGGCCCGG
Ngfr-sh2-F	GATCCCGGTCGAGAAGCTGCTCAATTTCAAGAGAATTGAGCAGCTTCTCGACCTTTTTT
Ngfr-sh2-R	CTAGAAAAAAGGTCGAGAAGCTGCTCAATTCTCTTGAAATTGAGCAGCTTCTCGACCGG
Ngfr-sh3-F	GATCCCGCATCCAGAGAGCTGACATTTCAAGAGAATGTCAGCTCTCTGGATGCTTTTTT
Ngfr-sh3-R	CTAGAAAAAAGCATCCAGAGAGCTGACATTCTCTTGAAATGTCAGCTCTCTGGATGCGG
shCON-F	GATCCCGTTCTCCGAACGTGTCACGTTTCAAGAGATGCACTGTGCAAGCCTCTTTTTT
shCON-R	CTAGAAAAAAGAGGCTTGCACAGTGCATCTCTTGAAACGTGACACGTTCGGAGAACGG

2.4.2. Construction of *Ngfr*-Overexpressed Vector

We used a vector, pCMV6-*Ngfr*-cDNA, containing a full-length mice *Ngfr* cDNA clone purchased from OriGene Technologies, Rockville, MD, USA. After plasmid transformation, extraction and sequencing, we constructed the *Ngfr* cDNA fragment into a pCAGIG plasmid. The restriction enzyme cutting sites we used were *Eco*RI and *Not*I.

2.4.3. Interference Efficiency Verification

We tested the interference efficiency of shRNA in vitro through co-transfecting the *Ngfr*-overexpressed pCAGIG plasmid with *Ngfr*-interference plasmid/control plasmid into HEK293T cells via Lipofectamine® 2000 (Invitrogen, Carlsbad, CA, USA) liposome. Forty-eight hours later, a Western blot was performed to measure the expression of *NGFR* protein (Western blotting system: Mini-protean® Tetra electrophoresis apparatus and matching transfer tank, Bio-Rad (Laboratories, Hercules, CA, USA); Simon automatic Western blot analysis system, ProteinSimple (San Jose, CA, USA). The cells were cultured in 6-well plates, and the transfection and following protein Western blot was repeated in 1 well for each group. The antibody we used was the *NGFR* antibody (Abcam, Cambridge, MA, USA), and the internal reference was β-actin.

2.4.4. Stereotaxic Surgeries and Microinjection

According to the interference efficiency verification experiment, *Ngfr*-sh2 decreased *NGFR* protein expression most effectively. Thereby, we used the virus vector, Hu6-MCS-CMV-EGFP (GV115), which encodes GFP as well, expressing an effective sequence of GGTCGAGAAGCTGCTCAAT, which was generated by the Genechem Co., Ltd. (Shanghai, China). Identified probe locations for animals used in assays were positioned to the dentate gyrus (DG) of the mouse brain hippocampus, according to its relative location with bregma and lambda on the mouse brain atlas by the stereo locator. The injections were administered in a bio-safety cabinet in a specific room.

Experimental equipment: 1 mL syringe, 10 mg/mL sodium pentobarbiturate, electronic balance, alcohol cotton, cotton swab, surgical scissors, forceps, skull drill (STRONG90), microinjector (Precision Instruments, Sarasota, FL, USA), 3% hydrogen peroxide, 75% alcohol, sterilizer PBS, wound stapler, animal antibiotics, electric blanket, etc.

We performed the intraperitoneal injection in the biosafety cabinet in a dedicated room (10 μL/g). We cut the hair on the top of the head of the mice under deep anesthesia and hung their upper jaw teeth on the positioning rod, which was fixed on the positioning instrument. Then the two sides of the positioning rod were pressed against the skull depression above the mouse's ear, and the three positioning rods were adjusted to fix the mouse's head. We cut a small hole lengthwise in the center of the mouse's head to expose the front funnel and the herringbone point and adjusted the height of the positioning rod to make the anterior fontanelle and the herringbone point consistent. In order to locate the dentate gyrus, we fixed two points with dye at 2 mm backward of anterior fontanelle, 1.6 mm to the left and the right, respectively, and drilled holes with skull surgery. We absorbed 2 μL PBS, 1 μL air and 4 μL virus successively and injected 2 μL virus into each hole at a rate of 500 nL/min. After 3 minutes of setting, we slowly withdraw the needle.

2.4.5. Behavioral Tests

Behavioral tests were performed after 2 weeks of recovery. After verifying by perfusion, we took the following behavioral tests on knockdown mice and controls in order, including an elevated plus maze, open field, novel object recognition, object–place recognition, social test, forced swimming, fear conditioning test and prepulse inhibition (PPI). The images were recorded with a video camera (JVC, TK-C9201EC), and behavior analysis was performed through EthoVision XT software (Version 8.0, Noldus Information Technology, Wageningen, The Netherlands). Data were collected from at least three independent experiments. *T*-tests and One-Way ANOVA for independent samples were performed.

In social tests, four juvenile mice (4–5 weeks) were prepared as social mice. The experimental device was a 3-chambered box system separated by 2 clapboards made of transparent acrylic materials, each with a small hole allowing the mice to enter and exit freely. A metal mesh cover was placed on each side chamber. The mice were placed in the middle chamber and allowed to move freely and adapt for 10 min. The mice were then removed and placed in a temporary empty chamber. A social mouse was then placed in a metal mesh chamber on one side, and the experimental mice were placed in the middle chamber. The activities of the mice were recorded within 5 min. Afterward, the experimental mice and social mice were taken out for the next round of experiments. We used the EthoVision XT software to calculate the time of the mice staying in the social mice chamber and the empty metal mesh chamber and the time of the mice exploring and smelling the social mice and the empty metal mesh chamber.

3. Results

3.1. Association-Study Analysis of the NGFR SNPs with Schizophrenia

We investigated four *NGFR* SNPs (rs1035050, rs575791, rs2072446, rs11466162) in 1010 patients with schizophrenia and 1034 healthy controls. The distributions of the allele and genotype frequencies of the SNPs in *NGFR* are shown in Table 3. The genotype distributions of all four *NGFR* SNPs conformed to the Hardy−Weinberg equilibrium (HWE, *p*-value > 0.05) in the control group, and only two of them (rs2072446 and rs11466162) appeared to deviate from HWE in the case group ($\chi^2 < 8$).

Table 3. Allele and genotype distribution of *NGFR* gene polymorphisms between SCZ patients and healthy participants.

SNP		Genotype Frequency			χ^2	*p* Value [a]	Allele Frequency		χ^2	*p* Value [a]	Odds Ratio (95%CI)	HWE *p* Value
		CC	CT	TT			C	T				
rs1035050	SCZ	570 (0.576)	362 (0.366)	57 (0.058)	1.01	0.605	1502 (0.759)	476 (0.241)	0.99	0.320	0.929 (0.803–1.07)	0.96
	Control	613 (0.596)	364 (0.354)	52 (0.051)			1590 (0.773)	468 (0.227)				0.83
		AA	AG	GG			A	G				
rs575791	SCZ	609 (0.604)	352 (0.349)	48 (0.048)	1.46	0.482	1570 (0.778)	448 (0.222)	1.40	0.237	0.914 (0.787–1.06)	0.75
	Control	645 (0.626)	344 (0.334)	41 (0.040)			1634 (0.793)	426 (0.207)				0.56
		CC	CT	TT			C	T				
rs2072446	SCZ	808 (0.813)	169 (0.170)	17 (0.017)	8.23	**0.016**	1785 (0.898)	203 (0.102)	4.17	0.041	1.23 (1.01–1.49)	0.022
	Control	784 (0.768)	224 (0.219)	13 (0.013)			1792 (0.878)	250 (0.122)				0.50
		AA	AG	GG			A	G				
rs11466162	SCZ	13 (0.013)	143 (0.142)	853 (0.845)	11.50	**0.0032**	169 (0.084)	1849 (0.916)	3.92	0.048	0.807 (0.652–0.998)	0.015
	Control	6 (0.006)	197 (0.192)	824 (0.802)			209 (0.102)	1845 (0.898)				0.11

[a] Pearson's *p*-value, significant *p* (<0.05) values are in bold. HWE: Hardy–Weinberg Equilibrium. SCZ: schizophrenia.

The results of association analyses showed that the difference in genotype frequency of rs2072446 and rs11466162 was significantly different between the two groups (*p*-value = 0.016, *p*-value = 0.0032). C allele of rs2072446 was associated with an increased schizophrenia risk (OR = 1.23, *p*-value = 0.041) and A allele of rs11466162 was associated with a decreased schizophrenia risk (OR = 0.807, *p*-value = 0.048). No significant linkage disequilibrium between the SNPs and no haplotype was found.

3.2. Correlation between NGFR Gene Polymorphisms and MDD

The distributions of the allele and genotype frequencies of four SNPs in *NGFR* among controls and MDD patients are shown in Table 4. We performed the association-study analysis. The genotype distributions of all four *NGFR* SNPs conformed to HWE in both the case and control groups. There was no statistically significant difference, and no haplotype was found to be significantly associated with MDD susceptibility in further analysis.

Table 4. Allele and genotype distribution of *NGFR* gene polymorphisms between MDD patients and healthy participants.

Scheme 2		Genotype Frequency			χ^2	*p* Value [a]	Allele Frequency		χ^2	*p* Value [a]	Odds Ratio (95%CI)	H-W *p* Value
		CC	CT	TT			C	T				
rs1035050	MDD	361 (0.605)	201 (0.337)	35 (0.059)	0.83	0.661	923 (0.773)	271 (0.227)	0.00082	0.977	1.00 (0.846–1.19)	0.32
	Control	613 (0.596)	364 (0.354)	52 (0.051)			1590 (0.773)	468 (0.227)				0.83
		AA	AG	GG			A	G				
rs575791	MDD	349 (0.589)	216 (0.364)	28 (0.047)	2.36	0.308	914 (0.771)	272 (0.229)	2.27	0.132	0.876 (0.737–1.04)	0.46
	Control	645 (0.626)	344 (0.334)	41 (0.040)			1634 (0.793)	426 (0.207)				0.56
		CC	CT	TT			C	T				
rs2072446	MDD	479 (0.800)	114 (0.190)	6 (0.010)	2.26	0.324	1072 (0.895)	126 (0.105)	2.19	0.139	1.19 (0.946–1.49)	0.79
	Control	784 (0.768)	224 (0.219)	13 (0.013)			1792 (0.878)	250 (0.122)				0.50
		AA	AG	GG			A	G				
rs11466162	MDD	7 (0.012)	107 (0.182)	475 (0.806)	1.91	0.3844	121 (0.103)	1057 (0.897)	0.0076	0.931	1.01 (0.798–1.28)	0.73
	Control	6 (0.006)	197 (0.192)	824 (0.802)			209 (0.102)	1845 (0.898)				0.11

[a] Pearson's *p*-value. HWE: Hardy–Weinberg Equilibrium. MDD: major depressive disorder.

3.3. Ngfr-Knockdown Mice Showing a Trend of Social Avoidance

We designed three shRNAs (sh1, sh2, sh3) and a control plasmid co-transfecting 293T cells with a pCAGIG-*Ngfr*-cDNA plasmid for 48 h, respectively. The expression level of *NGFR* protein is shown in Figure 1. As shown in Figure 2, in the interference efficiency verification system, both sh2 and sh3 can knock down the *NGFR* expression level, but sh2 shows a higher efficiency. Therefore, we chose sh2 for lentivirus packaging entrusting Shanghai Jikeiyin Chemical Technology Co., Ltd (Shanghai, China). We injected the virus into the hippocampal dentate gyrus region of the mouse brain. After 2 weeks, we took samples, extracted proteins from the hippocampal tissues of the mice, and performed Western blot detection (*NGFR* antibody: Cell Signaling Technology #8238S). As shown in Figure 3, injection of Lenti-*Ngfr*-sh2 virus successfully down-regulated *NGFR* in mouse hippocampus. We fixed mouse brain tissue by perfusion, performed frozen sections, and observed fluorescence under a microscope to confirm that the virus had been properly injected into the dentate gyrus region of the hippocampus (Figure 4).

pUEG-NGFR-shCON   pUEG-NGFR-sh2

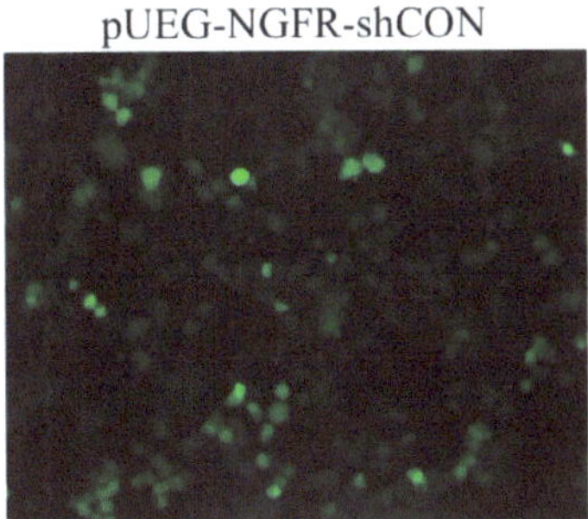
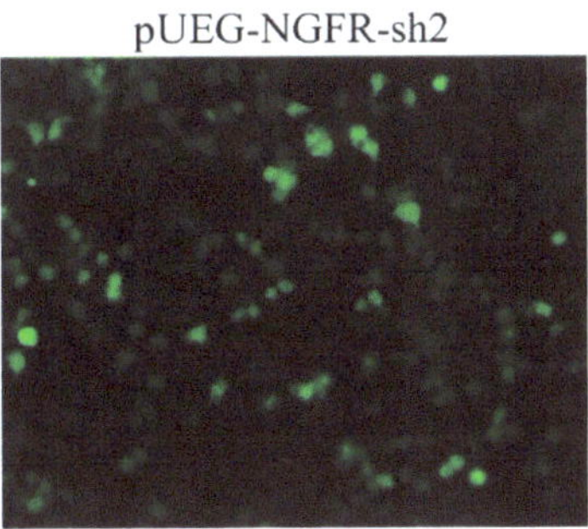

Figure 1. GFP expressed by the transfected cells. The 293T cells were co-transfected with shRNA plasmid and pCAGIG-*Ngfr*-cDNA.

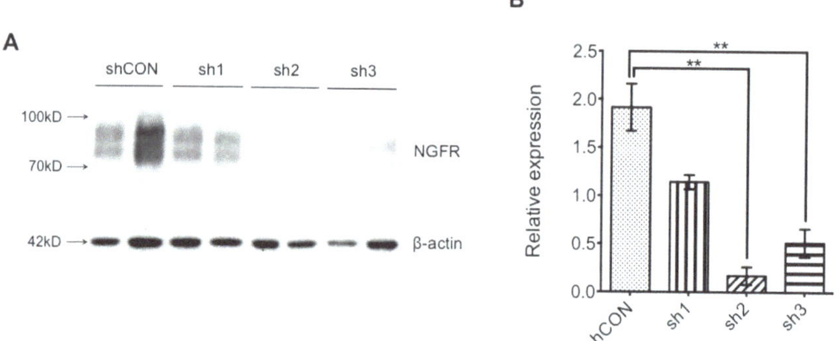

Figure 2. Down-regulation of *NGFR* in 293T cells. (**A**) Immunoblotting of protein extracted from 293T cells co-transfected with shRNA plasmid and pCAGIG-*Ngfr*-cDNA. (**B**) One-way ANOVA of the protein levels after normalization to β-actin; data showed as means ± SEM, ** $p < 0.01$.

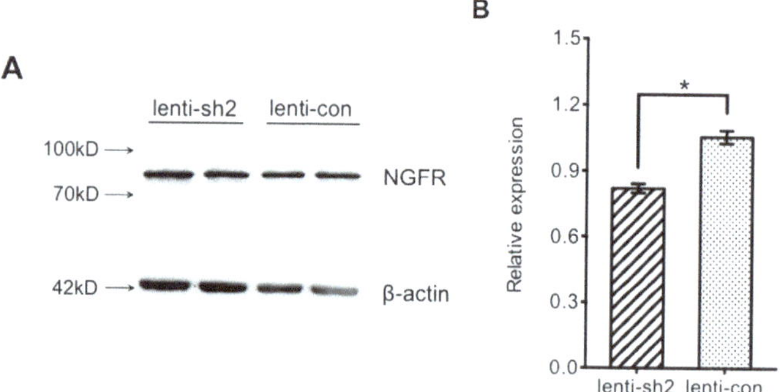

Figure 3. Down-regulation of *NGFR* in mouse hippocampus. (**A**) Lenti-*NGFR*-sh2 or control virus was injected into the dentate gyrus of adult mice. Two weeks after injection, extracts from hippocampus were immunoblotted. (**B**) One-way ANOVA of the protein levels after normalization to β-actin, data showed as means ± SEM, * $p < 0.05$.

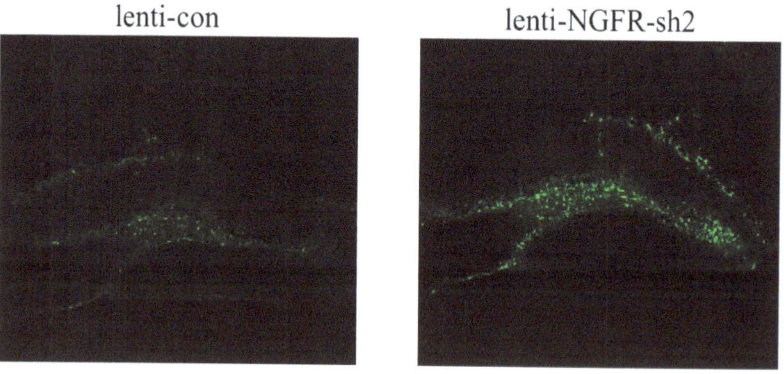

Figure 4. Fluorescence imaging showing the GFP expressed by infected cells in the dentate gyrus of adult mice.

In this study, 34 10-week-old C57BL/6 mice were injected with the virus in the hippocampal dentate gyrus. Seventeen mice in the experimental group were injected with Lenti-*Ngfr*- sh2 virus, and seventeen mice in the fossa control group were injected with the Lenti-con virus. All experiments and statistical analyses were conducted in a double-blind manner. In the eight behavioral tests we conducted, including the elevated cross maze experiment, open field experiment, new object recognition experiment, new position recognition experiment, social experiment, forced swimming experiment, conditioned fear experiment, and prepulse suppression experiment, we found significant differences in social behaviors between the experimental and control groups. As shown in Figure 5, we counted the time that mice stayed in the social mouse chamber and the empty chamber, respectively, during the 5-min test period; both groups of mice showed a longer time in the mouse chamber. Comparing the percentages of time spent in the mouse chamber between the two groups, we found that mice in the *Ngfr* shRNA knockdown group spent less time interacting with social mice, showing a trend of social avoidance (Figure 5). No significant differences in other behavioral phenotypes were found.

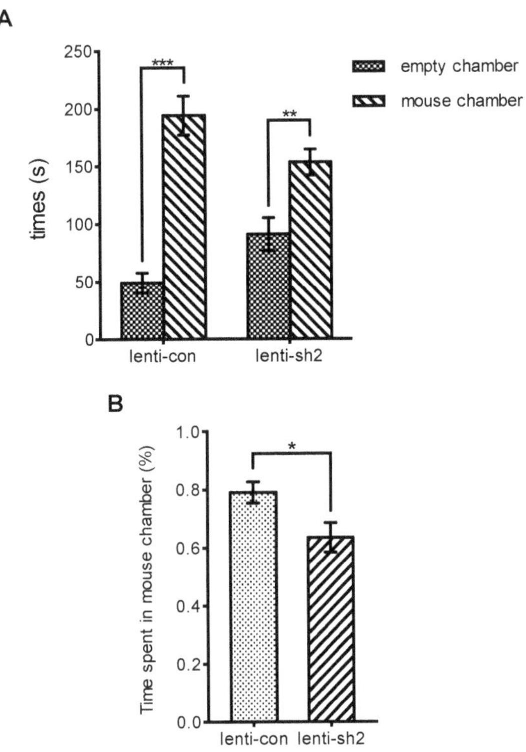

Figure 5. Results of the sociability test. (**A**) Time spent in the empty chamber and the mouse chamber for each mouse. (**B**) Percent of time spent in the mouse chamber was significantly lower for the shRNA mice than control. Results are presented as means $\pm$ SEM. *** $p < 0.001$, ** $p < 0.01$, * $p < 0.05$, one-way ANOVA.

4. Discussion

For the first time, our results indicate that the rs2072446 and rs11466162 polymorphisms in the *NGFR* gene are associated with schizophrenia in the Chinese Han population. We analyzed four *NGFR* SNPS (rS1035050, rs575791, rs2072446, rs11466162) in 1010 schizophrenic patients and 1034 control subjects and found that the genotype frequency of rs2072446 and rs11466162 were significantly different between the two groups (p-values were 0.016 and 0.0032, respectively). Our results suggest that the *NGFR* gene is associated with schizophrenia in Chinese Han populations.

The *NGFR* gene encodes the P75 neurotrophic factor receptor, which plays an important role in the regulation of neuronal growth, survival and plasticity. Previous research based on the Armenian Caucasian population [19] and Chinese population [20] revealed the level of *NGFR* in the plasma of schizophrenics was significantly lower than that of healthy controls, suggesting that *NGFR* is linked to the pathogenesis of schizophrenia, especially for the defect of synaptic plasticity. It was reported that the T allele of rs11466155 and T allele of rS2072446 might be risk factors, and the G allele of rs734194 may be a protective factor for schizophrenia in the Armenian Caucasian population [19]. Similar to these findings, in this study, we demonstrated that the genotype frequency distribution of rs2072446 and rs11466162 was significantly different between groups. T allele of rs2072446 may cause the conversion of serine, the 205th amino acid, to leucine in the *NGFR* polypeptide chain. This site is located in a genetically conserved region rich in intracellular serine/threonine, which may be involved in the development of schizophrenia by altering the function of proteins and thus has similar genetic analysis results in different populations. Rs11466162 is located at the 3′UTR of *NGFR,* which may affect psychiatric disorders by cooperating with other SNPs to form undetected haplotypes or regulate protein expressions.

Two recent articles reported higher levels of proBDNF and *NGFR* in depressed patients [17,18]. In the Japanese population, the T allele of rs2072446 was found to be a protective factor for major depression and suicidal behavior [21]. However, in the following association study in the Chinese population, five SNPs, including rs2072446, were found not to be significantly associated with MDD [22]. Rs2072446 and haplotypes rs2072446-rs11466155-rs734194 on the *NGFR* gene was reported to be associated with the efficacy of SSRI antidepressants in Chinese populations [22]. In this research, we also included 610 MDD patients but disproved that the *NGFR* gene was significantly associated with MDD in this Chinese population. These inconsistent results may be due to high heterogeneity among MDD patients and the different ethnic groups we investigated.

Our study suggests that the effects of *NGFR* polymorphisms are not consistent with each other among psychiatric disorders, including schizophrenia, bipolar disorder and depression. As mentioned above, researchers found patients with schizophrenia and bipolar mania have lower levels of serum *NGFR* than healthy people, while MDD patients have higher results, which was led by *NGFR* polymorphisms [17–20]. It is possible that the *NGFR* gene is directly involved in one or some behavioral phenotypes that take different proportions in different populations with psychiatric disorders, resulting in varying protein levels.

p75NTR knockdown in the hippocampus may affect cell death, survival, and differentiation by interacting with all kinds of TRK receptors, sorting proteins and NOGO receptors. For example, as a receptor for proBDNF, *NGFR* regulates neuronal death by co-acting with coreceptor sorting proteins to activate apoptotic signals. It was reported that chronic stress results in activation of neurodegenerative signaling of proBDNF/*NGFR*/sorting proteins (elevated levels of proBDNF, *NGFR,* and sorting proteins) and reduction in TrkB (inhibited BDNF/TrkB cell survival signals) in the neocortex and hippocampus of rodents. The imbalance between these two opposing pathways may be involved in the pathogenesis of depression and neurodegeneration during stress [23]. Clinical studies on serum *NGFR* protein levels in patients also suggest the potential role of proBDNF/*NGFR* signaling in the pathogenesis of depression [18].

Present behavioral research on *Ngfr* is mainly based on two gene knockout m models, *Ngfr*$^{exon\ III-/-}$ (carry ing exon III-targeted mutations but still expressing the *NGFR* short protein isoform receptor via alternative splicing [27]) and *Ngfr*$^{exon\ IV-/-}$ (complete inactivation of *NGFR* [28]). Neither of them has shown consistent behavioral results in *Ngfr*-knockout mice [25,29]. This may be due to the differences in the category background, gender, age and experimental environment of the experimental mice [30]. For example, considering gender factors, Puschban et al. studied the behavioral phenotypes of male and female *Ngfr*$^{exon\ IV-/-}$ mice and found Sex-specific behavior differences in three behavior tests [28]. The adult lentivirus injection knockdown method used in this paper did not affect the normal *NGFR* protein expression from mouse embryo to adult development. In contrast with previous studies, after eight behavioral tests, we only found that *Ngfr*-knockdown mice showed signs of social avoidance. Due to the different genetic models, whether there is a contradiction still needs to be discussed in the future. The phenotype of social disorder is associated with autism, negative symptoms of schizophrenia, depression, etc. Previous behavioral studies of the *Ngfr* gene in mice did not find similar situations. The results of this work also provide new evidence for the function of this candidate gene.

There are two main limitations in our study. One is the relatively small sample size, which may be the reason for the failure to completely repeat the result of previous research. The other is that we only tested the behavioral phenotypes of knockdown adult male mice and further refined the study by considering differences brought by genders and upregulated *NGFR* protein expression is in need. In the future, more SNPS involved in psychiatric disorders need to be included in the research. Inconsistent results of serum *NGFR* levels suggest that the overexpress test is also in need of further studies. Overexpress polymorphisms of the *NGFR* gene can be a potential way.

5. Conclusions

In this study, we demonstrated a significant association between the *NGFR* gene (rs2072446 and rs11466162) and schizophrenia. We investigated the role of *NGFR* in the behavioral phenotype of adult mice by injecting lentivirus in the hippocampal DG region to knock down the protein level and found the abnormal behavior of social avoidance in adult mice, which provided evidence for the *NGFR* gene to be involved in the pathogenesis of psychiatric disorders probably through proBDNF/*NGFR* signals.

Author Contributions: Conceptualization, L.Z. and X.L.; methodology, B.H. and F.Y. (Fan Yuan); software, Y.S. and L.L.; validation, L.J., D.R. and F.Y. (Fengping Yang); formal analysis, L.Z. and B.H.; investigation, T.Y., C.C. and C.L.; resources, B.B., Z.Y. and S.Y.; data curation, Y.B.; writing—original draft preparation, B.H. and L.Z.; writing—review and editing, G.H., S.W., Y.S. and X.L.; visualization, L.Z.; supervision, L.H.; project administration, S.W.; funding acquisition, G.H. and S.W. All authors have read and agreed to the published version of the manuscript.

Funding: This project is supported by the National Key Research and Development Program (2016YFC0906400), Innovation Funding in Shanghai (20JC1418600, 18JC1413100), the National Nature Science Foundation of China (82071262, 81671326), Natural Science Foundation of Shanghai (20ZR1427200, 20511101900), Shanghai Municipal Science and Technology Major Project (2017SHZDZX01), the Shanghai Key Laboratory of Psychotic Disorders (13dz2260500), the Shanghai Leading Academic Discipline Project (B205), Major Project of Lishui City (2017ZDYF15).

Institutional Review Board Statement: Informed consent was obtained from all subjects involved in the study.

Informed Consent Statement: Ethics Committee of the Bio-X center, Shanghai Jiao Tong University (M16033, 2016). Animal Care and Ethics Committee of Shanghai Jiao Tong University, China (202006003, 2020).

Conflicts of Interest: The authors declare no conflict of interest.

References

1. Hallmayer, J.; Cleveland, S.; Torres, A.; Phillips, J.; Cohen, B.; Torigoe, T.; Miller, J.; Fedele, A.; Collins, J.; Smith, K. Genetic heritability and shared environmental factors among twin pairs with autism. *Arch. Gen. Psychiatry* **2011**, *68*, 1095–1102. [CrossRef]
2. Tsuang, M.T.; Bar, J.L.; Stone, W.S.; Faraone, S.V. Gene-environment interactions in mental disorders. *World Psychiatry* **2004**, *3*, 73. [PubMed]
3. Shih, R.A.; Belmonte, P.L.; Zandi, P.P. A review of the evidence from family, twin and adoption studies for a genetic contribution to adult psychiatric disorders. *Int. Rev. Psychiatry* **2004**, *16*, 260–283. [CrossRef] [PubMed]
4. Nyatega, C.O.; Qiang, L.; Adamu, M.J.; Younis, A.; Kawuwa, H.B. Altered Dynamic Functional Connectivity of Cuneus in Schizophrenia Patients: A Resting-State fMRI Study. *Appl. Sci.* **2021**, *11*, 11392. [CrossRef]
5. Rog, J.; Błażewicz, A.; Juchnowicz, D.; Ludwiczuk, A.; Stelmach, E.; Kozioł, M.; Karakula, M.; Niziński, P.; Karakula-Juchnowicz, H. The Role of GPR120 Receptor in Essential Fatty Acids Metabolism in Schizophrenia. *Biomedicines* **2020**, *8*, 243. [CrossRef]
6. Silveira, S.; Hecht, M.; Adli, M.; Voelkle, M.C.; Singer, T. Exploring the Structure and Interrelations of Time-Stable Psychological Resilience, Psychological Vulnerability, and Social Cohesion. *Front. Psychiatry* **2022**, *13*, 804763. [CrossRef] [PubMed]

7.   Tanaka, M.; Szabó, Á.; Spekker, E.; Polyák, H.; Tóth, F.; Vécsei, L. Mitochondrial Impairment: A Common Motif in Neuropsychiatric Presentation? The Link to the Tryptophan—Kynurenine Metabolic System. *Cells* **2022**, *11*, 2607. [CrossRef]
8.   Gaebler, A.J.; Finner-Prével, M.; Sudar, F.P.; Langer, F.H.; Keskin, F.; Gebel, A.; Zweerings, J.; Mathiak, K. The Interplay between Vitamin D, Exposure of Anticholinergic Antipsychotics and Cognition in Schizophrenia. *Biomedicines* **2022**, *10*, 1096. [CrossRef]
9.   Tanaka, M.; Spekker, E.; Szabó, Á.; Polyák, H.; Vécsei, L. Modelling the neurodevelopmental pathogenesis in neuropsychiatric disorders. Bioactive kynurenines and their analogues as neuroprotective agents—In celebration of 80th birthday of Professor Peter Riederer. *J. Neural Transm.* **2022**, *129*, 627–642. [CrossRef]
10.   Avan, R.; Sahebnasagh, A.; Hashemi, J.; Monajati, M.; Faramarzi, F.; Henney, N.C.; Montecucco, F.; Jamialahmadi, T.; Sahebkar, A. Update on Statin Treatment in Patients with Neuropsychiatric Disorders. *Life* **2021**, *11*, 1365. [CrossRef] [PubMed]
11.   Wiescholleck, V.; Manahan-Vaughan, D. Persistent deficits in hippocampal synaptic plasticity accompany losses of hippocampus-dependent memory in a rodent model of psychosis. *Front. Integr. Neurosci.* **2013**, *7*, 12. [CrossRef] [PubMed]
12.   McAllister, A.K.; Katz, L.C.; Lo, D.C. Neurotrophins and synaptic plasticity. *Annu. Rev. Neurosci.* **1999**, *22*, 295–318. [CrossRef] [PubMed]
13.   Kraemer, B.R.; Yoon, S.O.; Carter, B.D. The Biological Functions and Signaling Mechanisms of the p75 Neurotrophin Receptor. *Handb. Exp. Pharmacol.* **2014**, *220*, 121–164. [PubMed]
14.   Zhou, X.F.; Rush, R.A.; McLachlan, E.M. Differential expression of the p75 nerve growth factor receptor in glia and neurons of the rat dorsal root ganglia after peripheral nerve transection. *J. Neurosci.* **1996**, *16*, 2901–2911. [CrossRef] [PubMed]
15.   VanGuilder Starkey, H.D.; Sonntag, W.E.; Freeman, W.M. Increased hippocampal NgR1 signaling machinery in aged rats with deficits of spatial cognition. *Eur. J. Neurosci.* **2013**, *37*, 1643–1658. [CrossRef] [PubMed]
16.   Malik, S.C.; Sozmen, E.G.; Baeza-Raja, B.; Le Moan, N.; Akassoglou, K.; Schachtrup, C. In vivo functions of p75(NTR): Challenges and opportunities for an emerging therapeutic target. *Trends Pharmacol. Sci.* **2021**, *42*, 772–788. [CrossRef]
17.   Zhou, L.; Xiong, J.; Lim, Y.; Ruan, Y.; Huang, C.; Zhu, Y.; Zhong, J.H.; Xiao, Z.; Zhou, X.F. Upregulation of blood proBDNF and its receptors in major depression. *J. Affect. Disord.* **2013**, *150*, 776–784. [CrossRef] [PubMed]
18.   Jiang, H.; Chen, S.; Li, C.; Lu, N.; Yue, Y.; Yin, Y.; Zhang, Y.; Zhi, X.; Zhang, D.; Yuan, Y. The serum protein levels of the tPA-BDNF pathway are implicated in depression and antidepressant treatment. *Transl. Psychiatry* **2017**, *7*, e1079. [CrossRef] [PubMed]
19.   Zakharyan, R.; Atshemyan, S.; Gevorgyan, A.; Boyajyan, A. Nerve growth factor and its receptor in schizophrenia. *BBA Clin.* **2014**, *1*, 24–29. [CrossRef]
20.   Chen, S.; Jiang, H.; Liu, Y.; Hou, Z.; Yue, Y.; Zhang, Y.; Zhao, F.; Xu, Z.; Li, Y.; Mou, X.; et al. Combined serum levels of multiple proteins in tPA-BDNF pathway may aid the diagnosis of five mental disorders. *Sci. Rep.* **2017**, *7*, 6871. [CrossRef] [PubMed]
21.   Kunugi, H.; Hashimoto, R.; Yoshida, M.; Tatsumi, M.; Kamijima, K. A missense polymorphism (S205L) of the low-affinity neurotrophin receptor p75NTR gene is associated with depressive disorder and attempted suicide. *Am. J. Med. Genet. Part B Neuropsychiatr. Genet.* **2004**, *129*, 44–46. [CrossRef] [PubMed]
22.   Gau, Y.T.; Liou, Y.J.; Yu, Y.W.; Chen, T.J.; Lin, M.W.; Tsai, S.J.; Hong, C.J. Evidence for association between genetic variants of p75 neurotrophin receptor (p75NTR) gene and antidepressant treatment response in Chinese major depressive disorder. *Am. J. Med. Genet. Part B Neuropsychiatr. Genet.* **2008**, *147*, 594–599. [CrossRef] [PubMed]
23.   Bai, Y.Y.; Ruan, C.S.; Yang, C.R.; Li, J.Y.; Kang, Z.L.; Zhou, L.; Liu, D.; Zeng, Y.Q.; Wang, T.H.; Tian, C.F.; et al. ProBDNF Signaling Regulates Depression-Like Behaviors in Rodents under Chronic Stress. *Neuropsychopharmacology* **2016**, *41*, 2882–2892. [CrossRef] [PubMed]
24.   Lin, L.; Zhou, X.F.; Bobrovskaya, L. Blockage of p75 NTR ameliorates depressive-like behaviours of mice under chronic unpredictable mild stress. *Behav. Brain Res.* **2020**, *396*, 112905. [CrossRef] [PubMed]
25.   Catts, V.S.; Al-Menhali, N.; Burne, T.H.J.; Colditz, M.J.; Coulson, E.J. The p75 neurotrophin receptor regulates hippocampal neurogenesis and related behaviours. *Eur. J. Neurosci.* **2008**, *28*, 883–892. [CrossRef] [PubMed]
26.   Olsen, D.; Kaas, M.; Schwartz, O.; Nykjaer, A.; Glerup, S. Loss of BDNF or its receptors in three mouse models has unpredictable consequences for anxiety and fear acquisition. *Learn. Mem.* **2013**, *20*, 499–504. [CrossRef] [PubMed]
27.   Lee, K.-F.; Li, E.; Huber, L.J.; Landis, S.C.; Sharpe, A.H.; Chao, M.V.; Jaenisch, R. Targeted mutation of the gene encoding the low affinity NGF receptor p75 leads to deficits in the peripheral sensory nervous system. *Cell* **1992**, *69*, 737–749. [CrossRef]
28.   Puschban, Z.; Sah, A.; Grutsch, I.; Singewald, N.; Dechant, G. Reduced Anxiety-Like Behavior and Altered Hippocampal Morphology in Female p75NTR(exon IV−/−) Mice. *Front. Behav. Neurosci.* **2016**, *10*, 103. [CrossRef] [PubMed]
29.   Martinowich, K.; Schloesser, R.J.; Lu, Y.; Jimenez, D.V.; Paredes, D.; Greene, J.S.; Greig, N.H.; Manji, H.K.; Lu, B. Roles of p75(NTR), long-term depression, and cholinergic transmission in anxiety and acute stress coping. *Biol. Psychiatry* **2012**, *71*, 75–83. [CrossRef] [PubMed]
30.   An, X.L.; Zou, J.X.; Wu, R.Y.; Yang, Y.; Tai, F.D.; Zeng, S.Y.; Jia, R.; Zhang, X.; Liu, E.Q.; Broders, H. Strain and sex differences in anxiety-like and social behaviors in C57BL/6J and BALB/cJ mice. *Exp. Anim.* **2011**, *60*, 111–123. [CrossRef] [PubMed]

Article

Memory Enhancement with Kynurenic Acid and Its Mechanisms in Neurotransmission

Diána Martos [1], Bernadett Tuka [1], Masaru Tanaka [1], László Vécsei [1,2,*] and Gyula Telegdy [3]

[1] MTA-SZTE Neuroscience Research Group, Hungarian Academy of Sciences, University of Szeged (MTA-SZTE), Semmelweis u. 6, H-6725 Szeged, Hungary; martos.diana@med.u-szeged.hu (D.M.); tuka.bernadett@med.u-szeged.hu (B.T.); tanaka.masaru.1@med.u-szeged.hu (M.T.)

[2] Department of Neurology, Albert Szent-Györgyi Medical School, University of Szeged, Semmelweis u. 6, H-6725 Szeged, Hungary

[3] Department of Pathophysiology, Albert Szent-Györgyi Medical School, University of Szeged, Semmelweis u. 5, H-6725 Szeged, Hungary; telegdy.gyula@med.u-szeged.hu

* Correspondence: vecsei.laszlo@med.u-szeged.hu; Tel.: +36-62-342-361

Abstract: Kynurenic acid (KYNA) is an endogenous tryptophan (Trp) metabolite known to possess neuroprotective property. KYNA plays critical roles in nociception, neurodegeneration, and neuroinflammation. A lower level of KYNA is observed in patients with neurodegenerative diseases such as Alzheimer's and Parkinson's diseases or psychiatric disorders such as depression and autism spectrum disorders, whereas a higher level of KYNA is associated with the pathogenesis of schizophrenia. Little is known about the optimal concentration for neuroprotection and the threshold for neurotoxicity. In this study the effects of KYNA on memory functions were investigated by passive avoidance test in mice. Six different doses of KYNA were administered intracerebroventricularly to previously trained CFLP mice and they were observed for 24 h. High doses of KYNA (i.e., 20–40 µg/2 µL) significantly decreased the avoidance latency, whereas a low dose of KYNA (0.5 µg/2 µL) significantly elevated it compared with controls, suggesting that the low dose of KYNA enhanced memory function. Furthermore, six different receptor blockers were applied to reveal the mechanisms underlying the memory enhancement induced by KYNA. The series of tests revealed the possible involvement of the serotonergic, dopaminergic, α and β adrenergic, and opiate systems in the nootropic effect. This study confirmed that a low dose of KYNA improved a memory component of cognitive domain, which was mediated by, at least in part, four systems of neurotransmission in an animal model of learning and memory.

Keywords: tryptophan; kynurenine; kynurenic acid; passive avoidance; cognitive domain; memory; cognitive enhancer; neurotransmission; receptor blockers; translational

Citation: Martos, D.; Tuka, B.; Tanaka, M.; Vécsei, L.; Telegdy, G. Memory Enhancement with Kynurenic Acid and Its Mechanisms in Neurotransmission. *Biomedicines* **2022**, *10*, 849. https://doi.org/10.3390/biomedicines10040849

Academic Editor: Giuseppe Tringali

Received: 20 March 2022
Accepted: 2 April 2022
Published: 5 April 2022

Publisher's Note: MDPI stays neutral with regard to jurisdictional claims in published maps and institutional affiliations.

1. Introduction

Worldwide, around 50 million people suffer from major neurocognitive disorders. Alzheimer's disease (AD) represents 60–70 percent of cases, imposing a physical, psychological, social, and economic burden on the elderly, their families, caregivers, and society [1]. Patients who develop AD first demonstrate a subtle decline in memory and learning, followed by changes in executive cognitive function and in language and visuospatial processing; indeed, recent evidence suggests that impairments in the ability to process contextual information and in the regulation of responses to threat are related to structural and physiological alterations in the prefrontal cortex (PFC) and medial temporal lobe, addressing how this progressive brain deterioration can eventually cause patterns of cognitive dysfunctions that might be observed in patients with AD [2]. The cause of major neurocognitive disorders remains unknown, but it is considered to be caused by convergence of multifactorial factors including genetic, environmental, infectious, and nutritional components, together with lifestyle, among others [3,4]. There is no remedy for neurodegenerative diseases. Disease-modifying and symptom-relieving measures are mainstays of treatment. Thus, a tremendous effort has been made to identify pathomechanisms, discover interventional targets, and design novel pharmaceutical agents [5].

KYNA is a metabolite of the Trp-kynurenine (KYN) metabolic system, known to possess a neuroprotective property [6]. The neuroprotective activities are considered to be attributed to the antagonism of the excitatory amino acid receptors (EAARs) such as the N-methyl-D-aspartate (NMDA) receptor, the α-amino-3-hydroxy-5-methyl-4-isoxazole propionic acid (AMPA) receptor, and the kainic acid receptor [7–10]. Furthermore, KYNA acts as an agonist of the G-protein-coupled receptor 35 (GPR35) and the aryl hydrocarbon receptor (AHR) [11–14]. In addition, opioid receptors are presumed to be interacting partners with KYNA [15,16].

It was previously postulated that the main component of KYNA-induced inhibition in glutamatergic neurotransmission may attribute to noncompetitive inhibition of α7-nicotinic acetylcholine receptors at glutamatergic presynaptic axon terminals [17], thereby regulating the release of glutamate. However, these results could not be subsequently reproduced by four different and independent groups. Thus, it is still questionable that KYNA may affect glutamate release via the mechanism [18–22]. KYNA plays crucial roles in the regulation of the intracellular Ca^{2+} and mitochondrial dysfunction-induced neuronal cell death in conditions associated with excitotoxicity (Figure 1).

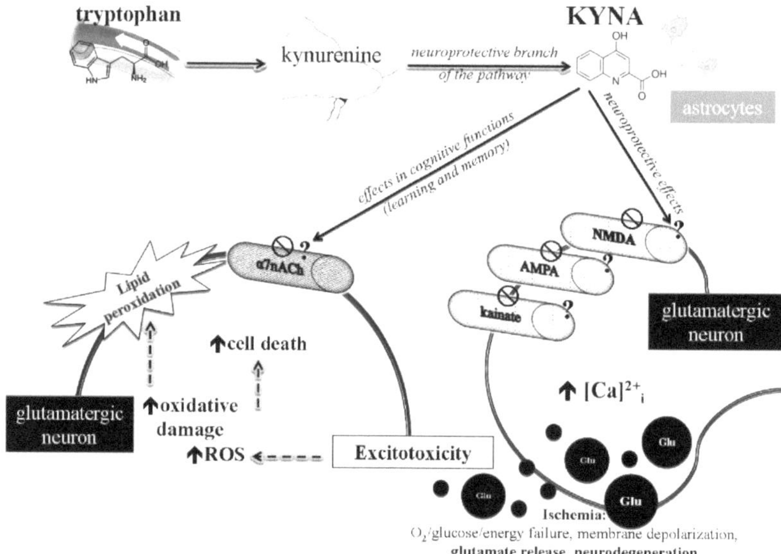

Figure 1. KYNA influences neuronal and glial glutamatergic neurotransmission.

Recently, KYNA and its novel pharmacokinetically favorable analogues demonstrated beneficial effects in animal models of neurologic diseases including pathologic pain sensation, migraine, ischemic stroke, and epilepsy, neurodegenerative diseases, and psychiatric disorder including depression, anxiety, and addiction [23–39]. Accordingly, neuroprotective KYN metabolites, their analogues, the inhibition of Trp-KYN enzymes that are responsible for production of toxic metabolites, their use for biomarkers, and their interaction with adjacent biosystems are under extensive research [40–48].

The beneficial effects were detected when these molecules were peripherally administered in an acute or semichronic manner with relatively high (millimolar) concentrations. Lower levels of KYNA were observed in patients with neurodegenerative diseases and psychiatric disorders [3,6,32,49]. Those illnesses are generally characterized by alterations in inflammatory mediators and mu-opioid receptor, and increased levels in neurotoxic

Try-KYN metabolites, which, furthermore, lead to changes in the amygdala [50]. However, manipulations to elevate KYNA levels have a potential risk of interfering with cognitive functions. Indeed, elevated levels of KYNA in the brain or its chronic application in higher doses are known to evoke cognitive impairment by inhibiting predominantly the glutamatergic system, a phenomenon having been linked to the pathophysiology of AD [51]. Furthermore, prenatal exposure of high levels of KYNA has also been experimentally shown to be associated with sustained cognitive deficits, with implications to schizophrenia [52,53]. Therefore, it is essential to identify the doses of KYNA and KYNA-related molecules to provide neuroprotection without any associated cognitive side effects.

In humans, KYNA is robustly synthesized in the endothelium and its serum levels correlate with homocysteine, a risk factor for cognitive decline; recent studies have suggested that a selective hippocampal increase in the KYNA level may be an important factor contributing to KYNA-related cognitive impairment. Identifying the mechanisms by which high KYNA levels in the hippocampal area may contribute to the deterioration of cognition would provide insight that might be used to manage inflammation-associated mental health disorders, including the discovery of new diagnostic and treatment therapies for depression. Recently, several studies have suggested the effectiveness of noninvasive brain simulation (NIBS) to interfere and modulate the abnormal activity of neural circuits including the amygdala-mPFC-hippocampus, involved in the acquisition and consolidation of memories, which are altered in psychiatric disorders, such as fear-related disorder, including anxiety disorder, phobias, posttraumatic stress disorder, and depression [54–56].

Our previous studies did not detect any behavior impairment of animals when they were treated intraperitoneally (i.p.) with millimolar doses of KYNA or its analogues [23,57]. The administration of KYNA and its analogues increased inducibility of long-term potentiation (LTP) in the CA1 region in rats, indicating better hippocampal function [58]. However, few data are available on the effects of a low dose KYNA. It was reported that KYNA has a dose-dependent dual action on AMPA receptors; the nanomolar and micromolar concentrations of KYNA could facilitate the responses of AMPA receptors via modulating their desensitization, whereas the millimolar doses of this compound antagonized these receptors [59].

It was demonstrated that KYNA was able to reduce the amplitudes of the field excitatory postsynaptic potentials (EPSPs) in hippocampal slices of young rats at micromolar concentrations, whereas the nanomolar concentrations evoked stimulation. Therefore, KYNA as a 'Janus-faced' molecule may display different effects according to its concentration by acting on different receptors and through mechanisms [60]. A lower endogenous formation of KYNA induces positive effects in cognition. Indeed, the role of the kynurenine aminotransferase II (KAT II), an enzyme responsible for the endogenous KYNA synthesis in the human brain, has been recently emphasized in the mechanisms of memory; activities of KAT I and II showed age-dependent increase with an exception for KAT II in the frontal cortex, which could be related to functional alterations in the PFC reported in psychiatric and brain-damaged patients' memory and learning abilities. Furthermore, recent studies revealed that naturally occurring bilateral lesions in the human ventromedial PFC compromise the capacity of associative learning [61–63], suggesting that PFC dysfunctions cause impairment of aversive learning and emotional memory circuits, which might be transversal across many psychiatric disorders in humans [64]. Pharmacological inhibition or genetic ablation of KAT II reduced KYNA levels in the brain and improved the performance in working/spatial memory and sustained attention tasks in different animal models [65–67]. The inhibition of KAT II, with a subsequent reduction in an endogenous KYNA level, restores normal cognitive function; thus, a manipulation of KYNA levels may be a promising therapeutic target in cognitive impairment associated with elevated concentrations of KYNA in the brain.

2. Materials and Methods

2.1. Experimental Animals and Ethics Statement

All animal experiments complied with the principles of animal care outlined in the instructions of the Ethical Committee for the Protection of Animals in Research of the University of Szeged (Szeged, Hungary), which specifically approved this study (XXIV/352/2012) and the protocol for animal care approved both by the Hungarian Health Committee (40/2013 (II.14.)) and by the European Communities Council Directive (2010/63/EU). CFLP male mice (body weight 25–28 g) were used. The animals were kept and handled during the experiments in accordance with the Regulations of the Faculty of Medicine, University of Szeged, Ethical Committee for the Protection of Animals in Research. Five animals per cage were housed under laboratory conditions with a 12 h dark/12 h light cycle in a temperature-controlled room (24–25 °C) in the Laboratory Animal House of the Department of Neurology in Szeged. Standard mouse chow and tap water were available ad libitum.

2.2. Surgery

The mice were anaesthetized with 40% Euthasol (in a dose of 60 mg/kg administered i.p.), and a plastic cannula was introduced into the lateral cerebral ventricle and fixed to the skull. The animals were allowed to recover for 5 days. The correct location of the cannula was controlled when dissecting the brain following the completion of the experiments. Only animals with the correct location of the cannula were used in the evaluation of the experiments. All experiments were performed during the morning period.

2.3. Materials

KYNA was purchased from Sigma-Aldrich Ltd. (Budapest, Hungary). The following receptor blockers were applied: cyproheptadine, a nonselective 5-HT2 serotonergic receptor antagonist, in a dose of 5 mg/kg (Tocris, Bristol, UK); phenoxybenzamine hydrochloride, a nonselective α-adrenergic receptor antagonist, in a dose of 2 mg/kg (Smith Kline and French, Hertz, UK); naloxone, a nonselective opioid receptor antagonist, in a dose of 0.3 mg/kg (Endo Lab Inc., Malvern, PA, USA), haloperidol, a D2, D3, D4 dopamine receptor antagonist, in a dose of 10 µg/kg (Richter Gedeon Plc., Budapest, Hungary), propranolol hydrochloride, a nonselective β-adrenergic receptor antagonist, in a dose of 2 mg/kg (ICI Ltd., Macclesfield, UK), atropine sulfate, the nonselective muscarinic acetylcholine receptor antagonist in a dose of 2 mg/kg (EGIS, Budapest, Hungary). The effective doses of the receptor antagonists have been determined based on the previous studies published and our previous work. The doses are calibrated in which no change in tested behaviors is observable [68–70]. KYNA was freshly dissolved in 0.9% aqueous saline solution and its pH was set to approximately 7.4 before use. The control animals received only 0.9% saline solution.

2.4. Experimental Groups and Treatments

Animals in the pilot study were divided into 4 groups (1 control and 3 for the different doses of KYNA applied). For the dose-effect examination, 7 groups were examined (1 control and 6 for the different doses of KYNA applied). Animals for further studies were divided into 24 groups (6 control, 6 KYNA, 6 for the different receptor blockers, and 6 combined groups) and the treatments were carried out following the training behavioral test (post-trial) on the second day, as presented in Table 1. KYNA was administered through a polyethylene tube with an external diameter of 1.09 mm (Becton Dickinson PE20) inserted stereotaxically into the right lateral brain ventricle in a volume of 2 µL i.c.v. The different receptor blockers were administered i.p. The dose of KYNA was selected based on the results of the dose–effect study (Figure 2); only the most effective dose was used during the different receptor blocker-testing experiments.

Table 1. Protocol of passive avoidance test and treatments.

	1th Day		2nd Day		3rd Day
Groups	**Trials**	**Trial**	**Post-Trial Treatments**		**Measure**
Control	3 × 2 min	Footshock in the dark part	i.p. saline	i.c.v. saline	300 s
KYNA	3 × 2 min	Footshock in the dark part	i.p. saline	i.c.v. KYNA	300 s
Receptor blockers	3 × 2 min	Footshock in the dark part	i.p. receptor blocker	i.c.v. saline	300 s
Combined	3 × 2 min	Footshock in the dark part	i.p. receptor blocker	i.c.v. KYNA	300 s

(30 min later)

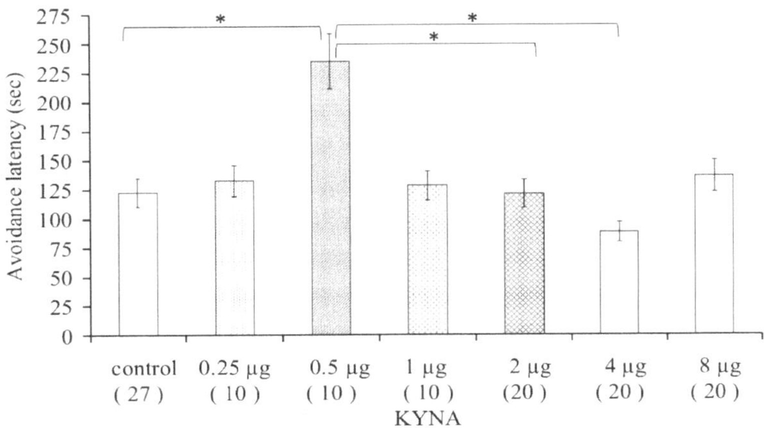

Figure 2. Dose–response examination of kynurenic acid in mice concerning the passive avoidance latency. * $p < 0.05$, the data in the plots are presented as means ± SEM. The exact subject numbers per group are indicated in brackets below the corresponding bar in the plots.

2.5. Behavioral Test: Passive Avoidance

The passive avoidance test was performed as previously described in Palotai et al. 2016 [71–74]. On the first day of testing, the mice were placed on an illuminated platform and were allowed to enter the dark compartment for 2 min. Since mice prefer the dark to the light, they normally entered within 5 s. This session was repeated 3 times with all animals, and an additional trial was performed on the following day. However, during this second trial, when the mice entered the dark part of the box, an unavoidable but not harmful mild electric footshock (0.75 mA, 2 s) was given through the grid floor. The gate between the light and dark compartments was closed and the animal could not escape. This learning trial was not repeated, but the mice were immediately removed from the apparatus and treated. The consolidation of passive avoidance behavior was tested 24 h later. Each animal was placed on the light platform and the latency to enter the dark compartment was measured up to a maximum of 300 secundum.

2.6. Statistical Analysis

Following the analyses of normality and variance, parametric tests were used in all cases of the receptor blocker measurements, but a nonparametric test was carried out in the KYNA dose–response investigation. The one-way analysis of variance (ANOVA) test was followed by Tukey post hoc test for multiple comparisons with unequal cell size. Kruskal–Wallis rank sum test was followed by pairwise comparisons using Tukey and Kramer (Nemenyi) test with Tukey-Dist approximation for independent samples.

Probability values (p) of less than 0.05 were considered significant. The data in the plots are presented as means $\pm$ SEM. The results (probability values) of treatments as presented in Table 2.

Table 2. The doses and binding affinity of receptor blockers and p-values.

Receptor Blockers (Doses)	Binding Affinity (Ki)	Control vs. Receptor Blocker	Control vs. KYNA	KYNA vs. Receptor Blocker	KYNA vs. Receptor Blocker Combined
Cyproheptadine (5 mg/kg)	1–9 nM [75]	$p < 0.384$	$p < 0.013$	$p < 0.001$	$p < 0.002$
Phenoxybenzamine (2 mg/kg)	108 nM [76]	$p < 0.739$	$p < 0.002$	$p < 0.001$	$p < 0.001$
Naloxone (0.3 mg/kg)	1 nM [77]	$p < 0.814$	$p < 0.022$	$p < 0.004$	$p < 0.006$
Haloperidol (10 µg/kg)	1.1 nM [78,79]	$p < 0.351$	$p < 0.014$	$p < 0.001$	$p < 0.003$
Propranolol (2 mg/kg)	8.7 nM [80]	$p < 0.711$	$p < 0.043$	$p < 0.003$	$p < 0.046$
Atropine (2 mg/kg)	0.5 nM [81]	$p < 0.998$	$p < 0.030$	$p < 0.041$	$p < 0.092$

3. Results

3.1. Passive Avoidance Tests

3.1.1. Pilot Study

To determine the most preferable effective dose of KYNA in the cognitive processes, 10, 20, and 40 µg of KYNA dissolved in 2 µL saline was administered i.c.v. to the mice ($n = 5$/group). In this preliminary experiment, we observed that 40 µg of KYNA substantially decreased the avoidance latency, whereas the lower doses did not significantly influence this parameter, as compared with the control animals. These results suggested that the positive cognitive effects of KYNA could be expected when administered in doses lower than 10 µg (data not shown).

3.1.2. Dose–Effect Examination

Male mice were used ($n = 10$–27/group) to determine the dose of KYNA that could significantly increase the avoidance latency. We investigated the effect of KYNA in doses of 0.25, 0.5, 1, 2, 4, and 8 µg in 2 µL saline. The 0.5 µg of KYNA prominently elevated the time until the animals entered the shock-associated dark part of the box, as compared with the control group ($p < 0.044$). We concluded that KYNA in a dose of 0.5 µg improved memory consolidation; therefore, this dose was used for further testing. Higher doses of KYNA were associated with significantly shorter avoidance latency as compared with the 0.5 µg KYNA-treated group (2 µg KYNA vs. 0.5 µg KYNA, $p < 0.013$; 4 µg KYNA vs. 0.5 µg KYNA, $p < 0.001$). Other doses did not significantly influence the avoidance behavior of mice (Figure 2).

3.1.3. Examination of Different Receptor Blockers

In all cases, the 0.5 µg/2 µL dose of KYNA significantly increased the avoidance latency of mice as compared with the healthy control group in the passive avoidance behavioral test. All groups of the tested receptor blockers were associated with significantly shorter avoidance latency as compared with the 0.5 µg KYNA-treated group. Furthermore, the groups receiving combined treatments (KYNA plus different receptor blocker compounds) were associated with significantly diminished time spent in the light part of the box, as compared with the group treated with 0.5 µg of KYNA alone, except for the one receiving atropine (Table 2, Figure 3). Compared to the control group, the applied receptor blockers did not influence remarkably the avoidance latency (in accordance with that previously reported in [71–74]), whereas the latency values observed in the combination groups did not differ significantly from those observed in the groups treated with the respective receptor blocker alone (Table 2, Figure 3).

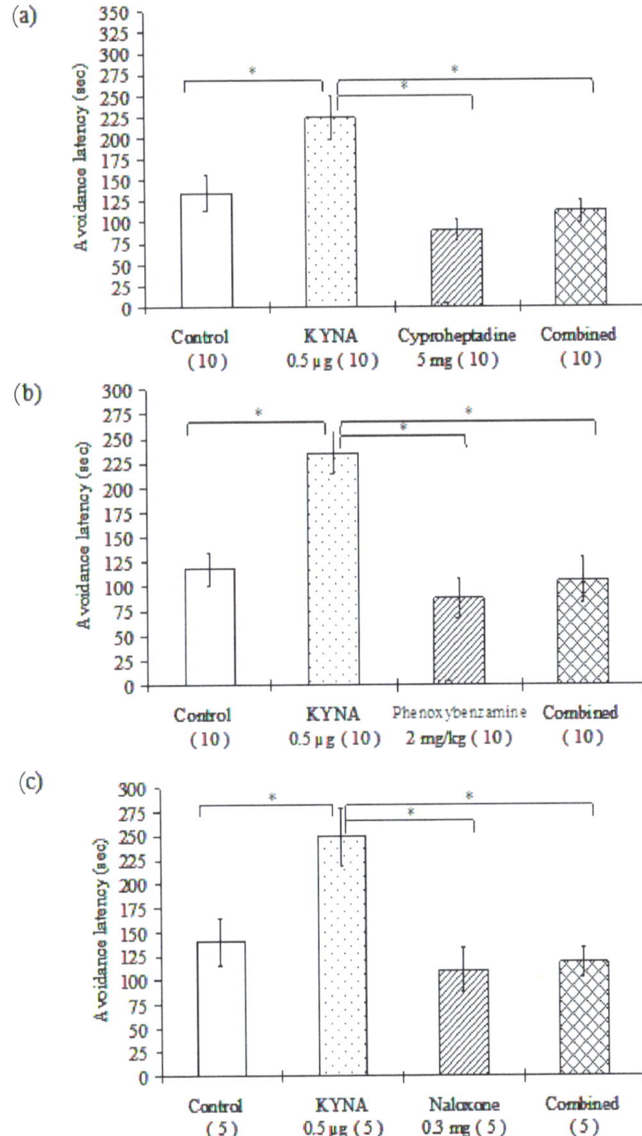

Figure 3. *Cont.*

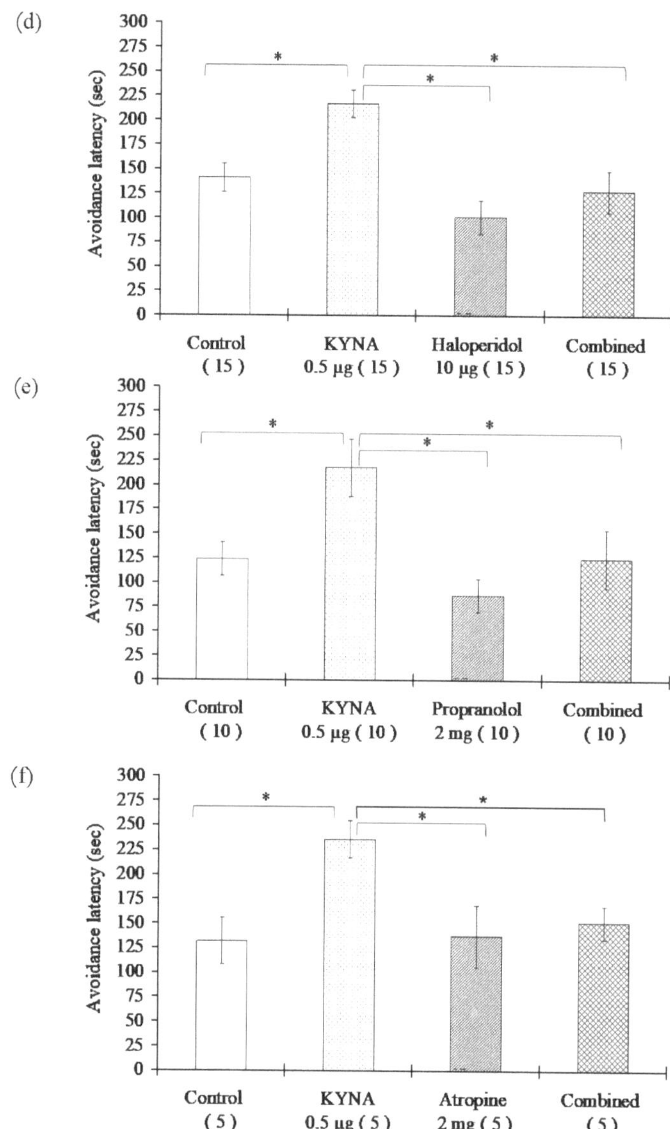

Figure 3. (**a–c**) The effects of different receptor blockers and their interaction with KYNA treatment in mice in the passive avoidance test: *Cyproheptadine*, a nonselective 5-HT2 serotonergic receptor antagonist (**a**); *phenoxybenzamine*, a nonselective α-adrenergic receptor antagonist (**b**); *naloxone*, a nonselective opioid receptor antagonist (**c**); * $p < 0.05$, the data in the plots are presented as means ⊥ SEM. The exact subject numbers per group are indicated in brackets below the corresponding bar in the plots. (**d–f**) The effects of different receptor blockers and their interaction with KYNA treatment in mice in the passive avoidance test. *Haloperidol*, a D2, D3, D4 dopamine receptor antagonist (**d**); *propranolol*, a nonselective β-adrenergic receptor antagonist (**e**); and *atropine*, a nonselective muscarinic acetylcholine receptor antagonist (**f**); * $p < 0.05$, the data in the plots are presented as means ± SEM. The exact subject numbers per group are indicated in brackets below the corresponding bar in the plots.

4. Discussion

Preclinical translational animal studies play a major role in neuroscience research to understand the roles of neuropeptides, neurohormones, and endogenous biomolecules in the normal function of human life such as cognition, emotion, and social interaction, and in pathological alterations developing into neurological and psychiatric disorders [82–97]. Various bioactive molecules are synthesized in the Try-KYN metabolic system. KYNA is generally described as a neuroprotective molecule, but it is also suspected of being a culprit of cognitive exacerbation in schizophrenia. Thus, the role of KYNA in cognitive function in the brain remains inconclusive [3,48].

This study attempts to determine whether KYNA influences the cognitive function positively in sufficiently low doses, to thus exhibit 'Janus-faced' property. The effects of low doses of exogenous KYNA administered by the intracerebroventricular (i.c.v.) route were examined in the passive avoidance cognitive test in mice, with special focus on memory consolidation, retention, and retrieval functions. The possible target(s) and transmitter system(s) involved in the observed effects of KYNA were evaluated by the application of different receptor blockers.

In a previous study, Chiamulera et al. detected that the KYNA treatment did not significantly change the avoidance latency in the passive avoidance tests in mice [98]. On the other hand, Potter et al. observed that the KAT II knockout mice performed better on the passive avoidance behavior test than their wild-type counterparts. The observation was linked to elevated levels of KYNA in the brain and cerebrospinal fluid patients with schizophrenia [67].

Our study confirms that KYNA influences the behavior of mice in the passive avoidance test. While high doses (i.e., 40 µg/2 µL) significantly decreased the memory performance of mice, a low dose of 0.5 µg/2 µL significantly enhanced the memory consolidation of mice by increasing in the avoidance latency.

To assess the mechanism of KYNA action in neurotransmission, we apply various receptor antagonists in combination (cyproheptadine for serotonergic neurotransmission; benzamine hydrochloride for α-adrenergic neurotransmission; naloxone for opioid neurotransmission; haloperidol for dopaminergic neurotransmission; propranolol hydrochloride for β-adrenergic neurotransmission; and atropine sulfate for muscarinic acetylcholine neurotransmission). The receptor blockers prevented the action of KYNA on passive avoidance learning, suggesting that the memory enhancement of KYNA is at least involved in serotoninergic, adrenergic, dopaminergic, and opiate systems, and implicating an indirect but functionally significant crosstalk between the kynurenine pathway and these systems of neurotransmission in the brain.

The glutamatergic synapse has decisive roles in cognitive brain functions (i.e., learning and memory); the role of NMDA receptors is important for triggering learning-related plasticity, whereas the AMPA receptors are essential for the expression of synaptic changes [58,99,100]. The activation of AMPA receptor-mediated neurotransmission ampakines was proposed as nootropics for mental disability, cognitive disturbances, and memory impairment [101].

Our presumption is that the applied doses of KYNA and its targets have crucial roles in the observed outcome effects. A shift in the balance of the Trp-KYN metabolic system toward the relative excess of neurotoxic molecules such as quinolinic acid (QUIN) has been implicated in the pathomechanisms of several neurological, neurodegenerative and psychiatric disorders, including epilepsy, Huntington's (HD), Parkinson's (PD), AD, and depressive disorder. Intervention to restore the balance or KYNA supplementation in the brain has been widely linked to neuroprotective actions in animal models of various diseases [102,103]. However, the influence of the KYN metabolites on certain diseases remains controversial. A potentially protective dose of KYNA may cause cognitive impairment via interfering with physiological NMDA- and AMPA-mediated currents [104,105]. In line with these findings, two concepts emerge regarding the role of an elevated KYNA levels in AD: a pathogenic factor in the development of memory impairment in AD and a compensatory mechanism against neurotoxicity [106]. Calibrating the equilibrium in the Trp-KYN metabolic system appears to be a complex maneuver.

In healthy subjects, the concentration of KYNA is in the nanomolar and micromolar ranges in the brain and the blood plasma, respectively; however, significant alterations were observed in the concentrations of KYN metabolites in neurodegenerative diseases associated with cognitive impairment [105,107]. Inhibitory effects of peripherally administered L-kynurenine (L-KYN) (single or daily repeated injections) were detected in rats in several behavioral tests; however, these treatments were applied in higher doses (100 and 200 mg/kg i.p.) [108]. These effects may be attributed to the inhibition of ionotropic glutamate receptors, for KYNA blocks both the AMPA and the kainate subtypes, and it has the highest dose-dependent affinity for the strychnine-insensitive glycine-binding site and the glutamate-binding site of NMDA receptors [7,109,110]. The antagonistic action can also induce neuroprotection via the prevention of glutamate excitotoxicity, predominantly through the inhibition of overactivated NMDA receptors localized extrasynaptically [105].

KYNA has dose-dependent dual effects on the AMPA receptors, for it exerts an inhibitory effect in the micromolar concentration range, whereas it evokes facilitation in low nanomolar concentrations [59,60]. The latter effect may be associated with a positive modulatory binding site at the AMPA receptors. The possible molecular mechanisms were detailed recently [111]. It can be hypothesized that the cognitive enhancing effect of KYNA may be attributed to this partial agonism at the AMPA receptors with a sufficient low dose of KYNA. It is suggested that a slight increase in the level of KYNA in the postsynaptic area may exert a preferential inhibition on the extrasynaptic NMDA receptors, thereby being able to protect against excitotoxic neuronal injury, while sparing or (in case of AMPA) even facilitating the physiological synaptic glutamate receptor-mediated currents without interfering with cognitive functions, or possibly even enhancing them (Figure 4).

The effects of cognitive enhancement by KYNA slightly resemble those of memantine, a molecule with a noncompetitive antagonistic low-to-moderate affinity to the NMDA receptors, which thereby has a modest beneficial effect on cognition [112–114]. Our results support that KYNA may have a cognitive enhancer effect when applied in low doses. However, KYNA is barely permeable to the blood–brain barrier (BBB) [115]. The injection procedure applied in our study is far from physiological circumstances, but at present this is the only method available to test the direct effects of KYNA. Our research group has attempted to package KYNA into core-shell nanoparticles to facilitate the penetration of KYNA through the BBB, thereby enhancing the concentration of KYNA in the brain [116]. The high doses of KYNA induced marked ataxia, stereotyped behavior, and muscular hypotonia in a dose-dependent manner. The effects can be alleviated by i.c.v. pretreatment with D-serine, a selective agonist at the strychnine-insensitive glycine binding site of the NMDA receptor complex [117].

L-KYN in combination with probenecid, an organic amino acid transporter inhibitor, improved the spatial memory in animal models of AD and PD [118,119]. The unwanted effects of KYNA and its analogues were tested in several behavioral tests such as spontaneous locomotor activity, working memory performance, and long-lasting, consolidated reference memory; however, the results showed that the higher concentration of KYNA in the brain via the administration of KYNA or its analogue does not cause a perturbation of working memory function or lead to impaired cognitive functions or any significant systemic side effect [23,57,120]. Additionally, an electrophysiological study revealed that one of the KYNA analogues did not decrease but rather increased the potentiation of field EPSPs. It can also be hypothesized that a partial agonistic effect of KYNA or its analogue on glutamate receptors accounts for the paradox effect [58,60]. There are data indicating a relationship between the adrenergic and KYN systems. Indeed, selective beta receptor agonists can increase the cortical endogenous level of KYNA in rat brain slices and mixed glial cultures, an effect that can be blocked by propranolol. This mechanism appears to be mediated by cyclic adenosine monophosphate- and protein kinase A-dependent processes [121].

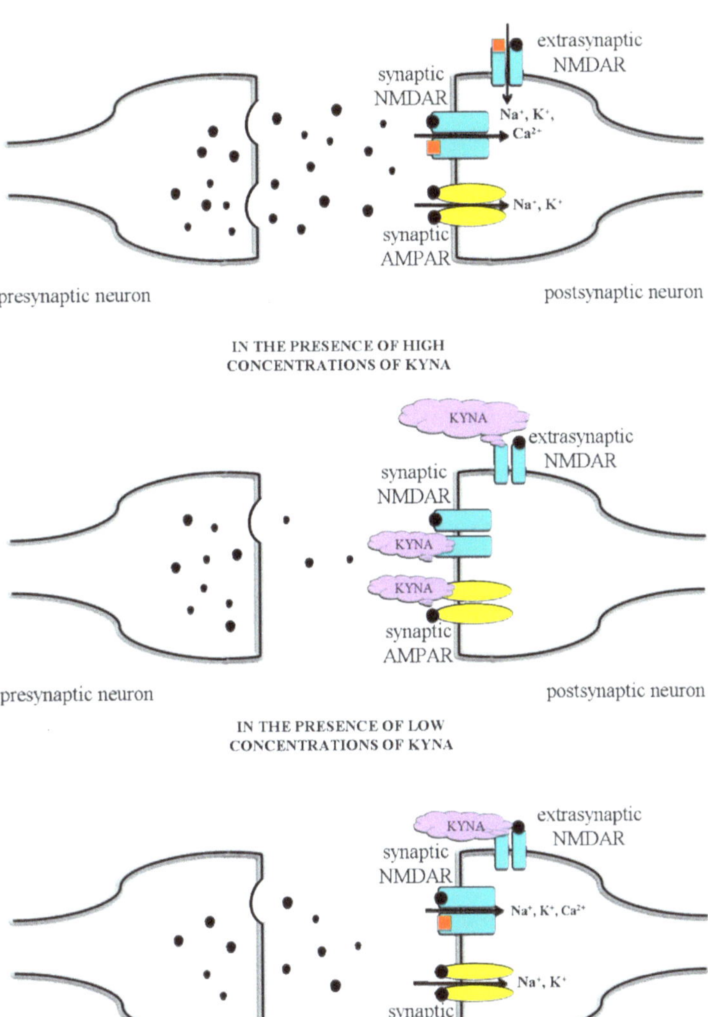

Figure 4. Hypothetical mechanisms, receptorial and current alterations in normal conditions of glutamatergic neurons and in the presence of KYNA in different dose. A slight increase in the level of KYNA in the postsynaptic area may exert a preferential inhibition on the extrasynaptic NMDA receptors, thereby being able to protect against excitotoxic neuronal injury.

Furthermore, the kynurenines and the dopaminergic systems are in a close relationship, for specific inhibition of KAT II markedly reduces the firing activity of dopaminergic neurons in the ventral tegmental area. The effect is proposed to be specifically carried out by NMDA-receptors and mediated indirectly via a γ-aminobutyric acidergic (GABA) disinhibition [122]. Trp is the common precursor for both serotonin and L-kynurenine. Thus, alteration in the activity of the rate-limiting step of the Trp-KYN metabolic system influences the serotonin pathway as well. This is suggested in the pathomechanisms of migraine, depression, and certain other psychiatric syndromes [123].

Finally, an indirect interaction may exist between the opioid and the KYN system. The activity of opioid receptor-mediated G-protein activity decreased after chronic systemic treatment with KYNA or its analogue in an animal study [16]. The widespread, complex molecular interactions of KYNA with different receptors may underlie its variable dose-dependent neuromodulatory effects and its significance in the processes of the central nervous system. It would be essential to unveil the effects of low doses of chronically administered KYNA by the systemic route. This would enable the identification the appropriate methods and doses that may be associated with both neuroprotective and cognitive enhancer effects without unwanted adverse effects.

5. Conclusions

Our results suggest that low doses of KYNA can facilitate learning and memory consolidation, as revealed by an experimental cognitive paradigm in healthy mice. Further, investigations are expected to reveal the potentially similar effects of low-dose KYNA in other memory tests, and longitudinal studies with extended follow-up are warranted to determine the effects of chronically administered KYNA in low doses. This approach may represent a potential therapeutic tool in neurodegenerative diseases and chronic conditions with associated cognitive impairments.

Author Contributions: Data collection, D.M. and B.T.; writing—original draft preparation, D.M. and B.T.; writing—review and editing, D.M., B.T. and M.T.; visualization, D.M. and B.T.; supervision, L.V. and G.T.; funding acquisition, L.V. All authors have read and agreed to the published version of the manuscript.

Funding: The current work was supported by National Scientific Research Fund OTKA138125, TUDFO/47138-1/2019-ITM, TKP2020 Thematic Excellence Programme 2020, University of Szeged Open Access Fund (5257).

Institutional Review Board Statement: Ethical Committee for the Protection of Animals in Research of the University of Szeged (Szeged, Hungary), which specifically approved this study (XXIV/352/2012), date of approval: 3 March 2012 and the protocol for animal care approved both by the Hungarian Health Committee (40/2013 (II.14.)) and by the European Communities Council Directive (2010/63/EU).

Informed Consent Statement: Not applicable.

Data Availability Statement: Not applicable.

Acknowledgments: The authors are grateful to Ilona Ungi for the invaluable help in the animal experiments.

Conflicts of Interest: The authors declare no conflict of interest.

Abbreviations

AD	Alzheimer's disease
AHR	aryl hydrocarbon receptor
AMPA	α-amino-3-hydroxy-5-methyl-4-isoxazole propionic acid
BBB	blood–brain barrier
EAARs	excitatory amino acid receptors
EPSPs	excitatory postsynaptic potentials
GABA	α-aminobutyric acid
GPR 35	G-protein-coupled receptor 35
HD	Huntington's disease
5-HT2	5-hydroxy-triptamin-2 receptor
KYNA	kynurenic acid

KYN	kynurenine
KAT II	kynurenine aminotransferase II enzyme
L-KYN	L-kynurenine
LTP	long-term potentiation
NMDA	N-methyl-D-aspartate receptor
PD	Parkinson's disease
PFC	prefrontal cortex
QUIN	quinolinic acid
Trp	Tryptophan

References

1. World Health Organization. Dementia, World Health Organization. 2020. Available online: https://www.who.int/news-room/fact-sheets/detail/dementia (accessed on 18 March 2021).
2. Battaglia, S.; Garofalo, S.; di Pellegrino, G. Context-dependent extinction of threat memories: Influences of healthy aging. *Sci. Rep.* **2018**, *8*, 12592. [CrossRef]
3. Tanaka, M.; Toldi, J.; Vécsei, L. Exploring the Etiological Links behind Neurodegenerative Diseases: Inflammatory Cytokines and Bioactive Kynurenines. *Int. J. Mol. Sci.* **2020**, *21*, 2431. [CrossRef] [PubMed]
4. Sini, P.; Dang, T.B.C.; Fais, M.; Galioto, M.; Padedda, B.M.; Lugliè, A.; Iaccarino, C.; Crosio, C. Cyanobacteria, Cyanotoxins, and Neurodegenerative Diseases: Dangerous Liaisons. *Int. J. Mol. Sci.* **2021**, *22*, 8726. [CrossRef]
5. Tanaka, M.; Török, N.; Vécsei, L. Novel Pharmaceutical Approaches in Dementia. In *NeuroPsychopharmacotherapy*; Riederer, P., Laux, G., Nagatsu, T., Le, W., Riederer, C., Eds.; Springer: Cham, Switzerland, 2021. [CrossRef]
6. Encyclopedia. The Tryptophan-Kynurenine Metabolic Pathway. Available online: https://encyclopedia.pub/8633 (accessed on 15 March 2022).
7. Kessler, M.; Terramani, T.; Lynch, G.; Baudry, M. A glycine site associated with N-methyl-D-aspartic acid receptors: Characterization and identification of a new class of antagonists. *J. Neurochem.* **1989**, *52*, 1319–1328. [CrossRef]
8. Birch, P.J.; Grossman, C.J.; Hayes, A.G. 6,7-Dinitro-quinoxaline-2,3-dion and 6-nitro,7-cyano-quinoxaline-2,3-dion antagonise responses to NMDA in the rat spinal cord via an action at the strychnine-insensitive glycine receptor. *Eur. J. Parmacol.* **1988**, *156*, 177–180. [CrossRef]
9. Perkins, M.N.; Stone, T.W. An iontophoretic investigation of the actions of convulsant kynurenines and their interaction with the endogenous excitant quinolinic acid. *Brain Res.* **1982**, *247*, 184–187. [CrossRef]
10. Alkondon, M.; Pereira, E.F.; Albuquerque, E.X. Endogenous activation of nAChRs and NMDA receptors contributes to the excitability of CA1 stratum radiatum interneurons in rat hippocampal slices: Effects of kynurenic acid. *Biochem. Pharmacol.* **2011**, *82*, 842–851. [CrossRef] [PubMed]
11. Cosi, C.; Mannaioni, G.; Cozzi, A.; Carla, V.; Sili, M.; Cavone, L.; Maratea, D.; Moroni, F. G-protein coupled receptor 35 (GPR35) activation and inflammatory pain: Studies on the antinociceptive effects of kynurenic acid and zaprinast. *Neuropharmacology* **2011**, *60*, 1227–1231. [CrossRef] [PubMed]
12. Wang, J.; Simonavicius, N.; Wu, X.; Swaminath, G.; Reagan, J.; Tian, H.; Ling, L. Kynurenic acid as a ligand for orphan G protein-coupled receptor GPR35. *J. Biol. Chem.* **2006**, *281*, 22021–22028. [CrossRef] [PubMed]
13. Moroni, F.; Cozzi, A.; Sili, M.; Mannaioni, G. Kynurenic acid: A metabolite with multiple actions and multiple targets in brain and periphery. *J. Neural Transm.* **2012**, *119*, 133–139. [CrossRef]
14. Stone, T.W.; Stoy, N.; Darlington, L.G. An expanding range of targets for kynurenine metabolites of tryptophan. *Trends Pharmacol. Sci.* **2013**, *34*, 136–143. [CrossRef] [PubMed]
15. Horvath, G.; Kekesi, G.; Tuboly, G.; Benedek, G. Antinociceptive interactions of triple and quadruple combinations of endogenous ligands at the spinal level. *Brain Res.* **2007**, *1155*, 42–48. [CrossRef]
16. Zador, F.; Samavati, R.; Szlavicz, E.; Tuka, B.; Bojnik, E.; Fulop, F.; Toldi, J.; Vecsei, L.; Borsodi, A. Inhibition of opioid receptor mediated G-protein activity after chronic administration of kynurenic acid and its derivative without direct binding to opioid receptors. *CNS Neurol. Dis. Drug Targets* **2014**, *13*, 1520–1529. [CrossRef] [PubMed]
17. Hilmas, C.; Pereira, E.F.; Alkondon, M.; Rassoulpour, A.; Schwarcz, R.; Albuquerque, E.X. The brain metabolite kynurenic acid inhibits alpha7 nicotinic receptor activity and increases non-alpha7 nicotinic receptor expression: Physiopathological implications. *J. Neurosci.* **2001**, *21*, 7463–7473. [CrossRef] [PubMed]
18. Zadori, D.; Veres, G.; Szalardy, L.; Klivenyi, P.; Fulop, F.; Toldi, J.; Vecsei, L. Inhibitors of the kynurenine pathway as neurotherapeutics: A patent review (2012–2015). *Expert Opin. Ther. Pat.* **2016**, *26*, 815–832. [CrossRef] [PubMed]
19. Stone, T.W. Kynurenic acid blocks nicotinic synaptic transmission to hippocampal interneurons in young rats. *Eur. J. Nneurosci.* **2007**, *25*, 2656–2665. [CrossRef] [PubMed]
20. Arnaiz-Cot, J.J.; Gonzalez, J.C.; Sobrado, M.; Baldelli, P.; Carbone, E.; Gandia, L.; Garcia, A.G.; Hernandez-Guijo, J.M. Allosteric modulation of alpha 7 nicotinic receptors selectively depolarizes hippocampal interneurons, enhancing spontaneous GABAergic transmission. *Eur. J. Neurosci.* **2008**, *27*, 1097–1110. [CrossRef] [PubMed]
21. Mok, M.H.; Fricker, A.C.; Weil, A.; Kew, J.N. Electrophysiological characterisation of the actions of kynurenic acid at ligand-gated ion channels. *Neuropharmacology* **2009**, *57*, 242–249. [CrossRef] [PubMed]

22. Dobelis, P.; Staley, K.J.; Cooper, D.C. Lack of modulation of nicotinic acetylcholine alpha-7 receptor currents by kynurenic acid in adult hippocampal interneurons. *PLoS ONE* **2012**, *7*, e41108. [CrossRef] [PubMed]

23. Nagy, K.; Plangar, I.; Tuka, B.; Gellert, L.; Varga, D.; Demeter, I.; Farkas, T.; Kis, Z.; Marosi, M.; Zadori, D.; et al. Synthesis and biological effects of some kynurenic acid analogs. *Bioorgan. Med. Chem.* **2011**, *19*, 7590–7596. [CrossRef]

24. Vamos, E.; Pardutz, A.; Varga, H.; Bohar, Z.; Tajti, J.; Fulop, F.; Toldi, J. Vecsei L: L-kynurenine combined with probenecid and the novel synthetic kynurenic acid derivative attenuate nitroglycerin-induced nNOS in the rat caudal trigeminal nucleus. *Neuropharmacology* **2009**, *57*, 425–429. [CrossRef]

25. Chauvel, V.; Vamos, E.; Pardutz, A.; Vecsei, L.; Schoenen, J.; Multon, S. Effect of systemic kynurenine on cortical spreading depression and its modulation by sex hormones in rat. *Exp. Neurol.* **2012**, *236*, 207–214. [CrossRef]

26. Tanaka, M.; Török, N.; Tóth, F.; Szabó, Á.; Vécsei, L. Co-Players in Chronic Pain: Neuroinflammation and the Tryptophan-Kynurenine Metabolic Pathway. *Biomedicines* **2021**, *9*, 897. [CrossRef]

27. Ciapała, K.; Mika, J.; Rojewska, E. The Kynurenine Pathway as a Potential Target for Neuropathic Pain Therapy Design: From Basic Research to Clinical Perspectives. *Int. J. Mol. Sci.* **2021**, *22*, 11055. [CrossRef]

28. Jovanovic, F.; Candido, K.D.; Knezevic, N.N. The Role of the Kynurenine Signaling Pathway in Different Chronic Pain Conditions and Potential Use of Therapeutic Agents. *Int. J. Mol. Sci.* **2020**, *21*, 6045. [CrossRef] [PubMed]

29. Spekker, E.; Tanaka, M.; Szabó, Á.; Vécsei, L. Neurogenic Inflammation: The Participant in Migraine and Recent Advancements in Translational Research. *Biomedicines* **2021**, *10*, 76. [CrossRef] [PubMed]

30. Fila, M.; Chojnacki, J.; Pawlowska, E.; Szczepanska, J.; Chojnacki, C.; Blasiak, J. Kynurenine Pathway of Tryptophan Metabolism in Migraine and Functional Gastrointestinal Disorders. *Int. J. Mol. Sci.* **2021**, *22*, 10134. [CrossRef] [PubMed]

31. Spekker, E.; Laborc, K.F.; Bohár, Z.; Nagy-Grócz, G.; Fejes-Szabó, A.; Szűcs, M.; Vécsei, L.; Párdutz, Á. Effect of dural inflammatory soup application on activation and sensitization markers in the caudal trigeminal nucleus of the rat and the modulatory effects of sumatriptan and kynurenic acid. *J. Headache Pain* **2021**, *22*, 17. [CrossRef]

32. Sas, K.; Robotka, H.; Rozsa, E.; Agoston, M.; Szenasi, G.; Gigler, G.; Marosi, M.; Kis, Z.; Farkas, T.; Vecsei, L.; et al. Kynurenine diminishes the ischemia-induced histological and electrophysiological deficits in the rat hippocampus. *Neurobiol. Dis.* **2008**, *32*, 302–308. [CrossRef] [PubMed]

33. Gellert, L.; Fuzik, J.; Goblos, A.; Sarkozi, K.; Marosi, M.; Kis, Z.; Farkas, T.; Szatmari, I.; Fulop, F.; Vecsei, L.; et al. Neuroprotection with a new kynurenic acid analog in the four-vessel occlusion model of ischemia. *Eur. J. Pharmacol.* **2001**, *667*, 182–187. [CrossRef] [PubMed]

34. Gellert, L.; Knapp, L.; Nemeth, K.; Heredi, J.; Varga, D.; Olah, G.; Kocsis, K.; Menyhart, A.; Kis, Z.; Farkas, T.; et al. Post-ischemic treatment with L-kynurenine sulfate exacerbates neuronal damage after transient middle cerebral artery occlusion. *Neuroscience* **2013**, *247*, 95–101. [CrossRef]

35. Demeter, I.; Nagy, K.; Gellert, L.; Vecsei, L.; Fulop, F.; Toldi, J. A novel kynurenic acid analog (SZR104) inhibits pentylenetetrazole-induced epileptiform seizures. An electrophysiological study: Special issue related to kynurenine. *J. Neural Transm.* **2012**, *119*, 151–154. [CrossRef] [PubMed]

36. Zadori, D.; Nyiri, G.; Szonyi, A.; Szatmari, I.; Fulop, F.; Toldi, J.; Freund, T.F.; Vecsei, L.; Klivenyi, P. Neuroprotective effects of a novel kynurenic acid analogue in a transgenic mouse model of Huntington's disease. *J. Neural Transm.* **2011**, *118*, 865–875. [CrossRef] [PubMed]

37. Hartai, Z.; Klivenyi, P.; Janaky, T.; Penke, B.; Dux, L.; Vecsei, L. Kynurenine metabolism in plasma and in red blood cells in Parkinson's disease. *J. Neurol. Sci.* **2005**, *239*, 31–35. [CrossRef] [PubMed]

38. Tanaka, M.; Bohár, Z.; Martos, D.; Telegdy, G.; Vécsei, L. Antidepressant-like effects of kynurenic acid in a modified forced swim test. *Pharmacol. Rep.* **2020**, *72*, 449–455. [CrossRef] [PubMed]

39. Morales-Puerto, N.; Giménez-Gómez, P.; Pérez-Hernández, M.; Abuin-Martínez, C.; Leticia, G.B.; Vidal, R.; Gutiérrez-López, M.D.; O'Shea, E.; Colado, M.I. Addiction and the kynurenine pathway: A new dancing couple? *Pharmacol. Ther.* **2021**, *223*, 107807. [CrossRef] [PubMed]

40. Tanaka, M.; Török, N.; Vécsei, L. Are 5-HT$_1$ receptor agonists effective anti-migraine drugs? *Expert Opin Pharmacother.* **2021**, *22*, 1221–1225. [CrossRef] [PubMed]

41. Erabi, H.; Okada, G.; Shibasaki, C.; Setoyama, D.; Kang, D.; Takamura, M.; Yoshino, A.; Fuchikami, M.; Kurata, A.; Kato, T.A.; et al. Kynurenic acid is a potential overlapped biomarker between diagnosis and treatment response for depression from metabolome analysis. *Sci. Rep.* **2020**, *10*, 16822. [CrossRef]

42. Balogh, L.; Tanaka, M.; Török, N.; Vécsei, L.; Taguchi, S. Crosstalk between Existential Phenomenological Psychotherapy and Neurological Sciences in Mood and Anxiety Disorders. *Biomedicines* **2021**, *9*, 340. [CrossRef]

43. Hunt, C.; Macedo e Cordeiro, T.; Suchting, R.; de Dios, C.; Cuellar Leal, V.A.; Soares, J.C.; Dantzer, R.; Teixeira, A.L.; Selvaraj, S. Effect of immune activation on the kynurenine pathway and depression symptoms—A systematic review and meta-analysis. *Neurosci. Biobehav. Rev.* **2020**, *118*, 514. [CrossRef]

44. Carrillo-Mora, P.; Pérez-De la Cruz, V.; Estrada-Cortés, B.; Toussaint-González, P.; Martínez-Cortéz, J.A.; Rodríguez-Barragán, M.; Quinzaños-Fresnedo, J.; Rangel-Caballero, F.; Gamboa-Coria, G.; Sánchez-Vázquez, I.; et al. Serum Kynurenines Correlate With Depressive Symptoms and Disability in Poststroke Patients: A Cross-sectional Study. *Neurorehabil. Neural Repair* **2020**, *34*, 936–944. [CrossRef]

45. Simonato, M.; Dall'Acqua, S.; Zilli, C.; Sut, S.; Tenconi, R.; Gallo, N.; Sfriso, P.; Sartori, L.; Cavallin, F.; Fiocco, U.; et al. Tryptophan Metabolites, Cytokines, and Fatty Acid Binding Protein 2 in Myalgic Encephalomyelitis/Chronic Fatigue Syndrome. *Biomedicines* **2021**, *9*, 1724. [CrossRef]

46. Ramírez Ortega, D.; Ugalde Muñiz, P.E.; Blanco Ayala, T.; Vázquez Cervantes, G.I.; Lugo Huitrón, R.; Pineda, B.; González Esquivel, D.F.; Pérez de la Cruz, G.; Pedraza Chaverrí, J.; Sánchez Chapul, L.; et al. On the Antioxidant Properties of L-Kynurenine: An Efficient ROS Scavenger and Enhancer of Rat Brain Antioxidant Defense. *Antioxidants* **2022**, *11*, 31. [CrossRef] [PubMed]

47. Marrugo-Ramírez, J.; Rodríguez-Núñez, M.; Marco, M.-P.; Mir, M.; Samitier, J. Kynurenic Acid Electrochemical Immunosensor: Blood-Based Diagnosis of Alzheimer's Disease. *Biosensors* **2021**, *11*, 20. [CrossRef] [PubMed]

48. Tanaka, M.; Vécsei, L. Monitoring the kynurenine system: Concentrations, ratios or what else? *Adv. Clin. Exp. Med.* **2021**, *30*, 775–778. [CrossRef] [PubMed]

49. Török, N.; Tanaka, M.; Vécsei, L. Searching for Peripheral Biomarkers in Neurodegenerative Diseases: The Tryptophan-Kynurenine Metabolic Pathway. *Int. J. Mol. Sci.* **2020**, *21*, 9338. [CrossRef] [PubMed]

50. Battaglia, S.; Fabius, J.H.; Moravkova, K.; Fracasso, A.; Borgomaneri, S. The Neurobiological Correlates of Gaze Perception in Healthy Individuals and Neurologic Patients. *Biomedicines* **2022**, *10*, 627. [CrossRef]

51. Baran, H.; Jellinger, K.; Deecke, L. Kynurenine metabolism in Alzheimer's disease. *J. Neural Transm.* **1999**, *106*, 165–181. [CrossRef] [PubMed]

52. Pershing, M.L.; Bortz, D.M.; Pocivavsek, A.; Fredericks, P.J.; Jorgensen, C.V.; Vunck, S.A.; Leuner, B.; Schwarcz, R.; Bruno, J.P. Elevated levels of kynurenic acid during gestation produce neurochemical, morphological, and cognitive deficits in adulthood: Implications for schizophrenia. *Neuropharmacology* **2015**, *90*, 33–41. [CrossRef] [PubMed]

53. Pocivavsek, A.; Thomas, M.A.; Elmer, G.I.; Bruno, J.P.; Schwarcz, R. Continuous kynurenine administration during the prenatal period, but not during adolescence, causes learning and memory deficits in adult rats. *Psychopharmacology* **2014**, *231*, 2799–2809. [CrossRef] [PubMed]

54. Borgomaneri, S.; Battaglia, S.; Sciamanna, G.; Tortora, F.; Laricchiuta, D. Memories are not written in stone: Re-writing fear memories by means of non-invasive brain stimulation and optogenetic manipulations. *Neurosci. Biobehav. Rev.* **2021**, *127*, 334–352. [CrossRef] [PubMed]

55. Borgomaneri, S.; Battaglia, S.; Avenanti, A.; Pellegrino, G.D. Don't Hurt Me No More: State-dependent Transcranial Magnetic Stimulation for the treatment of specific phobia. *J. Affect. Disord.* **2021**, *286*, 78–79. [CrossRef] [PubMed]

56. Borgomaneri, S.; Battaglia, S.; Garofalo, S.; Tortora, F.; Avenanti, A.; di Pellegrino, G. State-Dependent TMS over Prefrontal Cortex Disrupts Fear-Memory Reconsolidation and Prevents the Return of Fear. *Curr. Biol.* **2020**, *30*, 3672–3679.e4. [CrossRef]

57. Gellert, L.; Varga, D.; Ruszka, M.; Toldi, J.; Farkas, T.; Szatmari, I.; Fulop, F.; Vecsei, L.; Kis, Z. Behavioural studies with a newly developed neuroprotective KYNA-amide. *J. Neural Transm.* **2012**, *119*, 165–172. [CrossRef] [PubMed]

58. Demeter, I.; Nagy, K.; Farkas, T.; Kis, Z.; Kocsis, K.; Knapp, L.; Gellert, L.; Fulop, F.; Vecsei, L.; Toldi, J. Paradox effects of kynurenines on LTP induction in the Wistar rat. An in vivo study. *Neurosci. Let.* **2013**, *553*, 138–141. [CrossRef] [PubMed]

59. Prescott, C.; Weeks, A.M.; Staley, K.J.; Partin, K.M. Kynurenic acid has a dual action on AMPA receptor responses. *Neurosci. Lett.* **2006**, *402*, 108–112. [CrossRef] [PubMed]

60. Rozsa, E.; Robotka, H.; Vecsei, L.; Toldi, J. The Janus-face kynurenic acid. *J. Neural Transm.* **2008**, *115*, 1087–1091. [CrossRef]

61. Battaglia, S.; Harrison, B.J.; Fullana, M.A. Does the human ventromedial prefrontal cortex support fear learning, fear extinction or both? A commentary on subregional contributions. *Mol. Psychiatry* **2021**. [CrossRef] [PubMed]

62. Battaglia, S. Neurobiological advances of learned fear in humans. *Adv. Clin. Exp. Med.* **2022**, *31*, 217–221. [CrossRef] [PubMed]

63. Garofalo, S.; Timmermann, C.; Battaglia, S.; Maier, M.E.; di Pellegrino, G. Mediofrontal Negativity Signals Unexpected Timing of Salient Outcomes. *J. Cogn. Neurosci.* **2017**, *29*, 718–727. [CrossRef] [PubMed]

64. Battaglia, S.; Garofalo, S.; di Pellegrino, G.; Starita, F. Revaluing the Role of vmPFC in the Acquisition of Pavlovian Threat Conditioning in Humans. *J. Neurosci.* **2020**, *40*, 8491–8500. [CrossRef] [PubMed]

65. Koshy Cherian, A.; Gritton, H.; Johnson, D.E.; Young, D.; Kozak, R.; Sarter, M. A systemically-available kynurenine aminotransferase II (KAT II) inhibitor restores nicotine-evoked glutamatergic activity in the cortex of rats. *Neuropharmacology* **2014**, *82*, 41–48. [CrossRef] [PubMed]

66. Kozak, R.; Campbell, B.M.; Strick, C.A.; Horner, W.; Hoffmann, W.E.; Kiss, T.; Chapin, D.S.; McGinnis, D.; Abbott, A.L.; Roberts, B.M.; et al. Reduction of brain kynurenic acid improves cognitive function. *J. Neurosci.* **2014**, *34*, 10592–10602. [CrossRef] [PubMed]

67. Potter, M.C.; Elmer, G.I.; Bergeron, R.; Albuquerque, E.X.; Guidetti, P.; Wu, H.Q.; Schwarcz, R. Reduction of endogenous kynurenic acid formation enhances extracellular glutamate, hippocampal plasticity, and cognitive behavior. *Neuropsychopharmacology* **2010**, *35*, 1734–1742. [CrossRef] [PubMed]

68. Tanaka, M.; Schally, A.V.; Telegdy, G. Neurotransmission of the antidepressant-like effects of the growth hormone-releasing hormone antagonist MZ-4-71. *Behav. Brain Res.* **2012**, *228*, 388–391. [CrossRef] [PubMed]

69. Telegdy, G.; Schally, A.V. Involvement of neurotransmitters in the action of growth hormone-releasing hormone antagonist on passive avoidance learning. *Behav. Brain. Res.* **2012**, *233*, 326–330. [CrossRef] [PubMed]

70. Telegdy, G.; Tiricz, H.; Adamik, A. Involvement of neurotransmitters in urocortin-induced passive avoidance learning in mice. *Behav. Brain. Bull.* **2005**, *67*, 242–247. [CrossRef] [PubMed]

71. Palotai, M.; Telegdy, G.; Bagosi, Z.; Jaszberenyi, M. The action of neuropeptide AF on passive avoidance learning. Involvement of neurotransmitters. *Neurobiol. Learn Mem.* **2016**, *127*, 34–41. [CrossRef] [PubMed]
72. Palotai, M.; Telegdy, G.; Tanaka, M.; Bagosi, Z.; Jászberényi, M. Neuropeptide AF induces anxiety-like and antidepressant-like behavior in mice. *Behav. Brain Res.* **2014**, *274*, 264–269. [CrossRef]
73. Telegdy, G.; Tanaka, M.; Schally, A.V. Effects of the growth hormone-releasing hormone (GH-RH) antagonist on brain functions in mice. *Behav. Brain Res.* **2011**, *224*, 155–158. [CrossRef]
74. Telegdy, G.; Adamik, A.; Tanaka, M.; Schally, A.V. Effects of the LHRH antagonist Cetrorelix on affective and cognitive functions in rats. *Regul. Pept.* **2010**, *159*, 142–147. [CrossRef] [PubMed]
75. Petroianu, G.A. Hyperthermia and Serotonin: The Quest for a "Better Cyproheptadine". *Int. J. Mol. Sci.* **2022**, *23*, 3365. [CrossRef] [PubMed]
76. Regan, J.W.; DeMarinis, R.M.; Caron, M.G.; Lefkowitz, R.J. Identification of the subunit-binding site of alpha 2-adrenergic receptors using [3H]phenoxybenzamine. *J. Biol. Chem.* **1984**, *259*, 7864–7869. [CrossRef]
77. National Center for Biotechnology Information. PubChem Bioassay Record for Bioactivity AID 1135637—SID 103170037, Bioactivity for AID 1135637—SID 103170037, Source: ChEMBL. Available online: https://pubchem.ncbi.nlm.nih.gov/bioassay/1135637#sid=103170037 (accessed on 25 March 2022).
78. National Center for Biotechnology Information. PubChem Bioassay Record for Bioactivity AID 65111—SID 103167216, Bioactivity for AID 65111—SID 103167216, Source: ChEMBL. Available online: https://pubchem.ncbi.nlm.nih.gov/bioassay/65111#sid=103167216 (accessed on 25 March 2022).
79. Fan, L.; Tan, L.; Chen, Z.; Qi, J.; Nie, F.; Luo, Z.; Cheng, J.; Wang, S. Haloperidol bound D_2 dopamine receptor structure inspired the discovery of subtype selective ligands. *Nat. Commun.* **2020**, *11*, 1074. [CrossRef] [PubMed]
80. National Center for Biotechnology Information. PubChem Bioassay Record for Bioactivity AID 42040—SID 103164951, Bioactivity for AID 42040—SID 103164951, Source: ChEMBL. Available online: https://pubchem.ncbi.nlm.nih.gov/bioassay/42040#sid=103164951 (accessed on 25 March 2022).
81. Xu, J.; Chuang, D.M. Muscarinic acetylcholine receptor-mediated phosphoinositide turnover in cultured cerebellar granule cells: Desensitization by receptor agonists. *J. Pharmacol. Exp. Ther.* **1987**, *242*, 238–244.
82. Tanaka, M.; Telegdy, G. Neurotransmissions of antidepressant-like effects of neuromedin U-23 in mice. *Behav. Brain Res.* **2014**, *259*, 196–199. [CrossRef] [PubMed]
83. Thabault, M.; Turpin, V.; Maisterrena, A.; Jaber, M.; Egloff, M.; Galvan, L. Cerebellar and Striatal Implications in Autism Spectrum Disorders: From Clinical Observations to Animal Models. *Int. J. Mol. Sci.* **2022**, *23*, 2294. [CrossRef] [PubMed]
84. Tanaka, M.; Csabafi, K.; Telegdy, G. Neurotransmissions of antidepressant-like effects of kisspeptin-13. *Regul. Pept.* **2013**, *180*, 1–4. [CrossRef]
85. Correia, B.S.B.; Nani, J.V.; Waladares Ricardo, R.; Stanisic, D.; Costa, T.B.B.C.; Hayashi, M.A.F.; Tasic, L. Effects of Psychostimulants and Antipsychotics on Serum Lipids in an Animal Model for Schizophrenia. *Biomedicines* **2021**, *9*, 235. [CrossRef]
86. Swingler, T.E.; Niu, L.; Pontifex, M.G.; Vauzour, D.; Clark, I.M. The microRNA-455 Null Mouse Has Memory Deficit and Increased Anxiety, Targeting Key Genes Involved in Alzheimer's Disease. *Int. J. Mol. Sci.* **2022**, *23*, 554. [CrossRef] [PubMed]
87. Tanaka, M.; Kádár, K.; Tóth, G.; Telegdy, G. Antidepressant-like effects of urocortin 3 fragments. *Brain Res. Bull.* **2011**, *84*, 414–418. [CrossRef] [PubMed]
88. Giménez-Llort, L.; Marin-Pardo, D.; Marazuela, P.; Hernández-Guillamón, M. Survival Bias and Crosstalk between Chronological and Behavioral Age: Age- and Genotype-Sensitivity Tests Define Behavioral Signatures in Middle-Aged, Old, and Long-Lived Mice with Normal and AD-Associated Aging. *Biomedicines* **2021**, *9*, 636. [CrossRef]
89. Muntsant, A.; Giménez-Llort, L. Genotype Load Modulates Amyloid Burden and Anxiety-like Patterns in Male 3xTg-AD Survivors despite Similar Neuro-Immunoendocrine, Synaptic and Cognitive Impairments. *Biomedicines* **2021**, *9*, 715. [CrossRef]
90. Santana-Santana, M.; Bayascas, J.-R.; Giménez-Llort, L. Fine-Tuning the PI3K/Akt Signaling Pathway Intensity by Sex and Genotype-Load: Sex-Dependent Homozygotic Threshold for Somatic Growth but Feminization of Anxious Phenotype in Middle-Aged PDK1 K465E Knock-In and Heterozygous Mice. *Biomedicines* **2021**, *9*, 747. [CrossRef]
91. Vila-Merkle, H.; González-Martínez, A.; Campos-Jiménez, R.; Martínez-Ricós, J.; Teruel-Martí, V.; Blasco-Serra, A.; Lloret, A.; Celada, P.; Cervera-Ferri, A. The Oscillatory Profile Induced by the Anxiogenic Drug FG-7142 in the Amygdala–Hippocampal Network Is Reversed by Infralimbic Deep Brain Stimulation: Relevance for Mood Disorders. *Biomedicines* **2021**, *9*, 783. [CrossRef]
92. Smagin, D.A.; Kovalenko, I.L.; Galyamina, A.G.; Belozertseva, I.V.; Tamkovich, N.V.; Baranov, K.O.; Kudryavtseva, N.N. Chronic Lithium Treatment Affects Anxious Behaviors and theExpression of Serotonergic Genes in Midbrain Raphe Nuclei of Defeated Male Mice. *Biomedicines* **2021**, *9*, 1293. [CrossRef]
93. Lee, E.C.; Hong, D.-Y.; Lee, D.-H.; Park, S.-W.; Lee, J.Y.; Jeong, J.H.; Kim, E.-Y.; Chung, H.-M.; Hong, K.-S.; Park, S.-P.; et al. Inflammation and Rho-Associated Protein Kinase-Induced Brain Changes in Vascular Dementia. *Biomedicines* **2022**, *10*, 446. [CrossRef] [PubMed]
94. Castillo-Mariqueo, L.; Pérez-García, M.J.; Giménez-Llort, L. Modeling Functional Limitations, Gait Impairments, and Muscle Pathology in Alzheimer's Disease: Studies in the 3xTg-AD Mice. *Biomedicines* **2021**, *9*, 1365. [CrossRef] [PubMed]
95. Lamoine, S.; Cumenal, M.; Barriere, D.A.; Pereira, V.; Fereyrolles, M.; Prival, L.; Barbier, J.; Boudieu, L.; Brasset, E.; Bertin, B.; et al. The Class I HDAC Inhibitor, MS-275, Prevents Oxaliplatin-Induced Chronic Neuropathy and Potentiates Its Antiproliferative Activity in Mice. *Int. J. Mol. Sci.* **2022**, *23*, 98. [CrossRef]

96. Quirant-Sánchez, B.; Mansilla, M.J.; Navarro-Barriuso, J.; Presas-Rodríguez, S.; Teniente-Serra, A.; Fondelli, F.; Ramo-Tello, C.; Martínez-Cáceres, E. Combined Therapy of Vitamin D3-Tolerogenic Dendritic Cells and Interferon-β in a Preclinical Model of Multiple Sclerosis. *Biomedicines* **2021**, *9*, 1758. [CrossRef]

97. Jeong, W.-H.; Kim, W.-I.; Lee, J.-W.; Park, H.-K.; Song, M.-K.; Choi, I.-S.; Han, J.-Y. Modulation of Long-Term Potentiation by Gamma Frequency Transcranial Alternating Current Stimulation in Transgenic Mouse Models of Alzheimer's Disease. *Brain Sci.* **2021**, *11*, 1532. [CrossRef] [PubMed]

98. Chiamulera, C.; Costa, S.; Reggiani, A. Effect of NMDA- and strychnine-insensitive glycine site antagonists on NMDA-mediated convulsions and learning. *Psychopharmacology* **1990**, *102*, 551–552. [CrossRef] [PubMed]

99. Tocco, G.; Maren, S.; Shors, T.J.; Baudry, M.; Thompson, R.F. Long-term potentiation is associated with increased [3H]AMPA binding in rat hippocampus. *Brain Res.* **1992**, *573*, 228–234. [CrossRef]

100. Williams, J.M.; Guevremont, D.; Mason-Parker, S.E.; Luxmanan, C.; Tate, W.P.; Abraham, W.C. Differential trafficking of AMPA and NMDA receptors during long-term potentiation in awake adult animals. *J. Neurosci.* **2007**, *27*, 14171–14178. [CrossRef] [PubMed]

101. Galeotti, N.; Ghelardini, C.; Pittaluga, A.; Pugliese, A.M.; Bartolini, A.; Manetti, D.; Romanelli, M.N.; Gualtieri, F. AMPA-receptor activation is involved in the antiamnesic effect of DM 232 (unifiram) and DM 235 (sunifiram). *Naunyn-Schmiedeberg's Arch. Pharmacol.* **2003**, *368*, 538–545. [CrossRef] [PubMed]

102. Vecsei, L.; Szalardy, L.; Fulop, F.; Toldi, J. Kynurenines in the CNS: Recent advances and new questions. *Nat. Rev. Drug Discov.* **2013**, *12*, 64–82. [CrossRef] [PubMed]

103. Ostapiuk, A.; Urbanska, E.M. Kynurenic acid in neurodegenerative disorders-unique neuroprotection or double-edged sword? *CNS Neurosci. Ther.* **2022**, *28*, 19–35. [CrossRef] [PubMed]

104. Tanaka, M.; Bohár, Z.; Vécsei, L. Are Kynurenines Accomplices or Principal Villains in Dementia? Maintenance of Kynurenine Metabolism. *Molecules* **2020**, *25*, 564. [CrossRef] [PubMed]

105. Szalardy, L.; Zadori, D.; Toldi, J.; Fulop, F.; Klivenyi, P.; Vecsei, L. Manipulating kynurenic acid levels in the brain—On the edge between neuroprotection and cognitive dysfunction. *Curr. Top. Med. Chem.* **2012**, *12*, 1797–1806. [CrossRef]

106. Dezsi, L.; Tuka, B.; Martos, D.; Vecsei, L. Alzheimer's disease, astrocytes and kynurenines. *Cur. Alzheimer Res.* **2015**, *12*, 462–480. [CrossRef] [PubMed]

107. Vecsei, L. (Ed.) *Kynurenines in the Brain: From Experiments to Clinics*; Nova Science Publishers Inc.: New York, NY, USA, 2005; Available online: https://www.abebooks.com/9781594543654/Kynurenines-Brain-Experiments-Clinics-Vecsei-1594543658/plp (accessed on 18 March 2022).

108. Vecsei, L.; Beal, M.F. Influence of kynurenine treatment on open-field activity, elevated plus-maze, avoidance behaviors and seizures in rats. *Pharmacol. Biochem. Behav.* **1990**, *37*, 71–76. [CrossRef]

109. Swartz, K.J.; During, M.J.; Freese, A.; Beal, M.F. Cerebral synthesis and release of kynurenic acid: An endogenous antagonist of excitatory amino acid receptors. *J. Neurosci.* **1990**, *10*, 2965–2973. [CrossRef] [PubMed]

110. Birch, P.J.; Grossman, C.J.; Hayes, A.G. Kynurenic acid antagonises responses to NMDA via an action at the strychnine-insensitive glycine receptor. *Eur. J. Pharmacol.* **1988**, *154*, 85–87. [CrossRef]

111. Csapo, E.; Majlath, Z.; Juhasz, A.; Roosz, B.; Hetenyi, A.; Toth, G.K.; Tajti, J.; Vecsei, L.; Dekany, I. Determination of binding capacity and adsorption enthalpy between Human Glutamate Receptor (GluR1) peptide fragments and kynurenic acid by surface plasmon resonance experiments. Colloids and surfaces B. *Biointerfaces* **2014**, *123*, 924–929. [CrossRef] [PubMed]

112. Wilkinson, D. A review of the effects of memantine on clinical progression in Alzheimer's disease. *Int. J. Geriatr. Psychiatry* **2012**, *27*, 769–776. [CrossRef] [PubMed]

113. Herrmann, N.; Li, A.; Lanctot, K. Memantine in dementia: A review of the current evidence. *Expert Opin. Parmacother.* **2011**, *12*, 787–800. [CrossRef]

114. Majlath, Z.; Torok, N.; Toldi, J.; Vecsei, L. Memantine and Kynurenic Acid: Current Neuropharmacological Aspects. *Curr. Neuropharmacol.* **2016**, *14*, 200–209. [CrossRef] [PubMed]

115. Fukui, S.; Schwarcz, R.; Rapoport, S.I.; Takada, Y.; Smith, Q.R. Blood-brain barrier transport of kynurenines: Implications for brain synthesis and metabolism. *J. Neurochem.* **1991**, *56*, 2007–2017. [CrossRef] [PubMed]

116. Varga, N.; Csapo, E.; Majlath, Z.; Ilisz, I.; Krizbai, I.A.; Wilhelm, I.; Knapp, L.; Toldi, J.; Vecsei, L.; Dekany, I. Targeting of the kynurenic acid across the blood-brain barrier by core-shell nanoparticles. *Eur. J. Pharm. Sci.* **2016**, *86*, 67–74. [CrossRef]

117. Vecsei, L.; Beal, M.F. Intracerebroventricular injection of kynurenic acid, but not kynurenine, induces ataxia and stereotyped behavior in rats. *Brain Res. Bull.* **1990**, *25*, 623–627. [CrossRef]

118. Carrillo-Mora, P.; Mendez-Cuesta, L.A.; Perez-De La Cruz, V.; Fortoul-van Der Goes, T.I.; Santamaria, A. Protective effect of systemic L-kynurenine and probenecid administration on behavioural and morphological alterations induced by toxic soluble amyloid beta (25–35) in rat hippocampus. *Behav. Brain Res.* **2010**, *210*, 240–250. [CrossRef] [PubMed]

119. Silva-Adaya, D.; Perez-De La Cruz, V.; Villeda-Hernandez, J.; Carrillo-Mora, P.; Gonzalez-Herrera, I.G.; Garcia, E.; Colin-Barenque, L.; Pedraza-Chaverri, J.; Santamaria, A. Protective effect of L-kynurenine and probenecid on 6-hydroxydopamine-induced striatal toxicity in rats: Implications of modulating kynurenate as a protective strategy. *Neurotoxicol. Teratol.* **2011**, *33*, 303–312. [CrossRef] [PubMed]

120. Justinova, Z.; Mascia, P.; Wu, H.Q.; Secci, M.E.; Redhi, G.H.; Panlilio, L.V.; Scherma, M.; Barnes, C.; Parashos, A.; Zara, T.; et al. Reducing cannabinoid abuse and preventing relapse by enhancing endogenous brain levels of kynurenic acid. *Nat. Neurosci.* **2013**, *16*, 1652–1661. [CrossRef]
121. Luchowska, E.; Kloc, R.; Olajossy, B.; Wnuk, S.; Wielosz, M.; Owe-Larsson, B.; Urbanska, E.M. beta-adrenergic enhancement of brain kynurenic acid production mediated via cAMP-related protein kinase A signaling. *Prog. Neuropsychopharmacol. Biol. Psychiatry* **2009**, *33*, 519–529. [CrossRef] [PubMed]
122. Linderholm, K.R.; Alm, M.T.; Larsson, M.K.; Olsson, S.K.; Goiny, M.; Hajos, M.; Erhardt, S.; Engberg, G. Inhibition of kynurenine aminotransferase II reduces activity of midbrain dopamine neurons. *Neuropharmacology* **2016**, *102*, 42–47. [CrossRef] [PubMed]
123. Tajti, J.; Csati, A.; Vecsei, L. Novel strategies for the treatment of migraine attacks via the CGRP, serotonin, dopamine, PAC1, and NMDA receptors. *Expert Opin. Drug Metab. Toxicol.* **2014**, *10*, 1509–1520. [CrossRef]

 *pharmaceuticals*

Article

Morphofunctional Improvement of the Facial Nerve and Muscles with Repair Using Heterologous Fibrin Biopolymer and Photobiomodulation

Cleuber Rodrigo de Souza Bueno [1,2,3], Maria Clara Cassola Tonin [1], Daniela Vieira Buchaim [3,4], Benedito Barraviera [5,6], Rui Seabra Ferreira Junior [5,6], Paulo Sérgio da Silva Santos [7], Carlos Henrique Bertoni Reis [1,8], Cláudio Maldonado Pastori [2], Eliana de Souza Bastos Mazuqueli Pereira [4], Dayane Maria Braz Nogueira [9], Marcelo Augusto Cini [10], Geraldo Marco Rosa Junior [11] and Rogerio Leone Buchaim [1,12,*]

[1] Department of Biological Sciences, Bauru School of Dentistry (FOB/USP), University of São Paulo, Bauru 17012-901, Brazil; cleuberbueno@usp.br (C.R.d.S.B.); mariaclaratonin@usp.br (M.C.C.T.); dr.carloshenriquereis@usp.br (C.H.B.R.)

[2] Dentistry School, University Center of Adamantina (UNIFAI), Adamantina 17800-000, Brazil; claudiomaldonado@fai.com.br

[3] Medical School, University Center of Adamantina (UNIFAI), Adamantina 17800-000, Brazil; danibuchaim@alumni.usp.br

[4] Postgraduate Program in Structural and Functional Interactions in Rehabilitation, Postgraduate Department, University of Marilia (UNIMAR), Marília 17525-902, Brazil; elianabastos@unimar.br

[5] Center for the Study of Venoms and Venomous Animals (CEVAP), São Paulo State University (Universidade Estadual Paulista, UNESP), Botucatu 18610-307, Brazil; bbviera@gmail.com (B.B.); rui.seabra@unesp.br (R.S.F.J.)

[6] Graduate Program in Tropical Diseases, Botucatu Medical School (FMB), São Paulo State University (UNESP—Universidade Estadual Paulista), Botucatu 18618-687, Brazil

[7] Department of Surgery, Stomatology, Pathology and Radiology, Bauru School of Dentistry, University of São Paulo, Bauru 17012-901, Brazil; paulosss@fob.usp.br

[8] UNIMAR Beneficent Hospital (HBU), University of Marilia (UNIMAR), Marília 17525-160, Brazil

[9] Department of Prosthodontics and Periodontics, Bauru School of Dentistry (FOB/USP), University of São Paulo, Bauru 17012-901, Brazil; dayanenogueira@usp.br

[10] Medical School, University of West Paulista (UNOESTE), Guarujá 11441-225, Brazil; marcelo.cini2@gmail.com

[11] Dentistry School, Faculty of the Midwest Paulista (FACOP), Piratininga 17499-010, Brazil; geraldomrjr@yahoo.com.br

[12] Graduate Program in Anatomy of Domestic and Wild Animals, Faculty of Veterinary Medicine and Animal Science, University of São Paulo (FMVZ/USP), São Paulo 05508-270, Brazil

* Correspondence: rogerio@fob.usp.br; Tel.: +55-14-3235-8220

Citation: Bueno, C.R.d.S.; Tonin, M.C.C.; Buchaim, D.V.; Barraviera, B.; Ferreira Junior, R.S.; Santos, P.S.d.S.; Reis, C.H.B.; Pastori, C.M.; Pereira, E.d.S.B.M.; Nogueira, D.M.B.; et al. Morphofunctional Improvement of the Facial Nerve and Muscles with Repair Using Heterologous Fibrin Biopolymer and Photobiomodulation. *Pharmaceuticals* **2023**, *16*, 653. https://doi.org/10.3390/ph16050653

Academic Editors: Masaru Tanaka, Lydia Giménez-Llort, Simone Battaglia, Chong Chen and Piril Hepsomali

Received: 10 March 2023
Revised: 21 April 2023
Accepted: 23 April 2023
Published: 27 April 2023

Abstract: Peripheral nerve injuries impair the patient's functional capacity, including those occurring in the facial nerve, which require effective medical treatment. Thus, we investigated the use of heterologous fibrin biopolymer (HFB) in the repair of the buccal branch of the facial nerve (BBFN) associated with photobiomodulation (PBM), using a low-level laser (LLLT), analyzing the effects on axons, muscles facials, and functional recovery. This experimental study used twenty-one rats randomly divided into three groups of seven animals, using the BBFN bilaterally (the left nerve was used for LLLT): Control group—normal and laser (CGn and CGl); Denervated group—normal and laser (DGn and DGl); Experimental Repair Group—normal and laser (ERGn and ERGl). The photobiomodulation protocol began in the immediate postoperative period and continued for 5 weeks with a weekly application. After 6 weeks of the experiment, the BBFN and the perioral muscles were collected. A significant difference ($p < 0.05$) was observed in nerve fiber diameter (7.10 ± 0.25 μm and 8.00 ± 0.36 μm, respectively) and axon diameter (3.31 ± 0.19 μm and 4.07 ± 0.27 μm, respectively) between ERGn and ERGl. In the area of muscle fibers, ERGl was similar to GC. In the functional analysis, the ERGn and the ERGl (4.38 ± 0.10) and the ERGl (4.56 ± 0.11) showed parameters of normality. We show that HFB and PBM had positive effects on the morphological and functional stimulation of the buccal branch of the facial nerve, being an alternative and favorable for the regeneration of severe injuries.

Keywords: cranial nerve injuries; facial nerve; facial nerve diseases; low-level light therapy; fibrin tissue adhesive; biopolymers; transplantation; heterologous; facial muscles; facial expression

1. Introduction

The seventh pair of cranial nerves, the facial nerve, is responsible for the maintenance and dynamics of the muscles of facial expression [1,2], their injury can be caused by several factors such as facial trauma [3–5], tumors [6–8], iatrogenesis [2,9,10], viral infections [11,12] and metabolic diseases [13]. The impact of damage to this nerve includes facial aesthetic imbalance and loss of ophthalmic, nasal, and oral functions. Furthermore, there is a relevant psychological element involved, as it is directly related to the patient's ability to socially interact and society's negative perception of those with facial paralysis [14–17].

Peripheral nerve lesion levels define the extent of the prognosis for the patient. Pioneering, Seddon [18] developed a three-level classification involving the extent of damage to the axon and the connective tissue wrap of the nerve. The first degree, neuropraxia, is the milder degree and consists of a momentary functional decrease without direct nerve damage and support tissues, the next level is called axonotmesis, where we observe direct axon injury and local demyelination without loss of discontinuity of the axon structures. The most severe form that has an unfavorable prognosis is called neurotmesis, which involves the injury with total loss of axon discontinuity and connective tissue [19,20].

Given this, they would be important studies that propose to develop treatments or ways to accelerate axon regeneration to the target organ [21], especially neurotmesis, which necessarily surgical intervention is indicated [17,20,22]. The standard gold surgical treatment of neurotmesis without loss of tissue is end-to-end neurorrhaphy, with traditional suture coaptation [23–26]. However, sutures can trigger inflammatory processes, which can lead to neuroma formation and chronic neuropathic pain [27–30]. An alternative that has been tested is fibrin glue instead of traditional sutures with threads. However, they are expensive and contain fibrinogen and thrombin derived from human blood, which can transmit infectious and parasitic diseases [27,31,32].

Since 1990, the Center for the Study of Venoms and Venomous Animals (CEVAP/UNESP, Botucatu, Brazil) has been developing a new heterologous fibrin sealant. This is purified and extracted from snake venom (*Crotalus durissus terrificus*), and is biocompatible and biodegradable, hemostatic, adhesive, and does not produce adverse reactions [33]. Initially, fibrin sealant was used to glue nerves in preclinical studies and later used in clinical trials to treat chronic venous ulcers [34,35]. Due to its diverse biological properties, its production from animal products, and the possibility of use in different clinical situations, the name "heterologous fibrin sealant" was rethought, starting to be called "heterologous fibrin biopolymer" (HFB).

It proves to be a potentially risk-free method for the regenerative process, acting as a support and contributing to an axonal growth microenvironment in peripheral nerve injuries, obtaining promising results [36–38]. However, there is still no study with the coaptation of neurotmesis with the fibrin biopolymer in peripheral nerves analyzing the results in the innervated face muscles, so in this study, we perform the analysis of the cross-section area, which is a factor important for the functionality of the neuromuscular system [39–41].

However, in the search for faster and more effective morphological and functional recovery, the use of photobiomodulation (PBM) through the use of low-level laser therapy (LLLT) has been tested in peripheral nerves [10,42,43]. Effects such as increased mitotic activity and higher metabolic velocity, with a consequent increase in mitochondrial activity and transport of substances and oxygen, assist the activation of cell transcription factors, proliferation, survival and tissue repair, and nerve regeneration [44,45]. In addition, such benefits positively influence neuromuscular recovery when there are injured nerves, such as a decrease in the muscle degeneration process [46–49], but there is still no defined protocol in the literature for its use, and few studies with facial nerves.

Our group of researchers previously used a PBM protocol with the LLLT in the peripheral nerve defect repair process, obtaining promising results [37,38,45,50]. However, it is interesting to experiment with a new approach by modifying the frequency of treatment, aiming at better clinical adherence, patient convenience, and lower financial cost, increasing their use in clinical practice.

It can be hypothesized that HBF is effective for nerve repair and the PBM protocol used is capable of improving and accelerating the morphophysiological and functional recovery process. Therefore, the objective of this study was to investigate the repair of the buccal branch of the facial nerve using HFB as a means of coaptation of nerve stumps and LLLT with a new protocol with less frequency of applications, analyzing the effects on axonal and muscle regeneration and functional recovery.

2. Results

The results were distributed into topics according to the analyzes carried out in the studied groups: control (Control Group, normal and laser—CGn and CGl), denervated without repairing surgical treatment (Denervated Group—normal and laser, DGn and DGl); experimental in which we performed the lesion and repair with HFB (Experimental Repair Group—normal and laser, ERGn and ERGl) after the surgical procedures and photobiomodulation protocol described in the experimental design (Figure 1). We also present qualitative and quantitative analyzes of the distal stump of the buccal branch of the facial nerve and facial muscles. Finally, we present the results of the functional analysis of the animals' vibrissae.

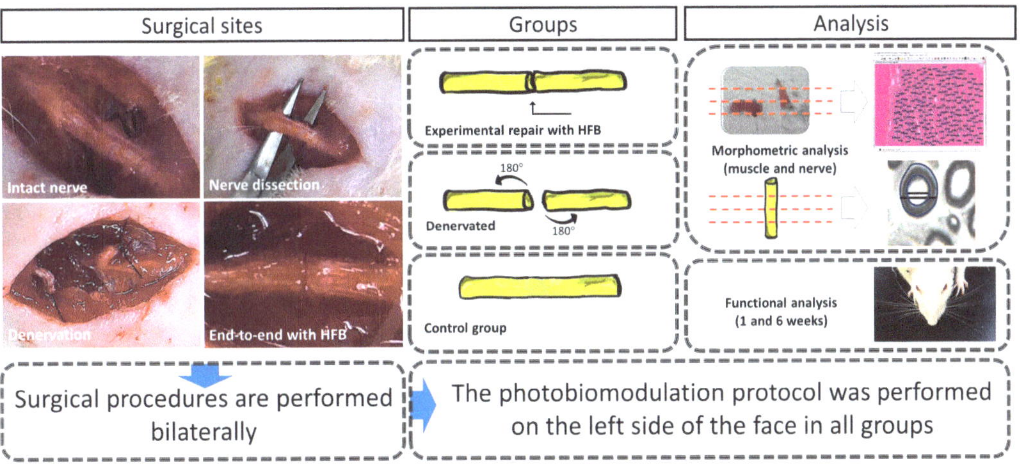

Figure 1. Surgical procedures and their treatments in the respective groups performed bilaterally. Morphological, morphometric, and functional evaluation.

2.1. Qualitative Nerve Analysis

It was observed in the groups CGn and CGl organized and myelinic axonal fibers, fascicular histological architecture organized by the presence of conjunctive wraps, with perineurium and endoneurium delimiting each fascicle and axonal fiber (Figure 2A–D).

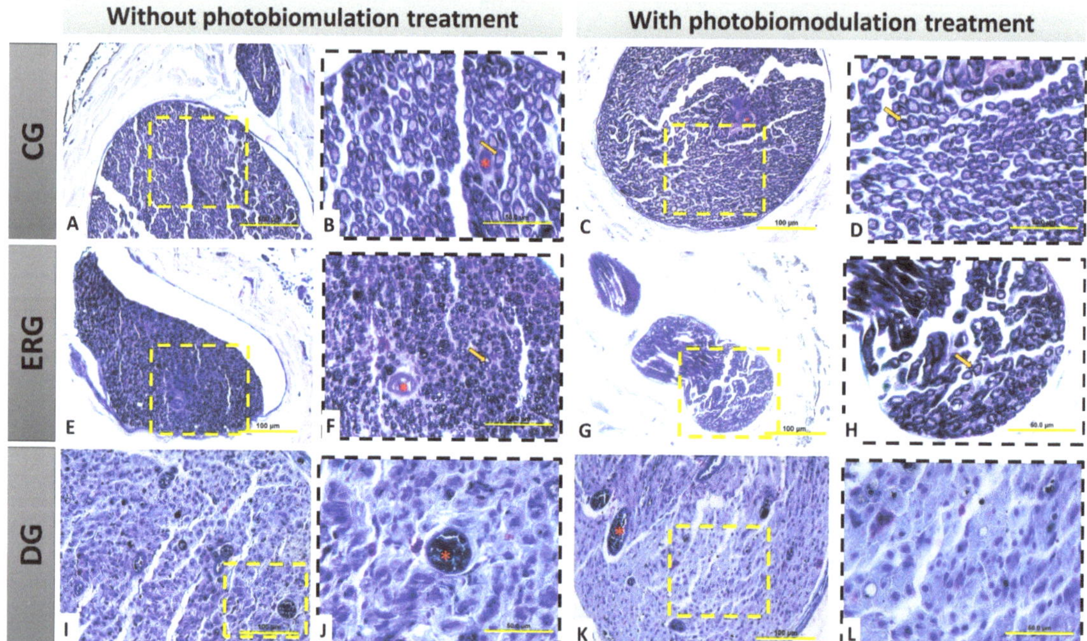

Figure 2. Histological view of the distal stump of the buccal branch of the facial nerve (BBFN) in cross-section, demonstrating the morphology in the different groups with or without photobiomodulation (PBM) treatment. In (**A–D**), seen at different magnifications of 40× (100 µm bar) and 100× (200 µm bar), there are myelin fibers with fascicular organization. In (**E–H**), seen at different magnifications (40 and 100×), it is observed that the groups with denervation and immediate repair with HFB also present myelin fibers, but in smaller size and less fascicular organization. In (**I–L**), seen at different magnifications of 40× (100 µm bar) and 100× (200 µm bar), the groups that underwent denervation and no surgical intervention was performed for repair, demonstrate severe morphological alterations of the distal stump of the nerve, resulting from this process, by observation the non-existence of myelin fibers, as well as a large invasion of scar tissue. CG = Control group, ERG= Experimental repair group, DG= Denervated group. Black arrow = myelin fiber; Asterisk (*) = blood vessel.

However, in the ERGn and ERGl groups, there was an invasion of dense connective tissue in connective envelopes, irregular axonal fibers, and visually smaller in relation to the CG, but regenerating reinnervated nerve fibers can be observed (Figure 2E–H). In addition, in the DGn and DGl histological slides, we observed the replacement of axonal fibers by dense connective tissue, highlighting the degenerated nervous tissue, demonstrating that the denervation was effective, and thus, in this group, it was not possible to perform the histomorphometric analysis (Figure 2I–L).

2.2. Histomorphometric Nerve Analysis

In the histomorphometry of the distal stump of the buccal branch of the facial nerve, it was observed, in the analysis of the areas of nerve fiber, axon, and myelin sheath, a significant difference between the control groups (GCn and CGl) and the experimental groups (ERGn and ERGl), and similarity between the experimental groups that were repaired with fibrin biopolymer, using or not the laser. Details of the mean and standard deviation values can be seen in Figure 3 and Table 1.

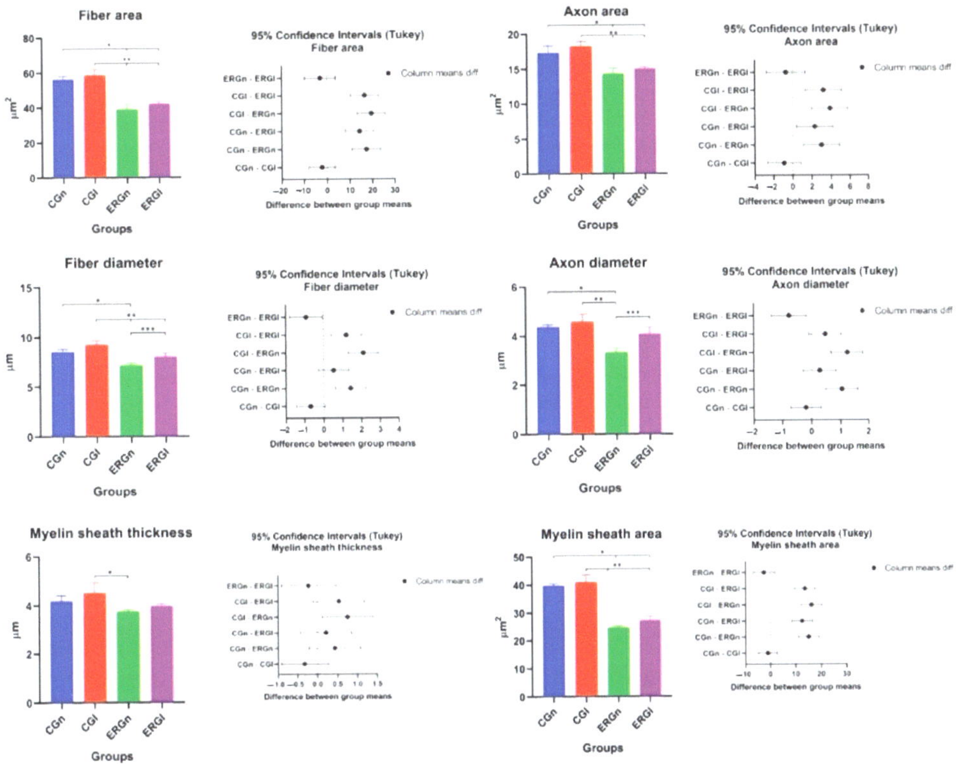

Figure 3. Histomorphometric results of the nerve (BBFN) of all studied groups demonstrated with mean and standard deviation column graph and standard deviation with the confidence intervals (Tukey). Asterisk (*, ** or ***) = significant difference between period/group (one-way ANOVA and Tukey, $p < 0.05$). CGn = Control group normal, CGl = Control group laser, ERGn = Experimental repair group normal, ERGl = Experimental repair group laser.

Table 1. Mean and standard deviation of histomorphometric analysis of the distal stump of the buccal branch of the facial nerve.

Groups	Fiber Area (μm^2)	Axon Area (μm^2)	Fiber Diameter (μm)	Axon Diameter (μm)	Myelin Sheath Area (μm^2)	Myelin Sheath Thickness (μm)
CGn	56.11 ± 1.93	17.27 ± 1.07	8.53 ± 0.28	4.36 ± 0,10	39.65 ± 0,70	4.16 ± 0.24
CGl	58.37 ± 3.66	18.17 ± 0.87	9.21 ± 0.43	4.55 ± 0.33	40,74 ± 2.81	4.49 ± 0.42
ERGn	38.77 ± 2.96	14.28 ± 0.77	7.10 ± 0.25	3.31 ± 0.19	24.57 ± 0.57	3.73 ± 0.08
ERGl	41.77 ± 1.50	14.97 ± 0.28	8.00 ± 0.36	4.07 ± 0.27	27.13 ± 1.46	3.95 ± 0.09

CGn = Control group normal, CGl = Control group laser, ERGn = Experimental repair group normal, ERGl = Experimental repair group laser.

Table 1 shows the areas of the nerve fibers, axons, and myelin sheath, as well as the diameters of the nerve, axon, and myelin sheath. In the morphometric analysis of the nerve, in the DGn and DGl groups, it was not possible to carry out the measurements due to the absence of myelin fibers (Figure 2I–L). Therefore, groups DGn and DGl from Figure 3 and Table 1 were removed (values = zero).

Regarding the diameter, CGn showed a significant difference with ERGn in nerve and axon fiber diameter measurements. In CGl we observe a significant difference in nervous fiber with experimental groups (ERGn and ERGl) and a significant difference with ERGN in axon diameter and myelin sheath thickness with ERGn.

We also observe a significant difference in nerve fiber parameters and axons between experimental groups (ERGn and ERGl) with the highest averages for the group treated with photobiomodulation. More details of the values of average and default diameter deviation can be observed in Figure 3 and Table 1.

2.3. Qualitative Muscle Analysis

In the control groups (CGn and CGl) were observed polygonal muscle fibers, with peripheral nuclei and organized histological architecture, visualized by the presence of conjunctive envelopments delimiting each fascicle and muscle fiber, highlighting the normal morphology of skeletal muscle tissue. In experimental groups (ERGn and ERGl), in some areas of the sample, we observe discreet invasion of connective tissue and muscle fibers with reduced cross-section area, permeated by larger section muscle fibers, however, continuous fascicular pattern and nuclei peripherals. Finally, in DGn and DGl groups, we observed the similarity of the qualitative pattern of experimental groups, however, with a greater number of muscle fibers with smaller area sections (Figure 4).

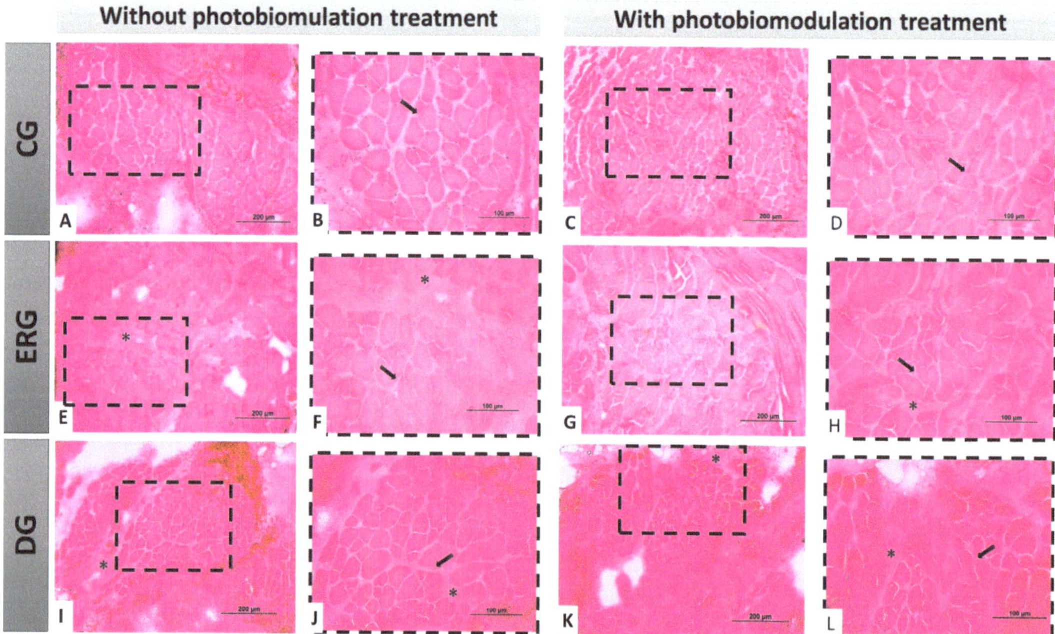

Figure 4. Histological view of the facial muscles in cross-section demonstrating the morphology of the different groups, with or without treatment with photobiomodulation. In (**A–D**), at different magnifications of 20× (200 μm bar) and 40× (100 μm bar), we observe polygonal muscle fibers, with peripheral nuclei and fascicular organization. In (**E–H**), at different magnifications (20 and 40×), the groups that underwent denervation and immediate repair with HFB, demonstrating a good histological pattern with few morphological changes. In (**I–L**), at different magnifications of 20× (200 μm bar) and 40× (100 μm bar), the groups that underwent denervation and no surgical intervention was performed for nerve repair, demonstrate some morphological alterations resulting from this situation, dimensional reduction of the muscle fibers and connective tissue invasion. CG = Control group, ERG = Experimental repair group, DG = Denervated group. Black arrow = muscle cell nucleus; Asterisk (*) intramuscular connective tissue.

2.4. Histomorphometric Muscle Analysis

In Table 2, the results related to the facial muscle fiber area are indicated. We observed a statistical difference between the control groups (CGn and CGl) and the groups that were denervated without treatment (DGn and DGl). Experimental groups (ERGn and ERGl) were statistically similar. ERGN showed a statistical difference with CGl, and ERGl was similar to control groups. Figure 5 shows the graphs of the mean values and standard deviation with the confidence intervals (Tukey) of the morphometric analysis of the cross-sections of the muscle fibers in the study.

Table 2. Median and standard deviation of histomorphometric analysis of the area of facial muscle fibers.

Groups	Fiber Muscle Area (μm^2)
CGn	1126.00 ± 250.50
CGl	1208.00 ± 249.60
DGn	603.60 ± 85.54
DGl	583.00 ± 44.36
ERGn	837.30 ± 113.80
ERGl	926.90 ± 183.60

CGn = Control group normal, CGl = Control group laser, DGn = Denervated group normal, DGl = Denervated group laser, ERGn = Experimental repair group normal, ERGl = Experimental repair group laser.

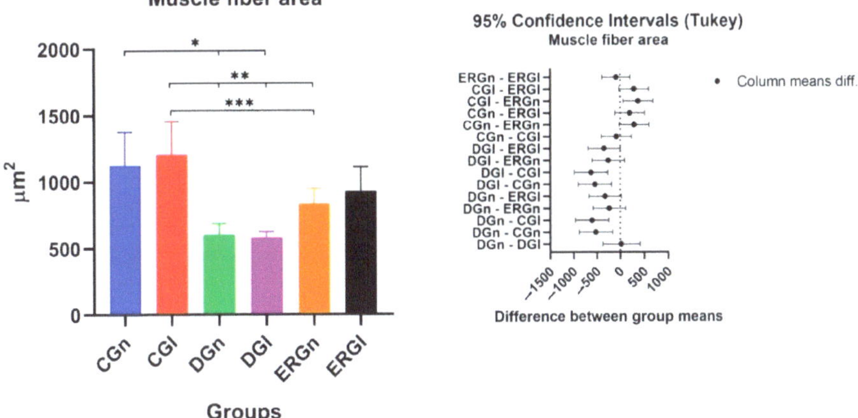

Figure 5. Mean and standard deviation of results of cross-section of muscle fibers and the difference between groups, demonstrated with mean and standard deviation column graph and standard deviation with the confidence intervals (Tukey). Asterisk (*, ** or ***) = significant difference between period/group (one-way ANOVA and Tukey, $p < 0.05$). CGn = Control group normal, CGl= Control group laser, DGn = Denervated group normal, DGl = Denervated group laser, ERGn = Experimental repair group normal, ERGl = Experimental repair group laser.

2.5. Functional Analysis of Whiskers Movements

Control groups (CGn and CGl) were used as a reference standard for the normality of whisker position and movement (score 5 for all animals), according to the parameters established by Faria et al. [51]. In the first week after surgery, we found similarities between experimental groups (ERGn and ERGl) and denervated groups (DGn and DGl), as well as significant differences with control groups.

The average score of ERGn was 3.90 ± 0.08 and the ERGl was 3.92 ± 0.12. In the sixth week after surgery, experimental groups showed similarities to each other and significant differences from denervated groups and controls. The average ERGn score was 4.38 ± 0.10 and the ERGl was 4.56 ± 0.11. The average values and standard deviation of the scores after functional analysis in 1 and 6 weeks after surgery are shown in Figure 6.

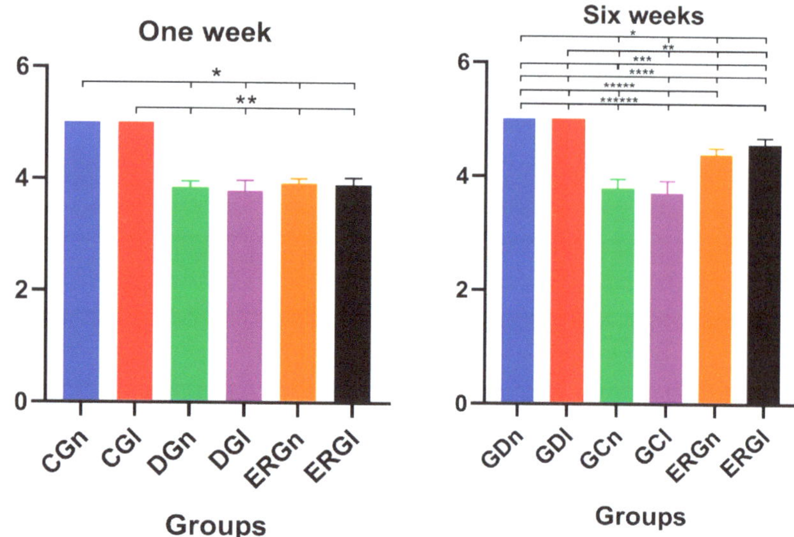

Figure 6. Functional results of vibrissae at 1 and 6 weeks postoperatively demonstrated with mean and standard deviation column graph. Asterisk (*, **, ***, ****, ***** or ******) = significant difference between period/group (one-way ANOVA and Tukey, $p < 0.05$). CGn = Control group normal, CGl = Control group laser, DGn = Denervated group normal, DGl = Denervated group laser, ERGn = Experimental repair group normal, ERGl = Experimental repair group laser.

In a general context, the morphological and morphometric evaluations of the distal stump of the BBFN demonstrated that the denervation was effective, with axonal degeneration (DGn and DGl) and, in the groups reinnervated by neurorrhaphy with HFB (ERGn and ERGl), axonal sprouting occurred, but with fibers of disorganized orientation in relation to the controls (CGn and CGl). All measurements showed lower mean values for reinnervated groups compared to controls. In the facial muscles, negative repercussions of denervation were observed, with connective tissue invasion, but like the nerve, PBM improved the repair process qualitatively, quantitatively, and functionally.

3. Discussion

In the field of tissue bioengineering, which studies effective alternatives for the treatment of peripheral nerve injuries, we evaluated the regeneration of a lesion (neurotmesis) of the facial nerve using the fibrin biopolymer (HFB) and the association of photobiomodulation (PBM) with only a weekly application of LLLT for 5 weeks, performing morphological, morphometric (nervous and muscular) and functional evaluations. We observed that the use of HFB was an efficient means of coaptation for the differentiation of BBFN and that PBM promoted positive changes in the morphological aspects of the nerve. In the functional aspect, the groups with HFB repaired with or without LLLT at the end of 6 weeks, results were considered normal for the movement of the animals' vibrissae.

HFB has been used for peripheral nerve repair, in order to replace or reduce the suture [52]. Studies comparing the use of fibrin sealants and sutures have shown less granulomatous inflammation, better axonal regeneration, and functional recovery in groups that used sealants for nerve coaptation [53–55]. With HFB we obtained the advantages of other sealants, but at a low cost, without adverse reactions or infections, due to the fact that it is produced with certified buffalo blood (*Bubalus bubalis*). Corroborating this fact, a study using HFB observed that its use reduced mechanical trauma to the nerve and reduced surgical time for peripheral nerve repairs, favoring the indication of this bioproduct [36].

The application of PBM in the regeneration of peripheral nerves has been reported in the literature with positive effects. Increased cellular metabolism, greater vascular budding, and collagen synthesis have been reported as indirect effects that aid in nerve regeneration. Added to this, increased Schwann cell proliferation, increased axonal growth velocity, and anti-inflammatory, analgesic, and anti-edema effects on the nerve have been described as positive effects directly related to nerve regeneration [45,56–59]. However, it is not yet clear which would be the most efficient protocols for each type of nerve (sensory, motor, or mixed), location (of the lesion and the effector organ), and classification of the lesion (neuropraxia, axonotmesis, or neurotmesis). In our study, it was thought of the use of a PBM protocol with LLLT using a single weekly application within 5 weeks after surgery, aiming at fewer clinical sessions, lower cost, and patient accessibility to treatment. The question would be whether therapy would maintain a positive photostimulant effect. Our group presents some previously published studies using an established parameter PBM protocol; however, using the application of therapy three times a week results in promising results in nerve regeneration of peripheral nerves [37,38,50].

Studies using 3 weekly applications for 4 weeks have been reported by other authors [59]. Lee et al. [60] demonstrated the increase in the diameter of axons using a wavelength of 604 nm. In our study, we also observed an increase in nerve fiber and axon diameter with 808 nm wavelength PBM. In another study performing axonotmesis, 980 nm PBM at three points along the nerve, three times a week for 5 weeks, had a beneficial effect on facial nerve regeneration, including the better functional capacity of the vibrissae and improvement in morphological nerve changes. Furthermore, there was a decrease and excessive suppression of apoptosis in Schwann cells induced by oxidative stress via activation of the PI3K/Akt signaling pathway [61].

However, PBM protocols with a single weekly dose are scarce in the literature. We aim to reduce the number of sessions, increasing the application time per point and, consequently, the energy per session. In the group that received PBM, after nerve repair, nerve, and axon diameters were similar to CG, suggesting that PBM with the once-weekly application protocol also accelerated nerve regeneration.

Regarding aspects of innervated muscles, after injury and during the period of nerve repair, neuromuscular changes occur, with consequent loss of trophic stimulus and, consequently, atrophy of muscle fibers and invasion of connective tissue [62–64]. Scarce scientific works have shown the effects on the target organ, being related to the success of neuromuscular regeneration [65], mainly in situations of neurotmesis. Thus, this study describes the effects of using HFB in the coaptation of neurotmesis, performing the histomorphometric analysis of the muscle innervated by BBFN. Thus, muscle preservation and reduction of the atrophic process in severe nerve injuries are challenges of regenerative medicine [66–68]. We observed that denervation led to a significant reduction in muscle fiber area in relation to the CG. However, the experimental groups showed less fiber atrophy, especially the group with PBM, which showed similarity in the area of muscle fibers with the GC, suggesting that the laser collaborated in reducing the post-repair atrophic process.

Therefore, regarding the functional analysis, BBFN is not the only one that provides motor innervation to rat vibrissae. When we only denervate this branch, we should not expect complete functional loss and total paralysis [69,70]. In the denervated group (DGn and DGl) there was loss of movement after 6 weeks, without total paralysis, probably due to having received innervation from another terminal branch of the facial nerve, such as the zygomatic. In ERGn and ERGl there was a normal movement score, according to the methodology of Faria et al. [51], not observing any difference between the groups with or without PBM.

The period of functional analysis the initial period to the postoperative (1 week) and 6 weeks postoperatively is described in studies in the literature [71–74]. Models such as the sciatic function index (SFI) and Basso, Beattie, and Bresnahan (BBB) are established in the literature for the sciatic nerve [71,72,75]. In a study that correlates these methods (SFI and BBB) for functional analysis, it is argued that the period of 6 weeks is adequate [72].

However, we must consider the different classifications of peripheral nerve injuries for the period of analysis, crush injuries (axoniotmesis), and neurotmesis without tissue loss. In a study that carried out the functional analysis of the facial nerve through the movement of the vibrissae using the total section of the nerve (neurotmesis) with and without tissue loss, a postoperative interval of four weeks was used. The authors noted that this was the time after surgery when the vibrissae began to demonstrate limited movement restitution in some animals. In the same study, they documented that during this period there was reinnervation of the motor plate in the levator labii superioris muscle, which is the largest muscle that innervates the vibrissae of animals, proving to be an interesting period for the analysis [76].

Another reason for choosing an initial and final period in our study was the fact that in 1 week after surgery, there is the presence of a post-trauma inflammatory process, making it possible to verify whether the use of LLLT in the initial phase could generate a functional gain. In the 6-week euthanasia period, the objective was to evaluate whether there was a gain in the functional speed of repair by comparing the groups that received or not LLLT. In addition, we used a shorter period of analysis of the results than in previous studies [38], which will allow a greater field of interpretation for future studies.

The denervated groups are important in this study because they allow us to observe the results of denervation in the facial muscles innervated by the buccal branch of the facial nerve. Added to this, we were able to compare the results of the surgical repair group with HFB both in the muscular and functional morphological scope with the vibrissae. The denervation technique used in the study, in which we performed a 180° rotation of the proximal and distal stump of the nerve and sutured it in the adjacent fascia and muscle, respectively, has already been proven to be effective in previous studies [40,75,77,78], not allowing spontaneous reinnervation. In this study, it was not possible to carry out the morphometric analysis of the distal stump of the buccal branch of the facial nerve in the DG groups, demonstrating the success of the technique, as there was a replacement of scar tissue and absence of myelin nerve fibers.

From future perspectives, the use of HFB associated with PBM in late repairs (not immediate to the injury) in neurotmesis can be investigated. An epidemiological study [79] revealed that 90% of the injuries caused were due to car accidents, an incidence that is increasing day by day compared to previous data [80]. Due to the morbidity of this type of trauma, nerve repair is often performed after days or weeks, so that there are stable conditions for the surgical procedure, which implies a longer period of convalescence. It is known that, in longer periods where the Schwann cells of the proximal stump of the injured nerve remain without contact with the axons of the distal stump after denervation, it is one of the main complicating factors, mainly in obtaining favorable results [62,81,82].

A possible limitation of this study was that we did not compare our previous protocol with three weekly sessions with the current protocol with fewer sessions; therefore, we cannot make any direct comparison between the protocols. In addition, in the present study, we used a device from another commercial brand, with greater power, which makes it difficult to directly associate the results. For the clinical translation of the association of HFB with PBM, it will be necessary to conclude phase III clinical studies of HFB, with relevant perspectives for use, as it has been shown to be a versatile and promising bioproduct in research in regenerative science.

4. Materials and Methods

The methodology used in this experimental protocol was based on previous studies by our group on peripheral nerve regeneration, causing an injury to the buccal branch of the facial nerve in rats [37,38,83]. In these studies, it was proven that HFB has the capacity to allow the repair of the lesion with results similar to the gold standard of end-to-end neurorrhaphy, which uses suture thread [84,85]. In an unprecedented way, in the present study, we evaluated a protocol with a smaller number of PBM sessions and also the effects on facial muscles, through morphofunctional analysis.

4.1. Experimental Design

This study was conducted in accordance with the Declaration of Helsinki, and approved by the Ethics Committee In Animal Use of the University of Marília (CEUA protocol code 033/2020 and date of approval 13 November 2020).

Twenty-one male Wistar rats (*Rattus norvegicus*) were used. The animals were 90 days old, weighing approximately 250–300 g at baseline. All animals were kept in appropriate boxes and received water and feed "ad libitum", with no restrictions on movement, respecting the 12 h light/dark regime and an approximate temperature of 22 °C. Throughout the experimental period, signs and symptoms of stress and unusual behavior of the animals were observed. The study was carried out according to the ARRIVE protocol (animal research: report of in vivo experiments) and based on the principles of the NC3Rs (National Center for Replacement, Refinement, and Reduction of Research Animals).

The animals were randomly divided into groups (controls and experimental) of 7 animals, with no inclusion and exclusion criteria. The experiment was conducted with the buccal branch of the right and left facial nerves of all animals of the experiment. In all groups, it was performed on the left side the photobiomodulation therapy with the proposed protocol. We performed the euthanasia of all animals after 6 postoperative weeks.

The groups were named as follows: Control group—normal and laser (CGn and CGl): where the incision and dissection of the buccal branch of the facial nerve was performed bilaterally without injury to it; Denervated group—normal and laser (DGn and DGl): We performed neurotmesis bilaterally in these animals and did not perform any type of surgical repair; Experimental Group Repair—normal and laser (ERGn and ERGl): We performed neurotmesis and immediate repair with heterologous fibrin biopolymer bilaterally.

4.2. Heterologous Fibrin Biopolymer (HFB)

In the groups with neurotmesis repair, the heterologous fibrin biopolymer (HFB) was used. This material was provided by the Center for the Study of Venoms and Venomous Animals (CEVAP) of the São Paulo State University (UNESP), Botucatu, Brazil. HFB has 3 components that are defrosted and homogenized prior to application. In sequence, with the aid of a micropipette, the substances were applied for the coaptation of the stumps of the injured nerve. The first was the thrombin-like enzyme fraction (5 µL), the second contains calcium chloride diluent (5 µL), and the last was fibrinogen extracted from buffalo blood (10 µL). After application, 1 min was allowed for the polymerization of the biopolymer and then a slight traction of the nerves was performed to certify its adhesion.

The fibrin biopolymer components and application formula are in accordance with patent number BR 102014011432-7 issued on 6 July 2022 by the National Institute of Industrial Property of Brazil (INPI). This material underwent a phase I/II clinical trial [86] which proved its safety for therapeutic use in humans, standing out as a promising therapeutic potential.

4.3. Surgical Procedures

For all surgical procedures, the animals underwent general anesthesia with an intramuscular injection of tiletamine hydrochloride and zolazepam hydrochloride (10 mg/kg-Telazol®; Fort Dodge Laboratories, IA, USA). Trichotomy was performed with the aid of a hair trimmer (Philips® Multigroom QG3250, São Paulo, Brazil) in the region of the bilateral face of the animals along the labial commissure to the tragus in order to obtain a smooth and hairless surface. Afterward, the animal was positioned in lateral decubitus in a surgical drip and antisepsis with 10% Polyvinyl Pyrrolidone Iodine PVPI (Povidine® Antiseptic, Vic Pharma Ind e Comércio Ltd., São Paulo, Brazil)

4.3.1. Denervation Surgery

A pre-auricular incision was made with blade 15 (Embramax®, São Paulo, Brazil) of approximately 3 cm with the aid of a surgical microscope (DF Vasconcelos®, São Paulo, Brazil). Subsequently, after division into planes, recognition, release, and sectioning of the buccal branch of the facial nerve (neurotmesis) were carried out in its central portion (point from the center of the line of the tragus to the labial commissure in lateral norm).

In DGn and DGl, in order to avoid spontaneous regeneration, the proximal stump was manipulated 180° degrees and sutured to the adjacent muscle fascia; the distal stump was manipulated 180° degrees and sutured to the adjacent musculature, both with 6-0 nylon thread [40,51,75]. The skin suture was performed with simple stitches using 4-0 Ethicon® nylon thread (Johnson & Johnson Ind e Comércio Ltd., São Paulo, Brazil).

4.3.2. Surgical Protocol for the Experimental Groups with Repair of the Buccal Branch of the Facial Nerve

After following the same steps described in 4.3.1 (Denervation surgery), the anatomical approximation of the sectioned nerve stumps was performed, without tension, and coaptation with fibrin biopolymer (see Figure 1) [37,51]. After restoring the neural continuity, the skin was sutured with simple stitches using 4-0 Ethicon® nylon thread (Johnson & Johnson Ind e Comércio Ltd., São Paulo, Brazil).

4.3.3. Post-Surgical Care

Immediately after the surgical procedures, the animals received a single dose of the antibiotic Flotril® 2.5% (Schering-Plough, Rio de Janeiro, Brazil), at a dosage of 0.2 mL/kg and analgesic Dipyrone Analgex V® (Agener União, São Paulo, Brazil) at a dose of 0.06 mL/kg in intramuscular applications. The application of the analgesic was maintained for 3 days, in addition to continuation with the analgesic Paracetamol® (Generic medication, Medley, São Paulo, Brazil) at a dose of 200 mg/Kg, 6 drops/animal dissolved in the water available in the drinker so far of euthanasia.

4.4. Photobiomodulation Protocol (PBM)

The treatment began in the immediate postoperative period and continued for 5 weeks with a weekly application, always occurring on the same day of the week throughout the protocol. The animals were manually immobilized (delicate restraint) and sedation was unnecessary during the application of photobiomodulation. All study groups received the PBM protocol in the buccal branch of the facial nerve on the left side using the protocol using the low-level laser of gallium aluminum arsenide (GaAlAs)—Therapy XT DMC® (São Carlos, Brazil), with three application points along the path of the injured nerve, each point received an energy dose of 4 J, corresponding to 40 s per point, more details of the parameters are shown in Figure 7. Prior to the applications, the device was calibrated and tested to certify the dose.

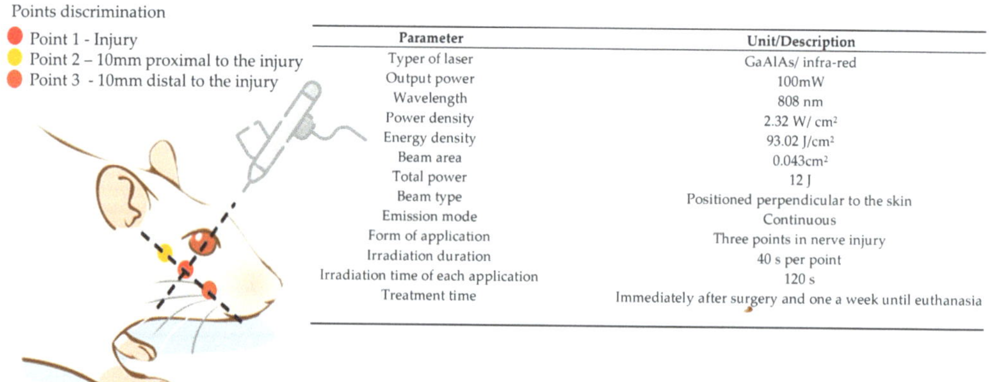

Parameter	Unit/Description
Typer of laser	GaAlAs/ infra-red
Output power	100mW
Wavelength	808 nm
Power density	2.32 W/ cm²
Energy density	93.02 J/cm²
Beam area	0.043cm²
Total power	12 J
Beam type	Positioned perpendicular to the skin
Emission mode	Continuous
Form of application	Three points in nerve injury
Irradiation duration	40 s per point
Irradiation time of each application	120 s
Treatment time	Immediately after surgery and one a week until euthanasia

Figure 7. PBM protocol (Therapy XT DMC® equipment, São Carlos, Brazil).

4.5. Functional Analysis

Observations of the animals' vibrissae movements were performed between 1 and 6 weeks postoperatively. The animals were placed in a box with a black background and the vibrissae were observed with spontaneous movements and when stimulated by the researcher (clap your hands 3 to 4 times), with the aim of triggering movements. The evaluator did not know which group was being evaluated (blind evaluation). After the observations, scores were assigned, following the methodology of Faria et al. [51]. During the observation, photographs of the animals were also taken.

4.6. Sample Collection and Euthanasia

After six weeks, the animals were anesthetized and the buccal branch of the facial nerve was carefully dissected and 10 mm of the distal stump from the neurotmesis site was collected in the experimental and denervated groups, as well as the intact nerve in the control group, under the magnified view of 16× of the surgical microscope (DF Vasconcelos®, São Paulo, Brazil). Next, the muscles of facial expression in the perioral region of all groups were dissected and carefully removed. Euthanasia was performed in a silent environment and away from the other animals, using an anesthetic overdose (triple dose—240 mg/kg of tiletamine hydrochloride + 30 mg/kg of zolazepam hydrochloride).

4.7. Histological Processing of Nerve and Muscle

The samples were fixed in a 10% buffered formaldehyde solution for 24 h and the historesin protocol was performed (Leica Mycrosistems®, Wetzlar, Germany) [38]. Sections were performed using a semiautomatic microtome (Model RM2245, Leica Microsystems®, Wetzlar, Germany) with a thickness of 5 μm. The slides were stained with Osmium Tetroxide and counterstained with 1% Toluidine Blue in distilled water. The sections were analyzed under an optical microscope.

Muscle samples were reduced to cylindrical fragments preserving the muscle belly, wrapped in surgical talc, immersed in liquid nitrogen, and included with an adhesive (Optimal Critical Temperature Tissue-Tek® (O.C.T., Sakura Finetek, Torrance, CA, USA)). Then, samples were kept in a freezer at $-80\,^\circ$C until the ten micrometer-thick histological sections were obtained in a cryostat (Model CM 1850, Leica Microsystems®, Wetzlar, Germany) at $-20\,^\circ$C, which were stained with hematoxylin and eosin (HE).

4.8. Histological Analysis of Nerve and Muscle

The morphometry of the distal region of the buccal branch of the facial nerve was performed with the measurement of 220 fibers of the nerve and muscle of all samples of each group using a microcomputer coupled to a photomicroscope (Olympus® BX50, Tokyo, Japan) and using a software of image capture and analysis (Image Pro-Plus® 6.2—Media Cybernetics, Bethesda, MD, USA). The morphometric variables studied in nerves were: the area of nerve fibers, the area of axons, the minimum diameter of nerve fibers, the minimum diameter of axons, myelin sheath area, and myelin sheath thickness [37,38,50,83], and in muscles the cross-sectional area of muscle fibers was measured [40,75,87].

4.9. Statistical Analysis

Data were organized into spreadsheets and tables in Excel format (Microsoft Office Excel®, Redmond, WA, USA) with means and standard deviation, which were subsequently submitted to statistical tests. We used the two-way variance test (ANOVA) and then Tukey's test for multiple comparisons between means. Statistical analyses and graphs were performed using the Graph Pad Prism version 8.0 program (GraphPad® Software, La Jolla, CA, USA). The level of statistical significance was set at $p < 0.05$ for all analyses.

5. Conclusions

In order to optimize the morphofunctional recovery of peripheral nerve injuries, we investigated the use of heterologous fibrin biopolymer (HFB) in the repair of the buccal branch of the facial nerve (BBFN) associated with photobiomodulation (PBM). Here we show that the use of this bioproduct (HFB) to reconnect nerves was effective in allowing axonal growth in the stump distal to the lesion and minimizing the effects on innervated muscles. There was an improvement in the functionality of the vibrissae, and the biostimulatory effects of PBM showed efficacy for nerve and muscle repair. Therefore, this study demonstrates a regenerative technique with the association of a versatile bioproduct and PBM with a single weekly application of LLLT, proving its translational potential for clinical studies in tissue bioengineering.

Author Contributions: Conceptualization, C.R.d.S.B., R.L.B. and G.M.R.J.; methodology, C.R.d.S.B., R.L.B., C.H.B.R. and P.S.d.S.S.; validation, B.B. and R.S.F.J.; formal analysis, C.R.d.S.B., D.V.B. and M.C.C.T.; investigation, D.M.B.N. and C.M.P.; data curation, E.d.S.B.M.P.; writing—original draft preparation, C.R.d.S.B.; writing—review and editing, R.L.B., C.R.d.S.B., B.B., R.S.F.J. and P.S.d.S.S.; visualization, M.A.C.; supervision, R.L.B.; project administration, R.L.B.; funding acquisition, R.S.F.J., B.B. and C.R.d.S.B. All authors have read and agreed to the published version of the manuscript.

Funding: The present study was financed in part by the Coordenação de Aperfeiçoamento de Pessoal de Nível Superior—Brasil (CAPES)—Finance Code 001; FAPESP 2021/11936-3 (B.B.); Rui Seabra Ferreira Júnior (R.S.F.J.) is a CNPq PQ1D research fellow No. 301608/2022-9; Benedito Barraviera (B.B.) is a CNPq PQ2 research fellow No. 306339/2020-0.

Institutional Review Board Statement: The study was conducted in accordance with the Declaration of Helsinki, and approved by the Institutional Review Board (or Ethics Committee) of Ethics Committee In Animal Use of the University of Marília (CEUA protocol code 033/2020 and date of approval 13 November 2020).

Informed Consent Statement: Not applicable.

Data Availability Statement: The data presented in this study are available on request from the corresponding author.

Acknowledgments: The authors thank the technical support of Cirilo Francisco Santos Neto for making the histological slides (University of Marilia, Marilia, Brazil).

Conflicts of Interest: The heterologous fibrin biopolymer (HFB) was provided by the Center for the Study of Venoms and Venomous Animals (CEVAP), São Paulo State University (UNESP), Botucatu, São Paulo, Brazil.

References

1. Yang, S.H.; Park, H.; Yoo, D.S.; Joo, W.; Rhoton, A. Microsurgical Anatomy of the Facial Nerve. *Clin. Anat.* **2021**, *34*, 90–102. [CrossRef]
2. Stuzin, J.M.; Rohrich, R.J. Facial Nerve Danger Zones. *Plast. Reconstr. Surg.* **2020**, *145*, 99–102. [CrossRef] [PubMed]
3. Lam, A.Q.; Tran Phan Chung, T.; Tran Viet, L.; Do Quang, H.; Tran Van, D.; Fox, A.J. The Anatomic Landmark Approach to Extratemporal Facial Nerve Repair in Facial Trauma. *Cureus* **2022**, *14*, e22787. [CrossRef] [PubMed]
4. Maxwell, A.K.; Lemoine, J.C.; Kahane, J.B.; Gary, C.C. Management of the Facial Nerve Following Temporal Bone Ballistic Injury. *Laryngoscope Investig. Otolaryngol.* **2022**, *7*, 1541–1548. [CrossRef] [PubMed]
5. Markiewicz, M.R.; Callahan, N.; Miloro, M. Management of Traumatic Trigeminal and Facial Nerve Injuries. *Oral Maxillofac. Surg. Clin. North. Am.* **2021**, *33*, 381–405. [CrossRef]
6. Cho, Y.S.; Choi, J.E.; Lim, J.H.; Cho, Y.-S. Management of Facial Nerve Schwannoma: When Is the Timing for Surgery. *Eur. Arch. Oto-Rhino-Laryngol.* **2022**, *279*, 1243–1249. [CrossRef]
7. Guntinas-Lichius, O.; Silver, C.E.; Thielker, J.; Bernal-Sprekelsen, M.; Bradford, C.R.; de Bree, R.; Kowalski, L.P.; Olsen, K.D.; Quer, M.; Rinaldo, A.; et al. Management of the Facial Nerve in Parotid Cancer: Preservation or Resection and Reconstruction. *Eur. Arch. Oto-Rhino-Laryngol.* **2018**, *275*, 2615–2626. [CrossRef]
8. Psillas, G.; Constantinidis, J. Facial Palsy Secondary to Cholesteatoma: A Case-Series of 14 Patients. *Audiol. Res.* **2023**, *13*, 86–93. [CrossRef] [PubMed]
9. Zourntou, S.-E.; Makridis, K.G.; Tsougos, C.-I.; Skoulakis, C.; Vlychou, M.; Vassiou, A. Facial Nerve: A Review of the Anatomical, Surgical Landmarks and Its Iatrogenic Injuries. *Injury* **2021**, *52*, 2038–2048. [CrossRef]
10. Green, J.D., Jr.; Shelton, C.; Brackmann, D.E. Iatrogenic facial nerve injury during otologic surgery. *Laryngoscope* **1994**, *104*, 922–926. [CrossRef]
11. Spencer, C.R.; Irving, R.M. Causes and Management of Facial Nerve Palsy. *Br. J. Hosp. Med.* **2016**, *77*, 686–691. [CrossRef]

12. Kim, Y.-H.; Kim, J.-E.; Yoon, B.-A.; Kim, J.-K.; Bae, J.-S. Bilateral Facial Weakness with Distal Paresthesia Following COVID-19 Vaccination: A Scoping Review for an Atypical Variant of Guillain–Barré Syndrome. *Brain Sci.* **2022**, *12*, 1046. [CrossRef]
13. Jung, S.Y.; Jung, J.; Byun, J.Y.; Park, M.S.; Kim, S.H.; Yeo, S.G. The Effect of Metabolic Syndrome on Bell's Palsy Recovery Rate. *Acta Otolaryngol.* **2018**, *138*, 670–674. [CrossRef] [PubMed]
14. Parsa, K.M.; Hancock, M.; Nguy, P.L.; Donalek, H.M.; Wang, H.; Barth, J.; Reilly, M.J. Association of Facial Paralysis with Perceptions of Personality and Physical Traits. *JAMA Netw. Open* **2020**, *3*, e205495. [CrossRef]
15. Okuma, H.; Nagano, R.; Takagi, S. Hemiplegic Peripheral Neuropathy Accompanied with Multiple Cranial Nerve Palsy. *Clin. Pract.* **2012**, *2*, e40. [CrossRef]
16. Li, M.K.K.; Niles, N.; Gore, S.; Ebrahimi, A.; McGuinness, J.; Clark, J.R. Social Perception of Morbidity in Facial Nerve Paralysis. *Head Neck* **2016**, *38*, 1158–1163. [CrossRef]
17. Fliss, E.; Yanko, R.; Zaretski, A.; Tulchinsky, R.; Arad, E.; Kedar, D.J.; Fliss, D.M.; Gur, E. Facial Nerve Repair Following Acute Nerve Injury. *Arch. Plast. Surg.* **2022**, *49*, 501–509. [CrossRef] [PubMed]
18. Seddon, H.J. Three Types of Nerve Injury. *Brain* **1943**, *66*, 237–288. [CrossRef]
19. Menorca, R.M.G.; Fussell, T.S.; Elfar, J.C. Nerve Physiology. *Hand Clin.* **2013**, *29*, 317–330. [CrossRef] [PubMed]
20. Krauss, E.M.; Weber, R.V.; Mackinnon, S.E. Nerve Injury, Repair, and Reconstruction. In *Plastic Surgery—Principles and Practice*; Elsevier: Amsterdam, The Netherlands, 2022; pp. 803–825.
21. Riccio, M.; Marchesini, A.; Pugliese, P.; Francesco, F. Nerve Repair and Regeneration: Biological Tubulization Limits and Future Perspectives. *J. Cell Physiol.* **2019**, *234*, 3362–3375. [CrossRef]
22. Grinsell, D.; Keating, C.P. Peripheral Nerve Reconstruction after Injury: A Review of Clinical and Experimental Therapies. *Biomed. Res. Int.* **2014**, *2014*, 698256. [CrossRef] [PubMed]
23. Panagopoulos, G.N.; Megaloikonomos, P.D.; Mavrogenis, A.F. The Present and Future for Peripheral Nerve Regeneration. *Orthopedics* **2017**, *40*, e141–e156. [CrossRef]
24. Lundborg, G. A 25-Year Perspective of Peripheral Nerve Surgery: Evolving Neuroscientific Concepts and Clinical Significance. *J. Hand Surg. Am.* **2000**, *25*, 391–414. [CrossRef]
25. Rönkkö, H.; Göransson, H.; Taskinen, H.-S.; Paavilainen, P.; Vahlberg, T.; Röyttä, M. Comparison of Peripheral Nerve Regeneration with Side-to-Side, End-to-Side, and End-to-End Repairs. *Plast. Reconstr. Surg. Glob. Open* **2016**, *4*, e1179. [CrossRef]
26. Battiston, B.; Artiaco, S.; Conforti, L.G.; Vasario, G.; Tos, P. End-to-Side Nerve Suture in Traumatic Injuries of Brachial Plexus: Review of the Literature and Personal Case Series. *J. Hand Surg. Eur. Vol.* **2009**, *34*, 656–659. [CrossRef]
27. Chimutengwende-Gordon, M.; Khan, W. Recent Advances and Developments in Neural Repair and Regeneration for Hand Surgery. *Open Orthop. J.* **2012**, *6*, 103–107. [CrossRef] [PubMed]
28. Wang, W.; Degrugillier, L.; Tremp, M.; Prautsch, K.; Sottaz, L.; Schaefer, D.J.; Madduri, S.; Kalbermatten, D. Nerve Repair With Fibrin Nerve Conduit and Modified Suture Placement. *Anat. Rec.* **2018**, *301*, 1690–1696. [CrossRef]
29. Chow, N.; Miears, H.; Cox, C.; MacKay, B. Fibrin Glue and Its Alternatives in Peripheral Nerve Repair. *Ann. Plast. Surg.* **2021**, *86*, 103–108. [CrossRef]
30. Bora, F.W.; Pleasure, D.E.; Didizian, N.A. A Study of Nerve Regeneration and Neuroma Formation after Nerve Suture by Various Techniques. *J. Hand Surg. Am.* **1976**, *1*, 138–143. [CrossRef] [PubMed]
31. Martins, R.S.; Siqueira, M.G.; da Silva, C.F.; Plese, J.P.P. Overall Assessment of Regeneration in Peripheral Nerve Lesion Repair Using Fibrin Glue, Suture, or a Combination of the 2 Techniques in a Rat Model. Which Is the Ideal Choice? *Surg. Neurol.* **2005**, *64*, S10–S16. [CrossRef] [PubMed]
32. Choi, B.-H.; Han, S.-G.; Kim, S.-H.; Zhu, S.-J.; Huh, J.-Y.; Jung, J.-H.; Lee, S.-H.; Kim, B.-Y. Autologous Fibrin Glue in Peripheral Nerve Regeneration in Vivo. *Microsurgery* **2005**, *25*, 495–499. [CrossRef]
33. Ferreira, R.S.; de Barros, L.C.; Abbade, L.P.F.; Barraviera, S.R.C.S.; Silvares, M.R.C.; de Pontes, L.G.; dos Santos, L.D.; Barraviera, B. Heterologous Fibrin Sealant Derived from Snake Venom: From Bench to Bedside—An Overview. *J. Venom. Anim. Toxins Incl. Trop. Dis.* **2017**, *23*, 21. [CrossRef] [PubMed]
34. Barros, L.C.; Ferreira, R.S.; Barraviera, S.R.C.S.; Stolf, H.O.; Thomazini-Santos, I.A.; Mendes-Giannini, M.J.S.; Toscano, E.; Barraviera, B. A New Fibrin Sealant From *Crotalus Durissus Terrificus* Venom: Applications in Medicine. *J. Toxicol. Environ. Health B Crit. Rev.* **2009**, *12*, 553–571. [CrossRef] [PubMed]
35. Gatti, M.; Vieira, L.; Barraviera, B.; Barraviera, S. Treatment of Venous Ulcers with Fibrin Sealant Derived from Snake Venom. *J. Venom. Anim. Toxins Incl. Trop. Dis.* **2011**, *17*, 226–229. [CrossRef]
36. Leite, A.P.S.; Pinto, C.G.; Tibúrcio, F.C.; Sartori, A.A.; de Castro Rodrigues, A.; Barraviera, B.; Ferreira, R.S.; Filadelpho, A.L.; Matheus, S.M.M. Heterologous Fibrin Sealant Potentiates Axonal Regeneration after Peripheral Nerve Injury with Reduction in the Number of Suture Points. *Injury* **2019**, *50*, 834–847. [CrossRef] [PubMed]
37. Buchaim, D.V.; Andreo, J.C.; Ferreira Junior, R.S.; Barraviera, B.; de Rodrigues, A.C.; de Macedo, M.C.; Rosa Junior, G.M.; Shinohara, A.L.; Santos German, I.J.; Pomini, K.T.; et al. Efficacy of Laser Photobiomodulation on Morphological and Functional Repair of the Facial Nerve. *Photomed. Laser Surg.* **2017**, *35*, 442–449. [CrossRef]
38. De Rosso, M.P.O.; Rosa Júnior, G.M.; Buchaim, D.V.; German, I.J.S.; Pomini, K.T.; de Souza, R.G.; Pereira, M.; Favaretto Júnior, I.A.; de Souza Bueno, C.R.; de Oliveira Gonçalves, J.B.; et al. Stimulation of Morphofunctional Repair of the Facial Nerve with Photobiomodulation, Using the End-to-Side Technique or a New Heterologous Fibrin Sealant. *J. Photochem. Photobiol. B* **2017**, *175*, 20–28. [CrossRef]

39. Lee, J.I.; Gurjar, A.A.; Talukder, M.A.H.; Rodenhouse, A.; Manto, K.; O'Brien, M.; Govindappa, P.K.; Elfar, J.C. A Novel Nerve Transection and Repair Method in Mice: Histomorphometric Analysis of Nerves, Blood Vessels, and Muscles with Functional Recovery. *Sci. Rep.* **2020**, *10*, 21637. [CrossRef] [PubMed]
40. de Bueno, C.R.S.; Pereira, M.; Favaretto Junior, I.A.; Bortoluci, C.H.F.; dos Santos, T.C.P.; Dias, D.V.; Daré, L.R.; Rosa Junior, G.M. Electrical Stimulation Attenuates Morphological Alterations and Prevents Atrophy of the Denervated Cranial Tibial Muscle. *Einstein* **2017**, *15*, 71–76. [CrossRef] [PubMed]
41. Bertin, J.S.F.; Marques, M.J.; Macedo, A.B.; de Carvalho, S.C.; Neto, H.S. Effect of Photobiomodulation on Denervation-Induced Skeletal Muscle Atrophy and Autophagy: A Study in Mice. *J. Manip. Physiol. Ther.* **2022**, *45*, 97–103. [CrossRef]
42. Pasquale, C.; Utyuzh, A.; Mikhailova, M.V.; Colombo, E.; Amaroli, A. Recovery from Idiopathic Facial Paralysis (Bell's Palsy) Using Photobiomodulation in Patients Non-Responsive to Standard Treatment: A Case Series Study. *Photonics* **2021**, *8*, 341. [CrossRef]
43. Hakimiha, N.; Rokn, A.R.; Younespour, S.; Moslemi, N. Photobiomodulation Therapy for the Management of Patients With Inferior Alveolar Neurosensory Disturbance Associated With Oral Surgical Procedures: An Interventional Case Series Study. *J. Lasers Med. Sci.* **2020**, *11*, S113–S118. [CrossRef] [PubMed]
44. Dias, F.J.; Fazan, V.P.S.; Cury, D.P.; de Almeida, S.R.Y.; Borie, E.; Fuentes, R.; Coutinho-Netto, J.; Watanabe, I. Growth Factors Expression and Ultrastructural Morphology after Application of Low-Level Laser and Natural Latex Protein on a Sciatic Nerve Crush-Type Injury. *PLoS ONE* **2019**, *14*, e0210211. [CrossRef] [PubMed]
45. Rosso, M.; Buchaim, D.; Kawano, N.; Furlanette, G.; Pomini, K.; Buchaim, R. Photobiomodulation Therapy (PBMT) in Peripheral Nerve Regeneration: A Systematic Review. *Bioengineering* **2018**, *5*, 44. [CrossRef] [PubMed]
46. Gigo-Benato, D.; Russo, T.L.; Tanaka, E.H.; Assis, L.; Salvini, T.F.; Parizotto, N.A. Effects of 660 and 780 Nm Low-Level Laser Therapy on Neuromuscular Recovery after Crush Injury in Rat Sciatic Nerve. *Lasers Surg. Med.* **2010**, *42*, 673–682. [CrossRef] [PubMed]
47. Andraus, R.A.C.; Maia, L.P.; de Souza Lino, A.D.; Fernandes, K.B.P.; de Matos Gomes, M.V.; de Jesus Guirro, R.R.; Barbieri, C.H. LLLT Actives MMP-2 and Increases Muscle Mechanical Resistance after Nerve Sciatic Rat Regeneration. *Lasers Med. Sci.* **2017**, *32*, 771–778. [CrossRef] [PubMed]
48. Mandelbaum-Livnat, M.M.; Almog, M.; Nissan, M.; Loeb, E.; Shapira, Y.; Rochkind, S. Photobiomodulation Triple Treatment in Peripheral Nerve Injury: Nerve and Muscle Response. *Photomed. Laser Surg.* **2016**, *34*, 638–645. [CrossRef] [PubMed]
49. Rochkind, S.; Geuna, S.; Shainberg, A. Phototherapy and nerve injury: Focus on muscle response. *Int. Rev. Neurobiol.* **2013**, *109*, 99–109. [CrossRef]
50. Buchaim, R.L.; Andreo, J.C.; Barraviera, B.; Ferreira Junior, R.S.; Buchaim, D.V.; Rosa Junior, G.M.; de Oliveira, A.L.R.; de Castro Rodrigues, A. Effect of Low-Level Laser Therapy (LLLT) on Peripheral Nerve Regeneration Using Fibrin Glue Derived from Snake Venom. *Injury* **2015**, *46*, 655–660. [CrossRef]
51. de Faria, S.D.; Testa, J.R.G.; Borin, A.; Toledo, R.N. Standardization of Techniques Used in Facial Nerve Section and Facial Movement Evaluation in Rats. *Braz. J. Otorhinolaryngol.* **2006**, *72*, 341–347. [CrossRef]
52. Koopman, J.E.; Duraku, L.S.; de Jong, T.; de Vries, R.B.M.; Michiel Zuidam, J.; Hundepool, C.A. A Systematic Review and Meta-Analysis on the Use of Fibrin Glue in Peripheral Nerve Repair: Can We Just Glue It? *J. Plast. Reconstr. Aesthet. Surg.* **2022**, *75*, 1018–1033. [CrossRef] [PubMed]
53. Sameem, M.; Wood, T.J.; Bain, J.R. A Systematic Review on the Use of Fibrin Glue for Peripheral Nerve Repair. *Plast. Reconstr. Surg.* **2011**, *127*, 2381–2390. [CrossRef] [PubMed]
54. Palazzi, S.; Vila-Torres, J.; Lorenzo, J. Fibrin Glue Is A Sealant and Not a Nerve Barrier. *J. Reconstr. Microsurg.* **1995**, *11*, 135–139. [CrossRef]
55. Rafijah, G.; Bowen, A.J.; Dolores, C.; Vitali, R.; Mozaffar, T.; Gupta, R. The Effects of Adjuvant Fibrin Sealant on the Surgical Repair of Segmental Nerve Defects in an Animal Model. *J. Hand Surg. Am.* **2013**, *38*, 847–855. [CrossRef]
56. Modrak, M.; Talukder, M.A.H.; Gurgenashvili, K.; Noble, M.; Elfar, J.C. Peripheral Nerve Injury and Myelination: Potential Therapeutic Strategies. *J. Neurosci. Res.* **2020**, *98*, 780–795. [CrossRef] [PubMed]
57. della Santa, G.M.L.; Ferreira, M.C.; Machado, T.P.G.; Oliveira, M.X.; Santos, A.P. Effects of Photobiomodulation Therapy (LED 630 Nm) on Muscle and Nerve Histomorphometry after Axonotmesis. *Photochem. Photobiol.* **2021**, *97*, 1116–1122. [CrossRef]
58. Faroni, A.; Mobasseri, S.A.; Kingham, P.J.; Reid, A.J. Peripheral Nerve Regeneration: Experimental Strategies and Future Perspectives. *Adv. Drug. Deliv. Rev.* **2015**, *82–83*, 160–167. [CrossRef]
59. Andreo, L.; Soldera, C.B.; Ribeiro, B.G.; de Matos, P.R.V.; Bussadori, S.K.; Fernandes, K.P.S.; Mesquita-Ferrari, R.A. Effects of Photobiomodulation on Experimental Models of Peripheral Nerve Injury. *Lasers Med. Sci.* **2017**, *32*, 2155–2165. [CrossRef]
60. Lee, J.; Carpena, N.T.; Kim, S.; Lee, M.Y.; Jung, J.Y.; Choi, J.E. Photobiomodulation at a Wavelength of 633 Nm Leads to Faster Functional Recovery than 804 Nm after Facial Nerve Injury. *J. Biophotonics* **2021**, *14*, e202100159. [CrossRef]
61. Li, B.; Wang, X. Photobiomodulation Enhances Facial Nerve Regeneration via Activation of PI3K/Akt Signaling Pathway–Mediated Antioxidant Response. *Lasers Med. Sci.* **2022**, *37*, 993–1006. [CrossRef]
62. Gordon, T. Peripheral Nerve Regeneration and Muscle Reinnervation. *Int. J. Mol. Sci.* **2020**, *21*, 8652. [CrossRef] [PubMed]
63. Geuna, S.; Raimondo, S.; Ronchi, G.; di Scipio, F.; Tos, P.; Czaja, K.; Fornaro, M. Chapter 3 Histology of the Peripheral Nerve and Changes Occurring During Nerve Regeneration. *Int. Rev. Neurobiol.* **2009**, *87*, 27–46. [CrossRef] [PubMed]

64. MacKinnon, S.E.; Dellon, A.L.; O'Brien, J.P. Changes in Nerve Fiber Numbers Distal to a Nerve Repair in the Rat Sciatic Nerve Model. *Muscle Nerve* **1991**, *14*, 1116–1222. [CrossRef]
65. Wang, M.L.; Rivlin, M.; Graham, J.G.; Beredjiklian, P.K. Peripheral Nerve Injury, Scarring, and Recovery. *Connect. Tissue. Res.* **2019**, *60*, 3–9. [CrossRef]
66. Fu, T.; Jiang, L.; Peng, Y.; Li, Z.; Liu, S.; Lu, J.; Zhang, F.; Zhang, J. Electrical Muscle Stimulation Accelerates Functional Recovery after Nerve Injury. *Neuroscience* **2020**, *426*, 179–188. [CrossRef]
67. Chu, X.-L.; Song, X.-Z.; Li, Q.; Li, Y.-R.; He, F.; Gu, X.-S.; Ming, D. Basic Mechanisms of Peripheral Nerve Injury and Treatment via Electrical Stimulation. *Neural Regen. Res.* **2022**, *17*, 2185–2193. [CrossRef]
68. Liu, M.; Zhang, D.; Shao, C.; Liu, J.; Ding, F.; Gu, X. Expression Pattern of Myostatin in Gastrocnemius Muscle of Rats after Sciatic Nerve Crush Injury. *Muscle Nerve* **2007**, *35*, 649–656. [CrossRef] [PubMed]
69. Angelov, D.N.; Ceynowa, M.; Guntinas-Lichius, O.; Streppel, M.; Grosheva, M.; Kiryakova, S.I.; Skouras, E.; Maegele, M.; Irintchev, A.; Neiss, W.F.; et al. Mechanical Stimulation of Paralyzed Vibrissal Muscles Following Facial Nerve Injury in Adult Rat Promotes Full Recovery of Whisking. *Neurobiol. Dis.* **2007**, *26*, 229–242. [CrossRef]
70. Pinto, M.M.R.; dos Santos, D.R.; de Barros Bentes, L.G.; Lemos, R.S.; de Almeida, N.R.C.; Fernandes, M.R.N.; Braga, J.P.; Somensi, D.N.; Barros, R.S.M. Anatomical Description of the Extratemporal Facial Nerve under High-Definition System: A Microsurgical Study in Rats. *Acta Cir. Bras.* **2022**, *37*, e370803. [CrossRef]
71. DeLeonibus, A.; Rezaei, M.; Fahradyan, V.; Silver, J.; Rampazzo, A.; Bassiri Gharb, B. A meta-analysis of Functional Outcomes in Rat Sciatic Nerve Injury Models. *Microsurgery* **2021**, *41*, 286–295. [CrossRef] [PubMed]
72. Dinh, P.; Hazel, A.; Palispis, W.; Suryadevara, S.; Gupta, R. Functional Assessment after Sciatic Nerve Injury in a Rat Model. *Microsurgery* **2009**, *29*, 644–649. [CrossRef]
73. Yian, C.H.; Paniello, R.C.; Gershon Spector, J. Inhibition of Motor Nerve Regeneration in a Rabbit Facial Nerve Model. *Laryngoscope* **2001**, *111*, 786–791. [CrossRef] [PubMed]
74. Liu, H.; Huang, H.; Bi, W.; Tan, X.; Li, R.; Wen, W.; Song, W.; Zhang, Y.; Zhang, F.; Hu, M. Effect of Chitosan Combined with Hyaluronate on Promoting the Recovery of Postoperative Facial Nerve Regeneration and Function in Rabbits. *Exp. Ther. Med.* **2018**, *16*, 739–745. [CrossRef]
75. de Bueno, C.R.S.; Pereira, M.; Favaretto-Júnior, I.A.; Buchaim, R.L.; Andreo, J.C.; Rodrigues, A.d.C.; Rosa-Júnior, G.M. Comparative Study between Standard and Inside-out Vein Graft Techniques on Sciatic Nerve Repair of Rats. Muscular and Functional Analysis. *Acta Cir. Bras.* **2017**, *32*, 287–296. [CrossRef] [PubMed]
76. Manthou, M.E.; Gencheva, D.; Sinis, N.; Rink, S.; Papamitsou, T.; Abdulla, D.; Bendella, H.; Sarikcioglu, L.; Angelov, D.N. Facial Nerve Repair by Muscle-Vein Conduit in Rats: Functional Recovery and Muscle Reinnervation. *Tissue Eng. Part A* **2021**, *27*, 351–361. [CrossRef]
77. Viterbo, F.; Brock, R.S.; Maciel, F.; Ayestaray, B.; Garbino, J.A.; Rodrigues, C.P. End-to-Side versus End-to-End Neurorrhaphy at the Peroneal Nerve in Rats. *Acta Cir. Bras.* **2017**, *32*, 697–705. [CrossRef]
78. Sulaiman, O.A.R.; Gordon, T. A Rat Study of the Use of End-to-Side Peripheral Nerve Repair as a "Babysitting" Technique to Reduce the Deleterious Effect of Chronic Denervation. *J. Neurosurg.* **2019**, *131*, 622–632. [CrossRef]
79. Kouyoumdjian, J.; Graç, C.; Ferreira, V.M. Peripheral Nerve Injuries: A Retrospective Survey of 1124 Cases. *Neurol. India* **2017**, *65*, 551–555. [CrossRef]
80. Kouyoumdjian, J.A. Peripheral Nerve Injuries: A Retrospective Survey of 456 Cases. *Muscle Nerve* **2006**, *34*, 785–788. [CrossRef]
81. Ronchi, G.; Cillino, M.; Gambarotta, G.; Fornasari, B.E.; Raimondo, S.; Pugliese, P.; Tos, P.; Cordova, A.; Moschella, F.; Geuna, S. Irreversible Changes Occurring in Long-Term Denervated Schwann Cells Affect Delayed Nerve Repair. *J. Neurosurg.* **2017**, *127*, 843–856. [CrossRef] [PubMed]
82. Jessen, K.R.; Mirsky, R. The Success and Failure of the Schwann Cell Response to Nerve Injury. *Front. Cell. Neurosci.* **2019**, *13*, 33. [CrossRef]
83. Buchaim, D.V.; de Rodrigues, A.C.; Buchaim, R.L.; Barraviera, B.; Junior, R.S.F.; Junior, G.M.R.; de Souza Bueno, C.R.; Roque, D.D.; Dias, D.V.; Dare, L.R.; et al. The New Heterologous Fibrin Sealant in Combination with Low-Level Laser Therapy (LLLT) in the Repair of the Buccal Branch of the Facial Nerve. *Lasers Med. Sci.* **2016**, *31*, 965–972. [CrossRef] [PubMed]
84. Tibúrcio, F.C.; Muller, K.S.; Leite, A.P.S.; de Oliveira, I.R.A.; Barraviera, B.; Ferreira, R.S., Jr.; Padovani, C.R.; Pinto, C.G.; Matheus, S.M.M. Neuroregeneration and immune response after neurorrhaphy are improved with the use of heterologous fibrin biopolymer in addition to suture repair alone. *Muscle Nerve* **2023**. ahead of print. [CrossRef]
85. Pinto, C.G.; Leite, A.P.S.; Sartori, A.A.; Tibúrcio, F.C.; Barraviera, B.; Junior, R.S.F.; Filadelpho, A.L.; de Carvalho, S.C.; Matheus, S.M.M. Heterologous fibrin biopolymer associated to a single suture stitch enables the return of neuromuscular junction to its mature pattern after peripheral nerve injury. *Injury* **2021**, *52*, 731–737. [CrossRef] [PubMed]

86. Abbade, L.P.F.; Barraviera, S.R.C.S.; Silvares, M.R.C.; de Lima, A.B.B.C.O.; Haddad, G.R.; Gatti, M.A.N.; Medolago, N.B.; Rigotto Carneiro, M.T.; dos Santos, L.D.; Ferreira, R.S.; et al. Treatment of Chronic Venous Ulcers With Heterologous Fibrin Sealant: A Phase I/II Clinical Trial. *Front. Immunol.* **2021**, *12*, 627541. [CrossRef]
87. Daré, L.R.; Dias, D.V.; Rosa Junior, G.M.; Bueno, C.R.S.; Buchaim, R.L.; Rodrigues, A.d.C.; Andreo, J.C. Effect of β-Hydroxy-β-Methylbutyrate in Masticatory Muscles of Rats. *J. Anat.* **2015**, *226*, 40–46. [CrossRef] [PubMed]

International Journal of
Molecular Sciences

Article

The Establishment of a Mouse Model of Recurrent Primary Dysmenorrhea

Fang Hong [1], Guiyan He [1], Manqi Zhang [2], Boyang Yu [1,3,*] and Chengzhi Chai [1,*]

[1] Jiangsu Provincial Key Laboratory of TCM Evaluation and Translational Research, School of Traditional Chinese Pharmacy, China Pharmaceutical University, Nanjing 211198, China; hongfang0221@126.com (F.H.); heguiyan0429@163.com (G.H.)

[2] Department of Medicine, Duke University, Durham, NC 27708, USA; manqi.zhang@duke.edu

[3] Research Center for Traceability and Standardization of TCMs, School of Traditional Chinese Pharmacy, China Pharmaceutical University, Nanjing 211198, China

* Correspondence: boyangyu59@163.com (B.Y.); chengzhichai@cpu.edu.cn (C.C.)

Abstract: Primary dysmenorrhea is one of the most common reasons for gynecologic visits, but due to the lack of suitable animal models, the pathologic mechanisms and related drug development are limited. Herein, we establish a new mouse model which can mimic the periodic occurrence of primary dysmenorrhea to solve this problem. Non-pregnant female mice were pretreated with estradiol benzoate for 3 consecutive days. After that, mice were injected with oxytocin to simulate menstrual pain on the 4th, 8th, 12th, and 16th days (four estrus cycles). Assessment of the cumulative writhing score, uterine tissue morphology, and uterine artery blood flow and biochemical analysis were performed at each time point. Oxytocin injection induced an equally severe writhing reaction and increased $PGF_{2\alpha}$ accompanied with upregulated expression of COX-2 on the 4th and 8th days. In addition, decreased uterine artery blood flow but increased resistive index (RI) and pulsatility index (PI) were also observed. Furthermore, the metabolomics analysis results indicated that arachidonic acid metabolism; linoleic acid metabolism; glycerophospholipid metabolism; valine, leucine, and isoleucine biosynthesis; alpha-linolenic acid metabolism; and biosynthesis of unsaturated fatty acids might play important roles in the recurrence of primary dysmenorrhea. This new mouse model is able to mimic the clinical characteristics of primary dysmenorrhea for up to two estrous cycles.

Keywords: primary dysmenorrhea; recurrent; mice model; two estrous cycles; writhing reaction; $PGF_{2\alpha}$; PGE_2; COX-2; uterine artery blood flow; metabonomic analysis

Citation: Hong, F.; He, G.; Zhang, M.; Yu, B.; Chai, C. The Establishment of a Mouse Model of Recurrent Primary Dysmenorrhea. *Int. J. Mol. Sci.* **2022**, *23*, 6128. https://doi.org/10.3390/ijms23116128

Academic Editors: Masaru Tanaka, Simone Battaglia, Lydia Giménez-Llort, Chong Chen and Piril Hepsomali

Received: 23 April 2022
Accepted: 28 May 2022
Published: 30 May 2022

Publisher's Note: MDPI stays neutral with regard to jurisdictional claims in published maps and institutional affiliations.

1. Introduction

Dysmenorrhea, defined as painful menstrual cramps of uterine origin, is the most common gynecological disorder in women [1–3]. The reported prevalence of dysmenorrhea is more than 50% [4–6]. Based on pathophysiology, dysmenorrhea is divided into primary dysmenorrhea (PD) and secondary dysmenorrhea [2]. PD refers to cyclic menstrual pain in the absence of pelvic anomalies [2,7,8] while secondary dysmenorrhea is caused by identifiable pathological conditions, such as adenomyosis, endometriosis, fibroids, and pelvic inflammatory disease [2,4,9].

Many women are lacking medical treatment because menstrual pain is usually regarded as a concomitant menstrual symptom [4]. However, PD not only affects the quality of life during menstrual periods but also has an effect on psychological changes, such as pain perception, trait empathy, anxiety, depression, etc. [10–12]. In addition, some studies even showed the altered white matter microarchitecture and altered posterior cerebellar lobule connectivity with the perigenual anterior cingulate cortex [13–15].

In recent years, many studies have been carried out focusing on the pathological mechanism of PD and tried to develop more drugs for PD treatment [16–22]. One acknowledged mechanism is that the overproduction of uterine PGs causes myometrial hypercontractility, resulting in ischemia and hypoxia of the uterine muscle and pain [2,4]. Additionally, elevated vasopressin and decreased progesterone are also accompanied by PD [23–26]. Nowadays, non-steroidal anti-inflammatory drugs (NSAIDs) and oral contraceptives are used to inhibit the production of $PGF_{2\alpha}$ [27,28]. Oxytocin receptor inhibitors, calcium channel inhibitors, and Chinese herbal medicine are also used for relieving menstrual pain [29–31]. Furthermore, encouragingly, there are studies which showed that some drugs and treatments had a long-term menstrual pain relief effect [32–34].

Though many drugs are used for menstrual pain in the clinic, most of them have common shortcomings of periodic medication and inevitable side effects [18]. Furthermore, due to the lack of an appropriate animal model, PD-related studies were only conducted on acute animal models or clinical studies [17,26,35,36], which hindered further research on PD pathological and the development of drugs for PD treatment. Unlike other animal models of chronic pain, an ideal PD model is very rare because the periodically occurring characteristics of menstrual pain are difficult to simulate [37].

Appropriate animal models, as replicas of human diseases, can recapitulate disease pathophysiology and clinical features [38]. Here, we describe a pioneering mouse model of recurrent PD induced by estrogen combined with oxytocin. This mouse model fully characterized the PD-related clinical features for two consecutive cycles. A metabolic analysis was performed in this study to validate the major features of the PD model and explore the mechanism of periodical PD attacks. The establishment of this mouse model could be used to investigate the pathological mechanism of recurrent PD and to screen radical treatment drugs for PD.

2. Results

2.1. Estrous Cycle Monitor

After three consecutive days of estradiol benzoate injection, 83.33% of the mice reached the estrus stage (fourth day). Interestingly, on the eighth day, vaginal smears demonstrated that 83.33% of the mice were in estrus again, but only 50% and 33.33% of them were in estrus on the 12th day and 16th day, respectively (Figure 1A,B). It suggested that the estrus cycle synchronization induced by estradiol benzoate was maintained for at least two estrous cycles (8 days).

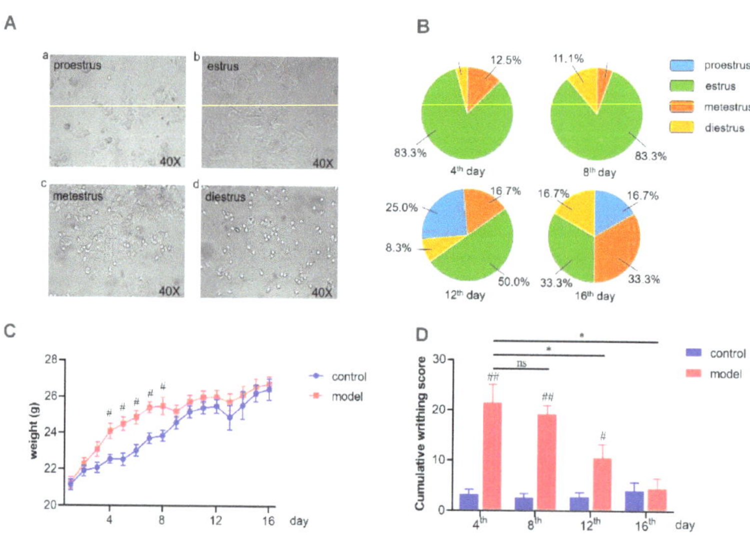

Figure 1. *Cont.*

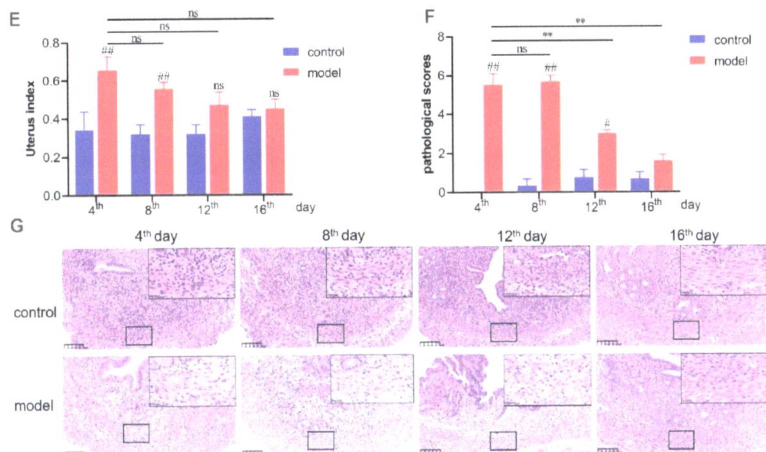

Figure 1. The regular detection indexes relative to the PD mice model. (**A**) Photomicrographs of vaginal smear from mice at proestrus (**a**), estrus (**b**), metestrus (**c**), and diestrus (**d**). (**B**) The percentage of each estrous stage from model group on 4th, 8th, 12th, and 16th days. (**C**) Body weight changes of the mice. (**D**) Oxytocin-induced writhing response score during the experiment. (**E**) Uterus index of control group and model group on 4th, 8th, 12th, and 16th days. (**F**) Pathological scores of uterus H&E staining. (**G**) Example pathological section of the uterus with H&E staining (20×). # Model group compared with corresponding control group. $^{\#}\, p < 0.05$; $^{\#\#}\, p < 0.01$. $^{*}\, p < 0.05$; $^{**}\, p < 0.01$. ns, no significance. (n = 6 per group). The statistical values are presented in Table S1.

2.2. Writhing Responses and Uterine Morphological Changes

Women's menstrual cycle, which is associated with menstrual distress, is always accompanied by water retention [39,40]. Therefore, the mouse body weight was recorded daily, and the average body weight of the model mice was significantly increased compared to the control mice from the 4th day to the 8th day (Figure 1C). The writhing response was similar in mice being treated with estradiol after the 4th and 8th days but decreased in mice treated after 12 days. The writhing score of the model group was even reduced compared to the control group on the 16th day (Figure 1D). Furthermore, the uterine tissue showed edema on both the 4th day and the 8th day but no significant difference on the 12th day and 16th day between the model group and the control group (Figure 1E).

2.3. Histomorphology Assessment of Uterus

Severe menstrual pain induces uterine histomorphological alterations, including discontinuous smooth muscle cells and disorganized arrangement of muscle fibers, i.e., disarray [16,21]. As shown in Figure 1F,G, compared with the control group, mice in the model group showed significant pathological changes on the 4th and 8th days. However, the uterine histomorphology started to recover from the 12th day. On the 16th day, there was no significant difference between the control group and the model group.

2.4. Characterization of Uterine Artery Blood Flow Features

By evaluating writhing response and histomorphology change, we showed that the PD-related indexes were significantly affected by the oxytocin injection in model mice, while no change was observed in the control group. We compared multiple blood flow indexes in the model mice and the control group before and after oxytocin injection on the 4th, 8th, 12th, and 16th days. Figure 2A shows a representative image of the uterine artery Doppler waveform from model mice. Oxytocin injection significantly decreased the maximum, minimum, and average flow velocity (V_{max}, V_{min}, and V_{mean}) of the uterine artery (Figure 2B–D) on the 4th and 8th days but was not able to do so on the 12th day and 16th day. The resistive index (RI) and the pulsatility index (PI) are frequently used to assess the resistance in the pulsatile vascular system. Here, after oxytocin injection, the RI was elevated on the 4th and 8th days, but no difference was detected on the 12th or 16th days (Figure 2E). Similarly, the PI was also elevated on the 4th and 8th days after oxytocin injection but was not changed on the 12th or 16th days (Figure 2F). The velocity time integral (VTI) declined on the 4th and 8th days but showed no change on the 12th or 16th days (Figure 2G). These data demonstrate that the model mice exhibited PD-induced uterine artery blood flow features on both the 4th and 8th days.

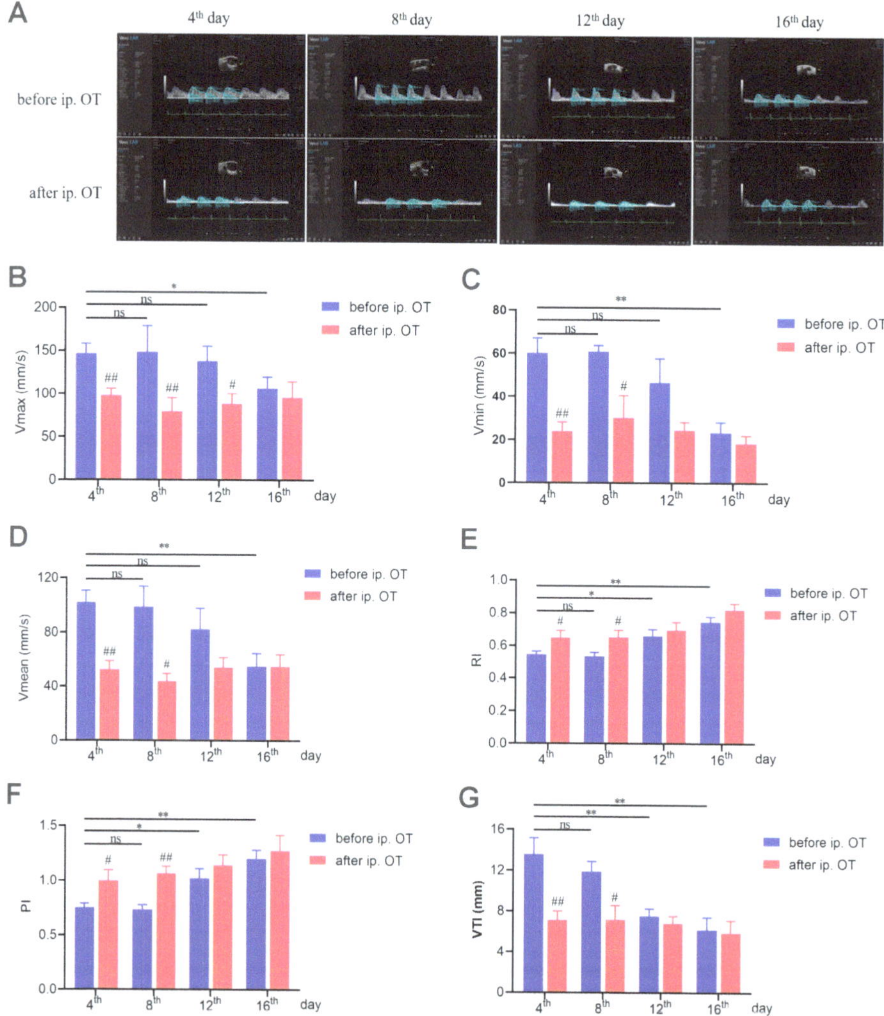

Figure 2. PD-related uterine artery blood flow features. (**A**) Typical images of uterine artery Doppler waveforms from model mice before and after oxytocin injection. (**B**) The maximum flow velocity (V_{max}) of the uterine artery. (**C**) The minimum flow velocity (V_{min}) of the uterine artery. (**D**) The average flow velocity (V_{mean}) of the uterine artery. (**E**) Resistive index (RI) of uterine artery blood flow. (**F**) Pulsatility index (PI) of the uterine artery. (**G**) Velocity time integral (VTI) of uterine artery blood flow. # Model group after oxytocin injection compared with the corresponding model group before oxytocin injection. # $p < 0.05$; ## $p < 0.01$. * $p < 0.05$; ** $p < 0.01$. ns, no significance. (n = 6 per group).

2.5. Biochemical Analysis of PD-Related Indicators

In the clinic and acute PD model mice, $PGF_{2\alpha}$ substantially increases in both serum and uterine tissue, but PGE_2 decreases [22,23]. Here, the levels of $PGF_{2\alpha}$ and PGE_2 were measured on the 4th, 8th, 12th, and 16th days. As shown in Figure 3A–F, $PGF_{2\alpha}$ was elevated but PGE_2 was reduced in serum and the uterus and the ratio of $PGF_{2\alpha}$ and PGE_2 was increased on both the 4th and 8th days. However, on the 12th day, a change in $PGF_{2\alpha}$ and PGE_2 was only observed in uterine tissue. On the 16th day, oxytocin injection was unable to induce a change in $PGF_{2\alpha}$ and PGE_2 in the serum, nor in the uterus.

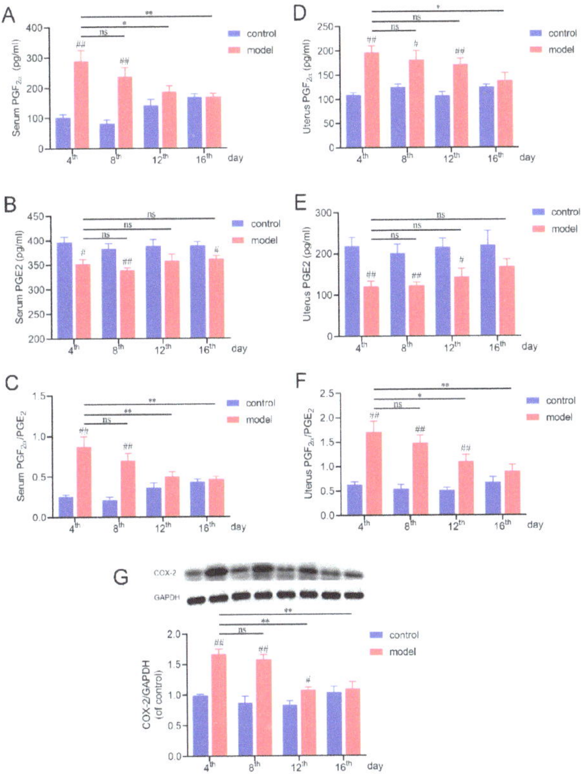

Figure 3. PD-related biochemical analysis. (**A–C**) $PGF_{2\alpha}$, PGE_2, and their ratio changes in serum. (**D–F**) $PGF_{2\alpha}$, PGE_2, and their ratio changes in uterine tissue. (**G**) COX-2 expression changes in uterine tissue during the experiment. # Model group compared with corresponding control group. # $p < 0.05$; ## $p < 0.01$; * $p < 0.05$; ** $p < 0.01$. ns, no significance (n = 6 per group). The statistical values are presented in Table S1.

2.6. COX-2 Expression

Cyclooxygenase 2 (COX-2), which can convert arachidonic acid (AA) into PGs, is an essential target in PD treatment [24]. Here, we showed that the expression of COX-2 was significantly upregulated on the 4th and 8th days in the model mice. The COX-2 induction was decreased from the 12th day and recovered to the basal level on the 16th day.

Our data showed that estradiol benzoate combined with oxytocin injection induced analogous menstrual pain in a similar context on the 4th and 8th days. Here, this PD mouse model is characterized as recurrent PD which was able to maintain its features for two estrous cycles.

2.7. Serum Metabolomics Analysis

We next performed a serum metabolomics analysis using this recurrent PD mouse model. The model mice were euthanized on the fourth day (M4) and the eighth day (M8). A clear separation was observed between the control (C) and M4/M8 groups in the principal component analysis (PCA) with both positive ion mode and negative ion mode (Figure 4A,B). We next compared the metabolic profiles of the control group vs. the M4 group and the control group vs. the M8 group using orthogonal partial least squares discriminant analysis (OPLS-DA) (Figure S1C–F). The overlapped metabolites might be involved during PD recurrence. We found a total of 61 metabolites that were significantly changed in the M4/M8 groups compared to the control (Table S2), and the relative expression levels are shown in Figure 4D. According to the enrichment of pathways (Figure 4E), arachidonic acid metabolism; biosynthesis of unsaturated fatty acids; linoleic metabolism; glycerophospholipid metabolism; and valine, leucine, and isoleucine biosynthesis are most affected.

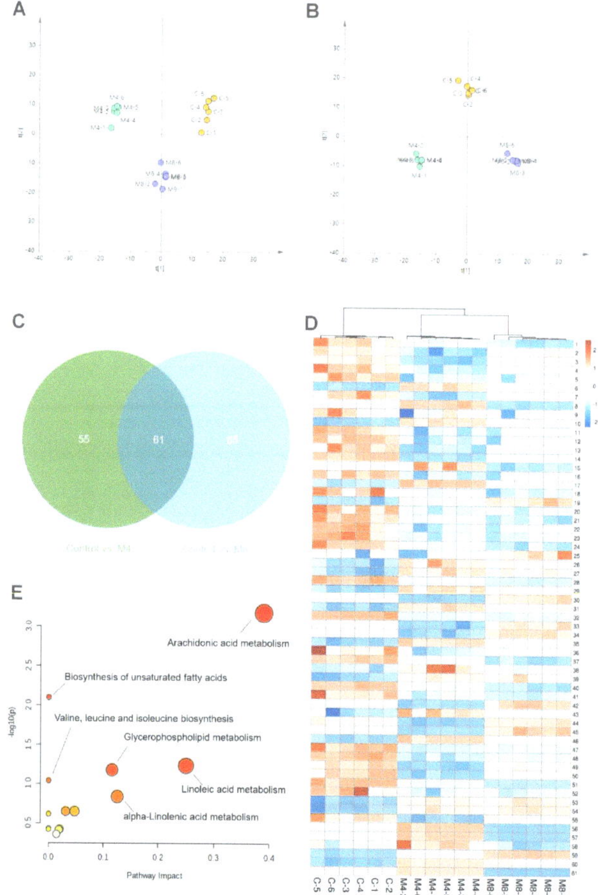

Figure 4. Serum metabolomics analysis of recurrent PD mouse model. (**A**) Principal component analysis (PCA) in positive mode. (**B**) PCA in negative mode. (**C**) Venn diagram of important metabolites relative to PD recurrence shared between control group vs. M4 group and control group vs. M8 group. (**D**) Heat map of important metabolites significantly changed in M4 group and M8 group compared to the control group. (**E**) Pathway analysis of important metabolites. n = 6 per group.

Furthermore, we performed a pathway and enrichment analysis of related regulatory enzymes. The protein interaction network was built up and is shown in Figure 5A. The node degree was ranked and the top 20 are shown in Figure 5B. The GO enrichment analysis suggested that the changed regulatory enzymes are mainly enriched in the lipid metabolic process, lipid catabolic process, and branched-chain acid catabolic process, which might play an important role in the development of PD. In addition, these enzymes are also enriched in the endoplasmic reticulum membrane and mitochondria (Figure 5C). The KEGG enrichment analysis indicated that these regulatory enzymes are involved in metabolic pathways, glycerophospholipid metabolism, glycerolipid metabolism, arachidonic acid metabolism, the phosphatidylinositol signaling system, etc. (Figure 5).

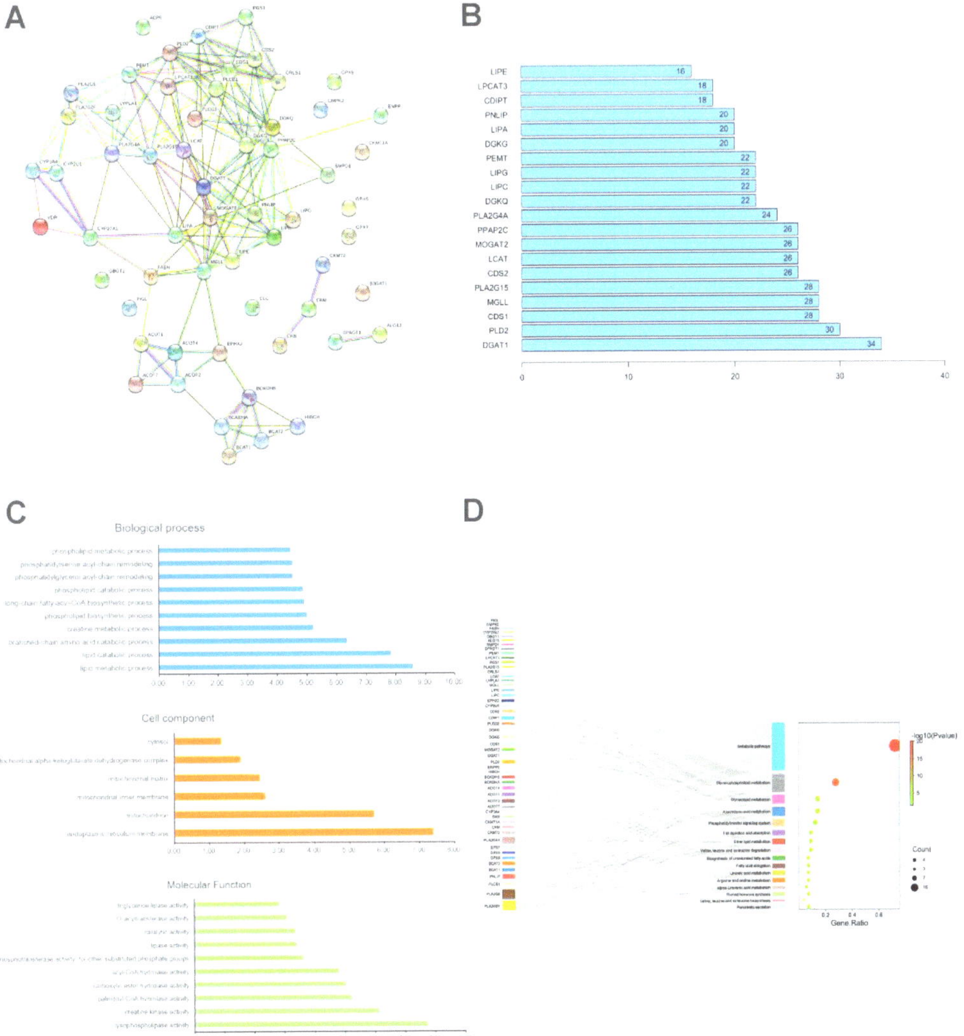

Figure 5. Protein–protein interactions and enrichment pathway analysis of regulatory enzymes. (**A**) Regulatory protein network map. (**B**) Top 20 proteins with high degree value in the PPI network. (**C**) GO enrichment analysis of the regulatory proteins. (**D**) KEGG enrichment analysis of the regulatory proteins.

3. Discussion

Primary dysmenorrhea (PD), characterized by recurrent crampy lower abdominal pain during menstruation but without pelvic pathology, is the most common complaint of women and affects more than 50% of women globally [1,4]. PD is a complex disorder involving hormone secretion, inflammation, metabolism, and sensory nerve conduction [41]. It is generally recognized that increased $PGF_{2\alpha}$ causes uterine contractions, which restricts blood flow and then produces hypoxia and ischemia in uterine tissue and perception of pain [1,2,4]. A clinical study showed that women with PD not only have elevated serum $PGF_{2\alpha}$ but also have higher serum ischemia-modified albumin (IMA) than those without menstrual pain during their menstrual period, indicating the role of ischemia in PD [42].

Though much PD-related research has been developed, it depends heavily on acute PD animal models or clinical trials. Animal model tests cannot achieve long-term monitoring of PD pathological processes, and clinical trials are not only time-consuming and expensive but also usually cannot meet the demand of basic research needs. Thus, developing an ideal PD animal model is quite necessary.

The animal models used at present are acute (single estrus cycle) PD models, which cannot present the characteristics of recurrent PD. Herein, we developed a mouse model of PD to mimic the periodic attacks of menstrual pain of women in the clinic. To the best of our knowledge, this is the first model of recurrent menstrual pain. As an improvement of the previously established acute PD model (one cycle, 4 days), this mouse model presented the same features of PD on the eighth day that we saw on the fourth day (Figure 1). In order to verify the consistency of PD indexes on both the 4th and 8th days, assessment of the cumulative writhing score, uterine tissue morphology, and uterine artery blood flow and biochemical analysis were performed. It is generally accepted that PD is mainly related to the abnormal synthesis and release of prostaglandins, resulting in hypoxia and ischemia of uterine muscle and the perception of pain [1,2,43]. $PGF_{2\alpha}$ has a negative effect as it induces potent vasoconstriction and myometrial contractions, whereas PGE_2 dilates blood vessels and increases blood flow. An increased ratio of $PGF_{2\alpha}/PGE_2$ leads to enhanced uterine smooth muscle contraction [44,45]. On both the 4th and 8th days, oxytocin injection induced a severe writhing response (Figure 1) and increased $PGF_{2\alpha}$ but reduced PGE_2 contents in serum and the uterus; furthermore, the ratio of $PGF_{2\alpha}/PGE_2$ also increased (Figure 3). In addition, COX-2, the major enzyme that induces $PGF_{2\alpha}$ production [42], was also significantly upregulated on the 4th and 8th days (Figure 3). What is more, the model mice showed decreased uterine artery blood flow and increased RI and PI on the 8th and 4th days. All these parameters in our research indicated that this PD mouse model could mimic the characteristics of the recurrence of PD.

As a miniature of human diseases, animal models make it convenient for researchers to take samples at any time according to the experimental purposes, which is difficult to do in clinical trials. In this regard, this recurrent PD model increases the window for sample collection, which is helpful for pathophysiology research on PD. In addition, this model is beneficial for screening drugs with long-term efficacy to address clinical needs.

Metabolomics analysis is a comprehensive examination of various metabolite reactions and is an effective technology for biomarker discovery, disease diagnosis, and therapeutic intervention [46,47]. To better understand the pathological mechanism of periodical PD attacks, serum samples of control mice and M4/M8 mice were analyzed by metabolomics. Finally, 61 differential metabolites were identified, which might be involved in the periodical occurrence of PD. The significantly changed metabolites were enriched in arachidonic acid metabolism; linoleic acid metabolism; glycerophospholipid metabolism; valine, leucine, and isoleucine biosynthesis; alpha-linolenic acid metabolism; and biosynthesis of unsaturated fatty acids.

Arachidonic acid metabolizes through different pathways, one of which is to generate $PGF_{2\alpha}$ through cyclooxygenase metabolism, and cyclooxygenase is a crucial target for alleviating menstrual pain [18,28,48,49]. Glycerophospholipid, an indispensable component of cell membranes and organelle membranes, can be hydrolyzed by phospholipase A2 to generate arachidonic acid [50,51]. Glycerophospholipid metabolism disorder leads to the excessive accumulation of arachidonic acid [52–54]. As a result, glycerophospholipid metabolism has an effect on the arachidonic acid pathway and influences $PGF_{2\alpha}$ and PGE_2 production. In addition, clinical studies have shown that reduced blood flow and an elevated level of oxidative stress were detected in PD-affected women [55–57]. Linoleic acid metabolism disorder is responsible for mitochondrial oxidative phosphorylation [58]. Studies showed that linoleic acid could stimulate ROS production by activating NADPH oxidase enzyme and thus disrupt mitochondrial function [59,60]. What is more, linoleic acid is the major fatty acid moiety of cardiolipin, which is a crucial component of the inner mitochondrial membrane and plays an important role in mitochondrial bioenergetics [61]. Serum cardiolipin significantly decreased in the model mice, which suggests that mitochondrial function might be impaired during menstrual pain, leading to oxidative stress. Branched-chain amino acids (BCAAs; valine, leucine, and isoleucine) are involved in a variety of biological processes, such as energy homeostasis, nutrient metabolism, and immunoreaction [62,63]. BCAA biosynthesis is related to the energy supply of mitochondria [64,65] and might take part in the energy metabolism in the uterus during PD.

Based on the above, the regulatory enzymes of these metabolites were analyzed (Figure 5). The GO enrichment analysis showed that these regulatory enzymes are mainly enriched in lipid metabolic and catabolic processes, the long-chain fatty-acyl-CoA biosynthetic process, and the branched-chain amino acid catabolic process, which is consistent with the metabolomics analysis. Furthermore, it also showed that lysophospholipase activity, creatine kinase activity, and palmitoyl-CoA hydrolase activity might participate in the pathological process of PD. Lysophospholipase, as the regulatory enzyme of lysophospholipids, mediates lipid homeostasis and lysophospholipid signaling [50,66]. Additionally, the regulatory enzymes of differentially expressed metabolites are also enriched in the cell components of reticulum membrane and mitochondria. As mentioned above, many metabolites are relevant to mitochondrial function, so it can be speculated that mitochondria are an important target in PD research. Studies also showed that intracellular calcium was significantly increased in uterine smooth muscles after oxytocin stimulation [67,68]. Endoplasmic reticulum membrane is essential for intracellular calcium homeostasis, and here, it might contribute to PD development through regulating intracellular calcium [69,70]. Consistently with this, the KEGG pathway enrichment analysis showed that these regulatory enzymes were included in glycerophospholipid metabolism, glycerolipid metabolism, arachidonic acid metabolism, etc. These metabolites and their corresponding regulatory enzymes might play important roles in PD recurrence. This study has provided a basis for subsequent research, especially the investigation of the mechanism by which the recurrence of PD is induced.

4. Conclusions

In summary, we established a mouse model of recurrent PD, whose characteristics were maintained for two estrous cycles. This model can provide an excellent platform for drug screening and pathological mechanism exploration of PD. Metabolism analysis is a great tool to help find important metabolites and pathways related to the recurrence of PD for further research and verification.

5. Materials and Methods

5.1. Animals

Female non-pregnant ICR mice (20 ± 2 g weight) were provided by the Experimental Animal Center of Yangzhou University, China (Animal quality certificate number: SCXK 2017–0007). Animals were housed in a 12-h light/dark cycle room at $23 \pm 1\ °C$ with adequate food and water. All animal welfare and experimental procedures were performed according to the National Institutes of Health Guide for the Care and Use of Laboratory Animals. Experiment protocols were approved by the Animal Ethics Committee of China Pharmaceutical University.

5.2. Reagents

Estradiol benzoate injections and oxytocin injections were purchased from Ningbo second hormone factory (China). $PGF_{2\alpha}$ and PGE_2 ELISA kits were obtained from Nanjing SenBeiJia Biological Technology Co., Ltd. (China). Bicinchoninic acid (BCA) assay kit and QuichBlockTM Western Bocking Buffer were purchased from Beyotime Biotechnology (China). COX-2 and GAPDH primary antibodies were obtained from Proteintech Group, Rosemont, IL, USA. Acetonitrile, methanol, and formic acid were obtained from Merck (Darmstadt, Germany). Ultrapure water was obtained using a Milli-Q purification system (Milford, MA, USA).

5.3. Animal Treatment

Forty-eight female mice were divided into a control group and a model group (24 per group). The estrous stage was monitored by vaginal smear for a total of 16 days (4 cycles) [71]. Model mice were intraperitoneally injected with estradiol benzoate ($10 \text{ mg·kg}^{-1} \text{·day}^{-1}$) for three consecutive days [41]. On the 4th, 8th, 12th, and 16th days, each mouse from the experimental group was intraperitoneally injected with oxytocin (0.4 U). The cumulative writhing score was recorded for each mouse within 30 min right after the injection. Mice in the control group were intraperitoneally injected with saline from the 1st day to the 3rd day. The criteria used for scoring the writhing were described in a previous study [72]. In brief, writhing was scored from 0 to 3, where 0 refers to normal body position and behaviors; 1 refers to body leaning to left or right; 2 refers to stretching of the hindlimbs and dorsiflexion of the hind paws, body stretched and flat on the bottom, the pelvis rotated sideward; and 3 refers to abdominal muscle contraction followed by body stretching and hind limbs extension.

5.4. Hematoxylin–Eosin Staining of Uterine Tissue

Uterine tissue was isolated and fixed in 4% paraformaldehyde for more than 24 h right after the writhing test. After dehydration, paraffin-embedded tissues were sliced to a thickness of 4 μm, hematoxylin and eosin staining were performed as previously described [35,73].

5.5. Uterine Artery Blood Flow Analysis

After anesthetization with 2% isoflurane, the uterine artery blood flow of the mice in the model group was monitored using Doppler ultrasound detection (Ultrasound biomicroscope, Vevo2100TM, VisualSonics, Toronto, ON, Canada) before and after oxytocin injection. Monitoring indicators included uterine artery blood flow maximum velocity (Vmax), uterine artery blood flow minimum velocity (Vmin), and average blood flow velocity (Vmean). The pulsatility index (PI) was calculated as follows: 2x (Vmax − Vmin)/(Vmax + Vmin); the resistance index (RI) was calculated as follows: (Vmax − Vmin)/Vmax.

5.6. $PGF_{2\alpha}$ and PGE_2 Measurement

After the writhing reaction, blood samples were collected and serum samples were obtained by centrifuging at $1000 \times g$ for 10 min. The uterine tissue was homogenized with PBS and then centrifuged at $15,000 \times g$ for 15 min to obtain the supernatant. The $PGF_{2\alpha}$ and PGE_2 contents in the serum and uterine tissue were tested using ELISA kits.

5.7. Western Blot Analysis

Total protein of uterine tissue was extracted using radioimmunoprecipitation assay buffer, and the protein concentration was quantified using the BCA assay kit. SDS-PAGE gels and PVDF membranes were used. The blots were blocked with QuichBlockTM Western Bocking Buffer (Beyotime Biotechnology, P0252), probed with COX-2 and GAPDH primary antibodies (Proteintech Group, 66351-1-Ig and 60004-1-Ig) overnight at 4 °C, and then incubated with the appropriate secondary antibodies. The values were expressed relative to the signal of GAPDH.

5.8. Untargeted Metabolomics Analysis

5.8.1. Serum Sample Pretreatment

At the end of each writhing experiment, serum was collected and stored at $-80\,^\circ$C for further metabolomics analysis. In brief, serum samples were thawed at room temperature and mixed with methanol in a ratio of 1:3. Next, the samples were centrifuged at $15,000\times g$ for 10 min (4 $^\circ$C), and the supernatant was transferred to a new tube. Subsequently, the supernatant was dried with nitrogen at room temperature and redissolved with methanol for further analysis.

5.8.2. LC-MS Analysis

Serum metabolite separation was performed on a SynergiTM Fusion-RP C18 column (50 mm $\times$ 2 mm internal diameter (i.d.), 2.5 μm) using an Agilent Technologies 6530 Accurate-Mass Q-TOF LC/MS system (Santa Clara, CA, USA). The mobile phase consisted of solvent A (0.1% aqueous formic acid solution) and solvent B (0.1% acetonitrile formic acid). The chromatographic conditions were as follows: 0–2 min, 0–5% B; 2–10 min, 5–60% B; 10–15 min, 60–70% B; 15–20 min, 70–80% B; 20–22 min, 80–95% B; 22–28 min, 95–60% B; 28–30 min, 60–30% B, and back to initial conditions (with 2 min for equilibration). The flow rate was 0.4 mL/min. Each sample injection volume was 5 μL. The MS was performed with an electrospray ionization (ESI) ion source in the positive (ESI+) and negative (ESI−) ion modes. The data were collected in both positive and negative ion modes with a mass range of 50–2000 Da. The MS parameters were set as follows: fragmental voltage, 120 V; nebulizer gas, 35 psig; capillary voltage, 4000 V; drying gas flow rate, 9 L/min; temperature, 325 $^\circ$C.

5.8.3. Method Assessment

The reproducibility and robustness of our experiment were tested using a quality control (QC) sample. In total, 10 μL of each serum sample was mixed and pretreated in the same way as the test samples to obtain a QC sample. During the analysis, the QC sample was analyzed every 6 injections. Next, the retention time and intensity of 5 different ions were extracted for calculating relative standard deviations (RSDs) to verify the repeatability of the method [44,53]. The typical TICs are shown in Figure S1A,B, and the RSD results of intensity and retention time are listed in Table S3.

5.8.4. Data Processing

The raw data were pretreated using an R package, including nonlinear retention time alignment, peak discrimination, filtering, alignment, and matching, etc. After that, the processed data were analyzed using SIMCA-P 14.1 (Umetrics, Umeå, Sweden) and MetaboAnalyst 3.0 on 22 November 2021 (https://www.metaboanalyst.ca/). Principal components analysis (PCA) and supervised orthogonal partial least squares discriminant analysis (OPLS-DA) were applied to analyze global metabolic profiles and detect the variables. The differential metabolites were identified with $p < 0.05$ (Student's t-test) and VIP > 1.0.

5.8.5. Metabolite Identification and Pathway Analysis

Differential metabolites' identification was reliant on online databases, including HMDB (http://www.hmdb.ca/ accessed on 22 April 2022), METLIN (http://metlin.scripps.edu/ accessed on 22 April 2022), and MassBank (http://www.massbank.jp/ accessed on 22 April 2022). Pathway enrichment analysis was carried out using MetaboAnalyst. The regulated enzymes were analyzed by STRING (https://cn.string-db.org/ accessed on 22 April 2022), DAVID (https://david.ncifcrf.gov/ accessed on 22 April 2022), and bioinformatics (http://www.bioinformatics.com.cn/en accessed on 22 April 2022).

5.9. Statistical Analysis

The experimental data were statistically analyzed using GraphPad Prism 7. The data were analyzed by Student's *t*-test or ANOVA and presented as mean $\pm$ SEM. A *p*-value < 0.05 was considered statistically significant.

Supplementary Materials: The following supporting information can be downloaded at: https://www.mdpi.com/article/10.3390/ijms23116128/s1.

Author Contributions: C.C. designed the experiments and revised the manuscript; F.H. and G.H. performed the experiments; F.H. and G.H. analyzed the data; M.Z. revised the manuscript; B.Y. contributed reagents/materials/analysis tools; F.H. and C.C. wrote the paper. All authors have read and agreed to the published version of the manuscript.

Funding: This work was supported by the "Double First-Class" University project (CPU2018GF06).

Institutional Review Board Statement: The animal study protocol was approved by the Ethics Committee of China Pharmaceutical University (No. 2020-12-004).

Informed Consent Statement: Not applicable.

Data Availability Statement: Not applicable.

Acknowledgments: This study acknowledges financial support from China Pharmaceutical University.

Conflicts of Interest: The authors declare no conflict of interest.

References

1. Ferries-Rowe, E.; Corey, E.; Archer, J.S. Primary Dysmenorrhea: Diagnosis and Therapy. *Obstet. Gynecol.* **2020**, *136*, 1047–1058. [CrossRef] [PubMed]
2. Iacovides, S.; Avidon, I.; Baker, F.C. What we know about primary dysmenorrhea today: A critical review. *Hum. Reprod. Update* **2015**, *21*, 762–778. [CrossRef] [PubMed]
3. Tu, F.; Hellman, K. Primary Dysmenorrhea: Diagnosis and Therapy. *Obstet. Gynecol.* **2021**, *137*, 752. [CrossRef]
4. Kho, K.A.; Shields, J.K. Diagnosis and Management of Primary Dysmenorrhea. *Jama* **2020**, *323*, 268–269. [CrossRef] [PubMed]
5. Carroquino-Garcia, P.; JimÃ©nez-Rejano, J.J.; Medrano-Sanchez, E.; de la Casa-Almeida, M.; Diaz-Mohedo, E.; Suarez-Serrano, C. Therapeutic Exercise in the Treatment of Primary Dysmenorrhea: A Systematic Review and Meta-Analysis. *Phys. Ther.* **2019**, *99*, 1371–1380. [CrossRef] [PubMed]
6. McKenna, K.A.; Fogleman, C.D. Dysmenorrhea. *Am. Fam. Physician* **2021**, *104*, 164–170. [PubMed]
7. Earl, R.A.; Grivell, R.M. Nifedipine for primary dysmenorrhoea. *Cochrane Database Syst. Rev.* **2021**, *12*, Cd012912. [CrossRef] [PubMed]
8. Guimarães, I.; Póvoa, A.M. Primary Dysmenorrhea: Assessment and Treatment. *Rev. Bras. Ginecol. Obstet.* **2020**, *42*, 501–507. [CrossRef]
9. ACOG Committee Opinion No. 760: Dysmenorrhea and Endometriosis in the Adolescent. *Obstet. Gynecol.* **2018**, *132*, e249–e258. [CrossRef]
10. Wang, C.; Liu, Y.; Dun, W.; Zhang, T.; Yang, J.; Wang, K.; Mu, J.; Zhang, M.; Liu, J. Effects of repeated menstrual pain on empathic neural responses in women with primary dysmenorrhea across the menstrual cycle. *Hum. Brain Mapp.* **2021**, *42*, 345–356. [CrossRef]
11. Dun, W.; Fan, T.; Wang, Q.; Wang, K.; Yang, J.; Li, H.; Liu, J.; Liu, H. Association Between Trait Empathy and Resting Brain Activity in Women With Primary Dysmenorrhea During the Pain and Pain-Free Phases. *Front. Psychiatry* **2020**, *11*, 608928. [CrossRef]
12. Bajalan, Z.; Moafi, F.; MoradiBaglooei, M.; Alimoradi, Z. Mental health and primary dysmenorrhea: A systematic review. *J. Psychosom. Obstet. Gynaecol.* **2019**, *40*, 185–194. [CrossRef]
13. Liu, J.; Liu, H.; Mu, J.; Xu, Q.; Chen, T.; Dun, W.; Yang, J.; Tian, J.; Hu, L.; Zhang, M. Altered white matter microarchitecture in the cingulum bundle in women with primary dysmenorrhea: A tract-based analysis study. *Hum. Brain Mapp.* **2017**, *38*, 4430–4443. [CrossRef]
14. He, J.; Dun, W.; Han, F.; Wang, K.; Yang, J.; Ma, S.; Zhang, M.; Liu, J.; Liu, H. Abnormal white matter microstructure along the thalamus fiber pathways in women with primary dysmenorrhea. *Brain Imaging Behav.* **2020**, *15*, 2061–2068. [CrossRef]
15. Wu, X.; Yu, W.; Tian, X.; Liang, Z.; Su, Y.; Wang, Z.; Li, X.; Yang, L.; Shen, J. Altered Posterior Cerebellar Lobule Connectivity With Perigenual Anterior Cingulate Cortex in Women With Primary Dysmenorrhea. *Front. Neurol.* **2021**, *12*, 645616. [CrossRef]
16. Du, Y.; Li, Y.; Fu, X.; Li, C.; Yanan, L. Efficacy of Guizhi Fuling Wan for primary dysmenorrhea: Protocol for a randomized controlled trial. *Trials* **2021**, *22*, 933. [CrossRef]
17. Chai, C.; Hong, F.; Yan, Y.; Yang, L.; Zong, H.; Wang, C.; Liu, Z.; Yu, B. Effect of traditional Chinese medicine formula GeGen decoction on primary dysmenorrhea: A randomized controlled trial study. *J. Ethnopharmacol.* **2020**, *261*, 113053. [CrossRef]

18. Oladosu, F.A.; Tu, F.F.; Hellman, K.M. Nonsteroidal antiinflammatory drug resistance in dysmenorrhea: Epidemiology, causes, and treatment. *Am. J. Obstet. Gynecol.* **2018**, *218*, 390–400. [CrossRef]

19. Allyn, K.; Evans, S.; Seidman, L.C.; Payne, L.A. "Tomorrow, I'll Be Fine": Impacts and coping mechanisms in adolescents and young adults with primary dysmenorrhoea. *J. Adv. Nurs.* **2020**, *76*, 2637–2647. [CrossRef]

20. Tang, B.; Liu, D.; Chen, L.; Liu, Y. NLRP3 inflammasome inhibitor MCC950 attenuates primary dysmenorrhea in mice via the NF-κB/COX-2/PG pathway. *J. Inflamm.* **2020**, *17*, 22. [CrossRef]

21. Osayande, A.S.; Mehulic, S. Diagnosis and initial management of dysmenorrhea. *Am. Fam. Physician* **2014**, *89*, 341–346. [PubMed]

22. Tong, H.; Yu, M.; Fei, C.; Ji, D.; Dong, J.; Su, L.; Gu, W.; Mao, C.; Li, L.; Bian, Z.; et al. Bioactive constituents and the molecular mechanism of Curcumae Rhizoma in the treatment of primary dysmenorrhea based on network pharmacology and molecular docking. *Phytomedicine Int. J. Phytother. Phytopharm.* **2021**, *86*, 153558. [CrossRef] [PubMed]

23. Akerlund, M. Vasopressin and oxytocin in normal reproduction and in the pathophysiology of preterm labour and primary dysmenorrhoea. Development of receptor antagonists for therapeutic use in these conditions. *Rocz. Akad. Med. Bialymst.* **2004**, *49*, 18–21. [PubMed]

24. Akerlund, M. Vascularization of human endometrium. Uterine blood flow in healthy condition and in primary dysmenorrhoea. *Ann. N. Y. Acad. Sci.* **1994**, *734*, 47–56. [CrossRef]

25. Hauksson, A.; Akerlund, M.; Forsling, M.L.; Kindahl, H. Plasma concentrations of vasopressin and a prostaglandin F2 alpha metabolite in women with primary dysmenorrhoea before and during treatment with a combined oral contraceptive. *J. Endocrinol.* **1987**, *115*, 355–361. [CrossRef]

26. Liedman, R.; Skillern, L.; James, I.; McLeod, A.; Grant, L.; Akerlund, M. Validation of a test model of induced dysmenorrhea. *Acta Obstet. Gynecol. Scand.* **2006**, *85*, 451–457. [CrossRef]

27. Wong, C.L.; Farquhar, C.; Roberts, H.; Proctor, M. Oral contraceptive pill as treatment for primary dysmenorrhoea. *Cochrane Database Syst. Rev.* **2009**, Cd002120. [CrossRef]

28. Marjoribanks, J.; Ayeleke, R.O.; Farquhar, C.; Proctor, M. Nonsteroidal anti-inflammatory drugs for dysmenorrhoea. *Cochrane Database Syst. Rev.* **2015**, *2015*, Cd001751. [CrossRef]

29. Zhu, X.; Proctor, M.; Bensoussan, A.; Wu, E.; Smith, C.A. Chinese herbal medicine for primary dysmenorrhoea. *Cochrane Database Syst. Rev.* **2008**, Cd005288. [CrossRef]

30. Akerlund, M. Targeting the oxytocin receptor to relax the myometrium. *Expert Opin. Ther. Targets* **2006**, *10*, 423–427. [CrossRef]

31. Earl, D.T.; Mercola, J.M. Calcium channel blockers and dysmenorrhea. *J. Adolesc. Health Off. Publ. Soc. Adolesc. Med.* **1992**, *13*, 107–108. [CrossRef]

32. Liu, Y.; Sun, J.; Wang, X.; Shi, L.; Yan, Y. Effect of herb-partitioned moxibustion for primary dysmenorrhea: A randomized clinical trial. *J. Tradit. Chin. Med. Chung I Tsa Chih Ying Wen Pan* **2019**, *39*, 237–245.

33. Witt, C.M.; Lüdtke, R.; Willich, S.N. Homeopathic treatment of patients with dysmenorrhea: A prospective observational study with 2 years follow-up. *Arch. Gynecol. Obstet.* **2009**, *280*, 603–611. [CrossRef]

34. Liu, Y.J.; Xiao, W.; Wang, Z.Z.; Zhao, B.J.; Zhou, Z.M.; Jiang, H.Z.; Jin, Z.; Wei, S.B.; Wang, Z.; Wang, D.M.; et al. Effects and safety of varying doses of guizhi fuling capsule in patients with primary dysmenorrhea: A multi-center, randomized, double-blind, placebo-controlled clinical study. *Zhongguo Zhong Yao Za Zhi Zhongguo Zhongyao Zazhi China J. Chin. Mater. Med.* **2013**, *38*, 2019–2022.

35. Yang, L.; Chai, C.Z.; Yan, Y.; Duan, Y.D.; Henz, A.; Zhang, B.L.; Backlund, A.; Yu, B.Y. Spasmolytic Mechanism of Aqueous Licorice Extract on Oxytocin-Induced Uterine Contraction through Inhibiting the Phosphorylation of Heat Shock Protein 27. *Molecules* **2017**, *22*, 1392. [CrossRef]

36. Cheng, Y.; Chu, Y.; Su, X.; Zhang, K.; Zhang, Y.; Wang, Z.; Xiao, W.; Zhao, L.; Chen, X. Pharmacokinetic-pharmacodynamic modeling to study the anti-dysmenorrhea effect of Guizhi Fuling capsule on primary dysmenorrhea rats. *Phytomedicine Int. J. Phytother. Phytopharm.* **2018**, *48*, 141–151. [CrossRef]

37. Pu, B.C.; Fang, L.; Gao, L.N.; Liu, R.; Li, A.Z. Animal study on primary dysmenorrhoea treatment at different administration times. *Evid. Based Complementary Altern. Med. Ecam* **2015**, *2015*, 367379. [CrossRef]

38. Robinson, N.B.; Krieger, K.; Khan, F.M.; Huffman, W.; Chang, M.; Naik, A.; Yongle, R.; Hameed, I.; Krieger, K.; Girardi, L.N.; et al. The current state of animal models in research: A review. *Int. J. Surg.* **2019**, *72*, 9–13. [CrossRef]

39. Carr-Nangle, R.E.; Johnson, W.G.; Bergeron, K.C.; Nangle, D.W. Body image changes over the menstrual cycle in normal women. *Int. J. Eat. Disord.* **1994**, *16*, 267–273. [CrossRef]

40. Haghighizadeh, M.H.; Karandish, M.; Ghoreishi, M.; Soroor, F.; Shirani, F. Body weight changes during the menstrual cycle among university students in Ahvaz, Iran. *Pak. J. Biol. Sci. PJBS* **2014**, *17*, 915–919. [CrossRef]

41. Yang, L.; Cao, Z.; Yu, B.; Chai, C. An in vivo mouse model of primary dysmenorrhea. *Exp. Anim.* **2015**, *64*, 295–303. [CrossRef]

42. Sen, E.; Ozdemir, O.; Ozdemir, S.; Atalay, C.R. The Relationship between Serum Ischemia-Modified Albumin Levels and Uterine Artery Doppler Parameters in Patients with Primary Dysmenorrhea. *Rev. Bras. De Ginecol. E Obstet. Rev. Da Fed. Bras. Das Soc. De Ginecol. E Obstet.* **2020**, *42*, 630–633. [CrossRef]

43. Zhang, Y.; Su, N.; Liu, W.; Wang, Q.; Sun, J.; Peng, Y. Metabolomics Study of Guizhi Fuling Capsules in Rats With Cold Coagulation Dysmenorrhea. *Front. Pharmacol.* **2021**, *12*, 764904. [CrossRef]

44. Zhang, K.; Su, J.; Huang, Y.; Wang, Y.; Meng, Q.; Guan, J.; Xu, S.; Wang, Y.; Fan, G. Untargeted metabolomics reveals the synergistic mechanisms of Yuanhu Zhitong oral liquid in the treatment of primary dysmenorrhea. *J. Chromatogr. B Anal. Technol. Biomed. Life Sci.* **2021**, *1165*, 122523. [CrossRef]

45. Barcikowska, Z.; Rajkowska-Labon, E.; Grzybowska, M.E.; Hansdorfer-Korzon, R.; Zorena, K. Inflammatory Markers in Dysmenorrhea and Therapeutic Options. *Int. J. Environ. Res. Public Health* **2020**, *17*, 1191. [CrossRef]

46. Moslehi, N.; Mirmiran, P.; Marzbani, R.; Rezadoost, H.; Mirzaie, M.; Azizi, F.; Tehrani, F.R. Serum metabolomics study of women with different annual decline rates of anti-MÃ$\frac{1}{4}$llerian hormone: An untargeted gas chromatography-mass spectrometry-based study. *Hum. Reprod.* **2021**, *36*, 721–733. [CrossRef]

47. Rinschen, M.M.; Ivanisevic, J.; Giera, M.; Siuzdak, G. Identification of bioactive metabolites using activity metabolomics. *Nat. Rev. Mol. Cell Biol.* **2019**, *20*, 353–367. [CrossRef]

48. Sun, Y.; Wu, D.; Zeng, W.; Chen, Y.; Guo, M.; Lu, B.; Li, H.; Sun, C.; Yang, L.; Jiang, X.; et al. The Role of Intestinal Dysbacteriosis Induced Arachidonic Acid Metabolism Disorder in Inflammaging in Atherosclerosis. *Front. Cell. Infect. Microbiol.* **2021**, *11*, 618265. [CrossRef]

49. Lee, K.; Lee, S.H.; Kim, T.H. The Biology of Prostaglandins and Their Role as a Target for Allergic Airway Disease Therapy. *Int. J. Mol. Sci.* **2020**, *21*, 1851. [CrossRef]

50. Law, S.H.; Chan, M.L.; Marathe, G.K.; Parveen, F.; Chen, C.H.; Ke, L.Y. An Updated Review of Lysophosphatidylcholine Metabolism in Human Diseases. *Int. J. Mol. Sci.* **2019**, *20*, 1149. [CrossRef]

51. Murakami, M.; Sato, H.; Taketomi, Y. Updating Phospholipase A(2) Biology. *Biomolecules* **2020**, *10*, 1457. [CrossRef] [PubMed]

52. Yuan, Z.; Yang, L.; Zhang, X.; Ji, P.; Hua, Y.; Wei, Y. Mechanism of Huang-lian-Jie-du decoction and its effective fraction in alleviating acute ulcerative colitis in mice: Regulating arachidonic acid metabolism and glycerophospholipid metabolism. *J. Ethnopharmacol.* **2020**, *259*, 112872. [CrossRef] [PubMed]

53. Ma, B.; Yang, S.; Tan, T.; Li, J.; Zhang, X.; Ouyang, H.; He, M.; Feng, Y. An integrated study of metabolomics and transcriptomics to reveal the anti-primary dysmenorrhea mechanism of Akebiae Fructus. *J. Ethnopharmacol.* **2021**, *270*, 113763. [CrossRef] [PubMed]

54. Huang, X.; Su, S.; Duan, J.A.; Sha, X.; Zhu, K.Y.; Guo, J.; Yu, L.; Liu, P.; Shang, E.; Qian, D. Effects and mechanisms of Shaofu-Zhuyu decoction and its major bioactive component for Cold—Stagnation and Blood—Stasis primary dysmenorrhea rats. *J. Ethnopharmacol.* **2016**, *186*, 234–243. [CrossRef]

55. Szmidt, M.K.; Granda, D.; Sicinska, E.; Kaluza, J. Primary Dysmenorrhea in Relation to Oxidative Stress and Antioxidant Status: A Systematic Review of Case-Control Studies. *Antioxidants* **2020**, *9*, 994. [CrossRef]

56. Kaplan, Ã.; NazÄ±roÄŸlu, M.; GÃ$\frac{1}{4}$ney, M.; Aykur, M. Non-steroidal anti-inflammatory drug modulates oxidative stress and calcium ion levels in the neutrophils of patients with primary dysmenorrhea. *J. Reprod. Immunol.* **2013**, *100*, 87–92. [CrossRef]

57. Dikensoy, E.; Balat, O.; PenÃŞe, S.; Balat, A.; Cekmen, M.; Yurekli, M. Malondialdehyde, nitric oxide and adrenomedullin levels in patients with primary dysmenorrhea. *J. Obstet. Gynaecol. Res.* **2008**, *34*, 1049–1053. [CrossRef]

58. Maekawa, S.; Takada, S.; Nambu, H.; Furihata, T.; Kakutani, N.; Setoyama, D.; Ueyanagi, Y.; Kang, D.; Sabe, H.; Kinugawa, S. Linoleic acid improves assembly of the CII subunit and CIII2/CIV complex of the mitochondrial oxidative phosphorylation system in heart failure. *Cell Commun. Signal CCS* **2019**, *17*, 128. [CrossRef]

59. Hatanaka, E.; Dermargos, A.; Hirata, A.E.; Vinolo, M.A.; Carpinelli, A.R.; Newsholme, P.; Armelin, H.A.; Curi, R. Oleic, linoleic and linolenic acids increase ros production by fibroblasts via NADPH oxidase activation. *PLoS ONE* **2013**, *8*, e58626. [CrossRef]

60. Ma, C.; Kesarwala, A.H.; Eggert, T.; Medina-Echeverz, J.; Kleiner, D.E.; Jin, P.; Stroncek, D.F.; Terabe, M.; Kapoor, V.; ElGindi, M.; et al. NAFLD causes selective CD4(+) T lymphocyte loss and promotes hepatocarcinogenesis. *Nature* **2016**, *531*, 253–257. [CrossRef]

61. Gasanoff, E.S.; Yaguzhinsky, L.S.; Garab, G. Cardiolipin, Non-Bilayer Structures and Mitochondrial Bioenergetics: Relevance to Cardiovascular Disease. *Cells* **2021**, *10*, 1721. [CrossRef]

62. Nie, C.; He, T.; Zhang, W.; Zhang, G.; Ma, X. Branched Chain Amino Acids: Beyond Nutrition Metabolism. *Int. J. Mol. Sci.* **2018**, *19*, 954. [CrossRef]

63. Yoneshiro, T.; Wang, Q.; Tajima, K.; Matsushita, M.; Maki, H.; Igarashi, K.; Dai, Z.; White, P.J.; McGarrah, R.W.; Ilkayeva, O.R.; et al. BCAA catabolism in brown fat controls energy homeostasis through SLC25A44. *Nature* **2019**, *572*, 614–619. [CrossRef]

64. Gannon, N.P.; Schnuck, J.K.; Vaughan, R.A. BCAA Metabolism and Insulin Sensitivity—Dysregulated by Metabolic Status? *Mol. Nutr. Food Res.* **2018**, *62*, e1700756. [CrossRef]

65. Biswas, D.; Duffley, L.; Pulinilkunnil, T. Role of branched-chain amino acid-catabolizing enzymes in intertissue signaling, metabolic remodeling, and energy homeostasis. *FASEB J. Off. Publ. Fed. Am. Soc. Exp. Biol.* **2019**, *33*, 8711–8731. [CrossRef]

66. Wepy, J.A.; Galligan, J.J.; Kingsley, P.J.; Xu, S.; Goodman, M.C.; Tallman, K.A.; Rouzer, C.A.; Marnett, L.J. Lysophospholipases cooperate to mediate lipid homeostasis and lysophospholipid signaling. *J. Lipid Res.* **2019**, *60*, 360–374. [CrossRef]

67. Aguilar, H.N.; Mitchell, B.F. Physiological pathways and molecular mechanisms regulating uterine contractility. *Hum. Reprod. Update* **2010**, *16*, 725–744. [CrossRef]

68. Huang, Y.J.; Chen, Y.C.; Chen, H.Y.; Chiang, Y.F.; Ali, M.; Chiang, W.; Chung, C.P.; Hsia, S.M. Ethanolic Extracts of Adlay Testa and Hull and Their Active Biomolecules Exert Relaxing Effect on Uterine Muscle Contraction through Blocking Extracellular Calcium Influx in Ex Vivo and In Vivo Studies. *Biomolecules* **2021**, *11*, 887. [CrossRef]

69. Marchi, S.; Patergnani, S.; Missiroli, S.; Morciano, G.; Rimessi, A.; Wieckowski, M.R.; Giorgi, C.; Pinton, P. Mitochondrial and endoplasmic reticulum calcium homeostasis and cell death. *Cell Calcium* **2018**, *69*, 62–72. [CrossRef]

70. Szymański, J.; Janikiewicz, J.; Michalska, B.; Patalas-Krawczyk, P.; Perrone, M.; Ziółkowski, W.; Duszyński, J.; Pinton, P.; Dobrzyń, A.; Więckowski, M.R. Interaction of Mitochondria with the Endoplasmic Reticulum and Plasma Membrane in Calcium Homeostasis, Lipid Trafficking and Mitochondrial Structure. *Int. J. Mol. Sci.* **2017**, *18*, 1576. [CrossRef]
71. Yuan, M.; Li, D.; Zhang, Z.; Sun, H.; An, M.; Wang, G. Endometriosis induces gut microbiota alterations in mice. *Hum. Reprod.* **2018**, *33*, 607–616. [CrossRef]
72. Schmauss, C.; Yaksh, T.L. In vivo studies on spinal opiate receptor systems mediating antinociception. II. Pharmacological profiles suggesting a differential association of mu, delta and kappa receptors with visceral chemical and cutaneous thermal stimuli in the rat. *J. Pharmacol. Exp. Ther.* **1984**, *228*, 1–12.
73. Maehara, T.; Fujimori, K. Inhibition of Prostaglandin F(2)(Î±) Receptors Exaggerates HCl-Induced Lung Inflammation in Mice. *Int. J. Mol. Sci.* **2021**, *22*, 12843. [CrossRef]

MDPI AG
Grosspeteranlage 5
4052 Basel
Switzerland
Tel.: +41 61 683 77 34

MDPI Books Editorial Office
E-mail: books@mdpi.com
www.mdpi.com/books